Lecture Notes in Civil Engineering

Volume 234

Lecture Notes in Civil Engineering (LNCE) publishes the latest developments in Civil Engineering - quickly, informally and in top quality. Though original research reported in proceedings and post-proceedings represents the core of LNCE, edited volumes of exceptionally high quality and interest may also be considered for publication. Volumes published in LNCE embrace all aspects and subfields of, as well as new challenges in, Civil Engineering. Topics in the series include:

- Construction and Structural Mechanics
- Building Materials
- Concrete, Steel and Timber Structures
- Geotechnical Engineering
- Earthquake Engineering
- Coastal Engineering
- Ocean and Offshore Engineering; Ships and Floating Structures
- Hydraulics, Hydrology and Water Resources Engineering
- Environmental Engineering and Sustainability
- Structural Health and Monitoring
- Surveying and Geographical Information Systems
- Indoor Environments
- Transportation and Traffic
- Risk Analysis
- Safety and Security

To submit a proposal or request further information, please contact the appropriate Springer Editor:

- Pierpaolo Riva at pierpaolo.riva@springer.com (Europe and Americas);
- Swati Meherishi at swati.meherishi@springer.com (Asia - except China, and Australia, New Zealand);
- Wayne Hu at wayne.hu@springer.com (China).

All books in the series now indexed by Scopus and EI Compendex database!

More information about this series at https://link.springer.com/bookseries/15087

Anil Kumar Dikshit · Balaji Narasimhan ·
Bimlesh Kumar · Ajey Kumar Patel
Editors

Innovative Trends in Hydrological and Environmental Systems

Select Proceedings of ITHES 2021

Springer

Editors
Anil Kumar Dikshit
Department of Environmental Science and Engineering
Indian Institute of Technology Bombay
Mumbai, Maharashtra, India

Bimlesh Kumar
Department of Civil Engineering
Indian Institute of Technology Guwahati
Guwahati, Assam, India

Balaji Narasimhan
Department of Civil Engineering
Indian Institute of Technology Madras
Chennai, Tamil Nadu, India

Ajey Kumar Patel
Department of Civil Engineering
National Institute of Technology Warangal
Warangal, Telangana, India

ISSN 2366-2557 ISSN 2366-2565 (electronic)
Lecture Notes in Civil Engineering
ISBN 978-981-19-0303-8 ISBN 978-981-19-0304-5 (eBook)
https://doi.org/10.1007/978-981-19-0304-5

This Springer imprint is published by the registered company Springer Nature Singapore Pte Ltd.
The registered company address is: 152 Beach Road, #21-01/04 Gateway East, Singapore 189721, Singapore

Contents

About the Editors

Dr. Anil Kumar Dikshit is Higher Administrative Grade (HAG) Professor in the Environmental Infrastructure & Clean Tech (EICT) Group and In-charge of GIS Laboratory for Environmental Management & Sustainability (GEMS) of the Environmental Science and Engineering (ESE) Department at Indian Institute of Technology, Bombay, India. He did his graduation and masters, in civil engineering in 1985 and 1987, respectively, and completed Ph.D. in civil and environmental engineering from Cornell University, USA in the year 1994. The areas of his teaching interest are environmental systems planning and management; environmental impact assessment; water supply and wastewater engineering; environmental engineering; and solid and biomedical waste management while the areas of research interest include environmental infrastructure; clean technology; urban, biomedical, hazardous and electronic solid waste management; environmental modeling, optimization, impact assessment, sustainability and life cycle assessment; and GIS applications to water and environment related problems. He has more than 30 years of experience in Teaching, Research and Consultancy at IIT Kanpur, Cornell University, IIT Kharagpur and IIT Bombay. He has also taken several short assignments as visiting professor at Kigali Institute of Science & Technology, Rwanda; University of Kwazulu-Natal, South Africa; Nanyang Technological University, and National University of Singapore, Singapore; Malardalen University, Sweden; and Asian Institute of Technology, Bangkok. He has guided 40 masters' and 22 doctoral scholars while five research scholars are working on their PhDs. He has also worked on more than 150 national and international research and consultancy projects. He is having research collaboration with German, French, US, Singapore, South Africa, Sweden, Norway and Thailand Universities and Institutions and has organized 10 International Conferences, eight Workshops and 30 Training Courses. He is prolific speaker and has delivered more than 300 invited talks at various national and international forums. The results of academic research work have provided one patent, seven books, 11 chapters in edited books, 100 reports, 135 publications in international & national journals, 140 in international & national conferences while many more papers are under review. In addition, more than 20 prototypes/equipment related to various environmental pollution management topics; academic versions of more than 10 environmental modeling and simulation software(s); and about seven GIS applications on environmental management problems have been developed under his guidance. He is associate editor, reviewer or editorial board member for many peer-reviewed environmental journals.

Dr. Balaji Narasimhan is currently a Professor at the Department of Civil Engineering, Indian Institute of Technology Madras, Chennai. He obtained his B.E. (Agricultural Engineering) from Tamil Nadu Agricultural University, Coimbatore; M.S. from University of Manitoba, Canada and Ph.D. from Texas A&M University, USA. Balaji Narasimhan is a Hydrologist with expertise in modelling the impact of landuse and climate change on hydrology using GIS, remote sensing and hydrologic/hydraulic/crop growth models with experience of over 20 years. He has published over 35 scholarly articles in reputed international journals. He received the Young scientist award at the 2013 International SWAT conference at Toulouse, France for his contribution in the development of algorithms for the SWAT model. He is one of the coordinators involved in the development of the Ganga River Basin Management Plan and is the lead investigator in the development of a hydrological flow health condition report for the entire Ganga River Basin. Balaji Narasimhan was also a Member of "Think Tank" for development of Water Resources Information System (WRIS), Ministry of Water Resources, Governtment of India. He was involved in a large team project to develop a flood forecasting system for Chennai city, following the floods of 2015, funded by the Principal Scientific Advisor, Government of India.

Dr. Bimlesh Kumar is currently a Professor at the Department of Civil Engineering, Indian Institute of Technology Guwahati, Guwahati. He obtained his B.E. (Civil Engineering) from Muzaffarpur Institute of Technology Muzaffarpur, Bihar, and M.E. and Ph.D. from the Indian Institute of Science, Bengaluru. His major areas of research interests include river hydraulics and sediment transport. He has published more than 100 papers in respected international journals. Dr. Bimlesh Kumar received the R. J. Garde research award in 2015.

Dr. Ajey Kumar Patel is currently an Assistant Professor at Department of Civil Engineering, National Institute of Technology, Warangal. He obtained his B.Tech. (Civil Engineering) from Harcourt Butler Technological Institute (HBTI), Kanpur; M.Tech. (Environmental Engineering) from Indian Institute of Technology Roorkee, Roorkee; Ph.D. from the Indian Institute of Science, Bangalore, Bangalore. Ajey Kumar Patel is an Environmental Engineer with expertise in modelling surface aeration system including both experimental and CFD models, urban flood modelling, water hammer analysis and water distribution network analysis. He has published over 30 papers in various peer reviewed journals and international conferences.

Hydrological Risk Assessment of the Coastal Megacity Chennai Using DRASTIC Method and SWAT

Preethi Vasudevan, M. A. Sherly, and Ranjana Ray Chaudhuri

Abstract Chennai is a coastal city located in the northern part of Tamil Nadu with an average annual rainfall of 1400 mm. It is among the major metropolis of India and has undergone rapid urbanization. Accommodating the growing population in the past few decades has been at the cost of an increased hydrological risk in the region. Chennai has been suffering from frequent floods and droughts due to poor land and water resources management practices. This study quantifies the hydrological risk in the region with respect to surface and groundwater using the DRASTIC method and SWAT model. Simulation results show that the run-off in 2018 was 141% of the run-off in 1998. This increase is heavily influenced by the land management practices of the study area. It has been found that 36.8% of the study area is under high vulnerability to groundwater contamination, while 58.4% has a medium vulnerability. Development plans need to include management practices both for surface and groundwater resources to ensure that the present level of risk is not further aggravated.

Keywords Chennai · Hydrological risk · SWAT · DRASTIC

1 Introduction

Water resources assessment and management are crucial for the sustainable development of water resources projects [1]. Run-off estimation is a significant parameter in watershed management, but often data on run-off is either partially or not available. Duvvuri [2] reports that an increase in urbanization has intensified water scarcity

P. Vasudevan · M. A. Sherly (✉) · R. R. Chaudhuri
Department of Regional Water Studies, TERI School of Advanced Studies, Delhi, India
e-mail: sherly.ma@terisas.ac.in

P. Vasudevan
e-mail: preethi.vasudevan@terisas.ac.in

R. R. Chaudhuri
e-mail: ranjana.chaudhuri@terisas.ac.in

A. K. Dikshit et al. (eds.), *Innovative Trends in Hydrological and Environmental Systems*, Lecture Notes in Civil Engineering 234,
https://doi.org/10.1007/978-981-19-0304-5_1

and run-off, especially in urban areas. With the increased demand for water, often households and communities turn to groundwater resources in the absence of reliable supply. This has led to the exploitation of groundwater resources in many parts of the country. Reducing the run-off and harvesting of this water are crucial to meet the water demand of the growing population. Rainfall run-off models have been used extensively to assess the impact of land-use change on run-off [3]. Understanding the relationship between land use and surface water hydrology will help in forming the baseline for policy interventions.

An increase in the number of impervious structures has led to a decrease in groundwater recharge with only extraction and limited recharge that has led to the rapid depletion [4]. Groundwater as a resource is relatively less vulnerable to pollution than surface water. But owing to the limited access, any pollution to the groundwater is difficult to detect and control which leads to the contamination, which may persist for years together. Prevention of contamination is easier than remediation of contaminated aquifers. Since groundwater monitoring is often expensive to adequately define the regional extent of contamination, assessing the groundwater vulnerability becomes an important step [5]. Groundwater vulnerability can be defined as the susceptibility of groundwater to contamination [6, 7]. It covers the tendency of contaminants to reach the groundwater when introduced at the surface. It is not an absolute measure but rather indicates the possibility of groundwater contamination [8].

Chennai has been experiencing rapid urbanization for the past three decades which has led to unregulated settlements on wetlands, watercourses, and small water bodies [9, 10]. The region experiences both flooding and droughts due to poor water management as evident from the 2015 flood catastrophe followed by a day zero in 2019 [11, 12]. Flood management is complicated as the city has a largely flat terrain and lacks a gradient for proper natural drainage [13]. The flooding in Chennai is also compounded by the fact that it lacks an effective sewerage and storm water drainage system. It was initially established in the 1980s, and the improvements over the years still do not cater to the growing population and city limits [14, 15]. The water supply in the city is inefficient and does not cover the entire population; most of the households depend on local groundwater sources for domestic purposes. Over-exploitation of the resource has brought about a drought-like situation in the city [14]. The presence of open water bodies alone does not prevent the occurrence of floods and droughts; increased catchment inflow and reduced groundwater recharge also impact the occurrence of these extreme events [14]. Along with the projected climate change, the management practices and land-use changes in the region have resulted in increased run-off generation and reduced groundwater recharge [13].

Chennai has been chosen as the focus of this study since it faces both surface water and groundwater issues due to expanding population growth. This study uses the DRASTIC method to assess the groundwater vulnerability to contamination, while the SWAT model to quantify the hydrological risk to the region from increased surface run-off.

2 Study Area

The study area (Fig. 1) is located in the south of Chennai city between 80.09 to 80.22 °E and 12.84 to 12.96 °N. It follows the catchment of the Pallikaranai marshland which extends to an area of 289.2 km^2 and covers parts of the Chennai and Kanchipuram districts. The average annual rainfall is about 1400 mm, with the majority of the rainfall occurring during the North-East monsoon from October to December. A tropical climate with a mean temperature of 24–32 °C prevails over the study area, and the region also experiences high humidity values in the range of 65–84% [10, 16].

Pallikaranai marshland is the only marsh wetland ecosystem in the city of Chennai. The wetland spanned an area of 50 km^2 in the 1960s, while currently only 5 km^2 of this area is preserved as a wetland. This loss in the wetland extent can be attributed to urbanization caused by the emergence of the city as a metropolis. The remaining wetland area still supports a huge amount of biodiversity including threatened species of birds such as spot-billed Pelican, woolly necked Stork, and black-headed Ibis. The wetland extends from Velachery to Karapakkam and connects to the Buckingham Canal through the Okkiyam Maduvu which acts as the outlet of the wetland [16, 17].

The second Chennai master plan drafted in 2006 for the year 2026 shows that parts of the wetland are planned to be replaced by industrial, institutional, and residential

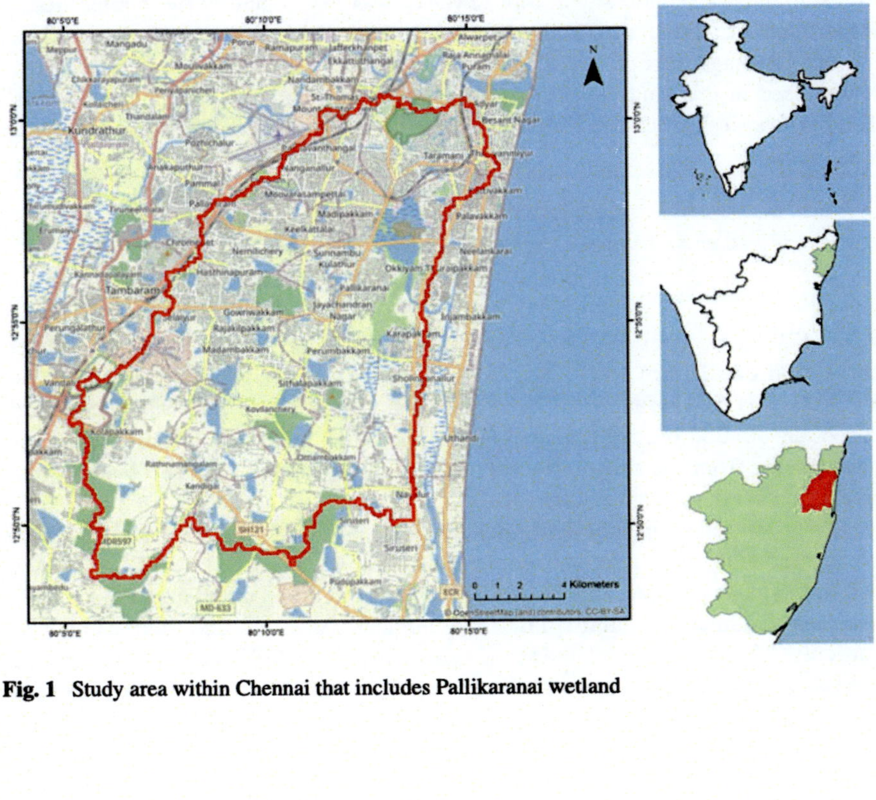

Fig. 1 Study area within Chennai that includes Pallikaranai wetland

set-ups to cater to this growing industry demand [18]. This region forms a part of the newly expanded Chennai city and has grown tremendously in the past few decades due to the presence of IT industries. The undeveloped parts of the wetland have since been declared a reserve forest [17]. Much of the existing settlements in this region lack a proper sewerage system for wastewater, and the residents are forced to adopt their own practices of disposal including dumping untreated wastewater in the nearby water bodies [14, 19]. This is the state of most of the water bodies in the region, contaminated with sewage and solid waste. Parts of the study area around the Pallikaranai marsh, like Velachery, Perungudi, Thoraipakkam, Pallikaranai, etc., are waterlogged almost every year during the monsoon. These areas are low-lying and heavily populated leaving little open green spaces making them at heavy risk for flooding [9, 10]. The population stress coupled with the lack of reliable water supply in certain places leads to over-exploitation of groundwater. The contaminated water bodies have the potential to contaminate the groundwater sources as well [14].

3 Materials and Methods

3.1 SWAT Model

The soil and water assessment tool (SWAT) was developed by the United States Department of Agriculture (USDA) for assessing and predicting the impact of land use, land management practices on water, sediment, and agriculture in watersheds. It was used in this study to estimate the run-off generated across two timescales. The hydrological processes simulated in the model are based on the water balance equation [20].

$$\mathrm{SW}_t = \mathrm{SW}_0 + \sum_{i=1}^{t}\left(R_{\mathrm{day}} - Q_{\mathrm{surf}} - E_a - w_{\mathrm{seep}} - Q_{\mathrm{gw}}\right). \quad (1)$$

where SW_t is the final soil water content (mm H_2O), SW_0 is the initial soil water content on day i (mm H_2O), R_{day} is the amount of precipitation on day i (mm H_2O), Q_{surf} is the amount of surface run-off on y i (mm H_2O), E_a is the amount of evapotranspiration on day i (mm H_2O), w_{seep} is the amount of water entering the vadose zone from the soil profile on day i (mm H_2O), and Q_{gw} is the amount of return flow on day i (mm H_2O).

The watershed delineator delineates the watershed into further sub-basins which are further subdivided into hydrological response units (HRUs) based on the land use, soil type, and slope characteristics. The land classified under a particular HRU has been assumed to exhibit similar hydrological properties, and the area under a particular HRU can be simulated as a single unit [20, 21]. Calibration of the model is necessary to match the results with the observed data and also to ensure the consistent

Table 1 SWAT model input data source

Parameter	Source
Digital elevation model (DEM)	NASA SRTM
Land use and land cover (LULC)	Landsat 5 and 8 images (1998 and 2018)
Soil data	FAO HWSD
Weather data	NASA POWER

performance of the model. Calibration of the model improves the reliability of the results obtained [22].

The input data used in the model is listed in Table 1. The watershed was divided into a total of six sub-basins and 35 HRUs. The SCS curve number method was used in the model for run-off calculation. The model was run from 1990 to 2018 in a monthly time step. An initial warm-up period of 5 years was provided. The watershed under consideration is ungauged; hence, for calibration, the spatial proximity regionalization method was used based on the Chengelpet catchment which is at an approximate distance of 52 km [23]. Calibration and validation of the model were carried out using the SUFI2 algorithm in SWAT-CUP. The calibration period considered was from 2000–2006, and the validation was done from the period 2007–2010.

3.2 DRASTIC Method

DRASTIC is a weighted index-based approach developed by the US Environmental Protection Agency (USEPA) to assess the groundwater vulnerability to pollution based on hydrogeologic parameters of the region [24]. It was used in this study to assess groundwater vulnerability to contamination. The seven parameters considered in the method are mentioned below:

- Depth to groundwater (*D*): The depth at which the groundwater is available from the surface. Shallower depth is more susceptible to pollution.
- Net recharge (*R*): Net recharge is the amount of water per unit area that adds to the groundwater. The recharge aids in the transport of water from the vadose zone to the saturated zone. The net recharge can determine the extent of the transport of contaminants. Higher recharge values could indicate more vulnerability to pollution.
- Aquifer media (*A*): The aquifer media is the saturated zone that stores water. Water gets collected in the pore spaces and fractures of the aquifer media. Certain aquifers help in the attenuation of pollutants due to the slower mobility of water [25].
- Soil media (*S*): The soil media influences the vertical movement of contaminants. If the soil has a higher permeability, the vulnerability is more as the contaminant is more likely to reach the water table.

- Topography (*T*): Areas with a higher slope retain the water for a lesser duration and are hence less vulnerable to contamination [25, 26]. The slope is defined as the steepness of the area.
- Impact on vadose zone (*I*): Vadose zone refers to the unsaturated or partially saturated soil layer above the water table. The transport of contaminants into the water table is dependent on the properties of the vadose zone [25]. A highly fractured vadose zone is more vulnerable to contamination.
- Hydraulic conductivity (*C*): Hydraulic conductivity governs the rate at which the groundwater flows. Higher conductivity corresponds with a higher vulnerability [27].

As shown in Table 2, the seven parameters have been evaluated with respect to each other to determine their relative significance [26]. Accordingly, they are assigned weights on a scale of 1–5 depending on their significance. Then each parameter has been divided into ranges or significant media types based on the impact on pollution potential. A rating varying from 1 to 10 is assigned to the various ranges for each DRASTIC parameter. Here, the rating assigned is after evaluation of its relative significance to pollution potential [26]. The equation for determining the DRASTIC

Table 2 DRASTIC parameters ratings and weights [26]

Parameters	Ranges	Ratings	Weights
Depth to groundwater	0–1.5	10	5
	1.5–4.5	9	
	4.5–9	7	
	9–15	5	
Net recharge	11–13	10	4
	9–11	8	
	7–9	5	
Aquifer media	Charnockite	4	3
	Alluvium	9	
Soil media	Sandy	9	2
	Sandy loam	6	
Topography (slope)	0–2	10	1
	2–6	9	
	6–12	5	
	12–18	3	
	18+	1	
Impact of the vadose zone	Alluvium	9	5
	Charnockite	4	
Hydraulic conductivity	4.71 E-7 to 4.71 E-5	1	3
	1.41 E-4 to 3.29 E-4	4	

Table 3 Sources of inputs used in DRASTIC method

Parameter	Source
Depth to groundwater (*D*)	The layer was derived using the inverse distance weighted tool in ArcGIS for an average water level. The average water level was obtained from the pre and post-monsoon water levels from 29 wells in the Chennai and Kanchipuram district. The well-log data was obtained from India WRIS and CGWB
Net recharge (*R*)	The net recharge was obtained from using the Piscopo [29] method which is considered a better field-based method [25]. The method considers the slope percentage, rainfall, and soil permeability to calculate the net discharge. The slope percentage was derived from the SRTM DEM, rainfall data was obtained from IMD, and soil permeability was obtained from the FAO soil properties
Aquifer media (*A*)	The aquifer media data was obtained from India WRIS
Soil media (*S*)	The soil characteristics have been obtained from FAO's harmonized world soil database
Topography (*T*)	The slope is derived from the SRTM DEM using the spatial analyst tool
Impact of vadose zone (*I*)	The vadose zone was obtained using Piscopo [29] method which uses soil permeability and the depth to groundwater
Hydraulic conductivity (*C*)	The hydraulic conductivity for the study area is determined from the standard values for the soil media cover [30]

vulnerability index (DVI) is

$$\mathrm{DVI} = D_r D_w + R_r R_w + A_r A_w + S_r S_w + T_r T_w + I_r I_w + C_r C_w \tag{2}$$

where r is the rating and w is the weight.

Once DVI map has been generated, it is possible to identify areas that are more likely to be susceptible to groundwater pollution relative to one another. The higher the DVI, the greater the groundwater pollution potential is. DVI provides only a relative evaluation and is not designed to provide absolute answers [26, 28]. Table 3 lists the input used for the DRASTIC assessment.

3.2.1 Sensitivity Analysis on DRASTIC Method

Sensitivity analysis is important to determine which of the parameters influence the vulnerability the most and to find the relative significance of the parameters [31]. Single-parameter sensitivity analysis was performed for the resultant DRASTIC map. A single-parameter sensitivity measure was developed to evaluate the impact of each DRASTIC parameter on the vulnerability index. It compares the effective weight with the theoretical weight to gauge the significance of a parameter [25, 32].

4 Results and Discussion

4.1 Surface Runoff Assessment

Figures 2 and 3 show the LULC of the study area in 1998 and 2018, respectively. The built-up area has increased from 13.46% to 44.69% within two decades as shown in Table 4. The percentage of water bodies is higher in 2018 than in 1998. This increase in water cover could be due to the fact that the 1998 image is from February and the 2018 image is from January. Another reason for this increase could be that most

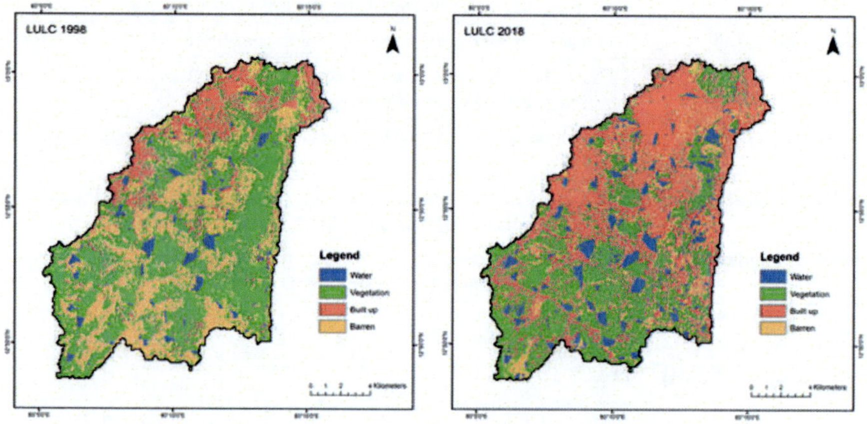

Fig. 2 LULC of the study area: Year 1998 (left) and Year 2018 (right)

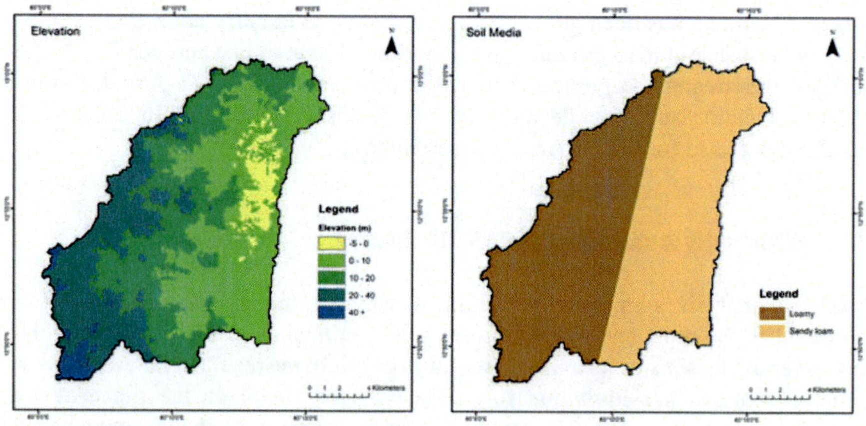

Fig. 3 Elevation and soil maps of the study area

Table 4 LULC changes in the study area during 1998–2018

Land use	% in 1998	% in 2018
Built-up	13.46	44.69
Vegetation	54.5	40.53
Water	3.49	7.09
Barren	28.55	7.69

of these water bodies are surrounded by urban settlements and receive untreated wastewater from them due to the lack of a proper sewerage system.

Figure 3 shows the elevation and soil maps where it is evident that the study area is on largely flat terrain with few small hills. The lowest region in the study area is the marshland itself. The soil in the region is mainly classified into two, namely sandy and sandy loam. Areas closer to the sea are sandy, and the soil turns loamy as the distance from the sea increases.

The simulated run-off in the year 2018 has become approximately 141% as compared to 1998. It is during the months of November and December that the region constantly experiences waterlogging and flood scenarios, whereas the drought-like conditions prevail during the hot summer months of April and May. The R^2 values of the calibrated and validated models are 0.66 and 0.64, respectively. The results could be improved further by using more accurate data and further calibration of parameters.

4.2 *Groundwater Vulnerability Assessment*

The raster calculator tool was used to assign ratings and weights and to overlay the multiple layers to form the DVI. The depth to groundwater has among the highest weight of 5. The average depth values range from 1.26 mbgl (metres below ground level) in Velachery to 11.55 mbgl in Tambaram.

The major part of the watershed falls in the range of 1.5–9 mbgl as shown in Fig. 4. The vulnerability decreases with an increase in groundwater depth. Most of the study area lies on a flat topography and receives rainfall of 1350 mm per year. Hence, the recharge is mainly governed by the infiltration capacity of the soil. It is observed that the coastal sandy soil has a greater recharge (Fig. 4). Higher recharge values indicate that the contaminant transport potential will be higher, and hence, the vulnerability is also higher.

The study area comprises two aquifer types: charnockite and alluvium. Alluvium covers the area near the coast, and the rest is dominated by charnockite. Permeability of the aquifer medium determines the mobility of the groundwater. Aquifers like alluvium are more vulnerable than charnockite. The soil media is mainly sandy and loamy sand. Sandy soil is concentrated near the coast. Soil with greater grain size and permeability are more vulnerable to pollution as they allow the percolation

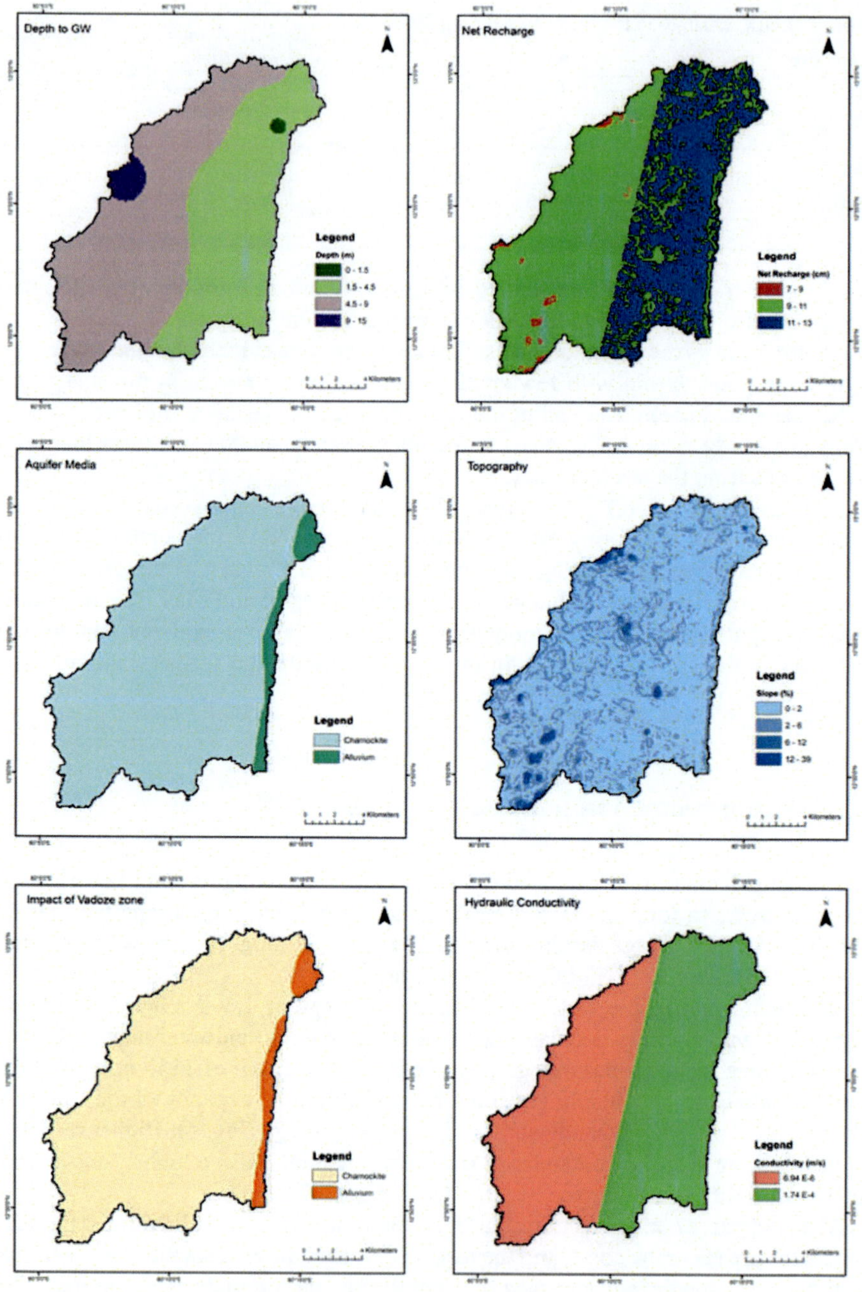

Fig. 4 Input layers used for generating the DRASTIC vulnerability index

of pollutants easily. The average slope is within 6% indicating that the terrain is relatively flat throughout. Flatter terrains retain water for longer, and hence, pollutants have a longer time to percolate into the soil, thereby increasing the vulnerability to pollution.

The impact of the vadose zone media is critical to assess the pollution vulnerability of the underlying aquifer. The impact of the vadose zone is dependent on the permeability of the unsaturated/partially saturated layer and the attenuation characteristics of the aquifer media. Lower ratings were given to materials having lower permeability. The impact of the vadose zone in the study area is evident from Fig. 4. Hydraulic conductivity in the study area ranges from 6.34 E-6 m/s in loamy sand to 1.76 E-4 m/s in sandy soil. A higher conductivity value means more transportation of pollutants, thereby increasing the vulnerability.

The resultant map (Fig. 5) indicating the vulnerability index of the study area was obtained by the weighted sum method. The DVI of the study area ranged from 101

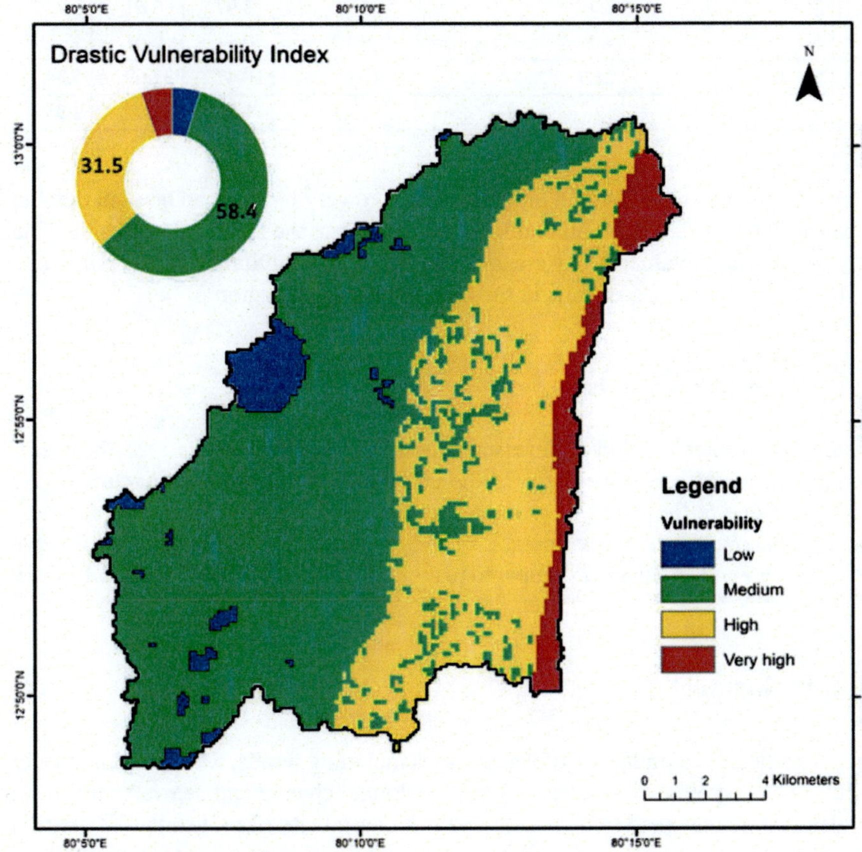

Fig. 5 Groundwater vulnerability map prepared for the study area using drastic vulnerability index

Table 5 DRASTIC vulnerability index classification

Range	Vulnerability	% area
101–120	Low	4.71
121–150	Medium	58.42
151–170	High	31.59
171–191	Very High	5.25

Table 6 Single-parameter sensitivity performed for DRASTIC method

	Theoretical weight	Theoretical weight %	Effective weight			
			Avg (%)	Min	Max	SD
D	5	21.7	28.85	19.34	34.65	4.56
R	4	17.4	24.19	16.39	32.79	4.61
A	3	13	10.57	6.28	19.01	3.2
S	2	8.7	5.19	3.97	6.61	0.73
T	1	4.3	5.19	0.74	9.01	2.42
I	5	21.7	17.61	10.47	31.69	5.34
C	3	13	5.61	1.99	9.92	2.95

to 196. The values were divided into classes low to very high. It can be seen that the area near the coast is more vulnerable to pollution than the area further inward. The high and very high vulnerability areas are underlain by alluvium. As shown in Table 5, 31.59% of the study area falls in the high vulnerability region.

4.2.1 Sensitivity Analysis

The single-parameter sensitivity measures the impact of each parameter on the index. It compares the effective weight with the theoretical weight of each parameter. As shown in Table 6, the depth to groundwater and net recharge are among the most effective parameters in the vulnerability index since they have a higher effective weight (28.85 and 24.19%) as compared to their theoretical weight (21.7 and 17.4%).

5 Conclusion

Chennai city faces a multitude of water issues including floods, water scarcity, water pollution, and groundwater contamination. Climate change and seawater intrusion are also factors that need to be considered in a coastal city like Chennai. The region has expanded in terms of area, infrastructure, and development tremendously in the past few decades. But the need of the hour is to consider the environmental impacts of any future development. Many of the water issues faced by the city have been brought about due to this rapid urbanization.

The results of the study also follow this pattern of urbanization. The increase in urban land use from 13.4% in 1998 to 44.6% in 2018 has impacted the run-off in all areas of the basin. The simulated run-off in 2018 has become approximately 141% as compared to the year 1998. Low-lying areas such as Velachery, Perungudi, Thoraipakkam, and Pallikaranai lie close to the Pallikaranai marsh and are water-logged and flooded every year. This could be attributed to the implementation of the second master plan of Chennai that planned to convert parts of the marsh into industrial, institutional, and residential areas. The importance of wetlands, especially in an urban metropolis like Chennai, is highlighted by this.

The obtained DVI was classified under four categories (low, medium, high, and very high) of vulnerability. It was estimated that about 5.3% of the area falls under very high followed by 31.6%, 58.4%, and 4.7% of the area under high, medium, and low vulnerability, respectively. Perungudi, Thoraipakkam, and Adyar are the most vulnerable to groundwater contamination as these localities are the nearest to the shore, and vulnerability decreases as the distance from shore increases. This can be attributed to the change in aquifer and soil properties as the distance from shore increases. St. Thomas Mount has a low vulnerability as it is at a higher elevation. The results of the study are meant to contribute towards identifying the vulnerable regions for future development.

Models like SWAT and DRASTIC can be useful in situations where it is not possible to physically measure the required parameters for hydrological risk assessment. Simulations can help identify the effect of land management practices in real time which can serve as an input to the decision-making processes. The results can be used by local regulatory authorities to formulate management action plans for surface and groundwater resources. Any further development of land use can also be regulated keeping in mind the vulnerable zones.

References

1. Jaiswal, R. K., Yadav, R. N., Lohani, A. K., Tiwari, H. L., Yadav, S. (2020). *Water balance modeling of Tandula (India) reservoir catchment using SWAT*.
2. Duvvuri, S. (2018). Hydrological modelling of Cooum river basin using GIS and SWAT Model. *International Journal for Research in Applied Science and Engineering Technology, 6*, 770–775. https://doi.org/10.22214/ijraset.2018.5127
3. Dutta, D., Das, R., Mazumdar, A. (2017). Assessment of runoff and sediment yield in the Tilaya reservoir, India using SWAT model. *Asian Journal of Water, Environment and Pollution, 14*, 9–18. https://doi.org/10.3233/AJW-170021
4. Murali, K.,Elangovan, R. (2013) Assessment of groundwater vulnerability in Coimbatore South Taluk, Tamilnadu, India using drastic approach. *International Journal of Scientific and Research Publications, 3*, 1–6.
5. Bai, L., Wang, Y., & Meng, F. (2012). Application of DRASTIC and extension theory in the groundwater vulnerability evaluation. *Water Environment Journal, 26*, 381–391. https://doi.org/10.1111/j.1747-6593.2011.00298.x

6. Kumar, S., Thirumalaivasan, D., Radhakrishnan, N., Kumar, S., Thirumalaivasan, D., Radhakrishnan, N. (2013). *Groundwater vulnerability assessment using SINTACS model,* 5705. https://doi.org/10.1080/19475705.2012.732119
7. Krogulec, E., Zabłocki, S., & Zadrozna, D. (2019). Variability of intrinsic groundwater vulnerability to pollution in river valley due to groundwater depth and recharge changes. *Applied Sciences, 9*, 9–12. https://doi.org/10.3390/app9061133
8. Durgadevagi, S., & Annadurai, R. (2016). Assessment of groundwater vulnerability using gis based drastic model: A case study of SIPCOT-Perundurai. *Erode Rasayan Journal of Chemistry, 9*, 465–470.
9. Suriya, S., Mudgal, B. V., & Nelliyat, P. (2012). Flood damage assessment of an urban area in Chennai, India, part I: Methodology. *Natural Hazards, 62*, 149–167. https://doi.org/10.1007/s11069-011-9985-3
10. Nithila Devi, N., Sridharan, B., & Kuiry, S. N. (2019). Impact of urban sprawl on future flooding in Chennai city. *Journal of Hydrology, 574*, 486–496. https://doi.org/10.1016/j.jhydrol.2019.04.041
11. Arabindoo, P. (2017). *Unprecedented natures? 20*, 800–821. https://doi.org/10.1080/13604813.2016.1239410
12. Krishnamurthy, L., Vecchi, G. A., Yang, X., van der Wiel, K., Balaji, V., Kapnick, S. B., Jia, L., Zeng, F., Paffendorf, K., & Underwood, S. (2018). Causes and probability of occurrence of extreme precipitation events like Chennai 2015. *Journal of Climate, 31*, 3831–3848. https://doi.org/10.1175/JCLI-D-17-0302.1
13. Dhamodaran, R., Suresh V. M. (2020) *An assessment of flood vulnerability zone in greater Chennai corporation, Tamil Nadu, 10*, 122–134.
14. Gupta, A. K., & Nair, S. S. (2010). Flood risk and context of land-uses: Chennai city case. *Journal of Geography and Regional Planning, 3*, 365–372.
15. Lavanya, A. K. (2013). Urban flood management—A case study of Chennai city. *Architecture Research, 2*, 115–121. https://doi.org/10.5923/j.arch.20120206.01
16. CGWB. (2008). *District groundwater brochure—Chennai.*
17. Tamil Nadu State Wetland Authority. http://tnswa.org/pallikaranai-marsh
18. CMDA: Chennai Second Master Plan. http://www.cmdachennai.gov.in/smp_main.html
19. CMWSSB: Sewerage System. https://chennaimetrowater.tn.gov.in/seweragesystem.html
20. Arnold, J. G., Moriasi, D. N., Gassman, P. W., Abbaspour, K. C., White, M. J., Srinivasan, R., Santhi, C., Harmel, R. D., Van Griensven, A., Liew, M. W. V., Kannan, N., Jha, M. K., Harmel, D., Member, A., Liew, M. W. V., & Arnold, J.-F.G. (2012). SWAT: Model use, calibration and validation. *Transactions of the ASABE, 55*, 1491–1508.
21. Shivhare, V., Goel, M. K., Singh, C. K. (2014) Simulation of surface runoff for upper tapi subcatchment area (burhanpur watershed) using swat. *International Archives of the Photogrammetry, Remote Sensing and Spatial Information Sciences*, ISPRS Arch. XL–8, 391–397. https://doi.org/10.5194/isprsarchives-XL-8-391-2014
22. Abbaspour, K. C., Vaghefi, S. A., Srinivasan, R. (2017) *A Guideline for Successful Calibration and Uncertainty Analysis for Soil and Water Assessment : A Review of Papers from the 2016 International SWAT Conference.* https://doi.org/10.3390/w10010006
23. Vandewiele, G. L., & Elias, A. (1995). Monthly water balance of ungauged catchments obtained by geographical regionalization. *Journal of Hydrology, 170*, 277–291. https://doi.org/10.1016/0022-1694(95)02681-E
24. Bataineh, S., Curtis, C., In, M., Alghwazi, Z. (2020). *Groundwater Resources, the DRASTIC Method and Applications in Jordan.*
25. Colins, J., Sashikkumar, M. C., Anas, P. A., Kirubakaran, M. (2016). GIS-based assessment of aquifer vulnerability using DRASTIC Model: A case study on Kodaganar basin. *Earth Sciences Research Journal, 20*, 1–8 (2016). https://doi.org/10.15446/esrj.v20n1.52469
26. Aller, L., Bennett, T., Lehr, J. H., Petty, R. J., Hackett, G. (1987). *DRASTIC : A standardized method for evaluating ground water pollution potential using hydrogeologic settings.* NWWA/Epa-600/2-87-035, 455.

27. Kumari, S., Jha, R., Singh, V., Baier, K., Sinha, M. K. (2016). Groundwater vulnerability assessment using SINTACS model and GIS in Raipur and Naya Raipur, Chhattisgarh, India. *Indian Journal of Science and Technology, 9*. https://doi.org/10.17485/ijst/2016/v9i41/99247
28. Wang, Y., Merkel, B. J., Li, Y., Ye, H., Fu, S., & Ihm, D. (2007). Vulnerability of groundwater in Quaternary aquifers to organic contaminants: A case study in Wuhan City. *China Environmental Geology, 53*, 479–484. https://doi.org/10.1007/s00254-007-0669-y
29. Piscopo, G. (2001). Groundwater vulnerability map explanatory notes groundwater vulnerability map explanatory notes Castlereagh catchment. *Centre for Natural Resources, NSW Department of Land and Water Conservation, 18.*
30. Duffield, G. *Hydraulic properties : Aquifer testing 101.* http://www.aqtesolv.com/aquifer-tests/aquifer_properties.htm
31. Lodwick, W. A., Monson, W., & Svoboda, L. (1990). Attribute error and sensitivity analysis of map operations in geographical informations systems: Suitability analysis. *International Journal of Geographical Information Systems, 4*, 413–428. https://doi.org/10.1080/02693799008941556
32. Sadat-Noori, M., & Ebrahimi, K. (2016). Groundwater vulnerability assessment in agricultural areas using a modified DRASTIC model. *Environmental Monitoring and Assessment, 188*, 1–18. https://doi.org/10.1007/s10661-015-4915-6

Analysis of Rainfall Data of Agartala Sadar, Tripura Using Statistical Approach

N. Vivekanandan, C. Srishailam, and R. G. Patil

Abstract Analysis of trend in rainfall is one of the important aspects to study about the temporal and spatial variations in nature, and assessment of extreme rainfall is a pre-requisite for planning, design and management of civil and hydraulic structures. For this purpose, Mann–Kendall (MK) test is applied for detecting a trend in rainfall data whereas Extreme Value Analysis (EVA) by fitting probability distributions such as Extreme Value Type-1, 2-parameter log normal and log Pearson type-3 (LP3) to the annual maximum rainfall series are adopted for rainfall estimation. Based on the variate under consideration, the method of moments and maximum likelihood method are used for determination of parameters of the distributions. The adequacy of fitting distributions adopted in EVA is evaluated by quantitative assessment through Goodness-of-Fit (viz., Chi-square and Kolmogorov-Smirnov), diagnostic (viz., D-index) tests and qualitative assessment with the fitted curves of the estimated rainfall. In this paper, a study on analysis of rainfall data for Agartala Sadar, Tripura using statistical approach is carried out. The MK test results witnessed that there is an increasing trend in annual maximum rainfall (AMR) and annual total rainfall (ATR) series with a rate of 5.51 mm/decade for AMR and 20.32 mm/decade for ATR. The results of quantitative and qualitative assessments showed that the LP3 (MLM) is a better suited amongst three distributions adopted in EVA for rainfall estimation.

Keywords Chi-square · D-index · Extreme Value Type-1 · Kolmogorov-Smirnov · Log Normal · Log Pearson Type-3 · Mann–Kendall · Rainfall · Trend analysis

1 Introduction

Rainfall is one of the most important natural input resources that help us to study about the temporal and spatial variations in nature. Also, it is witnessed that the analysis of trend in rainfall and determination of extreme (i.e. 1-day maximum) rainfall would enhance the management of water resources applications as well as the effective

N. Vivekanandan (✉) · C. Srishailam · R. G. Patil
Central Water and Power Research Station, Pune, Maharashtra, India

A. K. Dikshit et al. (eds.), *Innovative Trends in Hydrological and Environmental Systems*, Lecture Notes in Civil Engineering 234,
https://doi.org/10.1007/978-981-19-0304-5_2

utilization of water resources. Apart from this, estimation of extreme rainfall for a given return period is a pre-requisite for planning, design and management of various hydraulic structures, viz. dams, bridges, barrages, storm water drainage systems, etc. In this context, CWPRS carried out the study on analysis of rainfall data of Agartala Sadar, Tripura, using statistical approach for Agartala Smart City Limited (ASCL), which is constituted by Agartala Municipal Corporation (AMC) for implementation of Smart City Mission (SCM) for Agartala. The study carried out for ASCL consists of two parts, viz. (i) Trend analysis of Annual 1-day Maximum Rainfall (AMR) and Annual Total Rainfall (ATR) using Mann–Kendall (MK) test and (ii) Extreme Value Analysis (EVA) of rainfall using Extreme Value Type-1 (EV1), 2-parameter log-normal (LN2) and log Pearson Type-3 (LP3) distributions.

2 Literature Review

Trend analysis of a time series consists of the magnitude of trend and its statistical significance. In general, a non-parametric Mann–Kendall (MK) test is employed for identifying trends in time series data. The MK test compares the relative magnitudes of sample data rather than the data values themselves. The advantages of the MK test are (i) its low sensitivity to abrupt breaks due to in homogeneous time series and (ii) the data need not conform to any particular distribution [1]. Mann [2] originally used this test and Kendall [3] subsequently derived the distribution of MK test statistic. A number of studies have been carried out by different researchers to analyse the trends in rainfall and climatological variables using MK test. Immerzeel [4] studied the historical climate variation and future climate change for the entire Brahmaputra Basin. Joshi and Pandey [5] observed that no trend in annual precipitation over the entire Indian Territory and specific Indian regions for a study period of 100 years (1901–2000). Mondal et al. [6] have analysed the mean monthly precipitation data for the period 1971–2010 along Birupa River in Orissa, India, using MK test. They have found that there is either non-significant increase or decrease in precipitation trend in the various months of the year. Ganguly et al. [7] stated that there is a significant decreasing trend at 5% level of significance in the month of February for Kangra. They have expressed that there is a significant negative trend in annual rainfall data for Dharamshala. Study by Sharma [8] revealed that a statistically vital decreasing trend in annual mean rainfall of Punjab region. Li et al. [9] investigated the abstraction and temporal pattern of rain in reason of the Himalayas that was supported by four sets, viz. interpolated gridded data based on gauge observations, reanalysis data, and high-resolution simulation by a regional climate model. Praveen et al. [10] applied MK test and Sen's innovative trend analysis to analyse and forecast the long-term spatio-temporal changes in rainfall using the data from 1901 to 2015 across India at meteorological divisional level. Patel et al. [11] applied MK test to identify the trends in seasonal and annual rainfall data for river Bhogavo in Sabarmati lower basin, Gujarat. In light of the above, in the present study, MK test is applied for detecting the trends in AMR and ATR series.

In addition to trend analysis of rainfall data, the EVA of rainfall using AMR series is carried out. For which, Extreme Value Type-1 (EV1), 2-parameter log Normal (LN2) and log Pearson Type-3 (LP3) are widely adopted in EVA [12–14]. AlHassoun [15] carried out a study on developing empirical formula to estimate rainfall intensity in Riyadh region using Gumbel (also known as EV1), LN2 and LP3. He concluded that the LP3 distribution gives better accuracy amongst three distributions studied in estimation of rainfall intensity. Esteves [16] applied EV1 distribution to estimate the extreme rainfall depths at different rain-gauge stations in the southeast United Kingdom. Rasel and Hossain [17] applied EV1 distribution for development of intensity–duration–frequency curves for seven divisions in Bangladesh. Afungang and Bateira [18] applied EV1 distribution to estimate the maximum amount of rainfall for different periods in the Bamenda mountain region, Cameroon. Arvind et al. [19] stated that the EV1 is a better suited amongst five probability distributions (viz. normal, LN2, Pearson Type-3, LP3 and EV1) for analysing the annual and monthly rainfall data for Musiri region, Tiruchirappalli, Tamil Nadu. Esberto [20] determined the best-fit frequency distribution of rainfall patterns for event forecasting in order to address potential disasters using sixty probability distributions. Baghel et al. [21] applied eight probability distributions for frequency analysis of daily maximum rainfall data of Udaipur district. They have also found that LN2 and EV1 distributions are the best-fit distributions for a given sample population. Studies carried out by Sasireka et al. [22] indicated that the extreme rainfall for various return periods obtained from EV1 distribution could be used for design purposes by considering the risk involved in the operation and management of hydraulic structures in Tiruchirappalli region. Vivekanandan and Srishailam [23] compared the MoM and MLM estimators of EV1, LN2 and LP3 distributions applied in EVA of rainfall for Anakapalli, Atchutapuram, Kasimkota and Parvada rain-gauge sites. They have also found that the LP3 (MLM) is better suited distribution for EVA of rainfall for Anakapalli, whereas LN2 (MLM) for Kasimkota and EV1 (MLM) for Atchutapuram and Parvada. However, when number of probability distributions adopted in EVA of rainfall, a common problem that arises is how to determine which distribution model fits best for a given set of data. This possibly could be answered by quantitative and qualitative assessments, and the results are quantifiable and reliable. The adequacy of fitting of the probability distributions adopted in EVA of rainfall is evaluated by quantitative assessment through Goodness-of-Fit (GoF) (viz. Chi-Square (χ^2) and Kolmogorov-Smirnov (KS)) and diagnostic (viz. D-index) tests, and qualitative assessment with the fitted curves of the estimated extreme rainfall. This paper details a study on analysis of rainfall data of Agartala Sadar, Tripura, using statistical approach and the results obtained thereof.

3 Methodology

The methodology adopted in trend analysis of rainfall using MK test, EVA of rainfall using probability distributions, computation of GoF and diagnostic tests are briefly described in the following sections.

3.1 Trend Analysis

The Mann–Kendall (MK) test statistic (S) is defined by:

$$S = \sum_{i=1}^{N-1} \sum_{j=i+1}^{N} \text{sgn}(x(j) - x(i)) \tag{1}$$

where $x(i)$ and $x(j)$'s are the sequential data values and N is the total number of data points in the set. This statistic represents the number of positive differences minus the number of negative differences for all the differences considered. If the variable constantly decreases downward, then S is to be an indication of decreasing trend and the value (S) would be negative, whereas constantly increasing dataset shows a positive (S) value or increasing trend. For large samples ($N > 10$), the test is conducted using a normal distribution with the mean and the variance as follows:

$$E(S) = 0 \text{ and } \text{Var}(S) = (1/18)\left[N(N-1)(2N+5) - \sum_{k=1}^{n} t_k(t_k - 1)(2t_k + 5)\right] \tag{2}$$

where n is the number of tied (zero difference between compared values) groups and t_k the number of data points in the kth tied group. The standard normal deviate (Z-statistic) is then computed as:

$$Z = \begin{cases} \frac{S-1}{\sqrt{\text{Var}(S)}} & if \quad S > 0 \\ 0 & if \quad S = 0 \\ \frac{S+1}{\sqrt{\text{Var}(S)}} & if \quad S < 0 \end{cases} \tag{3}$$

If the computed value of $|Z| > Z_{\alpha/2}$, then the null hypothesis is rejected at the level of significance (α) in a two-sided test. Significance level (α) is used for testing either an increasing or decreasing monotonic trend (a two-tailed test).

3.2 Extreme Value Analysis

The cumulative function (CDF) and quantile estimator of EV1, LN2 and LP3 distributions adopted in EVA are presented in Table 1. The parameters of the distributions are determined by method of moments (MoM) and maximum likelihood method (MLM) [24] and also used for estimation of extreme (i.e. 1-day maximum) rainfall for different return periods. The procedures involved in determination of parameters of EV1 [25] and LN2 [26] are presented in Table 2.

Table 1 CDF and quantile estimator of EV1, LN2 and LP3 distributions

Distribution	CDF	Quantile estimator ($x(T)$)
EV1	$F(x) = \exp\left(-\exp -\left(\frac{x-c}{a}\right)\right)$	$x(T) = c\text{-}a\ln(\text{-}\ln(F(x))$
LN2	$F(y) = \varphi\left(\frac{y-\mu(y)}{\sigma(y)}\right)$ where in $y = \ln(x)$	$x(T) = \exp(\mu(y) + K(T)\sigma(y))$
LP3	$F(x) = \begin{cases} G\left(b, \frac{\ln(x)-c}{a}\right), & a > 0 \\ 1 - G\left(b, \frac{\ln(x)-c}{a}\right), & a < 0 \end{cases}$	No explicit expression of the quantile function is available

In Tables 1 and 2, c is the location parameter, a is the scale parameter, b is the shape parameter, $F(x)$ (or F) is the CDF of x (i.e. AMR), $F(y)$ is the CDF of y, ϕ is the standard normal distribution function, $G(\ldots)$ is the incomplete gamma integral, $\mu(x)$ and $\sigma(x)$ are the average and standard deviation of the observed data (x), $\mu(y)$ and $\sigma(y)$ are the average and standard deviation of the log-transformed series of the observed data (i.e. $y = \ln(x)$), $k(T)$ is the frequency factor of a return period (T) corresponding to coefficient of skewness (CS) (viz. CS = 0.0) and $x(T)$ is the estimated rainfall (x) for a return period (T). If $(1-p)$ is the probability of non-exceedance in a year then the return period (T) is defined by $T = 1/(1-p)$ and a relation between F and T is defined by F (or $F(x)$) $= 1-1/T$.

MoM of LP3 distribution

The MoM estimators of LP3 distribution [27] can be determined by solving the following system of equations:

$$\mu_r' = \frac{\exp(rc)}{(1-ra)^b} \text{ where in } 1 - ra > 0, \quad r = 1,\ 2,\ 3 \tag{4}$$

$$\mu(y) = \mu_1', \sigma(y)^2 = \mu_2' - \left(\mu_1'\right)^2 \text{ and } \gamma(y) = \mu_3' - 3\mu_2'\mu_1' + 2(\mu_1')^3 \tag{5}$$

where μ_r' is the rth moment of x about the origin. If $a > 0$, then $\gamma(y) > 0$; therefore, y must be positively skewed such that $f(y)$ is lower bounded ($c \le y < +\infty$). In this case, x must also be positively skewed; thus, x also possesses a lower bound [$\exp(c) \le x < +\infty$]. When $a < 0$, then $\gamma(y) < 0$, such that y is negatively skewed, such that $f(y)$ is upper bounded ($-\infty < y \le c$). In this case, x either can be positively or negatively skewed, depending on the values of the parameters a and b, but x is upper bounded [$0 < x < \exp(c)$]. For this case, the density function $f(x)$ may be defined as zero at $x = 0$.

MLM of LP3 distribution

The MLM estimators of LP3 distribution [27] can be determined by solving the following system of equations:

Table 2 Determination of MoM and MLM estimators of EV1 and LN2 distributions

Distribution	Estimators of the parameters of EV1 and LN2		
	Location	Scale	Shape
Method of Moments			
EV1	$c = \mu(x) - (0.5772157)a$	$a = \left(\sqrt{6}/\pi\right)\sigma(x)$	–
LN2	–	$\mu(y) = \left(\frac{1}{N}\sum_{i=1}^{N} y(i)\right) - \frac{1}{2}\left(\frac{1}{N}\sum_{i=1}^{N} y(i)^2\right) + \frac{1}{2}\left(\frac{1}{N}\sum_{i=1}^{N} y(i)\right)^2$ wherein $y(i) = \ln(x(i))$	$\sigma(y) = \left[\left(\frac{1}{N}\sum_{i=1}^{N} y(i)^2\right) - 2\left(\frac{1}{N}\sum_{i=1}^{N} y(i)\right)\right]^{1/2}$
Maximum Likelihood Method			
EV1	$c = -a\ln\left(\frac{\sum_{i=1}^{N}\exp(-x(i)/a)}{N}\right)$	$a = \mu(x) - \frac{\sum_{i=1}^{N} x(i)\exp(-x(i)/a)}{\sum_{i=1}^{N}\exp(-x(i)/a)}$	–
LN2	–	$\mu(y) = \frac{1}{N}\sum_{i=1}^{N} y(i)$	$\sigma(y) = \left(\frac{1}{N}\sum_{i=1}^{N}(y(i) - \mu(y))^2\right)^{1/2}$

$$\left.\begin{aligned}\sum_{i=1}^{N} y(i) - c &= Nab \\ \sum_{i=1}^{N} \frac{1}{a}(y(i) - c) &= N\psi(b) \\ \sum_{i=1}^{N} \frac{1}{y(i) - c} &= \frac{N}{a(b-1)}\end{aligned}\right\} \tag{6}$$

Here, $\psi(b)$ is the digamma function of estimator of the scale parameter (b).

3.3 Goodness-Of-Fit Tests

A number of GoF tests are available for checking the adequacy of probability distributions adopted in EVA. From this, the widely accepted GoF tests are χ^2 and *KS*, and hence, considered in the present study. Theoretical descriptions of GoF tests [28, 29] statistic are presented in Table 3.

Test criteria: If the computed values of GoF tests statistic given by the probability distribution are less than its theoretical values at the desired level of significance, then the distribution is acceptable for EVA at that level.

Table 3 Theoretical descriptions of GoF tests statistic

GoF test	Test statistic	Description of symbols
$\chi 2$	$\chi_c^2 = \sum_{j=1}^{NC} \frac{(O_j(x) - E_j(x))^2}{E_j(x)}$	m = Number of parameters of distribution
		χ_c^2 = Computed value of χ^2 statistic by probability distribution
		$O_j(x)$ = Observed frequency value (x) of jth class
		$E_j(x)$ = Expected frequency value (x) of jth class
		NC = Number of frequency classes
KS	$KS = \text{Max} \sum_{i=1}^{N} \lvert F_e(x(i)) - F_D(x(i)) \rvert$	$F_e(x(i))$ = Empirical CDF of $x(i)$ using Weibull plotting position formula ($P = r/(N + 1)$), wherein r is the rank assigned to the samples arranged in ascending order (i.e. $x(1) < x(2) < x(3).....x(N)$)
		$F_D(x(i))$ = Derived CDF of $x(i)$ by probability distribution
		N = Sample size

3.4 Diagnostic Test

A diagnostic test, viz. *D-index* is applied for the selection of a suitable probability distribution adopted in EVA for estimation of rainfall. The theoretical expression of *D-index* [30] is given as below:

$$D - \text{index} = (1/\mu(x)) \sum_{i=1}^{6} \left| x(i) - x(i)^* \right| \tag{7}$$

Here, $x(i)(i = 1\text{–}6)$ and $x(i)^*$ are the six highest observed and corresponding estimated rainfall by the probability distribution. The distribution with the lowest *D-index* is considered as a better suited distribution in comparison with the other distributions adopted in EVA for estimation of rainfall.

4 Application

Agartala is the capital city of the Indian state of Tripura, and is the second-largest city in northeast India after the State Guwahati. The city is governed by the Agartala Municipal Corporation. Agartala city is located on the banks of the Howrah, near the Bangladesh border, about 90 km east of Bangladesh's capital Dhaka. Agartala is being developed under the smart cities Mission, a flagship scheme of the Government of India.

The river Howrah is one of the major rivers in Tripura which flows through Agartala and Sadar sub-division of the western district of Tripura. This is also called as Saidra in Kokborok by the original inhabitants of the state. The Saidra or the Howrah originates from the Boromura hills in central Tripura, and flows through the foothills passing through important towns like Champaknagar, Jirania, Khumulwng, Khayerpur and the capital city Agartala before reaching Bangladesh. From the topography of Howrah river basin, it is observed that about 30% of the area falls in hilly terrain, 20% is under undulating and 50% is under plain land. The total flow length of the Howrah River in the state is about 61.2 km of which 52 km is flowing within Indian Territory. The catchment area of the river is about 457.92 km^2. The index map of the study area is presented in Fig. 1.

For the present study, the daily rainfall data of Agartala Sadar IMD rain-gauge station for the period from 1901 to 2014 is used. From the scrutiny of rainfall data, it is found that the rainfall data for the years 1952, 1954, 2004 and 2005 are missing. However, by considering the importance of the hydrological extremes, the data for the missing years are ignored and not considered in EVA.

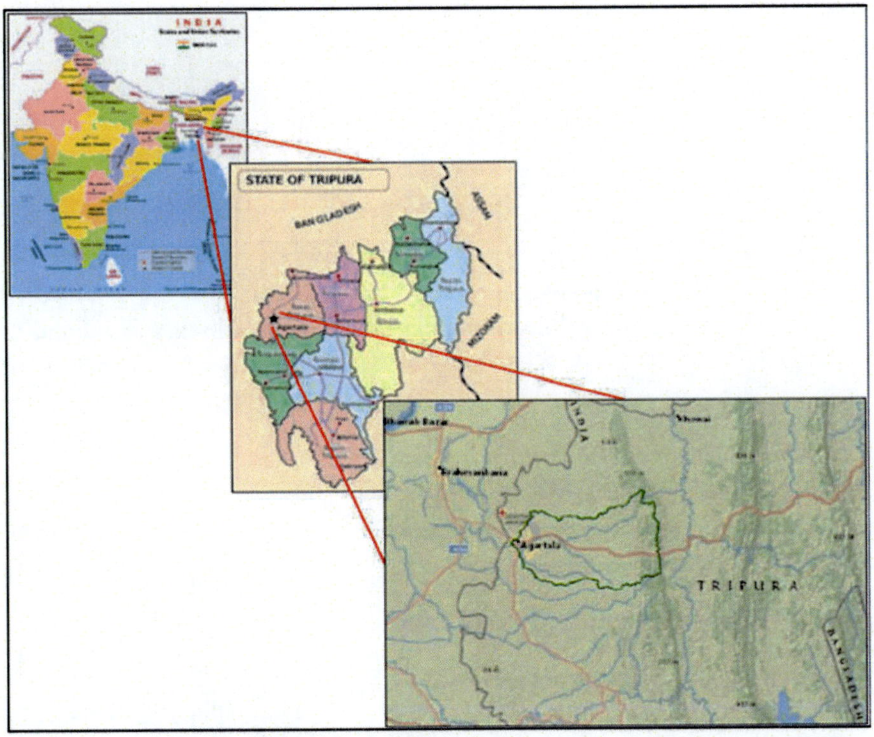

Fig. 1 Index map of the study area

5 Results and Discussions

By applying the procedures of trend analysis and EVA, as described above, a computer code was developed and used. The results obtained from trend analysis and EVA of rainfall are summarized and presented in the following sections.

5.1 Trend Analysis of Rainfall

The AMR and ATR series were extracted from the daily rainfall data and used in trend analysis for detecting the presence of statistically significant trend in rainfall data with reference to climate change. The statistical results indicated that there is a significant trend in AMR series at 5% level of significance, and also the graphical results (Fig. 2) showed that the trend is increasing at the rate of 5.51 mm/decade. Likewise, for ATR series, it was found that there is no significant trend at 5% level of significance. However, the graphical results (Fig. 3) witnessed that the trend is increasing at the rate of 20.32 mm/decade.

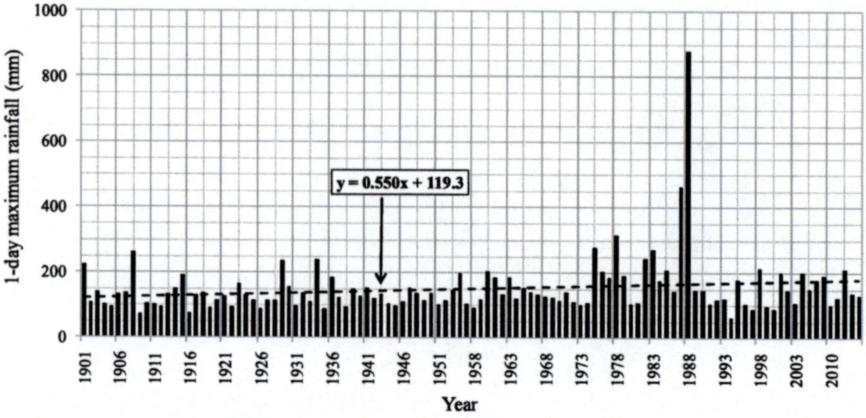

Fig. 2 Trend analysis of annual 1-day maximum rainfall

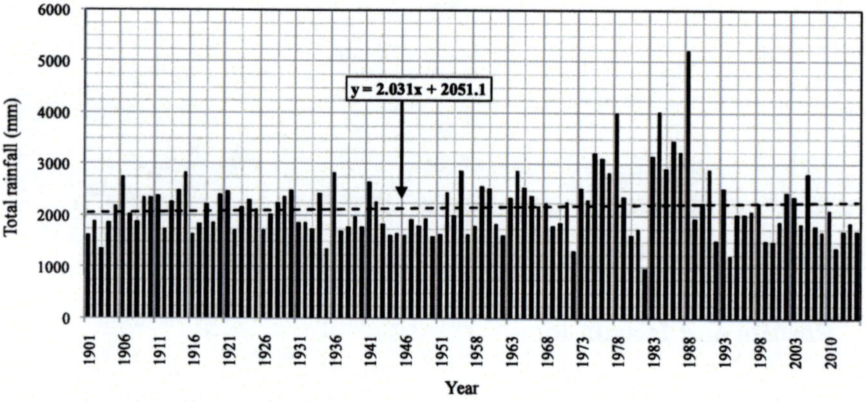

Fig. 3 Trend analysis of annual total rainfall

5.2 *EVA of Rainfall*

For EVA of rainfall, the AMR series for the period 1901–2014 (except for the year 1952, 1954, 2004 and 2005) were extracted from the daily rainfall data and used. For which, the parameters of the EV1, LN2 and LP3 distributions were determined by MoM and MLM, and also used for estimation of extreme (i.e. 1-day maximum) rainfall for different return periods. The EVA results of rainfall are presented in Table 4, while the plots are shown in Fig. 4. From EVA results of rainfall (Table 4), it is noted that (i) the rainfall estimates given by LP3 (MLM) distribution is higher than those values of EV1 and LN2 for the return periods from 10-year and above; and (ii) MoM gave higher estimates than MLM (for the return periods from 5-year and above), while the parameters of the EV1 and LN2 are determined by MoM and MLM.

Table 4 Estimated rainfall for different return periods using EV1, LN2 and LP3 distributions

Return period (year)	1-day maximum rainfall (mm)					
	EV1		LN2		LP3	
	MoM	MLM	MoM	MLM	MoM	MLM
2	135.4	137.6	128.8	137.4	125.4	145.5
5	215.0	183.7	205.4	188.5	178.1	208.7
10	267.8	214.1	262.1	222.4	227.2	276.3
20	318.4	243.4	320.6	254.9	287.1	366.9
25	334.5	252.7	340.0	265.2	309.2	402.2
50	383.9	281.3	402.3	297.2	387.9	535.7
100	433.0	309.6	467.9	329.3	484.9	714.2
200	481.9	337.9	537.4	361.6	604.4	953.3
250	497.6	347.0	560.6	372.1	648.5	1046.4
500	546.4	375.2	635.5	405.1	806.1	1398.0
1000	595.2	403.4	714.9	438.7	1000.3	1868.9

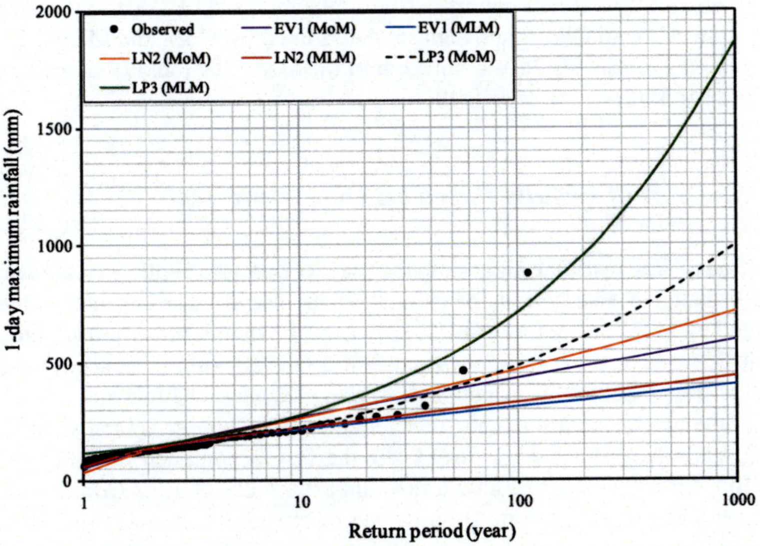

Fig. 4 Plots of estimated 1-day maximum rainfall using MoM and MLM estimators of EV1, LN2 and LP3 distributions with observed AMR data

From Fig. 4, it can be seen that the fitted lines of the estimated rainfall by EV1 and LN2 (using MoM and MLM) are in the form of linear curve, while the fitted lines using LP3 (using MoM and MLM) are in the form of exponentially growth curve.

Table 5 Computed values of GoF tests statistic by EV1, LN2 and LP3 with theoretical values

GoF tests	Theoretical value at 5% level			Computed values of GoF tests					
				EV1		LN2		LP3	
	EV1	LN2	LP3	MoM	MLM	MoM	MLM	MoM	MLM
χ^2	18.307	18.307	16.919	67.746	68.762	55.927	57.325	9.364	9.645
KS	0.130	0.130	0.130	0.216	0.219	0.205	0.210	0.057	0.059

5.3 Analysis of Results Based on GoF Tests

The adequacy of fitting probability distributions adopted in EVA is evaluated by GoF tests (viz. χ^2 and *KS*), as described earlier. In the present study, the number of frequency class (*NC*) is considered as thirteen while the degree of freedom (*NC-m-1* wherein *m* is the number of parameters of the distribution) was considered as nine for LP3, whereas ten for EV1 and LN2. The GoF tests results are presented in Table 5.

From the GoF tests results, the observations are drawn from the study:

(i) χ^2 and *KS* tests results confirmed the applicability of LP3 distribution for EVA of rainfall while the parameters are determined by MoM and MLM.
(ii) χ^2 and *KS* tests results did not support the use of EV1 and LN2 (using MoM and MLM) for EVA of rainfall.

5.4 Analysis of Results Based on Diagnostic Test

In addition to GoF tests, the diagnostic test (say, *D-index*) is applied for the selection of a best suitable distribution for estimation of rainfall. The *D-index* values are computed by MoM and MLM of EV1, LN2 and LP3 distributions and are presented in Table 6. Based on the *D-index* values, the rank (shown within brackets) is assigned to each of the probability distributions and also presented in Table 6.

From the diagnostic test results, it is noted that the *D-index* values given by LP3 (MoM), LN2 (MoM) and EV1 (MoM) are the first, second and third minimum in order. But, the GoF tests results did not support the use of LN2 (MoM) and EV1 (MoM) distributions though its D-index values are minimum next to LP3 (MoM). Moreover, the research reports indicated that the MoM estimators of the distributions are often less accurate than those values of MLM. Hence, by eliminating the *D-index*

Table 6 D-index values given by EV1, LN2 and LP3 distributions

Diagnostic test	EV1		LN2		LP3	
	MoM	MLM	MoM	MLM	MoM	MLM
D-index	4.868 (3)	5.682 (6)	4.643 (2)	5.100 (5)	3.768 (1)	5.032 (4)

values of LP3 (MoM), LN2 (MoM) and EV1 (MoM) from the selection, it is found that the *D-index* value of LP3 (MLM) is the next minimum in the array of *D-index* values. In view of the above, it is identified that the LP3 (MLM) is a better suited amongst three distributions adopted in EVA for rainfall estimation.

The estimated rainfall by LP3 (MLM) for different return periods with 95% confidence limits are presented in Table 7 while the plots are shown in Fig. 5. From Fig. 5, it can be seen that about 90% of the observed AMR data are within the confidence limits of the estimated rainfall using LP3 (MLM) distribution.

Table 7 Estimated rainfall with 95% confidence limits by LP3 (MLM) distribution

Return period (year)	Estimated rainfall (mm)	95% confidence limits (in mm)	
		Lower	Upper
2	145.5	120.1	170.9
5	208.7	160.0	257.4
10	276.3	200.3	352.2
25	402.2	254.5	550.0
50	535.7	300.4	770.9
100	714.2	350.0	1078.5
200	953.3	425.3	1481.4
500	1398.0	575.0	2221.0
1000	1868.9	750.2	2987.5

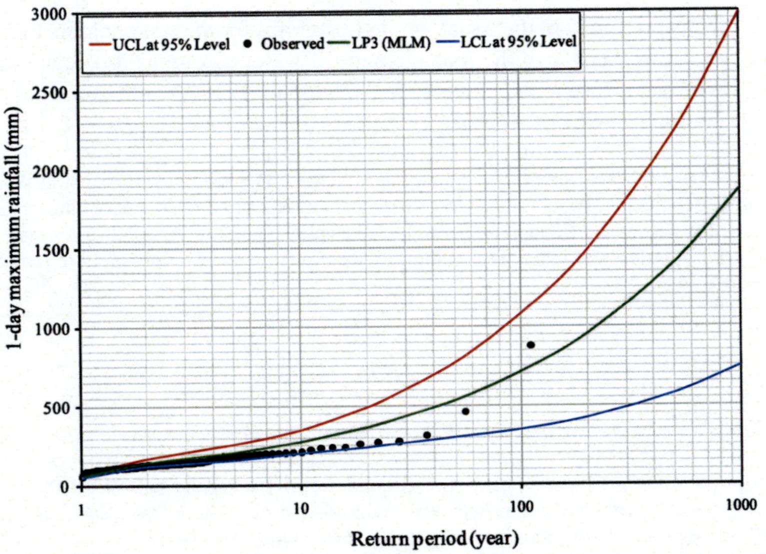

Fig. 5 Plots of estimated extreme rainfall with 95% confidence limits using LP3 (MLM) and observed AMR data

6 Conclusions

The paper presented a study on analysis of rainfall data of Agartala Sadar, Tripura, using statistical approach that includes trend analysis of AMR and ATR using Mann–Kendall (MK) test, and EVA of rainfall using EV1, LN2 and LP3 distributions. The parameters were determined by MoM and MLM, and also used for rainfall estimation. The adequacy of fitting probability distributions in EVA was evaluated by GoF (viz. χ^2 and *KS*) and diagnostic (viz. *D-index*) tests. On the basis of evaluation of the results through quantitative and qualitative assessments, some of the conclusions drawn from the study were summarized and are presented below:

(i) For AMR series, the results of trend analysis ascertained that there is a significant trend in AMR series at 5% level of significance, which is increasing at the rate of 5.51 mm/decade.
(ii) For ATR series, the results showed that there is no significant trend at 5% level of significance. But, the graphical results witnessed that the trend is increasing at the rate of 20.32 mm/decade.
(iii) χ^2 and *KS* tests results supported the use of LP3 distribution for EVA of rainfall while the parameters are determined by MoM and MLM.
(iv) The qualitative assessment (plots of EVA results) of the outcomes was weighed with *D-index* values indicated that the LP3 (MLM) is better suited amongst three distributions adopted in EVA for rainfall estimation.
(v) The fitted lines of the rainfall frequency curves indicated that about 90% of observed AMR data are within the confidence limits of the estimated rainfall by LP3 (MLM).

The results presented in the paper would be helpful for the stakeholders to resolve various issues, viz. enhance the management of water resources projects, effective utilization of water resources, planning and design of hydraulic structures regarding implementation of SCM of Agartala.

Acknowledgements The authors are grateful to Shri A.K. Agrawal, Director, Central Water and Power Research Station, Pune for encouragement and guidance during conduct of the studies and also according permission to publish this paper. The authors are also thankful to Agartala Smart City Limited, Agartala and India Meteorological Department, Pune for the supply of rainfall data used in the study.

References

1. Jaagus, J. (2006). Climatic changes in Estonia during the second half of the 20th century in relationship with changes in large-scale atmospheric circulation. *Theoretical and Applied Climatology, 83*, 77–88. https://doi.org/10.1007/s00704-005-0161-0
2. Mann, H. B. (1945). Nonparametric tests against trend. *Econometrica, 13*, 245–259. https://doi.org/10.2307/1907187
3. Kendall, M. G. (1975). Rank correlation methods, London

4. Immerzeel, W. (2008). Historical trends and future predictions of climate variability in the Brahmaputra basin. *International Journal of Climatology., 28*, 243–254. https://doi.org/10.1002/joc.1528
5. Joshi, M. K., & Pandey, A. C. (2011). Trend and spectral analysis of precipitation over India during 1901–2000. *Journal of Geophysical Research: Atmospheres, 116*, 1–9. https://doi.org/10.1029/2010JD014966
6. Mondal, A., Kundu, S., Mukhopadhyay, A.: Rainfall trend analysis by Mann-Kendall Test: A case study of North-Eastern part of Cuttack district, Orissa. *International Journal of Geology, Earth and Environmental Sciences, 2*, 70–81 (2012). http://www.cibtech.org/jgee.htm/Mondal etal
7. Ganguly A., Chaudhuri R. R., Sharma P. (2015). Analysis of trend of the precipitation data: A case study of Kangra district, Himachal Pradesh. *International Journal of Research-Granthaalayah, 3*, 87–95. https://doi.org/10.29121/granthaalayah.v3.i9.2015.2948
8. Sharma, M. R. (2017). Trend and variability of rainfall in Punjab: A statistical analysis, 1981–2010. *Research Guru Online Journal of Multidisciplinary subjects., 11*, 73–81.
9. Li, H., Haugen, J. E., & Xu, C. Y. (2018). Precipitation pattern in the Western Himalayas revealed by four datasets. *Hydrology and Earth System Science., 22*, 5097–5110. https://doi.org/10.5194/hess-22-5097-2018
10. Praveen, B., Talukdar, T., Shahfahad, Mahato, S., Mondal, J., Sharma, P., Islam, A. R. M. T., Atiqur Rahman, A. (2020). Analyzing trend and forecasting of rainfall changes in India using non-parametrical and machine learning approaches. *Scientific Reports (Nature Research), 10*, 10342. https://doi.org/10.1038/s41598-020-67228-7
11. Patel, P. S., Rana, S. C., & Joshi, G. S. (2021). Temporal and spatial trend analysis of rainfall on Bhogavo River watersheds in Sabarmati lower basin of Gujarat India. *Acta Geophysica, 69*, 353–364. https://doi.org/10.1007/s11600-020-00520-2
12. Mujere, N. (2011). Flood frequency analysis using the Gumbel distribution. *International Journal on Computer Science and Engineering., 3*, 2774–2778.
13. Sharma, N. K., Sharma, S. (2019). Frequency analysis of rainfall data of Dharamshala region. *International Journal of Science and Research, 8*, 886–892. https://www.ijsr.net/paperid.php?id=ART20195211
14. Tank, G., Dongre, P., Obi Reddy, G. P., & Sen, P. (2021). Rainfall trend analysis—A review. *International Research Journal of Engineering and Technology., 8*, 4028–4030.
15. AlHassoun, S. A. (2011). Developing empirical formulae to estimate rainfall intensity in Riyadh region. *Journal of King Saud University-Engineering Sciences., 23*, 81–88. https://doi.org/10.1016/j.jksues.2011.03.003
16. Esteves, L. S. (2013). Consequences to flood management of using different probability distributions to estimate extreme rainfall. *Journal of Environmental Management, 115*, 98–105. https://doi.org/10.1016/j.jenvman.2012.11.013
17. Rasel, M., & Hossain, S. M. (2015). Development of rainfall intensity duration frequency equations and curves for seven divisions in Bangladesh. *International Journal of Scientific and Engineering Research, 6*, 96–101.
18. Afungang, R., Bateira, C. (2016). Statistical modelling of extreme rainfall, return periods and associated hazards in the Bamenda Mountain. *NW, 1*, 5–19. https://doi.org/10.17127/got/2016.9.001
19. Arvind, G., Kumar, P. A., GirishKarthi, S., Suribabu, C. R. (2017). Statistical analysis of 30 years rainfall data: A Case Study. *Proceedings of IOP Conference. Series: Earth and Environmental Science*, Paper ID: 012067. https://doi.org/10.1088/1755-1315/80/1/012067
20. Esberto, M. D. P. (2018). Probability distribution fitting of rainfall patterns in Philippine regions for effective risk management. *Environment and Ecology Research, 6*, 178–186. https://doi.org/10.13189/eer.2018.060305
21. Baghel, H., Mittal, H. K., Singh, P. K., Yadav, K. K., Jain, S. (2019). Frequency analysis of rainfall data using probability distribution models. *International Journal of Current Microbiology and Applied Sciences, 8*, 1390–1396. https://doi.org/10.20546/ijcmas.2019.806.168

22. Sasireka, K., Suribabu, C. R., Neelakantan, T. R. (2019). Extreme rainfall return periods using Gumbel and Gamma Distributions. *International Journal of Recent Technology and Engineering, 8*, 27–29. https://doi.org/10.35940/ijrte.D1007.1284S219
23. Vivekanandan, N., & Srishailam, C. (2020). Selection of best fit probability distribution for extreme value analysis of rainfall. *CBIP Water and Energy International., 63*, 13–19.
24. Rao, A. R., & Hamed, K. H. (2000). *Flood frequency analysis.* CRC Publications.
25. Gumbel, E. J. (1985). *Statistic of extremes.* Columbia University Press.
26. Vivekanandan, N. (2020). Comparison of LN2 and WB2 probability distributions for low-flow frequency analysis. *International Journal of Emerging Engineering Research and Technology, 8*, 39–45.
27. Singh, V. P., & Singh, K. (1988). Parameter estimation for Log-Pearson Type-III distribution by POME. *Journal of Hydraulic Engineering, 114*, 112–122.
28. Zhang, J. (2002). Powerful goodness-of-fit tests based on the likelihood ratio. *Journal of Royal Statistical Society, 64*, 281–294. https://doi.org/10.1111/1467-9868.00337
29. Charles Annis, P.E. 2009. Goodness-of-Fit tests for statistical distributions.
30. United States Water Resources Council (USWRC). (1981). Guidelines for determining flood flow frequency, Bulletin 17A, U.S Geological Survey, Washington, D.C.

Geo-Spatial Analysis of October 2020 Hyderabad Flood

Vinay Ashok Rangari, C. M. Bhatt, Ajey Kumar Patel, and N. V. Umamahesh

Abstract Indian cities are exposed to increasing threat of urban flooding due to rapid urbanization and climate change. As a result, life in the affected city comes to a halt with power cuts, roads jammed, houses inundated, accidents and loss of life. A similar event occurred on October 2020 has paused the life in Hyderabad city affecting more than thousands of houses and 19 casualties. To capture the flooding extent and severity of this event, in present study we have employed HEC-RAS 2D simulations for urban catchment of Hyderabad city. The model simulation results are used to generate spatial flooding extent and develop flood risk map based on simulated flooding depth for Hyderabad city. The analysis of results accounts total areal flooding extent as 58% (658 km^2) for October 2020 event with 2–3 m deluge depth at some parts of the city. Further the model identifies respectively 47.44%, 38.37% and 13.69% areas falling under low, medium and high potential for risk. Therefore, to manage and reduce the impacts of urban floods it is must to regulate the urbanization identify the suitable places to divert additional storm water. Further, the natural lakes and flow streams need to bring back to its original state so that they can accommodate the excessive flood volumes.

Keywords Flood · Urbanization · Geo-spatial analysis · HEC-RAS · Modelling

---

V. A. Rangari (✉)
Civil Engineering Department, Nanasaheb Mahadik College of Engineering, Peth Naka, Sangli, Maharashtra 415407, India

C. M. Bhatt
Disaster Management Division, Indian Institute of Remote Sensing, Dehradun 248001, India

A. K. Patel · N. V. Umamahesh
Department of Civil Engineering, National Institute of Technology Warangal, Hanamkonda, Telangana 506004, India

A. K. Dikshit et al. (eds.), *Innovative Trends in Hydrological and Environmental Systems*, Lecture Notes in Civil Engineering 234,
https://doi.org/10.1007/978-981-19-0304-5_3

1 Introduction

Indian cities are exposed to increasing threat of urban flooding. Recent studies have admitted the fact of increase in flood frequencies with intense rainfall in past few years and it will continue to grow in future affecting the new cities [2, 4, 5, 14, 23]. The most known causes are increased and unregulated urban sprawl, rural to urban migration, encroachment of water bodies and flood plains, degraded storm water drainage system, and changed rainfall pattern due to climate change [3, 12, 17, 22]. With each flooding event, the associated damages are also going up due to limitations of available forecast systems, their geo-spatial accuracy, reliability, ignorance and our inability to learn from past [16, 14]. Present day we are having worse circumstance than ever but the hitches are same, i.e. a small rainfall heavy spell is enough to deluge a large area. The year 2020 alone experienced devastating urban floods with water rushing through the streets (Hyderabad, Bharuch, Jaipur, Dehradun, Solapur, Pune to name a few). The trend tails from Mumbai (2005), Ahmedabad (2005), Surat (2006), Kolkata (2007), Hyderabad (2008), Jamshedpur (2008), Delhi (2009), Leh (2010), Guwahati (2010), Haridwar–Rishikesh (2013), Srinagar (2014), Chennai (2015), Patna (2016), Chennai (2017), Mumbai (2017), Kerala (2018), Sangli (2019) and the most recent one Hyderabad 2020. Such events will enforce a great challenge that the city planners and governing authorities have to face towards disaster management and sustainable development. The provisions in standard operating procedure (SOP) to deal with urban floods recommends upgraded drainage system, advance forecasts, and simulation of numerical models to predict flood inundation extent and depth that will help in developing early warning systems and combat strategies [6, 9, 11]. Considering this approach, the present study aims at rapid geo-spatial analysis of October 2020 Hyderabad urban flood to assess spatial flooding extent and develop risk map based on 2D urban flood model simulation results.

Hyderabad, the age old city (approximately 400 years old) has seen many transformations in the landscape over past few decades and continues to grow by way of technological and infrastructural developments. As a result, the city has come forward as one of important engineering and technological hub in recent years. On the other hand, the alterations in natural topography have lessened the soil–water interactions making difficult to absorb and manage the chaotic rainfalls leading to urban floods. Persistent flooding diminishes the quality of life by posing environmental and public health concerns and huge damage to the properties. Table 1 enlists the most recent flood events in the history of Hyderabad city.

The natural topography of Hyderabad has distinctive gradient with some places at higher elevations and sloping down towards Musi River passing through the centre of city. The natural storm water flow from higher elevation to lower elevations was persistent through cascade system of interconnected lakes commonly known as 'Cheruvu' in local language. In monsoon, these Cheruvus were the natural distributers of storm water that protected the Hyderabad from floods in past [20]. Hyderabad has spread in all directions over the years with increase in impervious

Table 1 Important flash floods in recent time in Hyderabad [1, 15]

Event	Total rainfall (mm)	Duration of event (h)
October, 2020	322.2	24
September, 2019	133.0	24
September, 2016	167.2	30
August, 2008	220.7	36
August, 2000	312.1	33

areas more than 50% in last few decades [7, 8]. New roads, buildings and infrastructure have appeared. The persistent development has wiped out the system of natural flow paths connecting the Cheruvus in cascade system. Many Cheruvus went dry without incoming water flows over the years and few have disappeared from the map. Further, the climate change induced depressions and cyclones are filled with excessive rainfall and pouring more rain water over an area in a very short time span [10, 13, 18]. Thus, the altered landscapes, lass of natural water storages, climate change and excessive rainfall are conclusive evidences for the occurrence of urban flooding and the impacts are visible on ground in the form of October 2020 Hyderabad flood.

October 13, 2020: A low pressure region formed in the Bay of Bengal has given rise to exceedingly heavy precipitation that lead to massive floods killing over 70 people in Telangana, Andhra Pradesh and Maharashtra. Hyderabad recorded its rainiest day in 117 years, affecting thousands of houses and 19 casualties [21]. To capture the flooding extent and severity of this event, in present study we have employed HEC-RAS 2D simulations for urban catchment of Hyderabad city. The model simulation results are used to generate spatial flooding extent and develop flood risk map based on simulated flooding depth for Hyderabad city.

2 Study Area

Hyderabad is a metropolitan city of India located at 17.3700 °N, 78.4800 °E. The Greater Hyderabad Municipal Corporation (GHMC) has divided the natural terrain of Hyderabad Urban into sixteen storm water zones based on drainage outlet points and topography. Among these zones, Zone 12—Kukatpally and Zone 13—Alwal and Begumpet are major flood-prone areas. There are several ponds and lakes in the city. Among which Hussain Sagar is the largest lake, covering an area 4.4 km^2, with a maximum depth of 9.8 m. Musi is the only river passing through the centre of the city. Musi is a tributary of Krishna River. It divides the city into old city and new city. Figure 1 shows the location map of study area.

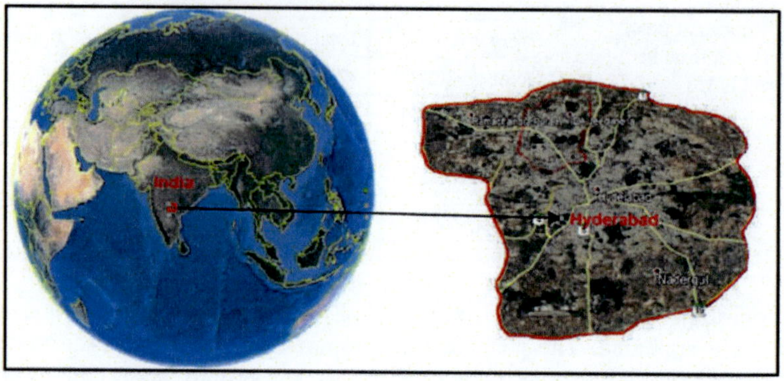

Fig. 1 Study area location map

3 Materials and Methods

A 2D urban flood modelling methodology described in present study is adopted from [15]. Figure 2 shows the flow chart of methodology. HEC-RAS 5.1 with ArcGIS integration is used to develop 2D flood simulation model for Hyderabad urban. The model performs the simulations based on digital terrain model (DTM) (Fig. 3a) that is developed from digital elevation model (DEM) as basic input for model development. The other inputs for the model development are the stream flow pattern,

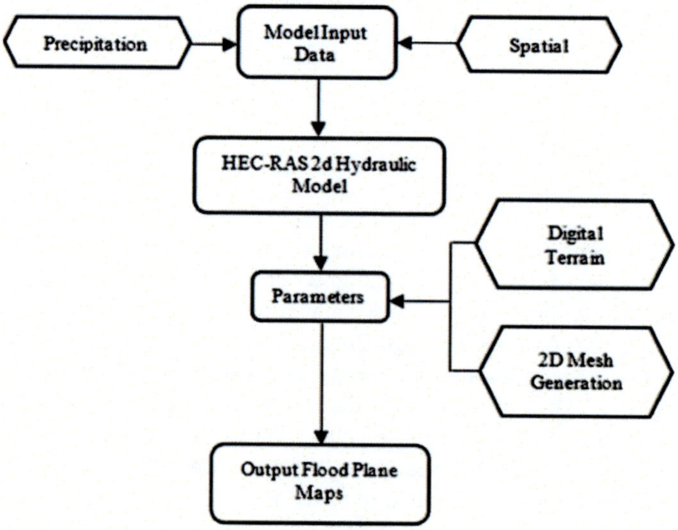

Fig. 2 Flow chart of methodology

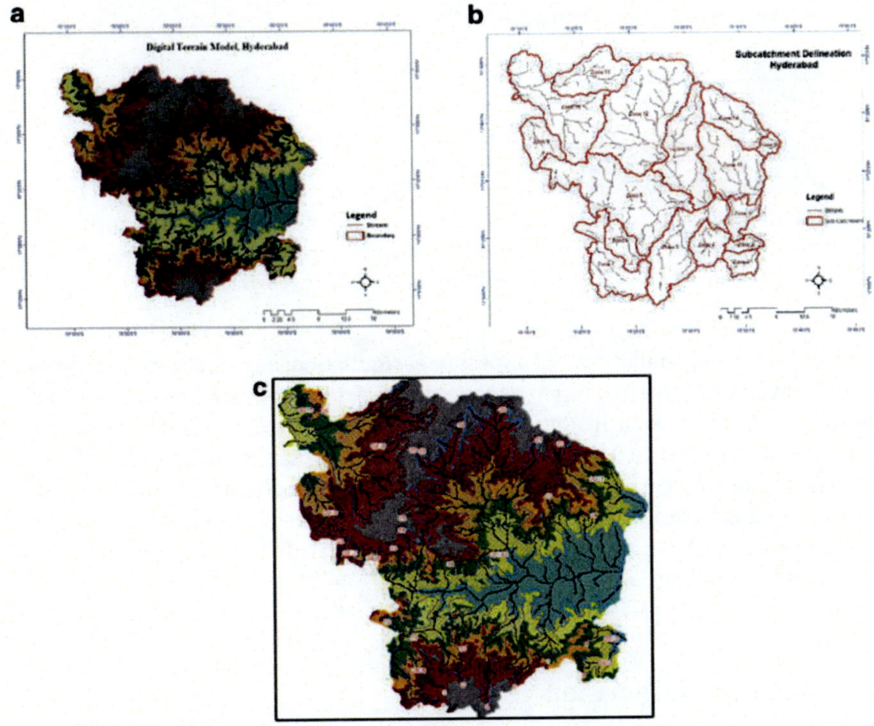

Fig. 3 **a** Digital terrain model (DTM), Hyderabad, and **b** Hyderabad catchment sub-basins, **c** RAS 2D flow area and computational mesh

basin properties (such as area, elevation, surface imperviousness, and Manning's roughness coefficient), 2D flow area, and 2D mesh.

The Cartosat DEM of 30 m resolution is obtained Bhuvan online data archive (www.bhuvan-app3.nrsc.gov.in). The DEM is processed in ArcGIS to generate natural stream flow pattern over the entire catchment. Based on natural stream flow pattern, the Hyderabad urban is divided into 16 sub-catchments (Fig. 3b). The basin properties (as mentioned above) for all the catchments are extracted in ArcGIS environment by DEM processing. Further the DEM is used as basic input to generate DTM model in RAS Mapper, a spatial processing component of HEC-RAS 2D. The DTM helps in pre-processing the 2D computational cells that generates the water surface profile over underlying terrain geometry from model simulation results. This is accomplished by marking 2D flow region along the stream flow pattern and establishes 2D computational mesh as presented in (Fig. 3c). The model thus developed is simulated for October 2020 flooding event to capture the spatial extent of flooding and examine the deluge severity. October 2020 event rainfall is attained from Indian meteorological department (IMD) as 3-h intensity distribution over the entire storm

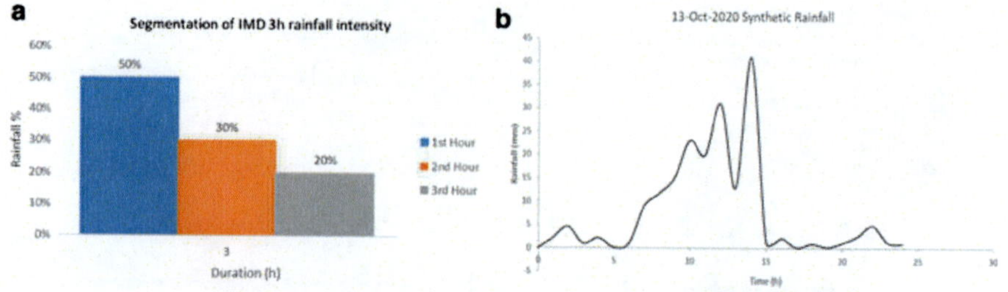

Fig. 4 **a** Segmentation of 3-h rainfall. **b** Synthetic event rainfall hydrograph

duration. To use the rainfall for real-time model simulation, a synthetic rainfall hyetograph is developed by estimating the total rainfall depth for every time interval. The estimated depth is shared in three consecutive parts of storm duration assuming 50% rainfall occurred in 1st part, 30% in second part and remaining 20% in 3rd part. The segmented data is then arranged by alternate block method to develop the synthetic rainfall hydrograph of event rainfall. Figure 4a, b shows condition employed for segmentation of 3-h rainfall and synthetic event rainfall hydrograph for model simulation respectively.

4 Results and Discussion

The RAS 2D model simulation results are presented in the form of time variant flooding extent and depth of inundation. Figure 5 signifies the progressive advancement of flood water acquired at 1st, 7th, 14th, and 21st hour after the commencement of 13 October 2020 event. A small amount of water as seen at 1 h after start of event has started accumulating at 7th hour in low laying areas along the natural streams and inundated more than 50% of area by the end of 14th hour after the commencement of event. The RAS 2D model simulation results are further processed in Arc GIS environment to analyse spatial extent of flooding and estimate deluge severity and risk. The analysis of results accounts total areal flooding extent as 58% (658 km^2) for October 2020 event. The flood inundation depth is used as a criterion to model the deluge severity and risk as presented in Table 2 and Fig. 6a presents the risk map for event simulation. The flood water started receding a week letter after the event that caused huge property loss and discomfort in many parts of the city.

Many localities, viz. Ramchandrapuram, Mallampet, Whisper valley, Laxmi Nagar, MIG Colony, Balanagar-Ida, Ramnathpur, Mallikarjun Nagar, Nagole, Krishna Nagar, Medbowli, Kismatpur, Toli Chowki, etc., have admitted heavy flooding by the end of 21st hour [19]. The Toli Chowki area, Mallikarjun Nagar, Begumpet, Malakpet, Ameerpet, and areas around Fox Sagar Lake, Hussain Sagar Lake experienced 2–3 m inundation depth and are visible on Google Earth projection of simulation results (Fig. 6b).

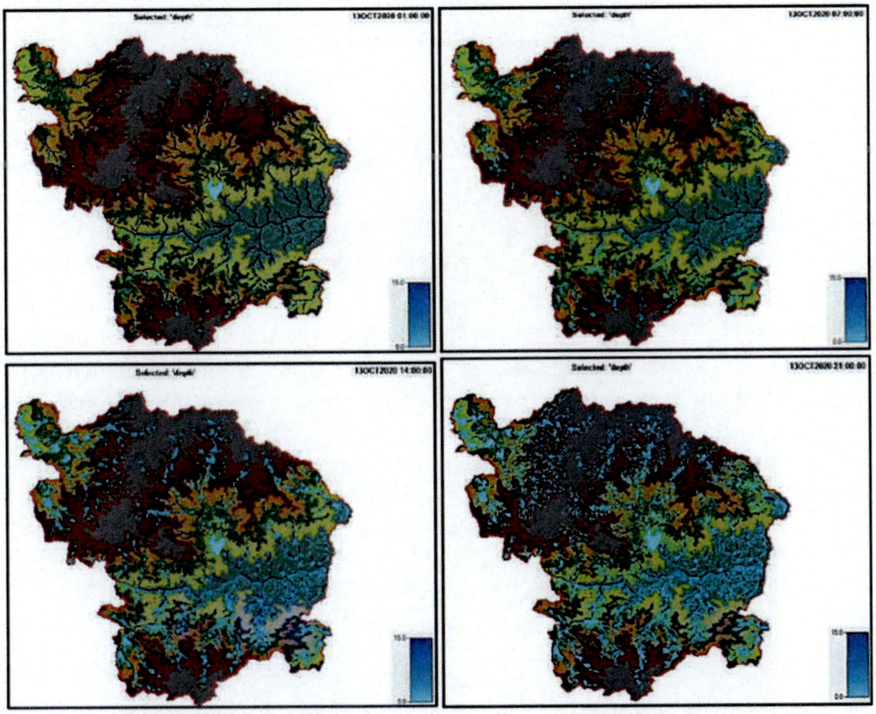

Fig. 5 Progressive advancement of flood water

Table 2 Deluge severity and risk analysis

Risk type	Inundation depth (m)	Spatial extend (Km^2)
High	>1.0	47.44
Medium	0.5–1.0	38.87
Low	<0.5	13.69

5 Conclusion

The paper presents a methodology for quick assessment (spatial extent and risk analysis) of urban floods. October 2020 Hyderabad flood is simulated using HEC-RAS 2D to analyse the spatial flooding extent and develop flood risk map based on the depth of flooding. A synthetic rainfall hyetograph is produced from rainfall distribution of 3-h intensity over Hyderabad city is used as input to carry out simulation run. The underlying Digital Terrain Model (DTM) works as base to generate the flood water sprawl and inundation depth with passing time. The inundation depth is used as parameter to describe potential for risk: low risk (<0.5 m), medium risk (0.5 m–1.0 m) and high risk (>1.0 m) of flooding. The simulation results identified spatial extent of October 2020 flood sprawl over 58% area. Further the model identifies

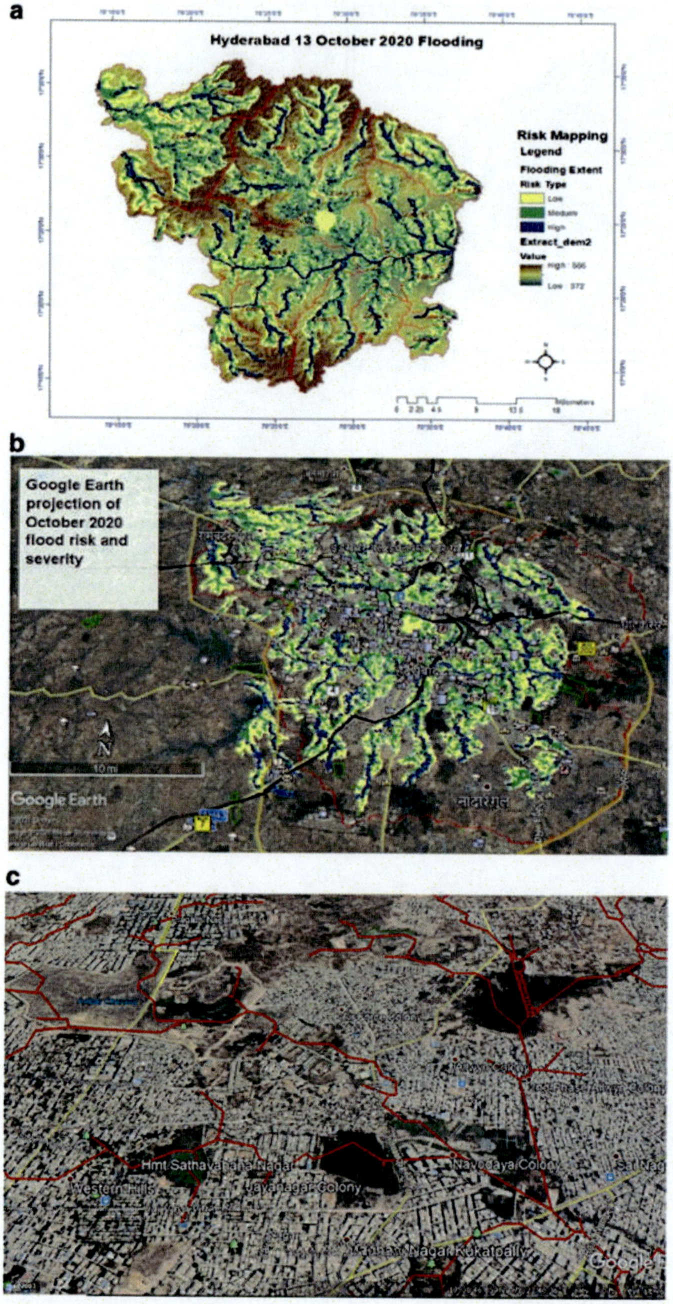

Fig. 6 **a** Risk map for event simulation, Hyderabad. **b** Google Earth projection of simulation results. **c** Present situation of natural lakes in Hyderabad city

respectively 47.44%, 38.37% and 13.69% areas falling under low, medium and high potential for risk. Further these risk maps can be used to establish an early warning system, prepare combat strategy and adopt relief measures as described in SOP.

Hyderabad has experienced chaotic floods in past few years and each event is becoming more severe with passing day. As mentioned earlier the changes in landscape, filling of natural depressions and lakes for high value land, vanishing of natural storm water flow paths and unavailability of space for water in urban environment are the key reasons for such intense flooding. Figure 6c points out the present situation of natural lakes in Hyderabad city. The natural flow paths of water (red colour line) undoubtedly have no presence on ground as these land has been transformed into urban sprawl over the period. Further, as the connection between the lake cascade is lost due settlements, many lakes dried with no water that resulted in encroachments in and around the lake areas. The encroachments have substantially reduced the lake capacity and spread area making these localities more vulnerable to flood risk. Therefore, to manage and reduce the impacts of urban floods it is must to regulate the urbanization identify the suitable places to divert additional storm water. Further, the natural lakes and flow streams need to bring back to its original state so that they can accommodate the excessive flood volumes.

Acknowledgements We thank IMD for sharing event rainfall data. We thank Greater Hyderabad Municipal Corporation (GHMC) and M/S Voyants Solutions Private Limited, Hyderabad, for sharing technical data. We also thank National Remote Sensing Centre (NRSC) for high-resolution dataset to take the study forward. The Landsat satellite images are downloaded from the United States Geological Survey. We also thank HEC-RAS technical team for their valuable suggestions and technical support.

References

1. Ahmed, Z., Rao, D. R. M., Reddy, K. R. M., & Raj, Y. E. (2013) Urban flooding—Case study of Hyderabad. *Global Journal of Engineering, Design & Technology, 2*(4), 63–66. https://www.longdom.org/articles/urban-flooding--case-study-of-hyderabad.pdf
2. Ali, H., Parth, M., & Mishra, V. (2019). Increased flood risk in Indian sub-continent under the warming climate. *Weather and Climate Extremes, 25*(100212). https://doi.org/10.1016/j.wace.2019.100212.
3. Awakimjan, I. (2015). *Urban Flood Modelling: Recommendations for Ciudad Del Plata.* BS thesis. University of Twente. http://purl.utwente.nl/essays/68990
4. Camarasa-Belmonte, A., & Soriano-Garcia, J. (2012) Flood risk assessment and mapping in peri-urban Mediterranean environments using hydrogeomorphology. Application to ephemeral streams in the Valencia region (eastern Spain). *Landscape and Urban Planning, 104*(2), 189–200. https://doi.org/10.1016/j.landurbplan.2011.10.009
5. Cavanaugh, N., Gershunov, A., Panorska, A., & Kozubowski, T. (2015). The probability distribution of intense daily precipitation. *Geophysical Research Letters, 42*(5), 1560–1567. https://doi.org/10.1002/2015GL063238
6. Chen, Y. (2017) *Draining the flood—A Combat against alert fatigue.*
7. Gumma, M. K., Irshad, M., Swamikannu, N., Anthony, W., & Carl, J. L. (2017). Urban sprawl and adverse impacts on agricultural Land: A Case Study on Hyderabad. *India. Remote Sensing, 9*(11), 1136. https://doi.org/10.3390/rs9111136

8. Kumar, N. N., Das, A., Mukerji, A., Mazumder, T., & Bera, S. (2020). Spatio-temporal dynamics of water resources of Hyderabad Metropolitan area and its relationship with urbanization. *Land Use Policy, 99*, 105010. https://doi.org/10.1016/j.landusepol.2020.105010
9. Kundzewicz, Z. W. (1999). Flood protection—sustainability issues. *Hydrological Sciences Journal, 44*(4), 559–571. https://doi.org/10.1080/02626669909492252
10. Ludwig, F., & Moench, M. (2012). *The impacts of climate change on water* (pp. 57–72). Routledge. https://doi.org/10.4324/9781849770743
11. Malik, U., & Kalra, A. M. (2017). Delineation of floodplains of streams to combat natural (flood) hazard in an urbanized watershed. *Water Resources, 44*(1), 16–22. https://doi.org/10.1134/S0097807817010195
12. NDMA (National Disaster Management Authority). (2010). GOI: Guidelines on Urban flooding in India, Government of India, New Delhi.
13. Patz, J. A., Maggie, L. G., & Limaye, V. S. (2014). When it rains, it pours: Future climate extremes and health. *Annals of Global Health, 80*(4), 332–344. https://doi.org/10.1016/j.aogh.2014.09.007
14. Pour, S., Abd Wahab, A., Shahid, S., & Dewan, A. (2020). Low impact development techniques to mitigate the impacts of climate-change-induced urban floods: Current trends, issues and challenges. *Sustainable Cities and Society, 62*, 102373. https://doi.org/10.1016/j.scs.2020.102373
15. Rangari, V. A., Prashanth, S. S., Umamahesh, N. V., & Patel, A. K. (2018). Simulation of urban drainage system using a storm water management model (SWMM). *Asian Journal of Engineering and Applied Technology, 7*(S1), 710. https://www.trp.org.in/issues/simulation-of-urban-drainage-system-using-a-storm-water-management-model-swmm-2
16. Sohn, W., Kim, J. H., Li, M. H., & Brown, R. (2019). The influence of climate on the effectiveness of low impact development: A systematic review. *Journal of Environmental Management, 236*, 365–379. https://doi.org/10.1016/j.jenvman.2018.11.041
17. SOPUF. (2017). Ministry of Urban Development, GOI: Urban flooding standard operating procedure. Government of India. https://amrut.gov.in/writereaddata/SOP_Urbanflooding_5May2017.pdf
18. Tabari, H. (2020). Climate change impact on flood and extreme precipitation increases with water availability. *Scientific Reports, 10*, 13768. https://doi.org/10.1038/s41598-020-70816-2
19. The New Indian Express. (2020). *Hyderabad floods: 100 colonies under sheet of water*. October 15, 2020. https://www.newindianexpress.com/states/telangana/2020/oct/15/hyderabad-floods-100-coloniesunder-sheet-of-water-2210468.html
20. Vani, M., & Kamraju, M. (2016). Impact of urbanization on lakes: A case study of Hyderabad. *Journal of Urban and Regional Studies, 5*(1).
21. Wikipedia, deep depression BOB 02. (2020). Accessed on October 2020. https://en.wikipedia.org/wiki/Deep_Depression_BOB_02_(2020)
22. Vorobevskii, I., Janabi, A., Schneebeck, F., Bellera, J., & Krebs, P. (2020). Urban floods: Linking the overloading of a storm water sewer system to precipitation parameters. *Hydrology 2020, 7*(2), 35. https://doi.org/10.3390/hydrology7020035
23. Xu, H., & Luo, Y. (2015). Climate change and its impacts on river discharge in two climate regions in China. *Hydrology and Earth System Sciences, 19*(11), 4609. https://doi.org/10.5194/hess-19-4609-2015

Flow Behavior in Surface Aeration System

Anurag Sharma and Bimlesh Kumar

Abstract The current work explores an experimental flume investigation of turbulence in surface aeration system. The laboratory arrangement is made of a rectangular tank, where flow is operated by uniformly placed aerators located on the fluid surface. The turbulent flow strictures at the inlet and outlet section of the tank were analyzed to observe the variation in their flow profiles. The results of turbulent statistics, such as time-averaged velocity, and Reynolds shear stress, are observed to be increased in flow outer layer with outlet section and decreasing in flow inner layer with inlet section of tank. Turbulent intensity increases with tank outlet section, and consequently, the momentum transfer also increases.

Keywords Velocity · Aerator · Inlet section · Outlet section

1 Introduction

The nature of turbulence on the free surface of fluid flows is essential in relation to mass transfer through the liquid–atmosphere interface. Re-aeration in streams, wastewater management, and further engineered methods depend on a mass transfer through a free liquid surface from one phase to another [1]. Stirred reactor planned for gas–liquid interaction is termed surface aerators. The critical purpose of surface aerators is to improve the oxygen exchange speed into the liquid, liquid-phase mixing to confirm oxygen accessibility in all portions of the contactor and microorganism suspension. Both liquid and gas film exchange parameters measure this transport system; the fluid side mass-exchange parameter measures the exchange speed for moderately soluble compounds like oxygen [2]. Orlins and Gulliver [1] investigated turbulence on the water-free surface in an oscillating grid tank and compared it with

A. Sharma (✉)
National Institute of Technology Rourkela, Rourkela, Odisha 769008, India
e-mail: sharmaan@nitrkl.ac.in

B. Kumar
Indian Institute of Technology Guwahati, Guwahati, Assam 781039, India

A. K. Dikshit et al. (eds.), *Innovative Trends in Hydrological and Environmental Systems*, Lecture Notes in Civil Engineering 234,
https://doi.org/10.1007/978-981-19-0304-5_4

that on the free fluid surface measured by particle image velocimetry. Kumar and Rao [3] observed the optimum geometric arrangement of continuous flow surface aeration systems and developed the design equations.

Nienow [4] has shown that circulation time is related to the basic turbulence model for a baffled mixing tank. Circulation due to impeller generates turbulence near the free surface, dissipates surface energy, and improves gas exchange through the air–water boundary [5]. Thus, the objective of the present work is to analyze the turbulence in continuous surface aeration systems.

2 Experimental Setup

To analyze the turbulent characteristics in surface aeration systems, a continuous flow rectangular tank of 40 × 60 × 400 cm was used. The impeller diameter (*D*) is 13.88 cm, which had six blades each of the Ruston type. The tank has a steel frame with sheet glass walls, with inlet and outlet connections for water recirculation. A 500 L tank is used to store water for the study. Water was re-circulated with a 370 W (0.5 hp) pump connected to the outlet and the water tank. The impellers are connected to a DC motor, which is further connected to the control panel. The latter has a digital display unit showing the current and voltage supplied to the impellers via the DC motor. The distance of measurement point to the inlet and outlet is around 25 cm from both ends of the tank so that the turbulence induced by the pump will not influence the experimental data measurements. A flow meter on the outlet pipe was used to check flow steadiness. The digital point gauge was used to measure the flow depth. The flow parameters are shown in Table 1, where N is the impeller rotation speed.

A four-beam down-looking acoustic Doppler velocimeter (VectrinoPlus) was used to measure three-dimensional instantaneous flow velocities. The velocity data sets were measured at the centerline of the channel cross section. The instantaneous velocities were measured by ADV at a position 50 mm beneath the probe emitter to reduce the effect of the probe on the experimental data. The data acquisition sampling frequency was 200 Hz. The sizes of the sampling cylinder were considered at 4×10^{-3} m in height and 6×10^{-3} m in diameter. The instantaneous ADV data were collected for the sampling duration of 300 s. The signal-to-noise ratio and correlation magnitude was higher than 10 and 60, respectively. The low-frequency noise in the ADV raw data was removed by a modified singular spectrum technique as suggested by Sharma et al. [6].

Table 1 Flow parameters at inlet and outlet

Median grain size, d_{50} (mm)	Discharge, Q (m^3/s)	Depth of flow, y (m)	Hydraulic radius, R (m)	Bed slope, S_0
0.418	0.0169	0.14	0.075	0.0015

3 Results and Discussions

The mean velocity (u) along the stream-wise (x direction) and vertically (z direction) were calculated using Eq. (1). To evaluate the coordinate point, the z-axis (z = 0) is considered the position of boundary measured positively in the upward direction. The x-axis was associated with centerline of the tank, therefore $x = 0$ was taken at the measurement point and positive along the flow course. The y-axis was considered positive in the rightward transverse direction. Thus, the measurement point had coordinates (0, 0, z).

$$u = \frac{1}{n}\sum_{i=1}^{n} u_i \tag{1}$$

where u_i is instantaneous velocities along the flow direction and n represents number of samples. Figure 1 presents the vertical distribution of the inlet and outlet section velocity profiles, where z is the vertical elevation measured from the boundary surface and h is the flow depth. The velocity decreases along the wall boundary because of resistance induced by the solid wall (Fig. 2).

Reynolds shear stress (RSS) was calculated using Eqs. (2) and (3). Figure 3 presents the vertical profile of Reynolds shear stress at the inlet and outlet section of the flume, where u' and w' are the fluctuating velocity components along the stream and vertically, respectively. Further, Fig. 3 pointed out that RSS advances toward the

Fig. 1 Experimental system (a-impeller, b-DC motor, c-impeller placing, d-storage tank, e-ADV, f-recirculating pump and g-display control panel)

Fig. 2 Vertical profile of mean velocity

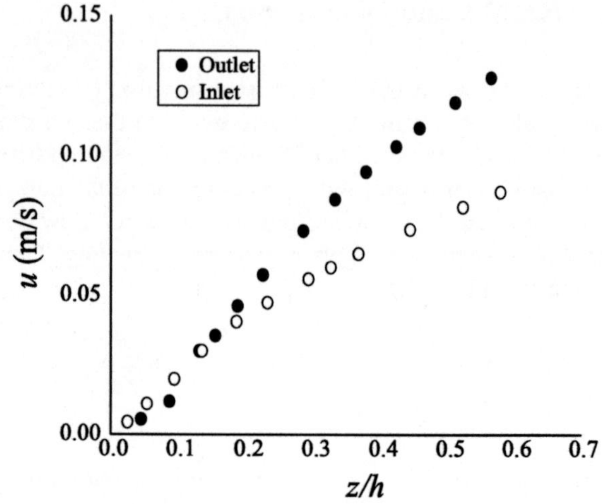

Fig. 3 Vertical profiles of Reynolds shear stress

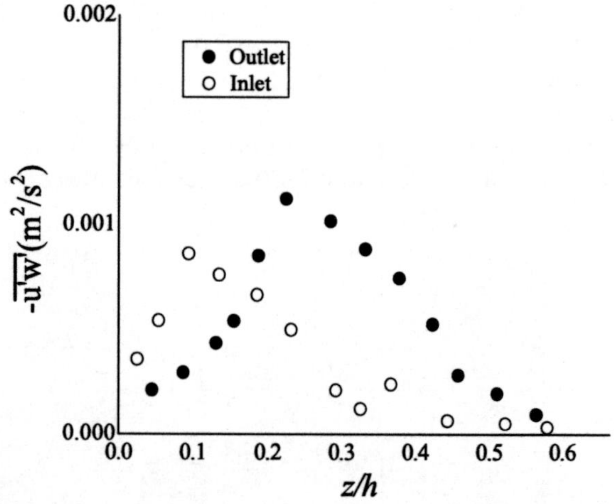

boundary surface and achieves a peak magnitude in the range $0.05 < z/h < 0.2$ and then again reduces very near to the bed surface due to the decrease in the fluctuating velocity components near the flow zone. The result also suggests that the pattern of Reynolds shear stress profiles is similar for both tank's inlet and outlet section. However, the greater value is observed in the inner layer with the inlet section, and the result is quite the opposite in the outer layer of flow.

$$\overline{u'w'} = \frac{1}{n}\sum_{i=1}^{n}(u_i - u)(w_i - w) \tag{2}$$

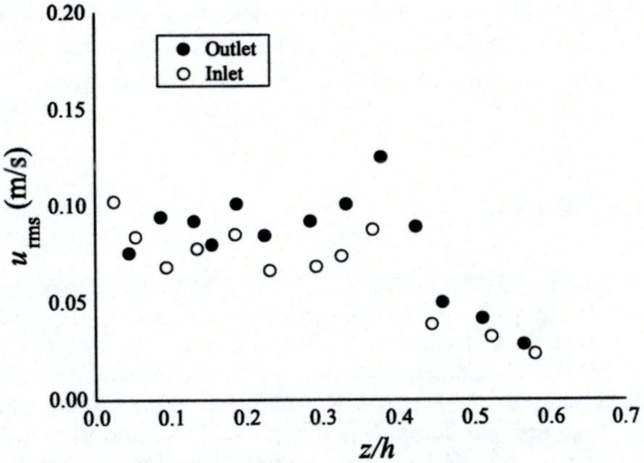

Fig. 4 Vertical profiles of Turbulence intensities

$$\tau_{uw} = -\rho_w \overline{u'w'} \tag{3}$$

The representation of fluctuating velocity components in terms of root-mean-square velocity is called turbulence intensity. Root-mean-square (RMS) velocities along the stream are represented by u_{rms} expressed by Eq. (4)

$$u_{rms} = \left[\frac{1}{n} \sum_{i=1}^{n} (u_i - u)^2 \right]^{0.5} \tag{4}$$

Turbulence intensity is a measure of the turbulence level of flow for rotational speed (Fig. 3). The distribution of turbulence intensity has followed uniform profiles implying high turbulence levels throughout the entire flow depth. Turbulence intensities display the peak magnitude between the flow depth 0.3–0.4 h from the boundary and, afterward, tends to be constant close to the free surface. In both the inlet and outlet section of a tank, it fluctuates higher in the vicinity of boundary than those of free surface due to boundary effect (Fig. 4).

4 Conclusions

This study is about the turbulence characteristics in continuous surface aeration systems. The turbulence characteristics are measured with an acoustic Doppler velocimetry. Various turbulent parameters are studied to achieve results of turbulent flow characteristics. The time-averaged stream-wise velocity decreases toward the wall boundary because of resistance induced by the solid wall. In general, the distributions of Reynolds shear stress are slightly scattered and advanced with an inlet section in the near boundary. The profiles of the Reynolds shear stress suggest a damping nature in the near boundary zone because of the decreased nature of velocity fluctuation in the corresponding area. Turbulence intensities increased with the outlet section, causing an increase in momentum transfer.

References

1. Orlins, J. J., & Gulliver, J. S. (2002). Measurements of free surface turbulence. *Geophysical Monograph Series*, pp. 247–253.https://doi.org/10.1029/gm127p0247
2. Jähne, B., & Haußecker, H. (1998). Air-water gas exchange. *Annual Review of Fluid Mechanics, 30*(1), 443–468. https://doi.org/10.1146/annurev.fluid.30.1.443
3. Kumar, B., & Rao, A. R. (2010). Continuous-flow surface aeration systems. *Chemical Engineering & Technology, 33*(2), 305–314. https://doi.org/10.1002/ceat.200900403
4. Nienow, A. W. (1997). On impeller circulation and mixing effectiveness in the turbulent flow regime. *Chemical Engineering Science, 52*(15), 2557–2565. https://doi.org/10.1016/s0009-2509(97)00072-9
5. Melville, W. K. (1996). The role of surface-wave breaking in air-sea interaction. *Annual Review of Fluid Mechanics, 28*(1), 279–321. https://doi.org/10.1146/annurev.fl.28.010196.001431
6. Sharma, A., Maddirala, A. K., & Kumar, B. (2018). Modified singular spectrum analysis for despiking acoustic Doppler velocimeter (ADV) data. *Measurement, 117*, 339–346. https://doi.org/10.1016/j.measurement.2017.12.025

Air Pollution Minimization of a Municipal Solid Waste Transport Sector: An Integrated Solid Waste Management Model Simulation on Kolkata

Samran Banerjee and Amit Dutta

Abstract With an average municipal solid waste generation of around 4000–4500 MT/day from 15 boroughs, the solid waste management system of Kolkata has various disparities to fulfill the need for a sustainable waste management. Waste under Kolkata Municipal Corporation (KMC) is carried by departmental (79%) and hired vehicles (21%). The major air pollutants for existing system are CO_2 (99%), NO_x (0.4%), CO (0.2%), SO_2 (0.02%), Particulate matter (0.04%) and Hydrocarbon (0.07%). A generic LP model has been developed considering central sorting station, incineration plant, composting plant, transfer station, and engineered landfill site to incorporate a proposed integrated solid waste management system where TATA LPT heavy-duty truck has been contemplated for waste transportation from transfer stations to three proposed engineered landfill sites. Different weighted average emission factors for respective pollutants have been calculated. The mathematical LP model was resolved using LINGO optimization software. In comparison with the proposed system, a decrease of 18% of total pollution from transport sector is possible. The estimated individual pollutant emission reductions are 18% CO_2, 6% SO_2, 8% NO_x, 13% PM, and 8% Hydrocarbon. Total vehicular pollutant emissions have been reduced by 19% from borough 9. Further implementation of Bharat Stage VI (BS 6) norms over the KMC vehicles shows 97–98% SO_2, NO_x, and PM reduction which is substantial.

Keywords Municipal solid waste management · Air pollution · LP model · Minimization · Kolkata

S. Banerjee (✉) · A. Dutta
Jadavpur University, Kolkata, India
e-mail: samran.civil@gmail.com

A. Dutta
e-mail: amit.dutta@jadavpuruniversity.in

A. K. Dikshit et al. (eds.), *Innovative Trends in Hydrological and Environmental Systems*, Lecture Notes in Civil Engineering 234,
https://doi.org/10.1007/978-981-19-0304-5_5

1 Introduction

Waste generation is an inherent outcome of human civilization in recent times. Expeditious growth of population density, sprawling urban cities, resource limitations, space constraints are leading causes of the inefficiency behind an effective sustainable Municipal Solid Waste Management (MSWM) system in developing countries like India. Waste transportation is a huge task to transfer the waste from various collection points, open vats, and dustbins to the transfer stations or disposal sites. Handcarts or tricycles with bins or containers, lightweight commercial vehicles like manually loaded trucks, hydraulic tipper trucks, dumper placer like mini trucks with garbage collection bins and payloaders are typically used for primary and secondary collection of waste. Most of the vehicles are aged with less efficiency and are more susceptible to pollution potential. Other than natural sources, littered waste in public spaces, gaseous emissions of different pollutants, and particulates from older conservancy vehicles used in waste management sectors followed by open dumping has a strong inter-relation with environmental degradation and global warming through air pollution. Mobile sources like automobiles are of highest concern in emission point of view [1]. Recently, aside from the criteria air pollutants, Carbon dioxide (CO_2) has been recognized as vulnerable to global warming by the climate scientists. Petrol-fueled vehicles emit Carbon monoxide (CO), unburnt Hydrocarbons (HC), Nitrogen oxides (NO_x), Benzene, and Butadiene as primary pollutants while diesel-fueled vehicles release Polycyclic Aromatic Hydrocarbons (PAH), Sulfur dioxide (SO_2), Nitrogen oxides (NO_x), and Particulate matters (PM) [2]. Not much research has been carried out worldwide linking the contribution of MSW conservancy transportation vehicles with air pollution. Jovičić et al. [3], in their research, had mainly focused on route optimization by GIS/GPS to increase fuel efficiency of conservancy vehicles. Vilms et al. [4] had investigated the amount of air pollutants (CO_2, CO, HC, NO_x, PM, and CO_2-eq) formed in the city center of Tallinn (Estonia) when the present waste collection system with conservancy trucks is applied.

1.1 Air pollution- Environmental and Health Hazards

Several studies showed the detrimental effects of various air pollutants coming from the automobile pollution causing both environmental degradation and hazards to the human health. Carbon monoxide (CO) lowers the oxygen-carrying capacity of blood, Nitrogen oxides (NO_x), PM10, and Sulfur dioxide (SO_2) causes respiratory illness, ultrafine particulates like PM1 and PM0.1 contribute to the overall toxicity of airborne particulate matter, while PAH's are carcinogenic [5]. Besides, Carbon dioxide (CO_2) promulgates greenhouse effect and climate change. About 30% of all CO_2 emissions in Canada come from automobile sectors. Vehicular emissions also contribute to smog formation and acid rain, producing 19% of nitrogen oxides (NO_x), 23% of volatile organic compounds (VOCs) which together create ground-level ozone, a major component of smog and 37% of the total carbon monoxide (CO) released each year by anthropogenic activity [6]. Long-term exposure to PM2.5 is associated with mortality and morbidity with shorter life expectancy. Around 87% of the total global population lives in areas exceeding the World Health Organization's (WHO) air quality guidelines for annual mean ambient standards. Recent assessments of the Global Burden of Disease (GBD) estimate that exposure to ambient air pollutants is a major contributing risk factor to regional and global burden of disease to the mankind [7, 8].

2 Present Solid Waste Management Scenario of Kolkata

In air pollution context, it affects enormously in densely populated metropolitan areas. Especially, cities like Mumbai (18.4 million), Delhi (16.3 million), and Kolkata (14 million) in developing countries like India where environmental regulation and monitoring of various pollutants are of greatest concern. High vehicular density, constraint in road space, low traffic speed, bad road conditions, rapid growth in vehicle population have led to a degradation of the atmospheric condition over Kolkata. 50% contribution toward air pollution is from automobile exhausts [9]. Kolkata generates about 4000–4500 MT/day of municipal solid waste which is equivalent to a per capita waste generation of 0.45–0.5 kg per day for residential population. In the existing SWM system of Kolkata with 15 no. of boroughs under Kolkata Municipal Corporation (KMC), lack of source segregation strategy, 60% house-to-house collection, 55% open vats, 50% operational efficiency of KMC transport system with 30–35% old vehicles, informal recycling system, uncontrolled waste disposal to major non-engineered landfill site Dhapa are contributing toward deterioration of environmental standards and human health hazards. In waste transportation system of KMC area, six different types of departmental vehicles carry 79% garbage and private hired vehicles carry 21% of garbage and entire silt/rubbish. For hired vehicles, hauling capacity is 7 MT for garbage transportation and 9 MT for silt transportation. KMC uses dumper placer Type-I (4.5 m^3), Dumper placer Type-II (7 m^3), Payloader tipper truck (11 m^3), Stationary compactor (10.5 m^3), Movable compactor Type-I (14 m^3), and Movable compactor Type-II (8 m^3). Distribution of different KMC vehicles with their total number, running vehicle and idle vehicles number is depicted in Fig. 1.

The municipal solid waste is to be transported to the landfill sites at the city's outskirts which generates a large amount of pollution in the urban atmosphere. CO, SO_2, PM, NO_x, and hydrocarbon emissions from waste conservancy vehicles are of increasing concern and probably possess the greatest long-term threat to the air quality of Kolkata. Absence of any transfer station leads the vehicles to travel longer distances from the borough centers. Usage of older KMC vehicles with BS-II, BS-III, and BS-IV engines that influence huge emission of air pollutants like CO_2, CO, NO_x, SO_2, HC, and PM.

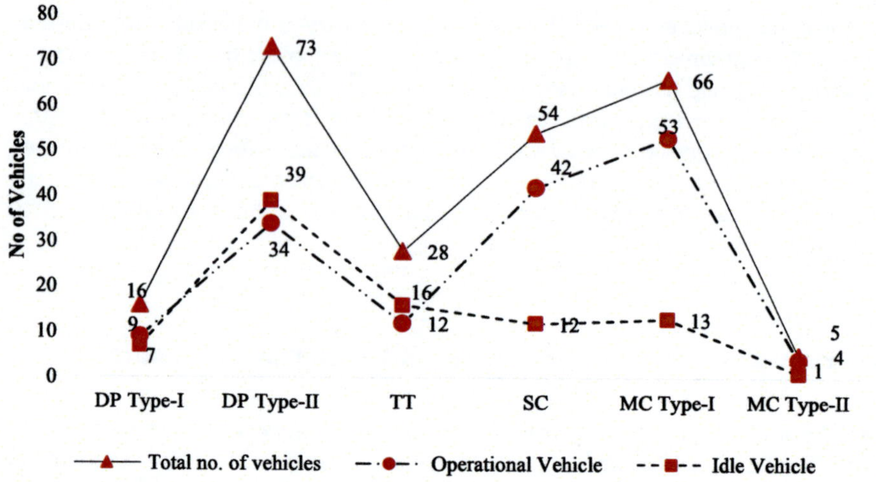

Fig. 1 Six different types of KMC vehicle and their distribution

2.1 *Effects of Vehicular Pollution*

PM pollution exacerbates both environmental standards and human health. Gasoline and diesel-operated vehicles produce PM10 during combustion of carbon and sulfur-based fuel and impact the climate adversely by accumulating in the upper atmosphere. Around 52% of total particulate resides in residential regions as PM10 in Kolkata [10]. PM affects the human respiratory system through bronchitis and chronic obstructive pulmonary disease (COPD) including serious effects like cancer, heart attacks, and strokes [11]. SO_x and NO_x emitted from the petrol and diesel combustion irritate the lungs and worsen the pre-existing lung diseases. Presence of considerable amount of PAHs in Kolkata air quality was found, some of which were supposed to be carcinogenic [12].

2.2 *Present Pollutant Emission from Solid Waste Transport Sector*

From detailed analysis, it is found that the major air pollutants emitting from existing solid waste transport sector are CO_2 (99%), NO_x (0.4%), CO (0.2%), SO_2 (0.02%), Particulate matter (0.04%) and Hydrocarbon (0.07%). In case of existing MSW management system, the total quantitative pollutant generation contributed by solid waste transport sector of Kolkata is around 4361 MT/year. The emission of Benzene and Butadiene is very negligible in comparison with the rest of pollutants. Emission of CO and NO_x from borough 10 is 12.3% and 13% whereas 1.5% and 1.5% are for borough 15 (Fig. 2).

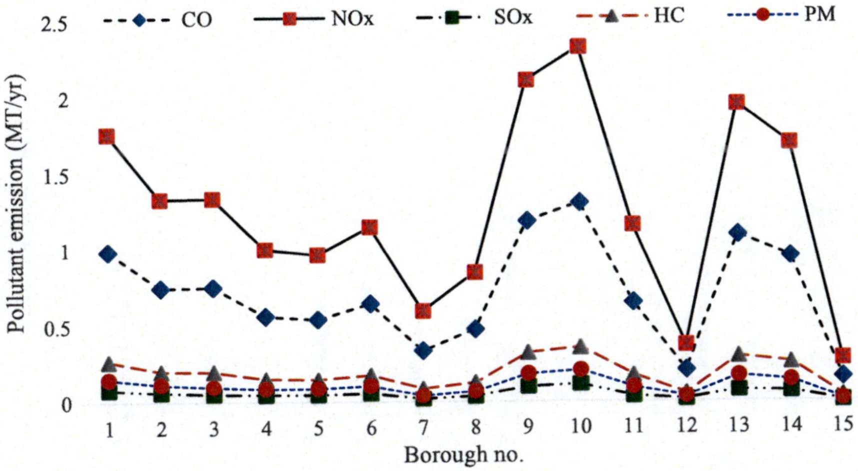

Fig. 2 Borough wise annual pollutant emission (MT/year)

Around 13% and 0.7% of CO_2 emission are recorded for borough 10 and borough 15, respectively. 99.2% of CO_2 emission is recorded among other pollutants from borough 10 (Fig. 3). The emission of CO_2 from the departmental vehicles is 89% (3879 MT/Year) and for private hired vehicles is 11% (447 MT/Year) because of constraint of disposal sites and higher distance covered compared to ISWM system.

Highest emission is recorded from the movable compactor type-I vehicles due to their maximum carriage of waste (36%) and higher distance covered among others. Emission of CO, HC, and NO_x from the departmental vehicles is 80%, 81%, and

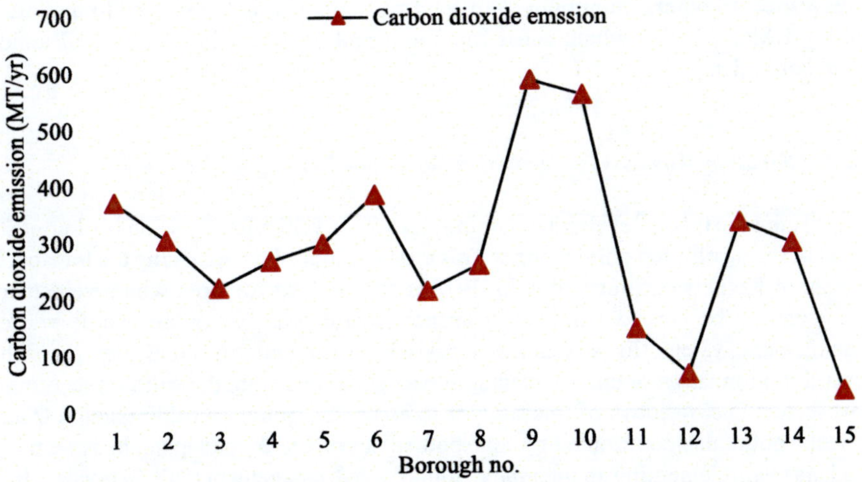

Fig. 3 Yearly CO_2 emission from different borough (MT/year)

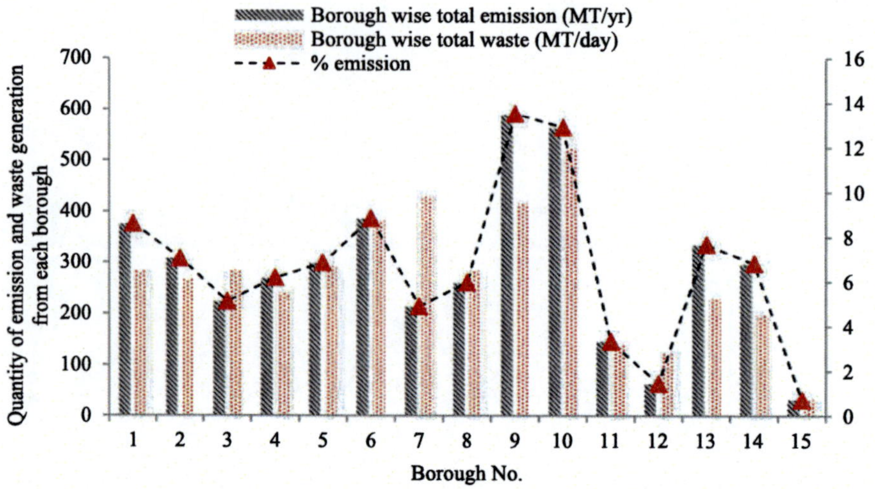

Fig. 4 Borough wise comparison of total emission (MT/year), total waste generation (MT/day), and percentage emission (%) for existing system

81%, respectively whereas respective emission from hired vehicles is 20%, 19%, and 19%. Movable compactor type-I type of vehicles show maximum coverage for the emission of these pollutants.

Total pollutant emission is higher in borough 9 and borough 10 out of 15 boroughs, which are around 13% that is influenced by waste generation, distance from landfill and sharing of waste fraction by departmental and private vehicles (Fig. 4). Contribution of total pollutants from borough 15 is negligible which is around 0.73% in comparison to others. As the distance of borough 15 from Dhapa landfill is maximum with respect to others, minimum % of total waste (0.8%) is transferred to disposal site by KMC vehicles which cause minimum pollutant contribution as well amid other boroughs.

2.2.1 Strategy Toward Integrated Solid Waste Management

For sustainable urban development, minimization of vehicular emission is of utmost concern from pollution point of view. This can be achieved by managing the transport system of Kolkata optimistically. MSWM is one such sector from which vehicular emission can be curbed. Although transport sector is a part of the whole waste management system, the present study envisages to encompass a realistic approach toward minimization of environmental impact generated from the solid waste transport sector. The objective of this study is to assess the pollution minimization from gaseous pollutants contributed by transport sector of MSW management system of Kolkata by implementing an Integrated Solid Waste Management (ISWM) system for better urban planning keeping into consideration that the minimization of air pollution should be sustained. The present study is an attempt to integrate the best feasible methods of different components of waste management along with socio-economic conditions in Kolkata considering systems approach. The integrated approach to SWM was a response to failure of the conventional approach in developed countries. Contemplating the hierarchy of integrated approach, i.e., waste prevention, minimization of waste generation by reuse, recycle and recovery, reduction of landfill volume by compost production and incineration. This sustainable approach personifies the development of an infrastructure in waste management sector not only to optimize the social aspects but also to minimize the environmental pollution.

3 Methodology

To suggest solutions for a better SWM, an extensive study is carried out on the various aspects of SWM in general for metropolitan cities of India and KMC's solid waste mechanism in particular. Model is something that predicts the effect of a proposed system that has not been developed yet. To implement the ISWM system over the existing SWM system, it is necessary to develop a model which can compare the effects between the proposed and the existing one. Presence of an exhausted landfill site Dhapa with surpassed carrying capacity needs to be replaced with engineered landfill sites for curbing the pollution problems. Longer distance traveling with littering waste leads to not only pollutant emission surrounding the city's atmosphere but also creating unhygienic conditions on the open road surfaces. Absence of source segregation system turns the volume of municipal solid waste (MSW) to a huge quantity consisting of more than 50% biodegradable fraction followed by recyclables (9–10%) and combustibles. Hence, proper implementation of an integrated system is of massive concern in pollution point of view. So, the methodology consists of developing linear programming model with basic assumptions and constraints, solving it by using linear programming equations developed in LINGO optimization software. Conventional analysis of pollutant emission with respect to existing waste transportation scenarios will give this model a realistic solution, which is the basic objective of this research work.

3.1 Development of Linear Programming Model

For the existing scenario, considering both socio-economic and environmental aspects, a generalized model is being developed using linear programming. To minimize all the potential conflicts arising in waste management sector can be overcome by proposing multi-objective decision techniques. This model is personified as multi-dimensional, however, the objective function solely depends on cost regulation. The current study envisages incorporating the integrated best feasible options in Kolkata to reduce the impact potential as well as the economy of the SWM system. ISWM includes waste prevention, source reduction, reusing, and recycling followed by processing to reduce the volume of materials sent for safe disposal. Development of a solid waste management mathematical model considering rate of waste generation, their composition, reuse and recycling, mode of transportation, revenues generating from waste processing, waste disposal to safe landfill sites have been accomplished. The constraints include those linking waste flows and mass balance, recycle amount, processing plants capacity, landfill capacity, transport vehicle capacity, number of trips, etc. The pollution generation from transport sector of this proposed ISWM system has been contemplated.

LP model representation

The integrated model assimilates different components of ISWM for realistic implementation approach. There are some assumptions made for developing the model. At household level, separation of organic and inorganic waste by means of two-bin system. Waste is presumed to be generated from the borough centers. A sorting station is considered for segregation of recyclables, further waste materials will be operated for thermal and biological processing. An incinerator-based plant for processing the combustibles with high calorific value and composting plant for organic processing have been incorporated. The whole waste is categorized into garbage and silt/rubbish. The garbage fraction will go through the whole processing whereas the silt will be directly disposed of at the disposal sites. Consideration of engineered landfill sites is of great concern due to enormous landfill gas emissions from the open dumping. Three engineered landfill sites have been contemplated in south, north, and east region of the city's outskirts. Engineered landfills will have proper leachate liner, cover, and gas collection system. Three transfer stations are considered in east, south, and northern region of the city. Processing will be done at the transfer station. All the garbage and silt will be delivered to the transfer station by the departmental and hired vehicles. Thereafter, silts will be directly loaded to the TATA LPT heavy-duty truck of capacity 20MT along with all the process rejects coming from the processing pre-sorters as well as post-processing. The city is distributed in two zones for each disposal site. Maximum and minimum number of trip limits is fixed as per the recent feasible data collected from KMC operation. Transfer stations are considered near KMC boundary which is found to be cost-effective [13]. The material balance flow diagram is interpreted considering total waste generation as 100 MT as shown in the following Fig. 5.

The objective function of the LP model is a cost optimization function. The cost incurred by different components is transportation cost, incentive cost, sorting cost, incineration cost, composting cost, landfilling cost. The total effective SWM cost will be given as cost incurred by different components deducted by revenue earned by recyclables, compost, and power generation from incineration. The LP model was solved using linear programming equations through LINGO optimization software. For example, pollutant generation from different vehicles for transporting the waste from borough center to the transfer stations/ disposal sites expressed by generic LP equations have been presented as follows.

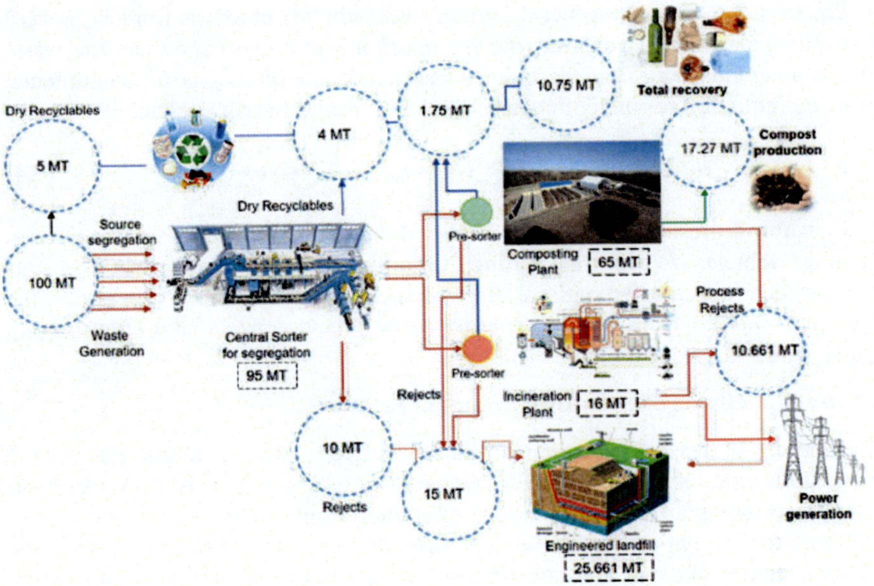

Fig. 5 Material balance flow chart of integrated solid waste management model

$$P\_emsn_{veh} - \sum_{br=1}^{BR} \sum_{veh=1}^{VEH} quanwaste\_veh_{br,veh,ds} \times veh\_ef\_P_{br,ds} = 0 \quad (1)$$

Equation 1 elucidates the quantitative generic form of all the pollutant (P_emsn_{dd}) emitting from the departmental and hired vehicles (*veh*) for transporting the garbage and silt (*quanwaste_veh*), respectively from the waste generation points (*br*) to the transfer stations/ disposal sites (*ds*). $veh\_ef\_P_{br,\, ds}$ is the vehicular emission factors for the respective pollutants considered from different vehicles.

$$P\_emsntruck - \sum_{ds=1}^{DS} truck\_ef_{ds} \times xlf_{ds} = 0 \quad (2)$$

Equation 2 depicts the generic form of all the pollutant (*P_emsntruck*) emitting from the TATA LPT heavy-duty truck for transporting the process rejects and silt (xlf_{ds}) from the transfer stations to the disposal sites (*ds*). $truck\_ef_{ds}$ is the truck emission factor for the respective pollutants under consideration.

$$total P\_emsn_{dv} - \sum_{dv=1}^{DV} P\_emsn_{dv} = 0 \quad (3)$$

Equation 3 express the generic form of total pollutant emission ($totalP\_emsn_{dv}$) from the departmental vehicles (dv) for transporting the garbage from the waste generation points to the transfer stations/ disposal sites. P_emsn_{dv} is the departmental vehicular emission for the pollutants considered from different vehicles.

$$totalP\_emsn_{hv} - P\_emsng_{hv} - P\_emsns_{hv} = 0 \tag{4}$$

Equation 4 presents the generic total pollutant emission ($totalP\_emsn_{hv}$) from the hired vehicles (hv) for transporting the garbage and silt from the waste generation points to the transfer stations/ disposal sites. P_emsng_{hv} and P_emsns_{hv} is the vehicular emission from hired vehicles for the pollutants considered for transporting garbage and silt, respectively.

Basis and methods of estimating pollutant emission

The purpose of this study is to analyze the gaseous pollutants generation from solid waste transportation sector of MSW management of Kolkata for better urban planning keeping in mind minimization of air pollution from it. There is number of steps followed for estimation of pollutant emission through model analysis. Calculation of vehicle wise average fuel consumption, estimation of weighted average emission factors for CO, NO_x, HC, and PM, weighted average pollutant emission rate from different boroughs in association with distance traveled by different vehicles, formulation and calculation of CO_2 and SO_2 depending upon departmental and hired vehicles are depicted as follows.

Vehicle wise average fuel consumption

The average weight carrying capacity per trip and fuel (diesel) consumption of departmental and private hired vehicles has been fed into the calculation of weighted average emission factors for different gaseous pollutants. Most of the privately hired vehicles are very old and their fuel consumption is considered by extrapolating the fuel consumption of Dumper Placer Type-II and that of Tipper truck. Average fuel consumption of DP Type-I (4.875 km/l), DP Type-II (4 km/l), TT (2 km/l), SC (1.55 km/l), MC Type-I (1 km/l), MC Type-II (1.46 km/l), hired vehicle (3.75 km/l) and LPT truck (6.25 km/l) are considered on basis of loaded and unloaded fuel consumption depending upon their waste carrying capacity.

Vehicle emission factors

The weighted average emission factors (g/km) for these respective pollutants are contemplated with the help of previous case studies, year-wise distribution of no. of vehicles and their respective emission factors [14] of the pollutants, and method of extrapolation. Density of the fuel is considered 820 kg/m^3. Percentage of carbon and sulfur assumed by weight is 87% and 0.035%, respectively. In the existing system, traveling distance is considered for single disposal due to the presence of only disposal ground Dhapa. However, in proposed model, distance traveled by vehicles has been segregated out into three different regions and emission rate is calculated as per the zone distribution and distance of engineered landfills from different borough centers.

Emission factors of NO_x (6.2–8.2 g/km) for different kinds of vehicles are more than CO and others because of diesel engines, although PM emission is less. Different ranges of emission factors for pollutants are CO (3.5–4.6 g/km), HC (0.9–1.3 g/km) and PM (0.3–0.8 g/km). CO_2 generation is calculated from fuel consumption, its carbon %, and utilization of carbon as CO and SO_2 emission is calculated from fuel consumption, its sulfur %. The weighted average pollutant emission rate is estimated incorporating the distance traveled by the vehicles from borough centers to disposal sites, vehicular capacity, and emission factors. This weighted average emission rate was fed as an input data for running the model to analyze various emission data of present and proposed waste management scenarios.

4 Result and Discussions

4.1 Air Pollution from Transport Sector of Proposed Model

In case of the proposed ISWM system, the total pollutant generation from transport sector of Kolkata is around 3590 MT/year. Total contribution of CO_2 emission is the highest (99%) due to high amount of carbon present in hydrocarbons which gets converted into CO_2 during burning. CO_2 is a major greenhouse gas that causes global warming. NO_x generation is the second largest (0.5%) which is higher than CO (0.3%) and HC (0.1%). This is because of high-temperature combustion in diesel engines also generates more thermal NO_x along with generation of fuel NO_x. It causes acid rain and photochemical oxidants. SO_2 emission (0.02%) is much less compared to NO_x due to stringent permissible sulfur percentage in diesel fuel. Though, the vehicles are under Bharat Stage IV norms. Implementation of BS-VI norms over these vehicles will enhance more reduction of SO_2 due to more stringent limit of percentage of sulfur content. The emission percentage of PM is less (0.04%) compared to other major pollutants but it has significant effect on respiratory system due to the presence of higher fraction of nanoparticles. The emission of Benzene and Butadiene is very negligible in comparison with the rest of pollutants. Total pollutant emission from solid waste transport sector is higher in borough 9 and 10 out of 15 boroughs, which is around 11% and 14%, respectively that is influenced by waste generation, distance from landfill and sharing of waste fraction by departmental and private vehicles. Emission of CO and NO_x from borough 10 is 12%, whereas 1.2% is from borough 15 which is minimum. 99.2% of CO_2 emission is recorded among other pollutants from borough 10. Emission result implies that the total emission of the departmental vehicle is 84%. The departmental vehicles carry 79% of the total waste due to incorporation of new compactor vehicles. In case of the private vehicles (hired ones), the total emission is 10% which carries 21% of the total waste. Pollutant emission of CO_2 from the departmental vehicles is 83% (2946 MT/Year) and for private hired vehicles is 9.5% (336 MT/Year) because of higher distance covered. The rest of 7% of CO_2 contribution is from TATA LPT truck for carriage of process rejects and silt from the transfer stations to their respective landfill sites. Highest emission is recorded from the movable compactor type-I type of vehicles due to their maximum carriage of waste (36%) and higher distance covered among others. Emission of CO, HC, and NO_x from the departmental vehicles is 69% (7.23 MT/yr), 75% (1.99 MT/yr), and 74% (12.92 MT/yr), respectively whereas respective emission from hired vehicles is 17%, 17.4%, and 18%. Movable compactor type-I departmental vehicles show maximum coverage for the emission of these pollutants. Emission of SO_2 and PM from the departmental vehicles is 83% and 87%, respectively whereas emission from hired vehicles is 9.4% and 12%.

4.2 Comparative Analysis

It has been possible to minimize every single pollutant emission by incorporating the integrated MSW management system. Detailed analysis of percentage reduction of these pollutant emissions is shown in the following Fig. 6. It is possible to reduce the total CO_2 emission by 18% in integrated SWM system over the existing system. Dumper placer type-I and movable compactor type-II departmental vehicles show maximum percentage of reduction among others (20%–29%), i.e., 41% and 37%, respectively. Though it has been possible to minimize the total CO emission by 1.25% only in which dumper placer type-I and movable compactor type-II vehicles show the maximum percentage of reduction, i.e., 41% and 37%, respectively. In case of SO_2 emission, the model shows a decrease of 6.5% over the existing SWM system. Dumper placer type-I types of departmental vehicles show 60% minimization over existing scenario. Dumper placer type-II and movable compactor type-II are also minimized by 26% and 37%, respectively which is noticeable. In case of NO_x, PM, and hydrocarbon emission, the percentage minimization from transport sector are 8.2%, 13%, and 8%, respectively.

5 Conclusions

Rapid urbanization, economic development, and massive rate of population growth are increasing the management process complex and in turn enhancing the quantity of solid waste for developing countries like India. Due to unplanned urbanization and heavy population demand, most of the Indian cities are experiencing such scarcity in a proper waste management venture. Previously, due to small population demand, the waste was simply acquired in nature. But waste management now a day have been very crucial for the megacities due to lack of proper planning of route selection, waste collection system, improper waste management, lack of proper Material Recovery Facilities (MRF), absence of proper processing facilities and open uncontrolled dumping of MSW is leading to consequent pollution generation both in air, water and soil. Generation of toxic gases, GHGs from older vehicles is causing global

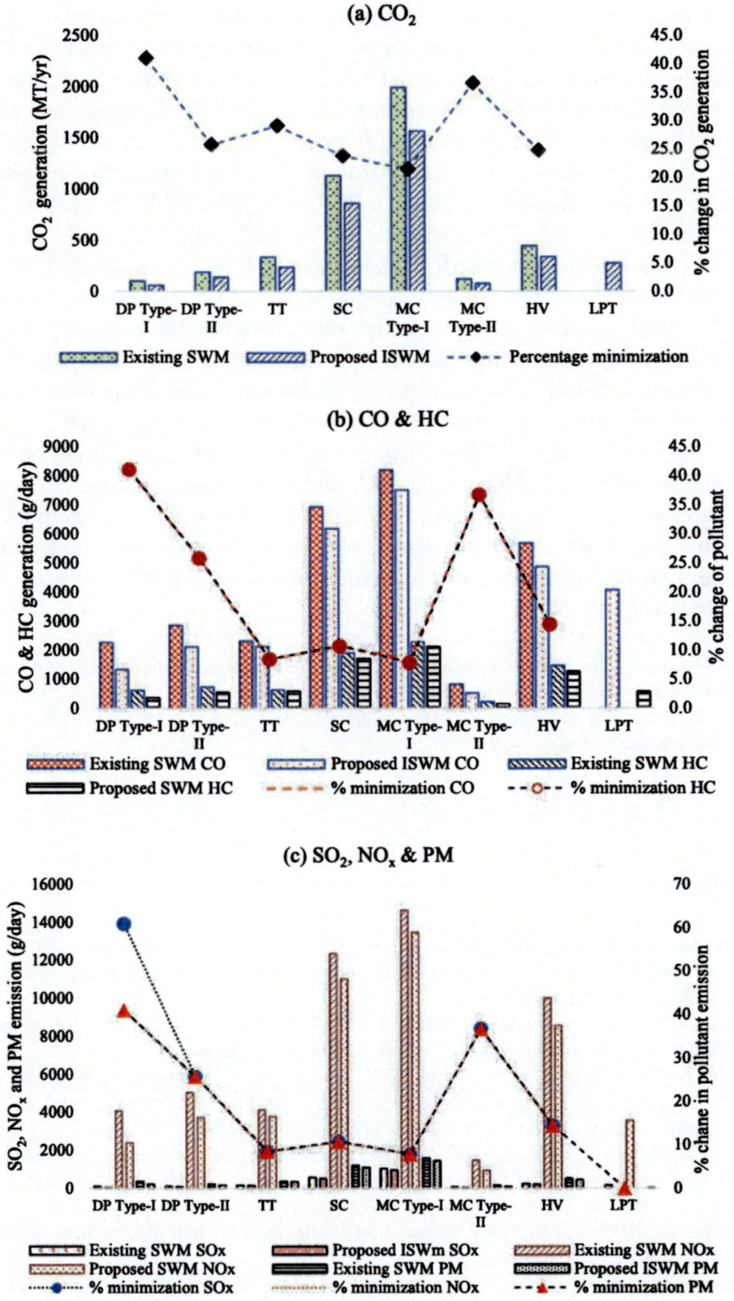

Fig. 6 Comparative analysis of pollutant emission from different vehicles **a** CO_2 (MT/year) **b** CO and HC (g/day) and (c) SO_2, NO_x, and PM (g/day)

warming which not only causes the destruction of resources but also the social and environmental sustainability. To meet the need for a sustainable environment, SWM rule 2016 implies the requirement of an Integrated SWM system to minimize the economic and pollution aspect in a sustainable manner. From the present study, it can be concluded that implementation of integrated approach reduces the pollution emission from the vehicular emission by around 18%. Total pollutant emission from solid waste transport sector of Kolkata in 2010 was 2988 MT/year [12] which has been increased to 4361 MT/year as seen in this study due to increase in number of departmental vehicles and waste compactors. But due to constraints in landfill site, the city's waste needs to dump to the single operating open landfill site Dhapa. Coverage of more distance with more trips requires more fuel consumption which in turn generates more pollutants. Lesser operational efficiency with older vehicles, constant running of engines of stationary vehicles emit more amount of pollutants and create odor problems. To curb this pollution strength, incorporation of modified engines with Bharat Stage VI (BS 6) norms following stringent emission criteria will ultimately reduce the SO_x, NO_x, HC, and PM emission by more than 95%. These strategic methodologies should be applied to the waste management sector not only to reduce the pollution but also to cope with the environmental degradation created through anthropogenic activities with direct and indirect measures toward sustainable development.

Acknowledgements The authors would like to acknowledge the Kolkata Municipal Corporation (KMC) that handles the whole Solid Waste Management system of city Kolkata, India, for their unconditional and enormous contribution to the development of this study and making it helpful for the authors to analyze such system.

References

1. JOAS (Journal of Applied Sciences). (2009). *Air pollution prevention applications for the transport sector by integrating urban area transport and vehicle emission models with the case study of Bangkok, Thailand*. Vol. 9.
2. Holman, C. (1999). Sources of air pollution. In S. T. Holgate , J. M. Samet , H. S. Koren & R. Maynard (Eds.), *Air pollution and health*. Academic Press, London.
3. Jovičić, N. M., Bošković, G. B., Vujić, G. V., Jovičić, G. R., Despotović, M. Z., Milanović, D. M., & Gordić, D. R. (2011). Route optimization to increase efficiency and reduce fuel consumption of communal vehicles. *Thermal Science, 14*(Suppl.), S67–S78.
4. Vilms, M., Voronova, V., & Loigu, E. (2015). Proceedings Sardinia 2015. In *15th International Waste Management and Landfill Symposium*. CISA Publisher, S. Margherita di Pula, Cagliari, Italy.
5. Marilena, K., & Elias, C. (2008). Human health effects of air pollution. *Journal of Environmental Pollution, 151*, 362–367.
6. Environment Canada. (2002). Canada's greenhouse gas inventory. 1990–2000, Greenhouse Gas Division, Ottawa, ON.
7. Forouzanfar, M. H., Alexander, L., et al. (2016). Global, regional, and national comparative risk assessment of 79 behavioural, environmental and occupational, and metabolic risks or clusters

of risks in 188 countries, 1990–2015: A systematic analysis for the Global Burden of Disease Study 2015. *Lancet, 388*, 1659–1724. https://doi.org/10.1016/S0140-6736(16)31679-8
8. Lim, S. S., Vos, T., Flaxman, A. D., et al. (2012). A comparative risk assessment of burden of disease and injury attributable to 67 risk factors and risk factor clusters in 21 regions, 1990–2010: A systematic analysis for the Global Burden of disease study 2010. *Lancet, 380*, 2224–2260. https://doi.org/10.1016/S0140-6736(12)61766-8
9. Gupta, A. K., Karar, K., Ayoob, S., & Kuruvilla, J. (2008). Spation-temporal characteristics of gaseous and particulate pollutants in an urban region of Kolkata, India. *Atmospheric Research*, 103–115.
10. Karar, K., Gupta, A. K., Kumar, A., et al. (2006). Characterization and Identification of the sources of chromium, zinc, lead, cadmium, nickel, manganese and Iron in Pm10 particulates at the two sites of Kolkata, India. *Environmental Monitoring and Assessment, 120*, 347–360. https://doi.org/10.1007/s10661-005-9067-7
11. Zoidis, J. D. (1999). *The impact of air pollution on COPD. RT: For decision makers in respiratory care.*
12. Chattopadhyay, S., Dutta, A., & Ray, S. (2010). Air pollution generation from municipal solid waste transport sector of Kolkata. *Indian Journal of Air Pollution Control, 10*(1), 1–8.
13. Paul, K., Chattopadhyay, S., Dutta, A., Krishna, A. P., & Ray, S. (2019). A comprehensive opti-misation model for integrated solid waste management system—A case study on Kolkata city India. *Environmental Engineering Research, 24*(2), 220–237. https://doi.org/10.4491/eer.2018.132
14. CPCB (Central Pollution Control Board). (2000). Environmental standards for ambient air, automobiles, fuels, industries and noise. CPCB, Govt. of India, New Delhi.

Estimation of soil hydraulic properties in flood-prone zone of NIT Manipur Campus using Mini Disk Infiltrometer

Yumnam Lanchenba Singh, Nameirakpam Momo Singh, and Thiyam Tamphasana Devi

Abstract Hydraulic conductivity is the depth of water that can penetrate into the soil with respect to time under some conditions and hydraulic gradient. Water can pass through the soil under saturated and unsaturated conditions. It is mainly affected by factors like water content, degree of saturation, density, void ration, grain size distribution, particle size, etc. NIT Manipur Langol campus is frequently submerged underwater during rainy season and thus it is imperative to understand the water storage capacity in this region by analyzing the soil behavior at various weather conditions through soil hydraulic conductivity. In this study, Mini Disk Infiltrometer (MDI) was used to obtain the cumulative infiltration curve at different time intervals for a particular location and then hydraulic conductivity value was determined from the field observation data. Eleven different locations were selected in the NIT Manipur Langol campus for the fieldwork and sample was collected from each location for estimating the moisture content and soil condition (soil types, soil weather condition). The study is also attempted to find out the estimated infiltration rate of each location and compare it with the infiltration rate obtain from empirical infiltration models (Hor- ton's and Philip's models). Results showed similar pattern in between the field observation and models of the published literature and Philip's model (considering in all the measurement locations) is the best fitting with high degree of coefficient of determination.

Keywords Mini Disk Infiltrometer · Cumulative infiltration curve · Hydraulic conductivity · Moisture content · Infiltration rate · Infiltration model

1 Introduction

Soil hydraulic properties (infiltration characteristics, soil hydraulic conductivity, etc.) provide an important insight into knowing the soil behavior, especially in flood/drought-prone zones. Hydraulic conductivity is the intensity at which water can

Y. L. Singh · N. M. Singh · T. T. Devi (✉)
Department of Civil Engineering, National Institute of Technology, Langol, Manipur 795004, India

A. K. Dikshit et al. (eds.), *Innovative Trends in Hydrological and Environmental Systems*, Lecture Notes in Civil Engineering 234,
https://doi.org/10.1007/978-981-19-0304-5_6

penetrate into the soil surface at specific soil conditions and hydraulic gradients. It can give you the ease with which pores of soil allow to pass the water through it. Water can pass through the soil under saturated and unsaturated conditions. It is mainly affected by factors like water content, degree of saturation, density, void ration, grain size distribution, particle size, etc. It is necessary to find out hydraulic conductivity in many geotechnical applications like design of drainage, seepage measurement, groundwater recharge, surface runoff, compaction of soil, soil erosion, monitoring of flood and drought, etc. Therefore, determination of soil hydraulic conductivity is required by collecting field observation and one of the commonly used instruments is Mini Disk Infiltrometer (MDI). Infiltration characteristics are mainly related to the hydraulic properties like saturated hydraulic conductivity, unsaturated hydraulic conductivity. Infiltration is the process of movement of water from ground surface to the soil pro- file and infiltration rate is the rate or speed at which water is entering into the soil. It measures how fast water is entering into the soil in certain time period. It is mainly measured in terms of millimeters per hour or centimeters per hour. When precipitation starts, the soil becomes wet and moves down into the soil surface by force of gravity. First, the soil grain becomes wet and extra water penetrates into the soil by gravitational force. Infiltration shows an important role in production of runoff volume, if the intensity of rainfall is more than the infiltration rate, then it consequences in either accumulation of water on surface or in runoff unless there is a physical barrier. The infiltration process is governed mainly by two factors, i.e., gravity which is a natural process and capillary action which is the ability of liquid to pass in narrow space. If there is a space available for holding additional water at soil surface, then the infiltration process will continue until the space is full. Through the larger pores of soil particles, water infiltrates downward by capillarity action and through the smaller pore of soil particles, water infiltrates downward by force of gravity. When the pores of soil particles are fully filled and intake capacity is decreasing, then infiltration rate will reduce. Infiltration rate is high at the beginning and decreases with time upto a nearly constant rate.

Describing of soil hydraulic properties is needed for consistent modeling of flow process and actual flood/drought modeling of the unsaturated zone. Infiltration characteristics of four different soils at different locations and soil hydraulic property such as hydraulic conductivity were calculated from the measured infiltration [1]. Evaluating the infiltration characteristics of sandy soil with different initial compaction states by using MDI and found that infiltration characteristics of soil are influenced by initial soil conditions like water content, dry density. The measured infiltration was used to calculate soil hydraulic properties such as hydraulic conductivity [2]. Estimation of infiltration rate of soils in NIT Kurushetra for the evaluation of performance of infiltration models at 10 different locations by using double-ring infiltrometer. By using two criteria, i.e., coefficient of correlation and root mean square error, the performance of models was determined [3]. Soil hydraulic conductivity is calculated by plotting quadratic equation graphs between cumulative infiltration depths with time at various suction rates of data measured using mini disk infiltrometer [4]. The infiltration rate of soil in NIT Rourkela at two different sites, i.e., garden and forest area was determined by using mini disk infiltrometer and found that infiltration rate is gradually decreased as compared to initial rate and also, it is affected due to cracks in plant root, earth movement, and desiccation of clay [5].

The main objective of this study is to find out the hydraulic conductivity of soil samples in NIT Manipur campus and to compare the estimated infiltration rate of the selected sites with the infiltration rate obtained by using empirical formulas, i.e., Hor- ton's and Philip's model. NIT Manipur Langol campus is frequently submerged underwater during rainy season and thus it is imperative to understand the water storage capacity in this region by analyzing the soil behavior (infiltration capacity, soil hydraulic conductivity, and moisture content and soil types) at various weather conditions.

2 Materials and Methods

2.1 Study Area

NIT Manipur, Langol Campus is in Imphal West District of Manipur was selected for the study. Geographically, it is on the easternmost part of the district which is shown in Fig. 1. The average latitude and longitude of the district are 25° 5 ′N–24° 39 ′N and 94° 8 ′E–93° 53 ′E. The general elevation of Imphal West is around 790 m

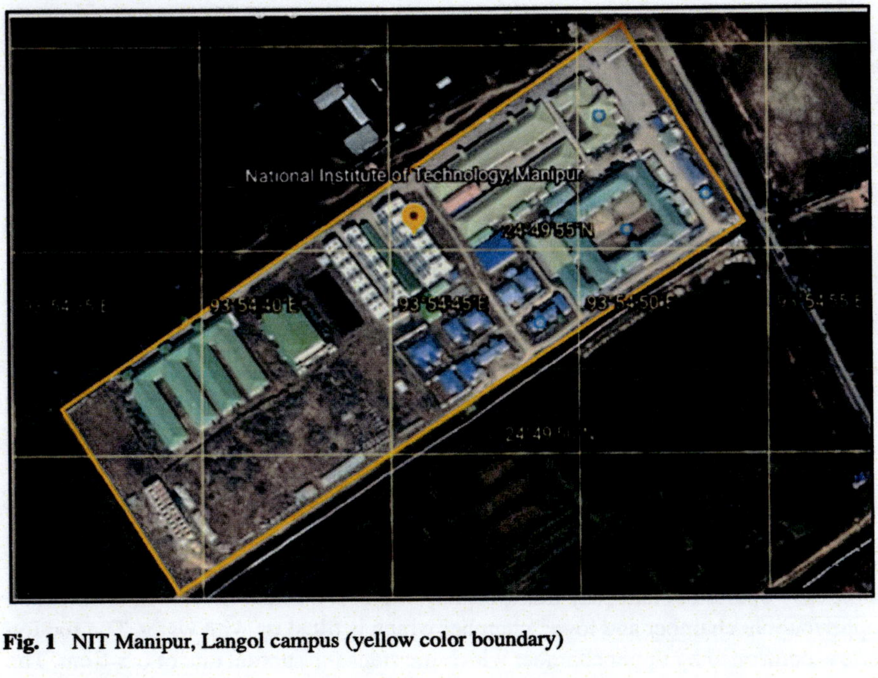

Fig. 1 NIT Manipur, Langol campus (yellow color boundary)

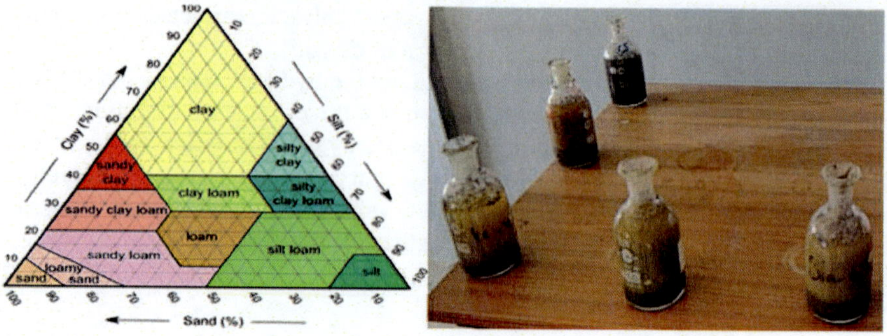

Fig. 2 Soil texture triangular (left) and Soil samples for jar test (right)

above mean sea level. The climate is moderate type in this region. During summer, the temperature ranges from 17 to 36 °C, and in winter, it may go upto sub-zero temperature. High rainfall is observed during the pre-monsoon (March to April) and monsoon season (May to July). Total eleven locations are selected within the study area.

2.2 *Determination of soil type and water content*

For understanding the variability in infiltration characteristics based on soil texture, eleven different locations are selected within the study area and their soil sample is collected and tested to identify the types of soil by using jar test with the help of soil texture triangle chart as shown in Fig. 2 (left). Five types of soil samples (Fig. 2 right side) are identified from the test, i.e., slay, silt loam, sandy loam, clay loam, and loam.

The water content of each soil sample is measured by standard method of oven drying. The sample collection should involve minimum disturbance to the soil. The soil sample is oven dried for 24 hours at 110^0C and calculates the difference of weight of soil sample before and after oven drying.

2.3 *Determination of hydraulic conductivity*

Mini Disk Infiltrometer (MDI) was used for measurement of hydraulic conductivity (K) of soil sample. The component of MDI is shown in Fig. 3a and is set up at the field measurement site in Fig. 3b. It is perfect for field measurement because of its compact size, easy operation, and easy transport anywhere. It consists of two chambers, i.e., upper/bubble chamber and lower chamber which is filled up with water. The suction rate is controlled by upper chamber which provides the suction rate of 0.5-6 cm. The

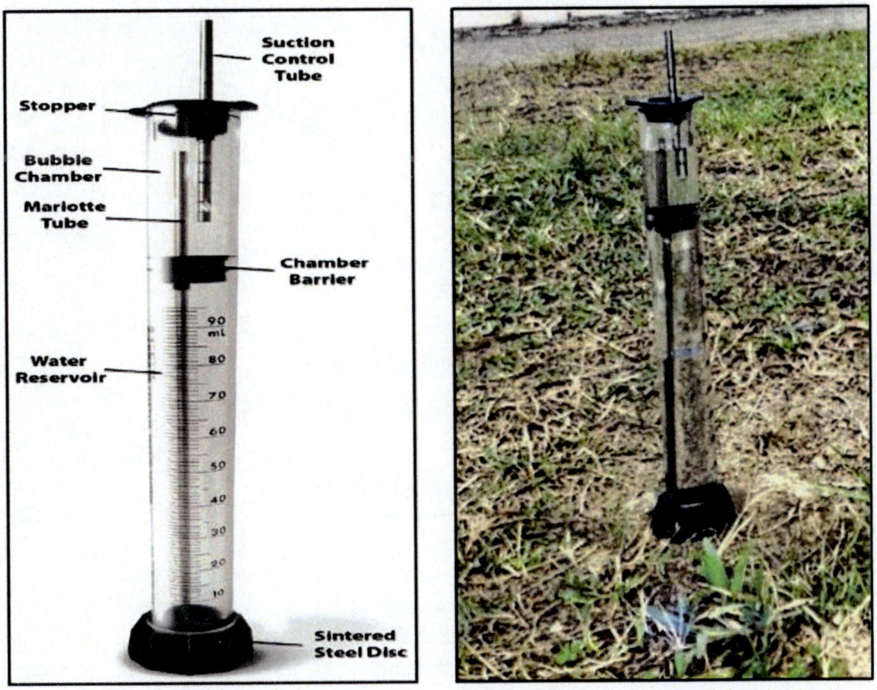

Fig. 3 **a** Components of Mini Disk Infiltrometer (MDI) and **b** its set up at field site

lower chamber provides the required water which is used for infiltrating into the soil. There is a suction control tube that is moved up and down for controlling the suction rate in upper chamber. The lower chamber consists of marriotte tube which is used to transfer suction head to the sintered steel disk through which water is infiltrating into the soil. Water starts to infiltrate into the soil from the lower chamber when it is placed on soil corresponding to time zero. As a function of time, the volume of water that infiltrates into the soil is obtained. For various time intervals, the volume of water infiltrates is obtained by repeating the process. From this, infiltration depth is obtained which is used for determination of hydraulic conductivity of the soil.

There are several methods of determining the soil hydraulic conductivity and a method proposed by Zhang [6], K is determined as in Eqn. (1):

$$K = \frac{C_{\mathrm{I}}}{A} \tag{1}$$

Where, C_{I} = slope of the quadratic equation, A = derived value from the van Genuchten parameters of α, n/h_0, and r_0. The value of A is calculated using Eqn. (2).

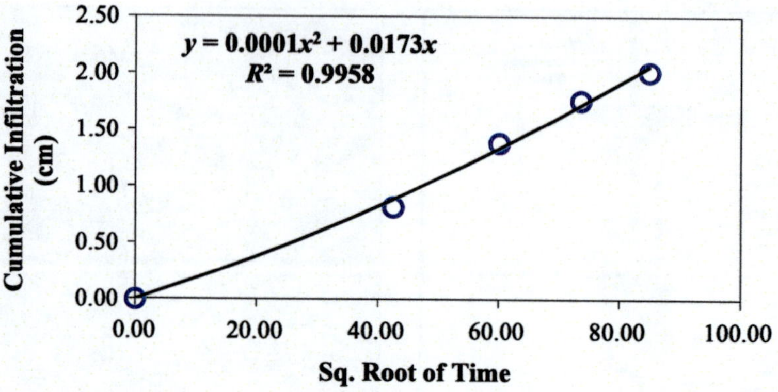

Fig. 4 Quadratic equation plot (Infiltrometer User's Manual [8])

$$A = \frac{11.65(n^{0.1} - 1)e^{[7.5(n-1.9)\alpha h_0]}}{(\alpha r_0)^{0.91}} \tag{2}$$

Where n and α are the van Genuchten parameters for the particular soil, r_0 is the radius of disk (cm) of the infiltrometer used in this study and its value is 2.25 cm. The h_0 is the suction head which depends on the soil types and its value is in the range of −0.5 to −6 cm. For the tabulated values of van Genuchten parameters (different soil types) can be referred to Carsel and Parrish [7]. A quadratic equation is plotted between square root of time in x-axis and cumulative infiltration depth (cm) in y-axis as shown in Fig. 4. The slope of the equation gives the value of C_I (cm/s) which is also known as infiltration slope and then K (cm/s) is calculated using Eqn. (1) for a particular value h_0 for the particular soil types.

2.4 Determination of infiltration rate by using empirical models

Two most commonly used empirical models were selected, i.e., Horton's and Philip's model for evaluating the estimating infiltration rate from the observed data.

2.4.1 Horton's model

Horton expressed the infiltration capacity with time as an exponential decay and is given as in Eqn. (3):

$$f_p = f_c + (f_0 - f_c)e^{-k_h t} \tag{3}$$

Where, f_0 = initial infiltration capacity, f_c = final steady-state infiltration capacity. Also f_c is known as constant infiltration capacity, k_h = Horton's decay coefficient

2.4.2 Philip's model

$$f_p = st^2 + k_p t \quad (4)$$

Where s is the sorptivity of soil and k_p is the Darcy's Hydraulic conductivity (cm/s) Then taking derivative of Eqn. (4), f_p could be expressed as in Eqn. (5):

$$f_p = \frac{1}{2} st^{-1/2} + k_p t \quad (5)$$

Where, f_p = infiltration capacity at any time, t

2.5 Statistical analysis

Coefficient of determination (R^2) is statistical measure that determines the amount of variance in the dependent variable that can be described by the independent variable. It gives the idea of how well the data fits the model. Calculating the coefficient of determination is convenient to define the relationship between observed data and calculated data of infiltration rate. As the coefficient of determination closer to 1 is measured to be the best-fitted model.

3 Results and Concluding Remarks

3.1 Types of soil and water content

The water content value of eleven different locations in the study area (NIT Manipur, Langol Campus) varies from 14.77 to 40.61% and five different soil types are identified, i.e., clay, loam, clay loam, silt loam, and sandy loam as shown in Table 1.

3.2 Hydraulic conductivity

The infiltration depth at the selected time interval and suction head was measured in all the selected locations under different soil conditions based on MDI observations.

Table 1 Types of soil and moisture content of soil for each location

Location	Soil composition	Types of soil	Moisture content (%)
Location 1 (in front of civil block)	Sand-17.14%, Silt-14.29% Clay-68.57%	Clay	16.35
Location 2 (right side of administrative block)	Sand-16.88%, Silt-64.94% Clay-18.18%	Silt loam	15.49
Location 3 (in front of administrative block)	Sand-10%, Silt-30% Clay-60%	Clay	24.91
Location 4 (in front of boys hostel)	Sand-63.49%, Silt-24.39% Clay-12.195%	Sandy loam	33.74
Location 5 (right side of boys hostel)	Sand-57.45%, Silt-31.91% Clay-10.64%	Sandy loam	18.5
Location 6(in front of barrack house)	Sand-34%, Silt-56% Clay-10%	Silt loam	15.106
Location 7 (in front of seminar hall)	Sand-38.88%, Silt-33.33% Clay-27.77%	Clay loam	38.73
Location 8 (in front of mechanical workshop)	Sand-34.78%, Silt-47.82% Clay-17.39%	Loam	40.61
Location 9 (in front of LH2)	Sand-40%, Silt-40% Clay-20%	Loam	21.74
Location 10 (in front of fluid mechanic lab)	Sand-37.93%, Silt-44.82% Clay-17.24%	Loam	25.77
Location 11 (in front of new boys hostel)	Sand-29.6%, Silt-37.03% Clay-33.33%	Clay loam	14.77

The detail of cumulative infiltration versus square root of time for all the locations are depicted and from the graph, we got the slope of curve which is used for determination of hydraulic conductivity as shown in Fig. 5 (for location 1: in front of civil block). The estimated hydraulic conductivity value of each location is shown in Table 2.

From Fig. 5, the observed hydraulic conductivity is maximum at location 11 having a value of 0.0029010 cm/hr, i.e., in front of new boys hostel and is minimum at location 3 having a value of 0.0000214 cm/hr, i.e., in front of administrative block.

3.3 *Comparison of Infiltration rate*

The observed infiltration rate with calculated values by Horton and Philip models are compared.

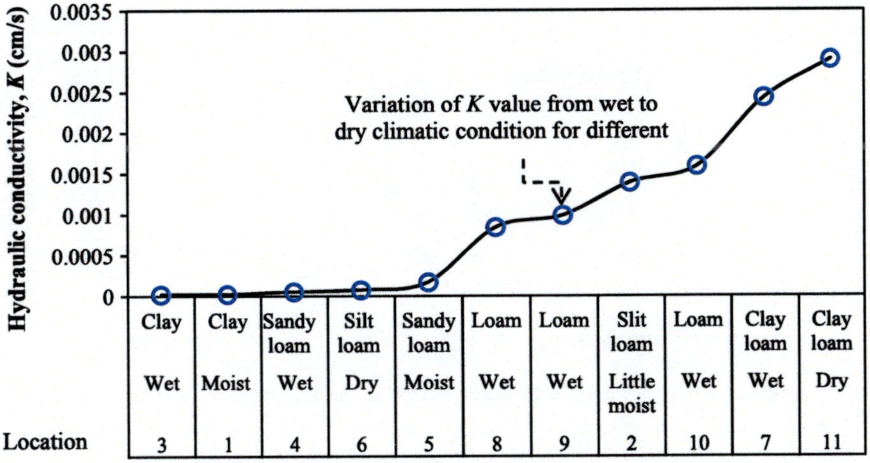

Fig. 5 Variation of hydraulic conductivity, *K* (cm/hr) at different locations

Table 2 Estimated hydraulic conductivity for all locations

Location	Soil condition	Soil type	Hydraulic conductivity, *K* (cm/s)
Location 1	Moist	Clay	0.0000239
Location 2	Little moist	Slit loam	0.001394
Location 3	Wet	Clay	0.0000214
Location 4	Wet	Sandy loam	0.0000538
Location 5	Moist	Sandy loam	0.00017
Location 6	Dry	Silt loam	0.0000751
Location 7	Wet	Clay loam	0.002433
Location 8	Wet	Loam	0.0008463
Location 9	Wet	Loam	0.0009921
Location 10	Wet	Loam	0.0015946
Location 11	Dry	Clay loam	0.0029010

3.4 Calculation of parametric value

The parametric (constant) value of empirical models (Horton and Philip) was obtained by plotting a graph (sample plot only for one location is shown in Fig. 6) and for Hor-ton's model, it is plotted between Ln (f_p–f_c) with time as shown in Fig. 6a for location 1 based on filed observed values which are shown in Table 3 (showing values only for one location).

For Philip's model, it is plotted between observed infiltration rates with square root of time as shown in Fig. 6b. From the graph Fig. 6a, we get the value of k_h = 0.936 and Ln (f_0–f_c) = 0.2707. So, (f_0–f_c) = e(0.2707) = 1.14 and from the graph

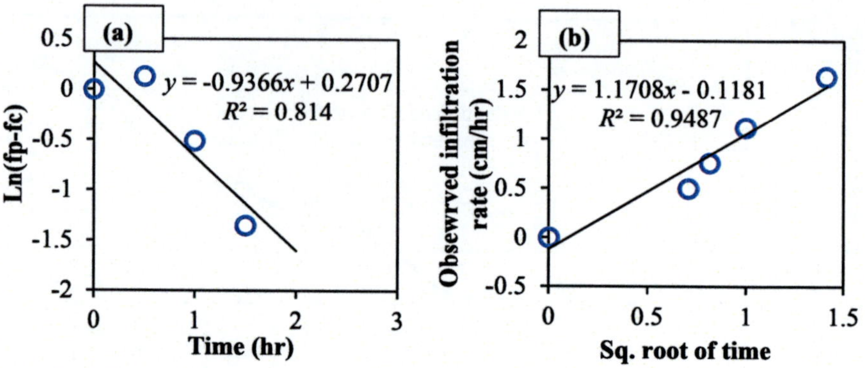

Fig. 6 **a** Plot of Ln(f_p−f_c) with time (Horton's model) and **b** infiltration rate with square root time (Philip's model)

Table 3 Observed infiltration rate (cm/hr) for location 1

Time (hr)	Infiltration depth (cm)	Time interval	Incremental depth (cm)	Observed Infiltration rate (cm/hr)
0	0	0	0	0
0.5	0.82	0.5	0.82	1.64
1	1.38	0.5	0.56	1.12
1.5	0.5	0.38	0.76	1.76
2	*0.5*	*0.25*	*0.5*	*2.01*

Fig. 6b, we get the value of $s/2 = 1.1708$. So, $s = 2.3416$ and $k_p = -0.118$. Likewise, the parametric value of the two empirical models of each location is calculated and is provided in Table 4.

3.5 *Calculation of infiltration rate by empirical models*

The calculated infiltration rate of eleven different locations by Horton and Philip is compared with observed value as shown in Fig. 7a–k. The field observed values of infiltration with the calculated values (infiltration rate by Horton's and Philip's model) for each location (eleven nos.) are also tabulated in Appendix-1. There is similar pattern observed when the field observed infiltration rate is compared with Horton's and Philip's model except in locations 7 and 8. Initially, (when time is zero) infiltration rate is zero then it reaches its peak (maximum) value then decreases because soil starts reaching its saturation point. Once the soil reaches its saturation point, its water holding capacity decreases and it takes longer time to infiltrate water into the soil. Between Horton's and Philip's models, Philip's model can predict in

Table 4 Parametric values of Horton's and Philip's models of each location

Location	Horton's model		Philip's model	
	kh	(f_0-f_c)	*s*	*kp*
Location 1	0.936	1.14	2.341	−0.118
Location 2	84.96	36.2	3.9	−4.89
Location 3	0.636	1.17	1.626	2.5
Location 4	18.633	4.908	1.318	−0.678
Location 5	7.854	3.055	0.758	−0.009
Location 6	35.465	22.51	3.344	−1.286
Location 7	6.013	0.85	0.252	−0.033
Location 8	5.864	0.837	0.408	−0.008
Location 9	3.95	0.85	0.293	−0.028
Location 10	5.593	1.04	0.736	−0.28
Location 11	4.8	1.13	0.618	−0.09

higher accuracy when compared with observed values. It is observed that at different locations, the magnitude of infiltration rate is varied and its maximum range reaches approximately 20cm/hr. But in most of the locations peak reaches upto 1 cm/hr (in five nos. of location) and 5cm/hr (in three nos. of location) on average.

4 Conclusions

Due to variation in soil type and moisture content, the value of hydraulic conductivity is different from one another. The parametric value of empirical models is varied due to variation in soil condition and place to place. It is observed that in most of the locations selected in the study area, Philip's model is best fitting with high degree of coefficient of determination (R^2) for a location and their soil condition. The result shows similar pattern between the field observation and models of the published literature. NIT Manipur Langol Campus is majorly covered by clay soil type which has lower rate of water absorbance capacity as the soil is in saturation state most of the time. Thus, if rainfall continues for few days, soil in this area exceeds its capacity to absorb the surface water and eventually accumulates in this region as the area is almost flat (valley area of the hilly state) which causes frequent floods during rainy season. There- fore, from this study, it can be understood that the soil behavior can provide detailed information for catchment management as preventive measures from floods or droughts. This study also helps in design of hydrological modeling, discharge adjustment, flood management, settlement and consolidation of soil, etc.

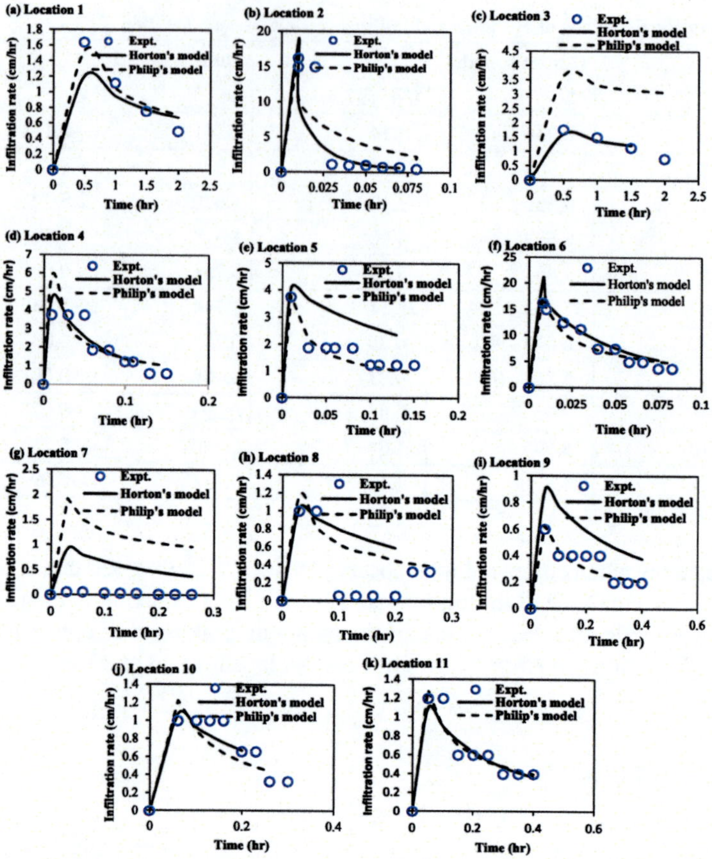

Fig. 7 Comparison of infiltration rate

Appendix

Field observed infiltration rate (cm/hr)

Site	Time (Hr)	Expt. (cm/ r)	Horton's model	Philip's model	Site	Time (Hr)	Expt. (cm/ hr)	Horton's model	Philip's model
Location 1	0	0	0	0	Location 5	0	0	0	0
	0.5	1.64	1.21	1.537		0.01	3.75	4.074	3.78
	1	1.12	0.947	1.052		0.03	1.87	3.664	2.18
	1.5	0.76	0.779	0.837		0.05	1.87	3.313	1.68
	2	0.5	0.675	0.709		0.06	1.87	3.158	1.54
Location 2	0	0	0	0		0.08	1.87	2.88	1.33
	0.01	16.25	18.845	16.911		0.1	1.25	2.644	1.19
	0.01	15	9.797	10.526		0.11	1.25	2.538	1.13
	0.02	15	4.827	7.442		0.13	1.25	2.351	1.04
	0.03	1.125	2.693	5.844		0.15	1.25	–	0.97
	0.04	1	1.611	4.74	Location 6	0	0	0	0
	0.05	1	1.017	3.83		0.008	16.25	20.699	17.13
	0.06	0.75	0.762	3.206		0.01	15	16.512	15.24
	0.07	0.75	0.632	2.7		0.02	12.5	13.025	10.54
	0.08	0.5	–	2.23		0.03	11.25	10.733	8.46
	0.08	0.5	–	1.878		0.04	7.5	9	7.22
Location 3	0	0	0	0		0.05	7.5	7.571	6.37
	0.5	1.76	1.61	3.649		0.058	5	6.627	5.86
	1	1.5	1.37	3.313		0.066	5	5.916	5.44
	1.5	1.14	1.21	3.16		0.075	3.75	5.324	5.05
	2	0.76	-	3.07		0.083	3.75	-	4.76
Location 4	0	0	0	0	Location 7	0	0	0	0

(continued)

(continued)

Site	Time (Hr)	Expt. (cm/ r)	Horton's model	Philip's model	Site	Time (Hr)	Expt. (cm/ hr)	Horton's model	Philip's model
	0.01	3.75	4.693	5.912		0.03	0.06	0.907	1.921
	0.03	3.75	3.426	3.126		0.06	0.06	0.788	1.495
	0.05	3.75	2.553	2.269		0.1	0.04	0.66	1.263
	0.06	1.87	2.224	2.012		0.13	0.04	0.583	1.165
	0.08	1.87	1.725	1.651		0.16	0.04	0.518	1.097
	0.1	1.25	1.381	1.405		0.2	0.02	0.449	1.03
	0.11	1.25	1.252	1.308		0.23	0.02	0.407	0.99
	0.13	0.62	–	1.149		0.26	0.02	0.372	0.961
	0.15	0.62	–	1.023					

References

1. Carsel, R. F., & Parrish, R. S. (1988). Developing joint probability distributions of soil water reten-tion characteristics. *Water Resources Research, 24*, 755–769. https://doi.org/10.1029/WR024i005p00755
2. Naik, A. P., Ghosh, B., & Pekkat, S. (2018). Estimating soil hydraulic properties using mini disk infiltrometer. *ISH Journal of Hydraulic Engineering (online)*. https://doi.org/10.1080/09715010.2018.1471363
3. Bhave, S., & Pekket, S. (2012). *Influence of initial soil condition on infiltration characteristics determined using a disk infiltrometer*. HYDRO, IIT Bombay.
4. Singh, B., Sihag, P., & Singh, K. (2018). Comparison of infiltration models in NIT Kurukshetra campus. *Applied Water Science, 8*(63), 1–8. https://doi.org/10.1007/s13201-018-0708-8
5. Fatehnia, M., & Abichou, K. (2014). Comparison of the method of hydraulic conductivity esti-mation from mini disk infiltrometer. *Electronic Journal of Geotechnical Engineering, 19*, 1047–1063.
6. Champatiray, A., Balmuri, V., Patra, K. C., & Sahoo, M. M. (2014). Standard test for determination of infiltration rate of soil using double ring Infiltrometer. In *Innovative trends in applied physical, chemical, mathematical science and emerging energy technology for sustainable development*, New Delhi, pp. 9–13. ISBN: 978-93-8303-83071-8.
7. Zhang, R. (1997). Infiltration models for the disk infiltrometer. *Soil Science Society of America Journal, 616*, 1597–1603. https://doi.org/10.2136/sssaj1997.03615995006100060008x
8. Infiltrometer User's Manual. (2007). Version 9, Decagon Devices, Inc. 2365 NE Hopkins Court Pullman, WA 99163.

Determination of Infiltration Rate in Imphal West District, Manipur: Field Study

Irungbam Jina, Nameirakpam Momo Singh, and Thiyam Tamphasana Devi

Abstract Present study is on determination of infiltration rate and soil hydraulic conductivity using Mini Disk Infiltrometer (MDI) at different site locations in Imphal West district which is majorly covered by three soil types (clay, clay loam, and marshyland). At each soil types, different land cover types (grass land, barren land, agricultural land, saturated land, etc.) are chosen for measurement. After analyzing the results, it is observed that grass-covered surfaces infiltrated more water than barren surfaces. It is also found that marshyland infiltrated more water than clay and clay loam soil types. Moreover, the area is mostly covered by clay and clay loam soils which has low infiltration rate, indicating that it has low absorbance of excess water and therefore accumulates on the soil surface during high-intensity rainfall. As the study area is in highly flood prone zone, such work will be beneficial in understanding the characteristics of natural grass cover and soil, planning a suitable artificial recharge of groundwater, surface run-off determination, achieving optimum crop production as well as prevention of flood and erosion.

Keywords Infiltration · Clay · Clay loam · Marshyland · Imphal West District · Mini Disk Infiltrometer · Infiltration rate

1 Introduction

Infiltration is the process of intake of water into the soil structure and a complex process with many factors contributing to the rate of infiltration. It is an integral process which adds in recharging groundwater and is an important parameter in the irrigation management, waste management, hydrological modeling of runoff, watershed modeling, etc. [1]. Generally, when the soil is dry, infiltration rate is high and slows down when it is wet as the water passes through a compacted layer or dense clay layer. The rate of infiltration decreases as water temperature reaches a freezing point, and therefore, no water infiltrates the frozen surface. Rashidi et al. [2] found

I. Jina · N. M. Singh · T. T. Devi (✉)
Department of Civil Engineering, National Institute of Technology, Manipur 795004, India

A. K. Dikshit et al. (eds.), *Innovative Trends in Hydrological and Environmental Systems*, Lecture Notes in Civil Engineering 234,
https://doi.org/10.1007/978-981-19-0304-5_7

that rate of infiltration of soil is the integral process which affects the efficiency and uniformity of surface irrigation because of the mechanism of transport and distribution of water to soil profile from the soil surface. Mini Disk Infiltrometer (MDI) also known as tension disk infiltrometer is used in various studies to find the hydraulic conductivity of soil sediments [3]. Angulo-Jaramillo et al. [4] found that tension disk infiltrometer, in spite of all the limitations, has shown to be convenient in determining hydraulic properties of soil in the field as the results obtained is quick and easy when compared to other methods. Clothier and White [5] MDI is a popular and useful instrument for obtaining soil permeability and infiltration characteristics. "MDI gives better results of hydraulic conductivity in comparison with other methods" [6]. Liu et al. [7] studied the rate of water infiltrated in cracked paddy fields and found that rate of infiltration increases significantly. Joshi and Tambe [8] estimated the influence of vegetation cover, slope, runoff on rate of infiltration, and yield of soil sediment at different predicted rainfall conditions in upper Pravara Basin, Maharastra. After which, it was concluded that bare surface with steep slope has minimum infiltration and maximum in those regions where vegetation cover is high and having gentle slope. They further reported the major influencing condition which induces infiltration with lowest runoff was due to vegetation cover which resulted to less sedimentation. Cerda [9] conducted a study in south Spain on infiltration rates considering contrasting slope by ponding method and simulated rainfall, and he reviewed that seasonal change plays a crucial role in differing infiltration rates; however, the steady state of infiltration rate is governed by slope, vegetation cover, and aspect. Adeniji et al. [10] stated that "the infiltration measurement of water in the ground is a key indication regarding improving the accessibility of water for development of plants and its metabolism, the efficiency of drainage and irrigation, enhancing the yield of crops and reducing erosion." Fox et al. [11] studied the influence of slope on last rate of infiltration considering inter-rill conditions and decreasing the infiltration intensity when slope angle increases were observed in their study. Although soil texture is the main factor while studying the varying hydraulic properties of soil using measured infiltration response, it may not be ignored. But, some research works were disclosed, and the differentiation of soil hydraulic characteristics with soil texture is studied [12]. Diamond and Shanley [13] used different techniques and various types of instruments for estimating the rate of infiltration, but the principal methods include sprinkling, flooding of basins or furrows, double or single-ring infiltrometers, and inversed auger hole. Ankeny et al. [14] determined unsaturated hydraulic conductivity using simple field method. Reynolds et al. [15] conducted a study using a tension disk infiltrometer for determining hydraulic conductivity. "The infiltration characteristics of soil under various land use practices in Owerri, South Eastern Nigeria" was conducted by Osuji et al. [16]. Various physical factors affecting infiltration rate for potential artificial recharge site selection were extensively discussed by Avinash [17].

In this study, infiltration rate is determined by using MDI in Imphal West district of Manipur in North Eastern Region of India. Infiltration rate on three soil types (clay, clay loam, and marshyland) was analyzed after taking measurement from field at several location in each soil types. And these results were compared under different climatic conditions.

2 Study Area

The study area selected is Imphal West District in Manipur. Lamphelpat city is the administrative headquarters of the district.

The district occupies an area of 558 km^2 and lies between 24° 30′ N to 25° 00′ N and 93° 45′ E to 94° 15′ E. Imphal West District has three foremost units, i.e., hilly areas in the extremes north, central parts, a tiny plain topography, and marshy land in the southern parts. Once upon a time, the valley was full of marshylands, mostly were Lamphalpat, Takyelpat, Sangaipat, Kakwapat, and Porompat (pat means Lake). Figure 1a–c shows the location map of the study area (Imphal West District) and Fig. 1d shows the location of sample points for the measurement in the study area.

3 Materials and Methods

MDI is used to measure the infiltration rate and soil hydraulic conductivity. Its basic components are shown in Fig. 2 (left) and soil surface profile at the site of measurement for the three types of soil such as (a) clay, (b) clay loam, and (c) marshyland.

The working principle of MDI is explained as: water is to be filled in both lower and upper chambers of the infiltrometer, and then, the upper chamber (or Mariotte chamber/bubble chamber) acts as suction controller. When the MDI is placed on the ground (site of measurement which is shown in Fig. 2 right side), already filled water in the reservoir (or lower chamber) started to infiltrate in to the soil. The rate of infiltrated water is indicated and controlled by a suction tube which is fixed in the bubble chamber of MDI. The other basic components of MDI are that it has a porous disk made up of stainless steel which is attached at the bottom and prevents water leaking. Such disk also helps to firmly hold the instrument on the ground while measurement is noted down. The level of water inside the reservoir could be understood from the graduated scale marked on the instrument. "Once the infiltrometer is on the soil, water begins to leave the lower chamber and infiltrate into the soil at a rate determined by the hydraulic properties of the soil and these measurements are used to calculate hydraulic conductivity" [18].

The field study has been conducted on different places of Imphal West District. After preparing the MDI, the starting water volume of the reservoir is recorded. Initially, the MDI is placed firmly on the ground soil surface, and then, the volume of water infiltrated in the soil is measured with respect to time. This process is performed at a specific suction head in the infiltrometer. For each measurement, the amount of water in the chamber (milliliters) is noted for various time intervals. Then, infiltration

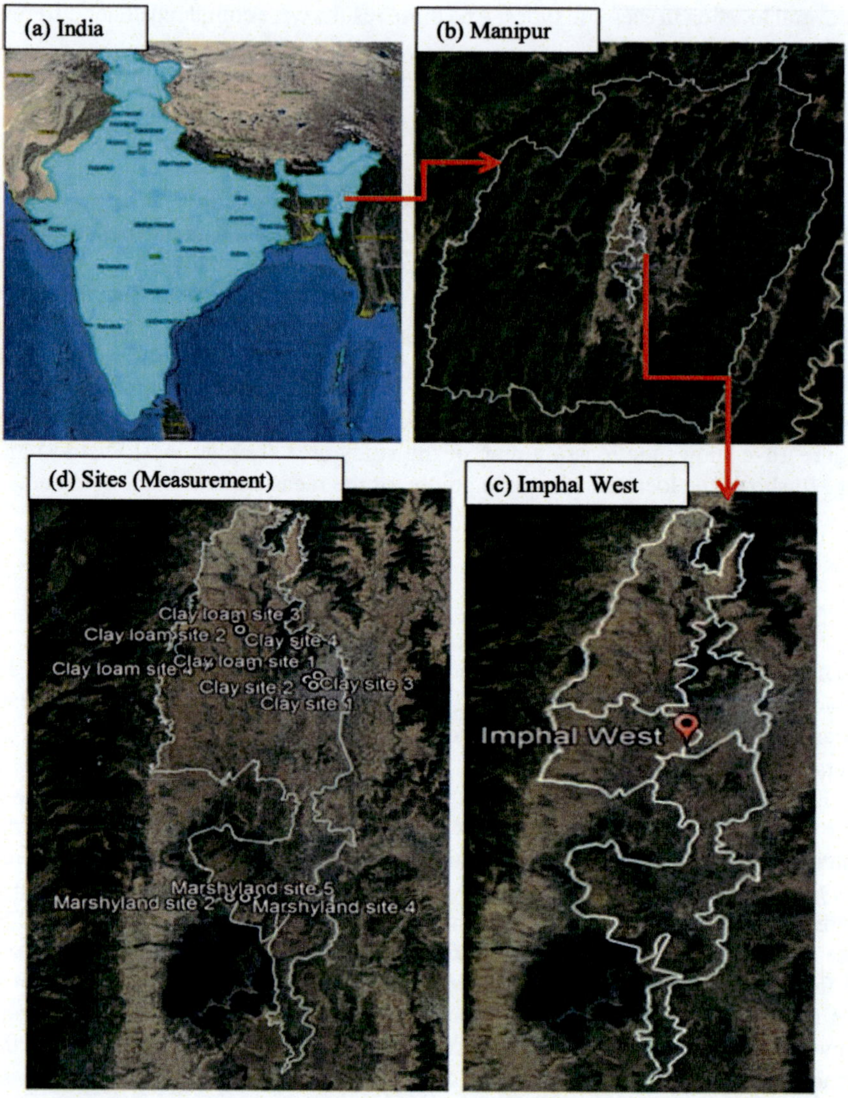

Fig. 1 **a–c** Location map of the study area (Imphal West District) and **d** locations of field measurement

depth (I_{depth}) is obtained. From the I_{depth} of different locations, rate of infiltration (I_{rate}) is calculated.

Soil hydraulic conductivity (K_{soil}) is a very important parameter to understand the nature of the soil types. There is a deep meaning between infiltration rate and hydraulic conductivity separately to understand the soil profile. Simply, infiltration is

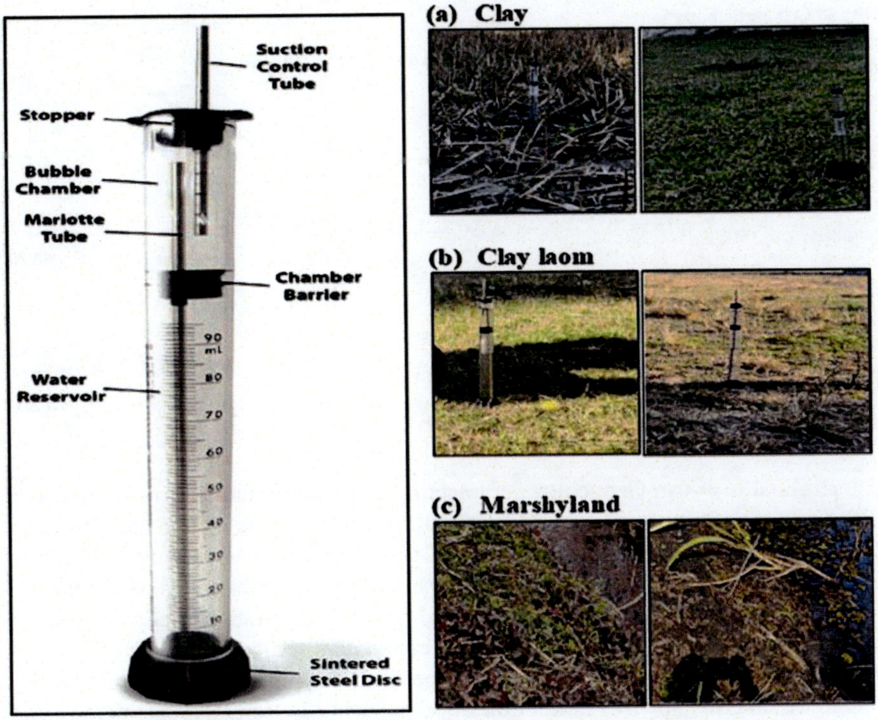

Fig. 2 Components of Mini Disk Infiltrometer (left) and at site for measurement (right) (present study) for different soil types—**a** Clay, **b** Clay loam, **c** Marshyland

the entry of water into the soil (ground surface of soil). For example, for a particular point or location, initial infiltration rate (rate of entry of water) is high, but afterward, it decreases; eventually, it becomes constant. However, hydraulic conductivity is the flow rate of water through the soil (layer or column or profile), and it could be vertical hydraulic conductivity or horizontal hydraulic conductivity base on direction of flow of water. There are several methods of determining the soil hydraulic conductivity, and a method proposed by Zhang [6], K_{soil} is determined as:

$$K_{\text{soil}} = \frac{C_{\text{I}}}{A} \tag{1}$$

where C_{I} = slope of the quadratic equation, A = derived value from the van Genuchten parameters of α, n/h_0, and r_0. The value of A is calculated using Eqs. (2a) and (2b).

$$A = \frac{11.65\left(n^{0.1} - 1\right) e^{[7.5(n-1.9)\alpha h_0]}}{(\alpha r_0)^{0.91}} \tag{2a}$$

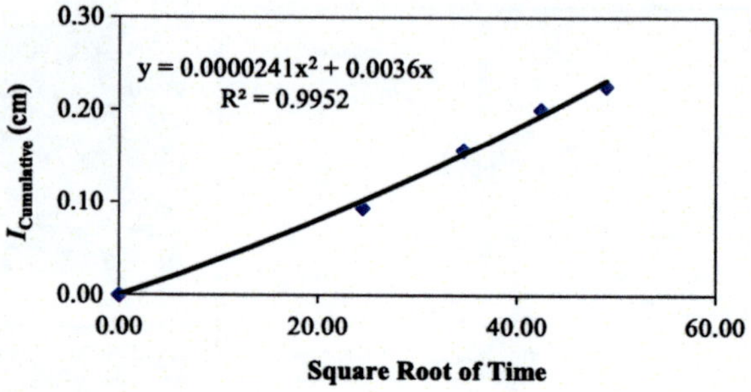

Fig. 3 Quadratic equation plot (Infiltrometer user's manual [18])

$$A = \frac{11.65\left(n^{0.1} - 1\right) e^{[2.92(n-1.9)\alpha h_0]}}{(\alpha r_0)^{0.91}} \quad (2b)$$

where n and $\alpha =$ van Genuchten parameters for the particular soil, $r_0 =$ radius of disk (cm) of the infiltrometer used in this study, and its value is 2.25 cm. The h_0 is the suction head which depends on the soil types and its value is in the range of −0.5 to −6 cm. For the tabulated values of van Genuchten parameters (different soil types) can be referred to Carsel and Parrish [19]. A quadratic equation is plotted between square root of time in x-axis and cumulative infiltration depth ($I_{\text{cumulative}}$) (cm) in y-axis as shown in Fig. 3. The slope of the equation gives the value of C_{I} (cm/s) which also known as infiltration slope and then K_{soil} (cm/s) is calculated using Eq. (1) for a particular value h_0 for the particular soil types.

For marshyland (van Genuchten parameters were not given for this type of site location in the standard table; therefore, particularly for this study, those parameters were derived using the standard table), the suction is considered as −0.5, and therefore, its corresponding values of α, n/h_0 and A are taken as 0.003, 1.06, and 6.4933, respectively. In this study, we defined marshyland as the region where transition from unsaturated to saturated land takes place.

4 Results and Discussions

The measurements are done on various places of Imphal West district with three different soil types, i.e., clay, clay loam, and marshyland between October 2020 and April 2021.

4.1 Infiltration Rate

From the infiltration depth of different locations, rate of infiltration is calculated and graphs of infiltration rate with time are represented for different soil types. The following are the results obtained after field study. The measured infiltration of different sites of clay (Table 1), clay loam (Table 2), and marshyland soil types (Table 3) is shown. The graphs of infiltration rate with time for different places under clay, clay loam, and marshyland are shown in Fig. 4a–c, respectively. And Fig. 5 represents the comparison for different soil types.

From the graphs of infiltration rate with time for different places at three soil conditions, i.e., clay, clay loam, and marshyland, results depict that soil types affect the infiltration rate. It is also observed that the initial infiltration rates were high, and its value decreases with time in all the measurements conducted on three types of soil at different climatic conditions. For clay soil type (Fig. 4a), infiltration rate was recorded highest at clay site 4 (peak was around 2.5 cm/hr achieved at time 0.18 h). For clay loam soil type (Fig. 4b), infiltration rate was recorded highest at clay loam site 4 (peak was around 8.5 cm/hr achieved at time 0.18 h). For marshyland soil type (Fig. 4c), infiltration rate was recorded highest at marshyland site 4 (peak was around at 16 cm/hr achieved at time 0.18 h). From (Fig. 5), comparison for different soil types, infiltration rate was highest for marshyland soil type.

4.2 Soil Hydraulic Conductivity

The K_{soil} different site location for each soil types is presented in Table 4 and in Fig. 6. The required values of infiltration slope C_{I} is determined using quadratic equation plotted in Fig. 3 (Sect. 3) to calculate the K_{soil} value at different site locations for each soil types. Then, K_{soil} is calculated using Eq. (1).

For clay soil, C_{I} and K_{soil} values are in the range of 2.41 to 9.126 × (10^{-05}) (cm/s) and 0.56 to 2.143 × (10^{-05}) (cm/s), respectively. Similarly, for clay loam, C_{I} and K_{soil} are in the range of 0.167 to 23.70 × (10^{-05}) (cm/s) and 0.072 to 3.563 × (10^{-05}) (cm/s), respectively, which is significantly varied in C_{I} values. And for marshyland, C_{I} and K_{soil} significantly vary in the range as 0.2165 to 82.0 × (10^{-05}) (cm/s) and 0.0334 to 12.30 × $(10^{-05)}$ (cm/s), respectively. As we observed different C_{I} and K_{soil} values for different sites location in each soil types, it is imperative to take as much as possible number of sites for the measurement for its accurate results. Such variation is a result of variations in climatic and temporal conditions. There were hot sunny days as well as rainy day during the measurement. It is observed from Fig. 6 that soil hydraulic conductivity is greatly varied in case of marshyland as compared with clay and clay loam. This may be because of considering the parameters for the calculation part and the defining criteria for marshyland as the study region involves such regions where there is transition between soil and water bodies apart from the well-defined clay and clay loam soil.

Table 1 Infiltration measurement result for clay soil

Clay (Site 1)			Clay (Site 2)			Clay (Site 3)			Clay (Site 4)		
Time (s)	I_{depth} (cm)	I_{rate} (cm/hr)	Time (s)	I_{depth} (cm)	I_{rate} (cm/hr)	Time (s)	I_{depth} (cm)	I_{rate} (cm/hr)	Time (s)	I_{depth} (cm)	I_{rate} (cm/hr)
0	0.00	0	0	0.00	0	0	0.00	0	0	0.00	0
600	0.09	0.54	600	0.19	1.14	600	0.25	1.5	600	0.38	2.28
1200	0.16	0.42	1200	0.31	0.72	1200	0.44	1.14	1200	0.69	1.86
1800	0.20	0.24	1800	0.41	0.6	1800	0.57	0.78	1800	0.82	0.78
2400	0.23	0.18	2400	0.47	0.36	2400	0.63	0.36	2400	0.94	0.72

Table 2 Infiltration measurement result for clay loam soil

Clay loam (Site 1)			Clay loam (Site 2)			Clay loam (Site 3)			Clay loam (Site 4)		
Time (s)	I_{depth} (cm)	I_{rate} (cm/hr)	Time (s)	I_{depth} (cm)	I_{rate} (cm/hr)	Time (s)	I_{depth} (cm)	I_{rate} (cm/hr)	Time (s)	I_{depth} (cm)	I_{rate} (cm/hr)
0	0.00	0	0	0.00	0	0	0.00	0	0	0.00	0
600	0.09	0.54	600	0.82	4.92	600	0.31	1.86	600	1.32	7.92
1200	0.16	0.42	1200	1.26	2.64	1200	0.44	0.78	1200	2.52	7.2
1800	0.19	0.18	1800	1.57	1.86	1800	0.57	0.78	1800	2.96	2.64
2400	0.22	0.18	2400	1.76	1.14	2400	0.63	0.36	2400	3.27	1.86

Table 3 Infiltration measurement result for marshyland soil

Marshyland (Site 1)			Marshyland (Site 2)			Marshyland (Site 3)			Marshyland (Site 4)		
Time (s)	I_{depth} (cm)	I_{rate} (cm/hr)	Time (s)	I_{depth} (cm)	I_{rate} (cm/hr)	Time (s)	I_{depth} (cm)	I_{rate} (cm/hr)	Time (s)	I_{depth} (cm)	I_{rate} (cm/hr)
0	0.00	0	0	0.00	0	0	0.00	0	0	0.00	0
600	0.25	1.5	600	2.33	13.98	600	2.64	15.84	600	0.57	3.42
1200	0.35	0.6	1200	3.77	8.64	1200	4.28	9.84	1200	0.82	1.5
1800	0.43	0.48	1800	4.72	5.7	1800	4.97	4.14	1800	1.01	1.14

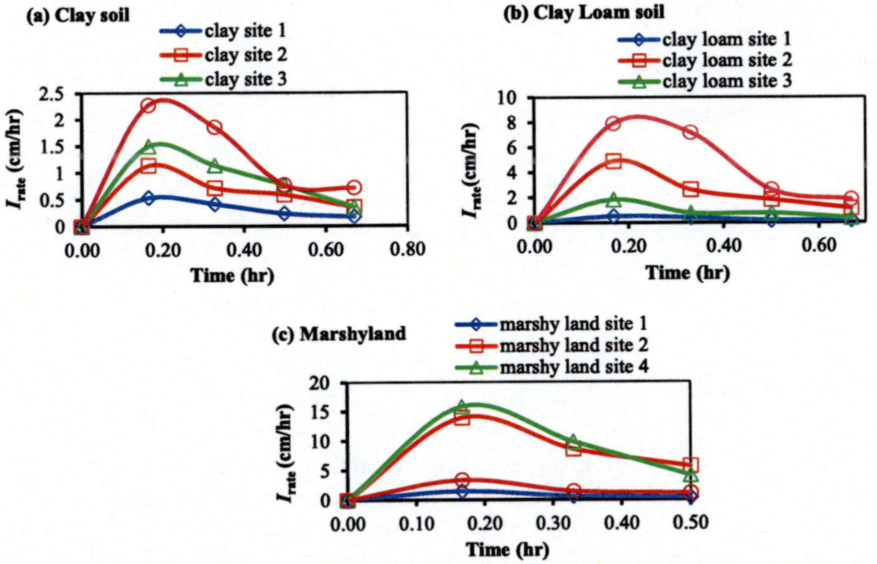

Fig. 4 Graph of infiltration rate with time for **a** clay soil, **b** clay loam soil, and **c** marshyland soil

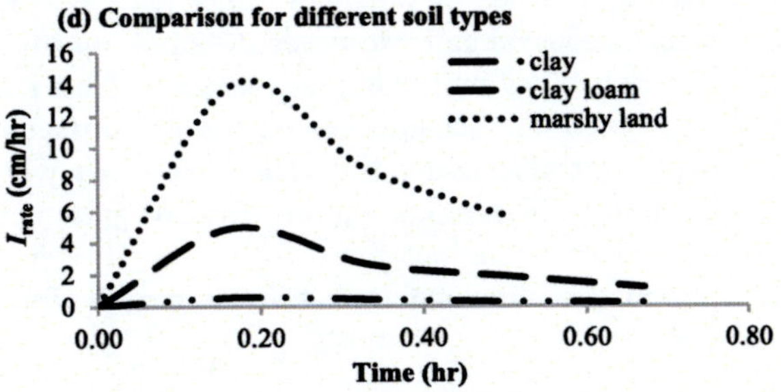

Fig. 5 Graph of infiltration rate with time for clay, clay loam, and marshyland soil types

Table 4 Infiltration slope (C_I) and soil hydraulic conductivity (K) of three soil types

Location of Sites	Clay $\times(10^{-05})$ (cm/s)		Clay loam $\times(10^{-05})$ (cm/s)		Marshyland $\times(10^{-05})$ (cm/s)	
	C_1	K_{soil}	C_1	K_{soil}	C_1	K_{soil}
Site 1	2.41	0.560	0.167	0.254	0.2165	0.0334
Site 2	6.71	1.680	6.795	1.023	82.00	12.30
Site 3	7.72	1.795	0.478	0.072	33.80	5.064
Site 4	9.216	2.143	23.70	3.563	3.145	0.484

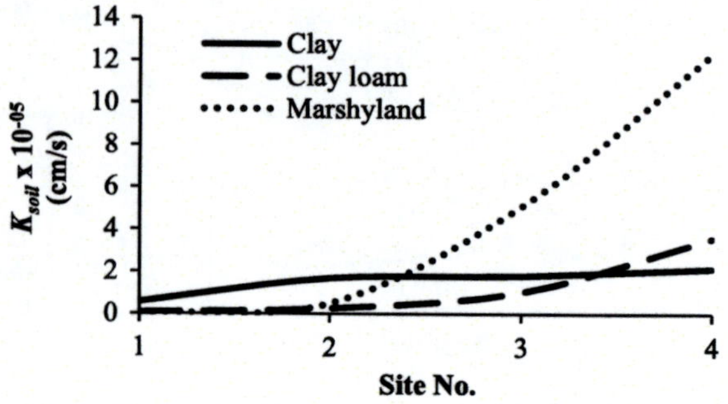

Fig. 6 Soil hydraulic conductivity (K_{soil}) at different site location of each three soil types

5 Conclusions

From the study, it is found that grass-covered surfaces have higher infiltration rate as compared to barren surfaces. It is also observed that marshyland soil types have higher infiltration rate as compared to clay and clay loam soils. This is because increase in plant material, dead or alive, generally improves infiltration. Soil hydraulic conductivity of marshyland also found higher as compared with clay and clay loam. The study area (Imphal West District) lies in the low-lying area of the state and is frequently affected by floods in recent years (2015, 2016, and 2017) because of heavy continuous rainfall. Moreover, the area is mostly covered by clay and clay loam soils which has low infiltration rate, indicating that it has low absorbance of excess water and therefore accumulates on the soil surface during high-intensity rainfall of both short and long duration. Thus, this study will help in combating floods as well as erosion in determining the effect of grass cover on soil, the nature of the soil, estimation of surface run off and groundwater recharge, achieving optimum crop production, etc., and could provide meaningful information to the concerned stakeholders.

References

1. Chow, T. V. (1988). *Applied hydrology*. Mc Graw-Hill Book Company.
2. Rashidi, M., Ahmadbeyki, A., & Hajiaghaei, A. (2014). Prediction of soil infiltration rate based on some physical properties of soil. *American-Eurasian J Agri Env Sci, 14*(12), 1359–1367. https://doi.org/10.5829/idosi.aejaes.2014.14.12.12461
3. Gadi, V. K., Tang, Y. R., Das, A., Monga, C., Garg, A., Berretta, C., & Sahoo, L. (2017). Spatial and temporal variation of hydraulic conductivity and vegetation growth in green infrastructures using infiltrometer and visual technique. *CATENA, 155*, 20–29. https://doi.org/10.1016/j.catena.2017.02.024
4. Angulo-Jaramillo, R., Vandervaere, J. P., Roulier, S., Thony, J. L., Gaudet, J. P., & Vauclin, M. (2000). Field measurement of soil surface hydraulic properties by disc and ring infiltrometers: A review and recent developments. *Soil Till Res, 551*, 1–29. https://doi.org/10.1016/S0167-1987(00)00098-2

5. Clothier, B., & White, I. (1981). Measurement of sorptivity and soil water diffusivity in the field. *Soil Science Society of America Journal, 45*, 241–245. https://doi.org/10.2136/SSSAJ1981.03615995004500020003X
6. Zhang, R. (1997). Infiltration models for the disk infiltrometer. *Soil Science Society of America Journal, 616*, 1597–1603. https://doi.org/10.2136/sssaj1997.03615995006100060008x
7. Liu, C. W., Cheng, S. W., Wen-Sheng, Y., & Chen, S. K. (2003). Water infiltration rate in cracked paddy soil. *Geoderma, 117*, 169–181. https://doi.org/10.1016/S0016-7061(03)00165-4
8. Joshi, V. U., & Tambe, D. T. (2010). Estimation of infiltration rate, run-off and sediment yield under simulated rainfall experiments in upper Pravara Basin, India, Effect of slope angle and grass-cover. *J Earth Sys Sci, 119*(6), 763–773. https://doi.org/10.1007/s12040-010-0055-0
9. Cerda, A. (1996). Seasonal variability of infiltration rate under contrasting slope conditions in South East Spain. *Geoderma, 69*, 217–232. https://doi.org/10.1016/0016-7061(95)00062-3
10. Adeniji, F. A., Umara, B. G., Dibal, J. M., & Amali, A. A. (2013). Variation of infiltration rates with soil texture—A laboratory study. *International Journal of Engineering and Innovative Technology, 3*(2), 454–459. IJEIT1412201308_86.pdf.
11. Fox, D. M., Bryan, R. B., & Price, A. G. (1997). The influence of slope angle on final infiltration rate for interrill conditions. *Geoderma, 80*, 181–194. https://doi.org/10.1016/S0016-7061(97)00075-X
12. Russo, D. (1988). Determining soil hydraulic properties by parameter estimation: On the selection of a model for the hydraulic properties. *Water Research, 243*, 453–459. https://doi.org/10.1029/WR024i003p00453
13. Diamond J, Shanley T (1998) *Infiltration rate assessment of some major soils*. End of Project Report, Armis 4102, Teagasc, Dublin. https://doi.org/10.1080/00750770309555810.
14. Ankeny, M. D., Ahmed, M., Kaspar, T. C., & Horton, R. (1991). Simple field method for determining unsaturated hydraulic conductivity. *Soil Science Society of America Journal, 552*, 467–470. https://doi.org/10.2136/sssaj1991.03615995005500020028x
15. Reynolds, W. D., & Elrick, D. E. (1991). Determination of hydraulic conductivity using a tension infiltrometer. *Soil Science Society of America Journal, 553*, 633–639. https://doi.org/10.2136/sssaj1991.03615995005500030001x
16. Osuji, G. E., Okon, M. A., Chukwuma, M. C., Nwarie, I, I. (2010). Infiltration charac-teristics of soil under selected land used practices in Owerri, South Eastern Nigeria. *World journal of Agricultural sciences, 6*(3), 322–326. https://citeseerx.ist.psu.edu/viewdoc/download?doi=10.1.1.415.2566&rep=rep1&type=pdf
17. Avinash, K. S. (2016). Determination of infiltration rate for site selection of artificial water recharge: An experimental study. *International Journal of Science and Research, 5*(6), 669–705. https://doi.org/10.21275/v5i6.NOV164327
18. Infiltrometer User's Manual. (2007). Version 9, Decagon Devices, Inc. 2365 NE Hopkins Court Pullman, WA 99163.
19. Carsel, R. F., & Parrish, R. S. (1988). Developing joint probability distributions of soil water retention characteristics. *Water Resources Research, 24*, 755–769. https://doi.org/10.1029/WR024i005p00755

Drag Measurement of Coral Reef Monitoring Robot with Various Submergence & Cross-Flow Conditions—A Case Study

Shri R. P. Gupta, K. G. Bhonde, H. R. Khandagale, and Shri Milankumar Someshwara

Abstract Drag is experienced by water due to wave formation when it is toed against water. Drag analysis is required to find out the hydrodynamic force acting on the object which will be useful for selection of propulsion system. The study envisages indigenously designing & developing technology for Coral Reef Monitoring Robot (CBOT). The approach and methodology adopted for evaluation of drag on an autonomous underwater vehicle (AUV) using Current Meter Rating Trolley (CMRT). The CBOT is held between two rigid rods with allowable flexibility in movement to cope with change in movement as well as fluid velocity. As the trolley travels in forward direction, the CBOT is exposed to backward movement. The experiments were conducted on CBOT for different depths of submergence for straight flow conditions and different analyses for cross-flow conditions. The experimental data analysis is indicated that drag force increases proportionately with increase in stream velocity. In straight flow condition, reduction in drag force experienced with increase in depth of submergence whereas during cross-flow condition increase in drag force experienced with increase in angle for various trolley velocity.

Keywords Current meter rating trolley · Coral reef monitoring robot · Drag force · Autonomous underwater vehicle

1 Introduction

Drag is experienced by moving body due to resistance offered by water when the body is towed against water stream. Resistance to motion of an object consists of air and water drag. The main components of resistance are friction drag and pressure drag (residual viscous). Drag analysis is required to find out hydrodynamic forces acting on the object, which is a prerequisite for selection of propulsion system. The study envisages indigenous designing & developing technology for Coral Reef Monitoring Robot (CBOT), an autonomous Underwater Vehicle (AUV) which may remain

S. R. P. Gupta (✉) · K. G. Bhonde · H. R. Khandagale · S. M. Someshwara
Central Water and Power Research Station, Khadakwasla, Pune, India

A. K. Dikshit et al. (eds.), *Innovative Trends in Hydrological and Environmental Systems*, Lecture Notes in Civil Engineering 234,
https://doi.org/10.1007/978-981-19-0304-5_8

submerged in coral reef environment for six months. The approach and methodology adopted for evaluation of the drag on CBOT using Current Meter Rating Trolley (CMRT) facility are described. Incorporation of electronic drive for CMRT system enables precise trolley speed ranging from 0.01 m/s to 7.5 m/s within +/- 0.1% accuracy and SCADA for data acquisition and analysis. Optical encoder gives precise measurement of distance traveled by the trolley resulting in accurate estimation of trolley speed.

2 Current Meter Calibration Facility, CWPRS

CW&PRS has a Rating Tank Facility, which was established in 1955 for testing of ship models as well as for calibration of current meters. The tank is 228 m long, 3.66 m wide, and has shallow water section, 0.84 m deep while deepwater section is 2.13 m deep. The rating trolley is self-propelled type which runs on two parallel straight steel rails 4.267 m apart, accurately aligned over the length of tank. The trolley is equipped with electrical servo motors to run along the track at various speeds.

The rating trolley facility was upgraded in 2003 and subsequently in 2020 incorporating electronic drive to achieve precise speed control and SCADA system for data acquisition and processing enabling fast and accurate data handling throughout the speed range, i.e., from 0.01 m/s to 7.5 m/s. Data acquisition software acquires simultaneous data for measurement of distance traveled by trolley through an optical encoder, time, and output of each instrument under test during each test run. Figure 1 shows Current Meter Rating Trolley (CMRT) facility at CWPRS.

The CMRT setup undertakes testing and calibration of contact type as well non-contact type of water stream velocity measuring instruments. This calibration of conventional current meters, i.e., most commonly used cup and propeller-type current meters is undertaken as per the recommendations of ISO 3455 / BIS 13371. The performance of Electromagnetic current meters, paddle wheels, and various velocity sensors working on the principle of Doppler Effect is also tested with this setup.

Fig.1 Current Meter Rating Trolley (CMRT) facility at CWPRS

The speed range for the test is determined by the type, size, and model of the sensor under test. Drag force measurement on CBOT is carried out using current meter rating trolley facility. The CBOT is held between two rigid rods, with allowable flexibility in movement to cope with change in movement as well fluid velocity.

3 Drag Force Evaluation

The drag is developed by a body to overcome the resistance offered to its motion by the flowing fluid. Generally, studies on the air drag and water drag are required to be undertaken extensively. Sometimes both the air as well water drag has to be given consideration simultaneously in a single amphibian body that is exposed to air as well water stream during its transit. It is observed that water drag is prominent in such cases. In case of water bodies fully submerged in water, only water drag is to be considered. Drag analysis is required to find out the hydrodynamic forces acting on the vehicle, which in turn determines the power required to drive the body. There are three different methods available for the drag force evaluations, which are as described below.

3.1 Theoretical Method

Empirical formulas are available for calculating the drag acting on vehicle based on its shape which is as given below:

Total Drag Force,

$$F_D = \frac{\rho S V^2 C_D}{2}$$

where

S	Frontal area of hull, m^2
ρ	Water mass density
CD	Total drag coefficient of similar vehicle
V	Vehicle velocity.

3.2 Model Testing Method

The physical model test is proved to be the most reliable as it derives the common resultant considering performance of all variable parameters. Conventional model testing involves, model fabrication and actual experimental testing in water stream.

Based on results of these tests, modifications in design of water body are made and again tests are carried out, till satisfactory results are obtained. Once the data is available for model, it can be applied to full-scale hull. In expansion of measured drag of the model to estimate the drag for full-scale hull, relationship between the components of resistance and Reynolds's number & Froude's number of model and full-scale hull are required to be considered.

3.3 Computational Fluid Dynamics (CFD) Method

Physical model testing involves a lot of valuable time, manpower, and money. In addition to this, setting up experimental facility is very difficult and expensive. To overcome this problem, Computational Fluid Dynamics (CFD) is very good tool. CFD is a science of finding numerical solutions of governing equations using high-speed digital computers. It is the science of predicting fluid flow, heat and mass transfer, chemical reactions, and related phenomena by solving numerically, set of governing mathematical equations. CFD uses a computer to solve the relevant mathematical equations, using information about parameters under consideration. Broadly, the strategy of CFD is to replace the continuous problem domain with a discrete domain using a grid. In the continuous domain, each flow variable is defined at every point in the domain. CFD has played an important role as a tool to help engineers and designers to achieve a better understanding of physical aspects of fluid flow, visualizing details of the flow in areas that are extremely difficult (or maybe impossible) to adopt using measuring techniques. In this paper, model testing experimental method adopted for the evaluation of the drag is discussed in detail.

4 Drag Measurement of "AUVs" in Straight Open Tank

Autonomous Underwater Vehicles (AUVs) are beneficial and provide substantial tools to achieve various actions and have increasing applications in discovering underwater environments. Due to the uncertainties in underwater environment and nonlinearity in fluid behavior, a control of underwater robots is difficult. Hence a number of parameters are required to monitor and control by the operator while operating AUVs thus becoming a challenge for engineers. Some parameters like position, drag, and direction of motion of the robot are controlled automatically by a control system facilitating substitution for eliminating risk of human life.

National Institute of Oceanography (NIO), Goa, has designed and fabricated a Coral Reef Monitoring Robot (CBOT) for long term monitoring of reefs with emphasis on high-quality photography along predetermined transects and were desirous of measuring the drag force exerted on CBOT due to flowing water when these vessels are deployed in water body. Drag force measurement on CBOT is carried out using Current meter rating trolley facility. The CBOT is held between two rigid rods, with allowable flexibility in movement to cope with change in movement as well fluid velocity. A flexible string connects the CBOT with digital load cell. One end of the string is tied at the CG of the CBOT and at another end, the load cell is connected. As the trolley travels in forward direction, the CBOT is exposed to backward movement initially. Then the model applies force to overcome the stream resistance and tries to attain steady speed same as that of trolley speed. The exerted drag can be directly shown by the digital load cell. The trolley tows the CBOT at different speeds and corresponding drag denoted by load cell is noted whenever steady movement of CBOT is achieved.

5 Autonomous Underwater Vehicle (AUV)

5.1 Coral Reef Monitoring Robot (CBOT)

Autonomous Underwater Vehicles (AUVs) are beneficial and provide substantial tools to achieve various actions and have increasing applications in discovering underwater environments. Figure 2 shows schematic view and Table 1 gives the dimensions of Coral Reef Monitoring Robot under test.

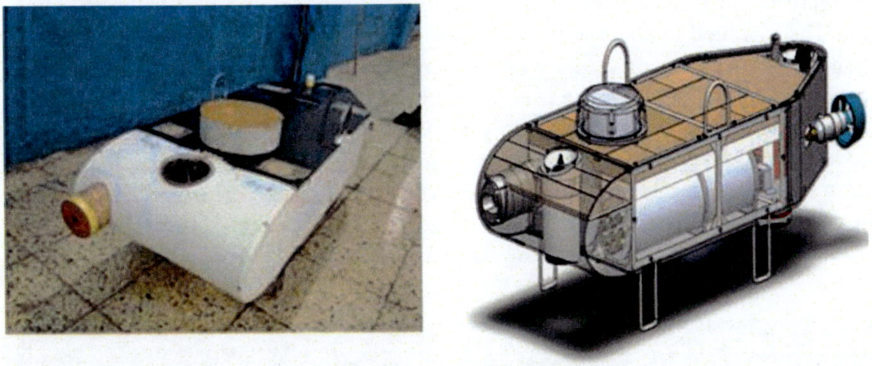

Fig. 2 Schematic view of coral reef monitoring robot (CBOT)

Table 1 Dimensions of CBOT

Name of the autonomous underwater vehicle	Coral reef monitoring robot (mm)
Length	1168
Width	426
Height	380

5.2 *Factors Affecting Underwater Vehicle Motion*

Buoyancy is the ability of an object to float which depends on weight of the body (W) and the buoyant force (B) exerted. Hydrodynamic damping forces act in the opposite direction to the motion of the body mainly due to drag and lift forces. Environmental forces also affect the motion and stability of the vehicle. The pressure drag is observed due to differential pressure between the front and rear exterior surfaces. Hence, it becomes necessary to evaluate the drag critically at various water velocities to consider all above-mentioned fluid dynamics effects.

6 Drag Force Measurement

6.1 *Experimental Setup*

Drag force measurement on CBOT is carried out using current meter rating trolley facility. The CBOT is held between two rigid rods, with allowable flexibility in lateral movement to absorb initial jerk with change in movement as well change in fluid velocity. A string connects the CBOT with digital load cell. One end of the string is tied at the CG of the CBOT and at another end, the load cell is connected as shown in Figure 3. The tension exerted by the string transmits drag from water body to load cell. The load cell is equipped with a spring to absorb the shocks developed while change in movement or velocity of water body is transmitted to load cell through string. The experimental setup is as shown in Figure 3.

As the trolley travels in forward direction, the CBOT is exposed to backward movement. The model applies drag to overcome the stream resistance and tries to attain steady speed same as that of trolley. The exerted drag can be directly shown by the digital load cell in kg. The trolley is run at different speeds and corresponding drag denoted by load cell is noted whenever steady movement of the CBOT is achieved.

Fig. 3 Experimental setup of CBOT on CMRT

Drag force measurement experiments were conducted with straight flow conditions, i.e., the CBOT is facing the streamflow at exactly 0° angles at different depths of submergence. The experiments were conducted at three different depths of submergence, i.e., 12 cm, 23 cm, and 29 cm. with various rating trolley speeds to assess the drag force.

6.2 Procedure

During the drag measurement studies, following procedure is followed:

- The CBOT is held between two rigid rods to position it at the central portion of rating trolley.
- One end of the string is tied to the hook at the CG position of the CBOT.
- Then the boat was lowered gradually to obtain the desired level of submergence.
- Dead weights were kept on the top surface of the CBOT to overcome the buoyancy effect and to achieve desired depth of submergence. The addition of weights should be such that the CBOT is exactly at the horizontal position below the water surface at required submergence.
- Then the string passes over three pulleys and the digital load cell is attached to other end of the string. The string should maintain sufficient tension to transmit the drag developed by the water body to load cell.
- Now the trolley is driven over the rails at minimum speed within predetermined speed range for the drag measurement.
- Once the steady velocity of rating trolley is achieved and the CBOT also attains steady movement same as that of rating trolley, then the digital reading given by the load cell shall be noted.
- In the same way, as mentioned in points f & g, repeat the same procedure for various speed and observed the value of the drag.

6.3 Cross-Flow Condition

Once the observed drag force readings were noted with straight flow conditions at different depths of submergence, a set of experiments were conducted with cross-flow conditions. In case of cross-flow condition, the angle of vertical axis of CBOT which is at 0° in straight flow condition is varied at 5° and 10° as shown in Fig. 4, and accordingly experiment is carried out as per the procedure mentioned above. The set of observations at various rating trolley speeds is conducted and the corresponding values of the drag were noted.

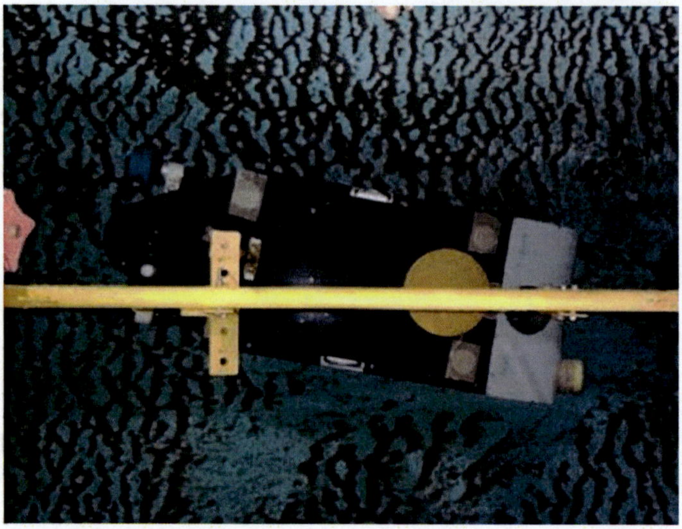

Fig. 4 Experimental setup of CBOT for cross flow on CMRT

7 Test Results

Drag force measurement experiments were conducted with straight flow conditions, i.e., the CBOT is facing the streamflow exactly at 0° angle to direction of flow at different depths of submergence. The experiments were conducted at three different submergence depths, i.e., 12 cm, 23 cm. and 29 cm at different rating trolley speeds, and the results are given in Tables 2, 3 and 4, respectively. During testing, it was observed that the drag force in straight flow condition varies directly with the velocity of rating trolley in turn velocity of water stream. This relationship pattern holds well at all the three submergence depths. In straight flow conditions, reduction in drag force is noticed with increase in submergence as shown in Fig. 5a.

During the cross-flow condition, the angle of vertical axis of CBOT is maintained at 5° and 10°, and accordingly set of observations at various rating trolley speeds were noted and the results are given in Tables 5 and 6, respectively. It was observed that the drag force increases with increase in angle of cross flow as shown in Fig. 5b.

8 Conclusions

The drag force measured data for various trolley speeds, straight flow with submergence condition, and cross-flow condition is analyzed and following conclusion drawn.

Table 2 Drag force with straight flow 12 cm submergence

S. No.	Trolley Velocity (m/s)	Drag force with 12 cm submergence (N)
1	0.234	2.763
2	0.401	4.823
3	0.519	5.939
4	0.609	8.213
5	0.715	14.033
6	0.806	20.658
7	0.923	31.272
8	1.078	43.982
9	1.125	49.869
10	1.228	61.524
11	1.313	71.430
12	1.424	82.585
13	1.536	91.474
14	1.618	106.761
15	1.732	142.286
16	1.828	174.538
17	1.911	201.253

Table 3 Drag force with straight flow 23 cm submergence

S. No.	Trolley velocity VT (m/s)	Drag force with 23 cm submergence
1	0.202	2.254
2	0.519	5.880
3	0.814	16.562
4	1.037	32.438
5	1.251	53.410
6	1.491	79.380
7	1.776	135.240
8	1.872	159.740
9	2.045	189.140

- The value of the drag force increases proportionately with the increase in stream velocity, i.e., trolley speed in case of experiments undertaken.
- In straight flow conditions as the depth of submergence is increased, the reduction of the drag force is observed for various stream velocities.
- Drag force values are observed to follow an increasing trend with increase in angle of cross flow for various stream velocities.

Table 4 Drag force with straight flow 29 cm submergence

S. No.	Trolley velocity VT (m/s)	Drag force with 23 cm submergence
1	0.401	1.225
2	0.592	8.820
3	0.794	16.660
4	1.034	33.320
5	1.213	50.470
6	1.485	79.380
7	1.703	110.740
8	1.890	142.100
9	2.610	244.020

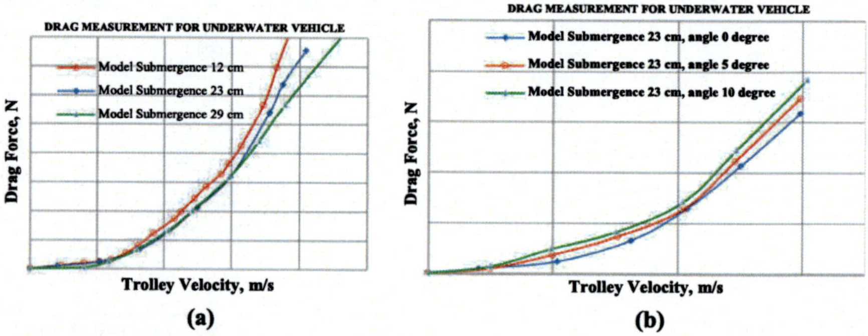

Fig. 5 Graphical presentation of drag at various submergence depth and cross-flow condition

Table 5 Drag force with Cross-flow condition 5°

S. No.	Trolley velocity VT (m/s)	Drag force at 5° crossflow
1	0.252	2.401
2	0.501	9.163
3	0.762	18.52
4	1.025	32.34
5	1.230	55.82
6	1.491	87.34

Table 6 Drag force with cross-flow condition 10°

S. No.	Trolley velocity VT (m/s)	Drag force at 10° crossflow
1	0.252	3.234
2	0.495	12.067
3	0.753	20.580
4	1.020	35.378
5	1.233	60.922
6	1.518	96.040

Bibliography

1. CWPRS Technical report No. 5879. (2020). Drag evaluation of coral reef monitoring robot (CBOT) for national institute of oceanography, Goa.
2. Amromin, E. (2016). A design of low- drag autonomous underwater vehicles and flow control, Trans RINA. *International Journal of Maritime Engineering, 158*(Part A1)
3. IS 13371: 2014 and ISO 3455. (2007). Calibration of rotating element current-meters in straight open tank.
4. Jiang, C. G., Xin, S. C., & Wu, C. W. (2011). Drag reduction of a miniature boat with super hydrophobic grille bottom. *AIP Advances, 1*, 032148.
5. Mohammad, A., & Majid, M. (2015). A review on the drag reduction methods of the ship hulls for improving the hydrodynamic performance. *International Journal of Maritime Technology, IJMT*, *4*, 51–64.
6. Sahoo, A., Dwivedy, S. K., & Robi, P. S. (2020). Dynamic modeling and control of a compact autonomous underwater vehicle. In *RITA 2018, proceedings of the 6th international conference on robot intelligence technology and applications* (pp. 303–321). https://doi.org/10.1007/978-981-13-8323-6_25.

A Remote Sensing, GIS Based Study on LULC Change Detection by Different Methods of Classifiers on Landsat Data

Krison Thokchom Singh, Nameirakpam Momo Singh, and Thiyam Tamphasana Devi

Abstract The present study was conducted to detect landuse/landcover (LULC) by using different methods of classifiers, i.e., (i) SAM (spectral angle mapper) algorithm, (ii) MLC (maximum likelihood classifier) and (iii) MDC (minimum distance classifier) algorithm. Landsat 5 and 8 were used to classify the decadal LULC for the past 31 years (from 1989 to 2020) in the Imphal West district of Manipur, which is a semi-urban region. This region has rapidly urbanized in the past few decades, consequently affecting the LULC pattern, which triggered the need to carry out this study by integrating GIS (Geographical Information System) tools with RS (Remote Sensing) data. The Decadal LULC change showed a significant difference in the areas of build-up (increased by 18.32%), forest (decreased by 4.19%) and agriculture (decreased by 10.17%). Thus, it concludes that there is a possibility that agriculture and forest areas will decrease due to development in the built-up areas as the population in this region has increased considerably. This study also demonstrates the potential of GIS and RS for identifying the LULC using Landsat data with the same resolution and for proper utilization of future land planning policies & natural resources.

Keywords LULC · GIS · RS · Change detection · SAM · MLC · MDC

1 Introduction

Landuse/landcover (LULC) reflects and often represents the overall environmental conditions of human activities as well as natural phenomena of a region. As LULC change informs the relationship between human activities and natural phenomena, a well-ahead and detailed knowledge of the change in LULC is imperative [1]. Assessment of global climate change and other major environmental disturbances like sustainable livelihood, ecosystem conservation & management etc. depends on the accurate information collected through LULC change [2, 3]. Therefore, Remote

K. T. Singh · N. M. Singh · T. T. Devi (✉)
Department of Civil Engineering, National Institute of Technology, Langol 795004, Manipur, India

A. K. Dikshit et al. (eds.), *Innovative Trends in Hydrological and Environmental Systems*, Lecture Notes in Civil Engineering 234,
https://doi.org/10.1007/978-981-19-0304-5_9

Sensing (RS) satellite data is the crucial source of input parameters as the accuracy of modelling and simulation of natural phenomena is based on input satellite data [4, 5]. And one of the major factors for LULC change is urbanization, as it is the most significant anthropogenic activity which destroys several habitats [6]. Urbanization directly impacts the natural vegetation cover as it replaces the natural covers with manmade covers. Though urbanization is seen as an economic advantage for a region, it has several limitations, such as it reduces natural resources and pollutes the environment. Once the natural habitat and environmental ecosystem are destroyed, it is unlikely to revert back to its normal condition.

Literature suggests [7–10] that the integrated approach of GIS and RS can be effectively used to provide information about LULC. Bisht and Kothyari [11] conducted a study on the Gurur Ganga watershed in Uttaranchal. They analyzed the LULC changes from 1963 to 1996 and 1986 to 1996 and observed that, due to urbanization, settlement areas increased whereas forest and barren land areas decreased. However, they also observed that agricultural activities had increased in the region, apparently due to various schemes of the government encouraging farmers to do agricultural activities. Sarma et al. [12] studied the impact of coal mining on LULC change in the Jaintia Hills in Meghalaya, India from 1975 to 2007 using RS and GIS tools. They found that a four-times increase in mining activities decreased the forest cover by three times in the region. Singh et al. [13] conducted a study based on an integrated approach of an advanced GIS tool with RS data to detect the decadal LULC changes in the urbanized district of Allahabad, UP, India. Their study observed that socio-economic and biophysical factors were significantly influenced by the growth of settlement and agricultural areas in the region. Prakasam [14] studied LULC change over a period of 40 years in Kodaikanal taluk, Tamil Nadu. A study found that the settlement and agricultural areas have increased, eventually decreasing the water bodies and forest land areas.

The overall accuracy of LULC change detection is highly influenced by the hypothesis and algorithms of the pixel classifier of the satellite data. There are generally three major LULC classifiers, such as SAM (spectral angle mapper) [15, 16]; MLC (maximum likelihood classifier) [17, 18] and (iii) MDC (minimum distance classifier) [1, 19] apart from ANN (artificial neural network) based models. Several studies have used these classifiers for LULC classification. Petropoulos et al. [20] compared SAM and ANN and found 90.28% and 83.82% overall accuracy in correctly predicting the LULC change. Shafri et al. [21] found that MLC gives 85.56% accuracy while SAM provides only 48.83%. Mishra et al. [18] conducted LULC classification based on MLC and MDC and overall accuracy was found to be better by MLC (79.21%) while for MDC, it was only 74.94%. Sharma et al. [19] have used MDC and found accuracy in the range of 85.97 to 95.03% when using different decomposition levels of satellite images.

The study region is a hub of different newly set-up government institutes, organizations, departments, as well as private properties which are significantly expended every year and significant changes in LULC are expected. Such an abrupt change in LULC always misses timely mapping and assessment of these changes. Rapid population growth has been witnessed in this study region over the past century and the population of Imphal West has dramatically increased from 140,989 in 1901 to 517,992 in 2011 as per Census [22] data. Therefore, in this study, the LULC change for the past three decades (1989–2020) has been studied using GIS tools with remotely sensed Landsat 5 and 8 satellite data in the Imphal West district of Manipur to understand the impact of urbanization. The major three classifiers (SAM, MLC and MDC) are used in LULC classification.

2 Study Area

Imphal West district is one of the semi-urban regions of Manipur, India, which has the highest population in the state. It is located between 24° 49′ N and 93° 54′ E, which is shown in Fig. 1. The total area of this district is 558 km^2. This district has a population of 517,992 as per Census 2011 report with a population density of 992 per km^2. Its population growth rate over the decade 2001–2011 was 15.82% (Fig. 2) and it is in the 545th rank in India out of a total of 640 districts.

Due to an increase in the population, this district is under stress in terms of conserving environmental ecosystems and maintaining the LULC pattern naturally. Figure 2 clearly depicts the rapid increase in the population every decade, which indirectly affects the socio-economic and natural livelihood in the region.

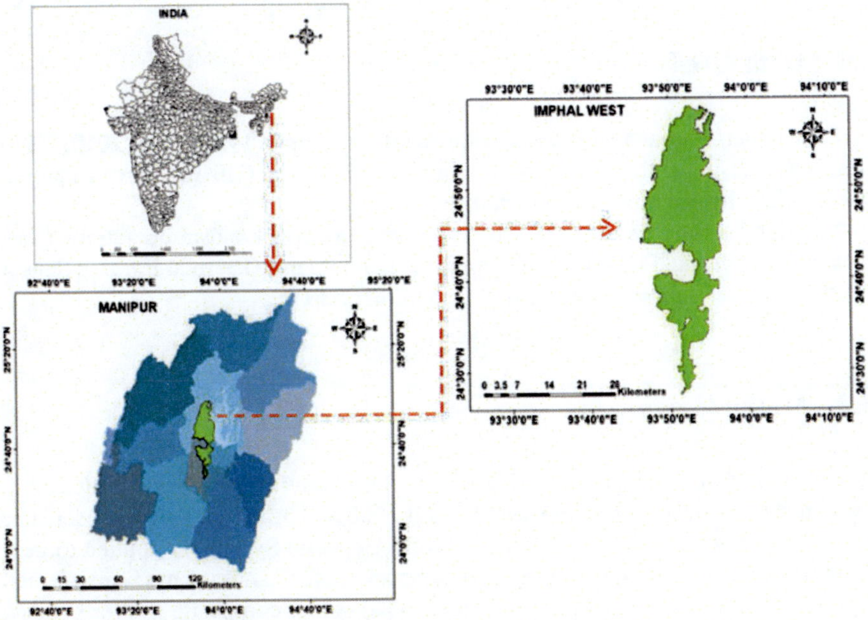

Fig. 1 Location of study area (Imphal West District)

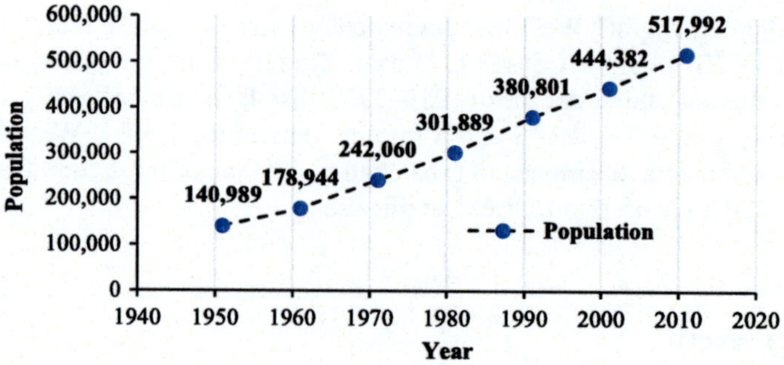

Fig. 2 The growth of population in study area (Imphal West district) (Census, 2011)

Table 1 Descriptions of Landsat satellite data

S. No.	Path/Row	Acquired date	Dataset	Satellite	Source
1	135/043	1989/03/27	TM_L1TP	LC05	USGS
2	135/043	2001/02/08	TM_L1TP	LC05	USGS
3	135/043	2010/02/01	TM_L1TP	LC05	USGS
4	135/043	2020/02/29	OLI_TIRS_L1TP	LC08	USGS

3 Data Used and Methodology

3.1 Data Used

Satellite data (Landsat 5 and 8) was collected (for the years 1989, 2001, 2010, 2020) from the website of the United States Geological Survey (USGS) Earth Explorer. Table 1 lists the description of the data used.

The Landsat images used for this study are from two Landsat operations (TM, thematic mapper, and OLI/TIRS, Operational Land Imager/Thermal Infrared Sensor bands). Landsat images have a spatial resolution of 30 m.

3.2 Methodology

A flowchart of the methodology followed in LULC classification for change detection using different classifiers is presented in Fig. 3. The Universal Traverse Mercator (UTM), zone 46 N with data WGS-84 (World Geodetic System) is applied to geo-reference the data. The colour composite composition for Landsat 5 has been selected as a 4-3-2 band for the red, green and blue (RGB) colour composite and for Landsat 8, it was selected as a 5-4-3 band combination.

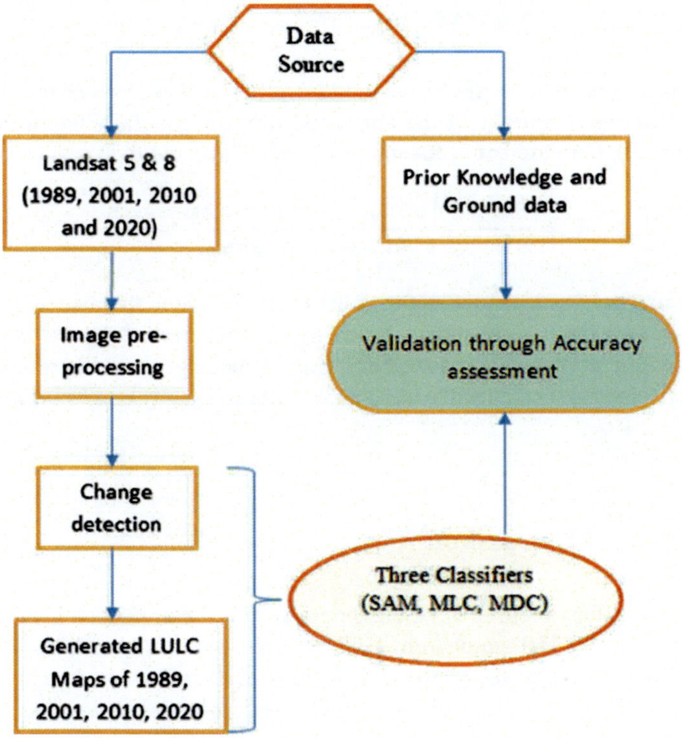

Fig. 3 Flowchart of Methodology used

The raw satellite data collected from a source has to undergo pre-processing (geometric correction, radiometric correction) before it is interpreted for classification analysis. After the satellite image is pre-processed, using three methods of classifier (SAM, MLC and MDC), LULC is generated and compared. The simulated results were validated with ground truth data through accuracy assessment.

The appropriate classifier for accuracy assessment is very important and three classifier algorithms were used in this paper. (i) The SAM (spectral angle mapper) algorithm is an automated method for comparing the direct imagery spectra to known spectra. SAM cares for both the vectors and spectra as vectors & computes spectral point of view between them [23]. (ii) MLC (maximum likelihood classifier) is based on the probability of a pixel belonging to a particular class [24]. The MLC equation assumes that each class in each band has normally distributed statistics and it calculates the probabilities of each pixel belonging to a specified class. (iii) The MDC (minimum distance classifier) algorithm calculates the distance between vector measurement for mean vector and pixel of each signature (Erdas 1999). The candidate pixel is assigned to the class with the closest mean [25].

3.3 Accuracy Assessment

The accuracy assessment for the classified LULC has been performed to validate the classified data. The accuracy of the simulated result is generally based on overall kappa value (K). Co-efficient of Kappa (K) is calculated in Eq. (1) as:

$$K = \frac{\text{Observed accuracy} - \text{Chance agreement}}{1 - \text{Chance agreement}} \quad (1)$$

This statistic serves as an indicator of the extent to which the percentage correct values of an error matrix are due to "true" agreement versus "chance" agreement. As true agreement observed approaches 1 and chance agreement approaches 0, K approaches 1 and this is the ideal case. The value of K usually ranges between 0 and 1 [26].

4 Results and Discussion

The LULC classification uses three classifiers and their comparison is presented in this section. The decadal study from 1989 to 2020 was conducted, but the classified LULC map only for the years of 1989 and 2020 is presented to avoid space consumption when a numerical analysis of all the decades is provided.

4.1 LULC Classification

For every LULC type, 20 samples were selected and accuracy assessments were performed. Table 2 shows the overall accuracy (OA) and Kappa coefficient or simply Kappa (K) for different classifiers, and it shows that the overall accuracy and kappa coefficient for MLC were better when compared to other classifiers.

The MLC method is the most widely used and is considered to be stable with good accuracy as compared with the SAM & MDC algorithms (Table 2) based on accuracy assessment. The comparison of different classifiers of LULC for the years 1989 and

Table 2 Overall accuracy and kappa coefficient for different classifiers

Classifier	1989		2001		2010		2020	
	OA (%)	Kappa	OA (%)	Kappa	OA (%)	Kappa	OA (%)	Kappa
MLC	85.734	0.780	90.352	0.847	88.247	0.814	86.480	0.788
SAM	88.247	0.814	89.412	0.833	87.526	0.803	79.811	0.689
MDC	84.521	0.774	86.524	0.794	89.476	0.834	81.311	0.709

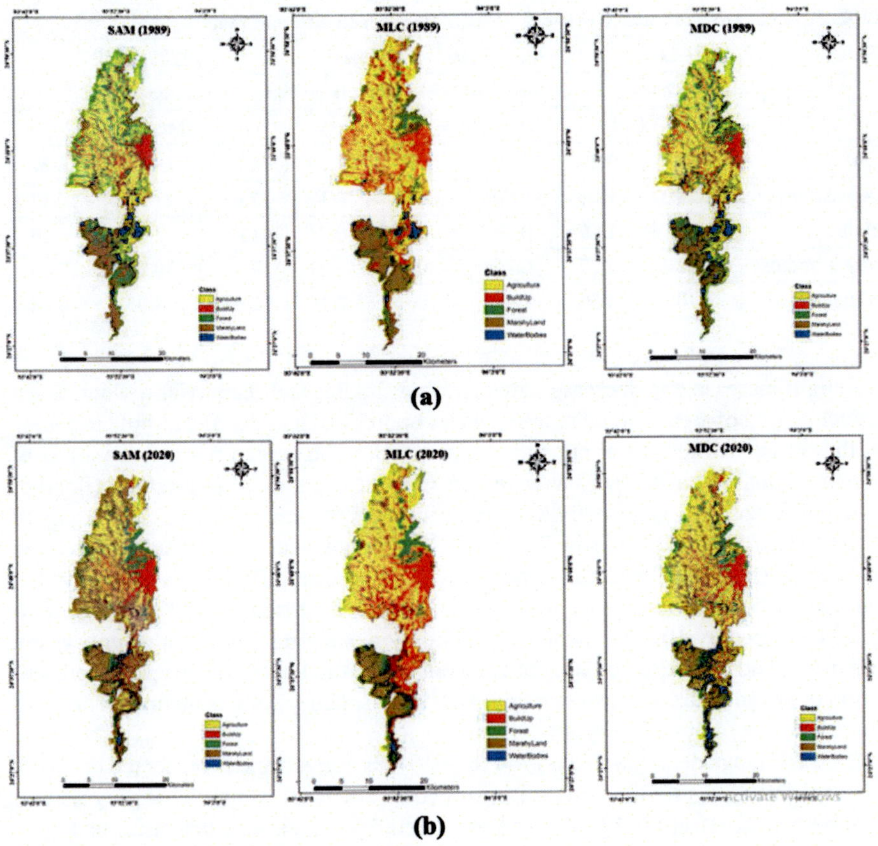

Fig. 4 **a** LULC map for the year 1989 using classifiers SAM, MLC and MDC, **b** LULC map for the year 2020 by using classifiers SAM, MLC and MDC

2020 is presented in Fig. 4a, b respectively. It is clearly visible the difference in classification of LULC among these three classifiers in both the years of calculation.

4.2 Change Detection

As MLC provides better accuracy in classification of LULC, this method is adopted for detecting LULC change. Change detection provides the introduction of human-knowledge from field experience to determine the amount of change over the study period. The difference in area of each land cover class for the entire study was compared in Table 3. In the first decade (1989–2001), the build-up and marshy-land areas increased while agriculture, forest and water bodies decreased. This trend continues in the other remaining decades (2001–2010 and 2010–2020), except there

Table 3 Area (km²) difference between LULC categories in different years

LULC	1989–2001		2001–2010		2010–2020		1989–2020	
	Area change		Area change		Area change		Area change	
	(km^2)	(%)	(km^2)	%	(km^2)	(%)	(km^2)	(%)
Build-Up	2.67	2.59	4.88	4.55	11.53	10.31	19.10	18.32
Agriculture	−2.28	−1.06	−3.50	−1.64	−16.12	−7.69	−21.90	−10.17
Forest	−0.50	−1.09	−0.58	−1.28	−0.84	−1.88	−1.93	−4.19
Water bodies	−3.32	−15.27	0.75	4.08	3.27	17.05	0.70	3.23
Marshyland	3.40	4.30	−1.53	−1.86	2.16	2.66	4.03	5.08

is a slight increase and decrease between water bodies and marshyland. Such slight interchanging of area covered between water bodies and marshyland is quite expected as these two entities are similar in characteristics. Marshyland are those areas of transition phase between land & water, so, in some decades, due to natural (flood or drought) or anthropogenic activities (encroachment of water banks or cleaning up of natural lakes, rivers, streams etc.). However, when decadal LULC change detection in 30 years (from 1989 to 2020) is observed, a significant difference was observed in the areas of build-up (increased by 18.32%), forest (decreased by 4.19%) and agriculture (decreased by 10.17%). And such a situation is alarming for natural ecosystem and livelihood protection as well as the wellbeing of habitats. The graphical presentation of decadal change is shown in Fig. 5, which clearly depicts the increasing and decreasing trend of LULC types.

LULC change detection from 1989 to 2020 was derived and is presented in Fig. 5. Change detection provides precise information about the slowly changing natural phenomena in terms of LULC. The interpretation of change detection in Fig. 6 is explained by the major five LULC types, such as agriculture, build-up, forest, water bodies and marshyland. It is a matrix of five LULC types. How much from agriculture to build-up and vice versa is changed is presented on this map. Similarly, for the other remaining LULC types, change is shown. Such a depiction demonstrates

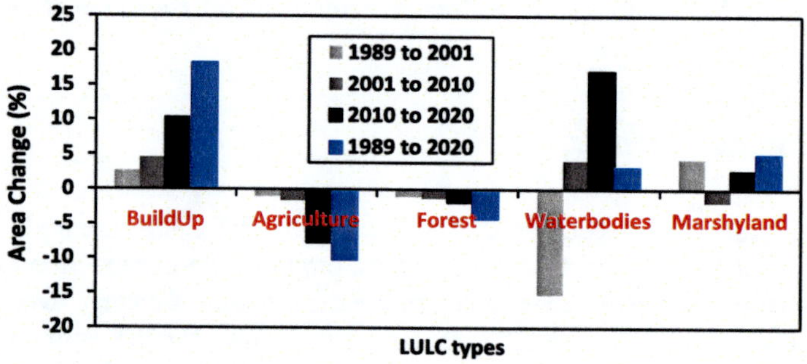

Fig. 5 Comparison of LULC change over the years

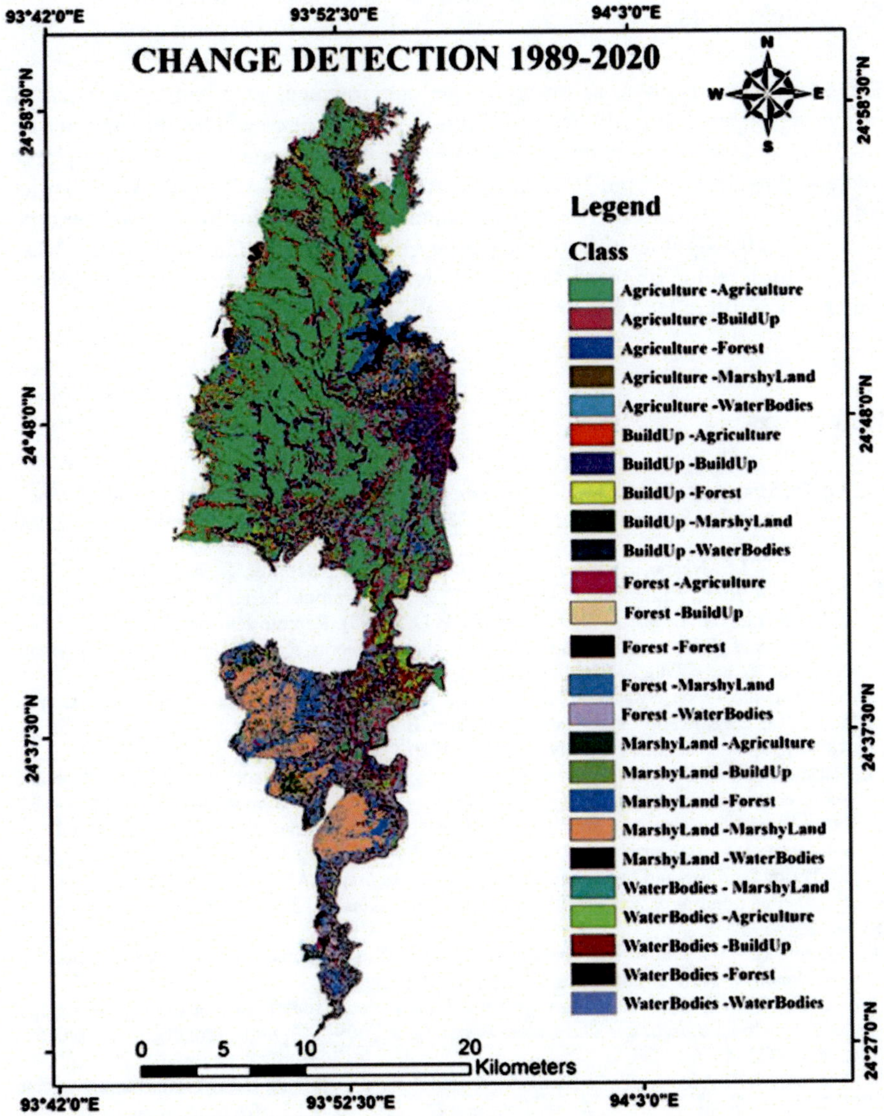

Fig. 6 LULC change detection from 1989 to 2020

the distribution of LULC while change is also shown simultaneously. It could be understood that the study region is majorly covered by agricultural areas and its areas are being converted into build-up areas quite significantly.

5 Conclusions

This study demonstrated, as evidence, the consequences of urbanization with an increasing population in the region. From the results, it concludes that agriculture (10.17%) and forest area decreased (4.19%) due to development in the built-up area (increased by 18.32% from 1989 to 2020) as the population of Imphal West district (study area) has witnessed remarkable population growth during the past century. This study also signifies the advantages of GIS and RS for identifying the LULC changes for a region by using Lansat data with the same resolution and for proper utilization of future land planning policies and natural resources.

References

1. Lu, D., Mausel, P., Brondizio, E., & Moran, E. (2004). Change detection techniques. *International Journal of Remote Sensing, 25*(12), 2365–2401. https://doi.org/10.1080/0143116031000139863
2. Sellers, P. J., Meeson, B. W., Asrar, G., Murphy, R. E., Schiffer, R. A., Bretherton, F. P., Dickinson, R. E., Ellingson, R. G., Field, C. B., Huemmrich, K. F., Justice, C. O., Melack, J. M., Roulet, N. T., Schimel, D. S., & Try, P. D. (1995). Remote sensing of the land surface for studies of global change: Models algorithms experiments. *Remote Sensing of Environment, 51*(1), 3–26. https://doi.org/10.1016/0034-4257(94)00061-Q
3. Vivekananda, G. N., Swathi, R., & Sujith, A. V. L. N. (2021). Multi-temporal image analysis for LULC classification and change detection. *European Journal of Remote Sensing, 5*(S2), 189–199. https://doi.org/10.1080/22797254.2020.1771215
4. Jensen, J. R., & Cowen, D. J. (1999). Remote sensing of urban/ suburban infrastructure and socio-economic attributes. *Photogrammetric Engineering and Remote Sensing, 65*, 611–622. https://www.asprs.org/wp-content/uploads/pers/1999journal/may/1999_may_611-622.pdf
5. Alawamy, J. S., Balasundram, S. K., Hanif, A. H. M., & Sung, C. T. B. (2020). Detecting and analyzing land use and land cover changes in the region of Al-Jabal Al-Akhdar, Libya using time-series Landsat data from 1985 to 2017. *Sustainability, 12*, 4490. https://doi.org/10.3390/su12114490
6. Alphan, H. (2003). Land-use change and urbanization of Adana, Turkey. *Land Degradation Development, 14*, 575–586. https://doi.org/10.1002/ldr.581
7. Aavikson, K. (1995). Simulating vegetation dynamics and land use in a mire landscape using a Markov Model. *Landscape and Urban Planning, 31*, 129–142. https://doi.org/10.1016/0169-2046(94)01045-A
8. Kuemmerle, T., Hostert, P., St-Louis, V., & Radeloff, V. C. (2009). Using image texture to map field size in Eastern Europe. *Journal of Land Use Science, 4*, 85–107. https://doi.org/10.1080/17474230802648786
9. Omar, N. Q., Ahamad, M. S. S., Hussin, W. M. A. W., Samat, N., & Ahmad, S. Z. B. (2014). Markov CA, multi regression, and multiple decision making for modeling historical changes in Kirkuk City. *Iraq Journal of Indian Society Remote Sensing, 42*(1), 165–178. https://doi.org/10.1007/s12524-013-0311-2
10. Tewabe, D., Abebe, A., Enyew, A., & Tsige, A. (2020). Determination of bed width on raised bed irrigation technique of wheat at Koga and Rib Irrigation Projects, North West, Ethiopia. *Cogent Food Agriculture, 6*(1). https://doi.org/10.1080/23311932.2020.1712767
11. Bisht, B. S., & Kothyari, B. P. (2001). Land-Cover Cchange analysis of Garur Ganga watershed using GIS/Remote Sensing technique. *Journal Indian Society Remote Sensing, 29*(3), 137–141. https://doi.org/10.1007/BF02989925

12. Sharma, K., Robeson, S. M., Thapa, P., & Saikia, A. (2017). Landuse/land-cover change and forest fragmentation in the Jigme Dorji National Park. *Bhutan Physical Geography, 38*(1), 18–35. https://doi.org/10.35940/ijitee.F3814.049620
13. Singh, S. K., Mustak, S. K., Srivastava, P. K., Szabó, S., & Islam, T. (2015). Predicting spatial and decadal LULC changes through cellular automata Markov Chain models using earth observation datasets and geo-information. *Environment Process, 2*, 61–78. https://doi.org/10.1007/s40710-015-0062-x
14. Prakasam. (2010). Land use and land cover change detection through remote sensing approach: A case study of Kodaikanal taluk Tamilnadu. *International Journal of Geomatics Geosciences, 1*(2), 150–158. http://www.ipublishing.co.in/jggsvol1no12010/EIJGGS1015.pdf
15. Talukdar, S., Eibek, K. U., Akhter, S., Ziaul, S., Islam, A. R. M. T., Mallick, J. (2021). Modeling fragmentation probability of land-use and land-cover using the bagging, random forest and random subspace in the Teesta River Basin, Bangladesh. Ecological Indicators.https://doi.org/10.1016/j.ecolind.2021.107612
16. Alshari, E. A., & Gawali, B. W. (2021). Development of classification system for LULC using remote sensing and GIS. *Global Transitions Proceedings, 2*, 8–17. https://doi.org/10.1016/j.gltp.2021.01.002
17. Manandhar, R., Odeh, I. O. A., & Ancev, T. (2009). Improving the accuracy of land use and land cover classification of Landsat data using post-classification enhancement. *Remote Sensing, 1*, 330–344. https://doi.org/10.3390/rs1030330
18. Mishra, P. K., Rai, A., & Rai, S. C. (2020). Land use and land cover change detection using geospatial techniques in the Sikkim Himalaya, India. *Egyptian Journal of Remote Sensing and Space Science, 23*, 133–143. https://doi.org/10.1016/j.ejrs.2019.02.001
19. Sharma, J., Prasad, R., Mishra, V. N., Yadav, V. P., & Bala, R. (2018). Land use and land cover classification of multispectral Landsat-8 satellite imagery using discrete wavelet transform. In *The international archives of the photogrammetry, remote sensing and spatial information sciences on geospatial technology—pixel to people, Dehradun, India*, (pp. 703–706). https://www.int-arch-photogramm-remote-sens-spatial-inf-sci.net/XLII-5/703/2018/
20. Petropoulos, G. P., Vadrevu, K. P., Xanthopoulos, G., Karantounias, G., & Scholze, M. (2010). A comparison of spectral angle mapper and artificial neural network classifiers combined with landsat TM imagery analysis for obtaining burnt area mapping. *Sensors, 10*(3), 1967–1985. https://doi.org/10.3390/s100301967
21. Shafri, H. Z. M., Suhaili, A., & Mansor, S. (2007). The performance of maximum likelihood, spectral angle mapper, neural network and decision tree classifier in hyperspectral image analysis. *Journal of Computational Science, 3*(6), 419–423. https://doi.org/10.3844/jcssp.2007.419.423
22. Census. (2011). https://censusindia.gov.in/2011census/dchb/DCHB_A/14/1406_PART_A_DCHB_IMPHAL%20WEST.pdf
23. Thirunavkkarsu, S., & Baboo, S. S. (2014) Performance of RGB and L Base supervised classification technique using multispectral satellite imagery. *Journal of Engineering Research and Applications, 4*(9), 44–49. https://core.ac.uk/download/pdf/25574405.pdf
24. Jensen, J. R. (2005) *Introductory digital image processing: A remote sensing perspective* (3rd ed.), Upper Saddle River, New Jersey: Pearson Prentice-Hall.
25. Erdas Inc. (1999). ERDAS Field Guide. *Imagine* (5th ed.). Atlanta, Georgia: ERDAS.
26. Lillesand, T. M., & Kiefer, R. W. (1999). *Remote sensing and image interpretation* (p. 576). Wiley.

A Comparative Study on Treatment of Industrial Wastewater Using Conventional Biomass and Chlorella Vulgaris

Khushbu Gohel, Nandini Moondra, and R. A. Christian

Abstract Microalgae growth at an industrial level has expanded and broadly diversified. Microalgae biomass production from textile effluent is a possible solution for the environmental effect of discharging the effluent into natural water bodies. This study was conducted to know the efficiency of microalgae in the bioremediation of industrial wastewater. The study comprises a comparison between the efficiency of bacterial biomass and microalgae for bioremediation of wastewater. *Chlorella Vulgaris*, a microalgal species treatability, was studied to efficiently reduce chemical oxygen demand (COD) from industrial effluent. The method involved the analysis of effluent from Common Effluent Treatment Plants (CETPs) and parameters such as pH, dissolved oxygen (DO), mixed liquor suspended solids (MLSS), chemical oxygen demand (COD) were measured. All setups were analyzed by providing specific HRT. This research work showed that microorganisms such as bacteria (conventional biomass) and microalgae enable the removal of pollutants from industrial wastewater. In the setup I, it was observed that the control reactor was able to remove maximum COD of 51%, whereas microalgae removed 56% of COD without adding any nutrition. And in setup II, maximum COD removal efficiency of control was obtained with 56% (with cow dung as a nutrient), whereas 65% was obtained using microalgae (no nutrient was added). The application of microalgae in effluent treatment has many advantages other than pollutants removal, such as CO_2 mitigation, aeration cost, and sludge quantity, etc. This study concludes that microalgae could be more effective by providing sufficient nutrients and favorable conditions to microorganisms.

Keywords Bacterial biomass · COD · Chlorella vulgaris · Industrial effluent · MLSS

K. Gohel · N. Moondra (✉) · R. A. Christian
Civil Engineering Department, Sardar Vallabhbhai National Institute of Technology, Surat, India

R. A. Christian
e-mail: rac@ced.svnit.ac.in

A. K. Dikshit et al. (eds.), *Innovative Trends in Hydrological and Environmental Systems*, Lecture Notes in Civil Engineering 234,
https://doi.org/10.1007/978-981-19-0304-5_10

1 Introduction

Industrialization in economic terms is the main key factor for developing countries. Environmental pollution poses a global threat to public health; certain environmental components have given rise to new environmental conservation efforts for economic and ecological reasons. The wastewater becomes the principal source of environmental pollution. The textile industry is one of the large industries that consume a massive quantity of water and several chemicals used for the dyeing process. The effluent produced from various textile production processes is a complex solution consisting of organic dyes, inorganic contaminants, phosphates, nitrates, and heavy metals. About 50% of the world's manufactured dyes are lost in the dyeing process and are usually released into textile wastewater [1]. Discharging this water directly into the environment contaminates the surface water, groundwater, soil, and air. The wastewater produced by chemical industries generally consists of organic and inorganic matter in different concentrations and a wide range of these substances that cannot be easily degraded. These contaminants are so harmful that they harm aquatic life, plants, animals and affect human health through the food chain. The colored wastewater released into the environment is the cause of a drastic source of environmental degradation. Eutrophication and disruption, including a decline in photosynthesis activity, also reduced DO and increased the biodegradable organic content in aquatic life [2]. Decolorization and detoxification are essential aspects of chemical dye effluents, significant concerns in complying with environmental legislation. Therefore, textile wastewater can potentially harm our environment, and proper treatment should be provided before discharge [3].

Nowadays, the critical concern is developing affordable and efficient methods for handling wastewater discharge from various industries to protect marine life in water bodies. Small-scale industries cannot afford to have their effluent treatment plant. Hence, the facility called CETP has been introduced in which combined effluent from many industries is to be brought to a centralized place for treatment. Many modifications have been introduced to traditional treatment processes as the performance of treatment can be improved. The new approaches could be physicochemical, biological, or both, producing effective technologies to remove wastewater contaminants from textile industries.

Phycoremediation can effectively biodegrade the dyes from textile wastewater effluent with minimum cost and optimum operational time can be considered a green technique. Phycoremediation refers to macroalgae or microalgae for pollutant removal or biotransformation, including nutrients and xenobiotics from wastewater CO_2 from waste air [4]. Wastewater consists of specific components that are essential for the growth of microalgae–bacteria. It can be considered a medium for growing them at a low-cost and new potential market [5]. Algae can combine with activated sludge to form an algae-bacterial symbiosis (ABS) [6]. Algae can use bacteria-exhaled CO_2 to produce O_2 and then contribute to the metabolism of bacteria. An algal–bacterial symbiosis system is known as a classic water self-purification process in natural water. Under light conditions, algae can remove pollutants through its degradation ability, such as biodegradable organic matter, ammoniacal-N, total nitrogen, and total phosphorus, and generate oxygen through photosynthesis utilized by bacteria to degrade pollutants in activated sludge. Therefore, the ABS reduces aeration cost and better removes contaminants than the conventional activated sludge process [7]. It is also effective in CO_2 mitigation and reduces eutrophication effects. Numerous studies have stated that microalgae are beneficial for removing pollutants from wastewater like domestic wastewater, bathroom greywater, dyeing, and printing effluents. Researchers reported that the microalgae strains such as *Nostoc Elepsosporum* and *Chlorella Vulgaris* successfully reduced COD by 98.00% and 97.63%, respectively, during textile effluent treatment. The application of *C. Vulgaris* in the bioremediation of textile wastewater for 28 days. They noted that the COD value decreased from 1700 mg/L to 243 mg/L [8].

The main objective of the study was to examine the treatability potential of *Chlorella Vulgaris* in removing COD from textile wastewater. To determine the effectiveness of the microalgal species, the removal efficiency of the algal system was compared using bacterial biomass (control system) for the primarily treated textile wastewater. This research work was conducted in January 2020. The reactor setup and analysis of samples were done at the institute campus.

2 Methodology

In this study, a microalgae strain named *Chlorella Vulgaris* was used. The microalgae culture was provided by an environmental consultant of Ahmedabad, Gujarat. The CETP selected for the collection of raw wastewater has contaminated wastewater from about 110 textile process houses and about 12 chemical industries and is one of the largest CETP in the city. Primary treated wastewater was collected from the inlet of the biological process at the CETP. Thus the raw wastewater for the study has passed primary screening, equalization, mixing with lime and ferrous sulfate. The initial characteristic of the primary treated industrial wastewater during the study (setup I and setup II) is shown in Table 1.

Table 1 Initial parameters of the wastewater sample of CETP

Parameters	Unit	Influent characteristics	
		Minimum	Maximum
pH		7.03	7.99
DO	mg/L	BDL	BDL
COD	mg/L	600	1200

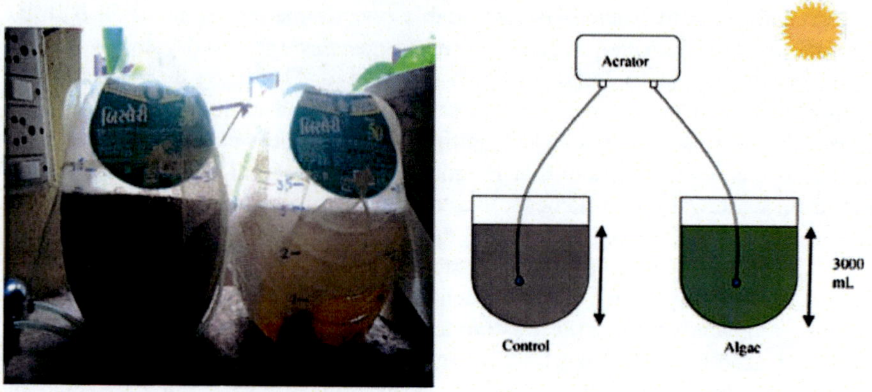

Fig. 1 Actual image and pictorial view of the reactor

2.1 Experimental Design

For the study, two reactors of volume 3L were set up. The study was conducted in an open space with ample sunlight, and external aeration was provided in the system. One reactor acted as a control in which bacterial biomass was provided, collected from CETP's biological reactor. In the second reactor, the influent was treated using microalgae. The microalgae content of 30% of the reactor's total volume was added, as shown in Fig. 1. Hydraulic Retention Time (HRT) of the study was 10 h (9 h aeration +1 h settling). For the acclimatization of algae, an aeration period of 24 h was provided. The supernatant was collected after the settling for the analysis from both the reactors. The study was conducted in two sets. The setup I was conducted without adding nutrients to both reactors, whereas in setup II, nutrients were added in the control reactor only.

3 Result and Discussion

3.1 Setup 1 (No Addition of Nutrient)

During the phase, the parameter studied was pH, MLSS, COD and DO as per the standard procedure mentioned in APHA 2012 [9]. The variation in all influent parameters and effluent in both the reactors for the setup I is shown in Table 2.

In the control system, the value of pH ranged between 6.8 and 7.6. Whereas for the algal system, an increase in pH was obtained due to the photosynthetic effect, ranging between 7.2 and 8.2. The variation of pH for the setup I is illustrated in Fig. 2a. An increase in DO concentration is contributed only due to external aeration 0.6 to 5.2 mg/L in the control reactor, whereas DO range between 1.0 and 5.6 mg/L

Table 2 Variations in parameters in influent and effluent during setup I

Parameters	Inlet	Control Reactor	Algal reactor
pH	7.50 ± 0.26	7.28 ± 0.25	7.67 ± 0.25
DO (mg/L)	0.00 ± 0.00	3.25 ± 1.43	3.55 ± 1.44
COD (mg/L)	931.75 ± 164.23	493.75 ± 78.56	538.13 ± 82.48
MLSS (mg/L)	–	3027.56 ± 767.66	–

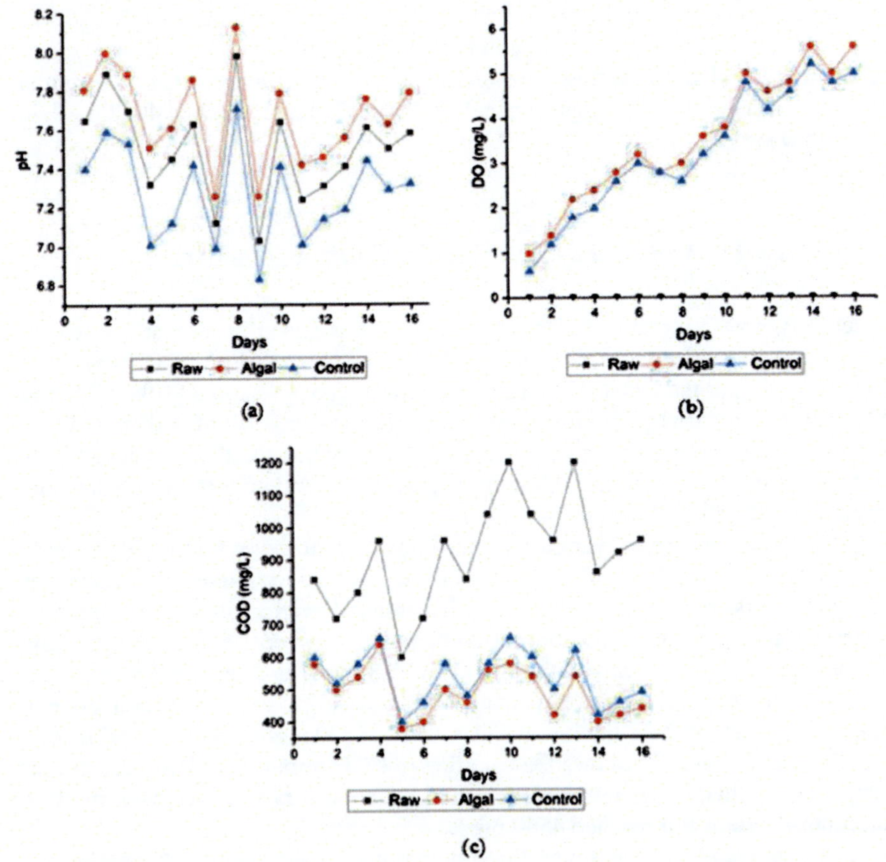

Fig. 2 Variation in **a** pH, **b** DO and **c** COD in algal and control reactor for the setup I

during microalgal treatment, as illustrated in Fig. 2b. An increase in DO for the algal system was the result of photosynthesis and external aeration. The higher the COD value, the higher the biodegradable organic compound in the wastewater, reducing dissolved oxygen [10]. COD removal of 51% had been achieved in the case of control, whereas 56% had been achieved with microalgal treatment. The variation in COD

Table 3 Variations in parameters in influent and effluent during setup II

Parameters	Inlet	Control reactor	Algal reactor
pH	7.55 ± 0.24	7.30 ± 0.26	7.77 ± 0.24
DO (mg/L)	0.00 ± 0.00	4.14 ± 1.21	3.85 ± 1.26
COD (mg/L)	1021.74 ± 146.09	487.83 ± 90.60	558.70 ± 87.30
MLSS (mg/L)	–	3372.35 ± 653.91	–

concentration for both reactors is shown in Fig. 2c. The amount of DO and low F/M ratio in the algal system results in higher COD removal. The concentration of MLSS was calculated to evaluate the microbial production in the system. The initial biomass concentration was approximately 1800 mg/L, and it reached 4072 mg/L in the control system.

3.2 *Setup II (Addition of Nutrient in Control Reactor)*

In setup II, cow dung as a nutrient for bacteria was added in the control reactor only; the rest was similar to the setup I. The raw wastewater had a BOD5 value of 600 mg/L. As per the BOD: N ratio, 30 mg/L nitrogen is required [10], and hence 90 mg of nitrogen is required for 3L of wastewater. The solution of 9000 mg of cow dung in 1L of water was prepared. Therefore, 30 mL of this solution was added daily to the reactor. The variation in all influent parameters and after treatment for both reactors in setup II is shown in Table 3.

In the control reactor, pH varied between 6.6 and 7.8; however, for the algal system, it lies between 7.3 and 8.3, for setup II as shown in Fig. 3a Initially, DO value was BDL for both the reactors, and after that, when influent treated with bacterial biomass (control reactor), effluent DO concentration varied between 1.8 and 5.4 mg/L and for the algal reactor DO vary between 2.0 and 5.8 mg/L as shown in Fig. 3b. COD removal efficiency varied between 56 and 65% when algae were used for treatment. The control reactor had a minimum COD value of 420 mg/L, and the algal reactor had 400 mg/L. Figure 3c shows the variation in COD removal for both reactors for setup 2. The initial biomass concentration in the control system was 2448 mg/L in the control system and reached 4560 mg/L.

During the study, it was found that influent wastewater had nearly neutral pH, high COD, and negligible DO depicting that CETP wastewater was highly polluted and favored the growth of microorganisms. It was noted that the pH value of the control decreases because bacteria release CO_2 gas that is acidic. In contrast, the pH of the algae reactor slightly increases than the inlet because of CO_2 consumption. Also, no significant change in the pH was observed after giving treatment to wastewater of both reactors. Several factors might explain the pH variation, such as microalgal growth and/or excretion of primary metabolites from biodegradation of organic matter [10]. The ratio of MLVSS to MLSS was obtained between 0.40–0.50.

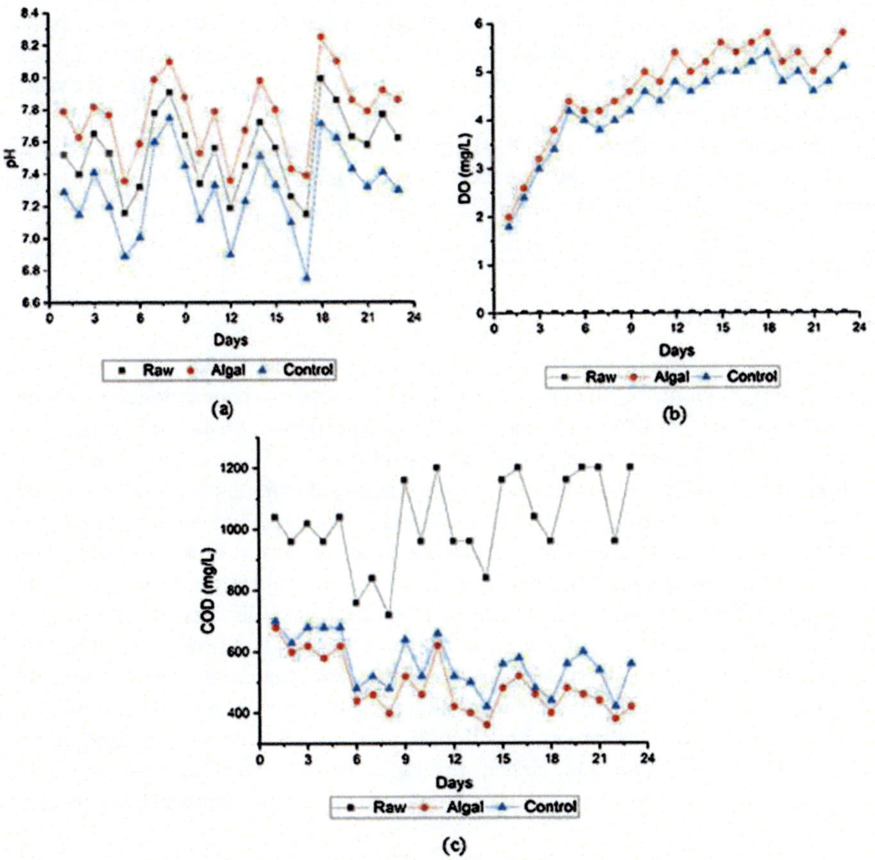

Fig. 3 Variation in **a** pH, **b** DO and **c** COD in algal and control reactor for setup II

Generally, it should be in the range of 0.70–0.80. The discharged limiting value for COD is 250 mg/L. But after giving treatment, these COD values were not obtained within the permissible limit. Hence the further treatment is required to meet the discharging limit. Removal of COD may not have been achieved exclusively by *C. Vulgaris,* but also by other causes, including chemical oxidation caused by aeration of medium and microorganisms present in wastewater, which may facilitate more reduction of COD of the medium [11].

Microalgae are known to remove dyes by bio-adsorption, biodegradation, and bio- conversion. Nitrogen and phosphorous are those nourishment sources that are necessary for the cultivation of microalgae. The efficiency of microalgae for bioremediation can be increased by adding nutrients such as cow dung which is rich in nitrogen and phosphorus. The essential cultural parameters that greatly influence and impact the overall yield of microalgae are the availability of nutrients, pH and salinity, mixing, mixotrophic cultivation, etc. It was also found that light is an essential operating factor in algal culture in terms of light intensity and cycle [12]. Light and temperature fluctuations affect the algal–bacterial wastewater treatment system by affecting the proportions of specific algal groups/species in biofilm [13].

Treatment of dye effluent with the help of microalgae got great interest because of the central role in carbon dioxide fixation [14]. Hence, it can be enclosed that the treatment of textile wastewater with microalgae can be adopted partially.

4 Conclusion

Algal-based effluent treatment is a viable alternative to conventional treatment processes to treat wastewater economically and sustainably. Microalgae have the ability for simultaneous bioremediation, reduction of CO_2, and the development of high added value molecules. During the study, the removal efficiency of the textile wastewater using bacterial biomass and microalgae had been conducted. It was noted that the maximum COD removal efficiency of control was obtained with 51%, whereas microalgae had 56% without adding any nutrition. When cow dung was added to the control reactor, maximum COD removal efficiency was 56%, and in the case of the algae reactor, it was 65% with no addition of nutrition. It can be seen that the efficiency of the algal system was higher than the control reactor for both setups. It can be concluded that this work addresses the application of the microalgae *C. Vulgaris* in textile wastewater, and it could be more effective by providing favorable conditions to microalgae. Hence, further research is required to obtain more efficiency of *C. Vulgaris* when used for the bioremediation of industrial wastewater.

References

1. El-Kassas, H. Y., & Mohamed, L. A. (2014). Bioremediation of the textile waste effluent by Chlorella Vulgaris. *Egyptian Journal of Aquatic Research, 40*, 301–308. https://doi.org/10.1016/j.ejar.2014.08.003
2. Andrade, C. J., & Andrade, L. M. (2018). Microalgae for bioremediation of textile wastewater: An overview. *MOJ Food Processing & Technology, 6*, 432–433. https://doi.org/10.15406/mojfpt.2018.06.00200
3. Elsadany, A. Y. (2018). The use of microalgae in bioremediation of the textile wastewater effluent. *Journal of Nature and Science, 16*, 98–104. https://doi.org/10.7537/marsnsj160318.11
4. Moondra, N., Jariwala, N. D., & Christian, R. A. (2021). Integrated approach of phy- coremediation in wastewater treatment: An insight. *WCM, 5*, 8–12. https://doi.org/10.26840/wcm.01.2021.08.12
5. Gou, Y., Yang, J., Fang, F., Guo, J., & Ma, H. (2020). Feasibility of using a novel al- gal-bacterial biofilm reactor for efficient domestic wastewater treatment. *Environmental Technology, 41*, 400–410. https://doi.org/10.1080/09593330.2018.1499812

6. Ji, X., Jiang, M., Zhang, J., Jiang, X., & Zheng, Z. (2018). The interactions of algae- bacteria symbiotic system and its effects on nutrients removal from synthetic wastewater. *Bioresource Technology, 247*, 44–50. https://doi.org/10.1016/j.biortech.2017.09.074
7. Moondra, N., Jariwala, N. D., & Christian, R. A. (2021). Microalgae based wastewater treatment: A shifting paradigm for the developing nations. *International of Jouranl Phytoremediation, 23*, 765–771. https://doi.org/10.1080/15226514.2020.1857333
8. Subashini, P. S., & Rajiv, P. (2018). An investigation of textile wastewater treatment using Chlorella vulgaris. *Oriental Journal of Chemistry, 34*, 2517–2524. https://doi.org/10.13005/ojc/340538
9. APHA. (2012). *Stardards methods for examination of water and wastewater*. (22nd ed.) Washington (DC). American Public Health Association.
10. Metcalf and Eddy Inc. (2003). *Wastewater Engineering: Treatment Disposal Reuse*. McGraw-Hill, Boston.
11. Hoh, D., Watson, S., & Kan, E. (2016). Algal biofilm reactors for integrated wastewater treatment and biofuel production: A review. *Chemical Engineering Journal, 287*, 466–473. https://doi.org/10.1016/j.cej.2015.11.062
12. Lin, C., Cao, P., Xu, X., & Ye, B. (2019). Algal-bacterial symbiosis system treating high-load printing and dyeing wastewater in continuous-flow reactors under natural light. *Water, 11*, 469–484. https://doi.org/10.3390/w11030469
13. Moondra, N., Jariwala, N. D., & Christian, R. A. (2020). Microalgal-bacterial consortia: An alluring and novel approach for domestic wastewater treatment. *WCM., 4*, 51–56. https://doi.org/10.26840/wcm.01.2020.51.56
14. Su, Y., Mennerich, A., & Urban, B. (2011). Municipal wastewater treatment and biomass accumulation with a wastewater-born and settleable algal-bacterial culture. *Water Research, 45*, 3351–3358. https://doi.org/10.1016/j.watres.2011.03.046

Modeling the Urban Growth of Kochi Using Remote Sensing and GIS

P. C. Afeefa, T. Anjali, D. B. Arunraj, K. Shilpi, Z. A. T. Nadha, and V. Agilan

Abstract The demographic data of India of the past few decades has illustrated a rapidly increasing urban population and a simultaneous growth of urban areas. The growth of urban areas has extensive socioeconomic and environmental effects, and hence, mapping and monitoring is vital for sustainable development, planning and management. This study is an attempt to understand the change in the extent of urban area in the city of Kochi and the immediate surrounding land area in the southern part of India, based on remote sensing (RS) and Geographical Information System (GIS) techniques. Suitable multispectral satellite images of the study area ranging between the years 1988 and 2017 were obtained from Landsat and used to extract land cover maps using QGIS and ArcGIS software. Supervised classification using maximum likelihood classifier (MLC) was applied to prepare land use land cover (LULC) maps. The accuracy of the classified maps was assessed by finding the kappa coefficient. Four major LULC classes—urban area, water body, barren land and vegetation—have been identified. The data obtained indicates a substantial increase in the extent of urban areas in the region.

Keywords Urban growth · Remote sensing · LULC · Kochi

1 Introduction

Urbanization can be defined in simple terms as the process by which an area or region becomes more urban. Some of the main components of urbanization include natural increase of population in the urban areas, net in-migration and widening of the city boundaries to include previously non-urban areas that have now been urbanized. Urban areas are characterized by concentrated population, industrial and commercial activities, buildings and infrastructure. In a way urbanization is desirable for human development. However, it also results in several problems including sub-standard living environment, acute problems of drinking water, noise and air

P. C. Afeefa · T. Anjali · D. B. Arunraj · K. Shilpi (✉) · Z. A. T. Nadha · V. Agilan
Department of Civil Engineering, National Institute of Technology, Calicut 673601, Kerala, India
e-mail: shilpi_b170457ce@nitc.ac.in

A. K. Dikshit et al. (eds.), *Innovative Trends in Hydrological and Environmental Systems*, Lecture Notes in Civil Engineering 234,
https://doi.org/10.1007/978-981-19-0304-5_11

pollution, disposal of waste, traffic congestion, etc. To minimize such environmental and socioeconomic problems in and around cities occurring as a result of rapid urbanization, the technological development in related fields has to address those problems. Such a broad and complex planning would require careful scrutiny of the spatial and temporal changes of the land use classes present in the area, particularly that of the urban area.

Remote sensing and GIS are essential tools in illustrating the nature and extent of urban area as well as other land use classes as they enable rapid and repetitive acquisition of data required for mapping, monitoring and modeling of urban expansion. Analyzing the extent of urban area using these tools coupled with spatial analysis serves as an effective tool for researchers and policy-makers for effective and sustainable land planning and management.

The scope of this paper is to use remote sensing and GIS tools to obtain and classify data on the changes in land use and land cover of Kochi and the immediate surrounding land area over the years 1988–2017. Satellite images obtained by Landsat are processed and classified by applying maximum likelihood classification (MLC), where four major land use classes—urban area, barren land, water body and vegetation—were identified. Maximum likelihood classification assigns each cell in the input raster to the class that it has the highest probability of belonging to. Several iterations were done to obtain a high value of kappa coefficient obtained in the accuracy assessment procedure. The LULC maps hence obtained denote the major shifts, and expansion of urban area and numerical data is also obtained that validates the increase in urban area.

2 Methodology

2.1 Study Area

Kochi is major port city located along the central coast of Kerala, a southern state in India, at an elevation of 0 m. It is one of the most populous metropolitan cities in the country and a major tourist attraction. The city of Kochi and its immediate surrounding land area was chosen as the study area as it has been a prominent center of rapid urbanization during the past two decades. Satellite images of the study area spanning 30 years were downloaded from the United States Geological Survey (USGS) Earth Explorer (http://earthexplorer.usgs.gov/). The study area map is shown in Fig. 1.

Four classes, namely urban area, waterbody, barren land and vegetation, are identified, and level-1 supervised classification was executed on the downloaded images. The satellite images were clipped by stacking the bands 2, 3 and 4 (Landsat 4-5 and 7) or bands 3, 4 and 5 (Landsat 8) to get a false-color composite. For the study, representative training areas were identified to denote the land use and land cover

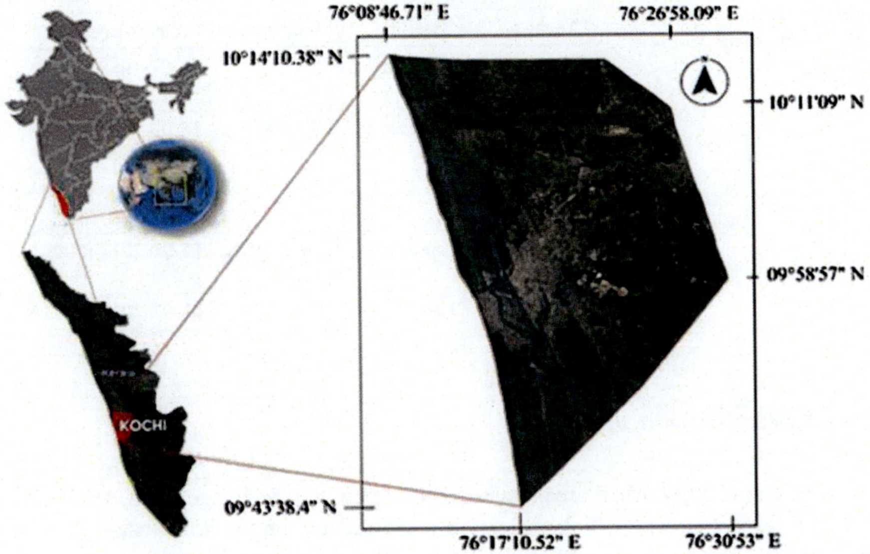

Fig. 1 Location map of the study area

classes such as water bodies, urban area, barren land and vegetation. Maximum likelihood classification (MLC) was selected because it can incorporate the statistics of the training samples before assigning the land covers to each pixel [1]. MLC requires sufficient spectral training sample data for each class to accurately estimate the statistics needed by the classification algorithm. In this 6-8, training samples of each land use and land cover classes were taken. The study area was classified into four classes, namely water bodies, urban area, barren land and vegetation.

2.2 *Data Collection*

Satellite images of the study area spanning 30 years (see Table 1) have been downloaded from the United States Geological Survey (USGS) Earth Explorer using the levels 1 and 2 of Landsat 4-5, 7 and 8. Additional criteria were given as (i) cloud cover less than 10% and (ii) day time. This is done to eliminate errors in classification.

Table 1 Details of satellite images used to model the urbanization

Year	Satellite	Sensor	Date of acquisition
1987–1990	Landsat 4-5	SAM	19-01-1988 (NASA, 1988)
2005–2008	Landsat 4-5	ETM+	14-03-2005 (NASA, 2005)
2008–2011	Landsat 8	OLI_TIRS	20-12-2008 (NASA, 2006)
2011–2014	Landsat 8	OLI_TIRS	11-01-2014 (NASA, 2014)
2014–2017	Landsat 8	OLI_TIRS	03-02-2017 (NASA, 2017)

2.3 *Classification of Images*

Four classes, namely urban area, waterbody, barren land and vegetation, are identified, and level-1 supervised classification was executed on the downloaded images using the software ArcGIS and QGIS 3.4.6.

Pre-processing. The satellite images were clipped by stacking the bands 2, 3 and 4 (Landsat 4–5 and 7) or bands 3, 4 and 5 (Landsat 8) to get false-color composite, FCC images and bands 3, 2 and 1(Landsat 4,5 and 7) or 2,3 and 4 (Landsat 8) true-color composite, TCC images obtained from the USGS site and the study area is masked out.

Supervised Classification

For the study, representative training areas were identified to denote the land use and land cover classes such as water bodies, urban area, barren land and vegetation. The GIS software develops a numerical description of the spectral attributes of each land cover type of interest in the scene, and each pixel in the image dataset is categorized into the land cover class; it most closely resembles. After classification, the results are presented in the output stage, where the classified image and its different attributes are obtained.

Maximum Likelihood Classification. Maximum likelihood classification (MLC) was selected because it can incorporate the statistics of the training samples before assigning the land covers to each pixel. MLC requires sufficient spectral training sample data for each class to accurately estimate the statistics needed by the classification algorithm [1]. For this purpose, 6-8 training samples of each land use and land cover classes were taken. The study area was classified into four classes, namely water bodies, urban area, barren land and vegetation. Using the signature files of each of the categories', supervised classification was carried out.

3 Result and Discussion

3.1 LULC Mapping

Land use and land cover maps covering four classes: water body, urban area, barren land and vegetation from interval 1988–2017 are shown in Fig. 2.

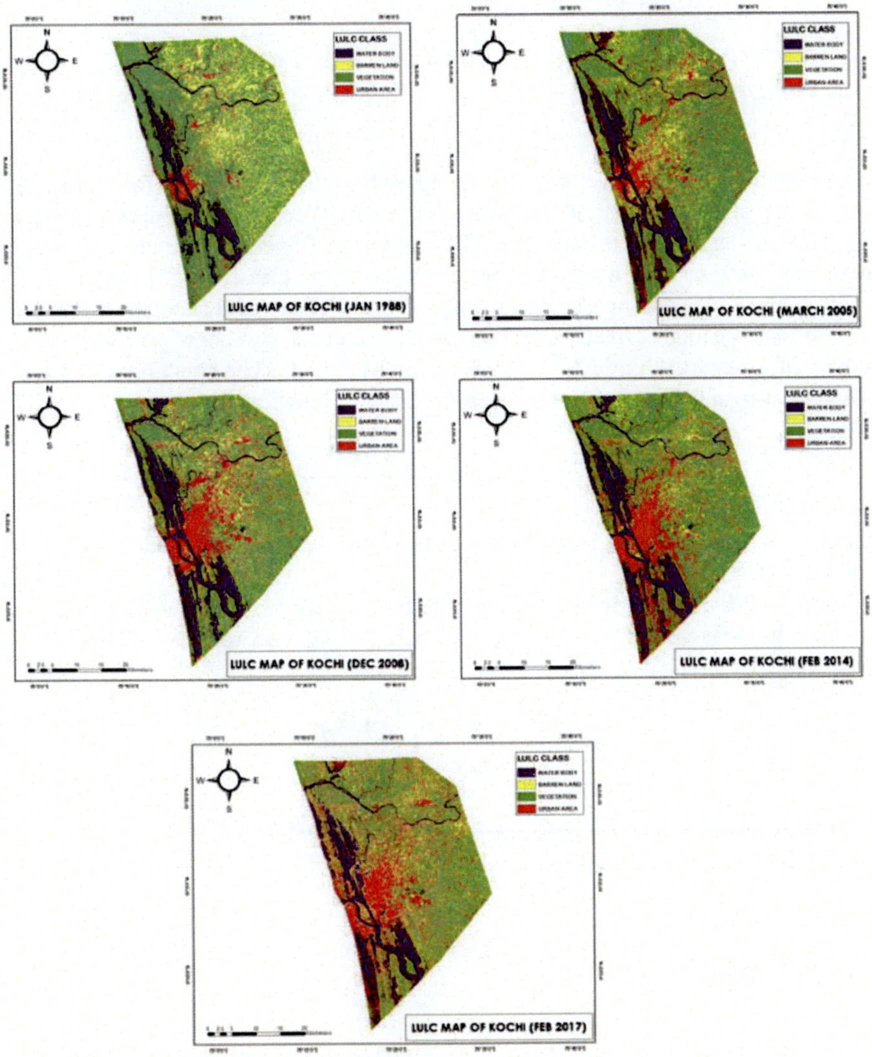

Fig. 2 Land use land cover maps of study area for different years

Table 2 Results of accuracy assessment

S. No.	Date of acquisition	Overall accuracy (%)	Kappa coefficient
1	19-01-1988	90	0.86
2	14-03-2005	98.92	0.98
3	20-12-2008	95.33	0.92
4	11-01-2014	95.23	0.63
5	03-02-2017	94.73	0.89

3.2 *Accuracy Assessment of Land Use and Land Cover Classification*

Every land use and land cover map was compared to the reference data for evaluating the accuracy of the designation. Reference data were collected by considering random sample points and Google Earth [2]. The measures of overall accuracy and kappa coefficient were used to assess the performance of the classifiers [3]. The overall accuracy and kappa coefficient are shown in Table 2. The overall accuracy is defined as the ratio of the total correctly classified pixels to the total number of pixels (the total number of ground truth reference pixels) given by Eq. (1). The area under different LULC classes in different years is given in Table 3. Equation (1):

$$\text{overall accuracy} = \frac{\text{TP} + \text{TN}}{\text{N}} \tag{1}$$

where TP is true positives and TN is true negatives is divided by total number of individuals tested.

Kappa coefficient (κ) is used to measure the accuracy of the classification, and it is used to measure the inter-relationship between the various class of the LULC maps, given by Eq. (2):

$$\kappa = \frac{p_o - p_e}{1 - p_e} = 1 - \frac{1 - p_e}{1 - p_e} \tag{2}$$

where p_o is the relative observed agreement among raters, and p_e is the hypothetical probability of chance agreement.

3.3 *Data Summary*

The spatial distribution pattern of LULC obtained from supervised classification was registered (Table 3). With this data, comparative graphs were created (Figs. 3 and 4).

Table 3 Areas covered by the LULC classes

S. No.	Date of acquisition	Area covered (km^2)			
		Water body	Urban area	Barren land	Vegetation
1	19-01-1988	142.72	65.76	287.89	820.74
2	14-03-2005	213.75	120.07	286.36	696.93
3	20-12-2008	215.73	184.08	216.85	700.43
4	11-01-2014	201.76	204.44	172.18	774.97
5	03-02-2017	186.50	260.30	293.13	577.16

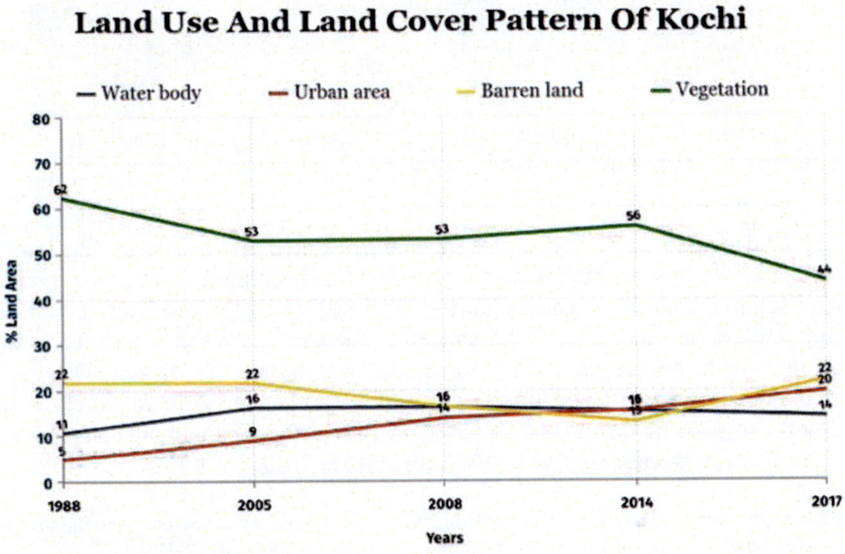

Fig. 3 LULC pattern of Kochi

Major changes in the urban area were observed in the interval 1999–2002, whereas it has increased on a comparatively higher note from 2002–2008. From 2008 onward, a considerable portion of barren land in the southwest region and vegetation regions in the northern parts of Kochi were converted into urban activities, as can be inferred from the graph (Fig. 4).

4 Summary

Land use and land cover change detection is a central component in current strategies for managing natural resources and monitoring environmental changes. In the present

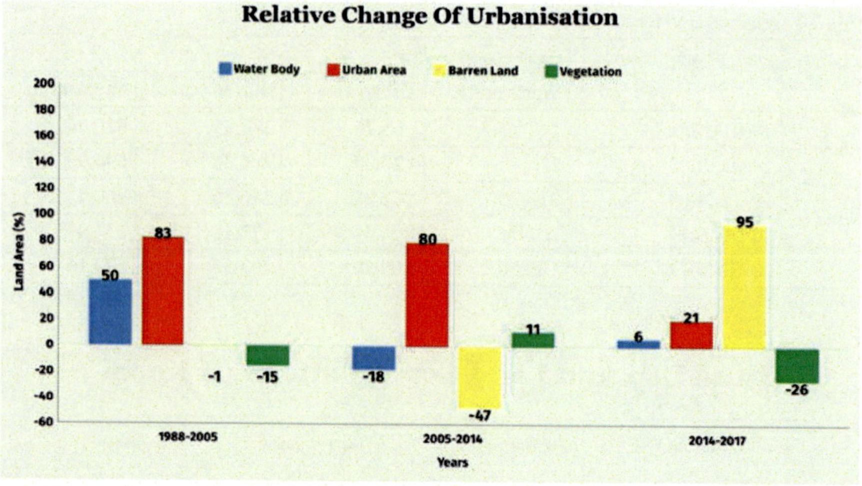

Fig. 4 Relative change in land use and land cover

study, land use and land cover (LULC) mapping of Kochi has been done using remote sensing. Multispectral satellite data of Landsat has been used to map and monitor urbanization changes that occurred during 1988–2017. LULC classification was done using ArcGIS 7.8 and QGIS 3.4.6 software. The maximum likelihood supervised classification technique was used to classify the images into four classes. The results obtained show considerable land use and land cover changes in the major class, viz. water bodies, urban area, barren land and vegetation. The results show that the urban area of Kochi city increased 116.79% from 2005 to 2017.

Acknowledgements The authors acknowledge the financial support from Science and Engineering Research Board, Government of India, through the Project No. SRG/2019/001424.

References

1. Agilan, V., & Umamahesh, N. (2015). Detection and attribution of non-stationarity in intensity and frequency of daily and 4-h extreme rainfall of Hyderabad, India. *Journal of Hydrology, 530*, 677–697.
2. Subramani, T., & Vishnumanoj, V., (2014). Land use and land cover change detection and urban sprawl analysis of Panamarathupatti Lake, Salem. *International Journal of Engineering Research and Applications, 4*(6) (Version 2). ISSN: 2248-9622.
3. Mishra, P. K., Rai, A., & Rai, S. C. (2019). Land use and land cover change detection and geospatial techniques in the Sikkim Himalaya, India. *The Egyptian Journal of Remote Sensing and Space Sciences, 23*(2020), 133–143.

Design of IOT-Based Flood Warning System

Nitish Kumar, Akhilesh S. Nair, and J. Indu

Abstract Flood is one of the most perilous natural disasters affecting mankind. This chapter highlights the effectiveness of Internet of Things (IOT) system and presents preliminary development of an in-house IOT system for data acquisition and transmission through telemetry. IOT provides big data with low latency by empowering physical resources through Internet connectivity. Through IOT system, we have discussed different types of sensors, communication devices that work with telemetry methods to effectively process data for monitoring water level. Data acquisition, data computation and data interpretation of water level will help us in preparation, response and recovery of water related natural disaster.

Keywords IOT · Flood · Water level

1 Introduction

Riverine flooding remains one of the natural disasters in India that causes loss of property and life. While disaster management is very crucial for a country to establish sustainable economic growth, the conventional approach of monitoring water levels in river by visual inspection of foreman using gauges is very difficult. As the foreman has to manually collect data to be sent to operational centers for generating warnings [1], this study demonstrates the potential of IOT (Internet of Things) for water level monitoring in rivers.

IOT is a network of physical objects that contain embedded technology to communicate and interact with external environment. An IOT system is defined by sensor, its connectivity and data processing capabilities. The sensor interacts with environment

N. Kumar · A. S. Nair · J. Indu (✉)
Department of Civil Engineering, IIT Bombay, Mumbai, India
e-mail: indusj@civil.iitb.ac.in

J. Indu
IDP in Climate Studies, IIT Bombay, Mumbai, India

A. K. Dikshit et al. (eds.), *Innovative Trends in Hydrological and Environmental Systems*, Lecture Notes in Civil Engineering 234,
https://doi.org/10.1007/978-981-19-0304-5_12

and collects the data, this acquired data is transmitted and made available to user in real time by implementing optimized telemetry methods [2].

Main causes of floods are heavy rainfall, inadequate capacity of rivers to carry the high flood discharge, inadequate drainage, etc., to name a few. According to World health organization, three types of floods exist namely, flash floods, river floods, coastal floods. Flash floods occur due to high rate of water flow and poor permeability of the soil. River floods are caused when consistent rain or snow melt forces a river to exceed capacity. Coastal floods are caused by storm surges associated with tropical cyclones and tsunami.

Extent of area affected by flood can be calculated based on increase in water level, therefore water level monitoring is required for flood analysis and to alert the authorities and community about the risk to avoid losses. Central water commission (CWC) of India has categorized various flood situations for monitoring floods in the country into four different categories namely Normal, above normal, Severe and extreme depending upon the water levels [3]. River is said to be in "normal flood" situation when water level of river is below warning level, if water level touches or crosses warning level but remains below danger level then the flood situation is called as "above normal," if water level touches or crosses danger level but is below the highest flood level then the flood situation is called "severe flood," the flood situation is said to be "extreme" if water level touches or crosses highest flood level. Authorities are informed based on flood situation above normal flood situation is sent up to secretaries of various ministers, while severe and extreme flood situations are sent up to Prime Minister office and cabinet secretariat [3].

This chapter discusses about how to use different set of sensors, communication devices and data acquisition methods along with telemetry methods to effectively process the data for monitoring water level of water bodies. This chapter outlines the hardware developments of an IOT system based on different electronic devices and transmission technologies. Use of IOT technology along with data analysis, can help monitor floods and provide useful information needed for predicting future floods. Along with IOT technology sensors and other technologies such as artificial intelligence techniques and machine leaning can be used for improving data acquisition and real time measurement. Here we also discuss about deployment of ultrasonic sensors and cellular transmission technologies for ensuring adequate transmission rates and prevention of data loss by using optimized telemetry methods and lightweight data structures. Main goal of this whole study is to design, develop and implement a system, based on different electronic devices and technologies, that can measure and monitor potential flooding in real time and generate data that can be used in future forecasting and prediction models.

2 Related Work

With development of wireless networking technologies and advancement in data analysis there have been many researches based on IOT, few of which have been summarized below.

Moreno et al. [4] have developed a RiverCore IOT device for river water level monitoring over cellular communications. RiverCore acquired data gets stored in Mongo DB database and is published to web platforms and devices of nodes can be operated from this web platform. Retrieved data can be downloaded in different formats for particular or all sensor variables. Feedback system ensures effective data transmission and optimum use of electrical energy.

Abubakr et al. [2] has used an ultrasonic sensor (HC-SR04) for monitoring the water level in a river. The author developed a wireless water level monitoring system with an ATMEGA32 microcontroller, an Ultrasonic sensor, RF transmitter, and LCD. Data measured was transmitted through Radiofrequency transmission technology which can transmit the data up to 10 m. Results show that accuracy for depth measurement was up to 2 cm.

Kruger et al. [5] developed and deployed BMRSS (Bridge Mounted River Stage Sensor) units on Iowa River, United States to monitor the water level. BMRSS units comprise ultrasonic sensor measuring module (Tough sonic 50 model), a solar panel, battery, charge controller, a GPS receiver, and cell modem which transmit data through the Internet to the server. Sensors can take measurements every 5 min and can make data available on the Internet. They have also discussed the lack of datasets due to insufficient gauge stations to monitor the water level. In 2016 USGS, the United States water level monitoring agency had 7000-gauge stations, and data was made available with the latency of 1 h at the earliest.

Panda et al. [6] discuss the accuracy of an ultrasonic sensor (HC-SR04). Temperature and humidity affect the speed of sound and hence the total time of travel is also affected. The author discusses the mathematical relationship between the speed of sound, temperature, and relative humidity. The author also concludes that millimeter accuracy was obtained and this sensor has been also used in medical operations.

Lo et al. [7] discuss four types of measurements of river state using pressure sensors, bubble gauges, float gauges, and non-contact gauges. There are several methods to sense levels, speeds, or distances in rivers or other flood situations, one of which uses ultrasonic waves. An ultrasonic sensing system does not require actual contact with the stream to gather information. Contact gauges potentially demand higher levels of maintenance and could be damaged or destroyed by environmental conditions. Using an ultrasonic sensor increases the longevity of the hardware system.

Mousa et al. [8] have used the MaxBotix ultrasonic sensor (MB 7066) and passive infrared sensors for flash flood detection in urban cities. Measurement using an ultrasonic sensor is easy when the environmental conditions are known. Temperature affects the speed of sound and hence sensor readings. Because of unknown boundary conditions and model parameters, a non-model-based artificial neural network approach is used for estimating the deviation in the readings. Results suggest that urban water levels can be reliably estimated with an error of less than 2 cm. This paper discusses how to use a different set of sensors, communication devices, and data acquisition methods along with telemetry methods to effectively process the data for monitoring water level of the river. This work also discusses the deployment of ultrasonic sensors and cellular transmission technologies for ensuring adequate transmission rates and prevention of data loss by using optimized telemetry methods and lightweight data structures.

3 IOT System

An IOT system is defined by sensors, its connectivity, data processing capability and user's command. Sensor interacts with the environment and collects the data; this data has to be transmitted to cloud for further processing through Internet connectivity [9]. There are two types of data which has to be transmitted to cloud, one is called as state information and other is called as telemetry. State information describes current status of device, whereas telemetry is the read-only data collected from the environment. Operational information for central processing units is also carried as telemetry data. After collection of data from sensor, device should provide data processing functionality before sending the data to cloud in order to avoid power loss due to unnecessary transmission of data, data processing includes converting data to another format, packaging the data in a way that is secure, validating data to ensure it meets a set of rules, sorting data to create preferred sequence, summarizing data to reduce volume and eliminating unwanted detail. Internet connectivity is required for sending data. Some sensors have connectivity and can directly transmit the data to other sensors, while some sensors do not contain network stack for Internet connectivity. Communication between those sensors takes place through Gateway. Gateway acts as a proxy and is used to provide connection and translation between devices and clouds. Gateway device also condenses data to maximize the amount that can be sent to cloud over single link [10]. These received data should be processed in order to convert raw data into useful data and user feedback or user command is required to manage the device and to establish the identity and control of device.

3.1 Data Transmission Technologies

There are many technologies by which interconnected devices can send data through the Internet. IOT system consists of set of sensors, Sensors collect data from environment, compute and integrate with the physical world. Data generated by sensors are transmitted to other electronic devices by different technologies depending on the following mentioned parameters: long or short range, high or low power, high or low bandwidth, wired or wireless and are stored in cloud platform for further processing.

Table 1 IOT telemetry technologies (medium to long range)

Technology	Governing body	Frequency band	Date rate (downlink)	Data rate (uplink)
NB-IOT	3GPP	7–900 MHz	150 kbps	150 kbps
Sigfox	Sigfox	902–928 MHz	32bps	100 bps
LoRa	LoRa alliance	780–915 MHz	100–900 kbps	100–900 kbps
LTE-M	3GPP	In-Band LTE	Up to 1 Mbps	Up to 1 Mbps

Table 2 IOT telemetry technologies (short range)

Technologies	Device used	Data rate	Distance Covered
RFID	Book tag, RFID passports, RFID badge	640 kbps	3–10 m
NFC	Smartphones, ticket stamping machine	106-424kbps	<10 cm
Bluetooth	Smartphones and many other mobile devices	1–24 mbps	<150 m

Communication technologies should be chosen with respect to distance and speed; these classifications are shown in Tables 1 and 2.

LTE-M (Long Term Evolution Machines) Cellular technology is most secured and widely used technology because of its transmission rate and range. LTE-M Cellular technology uses 2G/3G/4G frequency bandwidths for massive flow of data around the World [11]. REST (Representational State Transfer), MQTT (Message Queuing Telemetry Transport) are two protocols used for machine to machine communications because of its lightweight protocol, minimized data packets, low power usage [12]. MQTT uses publish/subscribe operations to exchange data between clients and the server [13]. REST is a communication protocol designed as request/response model which communicates over HTTP. REST protocol can be used to create channel, clear channel or create chart. Four HTTP methods are used in REST protocol namely GET, POST, DELETE, PUT for creating a record containing a new ID(Identity), to fetch list of records requested for, to update the record and to delete the record respectively.

3.2 *Ultrasonic Sensor*

Ultrasonic sensors are electronic devices that measure the distance of target by emitting ultrasonic sound waves (frequency exceeding 20 kHz) and it converts the reflected sound into an electrical signal. The ultrasonic sensor has two main components namely the transmitter (which emits ultrasonic waves) and receiver. Ultrasonic sensor measures the total time of travel of Ultrasonic waves. Transmitter and receiver are used to emit and collect the reflected ultrasonic pulses respectively. Distance is calculated by multiplying the velocity of sound with the time gap between emitting and receiving ultrasonic waves (Fig. 1).

$$D = V.\frac{R_x + T_x}{2} \tag{1}$$

where

D, is distance between water level and sensor.

V, is velocity of sound.

T_x, is time taken by ultrasonic waves from sensor to water surface.

R_x, is time taken by ultrasonic waves from water surface to sensor.

We have used HC-SR04 module (Ultrasonic sensor), which has a transmitter, a receiver and 4 pins. Trig and echo are input and output pins which are used for receiving and passing electronic signals through microcontroller units. Voltage common collector (VCC) and ground (GND) pins are used for power supply. Ultrasonic Sensor HC-SR04 (Hall Current Shock Resistance) measures distance through the total time of travel and a constant velocity of sound (Table 3). The velocity of sound does not remain constant with temperature, the US-100 (Ultrasonic-100) module calculates velocity of sound incorporating the temperature effect, using which distance is calculated. We need waterproof ultrasonic sensors to safeguard the electronics board when our application is to be installed outdoor. These types of sensors are available in two forms. The first one is with a probe and the other one is without a probe. A probe is a two-meter wire attached with a transmitter and is easy to handle.

MaxBotix is a US-based ultrasonic manufacturing industry. MaxBotix Company provides sensors with custom electronic specifications and the price varies from Rs. 1500/- to Rs. 18,000/- (Table 4). All sensors have four pins. VCC (Voltage Common Collector) and GND (Ground) pins are provided for the power supply to the sensor. VCC has a higher voltage compared to GND. Ultrasonic Sensor works on 5 V. Trig

Table 3 Specifications of ultrasonic sensors

Specifications	Waterproof ultrasonic sensor	US-100	HC-SR04
Operating Voltage	3–5.5 V	2.4–5 V	5 V
Average Current Consumption	30 milliampere	2 milliampere	2 milliampere
Frequency	40,000 Hz	40,000 Hz	40,000 Hz
Sensing angle	15°	15°	15°
Range	25–450 cm	2–450 cm	2–400 cm
Weight	25–50 gm	9 gm	10 gm
Size	4.13 cm × 2.9 cm	4.5 cm × 2 cm	4.5 cm × 2 cm
Accuracy	5 mm	3 mm	3 mm

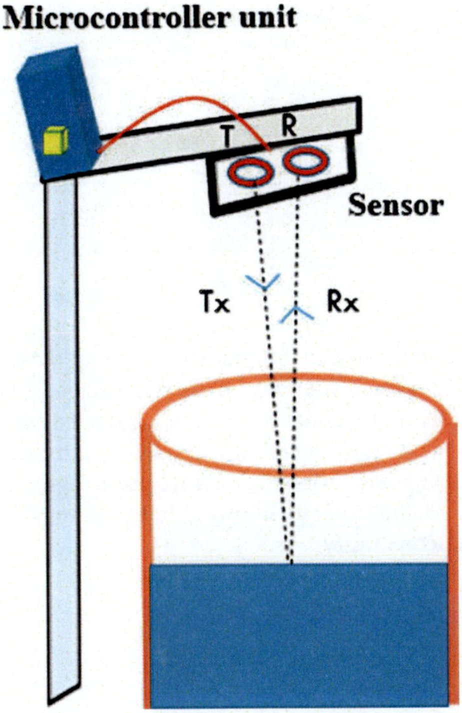

Fig. 1 Principle of IOT water level sensor

Table 4 Specifications of MaxBotix ultrasonic sensors

Sensor	Price	Feature	Precision (cm)	Range (cm)
MB1040	1790	Large object detection	2.5	15–645
MB1200	2899	Automatic calibration, Small object detection	1	25–765
MB1413	4449	Noise rejection	1	30–500
MB7052	6699	Weather resistant	1	20–765
MB7334	12,599	Snow level, weather station	1	50–500

is an input pin that triggers the sensor to produce ultrasonic pulses. It should be kept high for initializing the measurement. The echo is an output pin which remains high for short period, which is proportional to the time taken for the signal to return.

3.3 Microcontroller Units

Microcontroller unit are used for supplying power and communicating a set of instructions to the sensor. Some Microcontroller units are stacked with Internet

Table 5 Specifications of microcontroller units

Specification	Arduino Uno	Arduino Nano
Processor	ATmega328P	ATmega328P
Input voltage	5 V	5 V
Speed of CPU	16 MHz	16 MHz
USB	Regular	Mini
Size	68.6 × 53.4 mm	18 × 45 mm
Weight	25 g	7 g
Price	300	175

connectivity. In this study we have used Arduino board (Arduino Nano and Arduino Uno) for instructing ultrasonic sensor which does not have Internet stack Digital pins of Arduino boards are connected to trig and echo pins of ultrasonic sensor which notes the initializing and return time of electrical signal (Table 5). Arduino boards are programmed with open-source software Arduino IDE. We programmed Arduino to get total time of travel through digital pins to calculate distance and store it temporarily.

3.4 Telemetry Modules

Cellular communication is the most optimized and secured telemetry method for sending lightweight data [4]. In this study we have used Sim 800L and Sim 900A (Table 6). These modules are used for data transmission, sending and receiving messages. These modules are cellular GSM/ GPRS board that allows GPRS transmission, receiving and sending messages, and for making and receiving calls. This module supports quad-band GSM/GPRS network, which is available for SMS (Short Messaging Service) and GPRS data transmission. The module differs in current consumption and signal requirements. It has R_x (receive) and T_x (transmit) pins, power input pins and an antenna. Microcontroller unit transmits the data to sim module through T_x and R_x pins.

Table 6 IOT telemetry modules

Specifications	SIM 800L	SIM 900A
Operating voltage	3.8–4.2 V	9-12 V
Operating Temperature	−40 to + 85 °C	−40 to + 85 °C
Power consumption	Average: 0.25A, peak: 1A	Average: 0.45A, peak: 2A
Supported Frequencies	(850/950/1800/1900 MHz)	(850/950/1800/1900 MHz)
Dimensions	17 mm × 15 mm × 2 mm	24 mm × 24 mm × 3 mm
Weight	20 g	66 g

Table 7 Specification ESP8266

Operating voltage	3.3–5 V
Current Consumption	15μA-400 mA
Length	49 mm
Height	13 mm
Width	24 mm
Weight	50–60 gm

Esp8266 NodeMCU is a combination of microcontroller unit and telemetry module (Table 7). It has all the functions of microcontroller unit and can be used for transmitting data through Wi-Fi technology. It has a flash memory of 1–4 MB depending upon the series. It has 29 pins which are described as digital pins and power supply pins. For the power supply, we can use VU/VIN pin and GND pins both VU/VIN are 5-V pins to ground pins. It provides capabilities for 2.4 GHz Wi-Fi (802.11 b/g/n, supporting WPA/WPA2).

3.5 Arduino IDE

Arduino IDE is an open-source software which is used to program the microcontroller units. Microcontroller units include boards other than Arduino also can be programmed with Arduino IDE. It can be also used for displaying sensor outputs. Arduino web editor also allows to upload the program to board and save files in cloud. We define the functionality of digital pins and conversion of electrical signals to desired outputs through programming.

3.6 User Interface

Cloud platforms such as MathWorks cloud platform can be used to display the desired results. Data is temporarily stored in flash memory of microcontroller units and is transmitted using Rest API through Internet connectivity to cloud platform. Cloud platform allows to store, visualize and analyze live data acquired by sensor. Data can be also sent to Cellular mobile as cellular messages by using sim module. MathWorks cloud platform provides read and write API keys using this and REST API protocol data is sent by telemetry modules and are received at cloud platform.

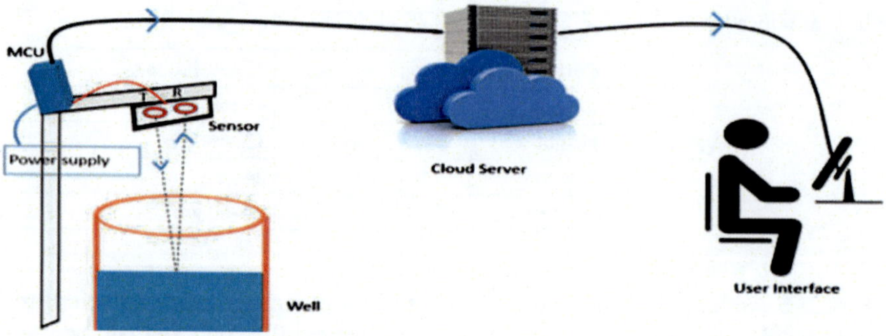

Fig. 2 Methodology water level monitoring

4 Methodology

The ultrasonic sensor transmits the ultrasonic sound waves; which get reflected by the water surface and are received by the transmitter. Time taken by waves for returning to the transmitter is multiplied by speed of sound to obtain distance (from the sensor to the water surface). The MCU is programmed with Arduino IDE, which sends commands to the sensor. Data is transmitted from MCU to cloud platform using Wi-Fi technology and from MCU to mobile handset through cellular messages (Fig. 2).

5 Flood Warning System

Flood warning system consists primarily of a hydrological model, supported by meteorological data obtained by monitoring, visual observation and weather forecasting. Water level can be monitored using different devices, such as sensors, seismographs and satellites [14]. In past information was made available on the Internet by volunteers and interpreted results were notified to people, now sensors are being used for collecting data.

IOT technology can be used along with data analysis for monitoring floods. Data acquired through sensors should be transmitted in real time to get analyzed in time, this can be ensured by implementing optimized telemetry methods and lightweight data structures [15]. Application of these data for generating a rule through machine learning has been discussed but not implemented in our present study.

IOT technologies provide benefits in respects of monitoring, tracking, controlling along with sensing the environmental data. Big data collected through IOT permits effective analysis of occurrence of flood. Big data comprises three stages data acquisition, data computation and data interpretation [16] this paper is primarily focused on data interpretation. It consists of three sections preprocessing of data, training phase and testing phase. Preprocessing of data involves computation of missing values and normalization of data. In training phase, we develop a classification rule, in testing phase based on input data it is classified into occurrence and non-occurrence of flood. Here we have discussed how artificial neural network, deep neural network and convolutional deep neural network to classify data acquired through IOT sensors.

5.1 Artificial and Deep Neural Networks

Artificial neural networks are the pieces of computing system designed to operate in a way human brain analyzes and processes information. An artificial neural network is comprised of a network of artificial neurons called as nodes. Nodes are connected to each other, and strength of their connections is assigned a value; if the value of connection is high, then it indicates that there is a strong connection. Input nodes, hidden nodes and output nodes are three nodes in an ANN that works on different algorithms that are feed-forward network and feedback network; if there is more than one hidden node, it is called as deep neural network (Fig. 3).

In feed-forward network input data is processed onto a layer of processing elements where it performs calculations. Each processing elements makes its computation based upon a weighted sum of its inputs. New calculated values are taken as input for next layers this process continues till it has gone through all layers and a final output is determined, while in feedback algorithm signals can be traveled in both direction using loops. Input nodes are provided with information which can be numerically expressed. Information is presented as activation values, where each node is given a number, higher the number greater the activation. Information is then

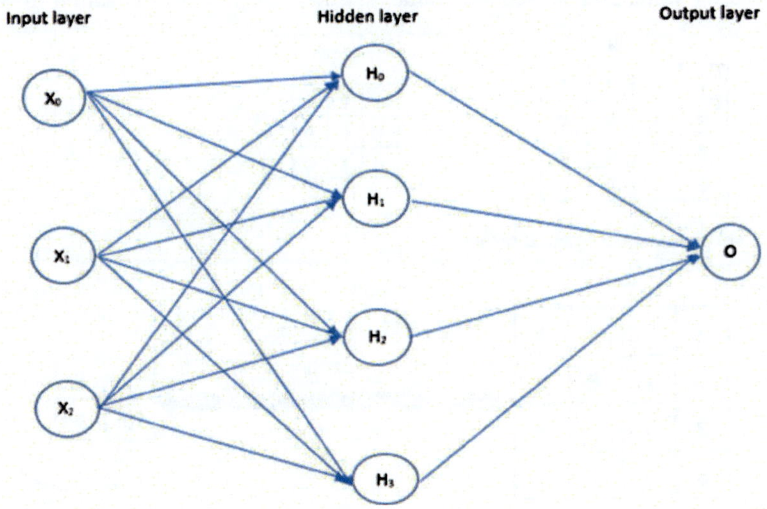

Fig. 3 Artificial neural network

passed throughout the network based on strength and activation function. Difference between predicted and actual output is back propagated and weights are reassigned and so with each passed information predicted value becomes more significant.

5.2 *Convolutional Deep Neural Network*

A convolutional neural network is a deep learning algorithm which takes image as input and assigns weights and biases to various aspects in the image and be able to differentiate one from the other. Convolutional deep neural network is to reduce the images into a form which is easier to process, without losing features which are essential for good prediction. Kernel is involved in carrying out the convolutional operation matrix multiplication takes between kernel and portion of the image it hops down from left of image with same stride till it transverses the whole image. First convolutional layer extracts low-level features such as color, gradient orientation, etc., with added layers, it extracts high-level features. In this paper, convolutional operation is followed by maximum pooling which is taking out of maximum value for defined size (Figs. 4 and 5).

Detection of flood is based upon Internet of things, big data and Machine learning. It has two phases, first phase is training phase, it starts with training of system for proposed CDNN to generate a rule for flood detection and other is testing phase [17]. Sensed information like water flow, water level, rain sensor, humidity, etc., are given as input to file system which has reduced function to sort the data with a set of rules. After reducing the repeated data preprocessing is done in two steps first is missing value imputation in which missing values in datasets can be replaced by mean value of preceding value and subsequent value, and next step is normalization in which

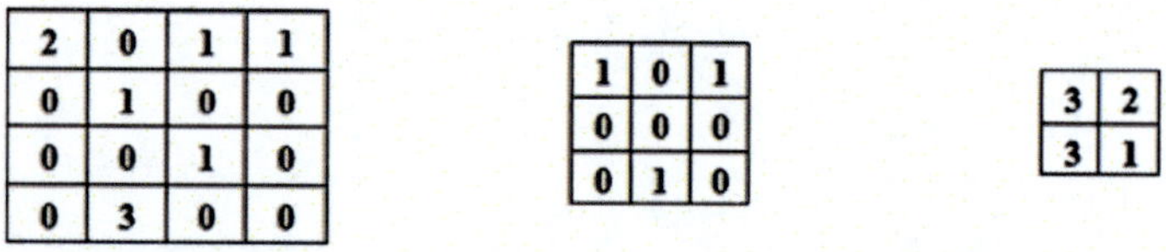

Fig. 4 Convolutional deep neural network

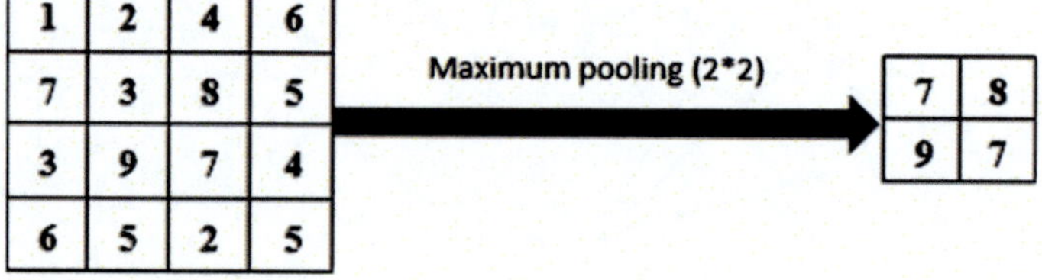

Fig. 5 Maximum pooling

Fig. 6 Training of dataset

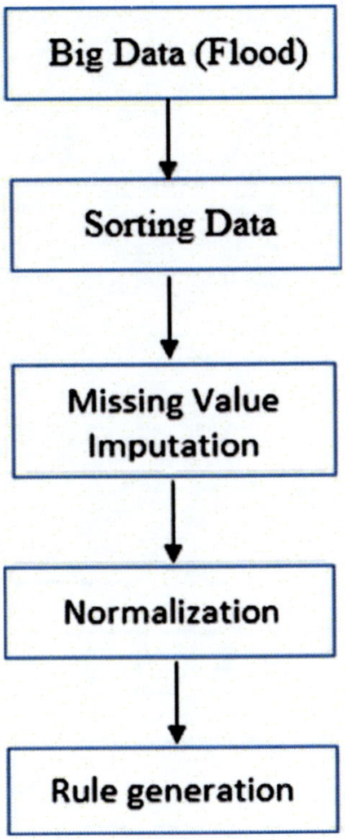

data is scaled to fit into a particular range by using maximum and minimum values of variables in datasets, after normalization these data are given as input to CDNN which can classify the data into chance of occurrence or non-occurrence of flood (Figs. 6 and 7).

6 Results

As a preliminary field experiment to validate the system, the device was kept at the top of a well located at (26°9′56″ N and 85°27′2″E) Mohanpur, Bihar. After setting up of device, leveling of sensor and sensor base was done to avoid false readings due to tilting of the sensor. Data received at cloud platform are displayed in below chart and table (Fig. 8).

Two sensors were set up for the experiment, wherein, sensor 2 was placed at 1.07 m downwards from sensor 1. Water level was measured from sensor to water surface with tape and was found to be 270 cm. Direct readings were taken from both sensors, then 1.07 m was added in sensor 2 readings as it was placed 1.07 m downwards from sensor 1. Water level was measured from tape to sensor with tape

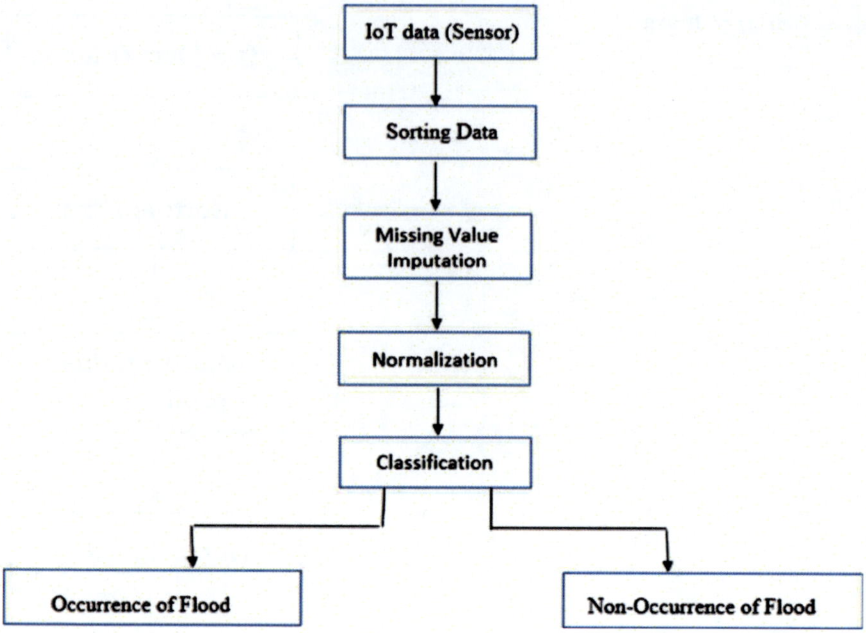

Fig. 7 Testing of dataset

and was found to be 270 cm, and average reading obtained from sensor 1 and sensor 2 are 266.526 cm and 267.728 cm respectively (Table 8).

7 Concluding Remarks

Accuracy in measured distance can be further improved by recording temperature and humidity as speed of sound changes with temperature and humidity [8]. Data pools and SD-card modules will be used to avoid data loss due to signal interferences and variances. Faster connection or more bandwidth is not required due to the use of small data packets and light communication protocols.

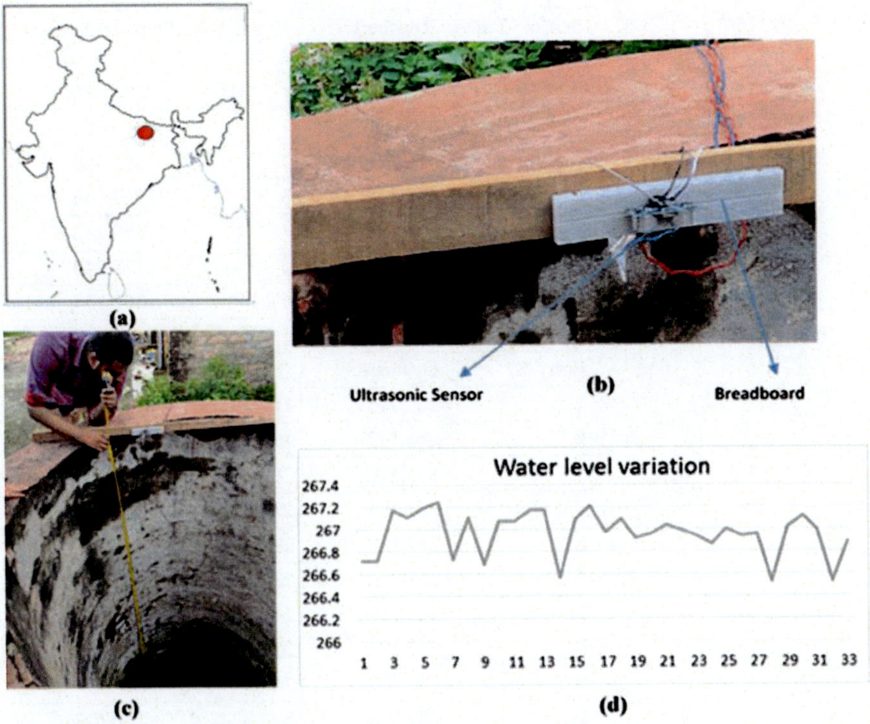

Fig. 8 **a** location of well experiment, **b** developed sensor setup, **c** collection of data for validation, **d** plot between distance and entry id

Table 8 Sensor readings

Sensor -2 direct Readings		Sensor 2 readings from sensor 1 reference	
Entry ID	Distance	Entry ID	Distance
1	158.865	1	265.865
2	160.565	2	267.565
3	162.146	3	269.146
4	159.732	4	266.732
5	165.240	5	272.240
6	158.134	6	265.134
Sensor 1 readings			
Entry ID	Distance	Entry ID	Distance
1	266.713	18	267.087
2	266.713	19	266.917
3	267.172	20	266.951
4	267.104	21	267.036
5	267.189	22	266.985
6	267.240	23	267.172
7	226.730	24	267.172
8	267.104	25	266.560
9	266.679	26	267.087
10	267.070	27	267.206
11	267.070	28	266.968
12	267.172	29	267.087
13	267.172	30	266.917
14	266.560	31	266.951
15	267.087	32	267.036
16	267.206	33	266.985
17	266.968		

References

1. Anbarasan, M., Muthu, B. A., Sivaparthipanc, C. B., Sundarasekar, R., Kadry, S., Krishnamurthy, S., Dinesh Jackson Samuel, R., & Dasel, A. A. (2020). Detection of flood disaster system based on IOT, big data and convolutional deep neural network. *Computer communications*, 150–157.
2. Abubakr, R. A. M., Wang, G. W. (2014). Real-Time wireless flood monitoring system using ultrasonic waves. *International Journal of Science and Research*, 100–103.
3. Flood Forecasting hydrological observation Available online: http://www.cwc.gov.in/flood-forecasting-hydrological-observation. Last accessed 25 Aug 2021.
4. Moreno, C., Aquino, R., Ibarreche, J., Pérez, I., Castellanos, E., Álvarez, E., Rentería, R., Anguiano, L., Edwards, A., Lepper, P., Edwards, R. M., & Clark, B. (2018) RiverCore: IOT device for river water level monitoring over cellular communications. *Sensors, 19*(1), 127.

5. Kruger, A., Krajewski, W. F., Niemeier, J. J., Ceynar, D. L., & Goska, R. (2016). Bridge-Mounted river stage sensors (BMRSS). *IEEE Access, 4*, 8948–8966.
6. Panda, K., Agrawal, D., Nshimiyimana, A., & Hossain, Ashraf. (2016). Effects of environment on accuracy of ultrasonic sensor operates in millimeter range. *Perspectives in Science, 8*, 574–576.
7. Lo, S.-W., Wu, J.-H., Lin, F.-P., & Hsu, C.-H. (2015). Cyber surveillance for flood disasters. *Sensors, 15*, 2369–2387.
8. Mousa, M., Zhang, X., & Claudel, C. (2016). Flash flood detection in urban cities using ultrasonic and infrared sensors. *IEEE Sensors Journal, 16*(19), 7204–7216.
9. IOT overview. Available online https://cloud.google.com/solutions/iot-overview. Last accessed 22 Aug 2021.
10. Mahmoud, M. S., & Mohamad, A. A. H. (2016). A study of efficient power consumption wireless communication techniques/modules for Internet of Things (IoT) applications. *Advanced Internet Things, 6*, 19–29.
11. Goudos, S., Dallas, P., Chatziefthymiou, S., & Kyriazakos, S. (2017). A survey of IoT key enabling and future technologies: 5G, mobile IoT, sematic web and applications. *Wireless Personal Communications, 97*(2), 1645–1675.
12. Eclipse Mosquitto™ Open Source Message Broker. Available online: https://mosquitto.org/. Last accessed 25 Aug 2021.
13. MQ Telemetry Transport. Available online: http://mqtt.org. Last accessed 25 Aug 2021.
14. Ghapar, A. A., Yussof, S., & Bakar, A. (2018). An Internet of things (IoT) architecture for flood data management. *International Journal of Future Generation Communication and Networking, 11*(1), 55–62 (2018).
15. Sakhardande, P., Hanagal, S., & Kulkarni, S. (2016). Design of disaster management system using IoT based interconnected network with smart city monitoring. In *2016 International Conference on Internet of Things and Applications* (pp. 185–190). IEEE.
16. Abdullah, M. F., Ibrahim, M., Zulkifli, H. (2017). Big data analytics framework for natural disaster management in Malaysia. In *International Conference on Internet of Things, Big Data and Security* (Vol. 2, 2017, pp. 406–411). Science and Technology.
17. Mitra, P., Ray, R., Chatterjee, R., Basu, R., Saha, P., Raha, S., Barman, R., Patra, S, Biswas, S. S, & Saha, S. (2016). Flood forecasting using internet of things and artificial neural networks. In *IEEE 7th Annual Information Technology, Electronics and Mobile Communication Conference*, (pp. 1–5). IEEE.

Impact of Climate Change on Precipitation Over India Using CMIP-6 Climate Models

Degavath Vinod and V. Agilan

Abstract Due to various changes in global climatic conditions and greenhouse gas emissions, it is critical to comprehend variations in precipitation patterns at higher spatial and temporal scales. To assess the impact of climate change on precipitation across India, the current study uses simulations from six Global climate models (GCMs) (ACCESS-CM2, ACCESS-ESM1-5, BCC-CSM2-MR, CanESM5, INM-CM4-8, and KACE-1-0-G) for the four SSP emission scenarios for the 1980–2014 historical period and 2022–2056, 2057–2091 future projections. The Indian Meteorological Department (IMD), Pune, provided observed data with a high spatial resolution of (0.25° × 0.25°) on a daily time scale. For downscaling the daily precipitation over India, a multiplicative change factor technique is used. The current study examines the change in daily average rainfall and daily maximum rainfall for the four climate emission scenarios (SSP126, SSP245, SSP370 and SSP585) with two distinct time scales (2022–2056 and 2057–2091) over India using six GCMs from the CMIP-6 simulations. Many portions of India saw an increase in daily average rainfall of 0–50 mm and a shift in maximum daily rainfall of 500–1000 mm as a result of the four climate emission scenarios.

Keywords Climate change · GCMs · Daily precipitation · Downscaling · Change factor

1 Introduction

Changes in global climatic conditions may have an impact on regional water availability [1]. Since a few decades, climate change and its impact on water supplies has picked-up the interest of the hydrologic research community. GCMs, the primary tools and exclusive means of identifying and assessing climate change, are gradually but steadily progressing to a more sophisticated formulation and more realistic depiction of the global climate system [2]. Climate change will inevitably accelerate

D. Vinod (✉) · V. Agilan
Department of Civil Engineering, National Institute of Technology Calicut, Calicut, Kerala, India

A. K. Dikshit et al. (eds.), *Innovative Trends in Hydrological and Environmental Systems*, Lecture Notes in Civil Engineering 234,
https://doi.org/10.1007/978-981-19-0304-5_13

redistribute water supplies and hydrological cycle and on both regional and temporal scales due to changes in temperature and air circulation. This will have an impact on the availability of water for agricultural, domestic, ecological, and hydropower generation purposes in a region and season, which will, in turn, have an impact on the region's social economy [3]. Climate change in future scenarios would have a wide range of consequences on watershed hydrologic regimes all around the world. Such effects on various sectors are complex, and methods for determining these changes are fraught with uncertainty [4]. In the last decade, numerous researches have been undertaken in various regions of the world on the application of various downscaling methodologies for projection of hydrological variables at river basin scale under climate change impacts. General circulation models (GCMs) are commonly used in climate impact studies to simulate current and future time series of climate variables for the entire planet [5].

Climate models are typically updated on a regular basis, as several modelling organisations throughout the world combine increasing spatial resolution, physical, biological, and chemical processes of the atmosphere, land, and ocean. These modelling groups time their updates to coincide with the publication of IPCC assessment reports, publishing "runs" of model results in the lead-up to each one. The IPCC's fifth assessment report (AR5) in 2013 used CMIP5 climate models, while the IPCC's sixth assessment report (AR6) in 2021 will use new state-of-the-art CMIP6 models. Approximately, 40 CMIP6 models have been published so far, and it is clear that they have a vastly greater climate sensitivity than CMIP5 models, contributing to projections of greater warming this century—around 0.4C warmer than similar scenarios run in CMIP5, though these warming estimates may change as more models become available. Global climate models (GCMs) are among the best instruments available for representing the main features of the global distribution of basic climate parameters such as precipitation, temperature, and so on. These models, on the other hand, are unable to accurately simulate regional climate conditions at temporal and spatial scales relevant to hydrological investigations. To put it another way, GCM outputs are typically at a resolution that is too coarse for many climate change impact studies. As a result, developing techniques for downscaling GCM climate change projections to regional, local, or station scales is critical [6]. Downscaling, according to the Intergovernmental Panel on Climate Change (IPCC), is a strategy for extracting global to small scale (100–25 km) information from larger-scale global climate models. The process of translating general circulation model (GCM) output to a finer spatial scale that is more relevant for understanding local and regional climate conditions is known as downscaling of climate projections. Downscaling is a technique for extracting high-resolution climate or climate change information from low-resolution GCMs. Downscaling strategies were created to bridge the gap between what the climate modelling community can now give and what the effects research community need. Changes in large-scale meteorological phenomena would signal the start of anthropogenic global climate change. Downscaling is essential despite the fact that the impact of large-scale feature changes on regional surface climate cannot be resolved in the current generation of GCMs [7].

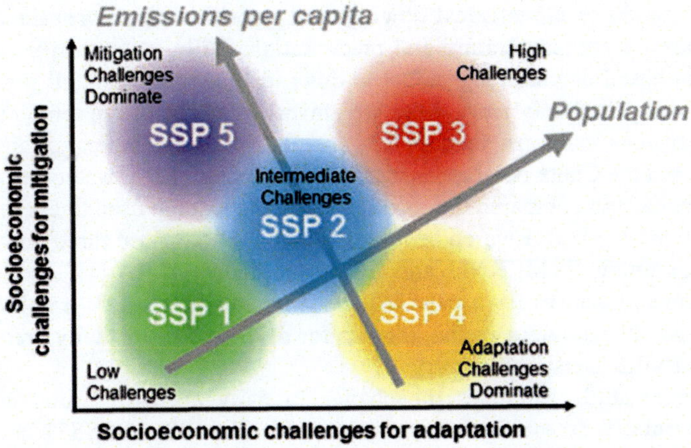

Fig. 1 Shared socioeconomic pathways (SSPs) under different challenges and mitigations. *Source* [9]

The energy modelling community established a new set of emissions scenarios termed "Shared Socioeconomic Pathways" in the run-up to the IPCC AR6, which are based on many socioeconomic assumptions (SSPs). A couple of these SSP scenarios were chosen to drive climate models in CMIP6. Four RCPs were included in the IPCC AR5 and looked at several scenarios for future greenhouse gas emissions. These RCP2.6, RCP4.5, RCP6.0, and RCP8.5 scenarios have been updated in CMIP6. The new scenarios include SSP1-2.6, SSP2-4.5, SSP3-7.0, SSP4-6.0, and SSP5-8.5, and they all result in identical 2100 radiative forcing levels as their predecessors in AR5 [8]. SSPs are projections of global socioeconomic transformation up to the year 2100. Alternative socioeconomic developments are described in the narratives (Fig. 1).

Global and regional climate models can be used to simulate the effects of climate change. GCM is often recognised as the most reliable technique for simulating climate time series [10]. GCMs are performed at too coarse a spatial resolution to generate trustworthy projections at the regional scale, hence downscaling is preferred. The two most widely utilised downscaling methods are dynamical and statistical downscaling; statistical downscaling has been widely employed to generate future climate scenarios at the local scale. Because of their simplicity and low processing cost, statistical downscaling tools are increasingly popular among scholars around the world. Devak et al. (2015) used dynamic coupling of the two well-known downscaling algorithms SVM and KNN in order to downscale the daily rainfall over Mahanadi river basin. ANN-based models were developed in order to obtain the mean, minimum and maximum temperature over chaliyar river basin [11]. Anandhi et al. [12] used SVM algorithm for downscaling monthly precipitation for Malaprabha river. Duhan and Pandey [13] created monthly maximum and minimum temperature forecasts for three sites in the Tons river basin: Satna, Rewa, and Allahabad, using three downscaling techniques: ANN, MLR, and LS-SVM. Goly et al.

[14] compared different statistical downscaling algorithms to downscale daily and monthly datasets for temperature and precipitation. Wilby et al. [15] used SDSM a regression-based downscaling model to downscale monthly rainfall for regional climate models. Coulibaly and Dibike [16] investigated the use of temporal neural networks as a downscaling method for the generation of daily temperature and precipitation series at the Chute-des-Passes station in the Serpent River watershed in north-eastern Canada, and compared the results to the most widely used linear regression model (SDSM) method. Raje and Mujumdar (2011) compared three downscaling algorithms, namely, SVM, KNN, and CRF. Mahmood and JIA [17] used additive change factor approach to downscale the daily temperature. Anandhi et al. [18] examined different change factor methodologies for downscaling the daily precipitation over cannonsville, Delaware country.

The current study examines the change in daily average rainfall and daily maximum rainfall for the four climate emission scenarios (SSP126, SSP245, SSP370, and SSP585) with two distinct time scales (2022–2056 and 2057–2091) over India using six GCMs from the CMIP-6 simulations. A multiplicative change factor methodology is applied to downscale the daily rainfall over India.

2 Data and Study Area

The present study considers the study area of all over the India with the latitude ranging from 6.5°N to 38.5°N and longitude ranging from 66.65°E to 100°E with 4641 grid points of 0.25° × 0.25° spatial resolution. The average annual rainfall for India is 120 cm, and majority of rainfall occurs in monsoon months, i.e. from June to September [19]. Indian Meteorological Department (IMD), Pune provides high spatial resolution data for both precipitation and temperature with 1.0° × 1.0°, 0.5° × 0.5°, and 0.25° × 0.25° spatial resolutions. The present study considers a high spatial resolution of 0.25° × 0.25° on the daily temporal scale as observed data.

The World Climate Research Programme (WCRP), in partnership with the Earth System Grid Federation (ESGF), is conducting research on climate change, provides Coupled Modelled Intercomparison Project (Phase 6), historical Global climate data by running multiple simulations based on different past emission criteria. Generally, many climate models have historical simulations from 1850–2014, i.e. past 115 years simulations. Each climate model provides the historical runs for more than a hundred climate variables, the present study focussed on precipitation (pr) variable for the ranking GCMs. The data collected from each global climate model is for 35 years, i.e. 1980–2014, on a daily temporal resolution. The spatial resolution of general circulation models varies with a different organization, as shown in Table 1. The general circulation models, organizations which developed the model and their respective spatial resolutions are presented in the Table1.

Table 1 List of GCMs, and spatial resolutions

S. No.	GCMs	Organization	Resolution (°)
1	ACCESS-CSM2	Australian Community Climate and Earth System Simulator	2.5 × 2.5
2	ACCESS-ESM1-5		
3	BCC-CSM2-MR	Beijing Climate Centre, China	2.5 × 2.5
4	CanESM5	Canadian Centre for Climate Modelling and Analysis, Environment and Climate Change Canada	5.0 × 5.0
5	INM-CM5-0	Institute of Numerical Mathematics, Russia	1.0X1.0
6	KACE-1.0-G	National Institute of Meteorological Sciences (NIMS) and Korea Meteorological Administration (KMA)	2.5 × 2.5

3 Methodology

To determine climate change, the change factor approach calculates the change in data for the baseline and forecast climate scenarios [4, 18]. Change Factor Methods (CFMs) come in a variety of shapes and sizes. Temporal scale, temporal resolution, mathematical formulation, and a variety of changing factors can all be used to classify these. The temporal scale and temporal domain from which CFMs are calculated determine the major type of CFM. The timeline (e.g. daily) used in the study is referred to as the temporal scale. At higher frequency temporal scales, the reliability of GCMs falls. Daily hydro—meteorological variables are also required in climate change-related hydrological and impact on the environment assessment studies. GCM simulations have been tested on daily timescales, and it has been determined that several of the GCMs (in the AR4 report) bestow significant skill at subcontinental sizes, even when examined using daily frequency distributions [20]. This increases the likelihood of utilising GCMs for regional evaluation and, in some situations, predicting extreme occurrences.

The mathematical formulation of the next type of CFM is used to classify it (additive or multiplicative). In an additive CF, the arithmetic difference between a GCM variable derived from a current climate simulation and a future climate scenario derived from the same GCM grid location is calculated. The resulting difference is then combined with observed local values to get the modelled future values. When it comes to temperature, rather than precipitation, the additive technique works better [21, 22], despite the accuracy of the GCM's present climate simulation, this assumption requires that the GCM delivers a reliable estimate of the change in the value of a given variable. A multiplicative change factor (CF) is identical to an additive CF except that the ratio between future and current GCM simulations is determined rather than the arithmetic difference; the observed values are then multiplied by the CF. This approach, which is commonly used for precipitation, assumes that the GCM gives a good estimate of the relative change in the value of a variable. There is no clear evidence in the literature to support the use of various forms of CFs. For meteorological variables like wind speed and solar radiation, CFs must be approximated additively or multiplicatively. However, these numbers are sometimes required in hydrology impact assessment studies. For improved simulation with regard to observed data, additive CF is used for temperature variables and multiplicative CF is used for precipitation variables (i.e. IMD). As a result, a methodology for applying CFs to a large range of meteorological variables is required [18].

The present study uses multiplicative change factor approach for downscaling the daily precipitation variable. The equation for the multiplicative change factor method is given by

$$\overline{GCMb} = \sum_{i=1}^{N_b} \frac{GCMb_i}{N_b} \tag{1}$$

$$\overline{GCMf} = \sum_{i=1}^{N_f} \frac{GCMf_i}{N_f} \tag{2}$$

where for a temporal domain, $\overline{GCMb}$ and $\overline{GCMf}$ show the values from a GCM historical or baseline and GCM future climatic scenarios, respectively. The average values from a GCM baseline and GCM future scenario for the given time domain are $GCMb$ and $GCMf$. The GCM baseline and GCM future scenarios, respectively, have Nb and Nf values in the time domain.

The next step is to evaluate the change factors (CF_{mul}) which are calculated as the ratio of average of every month for the future data set to the average of every month for the baseline data set which is shown in the following equation. To evaluate the average of each month the daily data set is converted to monthly and average of every month over the total time period is calculated. For example, let us consider all January months of the total time period (say 20 years). We will get 20 January monthly data. Now, apply the arithmetic mean for the January month. Finally, we get 12 monthly means for baseline and future scenarios, respectively.

$$CF_{\text{mul}} = \frac{\overline{GCMf}}{\overline{GCMb}} \tag{3}$$

Finally, multiply this multiplicative change factor with the observed IMD gridded data to get the downscaled data for that particular time period.

$$LSF_{\text{mul}} = LOb_i \times CF_{\text{mul}} \tag{4}$$

LOb_i are daily recorded values of the climatic variable precipitation (at the ith time step) at a particular weather station, or are the averaged climatic time series for a watershed for the defined time domain, where LSF_{mul} is local scaled future values.

4 Results and Discussion

The current study takes into account daily precipitation simulations from six GCMs from the Coupled Model Intercomparison Project-6 (CMIP-6) for 4641 grid locations with a high spatial resolution of 0.25° × 0.25°. The daily precipitation is downscaled using a multiplicative change factor algorithm. The data show that there is a huge difference in daily average rainfall and daily maximum rainfall across India.

Many portions of India saw an increase in daily average rainfall of 0 to 50 mm and a shift in maximum daily rainfall of 500 mm to 1000 mm as a result of the four climate emission scenarios. When compared to SSP126, SSP370, and SSP585, simulations from the SSP245 climate emission scenario revealed less. The heat maps in Fig. 2 depict the variations in daily average rainfall and daily maximum rainfall over two time periods. Table 2 shows the maximum and minimum rainfall amounts.

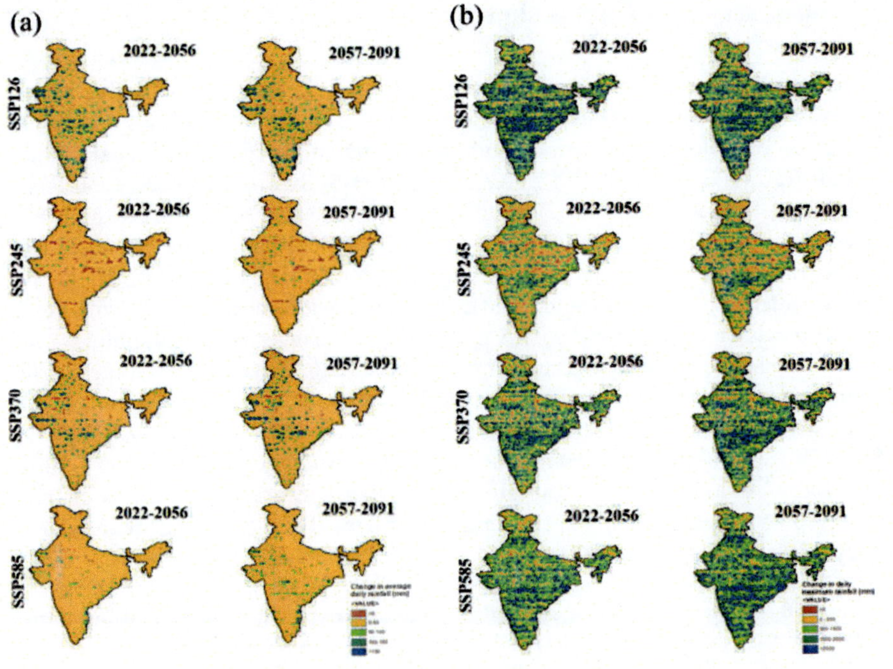

Fig. 2 Changes in daily rainfall across India. **a** Average daily rainfall, **b** maximum daily rainfall

Table 2 Changes in daily average and daily maximum rainfall over India

Scenario	Time period	Change in average daily rainfall (mm)		Change in maximum daily rainfall (mm)	
		Min. rainfall	Max. rainfall	Min. rainfall	Max. Rainfall
SSP126	2022–2056	−7.9	2996.8	−407.6	4934.8
	2057–2091	−7.2	4063.2	−364.9	4924.9
SSP245	2022–2056	−5.7	494.9	−301.8	4742.5
	2057–2091	−6.4	531.3	−357.9	4919.1
SSP370	2022–2056	−10.6	4236.4	−377.8	4868.9
	2057–2091	−10.2	6110.9	−376.0	4926.4
SSP585	2022–2056	−6.3	659.3	−374.5	4701.7
	2057–2091	−6.5	638.8	−377.2	4777.7

5 Summary and Conclusion

Understanding variations in precipitation patterns at greater spatial and temporal scales is crucial due to numerous changes in global climatic conditions and greenhouse gas emissions. The current study uses simulations from six Global climate models (GCMs) (ACCESS-CM2, ACCESS-ESM1-5, BCC-CSM2-MR, CanESM5, INM-CM4-8, and KACE-1-0-G) for the four SSP emission scenarios for the 1980–2014 historical period and 2022–2056, 2057–2091 future projections to assess the impact of climate change on precipitation across India. On a daily time frame, the Indian Meteorological Department (IMD) in Pune provided observed data with a high spatial precision of 0.25° × 0.25°. A multiplicative change factor technique is used to downscale the daily precipitation over India. Using six GCMs from the CMIP-6 simulations, the current work evaluates the change in daily average rainfall and daily maximum rainfall for the four climate emission scenarios (SSP126, SSP245, SSP370, and SSP585) with two separate time scales (2022–2056 and 2057–2091) over India. As a result of the four climate emission scenarios, many parts of India observed an increase in daily average rainfall of 0 to 50 mm and a shift in maximum daily rainfall from 500 to 1000 mm.

Acknowledgements The authors acknowledge the financial support from Science and Engineering Research Board, Government of India through the Project No. SRG/2019/001424.

References

1. Liu, J., Yuan, D., Zhang, L., Zou, X., & Song, X. (2016). Comparison of three statistical downscaling methods and ensemble downscaling method based on Bayesian model averaging in upper Hanjiang River Basin, China. *Advance in Meteorology, 2016*. https://doi.org/10.1155/2016/7463963

2. Devak, M., Dhanya, C. T., & Gosain, A. K. (2015). Dynamic coupling of support vector machine and K-nearest Neighbour for downscaling daily rainfall. *Journal of Hydrology, 525*, 286–301. https://doi.org/10.1016/j.jhydrol.2015.03.051
3. Duhan, D., & Pandey, A. (2015). Statistical downscaling of temperature using three techniques in the Tons River basin in Central India. *Theoretical and Applied Climatology, 121*, 605–622. https://doi.org/10.1007/s00704-014-1253-5
4. Vallam, P., & Qin, X. S. (2018). Projecting future precipitation and temperature at sites with diverse climate through multiple statistical downscaling schemes. *Theoretical and Applied Climatology, 134*, 669–688. https://doi.org/10.1007/s00704-017-2299-y
5. Raje, D., & Mujumdar, P. P. (2011). A comparison of three methods for downscaling daily precipitation in the Punjab region. *Hydrological Processes, 25*, 3575–3589. https://doi.org/10.1002/hyp.8083
6. Wilby, R. L., Wigley, T. M. L., Conway, D., Jones, P. D., Hewitson, B. C., Main, J., & Wilks, D. S. (1998). Statistical downscaling of general circulation model output: A comparison of methods. *Water Resources Research, 34*, 2995–3008. https://doi.org/10.1029/98WR02577
7. Gutiérrez, J. M., San-Martín, D., Brands, S., Manzanas, R., & Herrera, S. (2013). Reassessing statistical downscaling techniques for their robust application under climate change conditions. *Journal of Climate, 26*, 171–188. https://doi.org/10.1175/JCLI-D-11-00687.1
8. O'Neill, B. C., Tebaldi, C., Van Vuuren, D. P., Eyring, V., Friedlingstein, P., Hurtt, G., Knutti, R., Kriegler, E., Lamarque, J. F., Lowe, J., Meehl, G. A., Moss, R., Riahi, K., & Sanderson, B. M. (2016). The Scenario Model Intercomparison Project (ScenarioMIP) for CMIP6. *Geoscientific Model Development, 9*, 3461–3482. https://doi.org/10.5194/gmd-9-3461-2016
9. Hossain, M., Hales, R., & Sarker, T. (2017). Pathways to a sustainable economy: Bridging the gap between Paris climate change commitments and net zero emissions. In *Pathways to a Sustainable Economy Bridge Gap between Paris Climate Change Commitments Net Zero Emissions*, pp. 1–225. https://doi.org/10.1007/978-3-319-67702-6
10. Ghosh, S., & Mujumdar, P. P. (2006). Future rainfall scenario over Orissa with GCM projections by statistical downscaling. *Current Science, 90*, 396–404.
11. Chithra, N. R., Thampi, S. G., Surapaneni, S., Nannapaneni, R., Reddy, A. A. K., & Kumar, J. D. (2015). Prediction of the likely impact of climate change on monthly mean maximum and minimum temperature in the Chaliyar river basin, India, using ANN-based models. *Theoretical and Applied Climatology, 121*, 581–590. https://doi.org/10.1007/s00704-014-1257-1
12. Manatsa, D., Chingombe, W., & Matarira, C. H. (2008). The impact of the positive Indian Ocean dipole on Zimbabwe droughts Tropical climate is understood to be dominated by. *International Journal of Climatology, 2029*, 2011–2029. https://doi.org/10.1002/joc
13. Duhan, D., & Pandey, A. (2015). Statistical downscaling of temperature using three techniques in the Tons River basin in Central India. *Theoretical and Applied Climatology, 121*(3–4), 605–622. https://doi.org/10.1007/s00704-014-1253-5
14. Goly, A., Teegavarapu, R. S. V., & Mondal, A. (2014). Development and evaluation of statistical downscaling models for monthly precipitation. *Earth Interactions, 18*, 1–28. https://doi.org/10.1175/EI-D-14-0024.1
15. Wilby, R. L., Dawson, C. W., & Barrow, E. M. (2002). SDSM - A decision support tool for the assessment of regional climate change impacts. *Environmental Modelling and Software, 17*, 145–157. https://doi.org/10.1016/s1364-8152(01)00060-3
16. Coulibaly, P., Dibike, Y. B., & Anctil, F. (2005). Downscaling precipitation and temperature with temporal neural networks. *Journal of Hydrometeorology, 6*, 483–496. https://doi.org/10.1175/JHM409.1
17. Mahmood, R., & Jia, S. (2017). An extended linear scaling method for downscaling temperature and its implication in the Jhelum River basin, Pakistan, and India, using CMIP5 GCMs. *Theoretical and Applied Climatology, 130*, 725–734. https://doi.org/10.1007/s00704-016-1918-3
18. Anandhi, A., Frei, A., Pierson, D. C., Schneiderman, E. M., Zion, M. S., Lounsbury, D., & Matonse, A. H. (2011). Examination of change factor methodologies for climate change impact assessment. *Water Resources Research, 47*, 1–10. https://doi.org/10.1029/2010WR009104

19. Srinivasa Raju, K., & Nagesh Kumar, D. (2015). Ranking general circulation models for India using TOPSIS. *Journal of Water and Climate Change, 6*, 288–299. https://doi.org/10.2166/wcc.2014.074
20. Perkins, S. E., Pitman, A. J., Holbrook, N. J., & McAneney, J. (2007). Evaluation of the AR4 climate models' simulated daily maximum temperature, minimum temperature, and precipitation over Australia using probability density functions. *Journal of Climate, 20*, 4356–4376. https://doi.org/10.1175/JCLI4253.1
21. Akhtar, M., Ahmad, N., & Booij, M. J. (2008). The impact of climate change on the water resources of Hindukush-Karakorum-Himalaya region under different glacier coverage scenarios. *Journal of Hydrology, 355*, 148–163. https://doi.org/10.1016/j.jhydrol.2008.03.015
22. Hay, L. E., Wilby, R. L., & Leavesley, G. H. (2000). A comparison of delta change and downscaled GCM scenarios for three mountainous basins in the United States. *Journal of the American Water Resources Association, 36*, 387–397. https://doi.org/10.1111/j.1752-1688.2000.tb04276.x

On the Prediction of Extreme Wave Heights Under Cyclonic Events for the Design of Coastal Structures Situated at Remote Islands in Deep Sea

A. Basu and A. A. Purohit

Abstract Presently, due to change in the climatic scenario, India is witnessing increase in intensity/frequency of the extreme tropical cyclonic events and its associated disasters not only on the east coast of Indian peninsula but also on west coast, on the remote islands situated in the Arabian Sea. As such coastal structures situated on remote islands in Arabian Sea are also likely to get exposed to intensified extreme events in near future which may lead to damage/collapse of the structures, if they are not designed properly. Present study emphasizes in evolving the extreme significant wave heights (H_s) for various return periods (50/100 years) for design of coastal structures situated on remote islands (Kalpeni, Andrott, and Kiltan) in Arabian Sea. As the long-term historical measured wave data are not available near the remote islands, coupled hydrodynamic-wave model (TELEMAC-2D and TOMAWAC) with appropriate atmospheric forcing was applied to hindcast wave heights, storm surges at remote islands for past 40 years (1977–2017). Study reveals that due to the presence of steep slopes as well as deeper depths near the islands, storm surges are less (about 0.1 m–0.2 m) near the coastal structures, however, the extreme wave heights that reached over the structures are having higher magnitude (H_s about 6 m–6.5 m). The hindcast wave heights were fitted to generalized extreme value (GEV) distribution to forecast significant wave heights for various return periods (50/100 years). It will provide a basis to arrive at appropriate design wave condition for the coastal structures under extreme events.

Keywords Extreme wave heights · Generalized extreme value distribution · Hindcast · Storm surges

A. Basu (✉) · A. A. Purohit
Central Water and Power Research Station, Pune, India
e-mail: basuanimesh@cwprs.gov.in

A. K. Dikshit et al. (eds.), *Innovative Trends in Hydrological and Environmental Systems*, Lecture Notes in Civil Engineering 234,
https://doi.org/10.1007/978-981-19-0304-5_14

1 Introduction

Among the various natural hazards, tropical cyclone is one of the most severe event that poses great threat due to its associated high wind speed, strong waves, and extreme water levels to coastal habitats, marine structures [1]. Presently, due to the change in the climatic scenario, India is witnessing increase in intensity/frequency of the extreme events (tropical cyclone) and its associated disasters not only on the east coast of Indian peninsula but also on the west coast, remote islands situated in the Arabian Sea. Thus, the coastal structures built in the past on the remote islands in the Arabian Sea are also likely to get exposed to the intensified extreme events in the near future and it might also cause damage to the structure if they are not designed properly. This will not only affect the waterborne transport facility and in turn the lifeline to these islands. One of such coastal structure (breakwater) situated at Kalpeni island (as shown in Fig. 1) constructed during the period 1995–2002 to achieve safe berthing conditions at the wharf for the passenger's/fishing vessels for all weather seasons, got damaged extensively during the recent Ockhi cyclonic event (November–December 2017). In order to permanently restore the damaged breakwater, the design conditions need to be derived based on the historical extreme events like Ockhi. The prediction of extreme wave conditions depends primarily on long-term historical data, and as the measured data for limited period are available in the nearby area of the remote islands, hindcasting of wave data is the only way to get the historical wave data. Based on the information on fetch length, decay distance, duration of storm, geostrophic wind speed, etc., many researchers such as Sverdrup-Munk-Bretschneider (SMB) [2] had developed simplified semi-empirical relationships, nomograms to arrive at the hindcast wave heights, wave periods for the cyclonic events. The cyclonic wind fields were also described based on the simplified parametric model developed by Rankin Vortex [2], Holland [3], etc., and the distribution of atmospheric pressures during the extreme events was also formulated by Holland [3]. These parametric wind models

Fig. 1 Location of Kalpeni island and view of the breakwater before and after Ockhi cyclone. *Source* Google Earth

were also further improved by various researchers, and modified form of these models was published by Jelesnianski and Taylor [4], Holland [5], Carr and Elsberry [6], etc. In order to overcome the limitations of the simplified semi-empirical wave models in estimation of extreme wave conditions, the numerical hindcast simulations along the French coast by using TOMAWAC model forced by the wind field derived from ERA-40 meteorological reanalysis produced by the ECMWF were carried out by Lafon et al. [7]. Anand et al. [8] described simulation of Thane cyclonic sea state along east coast of India by using WAM model forced by ENCEP wind field. Murthy et al. [9] had applied coupled hydrodynamic-wave model (ADCIRC and SWAN) forced with a modified form of Jelesnianski and Taylor's parametric wind model to simulate extreme storm surges and wave characteristics for historical cyclonic events on the east coast of India. Cooper et al. [10] had applied TELEMAC-2D forced with parametric cyclonic wind field, atmospheric pressure to simulate storm surge due to the cyclone Yasi occurred in Australia, Queenland in 2011, and the study reveals that the simulated storm surge matches well with the measured data. Gourgue et al. [11] had also developed a storm surge model by using TELEMAC for Belgian coastal area.

In the present study, coupled hydrodynamic-wave model using TELEMAC-2D and TOMAWAC software was applied to determine hindcast wave heights, storm surges for past 40 years (1977–2017) for the coastal structures situated not only at Kalpeni but also for the nearby remote islands such as Andrott and Kiltan. The hindcast wave heights were fitted to GEV distribution function to predict the significant wave heights (Hs) for various return periods (50/100 years) so that it will be useful in deciding the design basis for coastal structures.

2 Data Used for the Study

The analysis of tracks of past cyclonic events as published by Indian Meteorological Department (IMD) shows that most of the tropical cyclonic events in the Indian Ocean generate primarily in the Bay of Bengal (BoB) due to its increased sea surface temperature (SST) while few are generated in the Arabian Sea. The past history also shows that in the Arabian Sea, very few storms are severe or extremely severe. However, presently due to the climate change, the SST is increasing and there is increase in intensity/frequency of the cyclonic event in the Arabian Sea also. Thus, design parameters such as extreme wave heights and storm surge levels need to be forecasted carefully for the safety of the coastal structures situated not only on the west coast of Indian peninsula but also on the remote islands in the Arabian Sea. The track of historical extreme events which are of relevance to the remote islands (Kalpeni, Andrott, and Kiltan) as published by IMD (for past 40 years) is shown in Fig. 2.

In order to hindcast the wave heights/storm surges for past 40 years for the remote islands, the numerical model was set up wherein the bathymetry for the model was

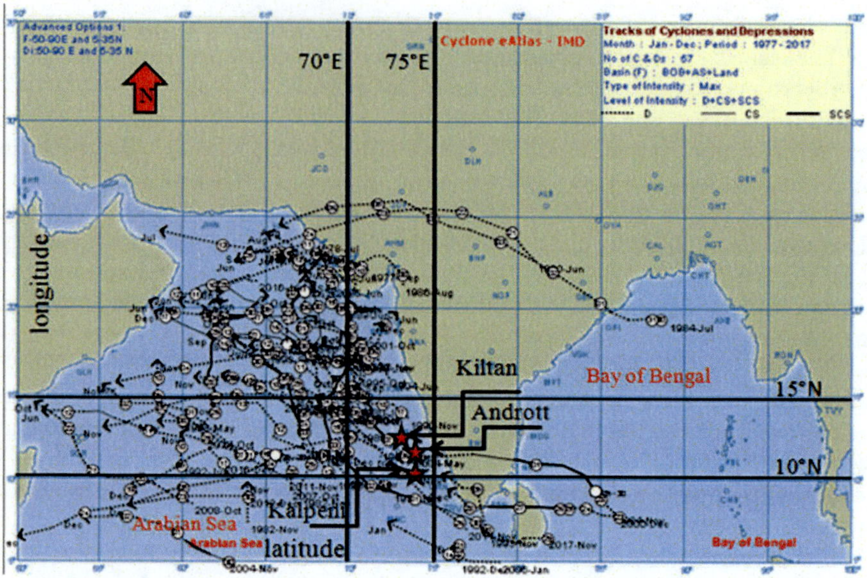

Fig. 2 Track of historical cyclonic events relevant for the coastal structures situated at Kalpeni, Andrott, Kiltan islands. *Source* IMD

obtained from CMAP (DHI) database. The track of cyclone, its forwards movement speed, distribution of atmospheric pressure, maximum sustained wind speed as published by (IMD) [12], etc., were also used.

3 Details of Forcing to Numerical Model

The measured historical tidal data, cyclonic wind field, and distribution of atmospheric pressure during cyclonic events near the islands were not available; as such the global inverse barotropic tidal database (TPXO atlas) [13], well established parametric cyclonic wind model developed by various researchers viz. Carr and Elsberry (1997), and the distribution of atmospheric pressure during the cyclonic event as published by IMD were used as forcing function for the coupled hydrodynamic-wave model. It enables transfer of appropriate energy into the system for the generation storm surges as well as extreme waves. Details of the atmospheric/tidal forcing are given in the following section.

3.1 Parametric Models of Tropical Cyclone

Cyclonic wind field and atmospheric pressure distribution are considered as the most important influencing parameters in generation of storm surges and extreme waves. Due to the non-availability of measured data on spatio-temporal varying wind field/atmospheric pressure distribution, simplified parametric models are normally used to represent the cyclonic wind field and distribution of atmospheric pressure. The parametric wind model is derived from the force balance at the geostrophic wind level, and the gradient wind speed is derived to the standard reference level (at +10 m above mean sea level) by considering boundary layer effects, wind inflow toward the vortex center, atmospheric pressure gradients, etc., [14]. These simplified wind models developed by various researchers such as Holland, Rankine Vortex, and Carr and Elsberry have been widely used by the engineers/practitioners, researchers to represent an equivalent cyclonic wind field. In the present study, the parametric wind model developed by Carr and Elsberry and validated by Purohit et al. [15] for the simulation of cyclonic waves for the Ockhi cyclone (2017) has been used to simulate the cyclonic wind field for past 40 years of historical cyclonic events. The information on maximum wind speed, track of cyclonic event as published by IMD at every 3 h interval for the tropical cyclonic events was also used to represent the cyclonic wind field near the radius of maximum wind (R_0). The tangential (v_c) and radial (u_c) wind components for the cyclonic wind field as per Carr and Elsberry's formulation can be computed as follows:

$$v_c = \frac{f_0}{2}\left[R_0\left(\frac{R_0}{r}\right)^X - r\right]\frac{\alpha^4}{(1+\alpha^4)} \tag{1}$$

$$u_c = \tan(\gamma)v_c(t) \tag{2}$$

where R_0 is the size of cyclone, r is the radial distance from center of cyclone, f_0 is Coriolis parameter, α is the scaling factor, X is a constant. The typical radial cyclonic wind profile (during Ockhi cyclone on 02.12.2017) derived from the Carr and Elsberry's parametric wind model and the distribution of atmospheric pressure as published IMD are shown in Fig. 3.

Similar to the parametric wind model, there are also simplified model developed by Holland [3] for representation of distribution of atmospheric pressure during the cyclonic event. However, for the present study, the spatio-temporal distributions of atmospheric pressure as published by IMD during each cyclonic event have been used.

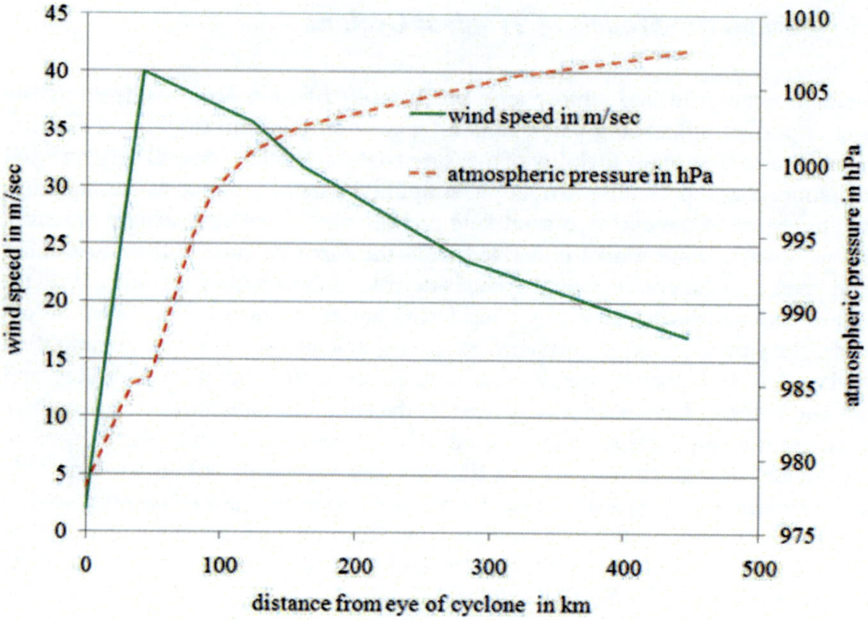

Fig. 3 The radial cyclonic wind profile and atmospheric pressure distribution for Ockhi cyclone

4 Details of Numerical Model

In the present study, coupled hydrodynamic (TELEMAC-2D) and wave model (TOMAWAC) were used to hindcast historical extreme waves and storm surges.

4.1 Hydrodynamic Model

The TELEMAC-2D solves Saint–Venant's shallow water equations of continuity and momentum, and these equations were derived by vertically averaging Navier-Stoke's equations. The governing equations that are solved in Cartesian co-ordinate system by finite element methods for hydrodynamic simulation in TELEMAC-2D are as follows:

$$\frac{\partial h}{\partial t} + \boldsymbol{u}.\nabla(h) + h\mathrm{div}(\boldsymbol{u}) = S_h \quad (3)$$

$$\frac{\partial u}{\partial t} + \boldsymbol{u}.\nabla(u) = -g\frac{\partial Z}{\partial x} + S_x + \frac{1}{h}\mathrm{div}(h\upsilon_t \nabla u) \quad (4)$$

$$\frac{\partial v}{\partial t} + \boldsymbol{u}.\nabla v = -g\frac{\partial Z}{\partial y} + S_y + \frac{1}{h}\mathrm{div}(hv_t\nabla v) \tag{5}$$

where h is depth of water; u and v are velocity components in x and y direction; Z is free surface elevation, S_h is source or sink of fluid; v_t is momentum diffusion coefficient; S_x and S_y are source terms representing the wind, Coriolis force, bottom friction, etc. [16]. TELEMAC-2D works both in Cartesian and spherical co-ordinate system.

The simulation of storm surge during the extreme events is primarily governed by (a) the energy/momentum imparted to the water column due to the passing of cyclonic wind over the water surface, (b) the change in water level due to variation of atmospheric pressure, as well as (c) by the tide-surge interaction. The interaction between air-sea interface in both spatial and temporal time scale being quite complex due to the complex behavior of air–water boundary layer, the drag force exerted by the cyclonic wind on the water surface is represented by simplified Flather's formulation [17] in TELEMAC-2D. TELEMAC also simulates spatio-temporal varying atmospheric pressure which enables simulation of the inverse barometric effect (rise in water level due to decrease in atmospheric pressure and decrease in water level due to increase in atmospheric pressure) during the cyclonic event. Apart from the atmospheric forcing, the storm surge simulation is also influenced by the interaction of storm surge and astronomical tide. The magnitude of the storm surge and the time of occurrence of resultant high/low water (due to storm and tide) are also influenced by the speed, phase of the two independent systems, i.e., astronomical tide and surge. In the present study to simulate the storm surges in TELEMAC-2D, the Coriolis parameter which is a function of the latitude, the tidal forcing from TPXO database, spatio-temporal varying cyclonic wind field (derived from parametric wind model) and atmospheric pressure (from IMD), wind drag force derived from Flather's formulation, bed friction coefficient of Chezy, etc., has been used.

4.2 Wind Wave Model

The TOMAWAC is a third generation spectral wind wave model, it solves the wave action density balance equation. The wave action balance equation is given as follows:

$$\frac{\partial N}{\partial t} + \frac{\partial\left(\frac{\partial\Omega}{\partial k_x}N\right)}{\partial x} + \frac{\partial\left(\frac{\partial\Omega}{\partial k_y}N\right)}{\partial y} + \frac{\partial\left(-\frac{\partial\Omega}{\partial x}N\right)}{\partial k_x} + \frac{\partial\left(-\frac{\partial\Omega}{\partial y}N\right)}{\partial k_y} = Q\left(k_x, k_y, x, y, t\right) \tag{6}$$

where N is wave action spectrum,

$$\Omega(\boldsymbol{k}, \boldsymbol{x}, t) = \sigma + \boldsymbol{k}.\boldsymbol{U} \tag{7}$$

where in **U** denotes depth-integrated current velocity; σ denotes the intrinsic or relative angular frequency; **k** is the wave number vector and k_x and k_y are its component along x, y; θ denotes the wave propagation direction; t is time [18]. The Q denotes source and sink terms and can be defined as follows:

$$Q = Q_{\text{in}} + Q_{\text{ds}} + Q_{\text{nl}} + Q_{\text{bf}} + Q_{\text{br}} + Q_{\text{tr}} + Q_{\text{veg}} + Q_{\text{ds,cur}} \tag{8}$$

Q_{in} is wind-driven wave generation; Q_{ds} is white-capping induced energy dissipation; Q_{nl} is nonlinear quadruplet interactions; Q_{bf} is bottom friction induced energy dissipation; Q_{br} is bathymetric breaking induced energy dissipation; Q_{tr} is nonlinear triad interactions; Q_{veg} is dissipation of energy due to vegetation; $Q_{\text{ds,cur}}$ is enhanced breaking dissipation of waves on a current. The computations in TOMAWAC are performed based on the finite element discretization method over unstructured mesh for infinite (deep water) or finite water (intermediate and shallow water) depths. TOMAWAC represents the sea state by splitting the directional wave spectrum into a finite number of frequencies and directions, and the wave action density balance equation is solved for each component of the frequency/direction. The energy per unit area of the wave field is obtained by integrating the wave energy spectrum along the discretized frequencies/directions, and the same is given in Eq. 9.

$$\sum_{f}^{f+df} \sum_{\theta}^{\theta+d\theta} \frac{1}{2}\rho g a_m^2 = E(f,\theta) df d\theta \tag{9}$$

where f and a_{m} are the frequency and the amplitude of each wave component, respectively. More details of governing equations for TELEMAC and TOMAWAC and their solution methods are described in the user manual of TELEMAC and TOMAWAC. The coupled TELEMAC-TOMAWAC model forced with spatio-temporal varying TPXO tidal base, cyclonic wind field derived from Carr and Elsberry's model, and the atmospheric pressure distribution (from IMD); simulates the physical processes such as input of energy from the wind, nonlinear wave-wave interactions, wave-current interaction, dissipation due to white-capping, bottom friction, and depth-induced breaking. The results of the coupled model are described in the following section.

5 Simulation Results and Discussion

The hindcast wave heights and storm surges were obtained from the numerical simulations for all the cyclonic events relevant to Kalpeni, Andrott, and Kiltan islands which had occurred in the past 40 years (1977–2017). These events also cover couple of extreme cyclonic events such as Gonu and Ockhi which had passed near these islands. The severe cyclonic storms (SCS) Gonu occurred during 5th to 10th May 2004 indicate that the maximum drop of central pressure near the eye of cyclone had reached up to 14 hPa and maximum sustained wind speed reached up to 55 knots. Also very severe cyclonic storm (VSCS) Ockhi occurred during 30th November 2017 to 5th of December 2017 had the maximum drop of central pressure near the eye of cyclone as 34 hPa and maximum sustained wind speed reached up to 85 knots. The Ockhi and Gonu cyclone being among the most severe extreme events in this region for these islands, their simulations, and results are discussed in this section.

The domain for the simulation of cyclonic events covers an area of about 8,33,000 km^2 (Fig. 4) and it is discretized with finite element mesh having the grid size varying from 2.5 km × 2.5 km in deep water to 50 m × 50 m in shallow water.

The hydrodynamic model developed has three liquid boundaries (north, west, and south) wherein the tidal forcing derived from TPXO tidal database was applied to simulate the tidal elevation as well as the tide induced currents. The spatio-temporal varying cyclonic wind field and atmospheric pressure were also applied as forcing function for the model. The Chezy's bed friction coefficient (C) was also varied with the water depth throughout the entire domain; wherein the water depth <30 m, the Chezy's (C) was kept to 70; for 30 m < water depth <90 m, the Chezy's (C) was kept to 60, and for the water depth >90 m, the Chezy's (C) was kept to 50. The time step was considered as 5 s for tidal hydrodynamic simulation as well as for the simulation

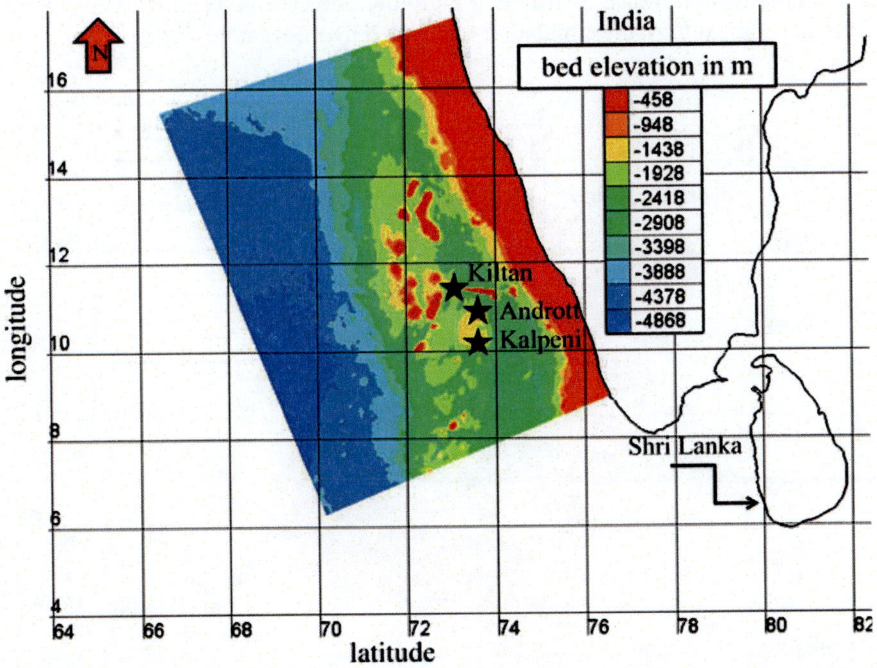

Fig. 4 Domain for the hindcast simulation indicating the locations of coastal structures at various islands

of spatio-temporal varying wind/atmospheric pressure as forcing function during its coupling.

Based on the information published by IMD (and available with Central Water and Power Research Station, Pune, India) about the distribution of atmospheric pressure, maximum sustained wind speed, track of cyclone, forward wind speed as well as the distribution of cyclonic wind field from the parametric wind model developed by Carr and Elsberry; the simulated atmospheric pressure distribution/cyclonic wind field for Gonu and Ockhi cyclonic events are shown in Figs. 5, 6, 7 and 8. The coordinate system for the simulation in the coupled model was in Mercator projection system.

The cyclonic events create storm tide which is the resultant of astronomical tide and the surge generated due to the cyclone. In order to estimate the magnitude of storm surge near the costal structures (at the water depth of 18 m to 22.5 m) located at Kalpeni, Andrott, and Kiltan islands; the comparison plots of simulated astronomical tide as well as storm tide are made, and the same are also shown in Figs. 9, 10, 11, 12, 13 and 14.

The study reveals that mainly due to the presence of steep slope, deep waters very near to the coastal structures; the storm surges are less and are about 0.1 m–0.2 m and as such are insignificant. Due to the interaction of storm tide, there is phase shift in the time of occurrence of resultant high/low tide. The wave heights for Ockhi cyclone obtained from the simulation of coupled hydrodynamic-wave.

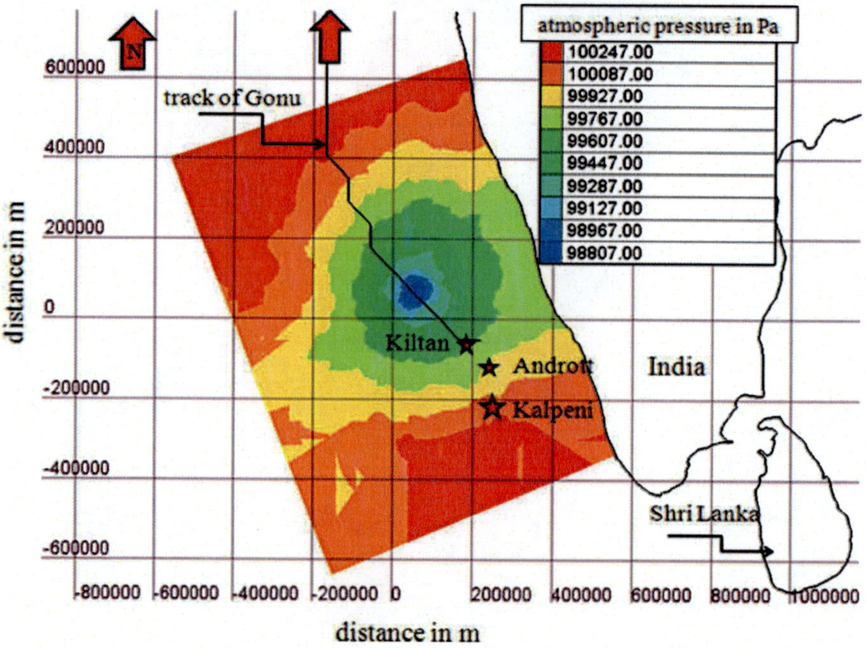

Fig. 5 Simulated atmospheric pressure during Gonu cyclone at 03.10 UTC of 7th May 2004

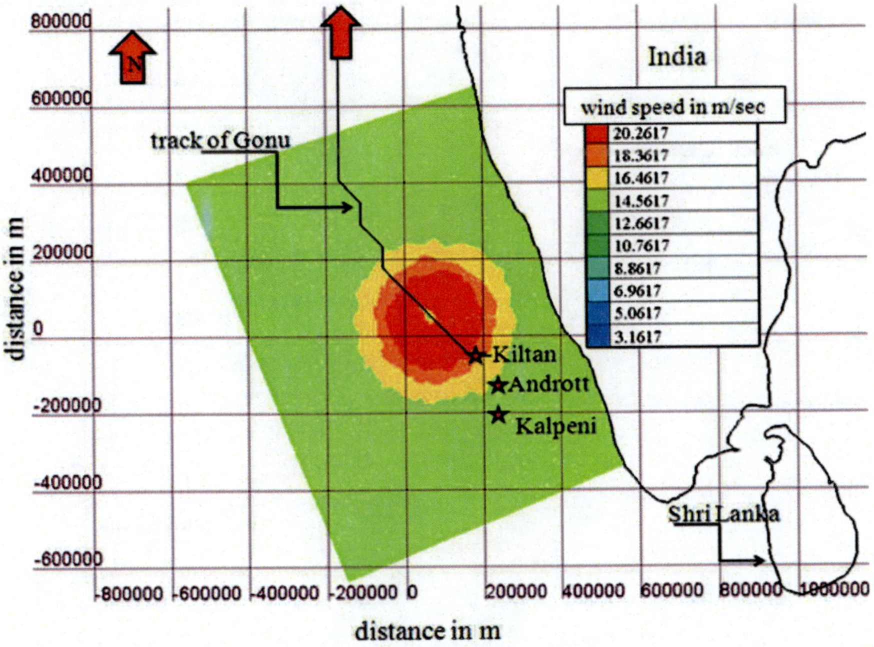

Fig. 6 Simulated cyclonic wind field during Gonu cyclone at 03.10 UTC of 7th May 2004

model for a typical time step are shown in Fig. 15. As the waves travel from deep water to shallow water, it undergoes various physical processes such as wave-wave interaction, wave-current interaction, and bottom friction induced energy dissipation and it results in reduction of wave energy. These dissipative processes of wave energy are expressed in the form of plot of spectral energy in frequency/directional domain. Such process is confirmed by extracting the information about spectral energy at two locations viz., deep water and at shallow water depths. The comparison plots of spectral energy at deep and shallow water near Kalpeni and Kiltan islands for the Ockhi cyclone (at 9.00 UTC of 1st December 2017) are shown in Fig. 16, Fig. 17, respectively.

The plots of spectral energy also confirm dissipative nature of wave energy as it travels from deep to shallow water. Some of the extreme significant wave heights and peak periods obtained from the hindcast study near the coastal structures (at water depth of 18–22.5 m) located at the islands are also given in Table 1.

The information given in the Table 2 indicates that the maximum significant wave height that had occurred during the Ockhi cyclone near the coastal structures at Kalpeni, Andrott, and Kiltan are of the order of 5.8–6.56 m having peak frequency (f_p) 0.09–0.18.

The hindcast wave heights and storm surges indicate that near the islands, and the extreme wave heights are having higher magnitude as compared to the storm surges. As such the impact of extreme waves will play a vital role in evolving the design

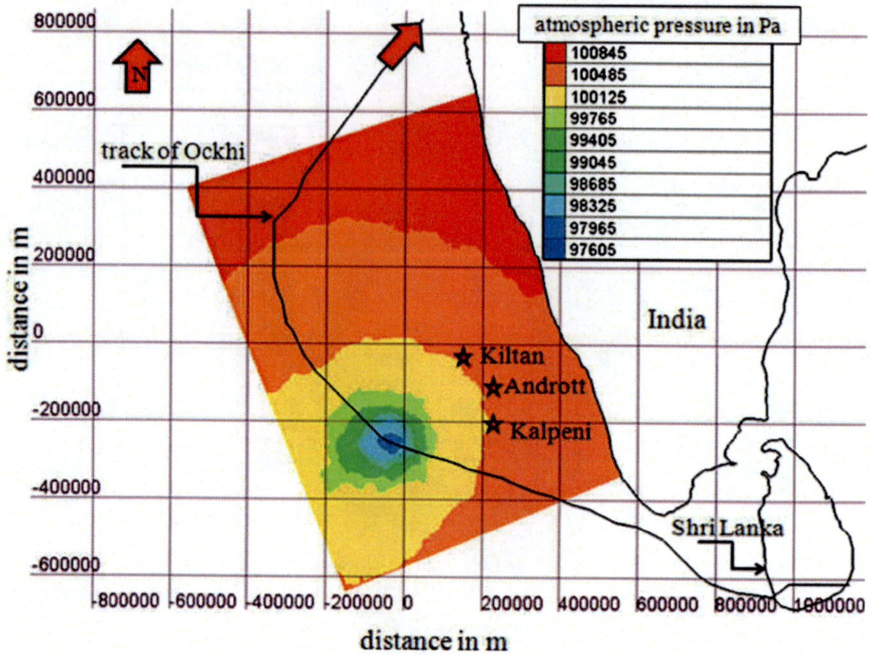

Fig. 7 Simulated atmospheric pressure during Ockhi cyclone at 03.10 UTC of 2nd Dec. 2017

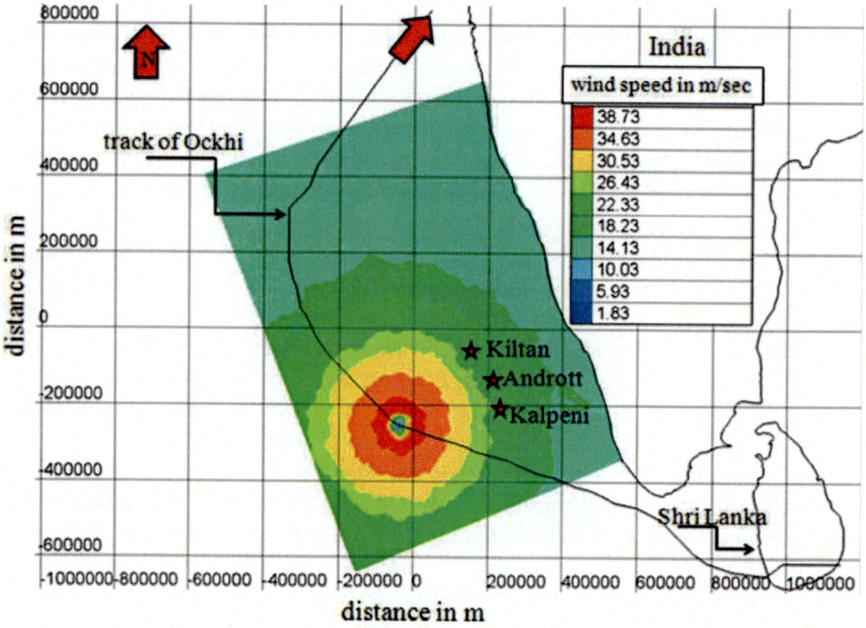

Fig. 8 Simulated cyclonic wind field during Ockhi cyclone at 03.10 UTC of 2nd Dec. 2017

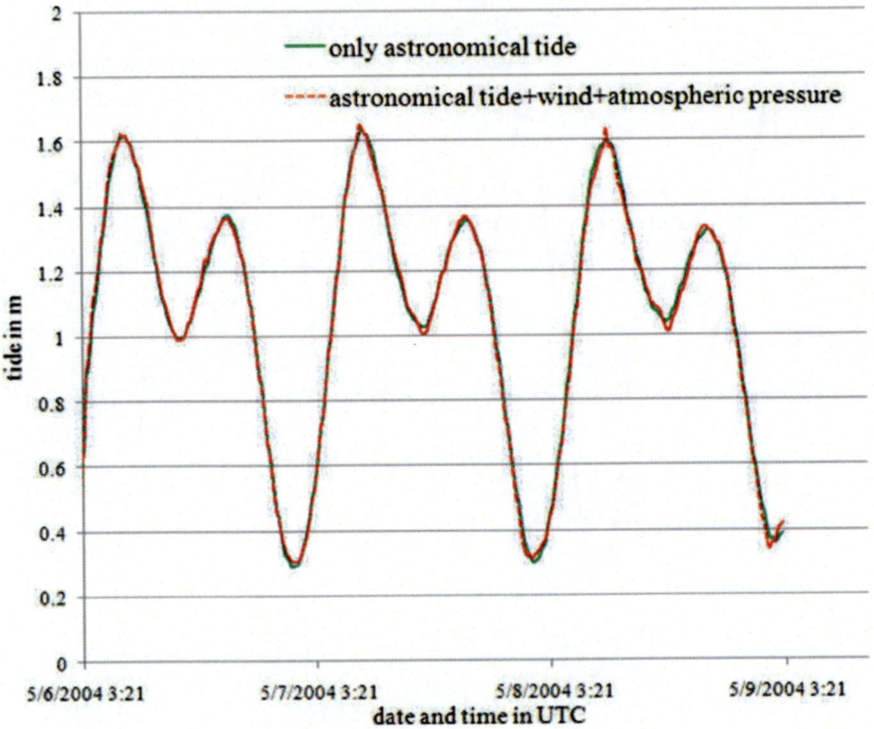

Fig. 9 Simulated storm tide/astronomical tide during Gonu cyclone at Kalpeni island

conditions for the coastal structures rather than storm surges. Hence, to arrive at the extreme wave heights for various return periods (25/50/100 years) for the design of coastal structures, the extreme value analysis of the hindcast wave heights was carried out and the same is described in details in the following section.

6 Extreme Value Analysis of Hindcast Wave Heights

The objective of extreme value analysis of hindcast wave heights is to arrive at the wave heights for various return periods so that the design basis for the coastal structures can be obtained. Goda [19] had described in details about the extreme value analysis of extreme wave heights. It was also explained about the superiority of peak value method, wherein the peak wave heights of statistically independent storms were selected rather than the total sample method, wherein all the regularly recorded wave heights which are mutually co-related are considered, and thereby total sample method violets the conditions of statistical independence. In peak value method, the dataset can be of either annual maximum series or partial duration

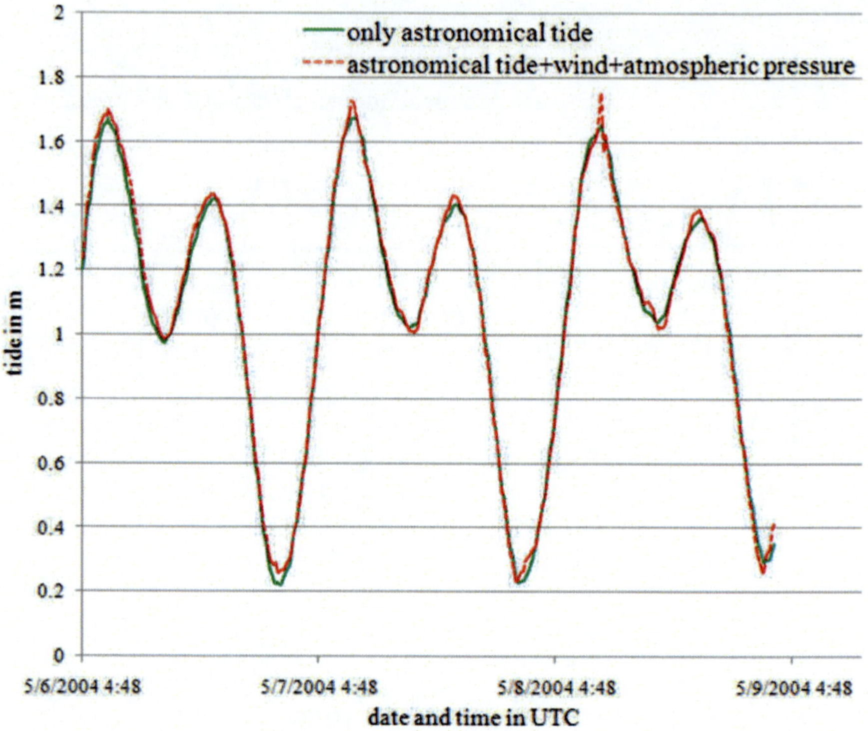

Fig. 10 Simulated storm tide/astronomical tide during Gonu cyclone at Andrott island

series. In the present study, as the hindcast database is derived during the period of storms, the dataset is of partial duration series type. In order to estimate the design wave heights for various return periods, mainly, two types of distribution families are used, i.e., either generalized extreme value distribution (GEV) or generalized Pareto distribution (GPD). In the present study, two types of distribution functions such as Gumbel distribution and Weibull distribution (from GEV distribution family) have been used to fit the hindcast wave heights. The general forms of cumulative distribution function for the GEV distribution are given in Eq. 10 and in Eq. 11.

$$F(X;k,\alpha,\beta)=\left\{\mathrm{e}^{-\left[1-k\left(\frac{x-\beta}{\alpha}\right)\right]^{\frac{1}{k}}}\text{for } k\neq 0\right\} \tag{10}$$

$$F(X;k,\alpha,\beta)=\left\{\mathrm{e}^{-e^{-\left(\frac{(x-\beta)}{\alpha}\right)}}\text{for } k=0\right\} \tag{11}$$

where k is the shape parameter, α is the scale parameter and β is the location parameter. Depending on the values of k, the GEV distribution is called Gumbel distribution

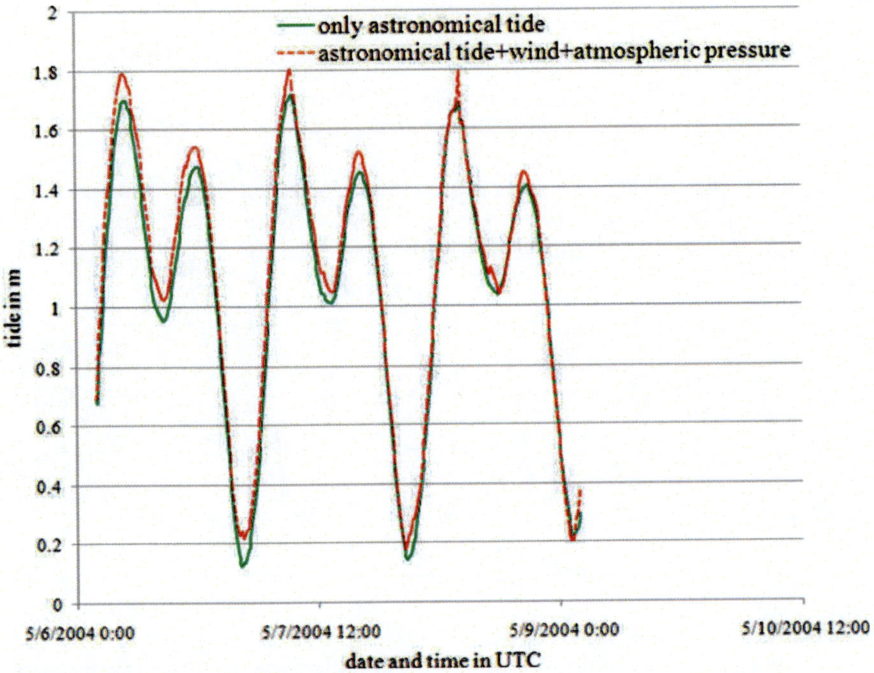

Fig. 11 Simulated storm tide/astronomical tide during Gonu cyclone at Kiltan island

(if $k = 0$), Frechet distribution (if $k > 0$), Weibull distribution (if $k < 0$). These parameters can be estimated by probability weighted method (PWM), maximum likelihood method (MLM). The detail descriptions about these methods are given in Coles [20]. The wave height (X_T) for a particular return period (T) for GEV distribution can be computed by using generalized expressions as given in Eq. 12 and in Eq. 13.

$$X_T = \left\{\beta + \frac{\alpha}{K}\left\{1 - [-\ln\left(1 - \frac{1}{T}\right)]^K\right\}\text{for } k \neq 0\right\} \tag{12}$$

$$X_T = \left\{\beta - \alpha \ln\left[-\ln\left(1 - \frac{1}{T}\right)\right]\text{for } k = 0\right\} \tag{13}$$

Goda [19] had summarized the expressions for Gumbel and Weibull distribution with the help some constants and is given in Eqs. 14 and 15, respectively.

$$F_{\mathrm{mI}} = 1 - \frac{(m - 0.44)}{(N_T + 0.12)} \tag{14}$$

$$F_{\mathrm{mII}} = 1 - \frac{(m - \alpha)}{(N_T + \beta)} \tag{15}$$

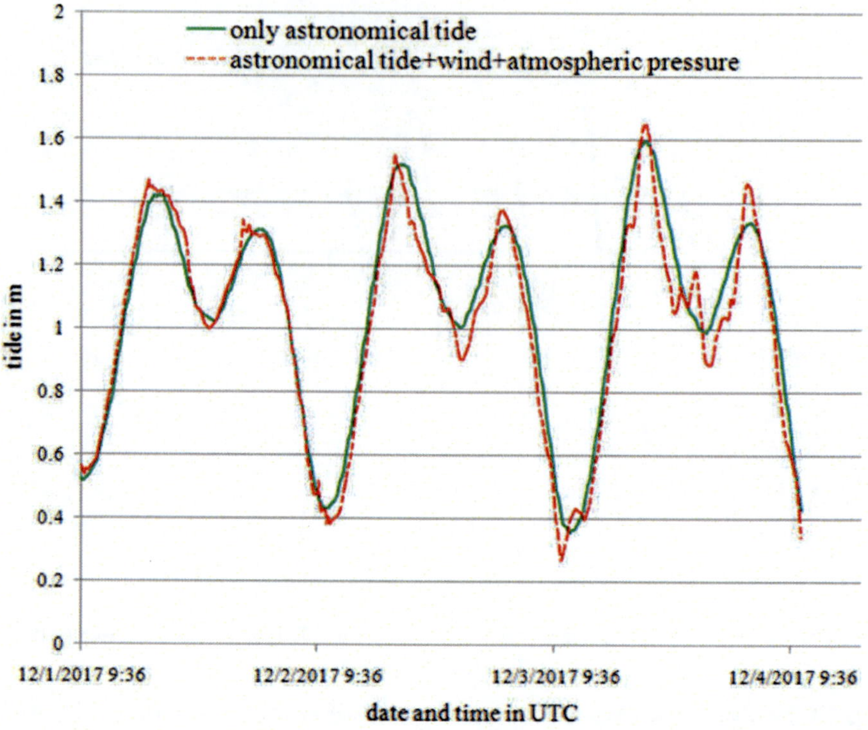

Fig. 12 Simulated storm tide/astronomical tide during Ockhi cyclone at Kalpeni island

where F_{mI} and F_{mII} are the distribution functions for Gumbel and Weibull, respectively, $\alpha = 0.2 + \frac{0.27}{\sqrt{k}}$ and $\beta = 0.2 + \frac{0.23}{\sqrt{k}}$, N_{T} is the total number of dataset considered for fitting in Gumbel or Weibull distribution, m varies from 1, 2, ... N and k can be 0.75 or 1 or 1.4 or 2. The wave height (X_{R}) for various return periods (R) is computed based on Eq. 16.

$$X_R = Ay_R + B \tag{16}$$

where in $y_R = -\ln\left\{-\ln\left[1 - \frac{1}{\lambda R}\right]\right\}$ for Gumbel distribution.

And $y_R = [\ln(\lambda R)]^{\frac{1}{k}}$ for Weibull distribution.

Where λ is the number of occurrence of the event per year, the scale (A) and location parameter (B) can be computed from the linear relationship (Eq. 17) between the ordered variate $X_{(m)}$ and its reduced variate $Y_{(m)}$.

$$X_m = AY_m + B \tag{17}$$

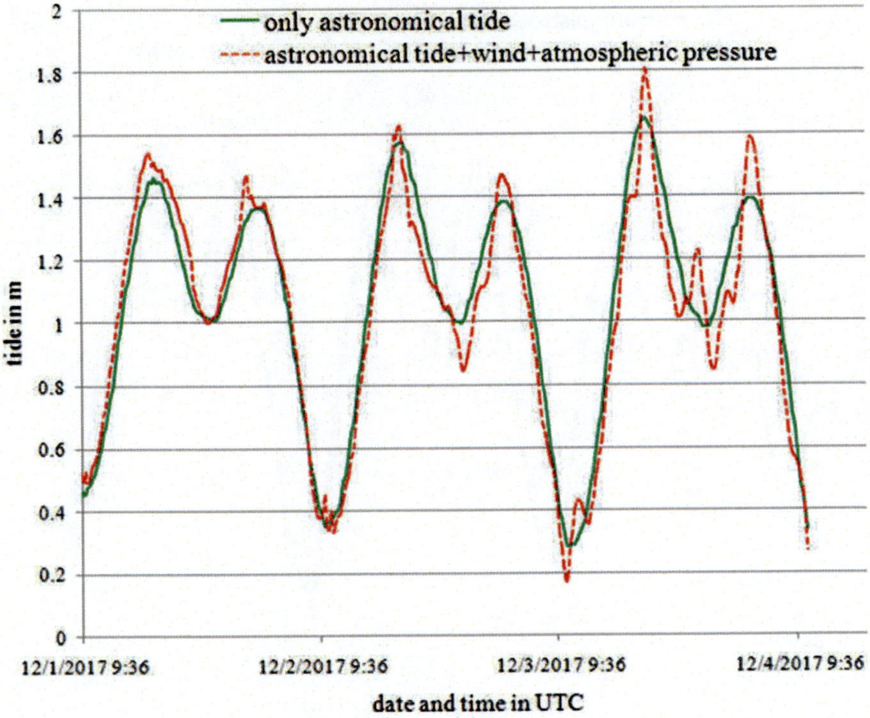

Fig. 13 Simulated storm tide/astronomical tide during Ockhi cyclone at Andrott island

More details about the procedure for extreme value analysis can be found in Goda [19]. In the present study, based on the shape, location and scale parameters, the hindcast wave heights at various islands were fitted to either Weibull distribution or Gumbel distribution function and depending on the goodness of fit/highest co-relation [19], the suitable distribution function was selected as well as the extreme wave heights for various return periods for the coastal structures were also calculated. The calculated extreme wave heights are given in Table 2, and the same is plotted in Fig. 18.

7 Conclusions

The global warming has resulted in increase in cyclonic events all over the world, and Indian Ocean is not an exception to this. The Bay of Bengal (BoB) and Arabian Sea have seen increase in cyclonic activity for the past decade or so with increase in trend in intensity of wind speed as well as change in monsoonal activities. Nowadays, there are also increase in occurrence of events like cloud bursting wherein there

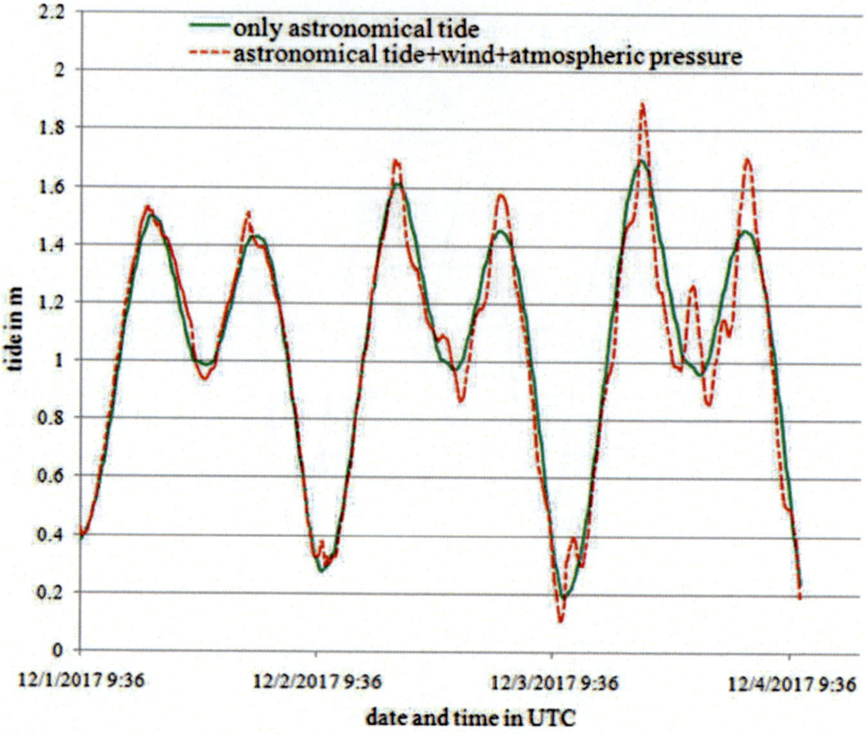

Fig. 14 Simulated storm tide/astronomical tide during Ockhi cyclone at Kiltan island

is heavy rain fall over short durations. Thus, the coastal structures in the form of breakwater/seawall/groin which used to protect harbors/beaches are subjected to severe attack of intensified sea waves. It may result in severe damages to the structures and also losing its functionality.

The coastal structures constructed on islands (Lakshadweep group of islands) in the Arabian Sea are subjected to attack of severe waves in the recent cyclonic events and thereby had resulted in damages to the structures at Kalpeni, Andrott, and Kiltan. The breakwater built at Kalpeni island to achieve safe berthing conditions at the wharf for the passengers, fishing vessels for all weather seasons got damaged extensively during Ockhi cyclone occurred in year 2017. This has not only affected the waterborne transport facility but also in turn the lifeline for the island. As such the design considerations for the structures situated at remote islands need to be evolved carefully by giving due considerations to the historical extreme cyclonic events. The long-term measured historical wave data (collected by wave rider buoy) being seldom available near these islands, hindcasting of extreme waves, storm surges are the only way to get the historical data. In the present study, coupled hydrodynamic-wave model developed for the part portion of Arabian Sea with appropriate forcing function (global inverse barotropic tidal database, validated parametric cyclonic wind

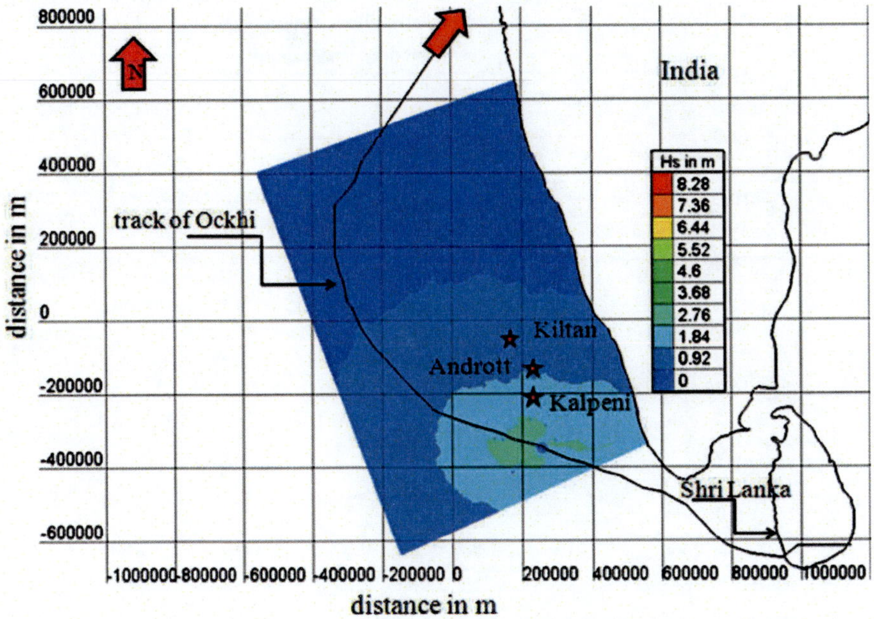

Fig. 15 Simulated significant wave height during Ockhi cyclone at 05.10 UTC of 1st Dec 2017

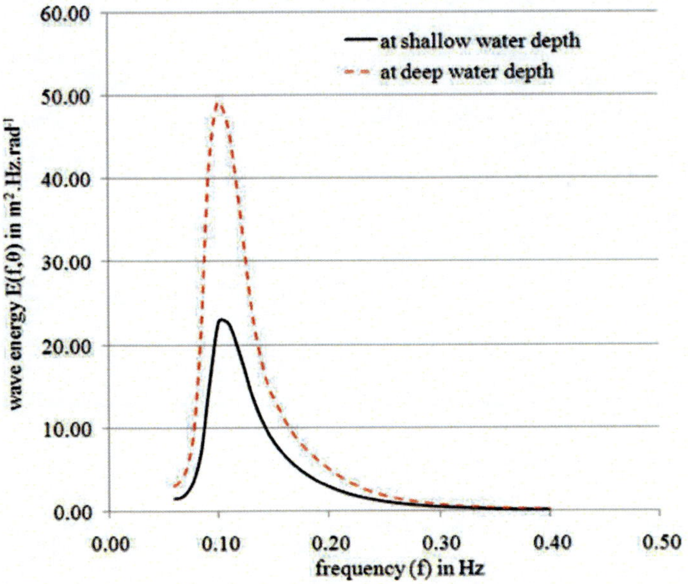

Fig. 16 Spectral density plot near Kalpeni island

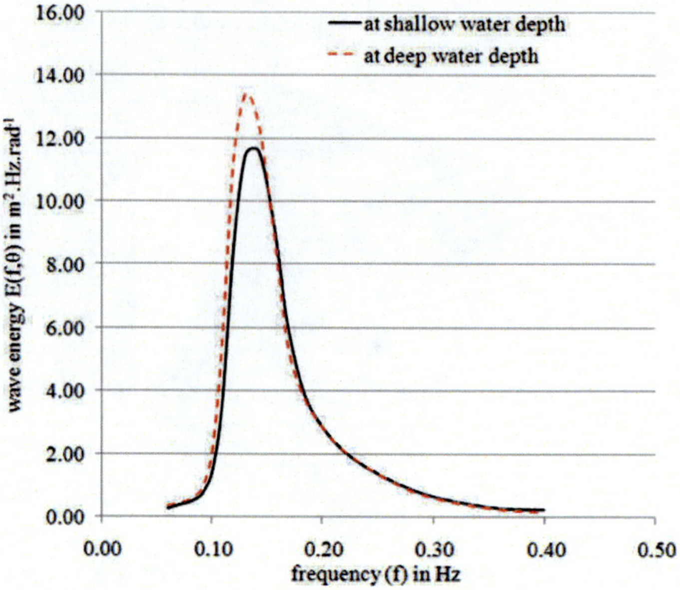

Fig. 17 Spectral density plot near Kiltan island

Table 1 Hindcast wave height during some of the storms occurred near the coastal structures of the islands

Date and time of occurrence of storms	Near the breakwater at Kalpeni island		Near the breakwater at Andrott island		Near the groin at Kiltan island	
	Hs in m	Peak frequency in Hz	Hs in m	Peak frequency in Hz	Hs in m	Peak frequency in Hz
21.11.1977 08.30 IST	2.4	0.16	4.74	0.12	3.49	0.135
06.05.2004 08.30 IST	4.9	0.10	5.4	0.1	5.49	0.1
07.05.2004 08.30 IST	2.2	0.17	2.2	0.16	2.94	0.15
01.12.2017 08.30 IST	4.7	0.098	3.68	0.135	2.95	0.148
01.12.2017 17.30 IST	5.8	0.094	6.23	0.1	5.34	0.11
02.12.2017 08.30 IST	4.7	0.12	5.47	0.11	5.29	0.11
02.12.2017 17.30IST	5.1	0.11	6.13	0.1	6.56	0.093
03.12.2017 17.30 IST	2.3	0.16	2	0.176	3.53	0.13

Table 2 Significant wave heights for various return periods for the coastal structures situated at various islands

Name of island	Distribution function and its co-relation coefficient (r)	Selected distribution function	Return period in years	Hs in m
Kalpeni	Gumbel (r = 0.93) Weibull (r = 0.95)	Weibull	50	5.96
			100	6.51
Andrott	Gumbel (r = 0.93) Weibull (r = 0.96)	Weibull	50	6.70
			100	7.36
			100	6.90

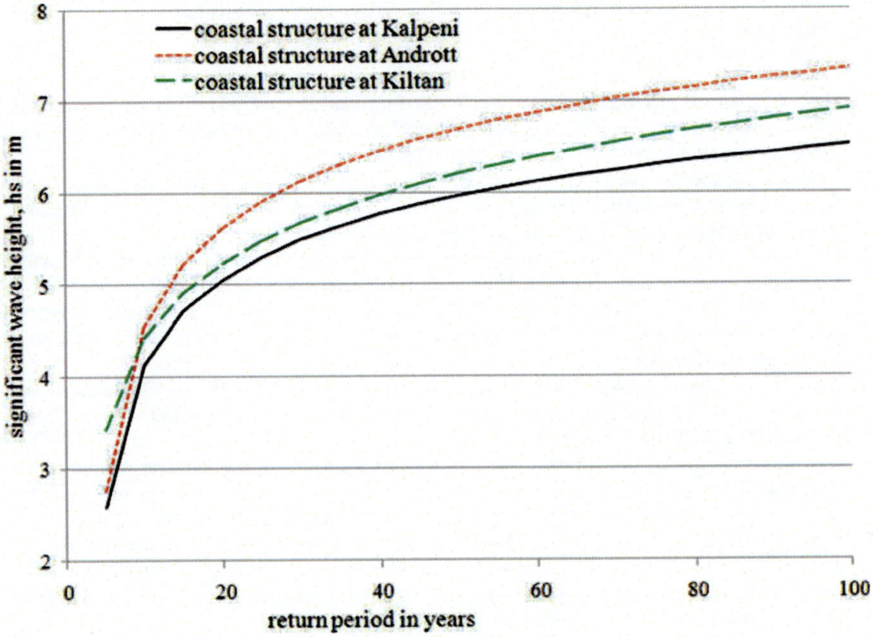

Fig. 18 Plot of hindcast wave heights for various return periods

model, atmospheric pressure distribution) was applied to hindcast the wave heights, storm surges for past 40 years. The study reveals that due to the presence of very steep slope near the islands, the storm surges are less (about 0.1 m–0.2 m) near the coastal structures. Also due to the tide-surge interaction, there is phase shift in the time of occurrence of high/low tide. However, during the cyclonic events, close to the islands, the wave heights are higher in magnitude (Hs about 6 m–6.5 m) with peak frequency of 0.09 Hz–0.18 Hz. As such the extreme wave height is having more significance as compared to the storm surge while assessing the hydraulic stability of the coastal structures. The hindcast significant wave heights being of partial duration series type, it was fitted in the GEV distribution (Gumbel, Weibull) to provide a basis for evolving design wave height for various return periods (50/100 years) for the coastal structures on the remote islands. Thus, extreme wave heights (Hs) predicted by analyzing the cyclones in the Arabian Sea reveals that the wave of Hs = 6.51 m, 7.36 m, and 6.90 m will be the design wave heights for the coastal structures at Kalpeni, Andrott, and Kiltan island, respectively.

Acknowledgements The authors are thankful to Shri A.K. Agrawal, Director, Central Water and Power Research Station, Pune (India) for his continuous encouragement for carrying out the research work.

References

1. Shao, Z., Liang, B., Li, H., Wu, G., & Wu, Z. (2018). Blended wind fields for wave modeling of tropical cyclones in the South China Sea and East China Sea. *Journal Applied Ocean Research, 71*:20–33.https://doi.org/10.1016/j.apor.2017.11.012
2. Coastal Engineering Research Centre, Department of the Army, Waterways Experiment Station, Crops of Engineers. (1984). Shore Protection Manual: Volume I. U.S. Army Engineer Waterways Experiment Station Coastal Engineering Research Centre.
3. Holland, G. J. (1980). An analytic model of the wind and pressure profiles in hurricanes. *Monthly Weather Review, American Meteorological Society, 108*, 1212–1218.
4. Jelesnianski, C.P., & Taylor, A. D. (1973) A preliminary view of storm surges before and after storm modifications. Technical memorandum, National Oceanic and Atmospheric Administration (NOAA).
5. Holland, G. J., Belanger, J. I., & Fritz, A. (2010). A revised model for radial profiles of hurricane winds. *Monthly Weather Review, American Meteorological Society, 138*, 4393–4401. https://doi.org/10.1175/2010MWR3317.1.
6. Das, Y., Mohanty, U. C., & Jain, I. (2016). Development of tropical cyclone wind field for simulation of storm surge/sea surface height using numerical ocean model. *Journal Modeling Earth Systems and Environment, 2*(13), 1–22. https://doi.org/10.1007/s40808-015-0067-5
7. Lafon, F., & Benoit, M. (2006). Estimation of extreme wave conditions from hindcast simulations with application to the wave climate along French coasts. In *30th international conference on coastal engineering* (pp. 739–751). World Scientific. https://doi.org/10.1142/9789812709554_0063
8. Anand, K. V., Sannasiraj, S. A., Sundar, V. (2014). Investigation on the cyclonic seastate along southeast coast of India. *Journal Marine Geodesy, 0*, 1–21https://doi.org/10.1080/01490419.2014.914111
9. Murty, P. L. N., Bhaskaran, P. K., Gayathri, R., Sahoo, B., Kumar, T. S., & Reddy, S. B. (2016). Numerical study of coastal hydrodynamics using a coupled model for Hudhud cyclone in the Bay of Bengal. *Journal Estuarine, Coastal and Shelf Science, 183*(part A):13–27. /https://doi.org/10.1016/j.ecss.2016.10.013.
10. Cooper, A., Turnbull, M., Grey, S., Loilier P (2013) Tropical cyclone modelling with TELEMAC-2D. In *20th Telemac Mascaret user conference*, (pp. 43–45). Karlsruhe.
11. Gourgue, O., Chen, M., & Sarhadi, E. (2014). Modeling storm surges in the Belgian coastal area: preliminary work. In *21st Telemac Mascaret user conference* (pp. 21–26). Artelia Eau & Environnement, France.
12. Indian Meteorological Department. https://rsmcnewdelhi.imd.gov.in/report.php?internal_menu=MzM=
13. TPXO Global Tidal Models. https://www.tpxo.net/global

14. Harper, B. A., & Holland, G. J. (1999). An updated parametric model of the tropical cyclone. In *AMS 23rd conference on hurricanes and tropical meteorology, Dallas*.
15. Purohit, A. A., Basu, A., & Agrawal, J. D. (2019). Importance of parametric wind model in simulation of extreme wave heights during cyclonic events. *24th Hydro 2019 international conference* (pp. 546–555). BS publications.
16. Ata, R. (2017). TELEMAC2d user manual version 7.2, EDF.
17. Gourgue, O., Sishah, B. B., Vanlede, J., Komijani, H., & Chen, M. (2015). Modelling tides and storm surges on the European continental shelf. In *22nd Telemac-Mascaret user conference* (pp. 154–159). STFC Daresbury Laboratory, UK.
18. Awk, T. (2017). TOMAWAC user manual version 7.2, EDF.
19. Goda, Y. (1988). On the methodology of selecting design wave height. In *21st International conference on coastal engineering* (pp. 899–913). Spain. https://doi.org/10.9753/icce.v21.67.
20. Coles, S. (2001). *An introduction to statistical modeling of extreme values*. Springer.

Forecast of Sea Surface Temperature and Chlorophyll-a Using ARIMA, South-Eastern Arabian Sea, Karnataka, India

Harshada, Satyanarayana, and K. S. Jayappa

Abstract Sea Surface Temperature (SST) and Chlorophyll-α (chl-a) concentrations are the two basic marine parameters used to monitor ocean productivity, ecosystem and global climate change. Thus, it plays an important role in managing marine ecosystems for sustainable development and conservation. SST and chl-a are closely coupled and linked to ocean physical and biogeochemical cycles. Satellite observations give a real-time synoptic view and regular repeated coverage of these variables, whereas time series machine learning helps to identify future trends. The time series modeling helps to identify trends, seasonality and to predict future values. In the present study, an attempt has been made to forecast SST and chl-a variations using the Seasonal Autoregressive Moving-Average (SARMA) model over the offshore waters of Karnataka, southeastern Arabian Sea. The accuracy measure results exhibit that SARIMA [1, 0, 1] $[1, 1, 2]_{12}$ and SARIMA [1, 0, 0] $[3, 1, 1]_{12}$ are effective models for forecasting SST and chl-a respectively. This model is used to forecast SST and chl-a for the next 24 months.

Keywords Sea surface temperature · Chlorophyll-α · Seasonal autoregressive integrated moving-average

1 Introduction

The sea surface temperature (SST) is one of the fundamental physical parameters extensively used in detecting global climate change signatures. Many studies have explored the relationship between increasing SST to increased intensity of the hurricane, typhoon and tropical cyclone activities [5, 35, 55, 57, 59]. It also gives insight into complex interactions between the ocean and the atmosphere [21, 36, 45]. SST anomalies are also linked to El Niño Southern Oscillation (ENSO) and Indian

Harshada (✉) · K. S. Jayappa
Department of Marine Geology, Mangalore University, Mangalore, Karnataka, India

Satyanarayana
Department of Statistics, Mangalore University, Mangalore, Karnataka, India

A. K. Dikshit et al. (eds.), *Innovative Trends in Hydrological and Environmental Systems*, Lecture Notes in Civil Engineering 234,
https://doi.org/10.1007/978-981-19-0304-5_15

Ocean Dipole (IOD) mode which could affect local weather and profoundly impact marine ecosystems by altering precipitation patterns and global ocean physical-biogeochemical cycles [3, 4, 14, 17, 20, 22, 29, 32, 46, 47, 58]. Chlorophyll-a (chl-a) is one of the crucial water quality parameters and is also used as a proxy for phytoplankton biomass [15]. Ocean phytoplankton contributes to net primary production [19] and play important role in the biological pump by sequestering carbon away from the atmosphere [8, 50]. In the literature, several studies have described the interrelationship between SST and chl-a [18, 30, 31, 33, 37–39, 56] which helps to study marine food web dynamics, also has the application in marine ecosystem monitoring and assessment, and fisheries as these two are the important parameters used in demarcating potential fishing zones (PFZ) [11, 44]. Satellite remote sensing can contribute to monitoring and acquiring time-series data along with real-time synoptic global coverage at a regular time interval. While the time series machine learning models help in identifying the future trend in the time series, Seasonal ARIMA (Auto-Regressive Integrated Moving Average) (SARIMA) model is one of such models [23, 34, 43, 49]. In several types of machine learning techniques, neural network and the ARIMA model have been extensively applied to predict future values of SST and chl-a in different parts of the ocean and sea [1, 2, 24, 41, 48, 51–53]. In these models out of several types of input, SST of previous/historical times (lagged SST values) were often adopted as input variable type (55%), whereas the remaining 45% inputs include fossil, marine sediment samples, spatio-temporal parameters such as net surface heat flux, wind stress, and dynamic wave height [25]. This study focuses on modelling of SST and chl-a time series data to predict their future values at two years lead over the offshore waters of Karnataka which covers both coastal and shelf regions of the Arabian Sea. For this purpose, we have used 18 years of SST and chl-a data derived from MODIS (Moderate-Resolution Imaging Spectroradiometer) Aqua and evaluated the accuracy of the model using statistical metrics over the study region.

2 Materials and Methods

2.1 Study Area

The area of interest is from 73 to 75° E and 12.45 to 15.17° N which mainly covers the coastal and shelf region of the Arabian Sea (Fig. 1). Area averaged SST and chl-a are used in this study. MODIS Aqua level-3 Standard Mapped Image (SMI) monthly products of SST (4 microns night) and chl-a (OCI algorithm), with a spatial resolution of 4 km, for the period of 18 years (July 2002 to July 2020) were obtained from the NASA data portal. (https://giovanni.gsfc.nasa.gov/giovanni). The methodology followed in this study to forecast SST and chl-a data are shown in Fig. 2.

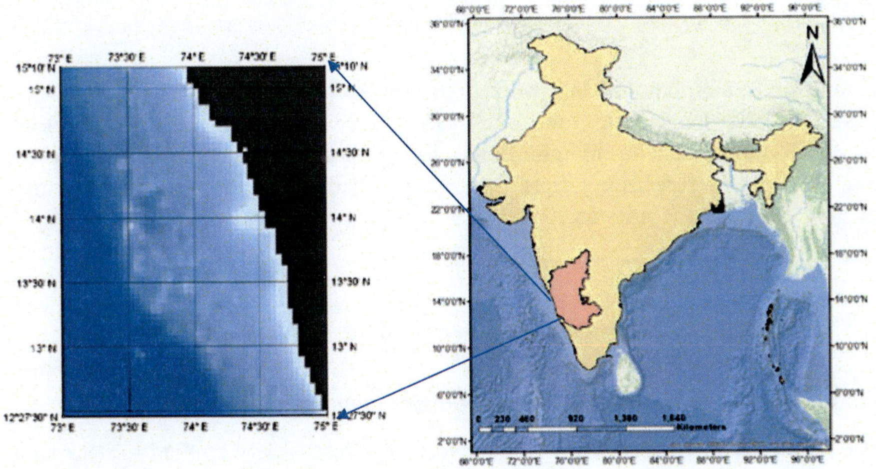

Fig. 1 Study area

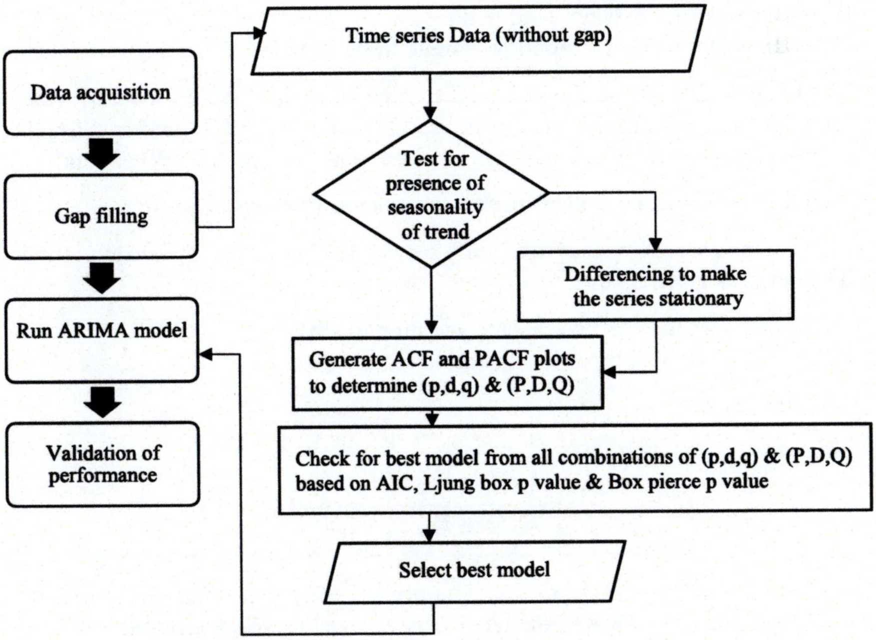

Fig. 2 Flowchart showing the steps followed to generate the forecast of SST and chl-a

2.2 Gap Filling of Missing Data

A total of 217 months of data were used, out of which, 13 months of monsoon chl-a data and one month (June 2019) SST data were missing due to the cloud cover. Missing of data during monsoon is a major problem in constructing complete time series and forecasting thus gaps were filled using the data interpolation method: *interp1* in R software.

2.3 Statistical Approach for Forecasting

Testing for the presence of trend component (Mann–Kendall trend test)

The Mann–Kendall trend test is a nonparametric test widely applied to detect the trend in a series even if a seasonal component exists [28, 40]. According to this test, the hypothesis is as follows.

H_0 → Indicates there is no trend in the data.
H_1 → Indicates there is a monotonic trend in the data.

The Mann–Kendall test is based on the calculation of Kendall's τ (tau), which measure the cross-correlation between two samples, which itself is based on the ranks of the samples. Here the main assumption is that observations are independent.

Testing for the presence of seasonality component (Rank-sum test)

The Rank-sum test is a test used to identify seasonality in a series [7]. The hypothesis for this test is as given below:

H_0 → Indicates there is no seasonal variation in the data.
H_1 → Indicates there is seasonal variation in the data.

Test statistic is

$$\chi_0^2 = \frac{12\sum_{j=1}^{D}\left(M_j - \frac{C(D-1)}{2}\right)^2}{CD(D+1)} \sim \chi^2(D-1)$$

where D—seasonality periods, C—total number of years, M_j—sum of the ranks for the jth period. If the chi-square calculated is more than the chi-square table value, we reject H_0 and conclude that there is seasonal variation in the data.

ARIMA model

An ARIMA model is used to predict future values based on the linear combination of previous values of the time series [42]. The seasonal ARIMA model [6, 26] is generally denoted as ARIMA $(p, d, q)(P, D, Q)_S$, where p is the trend autoregression order, d is trend difference order, q is the trend moving average, P represents seasonal autoregressive order, D is seasonal difference order, Q is seasonal moving average order and S is the period of repeating seasonal pattern.

If trend and seasonal components are present in the time series data, differencing is carried out to make the series stationary. The autocorrelation function (ACF) and partial autocorrelation function (PACF) plots of the stationary series are used to select the parameters (p, d, q) (P, D, Q) of the SARIMA model. All possible combinations using (p, d, q) (P, D, Q) were considered to select the best model for forecasting SST and chl-a. Then the best model is selected based on the Akaike Information Criterion (AIC), Ljung box (LB) p-value and Box pierce (BP) p-value [7, 9, 10, 12]. The univariate ARIMA model is run in R (4.0.3) software.

2.4 Model Validation

Performance of the SARIMA model predicted values against satellite-derived values were assessed using a set of cross-validation methods. In the literature Root Mean Square Error (RMSE), Mean Relative Percentage Error (MAPE) and Mean Absolute Relative Percentage Error (MRPE) and Goodness of fit (using the coefficient of determination (R^2), slope and intercept) are widely adopted as performance indicators [13, 16, 27, 54, 60, 61].

RMSE is used to determine the standard deviation of the residuals (i.e. errors in the prediction). As the residuals are the difference between the model-predicted and satellite-derived observations.

$$\mathrm{RMSE} = \sqrt{\frac{\sum_{i=1}^{n}(\mathrm{MP} - \mathrm{SD})^2}{n}} \tag{1}$$

MRPE is used to estimate the average bias percentage in the prediction. If the model predicted value is greater than the satellite measured value then the bias is positive (indicate overestimation) the inverse is negative bias indicate underestimation by the model.

$$\mathrm{MRPE} = \frac{1}{n}\sum_{i=1}^{n}\frac{\mathrm{MP} - \mathrm{SD}}{\mathrm{SD}} \times 100 \tag{2}$$

MAPE expresses the error percentage in predicted values. The higher MAPE indicates that the model predicted value is more than the actual/satellite-derived value and has less accuracy,

$$\mathrm{MAPE} = \frac{1}{n}\sum_{i=1}^{n}\frac{|\mathrm{MP} - \mathrm{SD}|}{\mathrm{SD}} \times 100 \tag{3}$$

where MP and SD represent model predicted values and satellite-derived data respectively.

The goodness of fit of the regression model is measured using the coefficient of determination (R^2) which lies between 0 and 1. This measures the relationship between the independent variable (satellite-derived) and the dependent (model-predicted) variable [54].

3 Results and Discussion

Figure 3a and b represents the time profile plots of the SST and chl-a time series data respectively. From the time profile plot, we observed that there is seasonality in the data. To confirm the same, tests for the presence of seasonality as well as trend were carried out. From the dataset, SST and chl-a data from 2002 to 2018 is considered as train-set and data from Jan 2019 to July 2020 is considered as test data.

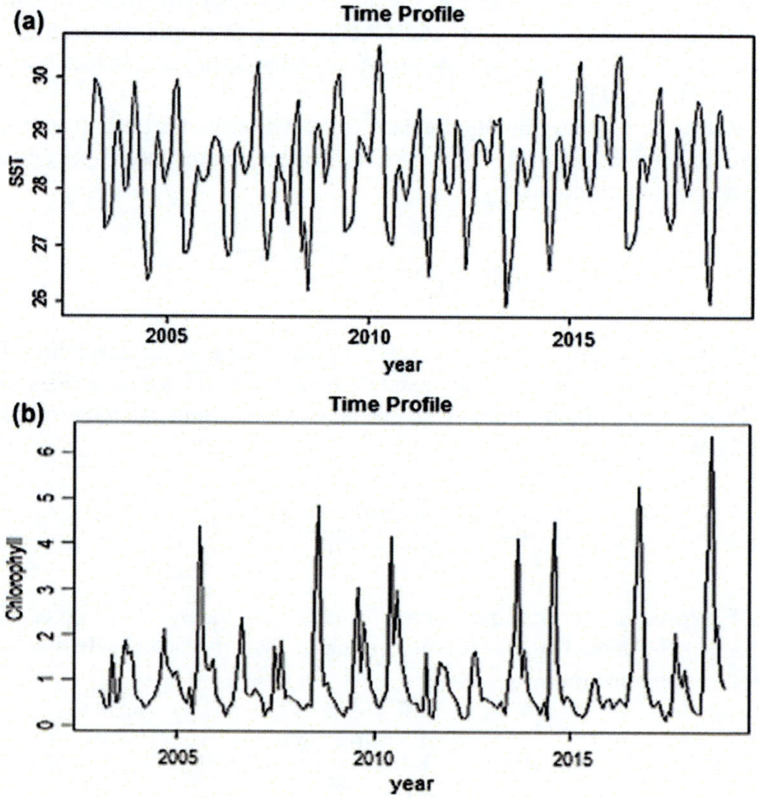

Fig. 3 **a** and **b** Profile plots of SST and chl-a time series

Table 1 Test result of the presence Rank Sum test and Mann Kendall test

	SST	Chl-a
Test for seasonality: Rank Sum test	Calculated value: 163.702 Table value: 9.675	Calculated value: 121.442 Table value: 19.675
	Since the calculated value > table value, we reject H_0 and conclude that there is seasonal variation in the data	
Test for the presence of trend: Mann Kendall test	*P* value: 0.352	*P* value: 0.426
	Since *P* value > 0.05, we accept H_0 and conclude that there is no trend component in the data	

As data (Table 1) contains only seasonal component, seasonal differencing was carried out to make the series stationary. From the seasonal differencing it is clear that after a one-time difference, the series is stationary and hence the value of $D = 1$ and due to the absence of trend component, trend difference order $d = 0$.

From the ACF and PACF plots of SST data (Fig. 4a and b), it is clear that $p = 2$, $q = 3$, $d = 0$, $P = 3$, $D = 1$, $Q = 2$. Since it is monthly data, the seasonal period is taken $S = 12$.

From the ACF and PACF plots (Fig. 5a and b) generated for chl-a $p = 1$, $q = 2$, $d = 0$, $P = 3$, $D = 1$, $Q = 2$. Since it is monthly data, the seasonal period is taken $S = 12$.

All possible combinations using (*p*, *d*, *q*) (*P*, *D*, *Q*) were considered to select the best model for forecasting SST and chl-a. Tables 2 and 3 show a few combinations of (*p*, *d*, *q*) (*P*, *D*, *Q*) for SST and chl-a.

Lower the AIC and higher the LB & BP indicate the better model. SARIMA [1, 0, 1] $[1, 1, 2]_{12}$ for SST and SARIMA [1, 0, 0] $[3, 1, 1]_{12}$ for chl-a had minimum AIC value and maximum p-value compared to all other models. Hence, these models are used for forecasting and compared with original data (test set).

The ACF of the residuals was plotted for both SST and chl-a to check for the model fit. The lags (Fig. 6a and b) in SST and chl-a residual plots lie within the sigma limits. It reflects that those forecasted values lie close to original data, SARIMA [1, 0, 1] $[1, 1, 2]_{12}$ of SST and SARIMA [1, 0, 0] $[3, 1, 1]_{12}$ are selected as the best models for forecasting the SST and chl-a respectively.

Forecasted time series (Fig. 7a and b) show that SST in the study area may increase in the coming two years and which may affect primary productivity as chl-a shows decreasing trend. Also, global warming and other accelerating factors may result in high SST in the future. Our results show in the coming two years on an average chl-a concentration over the study area may not exceed the concentration of 4 mg/m^3 and SST may drop to a minimum of about 26 °C and maximum may exceed 30 °C.

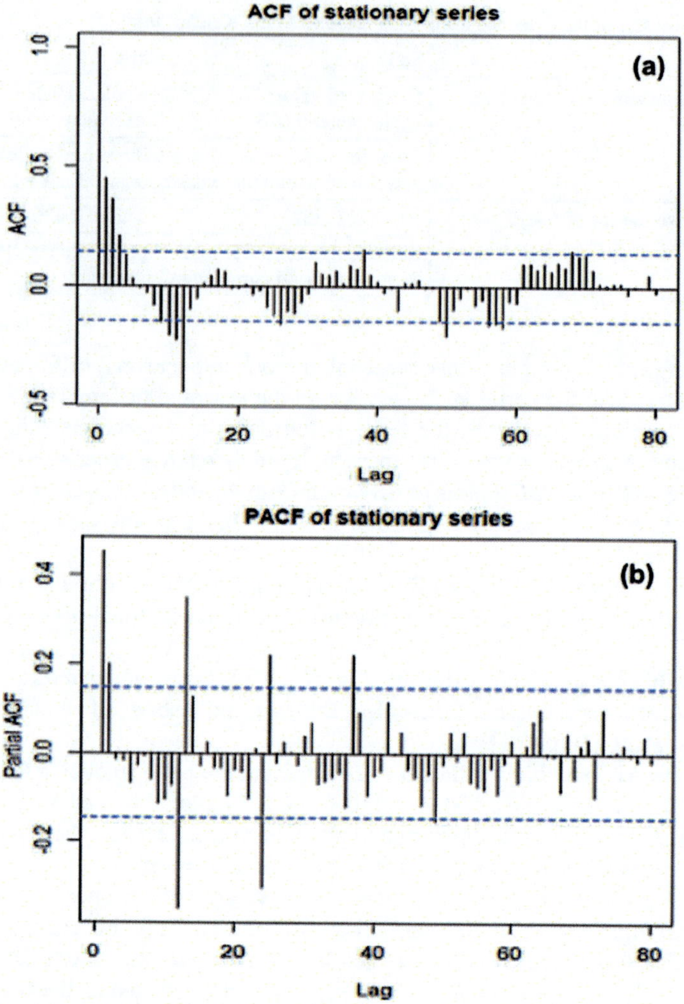

Fig. 4 **a** and **b** ACF and PACF plots of SST time series

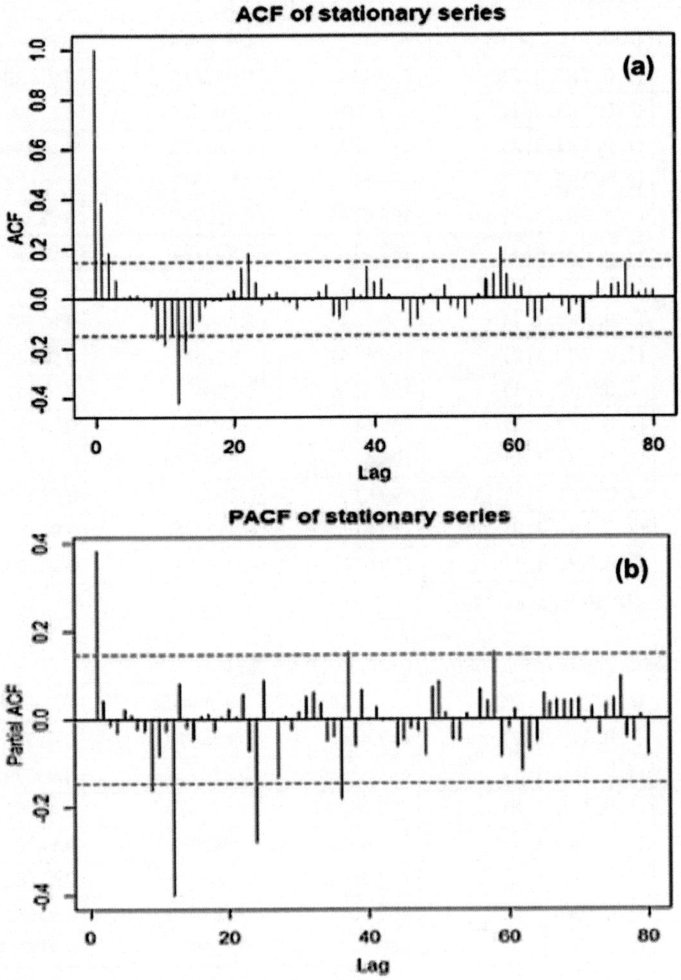

Fig. 5 **a** and **b** ACF and PACF plots of chl-a time series

Table 2 Different combinations of (*p*, *d*, *q*) (*P*, *D*, *Q*) for SST

Model	(*p*, *d*, *q*) (*P*, *D*, *Q*)	AIC1	LB: *p*-value	BP: *p*-value
1	(1, 0, 2)(3, 1, 2)	199.613	0.932645	0.961307
2	(1, 0, 1)(1, 1, 1)	200.3053	0.802639	0.879901
3	**(1, 0, 1)(1, 1, 2)**	**197.3801**	**0.934772**	**0.962344**
4	(2, 0, 1)(2, 1, 1)	196.6105	0.923512	0.956227
5	(2, 0, 3)(3, 1, 2)	199.613	0.880983	0.925875
6	(2, 0, 2)(1, 1, 1)	199.613	0.909721	0.945986
7	(1, 0, 3)(1, 1, 2)	199.613	0.909721	0.945986
8	(2, 0, 3)(3, 1, 1)	199.613	0.880983	0.925875
9	(1, 0, 2)(2, 1, 2)	199.613	0.932645	0.961307
10	(1, 0, 3)(2, 1, 1)	196.6105	0.898471	0.939431
11	(1, 0, 1)(3, 1, 1))	199.613	0.950556	0.972742
12	(1, 0, 1)(3, 1, 2)	199.613	0.950556	0.972742
13	(2, 0, 3)(1, 1, 1)	199.613	0.880983	0.925875
14	(2, 0, 1)(2, 1, 2)	199.613	0.932645	0.961307
15	(2, 0, 2)(2, 1, 1)	199.613	0.909721	0.945986

LB: Ljung box, BP: Box pierce

Table 3 Different combinations of (*p*, *d*, *q*) (*P*, *D*, *Q*) for chl-a

Model	(*p*, *d*, *q*) (*P*, *D*, *Q*)	AIC1	LB:*p*-value	BP:*p*-value
1	(1, 0, 1)(3, 1, 2)	447.4847	0.856071	0.915959
2	(1, 0, 1)(3, 1, 1)	445.615	0.877914	0.930798
3	(1, 0, 1)(3, 1, 0)	470.7897	0.273979	0.40953
4	**(1, 0, 0)(3, 1, 1)**	**443.6776**	**0.902791**	**0.947034**
5	(1, 0, 0)(3, 1, 2)	447.4847	0.887015	0.936829
6	(1, 0, 0)(3, 1, 0)	447.4847	0.887015	0.936829
7	(1, 0, 1)(2, 1, 2)	447.4847	0.856071	0.915959
8	(1, 0, 1)(1, 1, 1)	447.4847	0.856071	0.915959
9	(1, 0, 1)(2, 1, 1)	447.4847	0.856071	0.915959
10	(1, 0, 1)(2, 1, 0)	443.6776	0.874725	0.928648
11	(1, 0, 0)(2, 1, 2))	447.4847	0.887015	0.936829
12	(1, 0, 0)(2, 1, 1)	447.4847	0.887015	0.936829
13	(1, 0, 0)(2, 1, 0)	447.4847	0.887015	0.936829
14	(1, 0, 0)(1, 1, 1)	447.4847	0.887015	0.936829
15	(1, 0, 1)(1, 1, 2)	447.4847	0.856071	0.915959

LB: Ljung box, BP: Box pierce

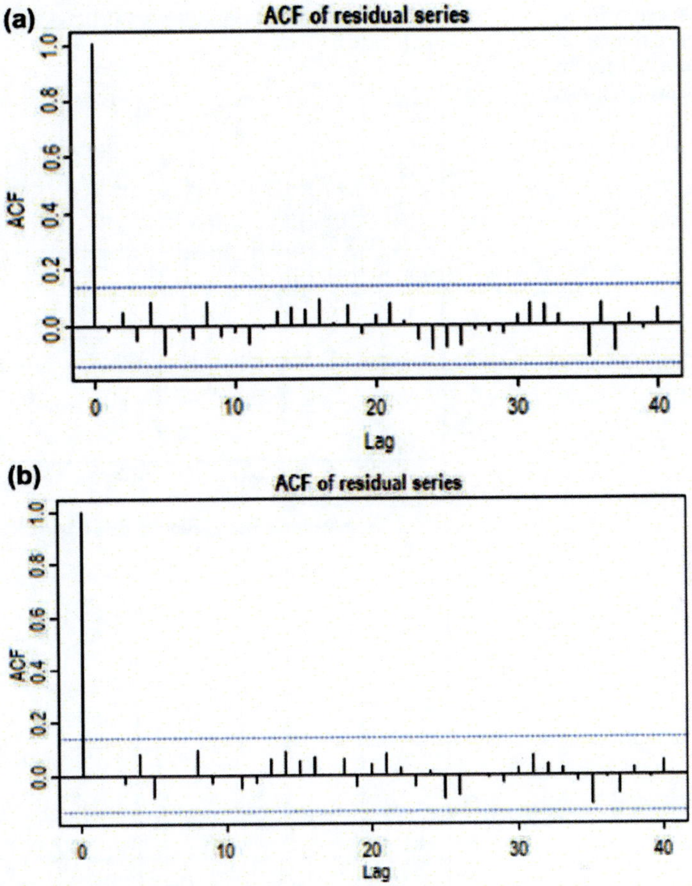

Fig. 6 **a** and **b** Residual plots of SST and chl-a time series data

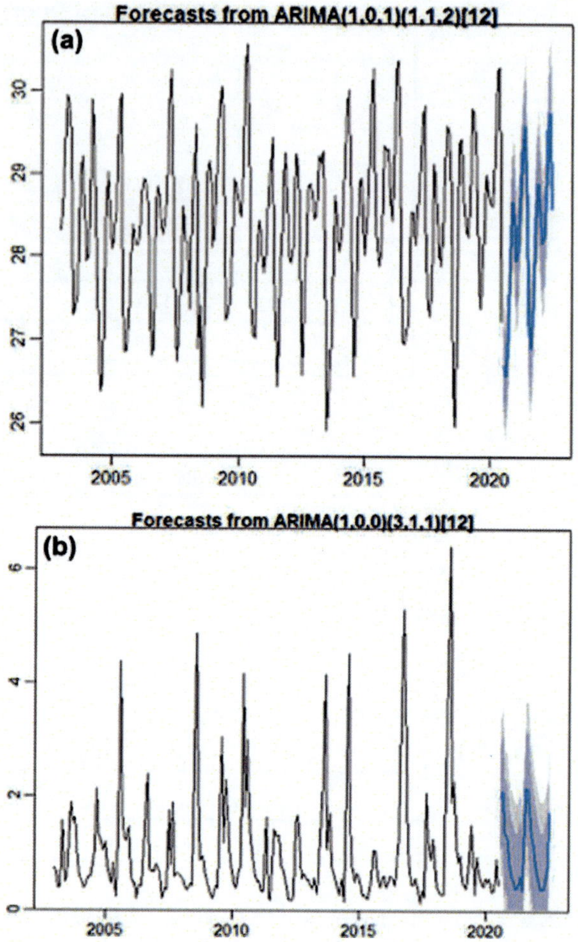

Figures 7 a and b Show time series and forecasted SST and chl-a data for the next 24 months (2021 and 2022)

Validation result of the model derived time-series data

The SARIMA model best fits SST data which is clear from the high value of the coefficient of determination ($R^2 = 0.87$) which is near to perfect fit value 1 and from the linear regression line obtained slope (0.876) is close to 1:1 line (Fig. 8a). Also, very less errors—RMSE (0.351), MAPE (0.905%) and negligible overestimation by the model revealed by MRPE = 0.047%. Hence SARIMA modelled SST shows good agreement with the original/satellite extracted SST data with an accuracy of about 99%.

The errors for modelled chl-a are found to be comparatively higher than modelled SST (RMSE = 0.671). The coefficient of regression R^2 (0.53) for the SARIMA model generated chl-a and satellite-derived data is found to be less and match-up points are found to be scattered around the trend line with a slope value of 0.50 (Fig. 8b). This shows the model can capture only about 50% of the variance of the original chl-a data set, which is also clear from the MAPE value of 51.266 (Table 4). The SARIMA model for chl-a is found to largely underestimate the satellite-derived chl-a which is clear from the MRPE value (−29.38%). The cross-validation results show that this method is more effective in modelling and predicting future values of SST than chl-a time series.

The validation results (Table 4) show that this method is effective in modelling SST and chl-a time series and predicting future values. The error for the modelled chl-a is higher than SST. Previous studies [52] also showed the performance of SARIMA was poor in estimating chl-a concentration due to the complexity of the chlorophyll-a in coastal water caused by the high level of spikes. However, this method was found to be more effective in modelling SST time series using historical/previous records.

The anomaly maps were generated from climatology to check for the spikes in the time series. Frequencies of high chl-a anomaly observed during monsoon and post-monsoon seasons are more than the inter-monsoon seasons (Fig. 9). Moreover, from the time series anomaly, it is clear that the magnitude of the chl-a anomaly is considerably higher than SST due to the bloom conditions during monsoon or post-monsoon seasons.

Table 4 Results of cross-validation used to measure the accuracy of the SARIMA model predicted SST and chl-a

	SST	Chl-a
y	$y = 0.8762x + 3.5027$	$y = 0.5033x + 0.4887$
R^2	0.868	0.526
RMSE	0.351	0.671
MAPE	0.905	51.266
MRPE	0.047	−29.376

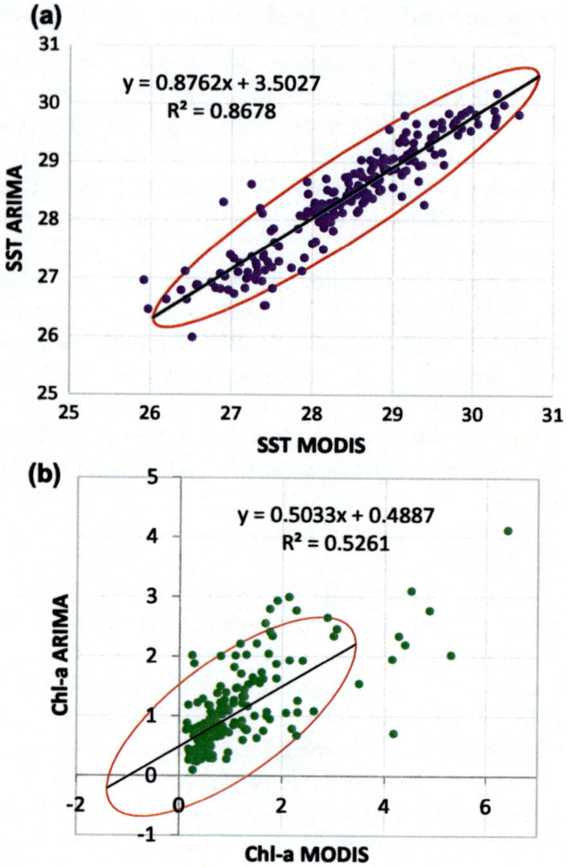

Fig. 8 **a** and **b** Scatter plots for MODIS data and ARIMA modelled data of SST and chl-a respectively

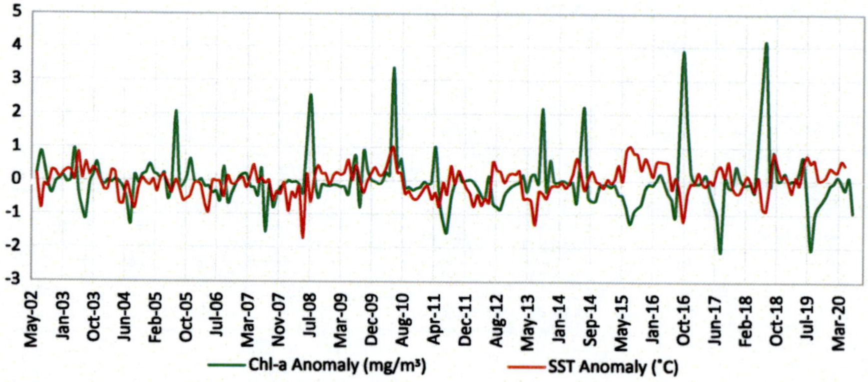

Fig. 9 Shows the anomaly of SST and chl-a in the study area

4 Concluding Remarks

Time series models are found to be helpful in characterizing temporal variation of oceanographic parameters and forecasting future values. In this study, SARIMA model is used to forecast the SST and chl-a values based on 18 years of satellite-derived SST and chl-a data (July 2002 to July 2020). The results show SARIMA [1, 0, 1] $[1, 1, 2]_{12}$ of SST and SARIMA [1, 0, 0] $[3, 1, 1]_{12}$ are the best models for forecasting the SST and chl-a respectively in the study area. The cross-validation results exhibit that this method is more effective in modelling SST than chl-a time series and predict future values, as the error for the modelled chl-a is higher than SST. The poor performance of SARIMA in estimating chl-a concentration is due to the complexity of the (Case-2) coastal waters and the high level of spikes caused due to the bloom conditions during monsoon or post-monsoon seasons which is clear from the chl-a anomaly plot.

Compliance with Ethical Standards

Conflict of Interest The authors declare no conflict of interest.

Data Availability MODIS Aqua data used in the present study were obtained from the NASA data portal (https://giovanni.gsfc.nasa.gov/giovanni).

References

1. de Amorim, F., Rick, J., Lohmann, G., & Wiltshire, K. H. (2021). Evaluation of machine learning predictions of a highly resolved time series of chlorophyll-a concentration. *Applied Sciences, 11*, 7208.https://doi.org/10.3390/app11167208
2. Aparna, S. G., D'Souza, S., & Arjun, N. B. (2018). Prediction of daily sea surface temperature using artificial neural networks. *International Journal of Remote Sensing, 39*, 4214–4231. https://doi.org/10.1080/01431161.2018.1454623
3. Barnston, A. G., & Smith, T. M. (1996). Specification and prediction of global surface temperature and precipitation from Global SST using CCA. *Journal of Climate, 9*, 2660–2697. https://doi.org/10.1175/1520-0442(1996)009%3c2660:sapogs%3e2.0.co;2
4. Behera, N., Swain, D., & Sil, S. (2020). Effect of Antarctic sea ice on chlorophyll concentration in the Southern Ocean. *Deep Sea Research Part II: Topical Studies in Oceanography, 178*, 104853. https://doi.org/10.1016/j.dsr2.2020.104853
5. Bogen, K. T., Jones, E. D., & Fischer, L. E. (2011). Hurricane intensity, sea surface temperature, and stochastic variation. *Recent Hurricane Research—Climate, Dynamics, and Societal Impacts*, 103–115
6. Box, G. E., & Jenkins, G. M. (1976). *Time analysis, forecasting and control*. Holden-Day.
7. Brockwell, P. J., & Davis, R. A. (2002). *Introduction to time series and forecasting* (2nd ed.). Springer.
8. Boyd, P. W., Claustre, H., Levy, M., Siegel, D. A., & Weber, T. (2019). Multi-faceted particle pumps drive carbon sequestration in the Ocean. *Nature, 568*, 327–335. https://doi.org/10.1038/s41586-019-1098-2
9. Bozdogan, H. (1987). Model selection and Akaike's information criterion (AIC): The general theory and its analytical extensions. *Psychometrika, 52*, 345–370. https://doi.org/10.1007/bf02294361

10. Burnham, K. P., & Anderson, D. R. (2004). Multimodel inference: Understanding AIC and BIC in model selection. *Sociological Methods & Research, 33*, 261–304. https://doi.org/10.1177/0049124104268644
11. Chassot, E., Bonhommeau, S., Dulvy, N. K., Mélin, F., Watson, R., Gascuel, D., & Le Pape, O. (2010). Global marine primary production constrains fisheries catches. *Ecology Letters, 13*, 495–505. https://doi.org/10.1111/j.1461-0248.2010.01443.x
12. Chatfield, C. (1980). *An introduction to the analysis of time series* (2nd ed.). Chapman-Hall.
13. Chen, C., Twycross, J., & Garibaldi, J. M. (2017). A new accuracy measure based on bounded relative error for time series forecasting. *PLoS ONE, 12*, e0174202. https://doi.org/10.1371/journal.pone.0174202
14. Chung, C. E., & Ramanathan, V. (2006). Weakening of North Indian SST gradients and the monsoon rainfall in India and the Sahel. *Journal of Climate, 19*, 2036–2045. https://doi.org/10.1175/jcli3820.1
15. Cullen, J. J. (1982). The deep chlorophyll maximum: Comparing vertical profiles of chlorophyll a. *Canadian Journal of Fisheries and Aquatic Sciences, 39*, 791–803. https://doi.org/10.1139/f82-108
16. Du, P., Wang, J., Guo, Z., & Yang, W. (2017). Research and application of a novel hybrid forecasting system based on multi-objective optimization for Wind Speed forecasting. *Energy Conversion and Management, 150*, 90–107. https://doi.org/10.1016/j.enconman.2017.07.065
17. Escribano, R., Daneri, G., Farías, L., Gallardo, V. A., González, H. E., Gutiérrez, D., Lange, C. B., Morales, C. E., Pizarro, O., Ulloa, O., & Braun, M. (2004). Biological and chemical consequences of the 1997–1998 El Niño in the Chilean Coastal Upwelling System: A synthesis. *Deep Sea Research Part II: Topical Studies in Oceanography, 51*, 2389–2411. https://doi.org/10.1016/j.dsr2.2004.08.011
18. Feng, J., Stige, L. C., Hessen, D. O., Zuo, Z., Zhu, L., & Stenseth, N. C. (2021). A threshold sea-surface temperature at 14 °C for phytoplankton nonlinear responses to Ocean Warming. *Global Biogeochemical Cycles, 35*, e2020GB006808. https://doi.org/10.1029/2020gb006808
19. Field, C. B. (1998). Primary production of the biosphere: Integrating terrestrial and oceanic components. *Science, 281*, 237–240. https://doi.org/10.1126/science.281.5374.237
20. Flanagan, P. X., Basara, J. B., Furtado, J. C., Martin, E. R., & Xiao, X. (2019). Role of sea surface temperatures in forcing circulation anomalies driving U.S. great plains pluvial years. *Journal of Climate, 32*, 7081–7100. https://doi.org/10.1175/jcli-d-18-0726.1
21. Frankignoul, C., & Sennéchael, N. (2007). Observed influence of North Pacific SST anomalies on the atmospheric circulation. *Journal of Climate, 20*, 592–606. https://doi.org/10.1175/jcli4021.1
22. Goswami, B. N., Madhusoodanan, M. S., Neema, C. P., & Sengupta, D. (2006). A physical mechanism for North Atlantic SST influence on the Indian summer monsoon. *Geophysical Research Letters, 33*.https://doi.org/10.1029/2005gl024803
23. Grigonytė, E., & Butkevičiūtė, E. (2016). Short-term wind speed forecasting using Arima model. *Energetika, 62*.https://doi.org/10.6001/energetika.v62i1-2.3313
24. Gupta, S. M., & Malmgren, B. A. (2009). Comparison of the accuracy of SST estimates by Artificial Neural Networks (ANN) and other quantitative methods using Radiolarian data from the Antarctic and Pacific Oceans. *e-Journal Earth Science 2*, 52–75. http://drs.nio.org/drs/handle/2264/3346
25. Haghbin, M., Sharafati, A., Motta, D., Al-Ansari, N., & Noghani, M. H. (2021). Applications of soft computing models for predicting sea surface temperature: A comprehensive review and assessment. *Progress in Earth and Planetary Science, 8*, 1–9. https://doi.org/10.1186/s40645-020-00400-9
26. Hipel, K. W., McLeod, A. I., & Lennox, W. C. (1977). Advances in Box-Jenkins modeling: 1. Model construction. *Water Resources Research, 13*, 567–575. https://doi.org/10.1029/wr013i003p00567
27. Hyndman, R. J., & Koehler, A. B. (2006). Another look at measures of forecast accuracy. *International Journal of Forecasting, 22*, 679–688. https://doi.org/10.1016/j.ijforecast.2006.03.001

28. Kendall, M. G. (1975). *Rank correlation methods* (4th ed.). Charles Griffin.
29. Khan, S., Piao, S., Zheng, G., Khan, I. U., Bradley, D., Khan, S., & Song, Y. (2021). Sea surface temperature variability over the tropical Indian Ocean during the ENSO and IOD events in 2016 and 2017. *Atmosphere, 12*, 587. https://doi.org/10.3390/atmos12050587
30. Kitsiou, D., & Topouzelis, K. (2014). Correlation between chlorophyll a concentration and sea surface temperature in the eastern Mediterranean Sea using GIS and satellite data. *Fresenius Environmental Bulletin, 23*, 2919–2925.
31. Kim, H. J., Miller, A. J., McGowan, J., & Carter, M. L. (2009). Coastal phytoplankton blooms in the Southern California bight. *Progress in Oceanography, 82*, 137–147. https://doi.org/10.1016/j.pocean.2009.05.002
32. Kim, S. U., & Kim, K. Y. (2021). Impact of climate change on the primary production and related biogeochemical cycles in the coastal and sea ice zone of the Southern Ocean. *Science of the Total Environment, 751*, 141678. https://doi.org/10.1016/j.scitotenv.2020.141678
33. Kumar, G. S., Prakash, S., Ravichandran, M., & Narayana, A. C. (2016). Trends and relationship between chlorophyll-AAND sea surface temperature in the central equatorial Indian Ocean. *Remote Sensing Letters, 7*, 1093–1101. https://doi.org/10.1080/2150704x.2016.1210835
34. Lai, Y., & Dzombak, D. A. (2020). Use of the autoregressive integrated moving average (ARIMA) model to forecast near-term regional temperature and precipitation. *Weather and Forecasting, 35*, 959–976. https://doi.org/10.1175/waf-d-19-0158.1
35. Lau, N. C. (1997). Interactions between global SST anomalies and the midlatitude atmospheric circulation. *Bulletin of the American Meteorological Society, 78*, 21–34. https://doi.org/10.1175/1520-0477(1997)078%3c0021:IBGSAA%3e2.0.CO;2
36. Lee, R. W., Woollings, T. J., Hoskins, B. J., Williams, K. D., O'Reilly, C. H., & Masato, G. (2018). Impact of gulf stream SST biases on the global atmospheric circulation. *Climate Dynamics, 51*, 3369–3387. https://doi.org/10.1007/s00382-018-4083-9
37. Li, W., El-Askary, H., Qurban, M., Proestakis, E., Garay, M., Kalashnikova, O., Amiridis, V., Gkikas, A., Marinou, E., Piechota, T., & Manikandan, K. (2018). An assessment of atmospheric and meteorological factors regulating Red Sea phytoplankton growth. *Remote Sensing, 10*, 673. https://doi.org/10.3390/rs10050673
38. Liu, C., Sun, Q., Xing, Q., Wang, S., Tang, D., Zhu, D., & Xing, X. (2019). Variability in phytoplankton biomass and effects of sea surface temperature based on satellite data from the Yellow Sea, China. *PLoS ONE, 14*, e0220058. https://doi.org/10.1371/journal.pone.0220058
39. Lotliker, A. A., Baliarsingh, S. K., Samanta, A., & Varaprasad, V. (2020). Growth and decay of high-biomass algal bloom in the northern Arabian Sea. *Journal of the Indian Society of Remote Sensing, 48*, 465–471. https://doi.org/10.1007/s12524-019-01094-3
40. Mann, H. B. (1945). Nonparametric tests against trend. *Econometrica, 13*, 245–259.
41. Mahongo, S. B., & Deo, M. C. (2013). Using Artificial Neural Networks to forecast monthly and seasonal sea surface temperature anomalies in the western Indian Ocean. *The International Journal of Ocean and Climate Systems, 4*, 133–150. https://doi.org/10.1260/1759-3131.4.2.133
42. Mudelsee, M. (2013). *Climate time series analysis* (Vol. 30). Springer.
43. Narayanan, P., Basistha, A., Sarkar, S., & Kamna, S. (2013). Trend analysis and Arima modelling of pre-monsoon rainfall data for Western India. *Comptes Rendus Geoscience, 345*, 22–27. https://doi.org/10.1016/j.crte.2012.12.001
44. Nurdin, S., Mustapha, M. A., & Lihan, T. (2013). The relationship between sea surface temperature and chlorophyll-a concentration in fisheries aggregation area in the archipelagic waters of Spermonde using satellite images. *AIP Conference Proceedings. American Institute of Physics, 1571*, 466–472. https://doi.org/10.1063/1.4858699
45. O'Carroll, A. G., Armstrong, E. M., Beggs, H. M., Bouali, M., Casey, K. S., Corlett, G. K., Dash, P., Donlon, C. J., Gentemann, C. L., Høyer, J. L., Ignatov, A., Kabobah, K., Kachi, M., Kurihara, Y., Karagali, I., Maturi, E., Merchant, C. J., Marullo, S., Minnett, P. J., … Wimmer, W. (2019). Observational needs of sea surface temperature. *Frontiers in Marine Science, 6*, 420. https://doi.org/10.3389/fmars.2019.00420
46. Oliver, E. C., Burrows, M. T., Donat, M. G., Sen Gupta, A., Alexander, L. V., Perkins-Kirkpatrick, S. E., Benthuysen, J. A., Hobday, A. J., Holbrook, N. J., Moore, P. J., Thomsen,

M. S., Wernberg, T., & Smale, D. A. (2019). Projected Marine Heatwaves in the 21st century and the potential for ecological impact. *Frontiers in Marine Science, 6*, 734. https://doi.org/10.3389/fmars.2019.00734
47. Park, J.-Y., & Kug, J.-S. (2013). Marine biological feedback associated with Indian Ocean Dipole in a coupled ocean/biogeochemical model. *Climate Dynamics, 42*, 329–343. https://doi.org/10.1007/s00382-012-1640-5
48. Patil, K., Deo, M. C., & Ravichandran, M. (2016). Prediction of sea surface temperature by combining numerical and neural techniques. *Journal of Atmospheric and Oceanic Technology, 33*, 1715–1726. https://doi.org/10.1175/jtech-d-15-0213.1
49. Rahman, M. A., Yunsheng, L., & Sultana, N. (2016). Analysis and prediction of rainfall trends over Bangladesh using Mann-Kendall, spearman's rho tests and Arima model. *Meteorology and Atmospheric Physics, 129*, 409–424. https://doi.org/10.1007/s00703-016-0479-4
50. Ryther, J. H., & Yentsch, C. S. (1957). The estimation of phytoplankton production in the ocean from chlorophyll and Light Data1. *Limnology and Oceanography, 2*, 281–286. https://doi.org/10.1002/lno.1957.2.3.0281
51. Sarkar, P. P., Janardhan, P., & Roy, P. (2020). Prediction of sea surface temperatures using Deep Learning Neural Networks. *SN Applied Sciences, 2*, 1–14. https://doi.org/10.1007/s42452-020-03239-3
52. Shehhi, M. R., & Kaya, A. (2020). Time series and machine learning to forecast the water quality from satellite data. arXiv preprint arXiv:2003.11923. https://arxiv.org/abs/2003.11923
53. Shirvani, A., Nazemosadat, S. M., & Kahya, E. (2015). Analyses of the Persian Gulf Sea surface temperature: Prediction and detection of climate change signals. *Arabian Journal of Geosciences, 8*, 2121–2130. https://doi.org/10.1007/s12517-014-1278-1
54. Sincich, T. (1996). *Business statistics by example* (5th ed.). Prentice Hall Englewood.
55. Thanh, N. T., Cuong, H. D., Hien, N. X., & Kieu, C. (2019). Relationship between sea surface temperature and the maximum intensity of tropical cyclones affecting Vietnam's coastline. *International Journal of Climatology, 40*, 2527–2538. https://doi.org/10.1002/joc.6348
56. Umbert, M., Guimbard, S., Ballabrera Poy, J., & Turiel, A. (2020). Synergy between ocean variables: Remotely sensed surface temperature and chlorophyll concentration coherence. *Remote Sensing, 12*, 1153. https://doi.org/10.3390/rs12071153
57. Wada, A., Usui, N., & Sato, K. (2012). Relationship of maximum tropical cyclone intensity to sea surface temperature and tropical cyclone heat potential in the North Pacific Ocean. *Journal of Geophysical Research: Atmospheres, 117*.https://doi.org/10.1029/2012jd017583
58. Wang, H., Schubert, S., Suarez, M., & Koster, R. (2010). The physical mechanisms by which the leading patterns of SST variability impact U.S. precipitation. *Journal of Climate, 23*, 1815–1836. https://doi.org/10.1175/2009jcli3188.1
59. Whitney, L. D., & Hobgood, J. S. (1997). The relationship between sea surface temperatures and maximum intensities of tropical cyclones in the eastern North Pacific Ocean. *Journal of Climate, 10*, 2921–2930. https://doi.org/10.1175/1520-0442(1997)010%3c2921:TRBSST%3e2.0.CO;2
60. Xu, Y., Yang, W., & Wang, J. (2017). Air Quality Early-warning system for cities in China. *Atmospheric Environment, 148*, 239–257. https://doi.org/10.1016/j.atmosenv.2016.10.046
61. Zurada, J., Levitan, A., & Guan, J. (2011). A comparison of regression and artificial intelligence methods in a mass appraisal context. *Journal of Real Estate Research, 33*, 349–388. https://doi.org/10.1080/10835547.2011.12091311

Experimental Comparison of Hydraulic Jump Characteristics and Energy Dissipation Between Sluice Gate and Radial Gate

Koroungamba Laishram, Thiyam Tamphasana Devi, and Ningombam Bishwajit Singh

Abstract A hydraulic jump is an abrupt change of flow when moving from higher velocity (supercritical flow) to lower velocity (subcritical flow). Also, the hydraulic jump is an energy dissipater for dissipating the excess energy of flow in the downstream through spillways, gates, weirs, dams, and other hydraulic structures, and it also helps to prevent scouring in gates and spillways. The precise selection of gates is very important for effective control of flow in the channel, to guarantee safety of the structure, easy to operate, economical, easy to maintain and to dissipate excess energy downstream. This study examines the characteristics of hydraulic jumps (a comparison of Sluice gate and Radial gate) and energy dissipation in hydraulic jumps experimentally. The experimentation was performed in a channel flume with a rectangular cross section (16 m long, 0.6 m wide and 0.8 m deep) in the Hydraulics Laboratory, Department of Civil Engineering, NIT Manipur. A sluice gate and radial gate are used in the channel to generate a hydraulic jump at the downstream of the flow. The characteristics of hydraulic jumps, such as length of jump, upstream and downstream head, velocity, sequent depth ratio, and amount of energy dissipated, etc., were measured with different heights of gate opening and flow rate. The results show that the Radial gate is much safer for the structure, dissipates more energy, and the control of the flow is much easier as compared to the Sluice gate, and can be considered as an appropriate model for dissipating energy.

Keywords Sluice gate · Radial gate · Channel flume · Hydraulic jump · Froude number · Energy dissipation

1 Introduction

When a flow moves from higher velocity (supercritical flow) to lower velocity (subcritical flow), the properties or direction of the flow change suddenly and this sudden change of flow is known as a hydraulic jump [1]. Hydraulic jump is an energy

K. Laishram · T. T. Devi (✉) · N. B. Singh
Department of Civil Engineering, National Institute of Technology, Langol, Manipur 795004, India

A. K. Dikshit et al. (eds.), *Innovative Trends in Hydrological and Environmental Systems*, Lecture Notes in Civil Engineering 234,
https://doi.org/10.1007/978-981-19-0304-5_16

dissipater for dissipating the excess energy of flow flowing downstream through spillways, gates, weirs, dams, and other hydraulic structures, and it also helps to prevent scouring in gates and spillways. Gates are the devices used for controlling the flow and flow measurement in an open channel. The flow downstream through tunnels, spillways, ducts, regulators, sluices, etc., is controlled by various kinds of hydraulic gates and hoists which work under different principles and mechanisms. In order to avoid the failure of structures downstream, which is mainly caused due to scouring at the downstream, it is of great importance to dissipate the water kinetic energy at the gate downstream. Radial gates, also known as Tainter gates, are a type of gate having a curved surface upstream in the form of an arc which is used for controlling the flow rate and head of water over spillways or canals. A sluice gate is another type of gate or barrier having a straight or perpendicular surface mainly made up of metal or wood, which is used for controlling the flow rate and water levels in rivers and canals. Gates are also used in mining operations and watermills for recovering minerals, and also in wastewater treatment plants. The precise selection of gates is very important for effective control of flow in the channel, to guarantee the safety of the structure, and to allow more energy dissipation. The selection of the gates depends upon numerous factors, such as safety, ease of operation, requirement for low maintenance, and cost-effective. According to the published documents [2–4], the hydraulic jump characteristics and amount of energy dissipated over a weir and a constricted channel were also studied. However, no study analyzing the hydraulic jump characteristics and the amount of energy dissipated between the sluice gate and radial gate has been conducted. Therefore, to get a better understanding and knowledge of this region, there is a requirement for conducting more studies regarding the hydraulic jump characteristics and energy dissipation between the sluice gate and radial gate.

Mohamed et al. [5] studied the influence of various arrangements of sill and without sill on the characteristics of flow at the radial gates downstream under the submerged condition of the new Nagaa Hammadi regulator in Egypt. The length of reverse flow, velocity, energy loss, hydraulic jump length, and formation of scour downstream were analyzed. Investigation shows that the downstream length of reverse flow is inversely proportional to the submergence ratio and Froude number. Also, the hydraulic jump length was reduced by 59% with the sill having right upstream and a slopping face downstream, and the stilling basin length was also decreased. Compared to the outcomes for flow with sill and without sill, the flow with sill reduces the local scour depth by 43% and increases the energy dissipation by 30%. Finally, empirical equations are developed using multiple linear regression analysis. Their study proves that the flow with sill gives good outcomes on flow characteristics and depth of local scour, particularly for sills with the right upstream and slopping face downstream. Kim et al. [6] conducted an experiment to examine the hydraulic phenomena of a flume using a weir which is fixed and a movable weir like a sluice gate. Different hydraulic characteristics were analyzed and compared for the two weir types. Additionally, based on different positions and heights of installation, the effectiveness of the two weirs in terms of energy dissipation was also studied. The result shows that even though sluice gates produced the same hydraulic jumps as fixed weirs, the overall hydraulic jumps were elongated due to an increase in the supercritical flow downstream. With reference to all the outcomes, energy dissipators help in reducing the energy and the sluice gate is the most operative in decreasing the hydraulic jumps and defending the riverbed. Abdelhaleem [7] examines the flow in submerged condition passing through the radial gates using a gate sill and without using it. The effect of various heights of gate on characteristics of hydraulic jump, hydraulic jump length, coefficient of contraction, velocity distribution, coefficient of discharge, and energy dissipation were analyzed. A new empirical equation was developed by using a combination of dimensional and regression analysis tools. Studies found that the highly turbulent flow was observed at the downstream of the gate for gates without a sill and for the radial gate using a sill. Other than turbulence for low submergences, the surface turbulence was increased. The hydraulic jump length was not affected by the sills under the submerged radial gate and the amount of energy dissipation increased with an increase in the height of the sill under the gate. Finally, the present investigation proves that the flow characteristics in submerged conditions passing through the radial gates without a gate sill are superior to the gates with a gate sill. Nangare and Kote [8] performed a study to build up a prototype of a stepped spillway with the profile of an ogee weir for Khadakwasla dam. An energy dissipator with a plain roller bucket and a slotted roller bucket were also constructed. An investigation was conducted into a design discharge with different heads using an energy dissipator of four arrangements, such as a stepped spillway and an ogee spillway with a plain roller bucket model, and a stepped spillway and an ogee spillway with a slotted roller bucket model. The capabilities of all four arrangements were equated with varying heads of 4 and 6 m. The result shows that out of the four arrangements, the arrangement at 6 m using a stepped spillway with a slotted roller bucket has the highest ability to lose 83.6% of specific energy and it is considered as the most suitable energy dissipating model for Khadakwasla dam. Ewah et al. [9] investigate and compare the ability to dissipate energy of hydraulic jumps using a weir and level bedded constricted channel. After the analysis and test run between the weir and constricted flume, the results obtained showed that the energy dissipated by the weir ranged from 0.013 to 0.020 m and from −0.001 to 0.001 m for level bedded constricted flume. From the results, it came to the conclusion that energy loss from weir is more than that of constricted channel. Also, the Froude number is in good agreement with the ratio of conjugate depth for the section before and after the hydraulic jump regardless of the instrument causing the jump.

This current study examines and compares the characteristics of hydraulic jumps produced using a Sluice gate and a Radial gate, and the energy dissipation capacity of the two gates. The characteristics of hydraulic jumps, such as changes in water head, length of jump, height of jump, velocities, sequent depth ratio, and amount of energy dissipated were analyzed.

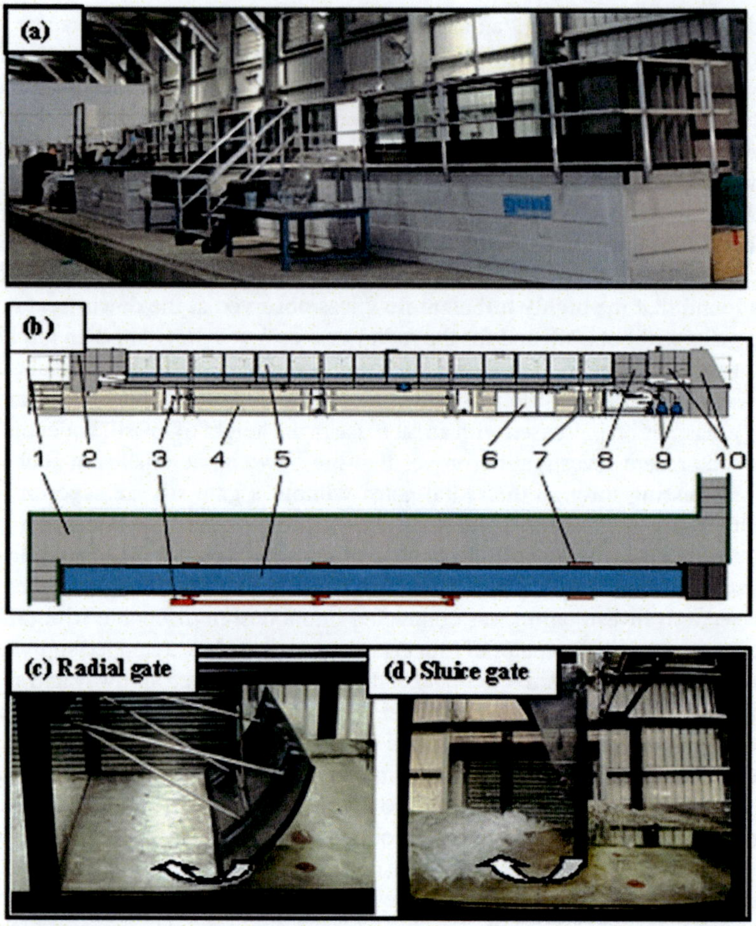

Fig. 1 **a** Experimental channel flume (16 m × 800 mm × 8600 mm). **b** Component of the channel flume and experimental set-up[1]. **c** Radial gate and **d** sluice gate used in present study

2 Materials and Methodology

2.1 Experimental Set-Up

The experiment was conducted in the Rectangular channel flume (16 m long, 0.6 m wide and 0.8 m high) in the Hydraulics Laboratory, Department of Civil Engineering, NIT Manipur, shown in Fig. 1a and b.

[1] Cross Sectional view with Details of the Channel Flume (Gunt Hm161): (1) Gallery (2) Inlet element (3) Jacking support with motorized inclination adjustment (4) Water tank (5) Experimental

A sluice gate (0.42 m long, 0.82 m wide and 0.7 m high) and a radial gate (0.31 m long, 0.16 m wide and 0.46 m high) shown in Fig. 1c and d, made by GUNT, Hamburg, Germany, are used in the channel to create a hydraulic jump. The channel flume has the facility to make it a sloping channel as well as horizontal, and it has smooth side glass walls to observe the flow patterns. The flume has a centrifugal pump which gives a maximum discharge of 412 m^3/h. The flow rate can be controlled by a control valve and the discharge can be noted from the control valve. The flume has 2 centrifugal pumps and it can be inclined from −0.75 to 2.1%. A velocity meter was used to measure the velocity downstream of the flow. A digital level gauge was used to find out the different heads in the channel.

2.2 *Procedure*

The channel flume and the instruments were set up in the required condition. For the first experimental run, the sluice gate was fixed on the channel flume and adjusted to the required height of gate opening. The pump was then turned on at some discharge to enable the flow and produce the hydraulic jump downstream while making a still water column upstream. For each experimental run, the characteristics of hydraulic jumps, such as initial depth, subsequent depth, the gate opening height, velocities, and the length of jump were recorded. The heights of gate openings were adjusted for different runs of experiments while forming a hydraulic jump downstream. A velocity meter was used for measuring the velocity of flow, a digital level gauge for depth of flow, and a meter scale for measuring the height of the gate. The Sluice gate is then replaced with a Radial gate at the same section of the channel. The same data was recorded with a similar height of gate opening and the same upstream head as the previous gate. A total of 6 experimental runs were conducted for different gates.

2.3 *Data Collection*

The experimentation work was carried out using Sluice and Radial gates to understand their characteristics in hydraulic jump occurrence.

(i) *Hydraulic jump by Sluice gate*

In this step, the data was collected to find the hydraulic jump characteristics using a sluice gate in a channel flume. The required depth of the hydraulic jump was achieved by adjusting the sluice gate opening at upstream and tail-water depth using the downstream gate of the flume. Measurements were taken on the upstream section and downstream section of the flume. The pre-jump

Section (6) Switch cabinet (7) Fixed support (8) Sediment trap HM 161.72, (9) Pump (10) Outlet element.

depths were taken at a location just before the formation of the surface roller and the post-jump depths were taken at a location where air bubbles cease to appear in the flow. A total of 3 experimental runs were conducted by adjusting the height of the gate, upstream head, and flow rate. The amounts of energy dissipated were calculated using data from hydraulic jump characteristics.

(ii) *Hydraulic jump by Radial gate*

The Sluice gate is then replaced by a Radial gate at the same section of the channel flume to find the hydraulic jump characteristics and amount of energy dissipated. The same procedure as in the previous scenario is followed to record the required data. A total of 3 experimental runs were conducted at the same height of gate opening, upstream head and flow rate as that of the Sluice gate for each run.

2.4 Basic Equation

Hydraulic jump is the pattern of flow between supercritical and subcritical flow. Figure 2 shows the typical presentation of a hydraulic jump.

By Belanger Equation [10], the sequent depth ratio for hydraulic jump (y_2/y_1) is given as:

$$\frac{y_2}{y_1} = \frac{1}{2}\left(\sqrt{1 + 8\mathrm{Fr}_1^2} - 1\right) \tag{1}$$

where, Fr_1 is the Froude number at upstream side (u/s) which is before the hydraulic jump occurs. And y_1 and y_2 are the (sequent depth) corresponding depth of flow before and after the hydraulic jump and is measured in cm. Fr_1 at u/s can be calculated as:

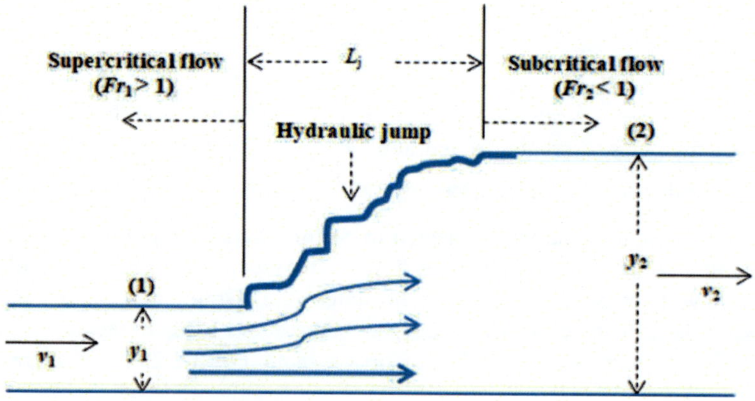

Fig. 2 Typical representation of hydraulic jump

$$\mathrm{Fr}_1 = \frac{v_1}{\sqrt{g y_1}} \quad (2)$$

Here, v_1 is the velocity of flow before the hydraulic jump. And the loss of energy (ΔE) by the hydraulic jump is calculated as:

$$\Delta E = E_1 - E_2 = \left(y_1 + \frac{v_1^2}{2g}\right) - \left(y_2 + \frac{v_2^2}{2g}\right) \quad (3)$$

where, v_2 is the velocity of flow after the hydraulic jump (cm/s) and g is the acceleration due to gravity which has constant value of 9.81 m/s_2. For a rectangular open channel, the energy loss (ΔE) is given as:

$$\Delta E = \frac{(y_2 - y_1)^3}{4 y_2 y_1} \quad (4)$$

where, E_1 is specific energy before the hydraulic and E_2 is specific energy after the hydraulic jump.

Length of jump (L_j) is given as:

$$L_j = 6(y_2 - y_1) \quad (5)$$

Height of jump (H_j) is given as:

$$H_j = y_2 - y_1 \quad (6)$$

3 Results and Discussion

The characteristics of hydraulic jump from Sluice gate and Radial gate were measured and analyzed at three different gate height openings (denoted by a) while maintaining the same flow rate (Q) and upstream head (y_0). The typical formation of hydraulic jumps in a channel flume when sluice gate is used is presented in Fig. 3.

The experimental data collected and other calculated values were tabulated in Table 1.

3.1 Sequent Depth Ratio (y_2/y_1)

To know the connection between the sequent depth ratio (y_2/y_1) and the Froude number (Fr_1) of the Sluice gate and Radial gate, a total of three experimental runs

Fig. 3 Hydraulic jump forming at downstream (present study)

Table 1 Experimental data collected for Sluice gate and Radial gate

Q (m^3/s)	v_1 (m/s)	a (m)	y_0 (m)	y_1 (m)	y_2 (m)	y_2/y_1	Fr_1	H_j/y_1	L_j/y_1	ΔE (m)
Sluice gate										
0.164	15.75	0.022	0.337	0.017	0.092	5.25	38	0.075	26.5	0.064
0.229	9.62	0.059	0.355	0.040	0.155	3.86	15	0.115	17.25	0.061
0.187	4.83	0.105	0.231	0.065	0.180	2.77	6	0.115	10.6	0.033
Radial gate										
0.165	10.76	0.022	0.337	0.026	0.129	4.96	21.5	0.103	23.77	0.081
0.23	7.73	0.059	0.355	0.050	0.195	3.91	11	0.145	17.4	0.078
0.187	4.3	0.105	0.231	0.073	0.208	2.85	5	0.135	11.09	0.040

each were conducted, analyzed, and represented graphically in Fig. 4. From the results, it can be proved that the sequent depth ratio rises with the rise in Froude number for both the Sluice gate and Radial gate. Figure 4 shows the hydraulic jump (represented by sequent depth ratio) from the Radial gate is safer for the structure.

3.2 Hydraulic Jump Height (H_j)

The data for Froude number and height of jump (H_j) for each experimental run were calculated using the data collected experimentally and the formulae mentioned for both the Sluice gate and Radial gate. The initial Froude number is then compared and analyzed with the hydraulic jump height for both the Sluice gate and Radial gate to know their relationship. Figure 5 shows the graphical representation that the height

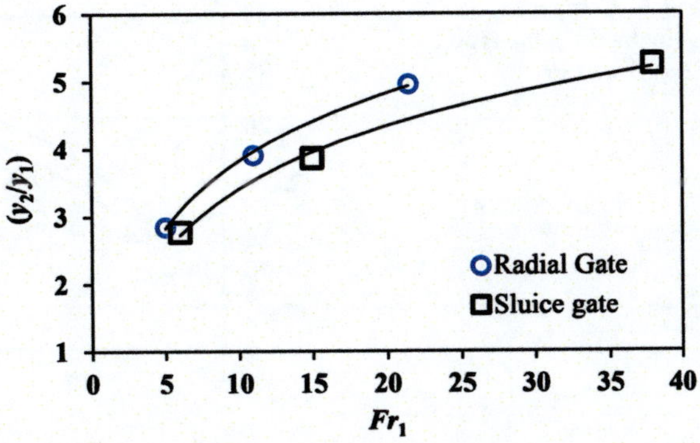

Fig. 4 Comparison of sequent depth ratio (y_2/y_1) w.r.t Fr_1

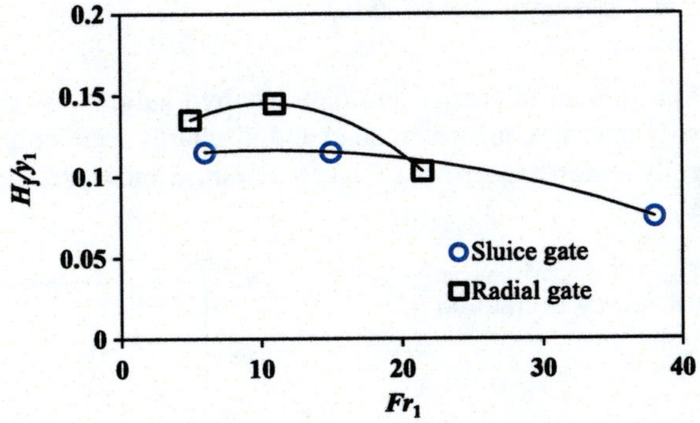

Fig. 5 Comparison of height of hydraulic jump ratio (H_j/y_1) w.r.t Fr_1

of the hydraulic jump is inversely related to the initial Froude number for both the gates.

3.3 *Length of Hydraulic Jump (L_j)*

The connection between the length of hydraulic jump ratio (L_j/y_1) and the initial Froude number was analyzed, compared, and graphically represented in Fig. 6. It shows that the length of jump ratio was in good agreement with the Froude number, i.e., the length of jump ratio rises with the rise in Froude number.

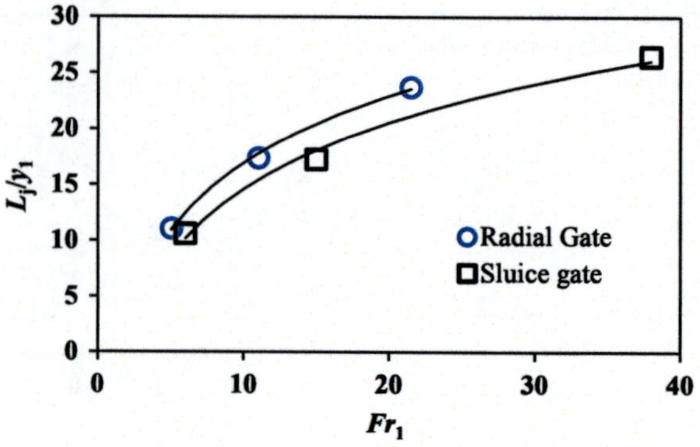

Fig. 6 Comparison length of hydraulic jump ratio (L_j/y_1) w.r.t Fr_1

3.4 Energy Loss (ΔE)

The amount of energy dissipated by hydraulic jump for Sluice gate and Radial gate with the same upstream head and discharge were compared, analyzed, and graphically represented in Figs. 7 and 8. For the Sluice gate, the amount of energy dissipated

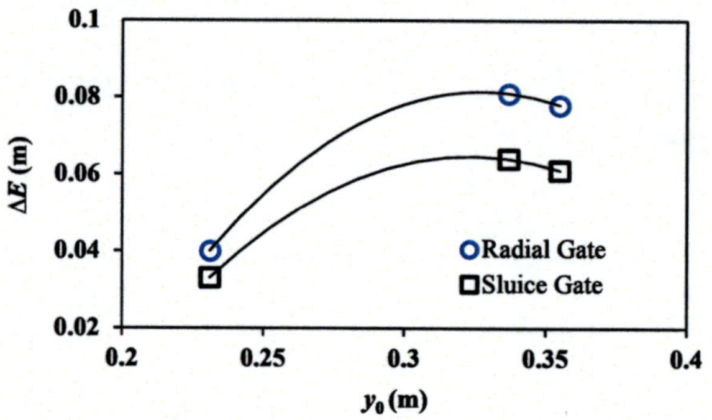

Fig. 7 Comparison energy loss (ΔE) w.r.t upstream head (y_0)

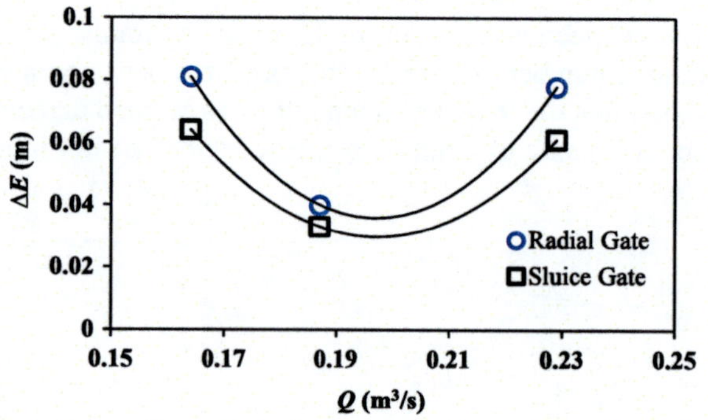

Fig. 8 Comparison of energy loss (ΔE) w.r.t flow rate (Q)

ranges from 0.033 to 0.064 m, whereas for the Radial gate, the amount of energy dissipated ranges from 0.040 to 0.081 m. Figures 7 and 8 show the graphical comparison between Radial and Sluice gates in terms of dissipated energy (energy loss) with respect to upstream head and flow rate. Experimental results and graphical representation depict that the hydraulic jump caused by a Radial gate dissipates higher energy than a Sluice gate in terms of the same upstream head and discharge. Also, the current study proves that the control of the flow is much easier as compared to the sluice gate.

4 Conclusions

The characteristics of hydraulic jump and the amount of energy dissipated from the Sluice gate and Radial gate were measured and analyzed at three different gate heights while maintaining the same flow rate and upstream head for each run. The experimental data for hydraulic jump characteristics and energy loss of the Sluice gate were then compared with the Radial gate data. From the comparison results, it is observed that for both the sluice and radial gates, the conjugate depth ratio rises with the rise of the initial Froude number, but the hydraulic jump height falls with the rise in Froude number. Furthermore, the length of jump ratio is found to be directly proportional to the Froude number for both the gates. At the same upstream head and discharge, it was found that the hydraulic jump caused by the Radial gate has the ability to dissipate more energy as compared to the energy loss by the Sluice gate. It is observed that at the same height of gate opening and same upstream head, the Radial gate dissipates 79% of specific energy more than the Sluice gate for discharge of 0.164 m^3/s, 78% for discharge of 0.229 m^3/s, and 82.5% for discharge of 0.187 m^3/s. Therefore, the hydraulic jump formed by the Radial gate has the ability to dissipate more energy than the Sluice gate and it can be concluded that the Radial gate is much safer for the structure, the control of the flow is much easier as compared to the Sluice gate, and it can be considered as an appropriate model for dissipating energy.

References

1. Chow, V. T. (1959). *Open channel hydraulics*. McGraw-Hill.
2. Jalil, S., Sarhan, S. A., & Yaseen, M. S. (2015). Hydraulic jump properties downstream a Sluice gate with prismatic sill. *Research Journal of Applied Sciences, 11*(4), 447–453.https://doi.org/10.19026/rjaset.11.1801
3. Basiouny, M. E., Nasrallah, T. H., Abdelhaleem, F. S., & Ibraheem, A. S. (2018). Roughened and corrugated aprons as scour countermerrged hydraulic jumps. In: *Twenty-first International Water Technology Conference, IWTC21*, Ismailia (pp. 200–214).
4. Mirzaei, H., & Tootoonchi, H. (2020). Experimental and numerical modeling of the simultaneous effect of sluice gate and bump on hydraulic jump. *Modeling Earth Systems and Environment, 6*, 1991–2002. https://doi.org/10.1007/s40808-020-00835-5

5. Mohamed, Y. A., Saleh, Y. K., & Ali, A. M. (2015). Studying the effect of different configurations and positions of sill over stilling basin on flow characteristics behind radial gate (case study: Nagaa Hammadi regulator). *Journal of Engineering Sciences, 43*(3), 311–329. https://doi.org/10.21608/JESAUN.2015.115186
6. Kim, Y., Choi, G., Park, H., & Byeon, S. (2015). Hydraulic jump and energy dissipation with sluice gate. *Water, 7*, 5115–5133. https://doi.org/10.3390/w7095115
7. Abdelhaleem, F. S. F. (2017). Hydraulics of submerged radial gates with a sill. *Journal of Hydraulic Engineering Division of the American Society of Civil Engineers, 23*(2), 1–10. https://doi.org/10.1080/09715010.2016.1273798
8. Nangare, P. B., & Kote, A. S. (2017). Experimental investigation of an ogee stepped spillway with plain and slotted roller bucket for energy dissipation. *International Journal of Civil Engineering and Technology, 8*(8), 1549–1555, Article ID: IJCIET_08_08_168.
9. Ewah, E. G., Nyah, E. E., Antigha, R. E. E., & Egbe, J. G. (2018). Experimental investigation of energy dissipation in hydraulic jump: A comparison of weir and level bedded constricted flume. *International Journal of Engineering Trends and Technology, 61*(1), 1–13. https://doi.org/10.14445/22315381/IJETT-V61P202
10. Bélanger, J. B. (1849). Notes sur le cours dhydraulique Notes on a course in hydraulics. Mém. Ecole Nat. Ponts et Chaussées, Paris, France, session 1849–1850, 222 pages in French.

Assessment of Soil Erosion in Karamana Watershed by RUSLE Model Using Remote Sensing and GIS

M. J. Pooja, Santra Thomas, U. B. Udayasurya, P. V. Praveej, and S. Minu

Abstract Soil erosion can be defined as the detachment of soil cover which is a serious land degradation problem today. The study estimates the average annual soil loss occurring in Karamana watershed, located in Thiruvananthapuram district of Kerala using the Revised Universal Soil Loss Equation (RUSLE) model integrated with remote sensing and Geographic Information System (GIS). Karamana watershed has a total catchment area of around 692 km^2. Factors such as rainfall, nature of soil, topographical features, land use, and land cover pattern were overlaid in ArcGIS software for the estimation of soil loss using RUSLE model. The average soil loss rate was estimated as 15.55 t-ha^{-1} yr^{-1} and the overall all soil loss was found as 1.07 million-t-yr^{-1}. It was observed that about 20.75% of the total study area is experiencing very high to severe erosion. The main parameters which are contributing to erosion are the topography factor and land-use pattern. The results of this study can be used in planning suitable land management approaches for minimizing the risk of soil loss within the watershed.

Keywords Remote sensing · GIS · RUSLE · Soil erosion · Watershed

1 Introduction

Soil erosion can be defined as the detachment of natural protective soil cover of land due to the action of rain, flowing water, and wind which causes its degradation. It happens because of the loss of canopy cover, improper land use, etc. In India, 45% of the total ground area is predominately affected by water and wind erosion and in Kerala, 67.1% of the total ground area is degraded by water erosion, waterlogging, and other complex problems [1]. Therefore, it is necessary to assess and map soil erosion-prone areas for soil conservation. The field measurements of erosion are there but they are time-consuming and usually done over restricted time and space

M. J. Pooja (✉) · S. Thomas · U. B. Udayasurya · P. V. Praveej · S. Minu
Department of Civil Engineering, Government Engineering College Barton Hill, APJ Abdul Kalam Technological University, Thiruvananthapuram, Kerala, India

A. K. Dikshit et al. (eds.), *Innovative Trends in Hydrological and Environmental Systems*, Lecture Notes in Civil Engineering 234,
https://doi.org/10.1007/978-981-19-0304-5_17

scales. Modeling soil erosion using remote sensing and GIS helps to identify, analyze and estimate soil erosion spatially.

Rainfall, spatial topography, vegetation, soil properties, and land use are the main factors controlling soil erosion. Several models are available to estimate soil erosion and compute the amount of sediment yield due to soil erosion. Soil erosion predictions by such models overcome the general limitations of the developing nations in their soil conservation research and planning efforts. Wischmeier and Smith [12] suggested the Universal Soil Loss Equation (USLE) model based on the concept of the separation and transport of particles from rainfall to calculate the amount of soil erosion in agricultural areas. The USLE has been enhanced later by several studies. Modified Universal Soil Loss Equation (MUSLE), RUSLE [10], Unit Stream Power-based Erosion Deposition (USPED) [7], etc. are proposed upon the improvement of USLE.

In this study, the RUSLE model integrated with geospatial technology is used to estimate the annual average soil loss by water occurring in the Karamana watershed in Thiruvananthapuram district Kerala. The soil erosion risk within the watershed is to be assessed and potential soil erosion zones are to be mapped. The soil loss estimation and identification of its spatial extent in the watershed may help in finding out the critical area for implementation of best management practices for the success of soil conservation in the Karamana watershed.

1.1 Study Area

Karamana watershed located in southern Kerala, India, has a catchment area of around 691.73 km^2 and extends between latitudes 8°21′43.2″ N and 8°41′45.6″ N and longitudes 77°14′13.2″ E and 76°51′14.4″ E. The Karamana river originates at Agastyarkoodam mountains located at the southern tip of the Western Ghats and flows 68 km westward, merging into the Arabian Sea at Thiruvallam-Karumam area near Kovalam. It has two important dams in its vicinity, namely, Aruvikkara Dam and Peppara Dam. Peppara Dam mainly supplies water for irrigation on the upper reach of the river, whereas Aruvikkara Dam supplies water for irrigation as well as drinking water to the city of Thiruvananthapuram. The main tributaries of the river are Kaviar, Attaiar, Vaiyapadiar, and Todiyar.

The rainfall over the area is uneven but the variation of rainfall from year to year is not significant. The average mean rainfall of the area is 2600 mm per year. Soil type found in the area is mainly gravelly clay. The study area mostly experiences a tropical humid climate.

Table 1 Particulars of data used

S. No.	Data type	Source	Description
1	Rainfall data	www.imdpune.gov.in	Gridded daily rainfall data for a period of 30 years (1991–2020)
2	Soil data	www.data.isric.org	Sand, silt, clay and SOC content maps at depths of 0, 5, 15 and 30 cm during the year 2017
3	Digital elevation model	www.earthexplorer.usgs.gov	SRTM DEM (resolution of 1 arc-second) published in the year 2014
4	Satellite image	www.earthexplorer.usgs.gov	Landsat 8 image of resolution 30 m (year 2021)

1.2 Data Sources

The study area is covered in the toposheets numbered 58H/2, 58H/3, 58D/14, and 58D/15 which is of scale 1:50000. Toposheets were downloaded from the official website of Survey of India (surveyofindia.gov.in). The toposheets were georeferenced and were used in the extraction of the study area using ArcGIS 10.2.2® software. To estimate the amount of soil loss occurring annually in the watershed, different data such as rainfall data, soil data, Digital Elevation Model (DEM), and satellite imagery of the study area of interest were collected from various sources which are given in Table 1.

2 Methodology

2.1 RUSLE Parameter Estimation

RUSLE model uses a set of mathematical equations to estimate sheet and rill erosion from fields subjected to different management practices [3]. RUSLE model calculates the average annual soil erosion per unit area (A) (t ha^{-1} yr^{-1}) using Eq. (1) [12].

$$A = R \times K \times \mathrm{LS} \times C \times P \tag{1}$$

where R is the rainfall erosivity factor (MJ-mm ha^{-1} h^{-1} yr^{-1}), K is the soil erodibility factor (t-ha-h MJ^{-1} mm^{-1}), LS is the topographic factor, C is the crop management factor, and P is the support practices factor.

2.2 Rainfall-Runoff Erosivity Factor (R)

The rainfall factor is the amount of the erosive power of rainfall energy. It depends on volume, intensity, and duration of the rainfall [2, 9]. Rainfall data of 30 years (1991–2020) for eight rain gauge stations were used for the estimation of R factor using Eq. (2) as per Singh et al. [11].

$$R = 79 + 0.363 \times \text{MAP} \tag{2}$$

where MAP = Mean Annual Precipitation in mm.

2.3 Soil Erodibility Factor (K)

K factor measures the ability of soil particles to withstand raindrop impact. It is the resistance shown by the soil to detachment of soil particles as well as to transportation of soil particles. *K* factor is determined through measurement of soil loss from a standard condition of 22.13 m long slope length with 9% of slope gradient. *K* factor was estimated using Eqs. (3)–(8) provided by Koirala et al. [5].

$$K = 0.1317 \times F_{\text{cs}} \times F_{\text{scl}} \times F_{\text{oc}} \times F_{\text{hs}} \tag{3}$$

where,

$$F_{\text{cs}} = 0.2 + 0.3 \times \text{e}^{(-0.0256 \times S \times 1 - M/100)} \tag{4}$$

$$F_{\text{scl}} = [M/(C + M)]^{0.3} \tag{5}$$

$$F_{\text{oc}} = \left[1 - 0.25 \times \text{OC}/\left(\text{OC} + \text{e}^{(3.72 - 2.95\text{OC})}\right)\right] \tag{6}$$

$$F_{\text{hs}} = [1 - 0.7 \times S_1/(S_1 + \text{e}^{(-5.51 + 22.9 \times S1)})] \tag{7}$$

$$S_1 = (1 - S/100) \tag{8}$$

where, *S*—percentage of sand content, *C*—percentage of clay content, *M*—percentage of silt content, and OC—percentage of soil organic carbon content.

Table 2 *C* factor values

S. No.	Class name	*C* factor
1	Agricultural area	0.63
2	Built-up land	0.09
3	Forest	0.003
4	Bare land	0.5
5	Waterbody	0

2.4 *Topographic Factor (LS)*

The topographic factor or the slope length and steepness (LS) factor, represents the ratio of soil loss under a given condition to that at a site with the standard slope steepness of 9% and slope length of 22.13 m. As the length and steepness of slope increase, the risk of erosion also increases. DEM data of the study area is used for calculating LS factor. LS is estimated using Eq. (9) as in Moore and Burch [8].

$$\mathrm{LS} = (\mathrm{FAC} \times 30/22.13)^{0.4} \times (\sin(\mathrm{slope} \times 0.01745)/0.09)^{1.3} \tag{9}$$

where FAC is flow accumulation.

2.5 *Land Management Factor (C)*

The land management factor (*C*) is the ratio of soil loss from a land cropped under specific conditions to the corresponding loss from clean, tilled continuous fallow land [6]. The Karamana watershed is classified on the basis of land-use pattern into five land useland cover (LULC) classes namely agriculture, forest, waterbody, bare land, and built-up area. LULC map was prepared using the supervised classification method of Landsat8 image of the study area acquired during the year 2021. Based on the land use land cover classes of the study area, the values of *C* factor are allotted as in Kim et al. [4] and are given in Table 2.

2.6 *Conservation Practice Factor (P)*

The practice factor (*P*) represents the ratio of soil loss from the land where conservation practices like contouring, strip cropping, terracing, etc. are adopted to those where they are not. Values ranging from 0 to 1 are assigned to various factors that help in reducing the rate of erosion, where a value of 1 indicates that there are no such conservation practices. Due to the lack of field data, the study on Karaman watershed assumes value of *P* factor as 1.

2.7 Annual Average Soil Erosion (A)

The annual erosion rate can be determined by using a raster overlay analysis of five factors in ArcGIS. Based upon the intensity of erosion, total annual soil erosion within the study area was categorized into six categories namely, slight, moderate, high, very high, severe, and very severe erosion.

3 Results and Discussion

3.1 R Factor

The rainfall erosivity map (Fig. 1) shows the resultant output of the effect of 30 years (1991–2020) rainfall on the erosion happening in the Karamana watershed. It indicates spatial variation in the rainfall across the region. The value of R factor ranges from 525.51 to 720.47 MJ-mm $ha^{-1\,h-1\,yr-1}$ with an average value of 647.68 MJ-mm $ha^{-1\,h-1\,yr-1}$. The rainfall erosivity map of the area (Fig. 1) shows that the northwestern part of the region experienced higher erosivity compared to the other portions of the study area. This is mainly because of the spatiotemporal variation of rainfall across the study area.

3.2 K Factor

The soil erodibility map obtained is shown in Fig. 2, it is evident that the *K* factor value varies from 0.014 to 0.021. *K* factor values are not defined for water bodies and hence those cells comprising water bodies are left as no value cells, which can be seen as white patches in the map.

3.3 LS Factor

The topographic factor map computed using DEM data is shown in Fig. 3. The value of the LS factor ranges between 0 and 820. The low values of LS factor are predominately observed in the middle portion of the study area which suggests that these areas could be better used for agriculture with proper support practice factors. Higher LS values are found along the northeastern mountainous region of the watershed and also along the channel length of the river. These higher LS values are due to the presence of higher gradients of the mountainous region and also due to high flow accumulation value along the path of flow of the river. These results

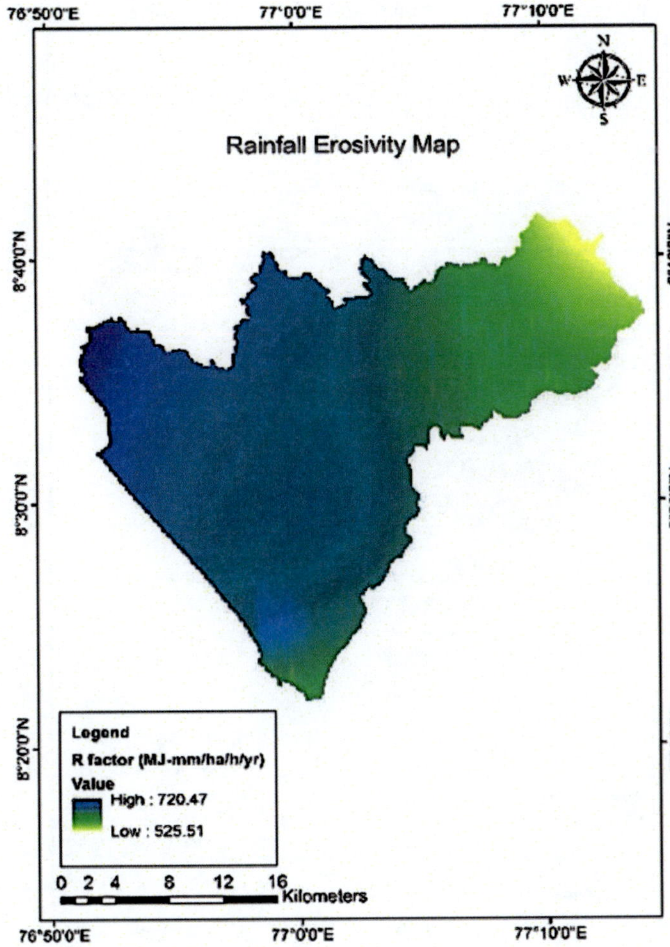

Fig. 1 Rainfall erosivity map of study area

indicate that LS factor increases with an increase in flow accumulation and slope steepness.

3.4 C Factor

The *C* factor map is shown in Fig. 4. Intense utilization of land using agriculture is evident and support practices against erosion are not much available. As a result, the highest value of the *C* factor is concentrated in a major portion of the watershed. Built-up land can be seen as a small patch in the western regions of the study area. Two

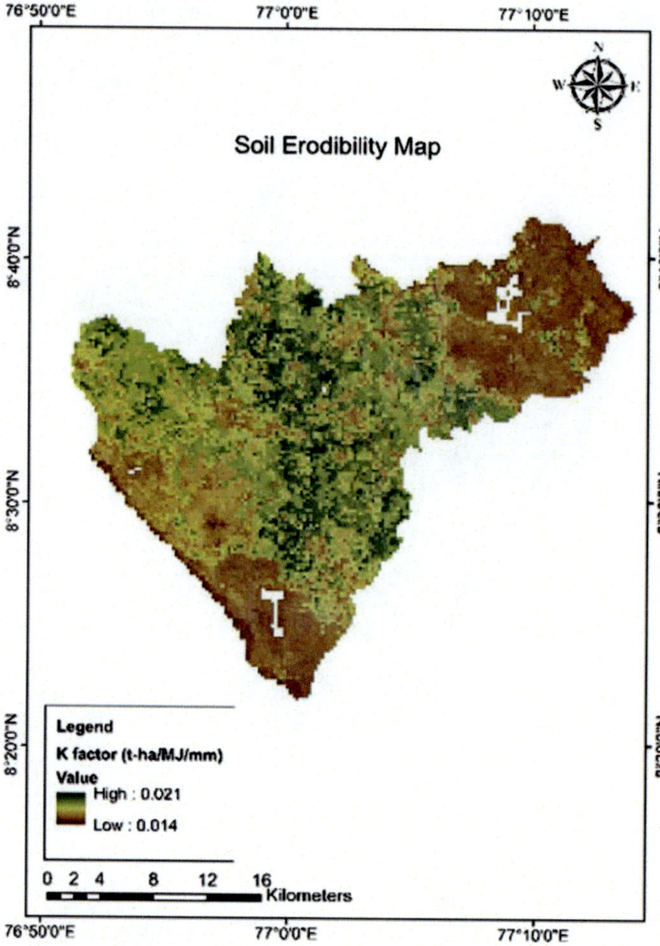

Fig. 2 Soil erodibility map

important lakes namely Akkulam and Veli are also found in the study area. Another salient feature is the presence of two reservoirs—Peppara Dam and Aruvikkara dam, in which Aruvikkara dam supplies the major water demand of Thiruvananthapuram city. Bare land is scattered in the watershed. The forest area is found in the north-eastern part with mountains including the southernmost tip of Western Ghats from where Karamana river originates.

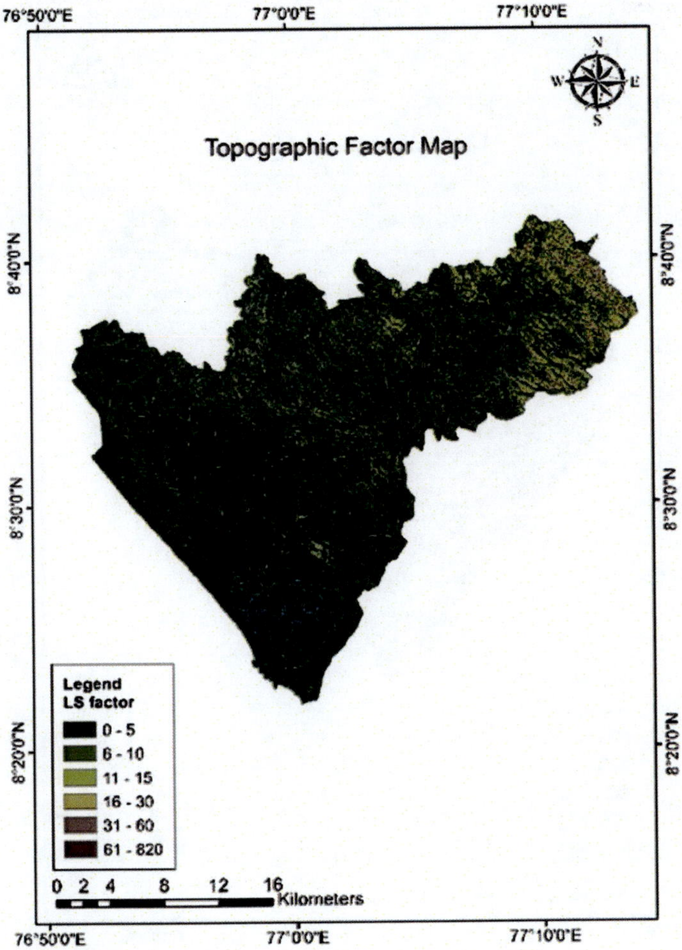

Fig. 3 Topographic factor map

3.5 *Annual Average Soil Erosion*

The annual average soil erosion map obtained by integrating the five factors is shown in Fig. 5. The estimated mean annual soil loss for the study area was found to be 15.55 t-ha$^{-1\,\mathrm{yr}-1}$ and the total erosion is 1.07 million-t-yr^{-1}. It was found that the main parameters which are contributing to erosion in the area are the topography factor and land-use pattern. Most of the region experiences erosion between 0 and 20 t-ha-1 yr^{-1}. However, in the central portion of the Karamana watershed, a soil loss of greater than 80 t-ha-1 yr^{-1} was noticed. This is due to different land-use patterns in that area characterized by the presence of scattered barren lands that are highly

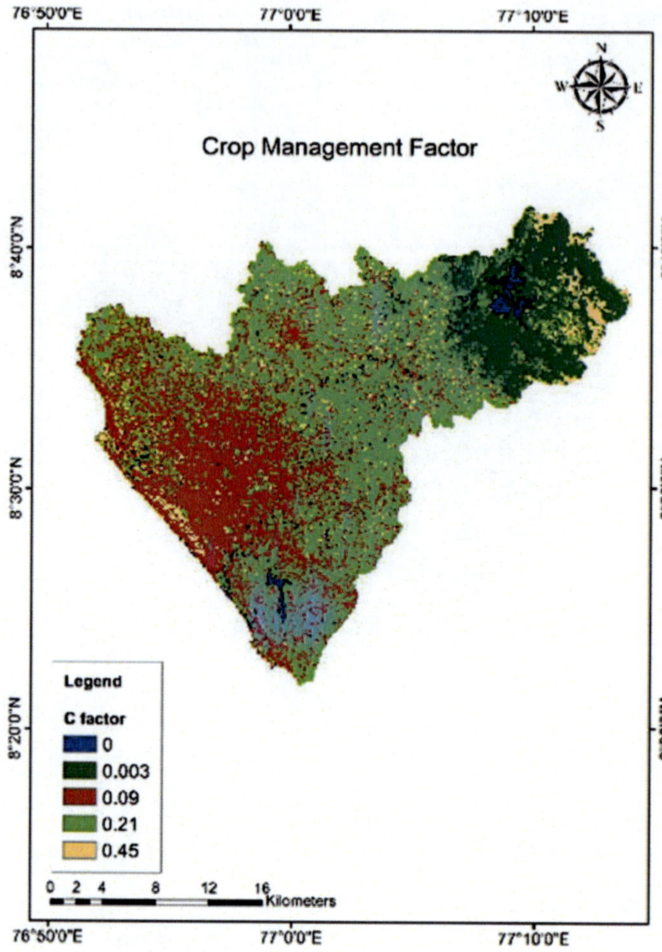

Fig. 4 Crop management factor map

vulnerable to erosion. The erosion severity map of the study area is prepared for the six classes of soil erosion as shown in Fig. 6. It is noticed that about 20.75% of the total area is experiencing very high to severe erosion as shown in Table 3.

Severe erosion was observed near the scattered bare lands near the Agasthyarkoodam Mountains which is located in the eastern portion of the study area. In addition to that, the illegal quarrying activity has changed the landscape of Mukkunnimala which is located in the central part of Karamana watershed where the risk due to erosion is severe. Mukkunnimala hills, which stretch from Edakkode to Vilavoorkal and Machel to Paamancode have disappeared over the past two decades due to illegal quarrying activity. The developmental activities and associated deforestation happening in the Karamana watershed have resulted in increased soil erosion.

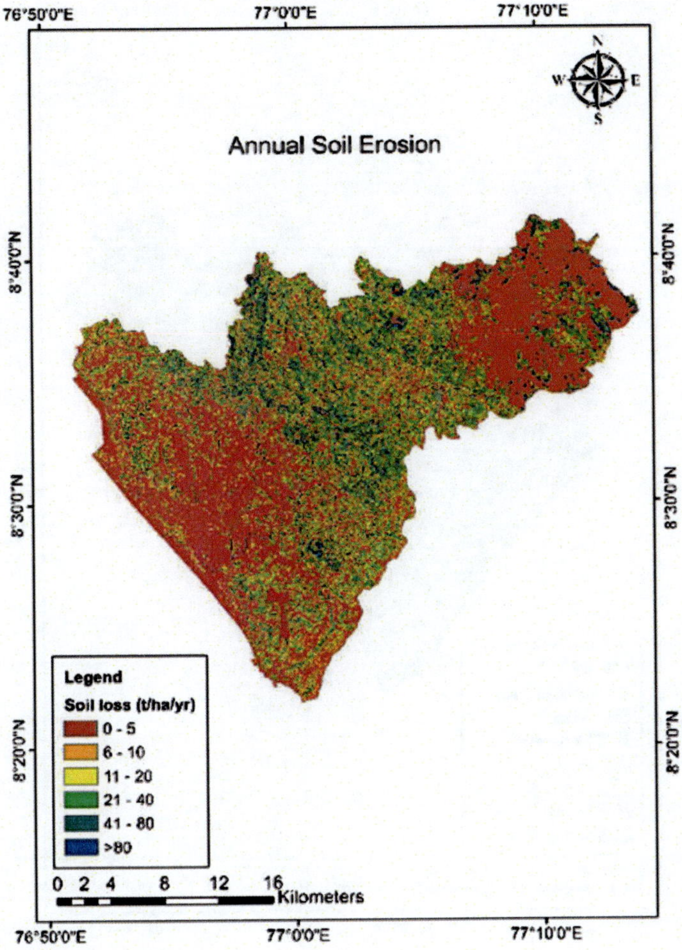

Fig. 5 Annual average soil erosion

Another important thing in the land-use pattern that contributed to very severe erosion is agriculture which prevailed in the major part of the area with fewer support practices against erosion.

The erosion happening in Karamana watershed can cause the siltation of dams located in the watershed which include Aruvikkara Dam and Peppara Dam in which Aruvikkara dam supplies the major water demand of Thiruvananthapuram city.

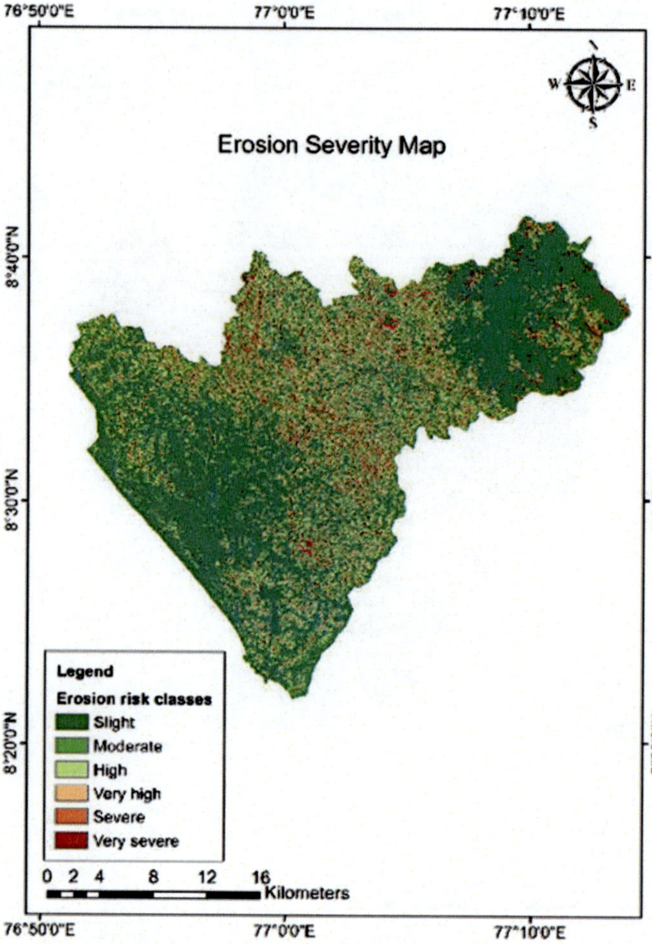

Fig. 6 Erosion severity map

Table 3 Percentage of area coming under each category

Erosion category	Area (km^2)	Area (%)
Slight	442.27	63.94
Moderate	41.30	5.97
High	64.35	9.30
Very high	75.73	10.95
Severe	45.82	6.62
Very severe	22.26	3.22
Total	691.73	100.00

3.6 *Suggestions to Reduce Soil Loss in Karamana Watershed*

The analysis of erosion rate suggests that there is a possibility for decreasing the erosion by adopting some changes in the land-use practices. The illegal quarrying activity in the Mukkunnimala region should be stopped to protect the soil cover of the region as well as to prevent the possible occurrence of landslides that might happen. Agriculture could be shifted from extreme erosion potential to less erosion risk areas with good management practices so that total erosion risk in the basin could be mitigated. Areas with less slope can be used for agricultural practices after ensuring proper support practices. Erosion was severe along the channel length of the flow of Karamana river where geotextiles can be used to prevent the wash off of the soil. The slopes can be riveted with stones or geotextiles to control soil wash-off. Sufficient cover should also be provided for the bare land scattered in the watershed so that the risk of erosion from these land classes can be reduced to an extent. Moreover, farmers should be educated on sustainable agricultural practices which conserve the soil.

4 Summary and Conclusion

The study exemplifies the application of the RUSLE equation using remote sensing and GIS to model soil erosion potential in the Karamana watershed. This study conducted in the Karamana watershed came up with the following conclusions:

- The average soil erosion rate was found out as 15.55 t-ha^{-1} yr^{-1} and the overall soil loss was estimated as 1.07 million-t-yr^{-1}.
- The results of the study show that LS factor and *C* factor were most prominent in governing soil loss of study area.
- A spatially distributed erosion severity class map was prepared for the watershed and it was observed that about 20.75% of the total study area is experiencing very high to severe erosion. It is required to plan appropriate land management approaches in severe erosion-prone areas.
- Stopping illegal quarrying activities, providing sufficient cover in the barren land, providing geotextiles along the bank, and adopting good agricultural practices can help in reducing the wearing away of soil in the Karamana watershed.
- The study demonstrates that the use of RUSLE model and geospatial technology is a beneficial tool in estimating the amount of soil loss, especially the soil loss due to sheet and rill erosion. But some limitations must be taken into account before modeling such as the quality of data and the spatial resolution used.

References

1. Bhattacharyya, R., Ghosh, B. N., Mishra, P. K., Mandal, B., Rao, C. S., Sarkar, D., & Franzluebbers, A. J. (2015). Soil degradation in India: Challenges and potential solutions. *Sustainability, 7*(4), 3528–3570. https://doi.org/10.3390/su7043528
2. Chinnasamy, P., Honap, V. U., & Maske, A. B. (2020). Impact of 2018 Kerala floods on soil erosion: Need for post-disaster soil management. *Journal of the Indian Society of Remote Sensing, 48*(10), 1373–1388.
3. Ganasri, B. P., & Ramesh, H. (2016). Assessment of soil erosion by RUSLE model using remote sensing and GIS—A case study of Nethravathi Basin. *Geoscience Frontiers, 7*(6), 953–961. https://doi.org/10.1016/j.gsf.2015.10.007
4. Kim, J. B., Saunders, P., & Finn, J. T. (2005). Rapid assessment of soil erosion in the Rio Lempa Basin, Central America, using the universal soil loss equation and geographic information systems. *Environmental Management, 36*(6), 872–885.
5. Koirala, P., Thakuri, S., Joshi, S., & Chauhan, R. (2019). Estimation of soil erosion in Nepal using a RUSLE modeling and geospatial tool. *Geosciences, 9*(4), 147. https://doi.org/10.3390/geosciences9040147
6. Mallick, J., Alashker, Y., Mohammad, S. A. D., Ahmed, M., & Hasan, M. A. (2014). Risk assessment of soil erosion in semi-arid mountainous watershed in Saudi Arabia by RUSLE model coupled with remote sensing and GIS. *Geocarto International, 29*(8), 915–940. https://doi.org/10.1080/10106049.2013.868044
7. Mitasova, H., Barton, C. M., Ullah, I., Hofierka, J., & Harmon, R. S. (2013). GIS based soil erosion modeling. In *Treatise on geomorphology* (pp. 228–258). Elsevier Inc. https://doi.org/10.1016/B978-0-12-374739-6.00052-X
8. Moore, I. D., & Burch, G. J. (1986). Physical basis of the length slope factor in the universal soil loss equation. *Soil Science Society of America Journal, 50*(5), 1294–1298. https://doi.org/10.2136/sssaj1986.03615995005000050042x
9. Prasannakumar, V., Vijith, H., Abinod, S., & Geetha, N. J. G. F. (2012). Estimation of soil erosion risk within a small mountainous sub-watershed in Kerala, India, using Revised Universal Soil Loss Equation (RUSLE) and geo-information technology. *Geoscience Frontiers, 3*(2), 209–215. https://doi.org/10.1016/j.gsf.2011.11.003
10. Renard, K. G., Foster, G. R., Weesies, G. A., McCool, D. K., & Yoder, D. C. (1997). Predicting soil erosion by water. *Agriculture handbook* (p. 703).
11. Singh, G., Babu, R., & Chandra, S. (1981). Soil loss prediction research in India. Bulletin Nos. T-12/D-9. Central Soil and Water Conservation Research and Training Institute.
12. Wischmeier, W. H., & Smith, D. D. (1978). *Predicting rainfall erosion losses: A guide to conservation planning* (No. 537). Department of Agriculture, Science and Education Administration.

Spatio-Temporal Analysis of Meteorological Drought Using IMD 0.25° Gridded Dataset for Marathwada Region

A. B. Pachore and R. Remesan

Abstract Drought is an extreme event and directly related water scarcity in the region. Marathwada region of Maharashtra state is worst affected by water stress. Gridded rainfall data of finer resolution, i.e., 0.25° * 0.25° from 1990 till 2019 (30 years) is downloaded from the webpage of the Indian Meteorological Department, Pune. Timescales selected for SPI calculations are 1, 3, 6, 9, and 12 months, and the index calculations are performed for every grid point. Percentage drought frequency is calculated for SPIs-1, 3, 6, 9, and 12. The average percentage drought frequency for 30 years is calculated, and spatial plots are prepared for mild, moderate, severe, and extreme droughts for SPI-1, 3, 6, 9, and 12. SPIs at 1-, 3-, and 6-month timescales refer to short-term precipitation deficiency along with agricultural drought. SPI at 6–12 months is correlated with hydrological drought. Spatial plots were used to view a spatial variation of percentage drought frequency which is further used to identify the worst drought-affected regions of the study area. P10, i.e., 10th percentile of percentage drought frequency over study area is calculated to know the threshold of percentage drought frequency for 90% of the area. Results show that mild drought is dominating over Marathwada region, and its percentage drought frequency is always between 57.7 and 70.4% for 90% of the study area. This study is helpful for policymakers to know the spatio-temporal variation of precipitation deficiency over the Marathwada region.

Keywords Precipitation deficiency · SPI · Percentage drought frequency · 10 percentiles

1 Introduction

There are major 22 drought events that occurred in India from the year 1871 to 2002. These drought years are 1873, 1877, 1899, 1901, 1904, 1905, 1911, 1918, 1920, 1941, 1951, 1965, 1966, 1968, 1972, 1974, 1979, 1982, 1985, 1986, 1987, and 2002

A. B. Pachore · R. Remesan (✉)
School of Water Resources, Indian Institute of Technology, Kharagpur, India
e-mail: renji.remesan@swr.iitkgp.ac.in

A. K. Dikshit et al. (eds.), *Innovative Trends in Hydrological and Environmental Systems*, Lecture Notes in Civil Engineering 234,
https://doi.org/10.1007/978-981-19-0304-5_18

[16]. Marathwada region of Maharashtra state is chronically affected by drought. Aurangabad, Beed, Nanded, Osmanabad, Parbhani are the most affected districts of the Marathwada region [8]. Maharashtra Economic Survey of 2015–2016 says 16,800 villages were drought-affected in 2014–2015, drought in the region directly affected 9 million farmers, and 3225 farmers committed suicide in 2015 alone.

Drought phenomenon is generally creeping in nature whose start and end are difficult to identify. Drought is generally characterized based on duration, severity, and interarrival time. Broad classification of drought is into four types which are: meteorological, hydrological, agricultural, and socio-economic drought. Meteorological droughts are generally defined depending on the number of dry days and also consider precipitation less than the certain threshold value. It takes account of both drought duration and magnitude. Meteorological drought is further correlated with agricultural drought. Agricultural drought occurred due to soil moisture deficiency during the initial growth stage of crops. When there is a prolonged deficiency of surface water and groundwater, it is termed as hydrological drought. Socio-economic drought is the combined effect of meteorological, hydrological, and agricultural drought. It has negative effect on the trade of goods [20].

The drought index is used to analyze the impact of drought to define drought characteristics such as drought severity, intensity, and frequency. Indices should be calculated at different timescales from months to year. A monthly timescale is appropriate for quantifying the scarcity of soil moisture, surface water, and groundwater [11]. World Meteorological Organization has led down guidelines in its handbook of drought indicators and indices [17] for selecting drought index for any situation which are as: (1) The drought index should facilitate on-time detection of dry event. (2) While determining start and end of drought condition, index should consider climate, space, and time element. (3) Index should be able to account for drought severity along with its impacts occurring on the field. (4) One should check for the need for integrated drought index. (5) Index selected for the area should be less complicated and easy to use. For many years, researchers are developing drought indices for analysis of drought events. Some of the popular indices for characterizing precipitation deficiency are Palmer drought severity index [15], rainfall anomaly index [18], deciles [5], standardized precipitation index [10], reclamation drought index [19]. Drought indices to quantify soil moisture scarcity are soil moisture anomaly [2], soil moisture deficit index [14]. Hydrological drought is characterized using Palmer hydrological drought index [15], standardized reservoir supply index [6], standardized streamflow index [12], streamflow drought index [13]. Remote sensing-based indices are enhanced vegetation index [7], normalized difference vegetation index [9].

The standardized precipitation index (SPI) is the most popular among different indices for meteorological drought characterization. SPI was originally computed at Colorado state university in 1993 for timescales of 3, 6, 12, 24, and 48 months by Mckee, Doesken, and Kleist. World Meteorological Organization in 2012 [21] has given a user guide for using SPI. Some advantages of SPI are: It is flexible to be computed at multiple timescales starting from 1 month till 48 months. Different timescales are referred for water scarcity of different types; e.g., 1-month to 6-month

SPI refers to meteorological and agricultural drought, and 6–12 months refer to surface and groundwater scarcity. SPI is consistent spatially, and hence, comparison of drought-hit areas with different climates is possible. SPI is normalized, and hence, it can quantify both wet and dry spells of precipitation. SPI is having many disadvantages as it only considers rainfall inputs and ignores temperature components; hence, SPI is not useful for analyzing climate change impact on precipitation and droughts. SPI did not consider soil moisture components as well and hence unable to quantify agricultural impacts of drought directly. The spatio-temporal assessment of meteorological drought is popularly done using SPI. SPI can be calculated at multiple timescales starting from 1 to 72 months, whereas the most suitable range is 1–24 months. SPI at different timescale refers to different types of water scarcity, i.e., precipitation, soil moisture, surface water, and groundwater deficiency. Timestep in SPI analysis is generally monthly moving average; e.g., SPI-6 at the end of July considers precipitation totals of February to July in that particular year and compares it with rainfall totals of February to July for all years in the record. When precipitation is above the median value, then SPI is positive and negative when below the median value.

Temporal variation of drought can be studied by calculating SPI at different timescales starting from a shorter timescales of 1 and 3 months to a longer timestep of 24 months. Spatial plots prepared using the ArcGIS platform are proved to be helpful to analyze the spatial variation of the dry event [1, 3]. This study focuses on the spatio-temporal analysis of meteorological drought in the Marathwada region of Maharashtra state using finer resolution (0.25° * 0.25°) precipitation data. Temporal angle is considered by calculating SPI at different timescales, and spatial plots are prepared in ArcGIS environment for studying the spatial extent of drought. Following are the specific objectives of the study: (1) to calculate SPI at different timescales, i.e., 1, 3, 6, 9, and 12 months using 0.25° * 0.25° gridded precipitation data of IMD for the period from 1990 till 2019. (2) To prepare spatial maps of percentage drought frequency for SPIs-1, 3, 6, 9, and 12. (3) To prepare spatial maps for different severity levels, i.e., mild, moderate, severe, and extreme. (4) To calculate P10, i.e., 10th percentile of percentage drought frequency for all severity levels at all timescales.

2 Materials and Methods

2.1 Study Area Details

Marathwada region of Maharashtra state, India, is selected for this study. Due to the lower amount of precipitation, drought situations are more likely in this region. The focused region comprises of total eight districts which are Aurangabad, Bid, Hingoli, Jalna, Latur, Nanded, Osmanabad, and Parbhani. Figure 1 shows the location map of the Marathwada region with respect to India. Longitudes of the area ranges between 70°5′ east and 78°5′ east, whereas latitude is between 17°5′ north and 20°5′ north.

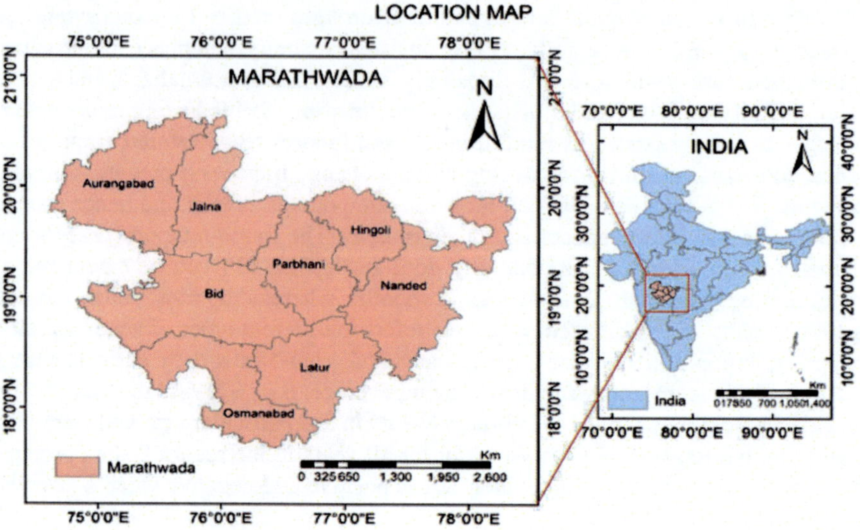

Fig. 1 Location of Marathwada in India

There are total 76 tehsils (administrative blocks) within eight districts of the study area with a total geographical area of 64,813 km^2. The region is having a maximum day temperature for summer at 43 °C, and the minimum temperature in the region is 6 °C, and due to this, climatic conditions in the region are dry. Relative humidity of region is extremely low, i.e., 35% but increases to 85% in the winter season (Source: IMD, Mumbai). Yearly rainfall in the region is 700 mm (Source: IITM, Pune). The Marathwada region is characterized by deep to very deep black cotton soils with good water-holding capacity. Mostly, rain-fed agriculture is adopted by farmers in the region. Major crops in the region are cotton, soybean, sugarcane, wheat, bajra.

2.2 *Dataset Used*

Monthly precipitation data having a finer resolution of 0.25° * 0.25° from 1990 till 2019 (30 years) is downloaded from the webpage of the Indian Meteorological Department in netCDF format which is further converted to .csv format using Python programming language.

2.3 Standardized Precipitation Index

Standardized precipitation index can be calculated at different timescales, and this flexibility in the calculation is the main advantage of this index. SPI calculations are based on the probability of rainfall. SPI is normalized, and hence, it gives values for both dry and wet conditions. Values of SPI will be positive if rainfall is above the median and negative if it is less than the median value. Standardized precipitation index values less than "1" are taken into account for drought characterization.

SPI is computed at timescale of "*k*" months for year "*i*" and month "*j*". For *j*th month and *i*th year, "*k*" months moving average is taken of rainfall time series Z_{ij}^{k} (*I* = 1, ..., *n*), where each "Z" term is summation of precipitation occurred in previous *k* months. Average monthly precipitation data series for timescales of 1, 3, 6, 9, and 12 months is fitted with gamma distribution [10]. Equation for gamma distribution function is given as below,

$$g_Z(Z) = \frac{1}{\beta^{\xi}\Gamma(\xi)} Z^{\xi-1} e^{-\frac{Z}{\beta}} \tag{1}$$

where ξ is the shape parameter and β is the scale parameter, $\Gamma(\xi)$ is gamma function at ξ. After integrating above Eq. (1), cumulative probability distribution is derived which is given as follows,

$$G_Z(Z) = \int_0^Z g_Z(Z)\mathrm{d}z = \frac{1}{\beta^{\xi}\Gamma(\xi)} \int_0^Z Z^{\xi-1} e^{-\frac{Z}{\beta}} \mathrm{d}z \tag{2}$$

Precipitation datasets mostly have zero values in it, but gamma distribution function does not take account of zero values; hence, cumulative distribution function (CDF) is derived with the help of mixed distribution function which is given as follow,

$$H_Z(Z) = q + (1 - q)G_Z(Z) \tag{3}$$

where *G*(*Z*) is cumulative distribution function with non-zero precipitation and *q* is the probability of null precipitation. Final equation of SPI is given as follows [4],

$$\mathrm{SPI} = \Psi^{-}(H(Z)) \tag{4}$$

CDF of the standard normal distribution is given by $\Psi()$, and $\Psi^{-}()$ is inverse of standard normal CDF.

Rainfall is the single input variable required for SPI calculation. Standardized precipitation index is having values on both positive and negative scales. Based on

Table 1 Drought classes based upon SPI values [10]

S. No.	SPI	Severity	Category
i	$\geq +2$	Extreme	Wet conditions
ii	≥ 1.5 and ≤ 1.99	Severe	
iii	≥ 1 and ≤ 1.49	Moderate	
iv	≥ -0.99 and ≤ 0.99	Mild	Dry conditions
v	≥ -1.49 and ≤ -1	Moderate	
vi	≥ -1.99 and ≤ 1.5	Severe	
vii	≤ -2.0	Extreme	

threshold value of SPI, dry events are identified and classified into different severity levels as shown in Table 1.

2.4 *P10 of Percentage Drought Frequency*

P10 is the 10th percentile value that exceeds 90% of the time in the given dataset. It means that 90% of data is below this value, and only 10% of data is above this value. The 10th percentile value is useful to know the nature of 90% of data. For calculating the 10th percentile, a dataset is arranged from smallest to largest and rank is given to each data point; after that, 0.10 is multiplied by the total number of values in a dataset which gives the rank of 10th percentile value from which one can get 10th percentile value. This can be calculated using the percentile function in excel.

3 Result and Discussion

The standardized precipitation index is calculated using monthly precipitation data of 30 years from 1990 to 2019. Gridded precipitation data (0.25°*0.25°) is used for this study, and SPI is calculated for rainfall record of 30 years at each grid point. SPI calculator which is a windows-based program developed by National Drought Mitigation Center (NDMC) is used for all SPI calculations. SPI values less than 1 are considered for drought analysis. SPI is calculated at every 3-month timescale starting from 1 month till 12 months.

3.1 *Drought Frequency Analysis*

Percentage drought frequency is the percentage of events having SPI value less than the threshold of "1". Percentage drought frequency is calculated for SPIs-1, 2, 3,

6, 9, and 12. Based on these percentage drought frequency values, spatial plots are prepared in ArcGIS 10.3 using the inverse distance weighted (IDW) method of interpolation. Spatial plots showing the spatial variation of percentage drought frequency at every SPI timescale are shown in Fig. 2.

For 30 years of monthly precipitation records from 1990 till 2019, as per SPI-1, almost the total study area is dry with percentage drought frequency ranging from 68.10 to 86.40%, whereas according to SPI-3, dry events are more frequent in some pockets of the northwest, eastern, and southern parts as compared to rest of the study area. SPI-6 shows that dry events are more frequent in Northern, Southern, and some parts of Eastern Marathwada. It is observed that dry events are more frequent in Northern, Western, and some parts of Eastern Marathwada when observed at SPI-9 and SPI-12.

3.2 *Drought Severity Analysis*

Spatial plots of percentage drought frequency for mild, moderate, severe, and extreme drought conditions are shown in Fig. 3. According to SPI-1, mild drought events are almost equally distributed from 1990 till 2019 with the same percentage drought frequency in the whole study area. Moderately dry events are more frequent in the southern and southeast parts of the study area. Severe dry events are more frequent for western and southwest part of the region. More frequent extreme dry events are in Northern Marathwada. When spatial plots of SPI-3 is observed, it is found that mild drought events are more frequent for the northern and eastern part of the Marathwada region. Central, Southern, and Southeast Marathwada show more frequent moderately dry events. A major part of Western Marathwada shows more frequent severely dry events. Extremely dry events are more frequent in the northern, southern, northeast, and central parts of the study area. As per SPI-6, mild dry events are more frequent in Northwest, Eastern, and Southern Marathwada. Northeast, eastern, and southern parts of Marathwada region show more frequent moderately dry events. Southern and Southwest Marathwada show more frequent severely dry events. Extreme dry events are more frequent in Northern and Central Marathwada. According to SPI-9, in the Northwest, Western, and Eastern Marathwada, mild drought is more frequent. Moderately dry events are more frequent for northeast, eastern, southeast, and southern parts of the study area. Western Marathwada shows more frequent severe dry events. The northern and northwest part of the Marathwada show more frequent extremely dry events. Spatial distribution of different drought classes as per SPI-12 shows that Northwest Marathwada is having more frequent mild drought events. Moderately dry events are more frequent in the southeast of the study area. Central, Southwest, and Southeast Marathwada show more frequent severe dry events. Northern and northwest part of Marathwada region face more frequent extreme dry events.

Box and whisker plot showed in Fig. 4 gives maximum value, minimum

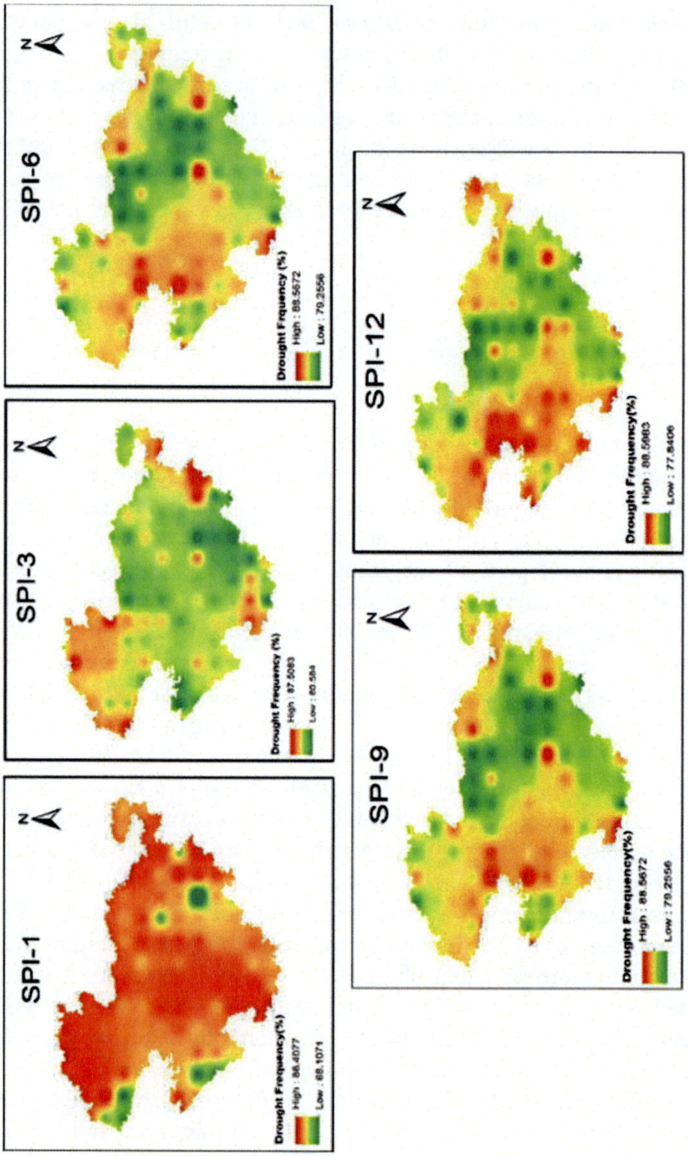

Fig. 2 Spatial plots of percentage drought frequency for all SPI timescales

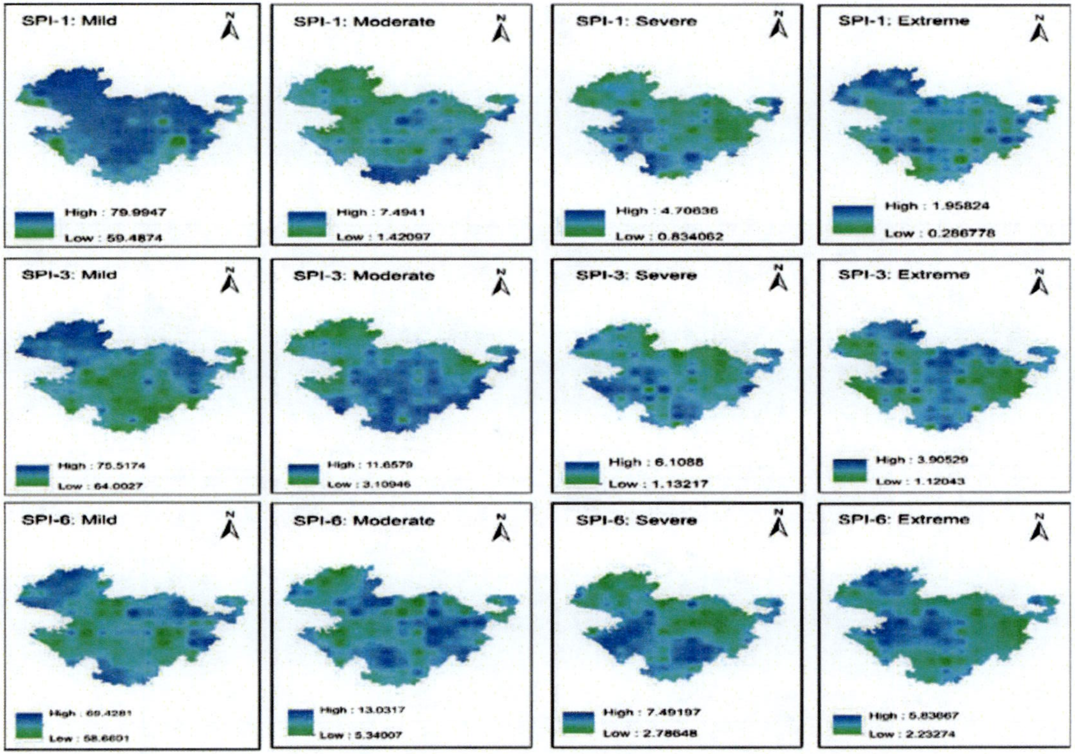

Fig. 3 Spatial plots of percentage drought frequency for all severity levels with multiple SPI timescales

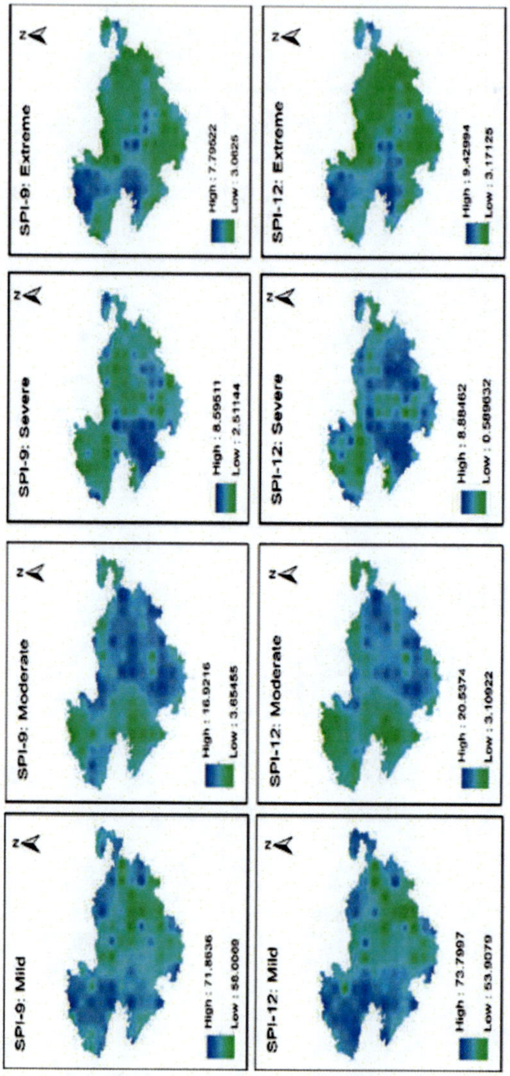

Fig. 3 (continued)

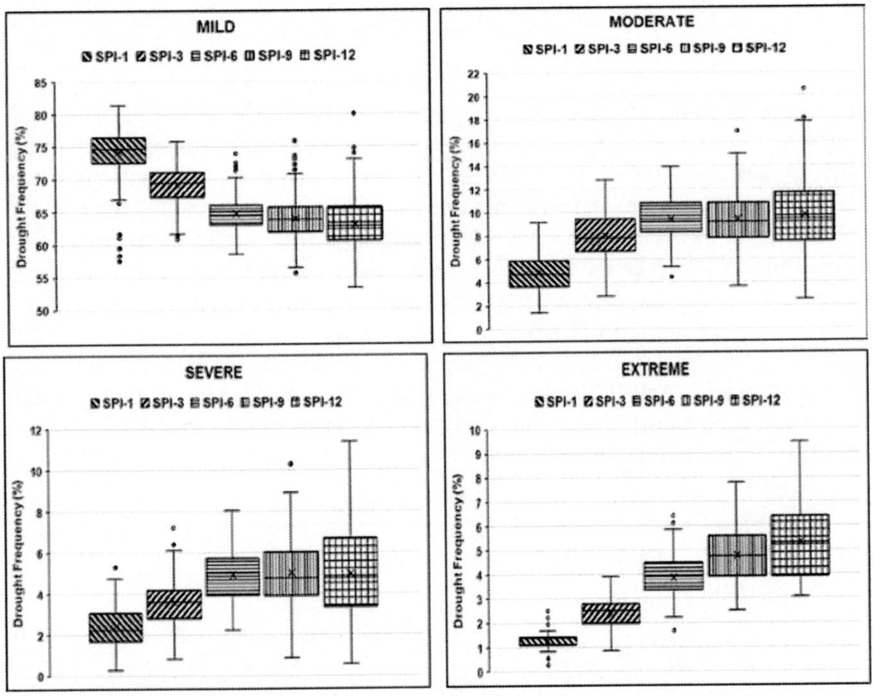

Fig. 4 Box and whisker plot of percentage drought frequency for all severity levels with multiple SPI timescales

value, 25th percentile, median, 75th percentile, interquartile range, and outliers of percentage drought frequency data for different severity at different timescales from 1990 to 2019. Median and 75 percentile values of percentage drought frequency at SPIs-6, 9, and 12 are almost consistent. In the case of extreme dry events, a large variation of percentage drought frequency is observed from SPI-1 to SPI-12. Mild drought events show a decreasing trend in percentage drought frequency when observed from SPI-1 to SPI-12, whereas moderate, severe, and extreme drought events show an increasing trend in percentage drought frequency when observed from SPI-1 to SPI-12. It is observed that for mild drought, the percentage drought frequency varies within 50–85%, whereas for moderate, severe, and extreme drought events, it is within 0–20% which is significantly less as compared to mild drought conditions.

3.3 P10 of Percentage Drought Frequency

P10 is the 10th percentile value of average percentage drought frequency from 1990 to 2019, which exceeds 90% of the area. As shown in Fig. 5, P10 of mild drought is always greater than 50% at any timescale, i.e., from 1 month till 12 months, whereas P10 of moderate, severe, and extreme drought is always less than 10%. The above observation shows that the Marathwada region is majorly experiencing more frequent mild dry events. P10 (percentage drought frequency covering 90% of the area) values of mild drought events show a decreasing trend as SPI timescale increases from 1 to 12 months. The highest P10 value of percentage drought frequency, i.e., 70.452% is for mild drought at SPI-1, and the lowest, i.e., 0.838% is for extreme drought at SPI-1 as shown in Table 2. While observing drought scenario under SPI at 1-month to 12-month timescale, the percentage drought frequency of mild drought decreases, and that of extreme drought increases.

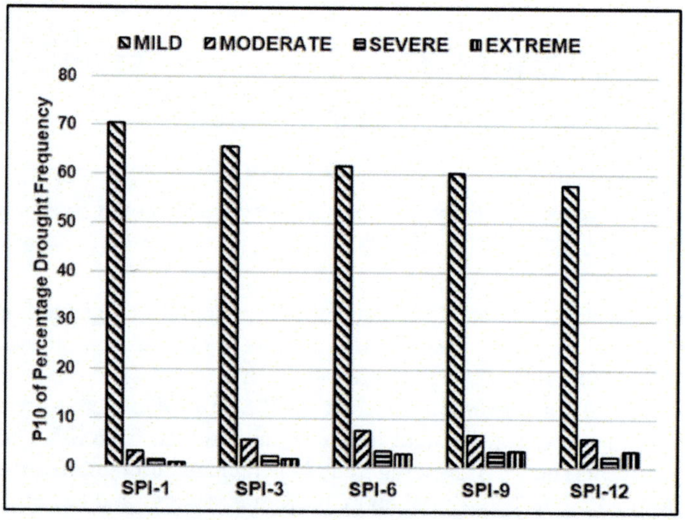

Fig. 5 Comparison of P10 of percentage drought frequency for multiple SPI timescales

Table 2 P10 values of percentage drought frequency

SPI	Mild	Moderate	Severe	Extreme
SPI-1	70.452	3.06	1.39	0.838
SPI-3	65.558	5.56	2.23	1.68
SPI-6	61.67	7.49	3.338	2.79
SPI-9	60.22	6.678	3.06	3.34
SPI-12	57.778	5.838	2.23	3.34

4 Conclusion

Spatial plots of percentage drought frequency at different severity levels and timescales are derived for the study area and helpful to identify worst drought-affected regions. Results show that median and 75th percentile values of percentage drought frequency from 1990 till 2019 for mild, moderate, and severe drought at 6-, 9-, and 12-month timescales are almost consistent. P10, i.e., 10th percentile of percentage drought frequency for mild drought varies from 57.7 to 70.4% for all timescales, whereas the maximum P10 value for moderate, severe, and extreme drought categories at all timescales is 7.49%; hence, it can be concluded that the Marathwada region is majorly affected by mild drought events. This study is helpful to policymakers to prepare drought mitigation and management policy by studying the frequency of meteorological drought of different severity in the Marathwada region from 1990 to 2019.

References

1. Awchi, T. A., & Kalyana, M. M. (2017). Meteorological drought analysis in northern Iraq using SPI and GIS. *Sustainable Water Resources Management, 3*(4), 451–463.
2. Bergman, K. H., Sabol, P., & Miskus, D. (1988, November). Experimental indices for monitoring global drought conditions. In *Proceedings of the 13th Annual Climate Diagnostics Workshop*, Cambridge, MA, USA (Vol. 31, pp. 190–197).
3. Das, J., Gayen, A., Saha, P., & Bhattacharya, S. K. (2020). Meteorological drought analysis using Standardized Precipitation Index over Luni River Basin in Rajasthan, India. *SN Applied Sciences, 2*(9), 1–17.
4. Ganguli, P., & Reddy, M. J. (2012). Risk assessment of droughts in Gujarat using bivariate copulas. *Water Resources Management, 26*(11), 3301–3327.
5. Gibbs, W. J., & Maher, J. V. (1967). Rainfall deciles drought inddicators.
6. Gusyev, M., Hasegawa, A., Magome, J., Kuribayashi, D., Sawano, H., & Lee, S. (2015, December). Drought assessment in the Pampanga River basin, the Philippines—Part 1: Characterizing a role of dams in historical droughts with standardized indices. In *Proceedings of the 21st International Congress on Modelling and Simulation (MODSIM 2015)*, November 29th to December 4th, Queensland, Australia.
7. Huete, A., Didan, K., Miura, T., Rodriguez, E. P., Gao, X., & Ferreira, L. G. (2002). Overview of the radiometric and biophysical performance of the MODIS vegetation indices. *Remote Sensing of Environment, 83*(1–2), 195–213.
8. India, M. O. A. (2009). *Manual for drought management*. Department of Agriculture and Cooperation, Ministry of Agriculture, Government of India.
9. Kogan, F. N. (1995). Droughts of the late 1980s in the United States as derived from NOAA polar-orbiting satellite data. *Bulletin of the American Meteorological Society, 76*(5), 655–668.
10. McKee, T. B., Doesken, N. J., & Kleist, J. (1993, January). The relationship of drought frequency and duration to time scales. In *Proceedings of the 8th Conference on Applied Climatology* (Vol. 17, No. 22, pp. 179–183).
11. Mishra, A. K., & Singh, V. P. (2010). A review of drought concepts. *Journal of Hydrology, 391*(1–2), 202–216.
12. Modarres, R. (2007). Streamflow drought time series forecasting. *Stochastic Environmental Research and Risk Assessment, 21*(3), 223–233.

13. Nalbantis, I., & Tsakiris, G. (2009). Assessment of hydrological drought revisited. *Water Resources Management, 23*(5), 881–897.
14. Narasimhan, B., & Srinivasan, R. (2005). Development and evaluation of Soil Moisture Deficit Index (SMDI) and Evapotranspiration Deficit Index (ETDI) for agricultural drought monitoring. *Agricultural and Forest Meteorology, 133*(1–4), 69–88.
15. Palmer, W. C. (1965). *Meteorological drought* (Vol. 30). US Department of Commerce, Weather Bureau.
16. Samra, J. S. (2004). *Review and analysis of drought monitoring, declaration and management in India* (Vol. 84). IWMI.
17. Svoboda, M., & Fuchs, B. (2016). *Handbook of drought indicators and indices.*
18. Van Rooy, M. P. (1965). A rainfall anomally index independent of time and space, notos.
19. Weghorst, K. M. (1996). *The Reclamation Drought Index: Guidelines and practical applications* (p. 6). Bureau of Reclamation.
20. Wilhite, D. A., & Glantz, M. H. (1985). Understanding: The drought phenomenon: The role of definitions. *Water International, 10*(3), 111–120.
21. World Meteorological Organization. (2012). Standardized precipitation index user guide. World Meteorological Organization (1090).

Evaluation of Rainfall Disaggregation Models for the Awash River Basin, Ethiopia

D. Ashenafi and S. Tripathi

Abstract Rainfall data at a fine temporal resolution (hourly or sub-hourly) are needed for most water resources assessment studies. It is possible to measure rainfall at high temporal resolution using recording rain gauges. The number and distribution of such instruments are still inadequate, particularly in developing countries such as Ethiopia where rainfall is usually measured manually at a daily time step. Hence, a scientific approach that disaggregates the available coarse temporal resolution rainfall data to finer resolution data is crucial. This study applied canonical and microcanonical multiplicative random cascade models to disaggregate daily rainfall into sub-hourly rainfall in Awash River basin, Ethiopia. The performance of the models in disaggregating daily rainfall to 45 min rainfall is analyzed for ten stations in the basin, for which the finer resolution rainfall data are available. The results show that two canonical models, namely the log-Poisson beta model and log-normal beta models outperformed other models in preserving the mean of 45-min rainfall. These two models also performed substantially better in capturing the variability in rainfall compared to other canonical models. The microcanonical model also performed well in capturing the variability in rainfall and representing the proportion of dry and wet periods. The results of this study demonstrate that multiplicative random cascade models can be a better alternative to ad hoc rainfall disaggregation models presently used in Ethiopia for design and assessment of water resources projects. The study also highlights some of the weaknesses of the multiplicative random cascade models that should be addressed in future studies.

Keywords Awash River basin · Rainfall disaggregation · Random cascade model · Intermittency

D. Ashenafi (✉) · S. Tripathi
Department of Civil Engineering, IIT Kanpur, Kanpur 208016, India
e-mail: ashenafi@iitk.ac.in

A. K. Dikshit et al. (eds.), *Innovative Trends in Hydrological and Environmental Systems*, Lecture Notes in Civil Engineering 234,
https://doi.org/10.1007/978-981-19-0304-5_19

1 Introduction

Water resources engineering projects, such as rainfall erosivity assessment and urban hydrological climate change assessment, require high resolution of temporal rainfall data. Precipitation data in Ethiopia are generally available at a daily or monthly time scale [1]. The number of meteorological stations with automatic gauges for recording high-resolution precipitation rates is relatively small. So, it requires a scientific approach to get the needed fine-resolution data by utilizing already available coarse-resolution data.

Many methods are proposed in the literature for disaggregating rainfall data into fine temporal scales from coarser scales. The uniform distribution is one of the simplest methods of disaggregation. The method assumes that the hourly rainfall intensity is constant throughout the day. Due to its limitation to represent rainfall variability, most researchers do not use this model to disaggregate rainfall. The other approach is the point process stochastic models that consider precipitation in time and space as random points. Models such as Neymann-Scott or Bartlett-Lewis [2] are good examples of point-rainfall simulations. The Poisson process controls the occurrence of rain events. The duration and intensity of the cells follow an exponential distribution. These models do not accurately describe physical processes since rain cells differ from the assumed randomly distributed state both spatially and temporally. Additionally, in low-level temporal aggregation, such models tend to produce unrealistic results that can negatively influence subsequent analyses [3, 4]. Also, Valdes et al. [4] noted that point process models have limitations in reproducing extreme value distributions and preserving correlation structures of rainfall sequences.

Among the various possible modeling approaches, the multiplicative cascade models, introduced first by Yaglom [5] appeared to be the most promising. Random multiplicative cascade models are derived from the multifractal theory. The model fragments the coarser timescale and the rainfall volume into smaller discrete time intervals and volumes until the desired time resolution is achieved [6]. Because of their link to the multifractal theory, multiplicative cascade models appear to be promising tools to simulate rainfall. Moreover, they are equally adapted for the simulation of rainfall in space and time [7]. This model is also parameter parsimonious and is theoretically easy to calibrate.

The rainfall disaggregation in the Ethiopian context has not been studied very well. Although few studies have been published, such as by Alem et al. [8], which generated maximum hourly rainfall data from existing daily rainfall records by regionalizing the rainfall stations of a river basin in Ethiopia. Without disaggregating the daily rainfall data, they extracted hourly maximum rainfall from maximum daily rainfall of annual duration. In this approach, the ratio of hourly maximum rainfall to daily maximum rainfall is modeled using site-specific parameters. Furthermore, Engida [1] used the modified Modified Bartlett–Lewis Rectangular Pulse Model (MBLRPM) to disaggregate daily rainfall data into hourly time series for the Nile basin Ethiopia. In his paper, Beyene et al. [9] utilized the Hyetos software package based on a point process approach to disaggregate the daily rainfall data in the Awash basin. Nevertheless, this approach has limitations in reproducing extreme events that are relevant to different water resource applications.

Due to lack of a scientifically rigorous method, the estimation of sub-daily rainfall data in Ethiopia is done through ad hoc techniques. The Ethiopian agricultural minister [10] has developed a design manual on drainage and irrigation systems which estimates the maximum hourly rainfall to be 50% of the daily rainfall for areas less than 5 sq. km. However, the accuracy of this method has been criticized. The Ethiopian Road Authority (ERA) has also developed a design manual for drainage structures. Nonetheless, the technique used in estimating the design flood for such infrastructures is dated and unreliable [11]. The ERA design manual divides Ethiopia into five regions based on intensity–duration–frequency (IDF) curves, which could be highly inaccurate due to the differences in rainfall across the country. The methods could overestimate or underestimate the maximum hourly rainfall magnitude [12]. In addition to updating existing methods, the effect of temporal rainfall variability also needs to be considered.

The aim of this study is to investigate the applicability of multiplicative random cascade models to disaggregate rainfall data for the Awash River basin in Ethiopia.

2 Research Methodology

2.1 Study Area

The Awash basin is the most vital river basin in Ethiopia. It covers an area of 110,000 km^2 with a population of 10.5 million [13]. The basin extents over Ethiopia's central highlands between the latitudes of 8° 30′ N and 12° 00′ N, and the longitudes of 38° 05′ E and 43° 25′ E (Fig. 1). The average annual rainfall ranges from 160 to 1600 mm, and the average annual temperature ranges from 20.8 to 29 °C [9]. Arid and semi-arid climate conditions prevail in most parts of the basin. In Fig. 1, the topography of the basin, which ranges from 259 up to 4055 m, is shown. According to Belete [14], the basin has a potential surface flow of 4.9 billion m^3.

2.2 Multiplicative Random Cascade Model

There are two types of multiplicative random cascade (RMC) model types. The first one is a canonical model which preserves mass on the average in disaggregation, and the second is a microcanonical model which preserves mass exactly in disaggregation. The basic theory of the random cascade models is briefly discussed here.

Multiple random cascade disaggregation methods are based on multifractal theory according to which an entity's structure can be derived from its parts. If objects can

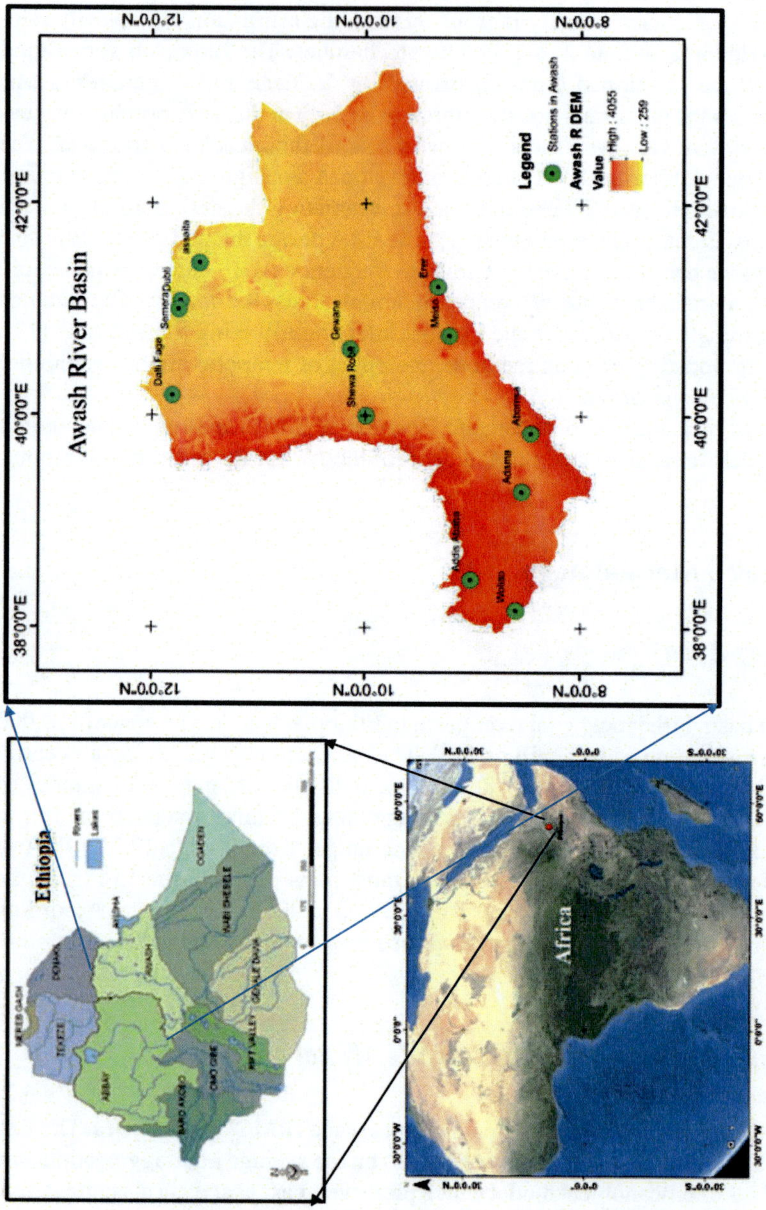

Fig. 1 Location and digital elevation model of the Awash River basin

be fragmented into minor copies of themselves, they are self-similar according to the scale-invariance theory [15]. Monofractal elements are homogeneous if they possess the same scaling properties, as defined by one exponent, called a singularity exponent. The cascade model belongs to the stochastic models' category [16]. Stochastic models are attractive in that they allow capturing the changes of rainfall over time scale.

2.2.1 The Canonical Model

The rainfall intensity at a coarser time scale is fragmented into two or more sub-intervals based on the branching number b through an iterative segmentation procedure. A set of weights W, called cascade generator, defines a partitioning mechanism for fragmenting data [17]. Considering the level n of the cascade, the number of available sub-intervals is b^n, and the corresponding cascade dimensionless spatial length scale is formulated as $\lambda_n = b^{-n}$. As described in Molnar [18], for the subdivided interval Δ_i in the cascade level n, the corresponding mass $\mu_n(\Delta_i)$ for i in range of 1 to n is:

$$\mu_n(\Delta_i) = m_0 \times \lambda_n \times \prod_{j-1}^{n} W_j(i) \quad (1)$$

Moment *S*caling

Determining W for a given rainfall data requires the estimation of sample moments. Moments of the order q are described as follows:

$$M_n(q) = \sum_{i=1}^{b^n} \mu_n^q(\Delta_n^i) \quad (2)$$

here q is the moment order ($q \geq 0$). As the value of n increases, the convergence of the sample moments to the ensemble moments is expected. Despite this $\lim_{n\to\infty} M_n(q) = 0$, the convergence rate of the moments is calculated as a function of the scale [18]. The Mandelbrot-Kahane-Peyriere (MKP) function describes how the scaling behavior follows a log–log–linear function [15, 19].

$$\chi_b(q) = 1 - q + \log_b E(W^q) \quad (3)$$

The scaling of the sample moments is used to calculate $\tau(q)$, and assuming a distribution for the cascade generator W, the parameters of the cascade model can then be calculated. Table 1 presents the canonical models used in this study: log-normal beta model, uniform model, log-Poisson model, log-Poisson beta model, and

Table 1 Canonical models and corresponding MKP function

Canonical models	$\chi(q)$ MKP function
Log-Poisson model $W = A\beta^N$ $P(N = M) = \frac{c^m e^{-c}}{m!}$	$\chi(q) = (1-q) + \frac{c\times(q\times(1-\beta)+)-(\beta^q-1)}{\ln b}$
Log-Poisson beta model $W = B \times Z$ $P(B = 0) = 1 - b^{-\beta}$ $\&P(B = b^\beta) = b^{-\beta}$ $Z = A\beta^N$ $P(N = M) = \frac{c^m e^{-c}}{m!}$	$\chi(q) = (\beta - 1) \times (1-q) + \frac{c\times(q\times(1-\beta)+)-(\beta^q-1)}{\ln b}$
Log-normal beta model $W = B \times Y$ $P(B = 0) = 1 - b^{-\beta}$ $\&P(B = b^\beta) = b^{-\beta}$ $Y = b^{\gamma+\sigma X}$ where X is normal $N(0, 1)$	$\chi(q) = (\beta - 1) \times (q-1) + \frac{\sigma^2\times\ln b\times(q^2-q)}{2}$
Uniform model $P(W = w) = \frac{1}{b}$ $b \in [0, b]$	$\chi(q) = 1 - q + \frac{\ln(\frac{b^q}{q+1})}{\ln b}$
Beta model $P(B = 0) = 1 - b^{-\beta}$ $\&P(B = b^\beta) = b^{-\beta}$	$\chi(q) = (\beta - 1) \times (q-1)$

beta model. These models differ in the methodology used to estimate the probability distribution functions representing W and in their mass conservation process [6].

Researchers have used different models for disaggregation of rainfall in various parts of the world. Kumar et al. [20] compared the performance of the log-normal beta model, the uniform model, and the log-Poisson model in the context of the South Indian climate. In another approach, two different distributions were used to represent intermittency and positive rainfall [21]. A similar approach was used in [18] where the beta and log-normal distributions were combined. They provide a detailed explanation of the log-normal beta model. No similar method has been used so far to enhance the performance of another model. The cascade generator generates a positive value for the log-Poisson model [22]. Thus, zero rainfall cannot be simulated using this model. Different researchers have attempted to solve this problem by forcing the generated rainfall threshold to zero [20, 22, 23]. However, the result obtained by this method was not satisfactory. Hence, we proposed improving the model's intermittency performance by combining it with the beta model. The following is a discussion for the proposed log-Poisson model.

Log-*P*oisson Beta Model

Log-Poisson beta model cascade generator W is composed of two independent random variables, $W = BY$, B being a random variable for determining the intermittency in rainfall and Y is the weight for assigning a rainfall amount in the interval. Based on the Bernoulli probability distribution, the interval between rain and no rain is determined for a rainfall process.

$$P(B = 0) = 1 - b^{-c} \quad \text{and} \quad P\left(B = b^{c}\right) = b^{-c} \tag{4}$$

If we take $Y = \mathcal{A}\beta^{\mathrm{N}}$ where $P(N = m) = \frac{c^m e^{-c}}{m!}$, the condition that the expectation of the random variable Y equal to 1,$E(Y) = 1$, gives $\mathcal{A} = \mathrm{e}^{c(1-\beta)}$.

The theoretical expression of $\chi(q)$ for the log-Poisson model will then be

$$\chi_b(q) = 1 - q + \frac{c[q(1-\beta) + (\beta^q - 1)]}{\ln b} \tag{5}$$

The derivation of the model is provided at the appendix part of the document.

2.2.2 Microcanonical Model

Multiplicative discrete random cascade (MDRC) models confirm that the disaggregated mass remains constant within each step of disaggregation. To each cascade level's weights, an independent probability distribution function can be fitted [18]. In each step of the disaggregation process, the weights are added to one to guarantee the conservation of mass. It is assumed that the weights are independent variables with the same distribution. The weight requires that at level n, at every subdivision into b subintervals, the following holds.

$$\sum_{k=1}^{b} W_n(b(i-1)+k) = 1 \quad \text{for } i = 1, 2, \ldots, b^{n-1} \tag{6}$$

These weights can also be referred to as breakdown coefficients. To account for rainfall intermittency, the model's weights are allowed to vary between 0 and 1. Based on each disaggregation level and each non-zero rainfall volume R, the corresponding rainfall volumes R_1 and R_2 in the lower disaggregation level are calculated, along with respective weights W_1 and W_2.

$$W_{i,1} = \frac{R_{\lambda,t}}{R_{\lambda-1,t-\Delta t}} \tag{7}$$

where.

$\lambda =$ the cascade level,

t = the measurement time in level λ, and

Δt = the time difference between level λ and $\lambda - 1$.

There are two parameters in the microcanonical model, P_{01} and α, related to the intermittency and variability of the generator W, respectively. P_{01} is the probability that the disaggregated mass will fall into one of the sub-intervals and is defined as follows.

$$P_{01} = p(W_1 = 0) + p(W_1 = 1) \tag{8}$$

A second possibility is when the rainfall volume is divided among the sub-intervals. In this scenario, the cascade generator is assumed to follow a certain probability distribution function. To represent the symmetric disaggregation of the weights, we used an asymmetric beta distribution function with a single parameter α [18].

$$f(w) = \frac{1}{B(\alpha)} w^{\alpha-1}(1-w)^{\alpha-1} \tag{9}$$

where

$B(\alpha)$ is the beta function.

The procedure of the microcanonical rainfall disaggregation is presented using a flowchart in Fig. 2.

Summary of the rainfall disaggregation process

1. Data on observed rainfall with a 15-min resolution was collected for the entire station.
2. Next, the observed rainfall data are aggregated to form 45 min rainfall data, which are then aggregated further to form daily rainfall data with $b = 2$.
3. We then calculated the probability of intermittency for the whole scale.
4. Using the moment scaling characteristics, the canonical models' parameters are estimated.
5. Using the intermittency and breakdown coefficients in the observed record, the parameters of the microcanonical random cascade model are estimated.
6. By disaggregating the complete record of observed daily rainfall, we generated 1000 series of 45-min rainfall based on these parameters.
7. Finally, the performance of the models was evaluated.

2.3 *Performance Evaluation Model*

Performance evaluation of the disaggregation models was carried out by utilizing three performance measures, namely root mean square error (RMSE), Nash–Sutcliffe efficiency (NSE), and percent bias (Pbias). The performance measures are evaluated by comparing observed (O_i) and model simulated (S_i) rainfall at 45 min interval.

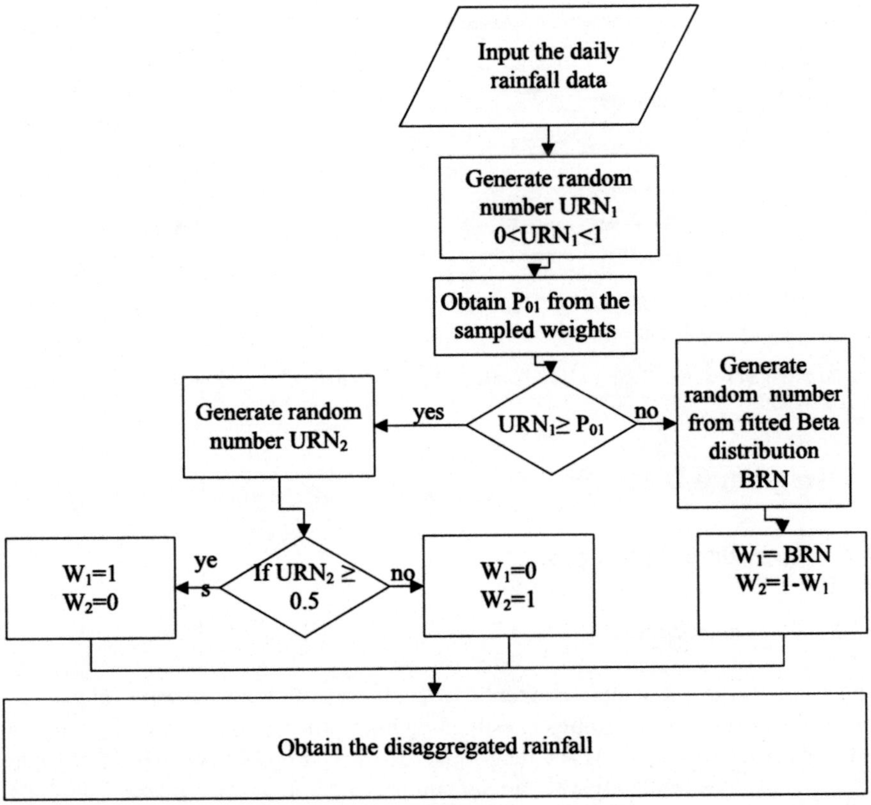

Fig. 2 Flow chart of microcanonical model. BRN: random number generated from the beta distribution; URN_1: random number generated at first stage from uniform distribution; URN_2: random number generated at second stage from uniform distribution; W_1: weight generator for the first half of the time interval; W_2: weight generator the second half of the time interval

$$\text{RMSE} = \sqrt{\frac{\sum_{i=1}^{n}(S_i - O_i)^2}{n}} \tag{10}$$

$$\text{NSE} = 1 - \frac{\sum_{i=1}^{n}(S_i - O_i)^2}{\sum_{i=1}^{n}(O_i - \overline{O})^2} \tag{11}$$

$$\text{where } \overline{O} = \frac{\sum_{i=1}^{n} O_i}{n} \tag{12}$$

$$\text{Pbias} = \frac{\sum_{i=1}^{n}(S_i - O_i)}{\sum_{i=1}^{n} O_i} \times 100 \tag{13}$$

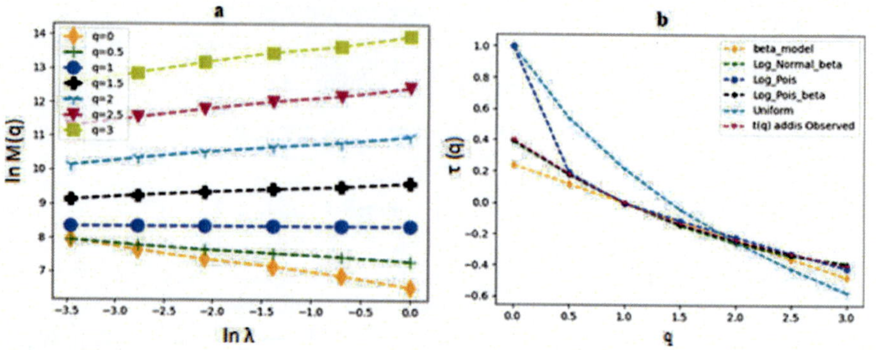

Fig. 3 Moment scaling relationships (**a**) and the MKP function (**b**) for the observed data at Addis Ababa station

3 Result and Discussion

3.1 Parameter Estimation

3.1.1 Canonical Model

As per Eq. 2, the qth sample moments, $(M_n(q)$, for each scale λ_n were calculated. To avoid statistical uncertainties in estimating high-order moments [24], only lower order moments $(0 < q < 3)$ are analyzed. Next, a relationship between $M_n(q)$ and λ_n was developed in the logarithmic domain for each qth order moment. The slope of each fitted line, $t(q)$, was determined by linear regression. Figure 3a displays an example of moment scaling relationships and Fig. 3b presents the MKP function for the observed data at Addis Ababa station.

The parameters of the canonical models are determined by a least square fit of the MKP function. These parameters are estimated using entire data and are considered constant. The model using these estimates will be referred to as the canonical model with constant parameters. Table 2 presents the parameters of log-Poisson beta model values for rainfall series for all the ten stations in the study area (Fig. 4; Table 3).

3.1.2 Microcanonical Model

Ethiopia's climate varies mostly with altitude. Accordingly, for this study, we chose two stations that will be representative of the high and low altitudes of the basin. Aside from the altitude, the length of available observational data in each station was also considered when selecting the stations. Based on these criteria, Addis Ababa and Meiso stations were chosen for model evaluation.

For the microcanonical model, the probability P_{01} was calculated as a frequency of occurrence. The results of the P_{01} at Meiso and Addis Ababa stations for five cascade

Table 2 Parameters of log-Poisson beta model for station in Awash River basin

Parameters	Shewa_Robit	Erere	Meiso	Addis Ababa	Woliso	Elidar	Dallifage	Dubuty	Gewane	Semara
Beta (β)	0.58	0.69	0.67	0.60	0.61	0.68	0.65	0.76	0.69	0.71
c	0.36	0.43	0.82	0.67	0.72	0.50	0.88	0.48	0.30	0.0001

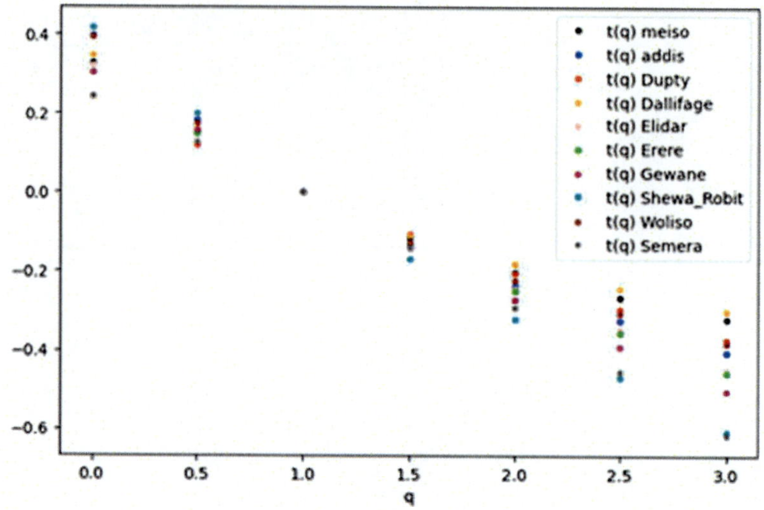

Fig. 4 Slope of moment scaling relationship at stations at Awash River Basin ($\tau(q)$)

Table 3 Parameters of canonical models at Addis Ababa station

Models	Beta model	Log-normal beta model		Log-Poisson model		Log-Poisson beta model		Uniform model
	Beta (β)	Beta (β)	c	Beta (β)	c	Beta (β)	c	b
Parameters	0.76	0.61	0.19	0.56	0.01	0.60	0.67	2.41

levels are presented in Table 4. The P_{01} parameter does not vary significantly between the two stations. As the temporal resolution decreases, P_{01} declines, except at the first level. The decline of P_{01} as we go to a finer time scale is in agreement with [18, 25]. The maximum likelihood estimates of α, which is the parameter of the model, for both the stations are also shown in Table 4. Based on the variation in the parameter

Table 4 Parameters of the microcanonical model at Meiso (left) and Addis Ababa (right) stations

Meiso station			Addis station		
Level	Parameters		Level	Parameters	
	P_{01}	Alpha (α)		P_{01}	Alpha (α)
90 min to 45 min (λ_5)	0.32	1.03	90 to 45 min (λ_5)	0.31	1.09
3 h to 90 min (λ_4)	0.36	0.74	3 h to 90 min (λ_4)	0.34	0.87
6 to 3 h (λ_3)	0.40	0.66	6 to 3 h (λ_3)	0.39	0.54
12 to 6 h (λ_2)	0.45	0.57	12 to 6 h (λ_2)	0.44	0.46
24 to 12 h (λ_1)	0.33	0.51	24 to 12 h (λ_1)	0.30	0.63

Table 5 Performance measures for microcanonical model at Addis Ababa and Meiso stations

Statistical parameters	Stations	
	Addis Ababa	Meiso
RMSE	0.54	0.75
Pbias	≈ 0	≈ 0
NSE	0.28	0.24

Table 6 Performance measures for various canonical models at Addis Ababa and Meiso stations

Statistical parameters	Addis Ababa				
	Beta model	Uniform model	Log-Poisson model	Log-Normal beta model	Log-Poisson model
RMSE (mm)	0.95	1.109	0.93	0.9	
Pbias	12.32	60.71	−0.06	−2.179	
NSE	−0.77	−0.456	0.93	−0.997	
Statistical parameters	Meiso				
	Beta model	Uniform model	Log-Poisson model	Log-normal beta model	Log-Poisson model
RMSE (mm)	1.29	1.56	1.17	1.16	1.29
Pbias	9.58	65.42	−6.09	−11.41	9.58
NSE	−0.60	−0.29	−0.99	−0.88	−0.60

alpha in Table 4, it appears that the breakdown coefficient distribution varies as we move from one timescale to another.

3.2 Evaluation of the Models

The performance of canonical models for Addis Ababa and Meiso stations was first evaluated using RMSE, NSE, and Pbias measures. The results are presented in Table 5 (Table 6).

The performance measures indicate that the simulated result does not fit the observed data. The model did well in representing the deviation (see Fig. 6), but its performance in terms of performance measures is poor.

3.3 Rainfall Depth and Distributions

The first test of the models in the generation of 45-min rainfall data from daily rainfall observations is whether they preserve the total rainfall depth at all temporal scales.

This is a test for the canonical models only because the microcanonical models, by virtue of their formulation, exactly preserve the total rainfall depth in disaggregation. The results for the log-Poisson beta and log-normal beta models were better compared to other models. The performance of the log-Poisson model in preserving the mean (Fig. 5) and standard deviation (Fig. 6) of rainfall series at 45-min time scale is better than all other models for Meiso Station.

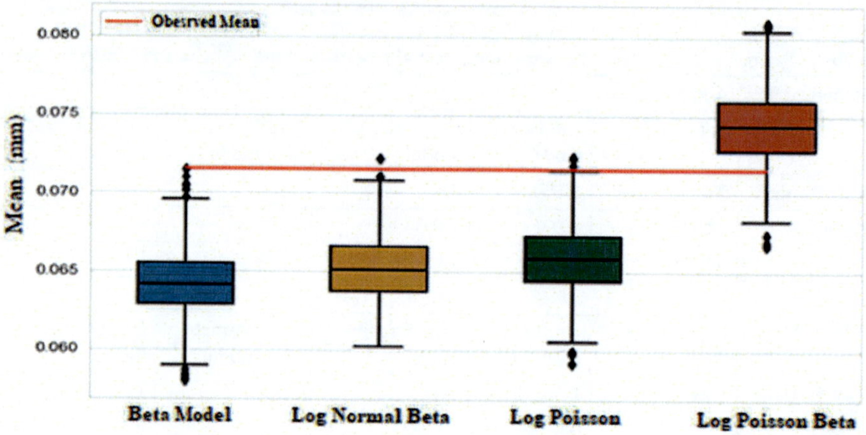

Fig. 5 Mean of 45-min disaggregated rainfall from 1000 simulations of different canonical models at Meiso station is shown using box plots. The mean of observed 45-min rainfall is represented by the solid red line

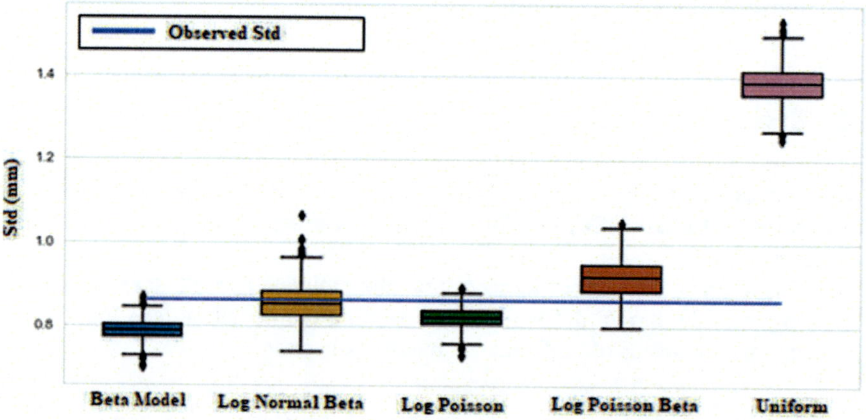

Fig. 6 Standard deviation (Std) of 45-min disaggregated rainfall from 1000 simulations of different canonical models at Meiso station is shown using box plots. The standard deviation of observed 45-min rainfall is represented by the solid red line

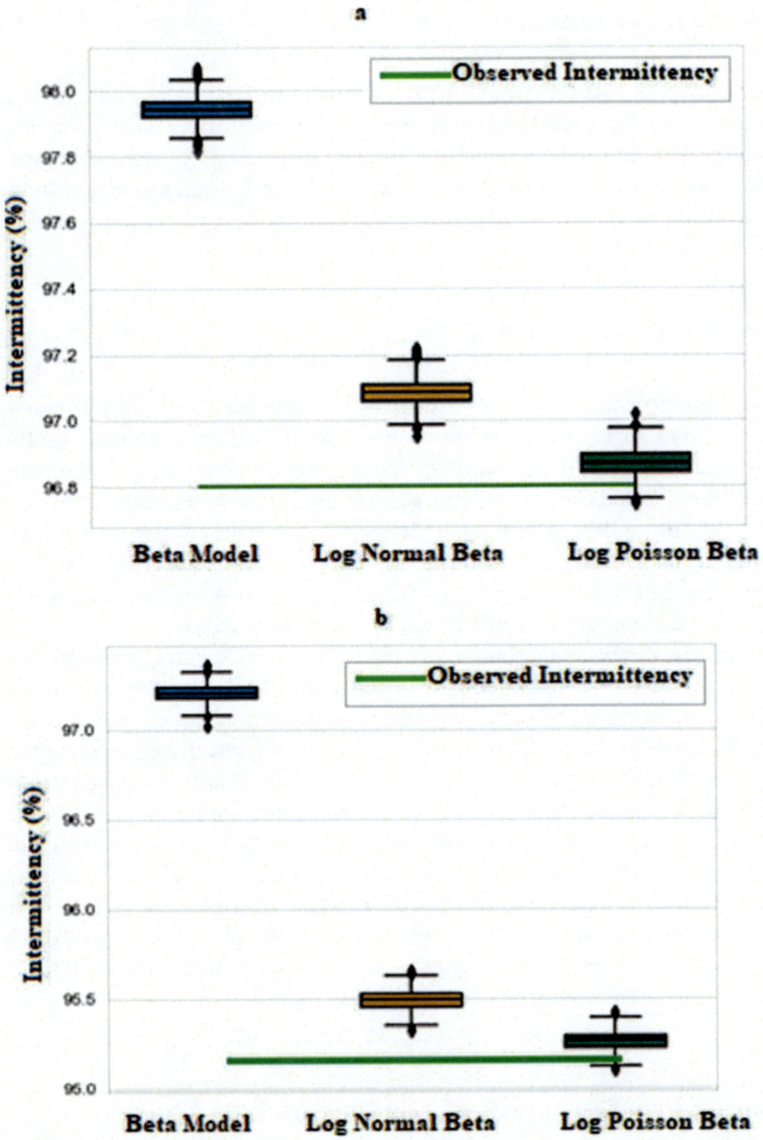

Fig. 7 Percentage intermittency of 45-min disaggregated rainfall from 1000 simulations of different canonical models at **a** Meiso station and **b** Addis Ababa stations is shown using box plot. The percentage intermittence of observed 45-min rainfall is represented by the solid green line

3.4 Dry Proportions

Figure 7 shows the performance of the canonical models in reproducing the dry proportion of the observed 45-min rainfall at Meiso and Addis Ababa stations. The log-Poisson beta and log-normal beta models, in general, produce disaggregated rainfall series with proper rainfall intermittency at the 45-min resolution for all the stations.

4 Summary and Conclusions

A discrete multiplicative random cascade model was examined for disaggregation of daily rainfall series into sub-hourly rainfall in Ethiopian climate conditions. Five canonical models—beta, log-normal beta, log-Poisson, log-Poisson beta, and uniform—were tested for ten stations, and the results for two stations are described in detail. In both Addis Ababa and Meiso, the log-Poisson beta and log-normal beta models performed better in preserving the mean and standard deviation of the observed 45-min rainfall. The two models also showed better performance in dry proportion of the observed 45-min rainfall at both the stations.

In this study, the beta distribution is used for the microcanonical model. The P_{01} parameter does not vary significantly between Addis Ababa and Meiso stations. The P_{01} values decreased with decreasing temporal resolution, except at the first level. The results showed that the microcanonical model with beta distributions performed well in capturing the variation in the sub-hourly rainfall and the dry proportion of the observed 45-min rainfall at Addis Ababa and Meiso station.

Overall, the study showed that the canonical models, log-Poisson beta and log-normal beta, and microcanonical model with beta distribution can be effective in disaggregating daily rainfall to sub-hourly rainfall. However, estimation of model parameters from limited observation was found challenging. Future studies should explore the utility of remote sensing data, which is available at sub-daily time step, to estimate the parameters of the disaggregation models.

Appendix: Proof for MKP of Log Poisson beta Model

The MKP function provides important information regarding the distribution of cascades generator W, determining the scaling properties of rainfall. It is given by

$$\chi_b(q) = 1 - q + \log E\left(W^q\right) \tag{14}$$

For log-Poisson model the $E(W^q)$ is derived as follows

$$W = A\beta^N \tag{15}$$

$$P(N = m) = \frac{c^m e^{-c}}{m!} \tag{16}$$

Therefore, $E[W^q]$ will be

$$E\left[W^q\right] = \sum_{N=0}^{\infty}\left(A\beta^N\right)^q \frac{c^N e^{-c}}{N!} \tag{17}$$

which is simplified to

$$E\left[W^q\right] = e^{-c}\left(A^q\right)\sum_{N=0}^{\infty}\frac{\left(\beta^q c\right)^N}{N!} \tag{18}$$

It is known that $\sum_{N=0}^{\infty}\frac{(\beta^q c)^N}{N!}$ is equal to $e^{(\beta^q)c}$.
Therefore, Eq. 18 simplifies to $E[W^q] = e^{-c}(A^q)e^{(\beta^q)c}$

or

$$E\left[W^q\right] = A^q e^{c((\beta^q)-1)} \tag{19}$$

We can use the condition that the expectation of W is equal to 1, $E[W] = 1$, to determine the expression for A.
Therefore, Eq. 18 simplifies to $E[W^q] = e^{-c}(A^q)e^{(\beta^q)c}$

$$E[W] = \sum_{N=0}^{\infty}\left(A\beta^N\right)\frac{c^N e^{-c}}{N!} \tag{20}$$

$$E\left[W^q\right] = 1 = e^{-c} \times A\sum_{N=0}^{\infty}\frac{\left(c\beta^N\right)}{N!}$$

$$\sum_{N=0}^{\infty})\frac{c\beta^N}{N!} \text{ is equal to } e^{(\beta)c}$$

Therefore, A is given by

$$A = e^{c(1-\beta)} \tag{21}$$

combining Eq. 19 with 21 and simplifying it yields

$$E\left[W^q\right] = e^{c[q(1-\beta)+(\beta^q-1)]} \tag{22}$$

Therefore, $\log_b E[W^q]$

$$\log_b E\left[W^q\right] = \frac{c[q(1-\beta)+(\beta^q-1)\}}{\ln b} \tag{23}$$

Therefore, the MKP function will be

$$\chi_b(q) = 1 - q + \frac{c[q(1-\beta)+(\beta^q-1)\}}{\ln b} \tag{24}$$

References

1. Engida, A. N., & Esteves, M. (2011). Characterization and disaggregation of daily rainfall in the Upper Blue Nile Basin in Ethiopia. *Journal of Hydrology, 399*, 226–234. https://doi.org/10.1016/j.jhydrol.2011.01.001
2. Rodriguez-Iturbe, I., Cox, D. R., & Isham, V. (1987). Some models for rainfall based on stochastic point processes. *Proceedings of the Royal Society of London. A. Mathematical and Physical Sciences. The Royal Society London*, 269–288.
3. Northrop, P. J., & Stone, T. M. (2005). A point process model for rainfall with truncated Gaussian rain cells (pp. 1–15). Dep Stat Sci Univ Coll London.
4. Valdes, J. B., Rodriguez-Iturbe, I., & Gupta, V. K. (1985). Approximations of temporal rainfall from a multidimensional model. *Water Resources Research, 21*, 1259–1270.
5. Yaglom, A. M. (1966). *The influence of fluctuations in energy dissipation on the shape of turbulence characteristics in the inertial interval* (pp. 26–29). Sov. Phys. Dokl.
6. McIntyre, N., Shi, M., & Onof, C. (2016). Incorporating parameter dependencies into temporal downscaling of extreme rainfall using a random cascade approach. *Journal of Hydrology, 542*, 896–912. https://doi.org/10.1016/j.jhydrol.2016.09.057
7. Thomas Mark, O. (1995). *Modeling space-time rainfall at the mesoscale using random cascades* (PhD Thesis).
8. Alem, A. M., Tilahun, S. A., Moges, M. A., Melesse, A. M., Beyene, T. D., Moges, M. A., Tilahun, S. A., Over, T. M., Yaglom, A. M., McIntyre, N., Shi, M., Onof, C., Valdes, J. B., Rodriguez-Iturbe, I., & Gupta, V. K. (1985). A regional hourly maximum rainfall extraction method for part of Upper Blue Nile Basin, Ethiopia. *Soviet Physics—Doklady, 542*, 26–29.
9. Beyene, T. D., Moges, M. A., & Tilahun, S. A. (2018). Development of rainfall disaggregation model in the Awash River basin, Ethiopia. In *International Conference on Advances of Science and Technology* (pp. 50–64).
10. Agriculture, M. O. (2001). *Irrigation and drainage design manual*. Ministry of Agriculture.
11. Ethiopian Road Authority. (2013). *Drainage design manual*. Federal Democratic Republic of Ethiopia
12. Pui, A., Sharma, A., Mehrotra, R., Sivakumar, B., & Jeremiah, E. (2012). A comparison of alternatives for daily to sub-daily rainfall disaggregation. *Journal of Hydrology, 470*, 138–157.
13. Taddese, G., Sonder, K., & Peden, D. (2009). *The water of the Awash River basin: A future challenge to Ethiopia*. ILRI.
14. Berhanu, B., Seleshi, Y., Melesse, A. M. (2014). Surface water and groundwater resources of Ethiopia: Potentials and challenges of water resources development. In: *Nile River Basin* (pp. 97–117). Springer
15. Mandelbrot, B. B. (1974). Intermittent turbulence in self-similar cascades: Divergence of high moments and dimension of the carrier. *Journal of Fluid Mechanics, 62*, 331–358.

16. McIntyre, N., & Bárdossy, A. (2017). Using probable maximum precipitation to bound the disaggregation of rainfall. *Water, 9*, 496.
17. Olsson, J. (1998). Evaluation of a scaling cascade model for temporal rainfall disaggregation. *Hydrology and Earth System Sciences, 2*, 19–30. https://doi.org/10.5194/hess-2-19-1998
18. Molnar, P., & Burlando, P. (2005). Preservation of rainfall properties in stochastic disaggregation by a simple random cascade model. *Atmospheric Research, 77*, 137–151. https://doi.org/10.1016/j.atmosres.2004.10.024
19. Kahane, J.-P., & Peyriere, J. (1976). Sur certaines martingales de Benoit Mandelbrot. *Advances in Mathematics (New York), 22*, 131–145.
20. Kumar, S., Sekhar, M., & Reddy, D. V. (2009). Improving the disaggregation of daily rainfall into hourly rainfall using hourly soil moisture. *IAHS Publication, 331*, 236.
21. Over, T. M., & Gupta, V. K. (1994). Statistical analysis of mesoscale rainfall: Dependence of a random cascade generator on large-scale forcing. *Journal of Applied Meteorology, 33*, 1526–1542.
22. Saad, S. M., Jemain, A. A., & Ismail, N. (2018). Temporal rainfall disaggregation by a simple random cascade model. *Jurnal Teknologi, 80*.
23. Onof, C., & Arnbjerg-Nielsen, K. (2009). Quantification of anticipated future changes in high resolution design rainfall for urban areas. *Atmospheric Research, 92*, 350–363.
24. de Lima, M. I. P. (1998). *Multifractals and the temporal structure of rainfall*. De Lima
25. El Hachem, A. (2019). *Application of a cascade model for temporal rainfall disaggregation* (p. 107). https://doi.org/10.13140/RG.2.2.33936.46084

Experimental Study on Pressure Flow Due to Vertical Contraction

Sofi Aamir Majid and Shivam Tripathi

Abstract Understanding the flow in open channels is important for the design of the bridges and other hydraulic structures. Flow in open channels is characterized by a constant pressure boundary. In extreme floods, the flow may also be subjected to a vertical contraction when the water surface rises above the bottom chord of the bridge. In the vertical contraction, a transition from atmospheric flow to the pressure flow occurs. The velocity gradient and the shear stresses are considerably modified in the contraction and can cause scouring or add to the already present scour in the contraction. Numerous studies have tried to establish empirical relations between the magnitude of scour depth and the independent variables such as approach flow depth, approach flow velocity, bridge opening, length of bridge and sediment size. No experimental or numerical study has thoroughly investigated the flow field inside and close to the contraction, which can be very useful in understanding the scouring process. In this paper, we present a laboratory experimental study on the vertical contraction using the particle image velocimetry (PIV). The methodology implemented in the acquisition and processing of the PIV data is reported. The results show that a non-uniform flow field is developed under the contraction and a stagnation zone is created under the bridge. The stagnation zone further decreases the vertical passage under the bridge. Besides, a recirculation zone is observed downstream of the contraction. The recirculation zone greatly modifies the velocity profile from the classical velocity profile observed in open channels.

Keywords Pressure flow · Vertical contraction · Stagnation zone · Recirculation zone · Particle image velocimetry · Open channel flow

1 Introduction

With the advancement in the numerical and experimental techniques, the understanding of the nature of the flow in open channels has improved significantly in the

S. A. Majid (✉) · S. Tripathi
Indian Institute of Technology Kanpur, Kanpur 208016, India
e-mail: aamirm@iitk.ac.in

A. K. Dikshit et al. (eds.), *Innovative Trends in Hydrological and Environmental Systems*, Lecture Notes in Civil Engineering 234,
https://doi.org/10.1007/978-981-19-0304-5_20

past century [1–3]. A good amount of literature is also available on the sediment transport in open channels [4]. Together they provide a scientific basis for hydraulic design of bridges. Nevertheless, most of the bridges have been designed assuming open channel flow condition, i.e., the flow remains under the bridge. However, with increase in magnitude and frequency of rainfall events due climate change scenarios, the existing bridges might face the condition of submergence and the flow may get subjected to the vertical contraction.

A schematic of the vertical contraction is shown in Fig. 1. The approach flow having a depth-averaged velocity of V_a and flow depth of H_a is subjected to a vertical contraction at the bridge site. The available vertical flow passage decreases to H_b, the depth-averaged flow velocity increases to V_b, and the flow itself changes from open channel flow to pressure flow. The velocity field is significantly modified, and the velocity gradients close to the bed increase. On mobile bed channels, this condition can initiate scouring on previously uneroded bed or can add to the scouring already present in the contraction.

Shan et al. studied the velocity field inside the vertical contraction with the help of particle image velocimetry and numerical simulations [5]. They reported that a stagnation zone is formed under the bridge. The depth of the stagnation zone increases from the upstream edge of the contraction toward the downstream. They developed an empirical model to estimate the maximum stagnation zone thickness as a function of the flow geometry and flow properties. Carnacina et al. studied the characteristics of flow structure around cylindrical bridge piers in pressure flow conditions using acoustic Doppler velocimetry (ADV) [6]. They reported that the maximum average velocity components are accelerated due to vertical contraction and this acceleration is further increased by the development of boundary layer under the bridge deck. They also reported a significant increase in the magnitude of bed shear stresses inside the vertical contraction. Yoon et al. studied the time-averaged turbulent velocity field near the bridge abutments in pressure flow conditions due to vertical contraction using an ADV [7]. They reported that pressure flow imparts additional flow acceleration close to the abutment, resulting in the increased scour near it.

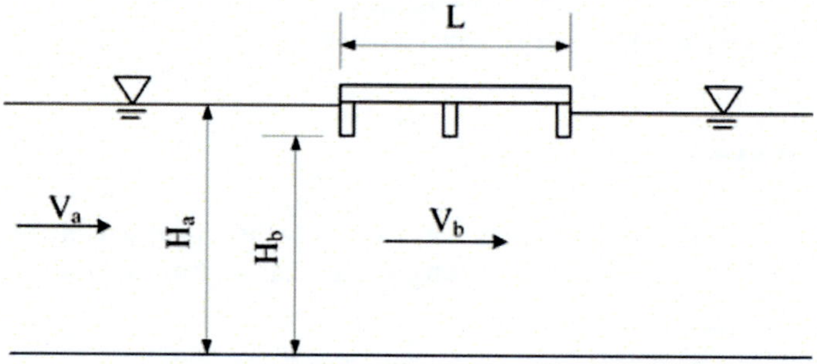

Fig. 1 Schematic of vertical contraction to the flow

In the present study, flow over a rigid bed with a vertical contraction is studied using particle image velocimetry (PIV). The various steps in the PIV data acquisition and processing are described. The data is validated against the classical velocity profile model for open channel flows. Next, the results of the vertical contraction are presented and discussed. The present study explores the kinematics and dynamics of the bulk flow as well as flow close to the bed. The study may be extended to understand the effects of the approach flow variables and flow geometry on the flow properties inside the contraction.

2 Experimental Setup

2.1 Laboratory Flume and Model Bridges

The experiments are carried in the hydraulics laboratory of IIT Kanpur on a flume of length 15 m. The width and depth of the flume are respectively 0.3 m and 0.45 m. A schematic of the flume is shown in Fig. 2a. The flow is recirculated with the help of a pump. An electromagnetic flow meter is used to measure the discharge and can be accordingly controlled using a control panel. The flow depth inside the flume is adjusted using the downstream gate. The side walls of the flume are made of hardened glass to ensure visibility.

A model bridge is used to create the vertical contraction. It is also made of glass and measures 0.15 m (bridge length, L) in the flow direction, i.e., along the length of the flume. The bridge is inserted from the top of the flume and is lowered to the required position such that the bottom of the bridge is 0.072 m above the flume bed. The sides of the model bridge touching the flume wall are covered with rubber. Once the bridge is at the required position, the side walls are tightened against the bridge to fix the bridge in place.

In addition to the vertical contraction experiment, one simple open channel flow experiment with the same flow parameters as that of vertical contraction is performed. In both the experiments, the approach flow depth, H_a, is kept as 0.10 m and the approach flow velocity, V_a, 0.20 m/s.

2.2 Components of PIV

The PIV setup consists of a laser, a camera, and a synchronizer. Nd:YAG laser of power 200 mJ and frequency 532 nm is used to illuminate the flow. The laser optics placed at the top of the flume, is used to spread the laser beam along the length of the flume. A laser arm is used to transmit the laser beam from the laser to the laser optics. A high-speed camera is placed in front of the flume to capture the PIV images.

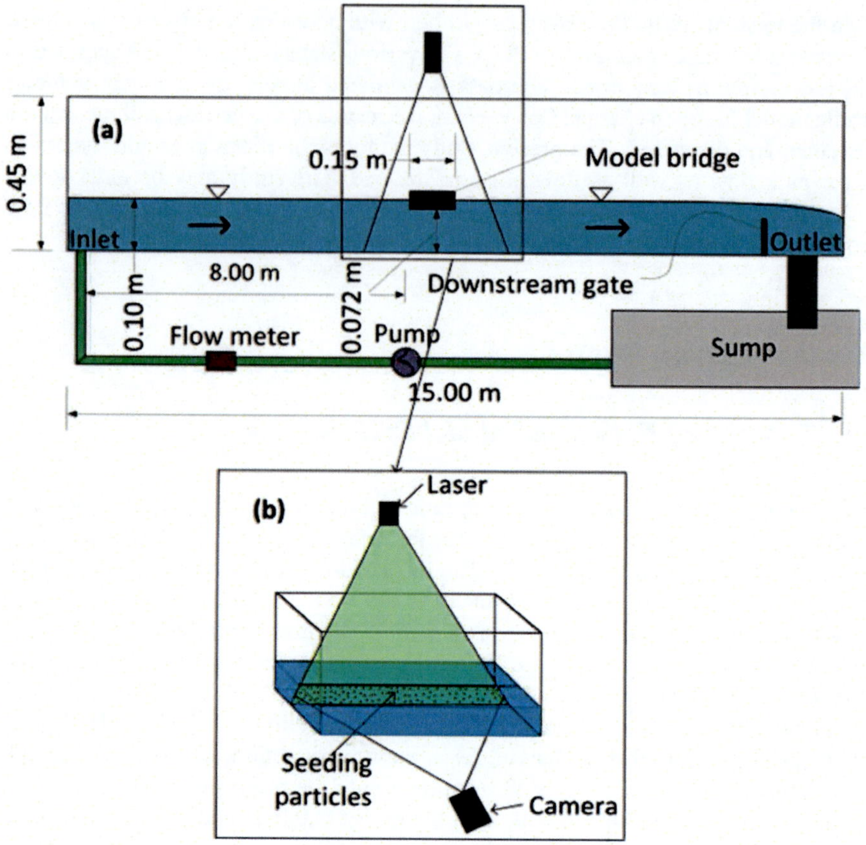

Fig. 2 **a** Line diagram of the flume and the model bridge used in the experiment, **b** image acquisition process

The images are dumped to a local computer. A six-channel delay generator is used to synchronize the laser and the camera.

2.3 *Image Acquisition and Calibration*

The flow was seeded with polyamide particles of diameter 10 μm and specific gravity 1.05. For each experiment, 500 pairs of images are acquired (Fig. 2b). The time delay between two successive pulses is set as 2 ms and 10 pairs of images are recorded per second. A calibration image is taken by putting a metric scale in the flow. The calibration image is used to convert the photographic pixels into real world distances.

3 Processing of PIV Images

A MATLAB-based open-source toolbox PIVlab [8] is used for processing the images. The PIV images are processed in the following steps to obtain the instantaneous velocity fields.

- Initially a region of interest (ROI) as per the problem is chosen. The subsequent processing is applied to the ROI only.
- The ROI is divided into small interrogation windows. The size of the window determines the spatial resolution of the velocity vectors. For each window a velocity vector is obtained. Thus, the velocity vector represented the average motion of the fluid in that window. Besides, the seeding particle density is also important in the selection of the window size. Ideally, the window size should be such that the number of particles in the window is around 10 [8]. In the present study, the interrogation windows overlap 50% with each other. This is done to increase the spatial resolution of the velocity vector field.
- The images are preprocessed to enhance the image quality. PIVlab offers multiple preprocessing filters which include contrast limited adaptive histogram equalization (CLAHE), high pass filter, intensity capping filter and Gaussian low pass filter. For the detailed applicability of the filters for PIV images, readers are referred to [8]. In the present study, CLAHE filter is applied to the images.
- PIV works on the principle of cross-correlation. For two successive frames, a cross-correlation algorithm (CCA) is used to find the average movement of the particles. The algorithm is applied individually for each interrogation window and generates a single velocity vector for each window (Fig. 3). This step is computationally most exhaustive step in the PIV analysis. Two algorithms are available in the PIVlab to perform the cross-correlation. Direct cross-correlation (DCC) and fast Fourier transform (FFT) cross-correlation. The later reduces the computation time and is performed in multiple passes to improve the results. A detailed discussion on the cross-correlation in PIV is available in [9]. In the present study, FFT cross-correlation is done in 3 passes starting with a window size 128 $\times$ 128 pixels. The final size of the interrogation was 32 $\times$ 32 pixels. For a flow depth of 0.1 m, approximately 50 vertical vectors are obtained, the lowest vector about 2 mm from the bed.

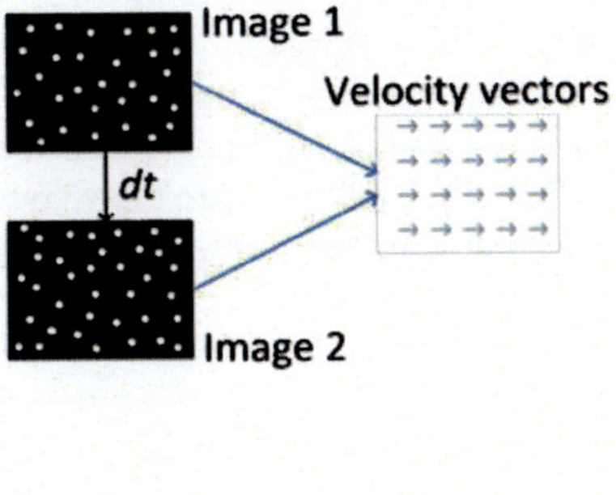

Fig. 3 Processing of PIV images

- The pixel location of the cross-correlation peak gives the mean displacement in the window and thus the accuracy of the results is 0.5 pixels. To improve the accuracy, sub-pixel interpolation is carried out. PIVlab offers to choose between Gauss 3-point estimation and Gaussian 2D estimation [8]. In the present study, Gauss 2D is used. In this method, a 2D surface is fitted to the neighborhood of the correlation peak and the maximum of the surface gives the displacement in the window.
- Once all the frames are analyzed and the instantaneous vector fields are obtained, the vector fields are checked for possible outliers. Classical methods of outlier detection can be used. PIVlab has two inbuilt functions for this process and only spatial filtering can be performed. Standard deviation (SD) filter, which operates on the principle that any vector whose standard deviation from the mean velocity in the frame is more than some user-specified standard deviations is an outlier. The median filter on the other hand operates locally and the vector is compared with the local vectors in the neighborhood [10]. This process of outlier detection is also called vector validation.
- After the validation of the vectors, the calibration is done. For the calibration, the time step between the pair of images is required in addition to the calibration image. The calibration converts the photographic pixels into distances.

At the end of this step instantaneous vector field for each frame in world coordinates (in dimensions of length and time) is obtained. The vectors fields can now be post processed to obtain quantities of interest which is discussed in the next section.

4 Postprocessing of the Data

4.1 Temporal Averaging

The data is postprocessed to obtain the averaged velocity fields. For uniform flows, double averaging (temporal and spatial averaging) can be performed, while for non-uniform flows like vertical contraction case, only temporal averaging can be done. In the temporal averaging, average velocity field is obtained by taking the mean of the velocity fields for all 500 frames. PIVlab outputs the velocity data in two components, x component, u_{ij}, and y component, v_{ij}. u_{ij} corresponds to the velocity component along the flow direction at the point having coordinate (i, j) and v_{ij} corresponds to vertical velocity component. The process of temporal averaging is depicted in Eqs. (1) and (2). U_{ij} and V_{ij} are the temporally averaged velocity components at (i, j). $\left(u_{ij}\right)_n$ and $\left(v_{ij}\right)_n$ are the velocity components at (i, j) in the nth frame and N is the total number of frames. In the present study, N equals 500.

$$U_{ij} = \frac{1}{N}\sum_{n=1}^{N}\left(u_{ij}\right)_n \tag{1}$$

$$V_{ij} = \frac{1}{N}\sum_{n=1}^{N}(v_{ij})_n \tag{2}$$

4.2 Spatial Averaging

For the open channel flow case, spatial averaging follows temporal averaging referred to as double averaging. In the spatial averaging, the temporal averaged results are averaged again along the longitudinal direction with the assumption that the flow is uniform in the longitudinal direction. Besides, the vertical component in open channel flow is negligible. Equation (3) shows the procedure of spatial averaging. U_j is the spatially averaged longitudinal velocity, and I is the total number of velocity vectors in longitudinal direction.

$$U_j = \frac{1}{I}\sum_{i=1}^{I}U_{ij} \tag{3}$$

5 Results and Discussions

5.1 Validation of the Experimental Data

The performance of the PIV is validated by comparing the experimental velocity profile with theoretical velocity profile models proposed by Coles called as Coles wake law (Eq. 4) [11]. In Eq. (4) u is the double averaged longitudinal velocity at height y from the bed, u_* is the shear velocity given as $u_* = \sqrt{(\tau_b/\rho)}$, where τ_b is the bed shear stress and ρ is the density of fluid. κ is equal to 0.41, B is equal to 5.3 [2], H_a is the total flow depth and Ω is the Coles wake parameter and has been reported to be equal to 0.55 by Coles [11]. Nezu and Rodi obtained the value for Ω between 0 and 0.20 [2]. In the present study, Ω has been taken as 0.20.

$$\frac{u}{u_*} = \frac{1}{\kappa}\ln\frac{yu_*}{\nu} + B + \frac{2\Omega}{\kappa}\sin^2\frac{\pi y}{2H_a} \tag{4}$$

Figure 4 shows the comparison of the experimental data obtained using PIV with the Coles wake law. In the figure, the vertical distance and the velocity have been non dimensionalized using the shear velocity, u_*. u_* is obtained by fitting Eq. (4) to the experimental data within the range of $0.1 < y/H_a < 0.9$. Close to the bed, in the region $y/H_a < 0.1$, viscous effects are dominant, and the velocity profile deviates

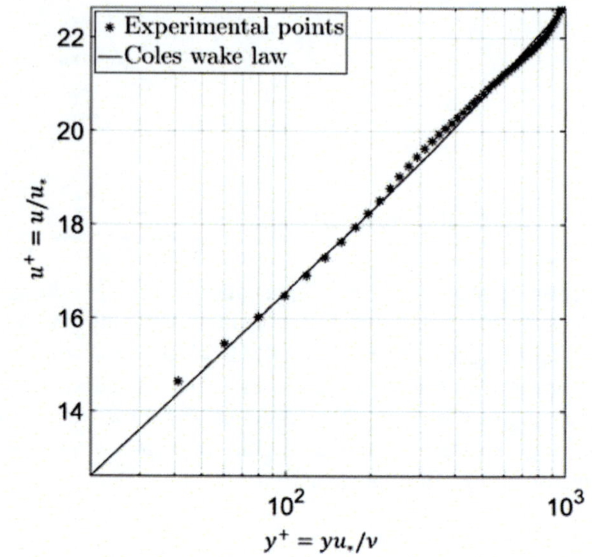

Fig. 4 Comparison of experimental velocity profile with the Coles wake law

from the logarithmic profile. Similarly, for the regions close to the free surface, the secondary or wall effects come into play especially for narrow channels. For turbulent boundary layers, viscous length scale, ν/u_* has been widely used to non dimensionalize the vertical distance, y, $y^+ = yu_*/\nu$ [2]. For $y^+ < 5$, called as the wall region, the flow is laminar and is called as the viscous sublayer, $5 < y^+ < 30$, a transition from laminar to turbulent flow occurs and for $y^+ > 30$, the flow is fully turbulent. In Fig. 4, a good agreement between the experimental points and the model is observed for $y^+ > 30$.

5.2 *Vertical Contraction Experiment*

In the vertical contraction experiment, the velocity field and stagnation zones are investigated. The stagnation zone is defined as the region from the top end of the flow field up to which the cumulative discharge is less than 1% of the total discharge. Ideally, the threshold should be zero, but a slightly conservative approach is used here to account for the uncertainty associated with PIV measurement. Besides, the stagnation zone is also evaluated for the temporally averaged velocity field. Mathematically, discharge is obtained by the vertical integration of the velocity profile. The cumulative discharge from the top up to a height h from the bottom is given as Eq. (5).

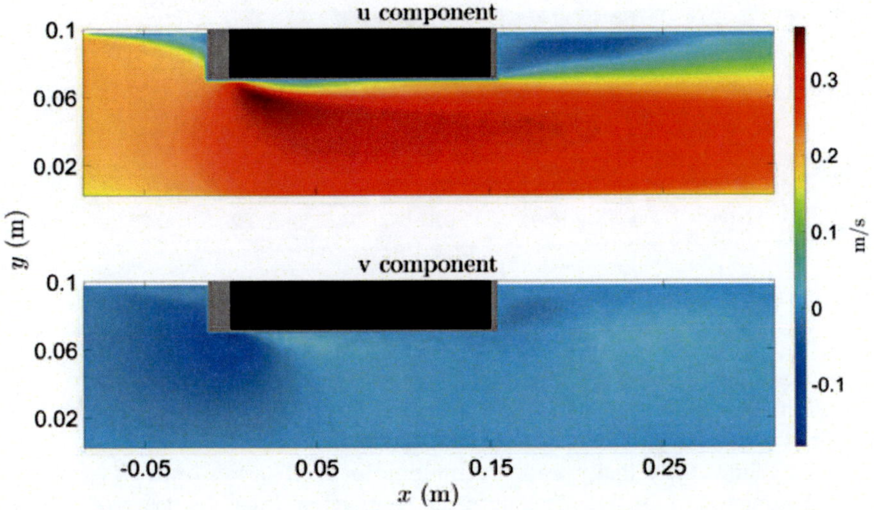

Fig. 5 Longitudinal and vertical velocity color contours. The black color is the bridge, and the gray color is the masked region during the processing of PIV images

$$Q_h = \int_{H_a}^{H_a - h} u \, \mathrm{d}y \tag{5}$$

Figure 5 shows the color maps of the longitudinal and the vertical velocity components of the flow during the vertical contraction. It can be observed that the flow is accelerated in the longitudinal direction and the peak velocity is attained inside the contraction. A low velocity zone is created under the bridge starting from the entrance and increases in size and then as the flow develops inside the contraction, the low velocity zone also disappears. At the downstream end of the contraction, a recirculation zone is observed at the top. The recirculation occurs because of the flow expansion. The recirculation zone does not allow the deceleration of the flow as the flow leaves the contraction. It appears as if the apparent length of the contraction is increased by the recirculation zone.

At the entrance section of the contraction, significant vertically downward velocity components are produced due to the contraction of the flow and a small stagnation region is also observed upstream of the contraction. The vertical velocity components can greatly influence the bottom bed shear stresses. Besides, the vertical components also decrease as the flow stabilizes inside the contraction. Vertical velocity components are also observed in the flow expansion downstream of the contraction.

Figure 6 shows the streamlines and the stagnation region. It is observed that the streamlines are significantly curved at the entrance section and a small stagnation zone is observed. The streamlines expand downstream of the contraction and a relatively large recirculation zone is observed. The stagnation zone downstream of the

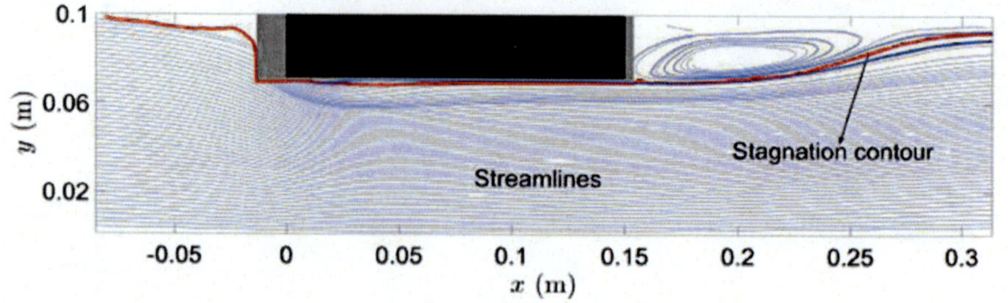

Fig. 6 Streamlines and stagnation zone

contraction slowly decays and after around one bridge length, L, becomes zero. The stagnation zone under the contraction decreases the flow passage further and thus modifies the velocity profile inside the contraction.

Figure 7 shows the velocity profiles at multiple location starting from the upstream of the contraction to the downstream of the contraction. It can be observed that at $x = -0.075$ m, i.e., $x = -L/2$, the velocity profile is like the one observed in open channels but as the flow approaches the contraction, the upper layers are brought to rest. As the flow enters the contraction, $x = 0.05$ m, the velocity gradients close to the channel bottom are small as compared to the gradients underneath the contraction. This implies that shear stresses are generated under the bridge surface. As the flow crosses the center of the contraction, the gradients stabilize and the velocity profile assumes a fuller shape. Just before the expansion of the flow, at $x = 0.15$ m, the velocity profile has become more symmetric, and appears like the velocity profile

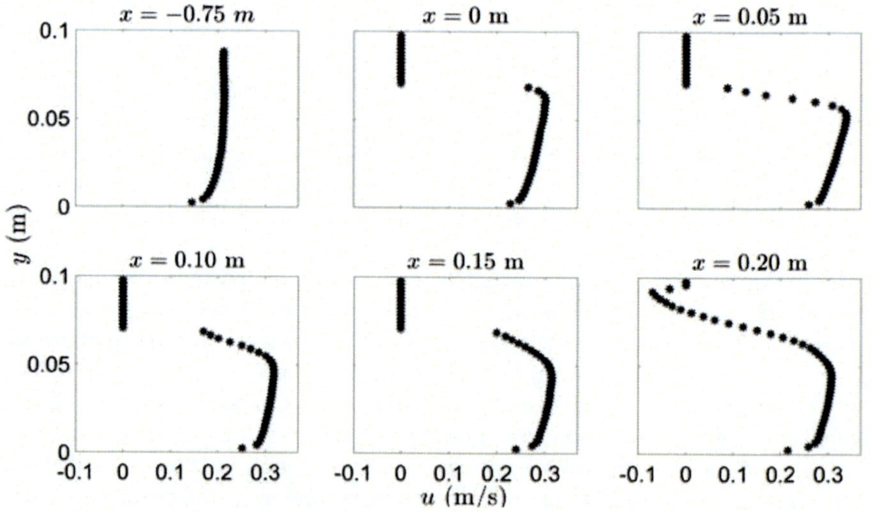

Fig. 7 Velocity profiles at multiple sections. $x = 0$m, $x = 0.15$m are the start and end of the contraction respectively. Refer to Fig. 5

observed in pipes. Downstream of the contraction, at $x = 0.20$ m, the velocity profile is completely different from open channels flows. Significant negative velocities are observed toward the free surface. It is also observed that the velocity close to the bed is significantly larger inside the contraction and the downstream of the contraction as compared to that at the upstream. This observation can be useful in studying the pressure flow scour process and designing bridges for submerged conditions.

6 Conclusions

In this study, the particle image velocimetry (PIV) is used to study open channel flow and vertical contraction. The velocity profile reported in the literature matches well with the velocity profile obtained using the PIV for open channel flow. The main conclusions drawn for the flow with vertical contraction are listed below.

- The flow is accelerated inside the contraction which is further increased by the development of the stagnation zone under the contraction. After some length, the flow stabilizes, and the stagnation zone disappears.
- A recirculation zone is formed at the downstream of the contraction which apparently increases the effective length of the bridge.
- The velocity profile gets modified inside the contraction and tends to become more symmetric about the centerline as the flow moves inside the contraction.
- Negative velocity components are observed downstream of the contraction inside the recirculation zone.
- Higher velocity components and velocity gradients are observed close to the bed inside and downstream of the contraction.

This study can be extended to investigate the effects of the flow properties and the flow geometry on the velocity fields, stagnation, and recirculation zones. Besides, the shear stress and energy fields can be studied which can provide vital information for designing bridges against the pressure flow scour.

Acknowledgements We would like to extend our thanks to Profs. Debopam Das and Kamal Poddar (Department of Aerospace engineering, IIT Kanpur) for providing the core components of PIV setup and helping in PIV measurements. The support provided by the Science and Engineering Research Board (File No. YSS/2015/000574-SERB) to the second author is also duly acknowledged.

References

1. Bigillon, F., Niño, Y., & Garcia, M. H. (2006). Measurements of turbulence characteristics in an open-channel flow over a transitionally-rough bed using particle image velocimetry. *Experiments in Fluids, 41*, 857–867. https://doi.org/10.1007/s00348-006-0201-2

2. Nezu, I., & Rodi, W. (1986). Open-channel flow measurements with a laser Doppler anemometer. *Journal of Hydraulic Engineering Division of the American Society of Civil Engineers, 112*, 335–355. https://doi.org/10.1061/(asce)0733-9429(1986)112:5(335)
3. Wu, W., Rodi, W., & Weaka, T. (2000). 3D numerical modeling of flow and sediment transport in open channels. *Journal of Hydraulic Engineering Division of the American Society of Civil Engineers, 126*, 4–15. https://doi.org/10.1061/(ASCE)0733-9429(2000)126:1(4)
4. Chien, N., & Wan, Z. (1999). Mechanics of sediment. *Transport*. https://doi.org/10.1061/9780784404003
5. Shan, H., Zhaoding Xie, C. B., Suaznabar, O., Steven Lottes, J. S., & Kerenyi, K. (2012). Submerged flow bridge scour under clear water conditions. Fed. Highw. Adm. Publ. NO. FHWA-HRT-12-034. 51.
6. Carnacina, I., Leonardi, N., & Pagliara, S. (2019). Characteristics of flow structure around cylindrical bridge piers in pressure-flow conditions. *Water (Switzerland), 11*. https://doi.org/10.3390/w11112240
7. Yoon, K. S., Lee, S. O., & Hong, S. H.: Time-averaged turbulent velocity flow field through the various bridge contractions during large flooding. *Water (Switzerland), 11* (2019). https://doi.org/10.3390/w11010143
8. Thielicke, W., & Stamhuis, E. J. (2014). PIVlab—Towards user-friendly, affordable and accurate digital particle image velocimetry in MATLAB. *Journal of Open Source Software, 2*. https://doi.org/10.5334/jors.bl
9. Keane, R. D., & Adrian, R. J. (1992). Theory of cross-correlation analysis of PIV images. *Applied Scientific Research, 49*, 191–215. https://doi.org/10.1007/BF00384623
10. Westerweel, J., & Scarano, F. (2005). Universal outlier detection for PIV data. *Experiments in Fluids, 39*, 1096–1100. https://doi.org/10.1007/S00348-005-0016-6
11. Coles, D. (1959). The law of the wake in compressible turbulent boundary layers. *Journal of the Aerospace Sciences, 26*, 541–542. https://doi.org/10.2514/8.8191

Impact Assessment of Land-Use/Land-Cover Changes on Hydrology and Storm Water Drain Network in Yelhanka Watershed, Bangalore

R. L. Gouri, V. V. Srinivas, S. N. Soumya, and M. Amulya

Abstract Globally, urban areas are witnessing rapid development in the recent decades, and there is growth in flash flood incidents following even short-duration storm events of less rarity (return period). Stormwater drains (SWDs) are an integral part of urban infrastructure. Owing to this, the existing natural drainage system gets affected, as the SWD network might replace only a part of the natural system. Further, changes in land-use/land-cover (LULC) associated with urbanization of catchments have implications on generated runoff in terms of increase in peak discharge, runoff volume, and velocity. An increase in runoff velocity causes a reduction in time to peak discharge, resulting in flash floods. In addition, encroachment into storm sewers limits their capacity to convey runoff, causing more devastation during floods due to an increase in stage for a given discharge. The present study is aimed at assessing LULC change in an urban catchment located in Bangalore (North) and evaluating its effect on runoff generation. Remote sensing satellite images corresponding to the years 1996, 2002, 2006, and 2012 have been analyzed to identify changes in LULC. The impact of the changes on runoff is investigated by inputting design hyetographs to Storm Water Management Model (SWMM) developed for the urban catchment for the various LULC scenarios and analyzing the runoff generated by the model. Vulnerable reaches in the existing stormwater drainage network of the studied urban catchment are identified, and the failure state of those drains is marked.

Keywords Storm Water Management Model (SWMM) · Land-use/land-cover (LULC) · Stormwater drains (SWDs) · Urbanization and vulnerability

---

R. L. Gouri · V. V. Srinivas (✉) · M. Amulya
Interdisciplinary Centre for Water Research (ICWaR), Indian Institute of Science, Bangalore 560012, India
e-mail: vvs@iisc.ac.in

V. V. Srinivas · S. N. Soumya
Department of Civil Engineering, Indian Institute of Science, Bangalore 560012, India

A. K. Dikshit et al. (eds.), *Innovative Trends in Hydrological and Environmental Systems*, Lecture Notes in Civil Engineering 234,
https://doi.org/10.1007/978-981-19-0304-5_21

1 Introduction

The process of urbanization is often rapid and disorganized in developing countries like India. Performance of stormwater drains (SWDs) is severely affected by an increase in inflow caused by significant changes in land-use and land-cover (LULC) and the associated changes in climate, storm patterns, and hydrological processes (e.g., evapotranspiration, infiltration, and runoff). A surge in peak discharge, volume and velocity of flow, and reduction in time taken for the flood discharge to attain its peak are some of the typical effects of urbanization. Furthermore, the SWD networks demand substantial financial investment for construction/installation, maintenance, and retrofitting. Financial constraints pose a challenge for judiciously allocating the resources toward these activities. In this perspective, there is a need to understand how the performance of the SWDs is affected by LULC changes so that flood mitigation and management strategies can be devised to minimize the consequent direct and indirect damages to the urban environment and population. Most studies in the past have assessed the effect/impact of LULC change on (i) water balance components in agricultural watersheds [14, 16]; and (ii) runoff generation in river basins [6, 7, 9, 12, 15]. They have concluded that reduction of vegetation cover leads to an increase in runoff and vice versa [8, 10] which is the case in urban catchments. There is a stressing need for studies on impact assessment of LULC change on runoff generation in urban catchments in India, as most cities are witnessing rapid development.

The objectives of the present study are to develop a hydrological model, Storm Water Management Model (SWMM), for a rapidly urbanizing Yelahanka watershed in Bangalore and estimate overland flow and runoff entering an existing SWD network in the area for different LULC scenarios. The LULC changes were analyzed based on remote sensing products corresponding to four typical years (1996, 2000, 2006, and 2012) using Geographic Information System (GIS) tools. The changes were found to be significant, especially during recent years. The SWMM developed corresponding to each chosen historical LULC scenario was forced with synthetic design storm hyetographs (DSHs), and watershed responses were recorded, which reflected the implications of the LULC change. Discharge generated from SWMM models was used to assess the performance of each of the drains in the SWD network to identify those that are susceptible to failure. Furthermore, suggestions are provided on possible mitigation measures which could be implemented to enhance/improve the performance of the drains for alleviating/minimizing the flood-damage potential from critical storm events.

The remainder of the chapter is structured as follows. A detailed description of the methodology utilized for the analysis is presented in Sect. 2. Following this, a case study is presented in Sect. 3 on the stormwater distribution system in the Yelahanka watershed of Bangalore city. Subsequently, conclusions drawn based on the study are provided in Sect. 4.

2 Methodology

2.1 *Methodology for Reliability Analysis of SWDs*

In the study, the reliability of the drains was determined under two failure modes. A drain is said to have failed when (i) runoff in the drain exceeds its flow capacity and (ii) the actual flow velocity in the drain exceeds the maximum allowable velocity for scouring control. The reliability of any system generally depends on the interaction between loads on the system and resistance by the system. Loadings are the external stresses that act on the system, and resistance is the capacity of the system to sustain the load. In the context of SWDs, the load is actual discharge generated from urban area (Q_D) into the channel, and resistance is flow capacity (Q_C) of the drain under consideration. In this study Q_C is calculated using Manning's equation (Eq. 1) and Q_D is extracted from the SWMM output.

$$Q_C = \frac{1}{n} b^{5/3} d^{5/3} \left(\frac{1}{b + 2d}\right)^{2/3} s^{1/2} \quad (1)$$

where *b, d,* and *s* denote base width, depth, and slope of the drain, respectively, and *n* represents Manning's roughness coefficient. Failure of the drain is said to have occurred when the actual discharge generated in the drain from the urban watershed (Q_D) exceeds the flow capacity (Q_C) of the drain [5]. In this perspective, discharge-based reliability estimate of a drain (R_d) can be expressed as,

$$R_d = \begin{cases} 1 \text{ if } Q_C \geq Q_D \\ 0 \quad \text{otherwise} \end{cases} \quad (2)$$

Analogous to discharge-based reliability (R_d), velocity-based reliability (R_v) can be defined for the failure of a drain due to maximum velocity (V_D) exceeding maximum allowable velocity (V_{Max}) of the drain. The value for V_D is extracted from SWMM output, whereas the value of V_{Max} is considered to be 3 m/s [2, 4]. Mathematically, this can be expressed as,

$$R_v = \begin{cases} 1 \text{ if } V_{\text{D}} \leq V_{\text{Max}} \\ 0 \quad \text{otherwise} \end{cases} \quad (3)$$

2.2 *Storm Water Management Model (SWMM)*

The SWMM, used for the case study, is a dynamic rainfall-runoff simulation model widely used to quantify runoff generated from urban areas. It has two modules, the first being the rainfall-runoff module and the second being the transportation module [13]. The rainfall-runoff module calculates the runoff generated from a collection of sub-catchments that receive precipitation. The transportation module routes the runoff through the stormwater drains or pipes during the simulation period. Spatial variability is accounted for by dividing the study area into a collection of smaller sub-catchments, each containing some fraction of pervious and impervious sub-areas. Surface runoff from each sub-watershed is calculated after accounting for initial abstractions (evaporation and infiltration). Inputs to SWMM include design rainfall hyetograph for the duration and return period of interest, LULC corresponding to different years, hydrologic soil groups for the study area identified based on the soil map.

3 Case Study

The methodology presented in Sect. 2 was applied to an existing SWD network in the Yelahanka watershed (Fig. 1) located in the northern part of Bangalore, India. The reliability of the network is evaluated by considering its discharge carrying capacity and maximum flow velocity criteria. The area of the watershed contributing flow to drains in the network is about 355 km^2. The total length of the drain is 176 km, which was further divided into 500 smaller drains for ease of analysis. The watershed was divided into 180 sub-watersheds, which are shown as darkened rectangles in Fig. 2.

For use in the analysis, daily rainfall records corresponding to five rain gauges (base stations) located at GKVK (Gandhi Krishi Vignan Kendra), Yelahanka,

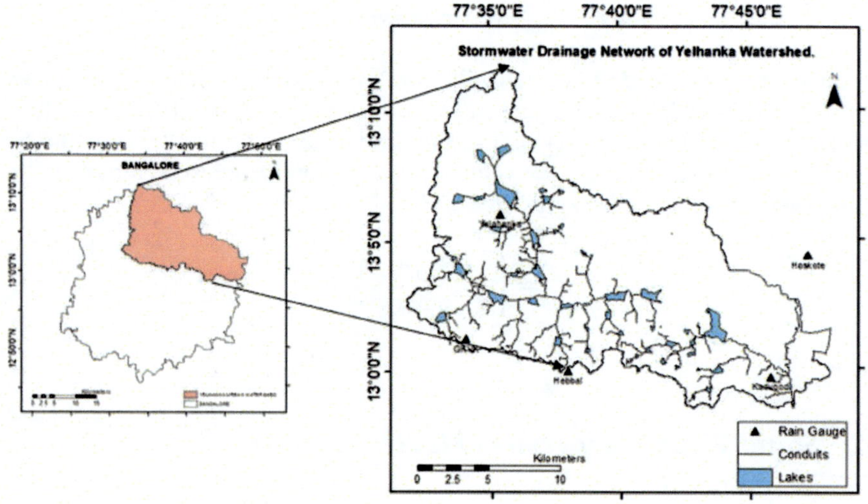

Fig. 1 Location map of Yelahanka watershed

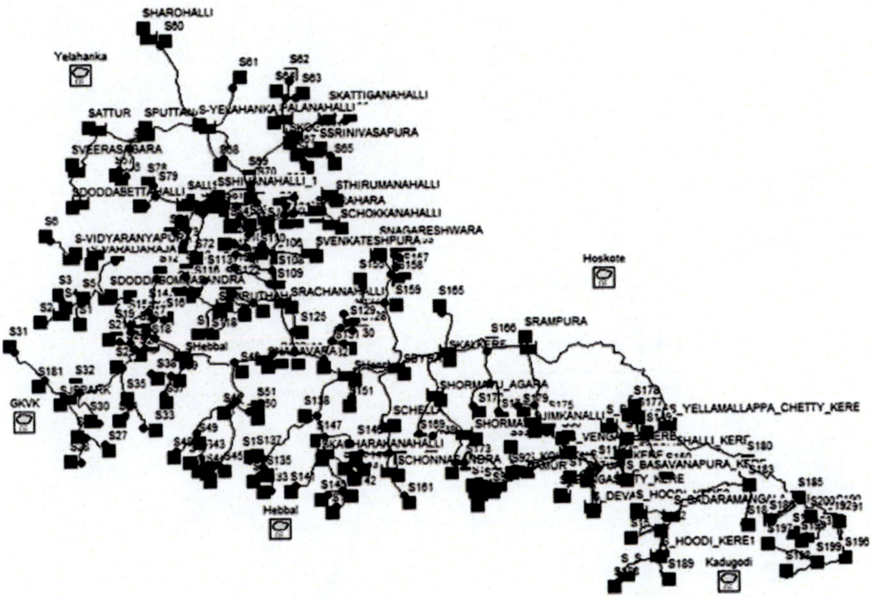

Fig. 2 Representation of the sub-watersheds and rain gauges in SWMM

Hoskote, Hebbal, and Kadugodi were considered for the period 1988–2010. Records of the latter four gauges were collected from the Directorate of Economics and Statistics, Bangalore, whereas those of GKVK were collected from the GKVK Agricultural college campus. Locations of the gauges are shown as "square enclosing clouds" in Fig. 2. Existing SWD network details were collected from the SWD department, Bruhat Bengaluru Mahanagara Palike (BBMP).

Discharge from the Yelahanka watershed into the SWD network was estimated corresponding to various rainfall durations and return periods using SWMM. Inputs to SWMM include design rainfall hyetograph for the duration and return period of interest, LULC corresponding to different years, and curve number for the sub-watersheds. The design hyetographs were constructed based on Intensity-duration-frequency (IDF) curves developed for the study area (Fig. 3). As the records at base stations were available at a daily scale, they had to be disaggregated to hourly scale for the construction of design hyetographs corresponding to sub-daily durations.

The disaggregation was carried out using the *k*-nearest neighbor method [1]. For this purpose, relationships between daily and hourly rainfall records corresponding to a rain gauge located at the Indian Institute of Science (IISc) Campus were considered. The gauge at IISc had 10-min scale rainfall records over the period 2003–2016, and they were found to be reasonably well correlated at daily scale with records corresponding to gauges located at the base stations. IDF curves were developed [3] for each of those stations using the derived hourly rainfall by frequency analysis using Extreme value type-I distribution. Subsequently, those curves were considered as the

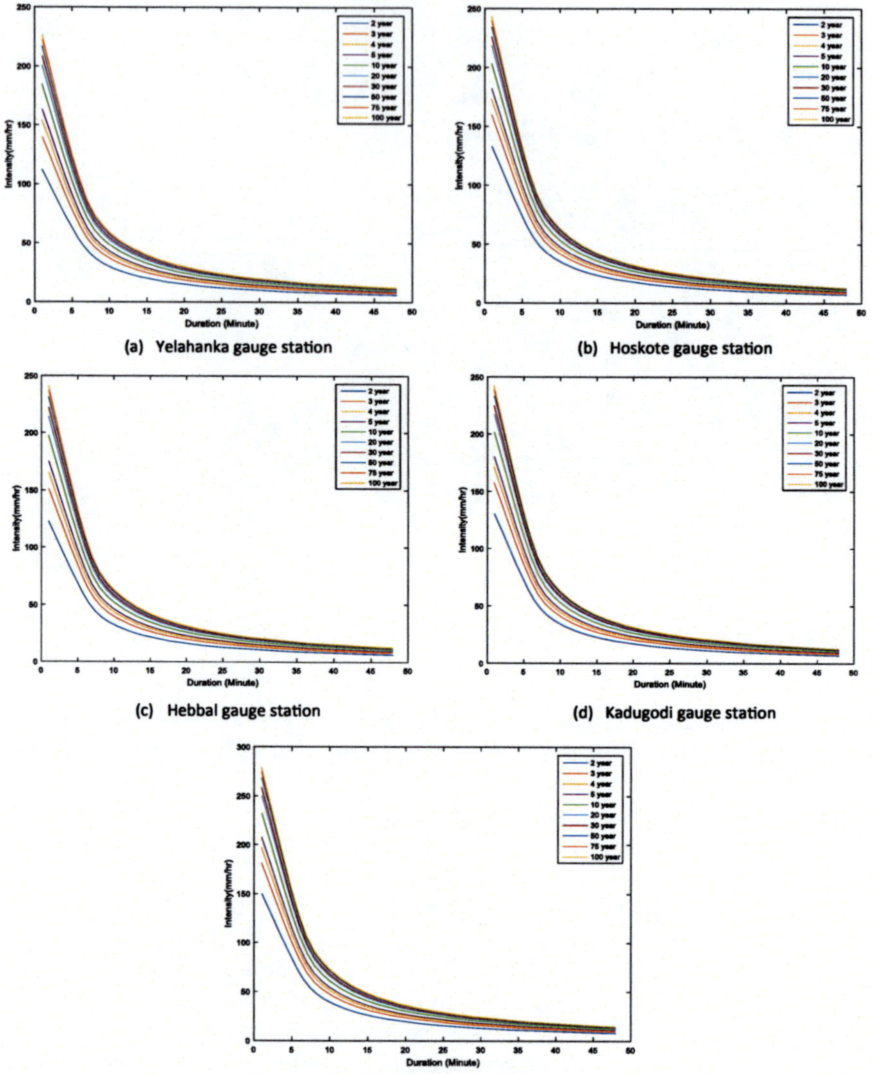

(a) Yelahanka gauge station

(b) Hoskote gauge station

(c) Hebbal gauge station

(d) Kadugodi gauge station

(e) GKVK gauge station

Fig. 3 The IDF curves plotted for the base stations

basis to construct design hyetographs corresponding to various sub-daily durations (1, 2, 3, 6, and 12 h) and return periods (2, 5, 10, 25, and 50 years) for the base stations locations. Herein, it is to be mentioned that records corresponding to gauge at IISc were short, and hence those were not considered for frequency analysis to construct IDF curves. Typical hyetographs plotted for the base stations are presented in Fig. 4.

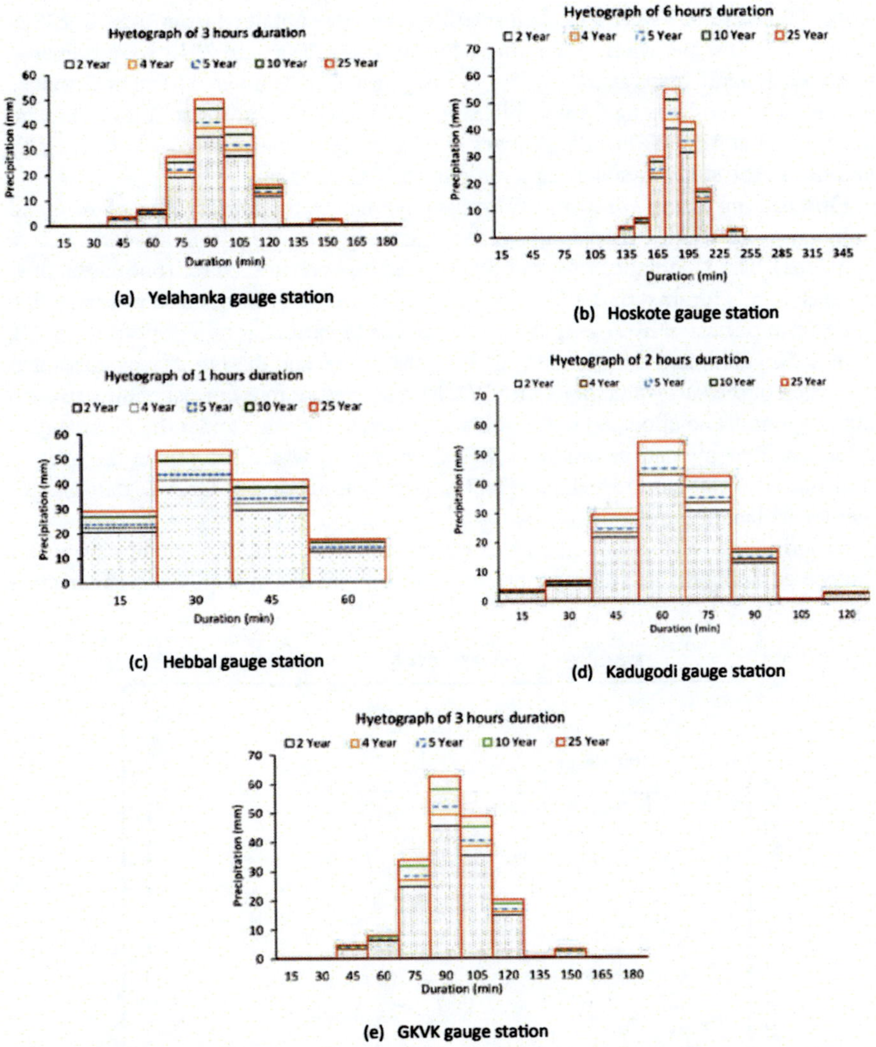

Fig. 4 Typical hyetographs constructed for the base stations

Initial abstractions were calculated using the NRCS curve number method [16]. Implementation of the method involved assignment of curve numbers to each sub-watershed based on LULC information corresponding to different years and hydrological soil groups. The hydrologic soil groups for the study area were identified based on a soil map obtained from the National Bureau of soil survey and land-use planning (NBSS & LUP [11]). For the study area, LULC maps were prepared corresponding to the years 1996, 2002, 2006, and 2012. Among those, maps corresponding to the years 1996 and 2002 were based on satellite imageries obtained from IRS-IC-LISS-III at 1:50,000 scale, whereas the maps for the years 2006 and 2012 were prepared based on thematic maps available for the study area from Bhuvan-Thematic Services National Remote Sensing Center (NRSC), ISRO, Hyderabad, India, (http://bhuvan.nrsc.gov.in) at 1:50,000 scale. A level II classification was carried out to classify land-use in the study area corresponding to the LULC maps.

Flow routing within a drain in SWMM is governed by the conservation of mass and momentum equations for gradually varied, unsteady flow (i.e., the Saint–Venant flow equations). The kinematic wave routing method was considered for routing the flow in the channels (drains) of the SWD network. The rainfall information was provided in the form of design hyetographs corresponding to various return periods (2, 5, 10, 25, and 50 years) and durations (1, 2, 3, 6, and 12 h), and the runoff was generated from each of the sub-watersheds. In SWMM, the hyetographs corresponding to each rain gauge must be allocated to the sub-watershed that comes under the rain gauge's influence. This allocation was done based on Fig. 5, which represents the area of influence of Yelahanka, Hoskote, Hebbal, and Kadugodi and GKVK rain gauges over the Yelahanka watershed.

Experiments were conducted to examine the effect of LULC change on water balance components such as evaporation, infiltration, and runoff. Each experiment

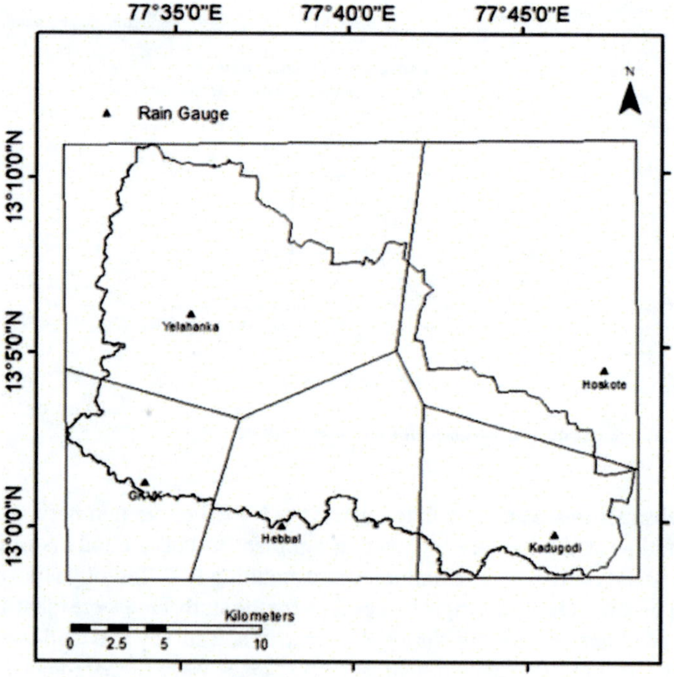

Fig. 5 Area of Influence of rain gauges over the Yelahanka watershed

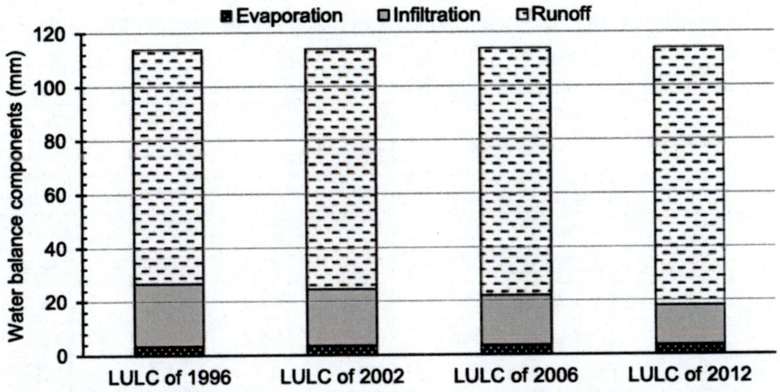

Fig. 6 Water balance components quantified (in-depth units) corresponding to 2 h duration 10-year return period design hyetograph

involved providing a design hyetograph as input to SWMM along with LULC information corresponding to each of the four years (1996, 2002, 2006, and 2012) and analyzing output from the model to quantify changes in each of the water balance components due to LULC change. To quantify the effect of design hyetograph on the results, hyetographs corresponding to various durations and return periods were developed. One of those hyetographs was considered at a time for experimenting. Changes in water balance components due to LULC change were quantified corresponding to each of the hyetographs. Typical results obtained from these experiments are presented in Fig. 6. Comparison of runoff generated for different years shows an increase in runoff and consequent reduction in infiltration. Reduction in infiltration and increase in runoff could be attributed to an increase in imperiousness associated with an increase in the built-up area. On the other hand, change in evaporation was found to be insignificant, possibly due to insignificant changes to factors (e.g., temperature, humidity, wind speed) affecting the component. For brevity, the distribution of catchment response to a design hyetograph corresponding to 3-h duration and 5-year return period is shown in Fig. 6 (Table 1).

Peak flow discharge (Q_D) and velocity of flow (V_{Max}) corresponding to peak flow in the drain for each specified return period and duration were extracted from "Link Flow summary" from the output file generated by the SWMM model at 15-minute scale. The values for discharge reliability R_d and velocity reliability R_v is estimated based on Equations 2 and 3 where the value of Q_C is calculated using Equation 1. For estimating R_v the value of maximum allowable velocity (V_{Max}) for a lined drain is considered to be 3m/s (e.g., [4]). The results are presented in the pictorial format in Figures 7, 8, 9 and 10 for each of the LULC scenarios (1996, 2002, 2006, and 2012) considered. In the figures, the drains that are safe/reliable are shown in green, whereas the drains that are failing/unreliable are shown in red. The effect of LULC change on the SWDs is evident and can be observed in the figures. From the analysis based on LULC for the year 1996, it can be observed that R_d for most of the drains is

Table 1 Water balance components corresponding to LULC for the year 2012 for various return periods and durations

Duration in hours		Return period in years			
		2	5	10	25
		Depth in mm			
1	Total precipitation	66.7	94.0	105.6	118.2
	Evaporation loss	1.2	2.0	2.1	2.2
	Infiltration loss	37.9	43.3	44.9	46.2
	Surface runoff	27.9	48.2	57.8	68.5
2	Total precipitation	68.2	96.2	108.1	120.9
	Evaporation loss	2.3	2.5	2.5	2.5
	Infiltration loss	31.8	37.2	38.6	39.7
	Surface runoff	34.4	56.6	66.7	77.9
3	Total precipitation	68.4	96.4	108.2	121.1
	Evaporation loss	3.1	3.2	3.2	3.2
	Infiltration loss	26.0	29.9	31.2	32.3
	Surface runoff	38.1	61.3	71.6	83.0
6	Total precipitation	68.6	96.5	108.3	121.2
	Evaporation loss	3.1	3.1	3.1	3.1
	Infiltration loss	25.4	29.6	31.0	32.2
	Surface runoff	38.5	61.4	71.6	82.9
12	Total precipitation	68.9	96.9	108.7	121.7
	Evaporation loss	2.9	2.9	2.9	2.9
	Infiltration loss	25.9	30.2	31.6	32.9
	Surface runoff	38.2	61.1	71.3	82.6

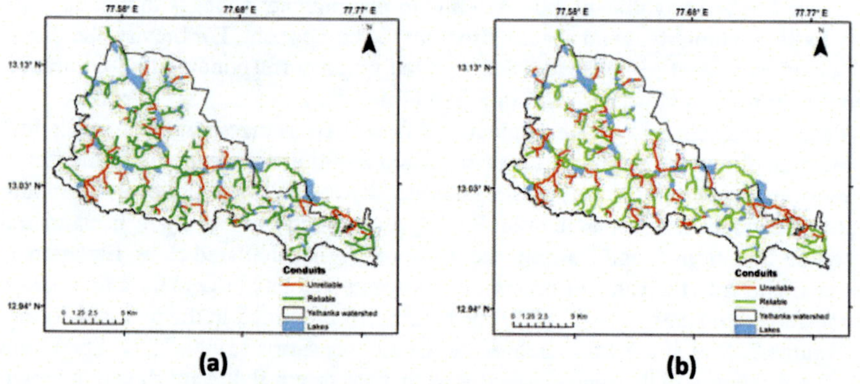

Fig. 7 **a** R_d values for LULC 1996, for 3-year return period and 2-h duration hyetograph. **b** R_v values for LULC 1996, for 3-year return period and 2-h duration hyetograph

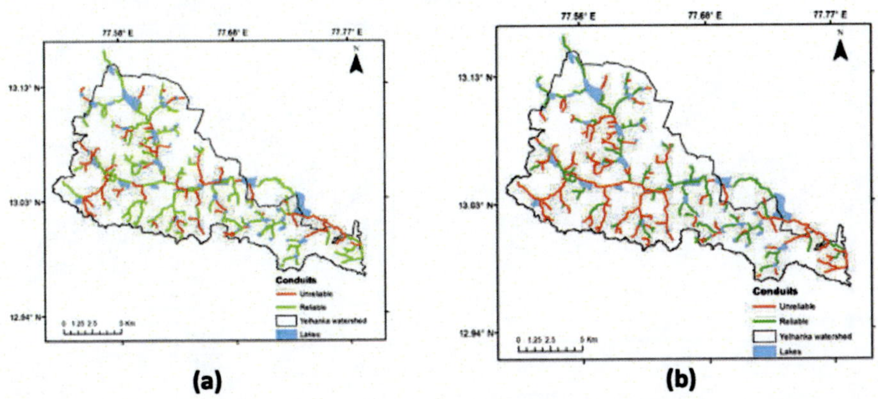

Fig. 8 **a** R_d values for LULC 2002, for 3-year return period and 2-h duration hyetograph. **b** R_v values for LULC 2002, for 3-year return period and 2-h duration hyetograph

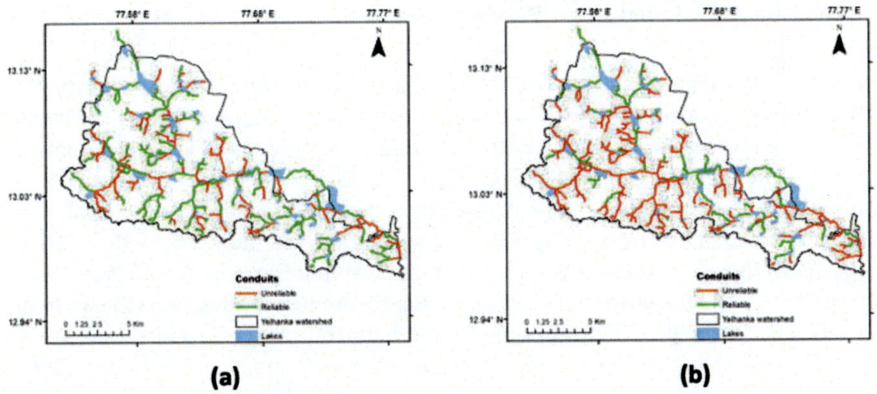

Fig. 9 **a** R_d values for LULC 2006, for 3-year return period and 2-h duration hyetograph. **b** R_v values for LULC 2006, for 3-year return period and 2-h duration hyetograph

positive, whereas from analysis based on LULC for the year 2012, R_d for most of the drains is negative and such drains are identified as vulnerable reaches in the SWD network being studied. These drains are found to be inadequate to convey discharge corresponding to various design hydrographs (durations greater than 3 hours, return period higher than 5 years), and the velocity of flow is exceeding the maximum allowable velocity, which may cause scouring. More drains were in failure mode in reliability analysis with maximum velocity than with maximum discharge for each of the LULC scenario. Furthermore, the count of drains in failure mode increased with the change in LULC from 1996 to 2012.

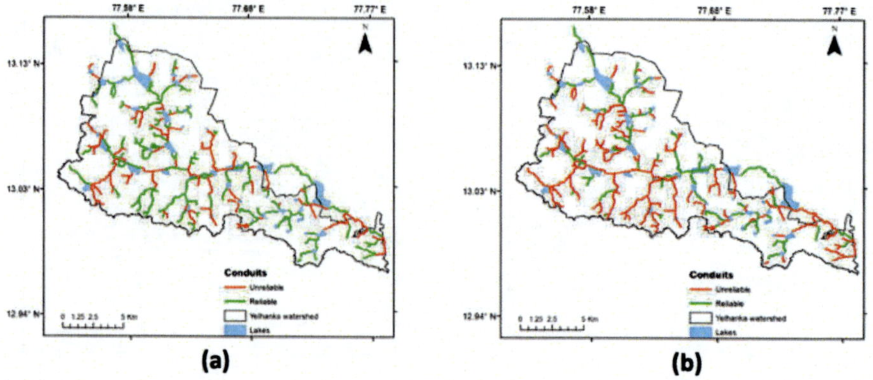

Fig. 10 **a** R_d values for LULC 2012, for 3-year return period and 2-h duration hyetograph. **b** R_v values for LULC 2012, for 3-year return period and 2-h duration hyetograph

4 Results and Concluding Remarks

SWMM models were developed for the Yelahanka watershed in Bangalore city for LULC scenarios corresponding to four years (1996, 2002, 2006, and 2012). The overland flow/runoff corresponding to synthetic design hyetographs of various durations (1, 2, 3, 6, and 12 h) and return periods (2, 5, 10, 25, and 50 years) was estimated. Furthermore, its characteristics (peak discharge, maximum velocity) were analyzed in each of the drains of an existing SWD network. Between the years 1996 to 2012, a significant increase in runoff volume and reduction in infiltration was observed for the various LULC scenarios considered, owing to an increase in imperviousness in the Yelahanka watershed. Furthermore, the discharge reliability R_d and velocity reliability R_v was calculated by extracting conduits flow summary from SWMM outputs. The results indicated that for storms of higher return periods (> 5 years) and durations (> 3 h): (i) the discharge exceeds capacity in several drains, and (ii) flow velocity is higher, which causes scouring in a considerable number of drains. Comparison of discharge generated corresponding to various LULC scenarios indicated that the number of surcharged drains and the duration of flooding in those have increased owing to urbanization. More drains were in failure mode in reliability analysis with maximum velocity than with maximum discharge.

Acknowledgements The authors acknowledge BBMP Bangalore and the directorate of economics and statistics (DES), Karnataka, for providing data for this study. They express their gratitude to the Department of Science and Technology (DST) for grants provided through Project No. DST/CCP/NUC/146/2018 (G), and to Ministry of Earth Sciences for the support provided to ICWaR of IISc through Project No. MOES/PAMC/H&C/41/2013-PC-II.

References

1. Anandhi, A., Srinivas, V. V., Kumar, D. N., & Nanjundiah, R. S. (2012). Daily relative humidity projections in an Indian river basin for IPCC SRES scenarios. *Theoretical and Applied Climatology, 108*(1–2), 85–104.
2. Central Public Health and Environment Engineering Organization (CPHEEO). (2019). *Manual on storm water drainage system, Volume I, Part-A: Engineering design* (1st ed.).
3. Chow, V. T., Maidment, D. R., & Mays, L. W. (1988). Applied hydrology. McGraw-Hill.
4. Chaudhry, M. H. (2007). *Open-channel flow*. Springer
5. Gouri, R. L., & Srinivas, V. V. (2015). Reliability assessment of a storm water drain network. *Aquatic Procedia, 4*, 772–779. https://doi.org/10.1016/j.aqpro.2015.02.160
6. Hundecha, Y., & Bárdossy, A. (2004). Modeling of the effect of land use changes on the runoff generation of a river basin through parameter regionalization of a watershed model. *Journal of Hydrology, 292*(1), 281–295.
7. Liu, Z., Yao, Z., Huang, H., Wu, S., & Liu, G. (2012). Land use and climate changes and their impacts on runoff in the Yarlung Zangbo river basin, China. *Land Degradation & Development*.
8. Miller, J. D., Kim, H., Kjeldsen, T. R., Packman, J., Grebby, S., & Dearden, R. (2014). Assessing the impact of urbanization on storm runoff in a peri-urban catchment using historical change in impervious cover. *Journal of Hydrology, 515*, 59–70. https://doi.org/10.1016/j.jhydrol.2014.04.011
9. Niehoff, D., Fritsch, U., & Bronstert, A. (2002). Land-use impacts on storm-runoff generation: Scenarios of land-use change and simulation of hydrological response in a meso-scale catchment in SW-Germany. *Journal of Hydrology, 267*(1), 80–93.
10. Paule-Mercado, M. A., Lee, B. Y., Memon, S. A., Umer, S. R., Salim, I., & Lee, C. H. (2017). Influence of land development on stormwater runoff from a mixed land use and land cover catchment. *Science of the Total Environment, 599–600*, 2142–2155. https://doi.org/10.1016/j.scitotenv.2017.05.081
11. Prasad, C. R. S., Reddy, R. S., Seghal, J., & Velayutham, M. (1998). Soils of Karnataka for optimising land use. NBSS Publ. 47b (Soils of India series), National Bureau of Soil Survey and Land Use Planning.
12. Rongrong, W., & Guishan, Y. (2007). Influence of land use/cover change on storm runoff—A case study of Xitiaoxi River Basin in upstream of Taihu Lake Watershed. *Journal of Chinese Geographical Science, 17*(4), 349–356.
13. Rossman, L. A. (2009). Storm Water Management Model User's Manual Version 5.0. EPA/600/R-05/040. National Risk Management Research Laboratory, United States Environmental Protection Agency.
14. Schilling, K. E., Jha, M. K., Zhang, Y. K., Gassman, P. W., & Wolter, C. F. (2008). Impact of land use and land cover change on the water balance of a large agricultural watershed: Historical effects and future directions. *Water Resources Research, 44*(7), 1–12.
15. Tang, Z., Engel, B. A., Pijanowski, B. C., & Lim, K. J. (2005). Forecasting land use change and its environmental impact at a watershed scale. *Journal of Environmental Management, 76*(1), 35–45.
16. USDA. (1986). *Soil conservation service national engineering handbook*. USA.

Evaluating Different Approaches to Model Land Use and Land Cover Change—A Case Study on Yelahanka Urban Watershed, Bangalore

M. Amulya, R. L. Gouri, V. V. Srinivas, and S. N. Soumya

Abstract Most of the cities in India have been undergoing rapid development in recent decades, and many rural localities are undergoing transformation to urban hotspots. These developments have associated land use/land cover (LULC) changes that affect hydrological response (e.g., evapotranspiration, infiltration, and runoff). There is a need to account for LULC change information in studies related to urban watershed planning and development. In recent years, studies on the impact of LULC change on the landscape and hydrology have gained prominence due to the realization that changes in LULC also influence climate, ecosystem, and hydrological processes. The study presented in this chapter is motivated to model LULC change in the Yelahanka urban watershed of Bangalore, which has witnessed rapid urbanization in the recent past. For the study area, historical LULC maps were prepared corresponding to the years 1996, 2002, 2006, 2012, and 2016. Models based on Markov chain and cellular automata Markov analysis are developed to identify and model changes in LULC patterns. Change detection analysis indicated significant temporal changes in the LULC classes. The developed models are further used for obtaining future projections of LULC. It is observed that among the models considered, the cellular automata Markov model exhibited better performance in yielding projections of the land use classes.

Keywords LULC change · Future projection of LULC · Markov chain model · Cellular automata Markov chain model · TerrSet software

M. Amulya · R. L. Gouri · V. V. Srinivas (✉)
Interdisciplinary Centre for Water Research (ICWaR), Indian Institute of Science, Bangalore 560012, India
e-mail: vvs@iisc.ac.in

V. V. Srinivas · S. N. Soumya
Department of Civil Engineering, Indian Institute of Science, Bangalore 560012, India

A. K. Dikshit et al. (eds.), *Innovative Trends in Hydrological and Environmental Systems*, Lecture Notes in Civil Engineering 234,
https://doi.org/10.1007/978-981-19-0304-5_22

1 Introduction

Urbanization is a way of developing a locality into a metropolitan area. The development often has disorienting effects on social, economic, and political growth, leading to chaotic changes in the area. These developments have associated land use/land cover (LULC) changes that influence climate, ecosystem, and hydrological processes (e.g., evapotranspiration, infiltration, and runoff) in the watershed. Hence, there is an impetus on studies to monitor LULC change and assess its associated impacts. Water resources are influenced by different land uses, such as industrialization, urbanization, forestry, and agriculture. Hence, spatial-temporal changes in LULC may significantly affect the water resources if they are improperly managed [30]. Several techniques are developed to identify, model, and project future changes in LULC patterns. Literature review indicates that the Markov chain model and hybrid cellular automata (CA) Markov chain model are widely used by practitioners. Table 1 provides a summary of some recent studies.

The utility of LULC models is mainly in studies focused on aspects such as land use planning (e.g., [5]), evaluation of urban growth (e.g., [15, 27, 36]), and assessing the impact of LULC change on runoff generation (e.g., [22, 32, 33]). Detection of LULC changes in a specific region is possible by comparing remote sensing imageries corresponding to different times. Remote sensing data are considered reliable for identifying/detecting LULC changes due to their high temporal frequency on global land cover information, accessibility, and ease of use for calculations owing to digital format [13, 19, 31, 33]. Recent developments in remote sensing technology and geographic information system (GIS) framework have provided an impetus for undertaking studies on modeling LULC changes.

This chapter presents a study focused on assessing and modeling LULC changes in the Yelahanka urban watershed (Bangalore), using the Markov Chain model and CA-Markov chain model. These models are suitable for LULC change detection and future LULC projection, as they consider the spatial and temporal components of land cover dynamics. In the study area, there has been an increase in shortage of water resources, and exploitation of groundwater resources in recent decades, accompanied by low water quality and soil erosion. This study is undertaken to verify if there are significant changes in LULC, so that extended work could be envisaged to investigate if the LULC changes led to an alteration in hydrological processes such as evapotranspiration and infiltration. The remainder of the chapter is structured as follows. The methodology followed for obtaining future projections of LULC is presented in Sect. 2. Description of the study area and its data is presented in Sect. 3, along with results pertaining to the investigation of the objective set for this study. Following this, a summary of the results and conclusions drawn based on the study is presented in Sect. 4.

Table 1 Some recent studies focused on obtaining future projections of LULC

Study	Study area	Purpose of the study	Model used
Rahnama [24]	Mashhad metropolis, Iran	Investigating trend of land use change from 2016 to 2020	Cellular automata (CA)-Markov chain model
Wang et al. [34]	Thimphu city, Bhutan	Detection and prediction of LULC change	CA-Markov chain model
Babbar et al. [3]	Sariska Tiger Reserve, India	To predict carbon sequestration	Markov chain model
Mustafa et al. [20]	New York City, USA	To simulate multi-density urban expansion	Markov chain cellular automata model
Mondal et al. [19]	Brahmaputra River Basin	Modeling of LULC change prediction results	CA-Markov chain model
Shen et al. [28]	Stoney Creek Watershed, British Columbia, Canada	Object-based image analysis to map and predict LULC changes	Markov chain model
Bose et al. [4]	Siliguri, West Bengal, India	Modeling of spatio-temporal urban expansion and LULC change	Markov chain model
Mansour et al. [17]	Oman	Monitoring LULC changes and predicting the urban growth	CA-Markov model
Kafy et al. [12]	Rajshahi, Bangladesh	Modeling future LULC changes and their impacts on land surface temperatures	Multi-layer perceptron-Markov chain
Nasiri et al. [21]	Arasbaran, Iran	To detect and predict LULC changes	Markov chain analysis
Hamad et al. [10]	Kurdistan, Iraq	Predicting LULC changes under two different scenarios	CA-Markov chain model
Paul et al. [23]	Kiskatinaw River Watershed, British Columbia, Canada	Predicting LULC change	Markov chain model
Lu et al. [16]	London	To simulate and predict LULC changes for future	CA-Markov chain model
Ahmad et al. [1]	Ranchi, India	Developing LULC maps	Markov chain model

(continued)

Table 1 (continued)

Study	Study area	Purpose of the study	Model used
Rimal et al. [26]	Jhapa District of Nepal	Spatial-temporal analysis of LULC change as well as the monitoring and modeling of urban expansion	CA-Markov chain model
Gidey et al. [9]	Raya, Northern Ethiopia	Prediction of future LULC scenarios	CA-Markov model

2 Methodology

The methodology involves (i) evaluating historical LULC changes in the Yelahanka urban watershed,

(ii) modeling the changes using Markov chain and CA-Markov chain models, and (iii) utilizing the developed models to arrive at the future projections of LULC to assess implications of the same on hydrological processes.

2.1 *Markov Chain Model*

Markov chain model for LULC projection is a stochastic model which utilizes knowledge on the evolution of a particular land use class from a time step $t-1$ to step t to project the LULC corresponding to the time step $t+1$. Transition probabilities for changes in land use classes from the past $(t-1)$-th step to t-th step are considered as the basis to detect changes to each of the land use classes considered. Each diagonal cell in the transition probability matrix represents the transition probability for a particular land use class between the chosen time steps. In contrast, the rest of the cells represent the probability of transition from one land use class to another. The stochastic nature of the approach does not allow for a thorough explanation of the change processes involved. Furthermore, since land cover change is described in aggregated terms, the model is not spatially explicit [14]. A limitation of the Markov chain approach in deciding the land use class for each pixel in the study area is that it does not consider knowledge on land use classes of its neighboring pixels. To address this, modelers are using cellular automata (CA) in conjunction with the Markov chain approach to arrive at reliable future LULC projections (e.g., [11, 18]). It is henceforth referred to as the CA-Markov model, which aids in detecting LULC changes and future projections effectively by providing a spatial sense to the transitional probabilities.

2.2 CA-Markov Model

The CA-Markov model is a combined cellular automaton, Markov chain, multi-criteria evaluation, multi-objective land allocation LULC prediction procedure that adds an element of spatial contiguity as well as knowledge of the likely spatial distribution of transitions to Markov chain analysis. In the application of CA, each pixel (cellular entity) varies its state independently based on its previous state and that of its immediate neighbors (pixels) based on a specified set of user-defined rules (e.g., [7]). The rules aid in associating an extent to the land use change spatially. Owing to the advantages of the CA-Markov approach, the CA-Markov module available with IDRISI software v17.02 [8] was considered in the present study for modeling LULC. The module allows simultaneous simulation of different land use and land cover classes. A detailed description of CA can be found in Schiff [27].

The steps involved in the methodology considered to determine future projections of LULC are as follows.

(1) Data collection: Procure DEM data and satellite imageries corresponding to different time instances for the study area. Description of the data utilized to prepare LULC maps for the study area is provided in Sect. 3.
(2) Preparation of LULC maps: Classify the LULC in the study area into different classes and prepare LULC maps corresponding to different time instances. In this study, land use and land cover were classified into eight classes by considering the level II classification system [2]. The LULC classes include (i) built-up urban, (ii) built-up rural, (iii) agricultural cropland, (iv) agricultural plantation, (v) barren land, (vi) forest, (vii) wetland, and (viii) lakes/ponds.
(3) *Change detection analysis*: Visually inspect the LULC maps corresponding to different time instances to detect any changes. In addition, quantify the changes in terms of transition probability matrix, each of whose elements describes the probability for the transition from one class of LULC to another/same class.
(4) *Prediction of future LULC maps and their validation*: In this study, the Markov chain module and CA-Markov module available with IDRISI, v17.02 software [6] were utilized to model LULC in the study area for change detection and predict future changes to LULC. The module allows simultaneous simulation of different LULC classes. LULC images corresponding to two different time steps are provided as input to the Markovian transition estimator of IDRISI. The images are referred to as the first (earlier) LULC image and second (later) LULC image. The initial Markov transition probabilities matrix is determined based on the observed frequencies of transition between the first and the second LULC images using the Markovian transition estimator. The matrix establishes the frequency with which each cell within a specific LULC class will undergo a transition to any other land cover class. The Markov transition area files contain information on how many pixels of each LULC class undergo a transition to each of the other LULC classes. The transition probability matrix from a Markov chain analysis of two prior LULC maps establishes the quantity of expected LULC change from each existing category to each other category

in the next time step for prediction of future LULC map. Since the Markov process does not establish where (in the study area) the land use transitions would occur, the predicted future LULC map will not be a smooth prediction. To address this issue, a group of suitability images (also called conditional probability images) is prepared for each of the LULC classes. The conditional probability image for a LULC class provides the probability associated with each pixel in the image for transition to the LULC class. The time (in years) between the second LULC image (later image) and the future LULC image determines the number of iterations that must be considered in the simulation to arrive at Markov transition probabilities and Markov transition areas file for the future year.

The Markov transition probabilities and conditional probability images form the basis for the prediction of LULC based on the CA-Markov model. Along with these, a contiguity filter helps determine the suitability of pixels under consideration for deciding the LULC class by giving preference to neighboring pixels. The filter is an integral step to the action of the CA component. This filter provides each cell a certain chance to be assigned to a LULC class if it is near an existing cluster of cells of the same LULC class. To check if LULC class of the target cell undergoes a transition to urban, the filter is placed on the target cell in the suitability image corresponding to the LULC class under consideration (Urban class suitability image). The CA filter helps to develop a LULC map that is spatially explicit to the change of state of cells based on its neighbors, thus giving geography more importance in the solution [6]. For the target cell, the surrounding cells must belong to the same state (urban LULC class). Hence, the target cell's tentative LULC class (urban) could be declared as the true class. The model's performance is validated by verifying if the LULC class corresponding to each of the cells (in the true future image) is predicted correctly by the projected image. If the model performance is fairly good, the validated model could be used to predict LULC maps corresponding to years far into the future. The methodology flow chart is presented in (Fig. 1).

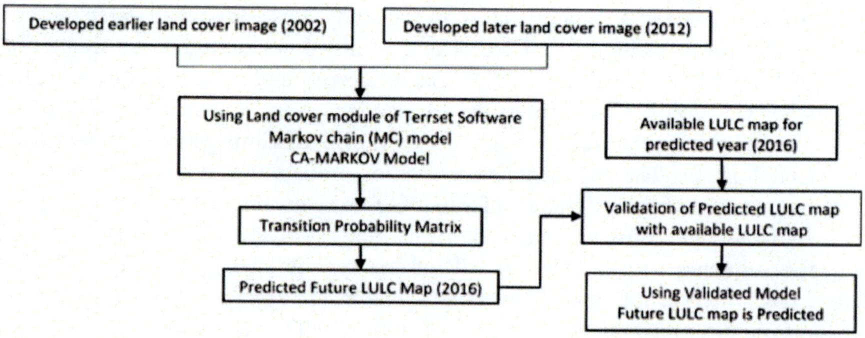

Fig. 1 Methodology flowchart

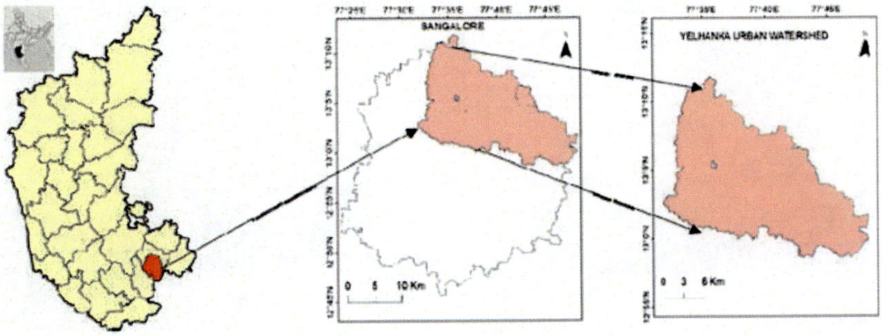

Fig. 2 Location map of Yelahanka watershed considered for the study

3 Case Study

3.1 Study Area and Data

The study is conducted on the Yelahanka urban watershed located within Bangalore city in the Karnataka state of India. The watershed covers an area of 356 km^2 and is a part of the Arkavathi river basin. As mentioned earlier, this study is envisaged to evaluate the performance of different models in simulating LULC changes witnessed over the Yelahanka urban watershed, Bangalore. The watershed has witnessed rapid urbanization in the recent past, engulfing its neighboring villages/towns. Location map for Yelahanka watershed is presented (Fig. 2).

Historical LULC maps were prepared for the study area corresponding to five years (1996, 2002, 2006, 2012, and 2016). The maps related to the years 1996 and 2002 (Fig. 3a, b) were extracted from 1:50,000 scale imageries taken by Indian remote sensing satellite IRS-1C with linear imaging self scanner (LISS)-III sensor. The maps for the latter years 2006, 2012 (Fig. 3c, d), and 2016 were prepared based on thematic maps available at 1:50,000 scale for the study area from Bhuvan-Thematic Services of National Remote Sensing Center (NRSC), ISRO, Hyderabad, India. A level II classification was carried out to classify LULC in each of the five maps into eight classes: (i) built-up urban, (ii) built-up rural, (iii) agricultural cropland, (iv) agricultural plantation, (v) barren land, (vi) forest, (vii) wetland, and (viii) lakes/ponds.

3.2 Prediction of Future Lulc Maps and Their Validation

Change detection analysis is very helpful in identifying and locating various geospatial changes occurring in different classes of land use. Post-LULC classification comparison method was considered to analyze the prepared LULC maps and detect

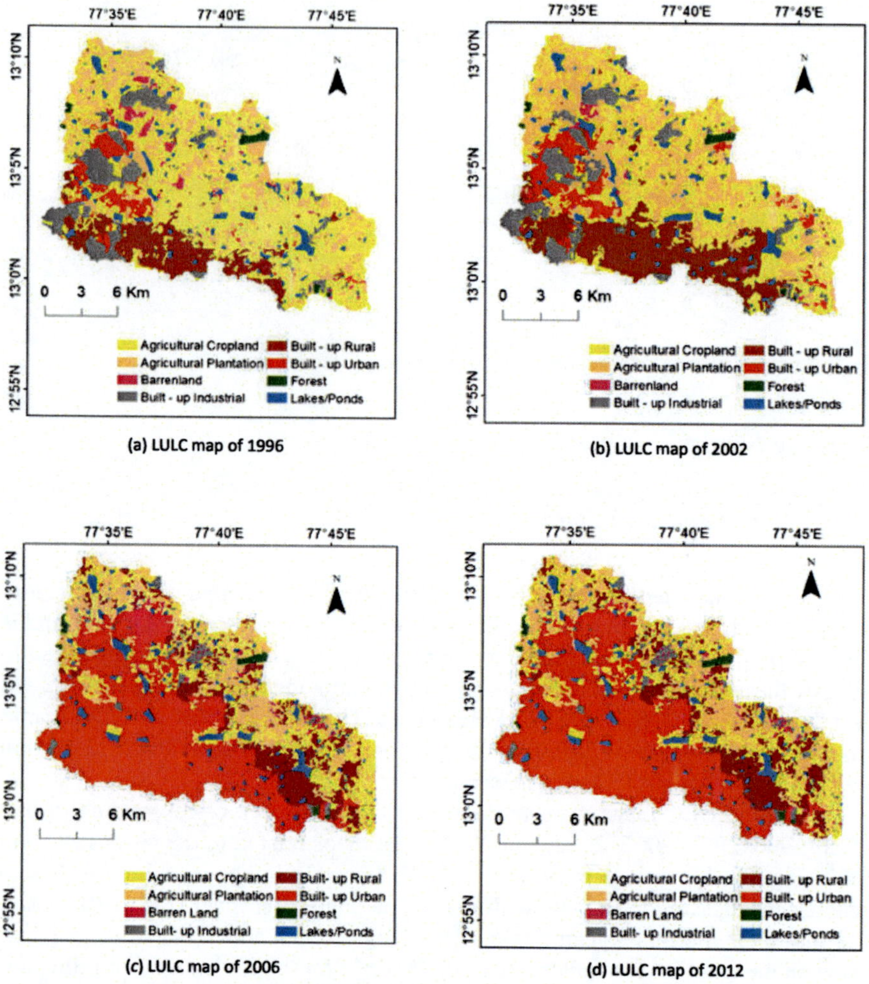

Fig. 3 LULC maps of Yelahanka from the years **a** 1996, **b** 2002, **c** 2006, and **d** 2012

temporal changes in various land use classes for 30 m resolution pixels in the study area. It involved a comparison of calculated areas corresponding to each of the eight LULC classes across the years 1996–2016 to ascertain the changes which have taken place (Fig. 4).

It can be observed from Fig. 4 that LULC has changed in the Yelahanka watershed during the period 1996–2016, and the changes are significant between the years 1996 and 2006, during which area under agricultural cropland has decreased from approximately 160–60 km^2. Reduction is also evident in the area classified as built-up industrial. In contrast, the built-up urban area has significantly increased (from about 15 km^2 to about 175 km^2) during the period 1996–2016. The increase could

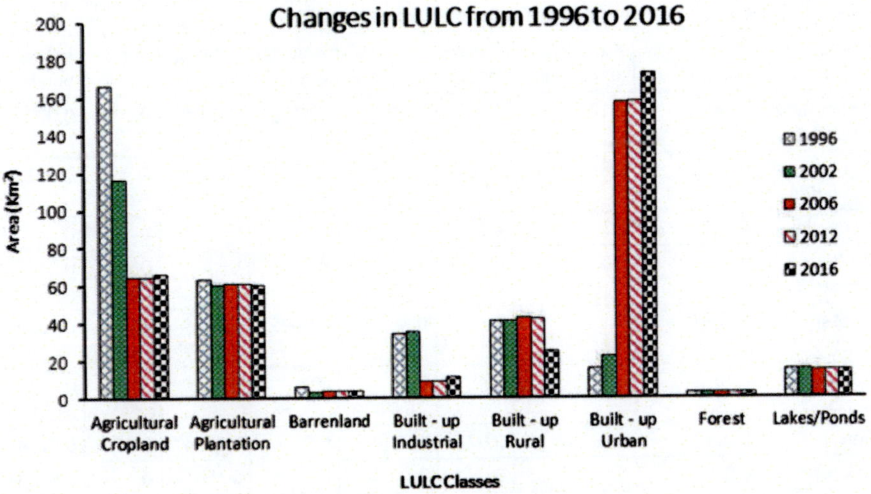

Fig. 4 Changes in different LULC classes for the Yelahanka watershed between the years 1996 and 2016

be primarily attributed to a reduction in (i) agricultural cropland during 1996–2006, (ii) built-up industrial area during the period 2002–2006, and (iii) built-up rural area during the period 2012–2016. The changes appear insignificant for areas classified as agricultural plantation, barren land, forest, and lakes/ponds.

To obtain future projections of LULC, Markov chain and CA-Markov models were developed for several map pairs (e.g., 1996–2002, 2002–2006, 2006–2012, 2012–2016, and 2002–2012) using IDRISI software. The model utilizes suitability maps corresponding to each LULC class and LULC transition probability matrix across the several map pairs to arrive at future projections of LULC. The transition probability matrices were subsequently analyzed to draw inferences on LULC changes across different years. The matrices indicated that changes in LULC are significant across several map pairs, suggesting that modeling of LULC in the study area is necessary to assess possible changes in runoff generated from the watershed. For brevity, the transition probability matrix for the 2002–2012 map pair is shown in Table 2.

The CA-Markov model/Markov chain model developed for each map pair was validated by comparing the LULC map projected by it for the year 2016 with the true LULC image prepared for the year using a 1:50,000 thematic map from NRSC. The comparison was in terms of (i) correlation coefficient R and (ii) the percentages of area and pixels under each LULC class correctly predicted by the model. The validation tool REGRESS available in TerrSet software was used to determine the *R*-value. Table 3 presents the *R*-values obtained for the models (Markov chain and CA-Markov) developed corresponding to several map pairs considered. The table shows that the *R*-value for the CA-Markov model is higher than that for the Markov chain model for each of the map pairs considered. Furthermore, the *R*-value is highest for the CA-Markov model corresponding to the 2002–2012 map pair.

Table 2 Transition probability matrix for the map pair 2002–2012

Classes	Agricultural cropland	Agricultural plantation	Barren land	Built-up industrial	Built-up rural	Built-up urban	Forest	Lakes/ponds
Agricultural cropland	0.8494	0.1221	0.0012	0.0015	0.0194	0.0019	0.0023	0.0023
Agricultural plantation	0.1131	0.8496	0.0018	0.0039	0.0295	0.0001	0	0.002
Barren land	0.1503	0	0.8497	0	0	0	0	0
Built-up industrial	0	0	0	0.8498	0.1502	0	0	0
Built-up rural	0.0084	0	0	0	0.8495	0.1419	0	0.0002
Built-up urban	0.15	0	0	0	0	0.85	0	0
Forest	0.0214	0.0214	0.0214	0.0214	0.0214	0.0214	0.85	0.0214
Lakes/ponds	0.0002	0	0	0	0	0.154	0	0.8458

Table 3 Estimates of correlation coefficient (*R*) for the models developed corresponding to different map pairs. The optimal value of R is highlighted in bold font

Year	*R*-value for Markov chain model	*R*-value for CA-Markov model
2006–2012	0.92	0.93
2002–2012	0.90	**0.94**
1996–2006	0.90	0.92
2002–2006	0.82	0.89
1996–2002	0.74	0.76

The area corresponding to each LULC class in the image projected by the developed Markov chain and CA-Markov models for the year 2016 in analysis with each of the map pairs was expressed as a percentage of the corresponding area in the true LULC image of the 2016 image (Tables 4 and 5). If the value corresponding to a LULC class is greater than 100%, it indicates that the model overpredicts the area belonging to that class and vice-versa. For the best-fit CA-Markov model developed based on the maps pair 2002–2012, the values for most classes were around 100%, except for the built-up rural class that covers about 10% of the study area. The number of pixels corresponding to each LULC class predicted correctly in the true image of the year 2016 was assessed, and errors were quantified (Table 6). From the table, it can be noted that the errors are fairly low for all the LULC classes, except for the built-up rural class, as expected. The LULC maps projected by Markov chain and CA-Markov models for the year 2016 (based on the 2002 and 2012 maps pair) are compared with the true LULC image of the year (Fig. 5). The best-fit model (i.e., CA-Markov) is used to arrive at the projected LULC map for the future year 2026, which is presented in Fig. 6.

Table 4 Area corresponding to each LULC class in the image projected by Markov chain model for the year 2016 (expressed as a percentage of the corresponding area in the true LULC image)

LULC classes	Area predicted using Markov chain model (%)				
	1996–2002	1996–2006	2002–2006	2002–2012	2012–2016
Agricultural cropland	141.56	70.03	25.31	133.57	97.55
Agricultural plantation	133.64	57.01	29.42	98.90	100.54
Barren land	69.82	70.13	104.02	90.96	105.20
Built-up industrial	321.61	40.29	20.73	67.82	79.29
Built-up rural	380.54	118.77	87.56	160.81	168.70
Built-up urban	17.44	125.52	155.40	82.75	91.73
Forest	67.91	88.88	100.43	92.06	100.49
Lakes/ponds	101.33	94.66	103.21	86.80	102.15

Table 5 Area corresponding to each LULC class in the image projected by the CA-Markov chain model for the year 2016 (expressed as a percentage of the corresponding area in the true LULC image)

LULC classes	Area predicted using CA-Markov model (%)				
	1996–2002	1996–2006	2002–2006	2002–2012	2012–2016
Agricultural cropland	141.56	73.45	38.00	**97.42**	162.66
Agricultural plantation	133.64	67.23	37.17	**100.39**	143.25
Barren land	69.82	74.21	117.83	**104.59**	88.91
Built-up industrial	321.61	48.66	63.52	**91.60**	312.94
Built-up rural	380.54	109.56	184.14	**163.55**	315.39
Built-up urban	17.44	129.27	132.12	**91.64**	15.56
Forest	67.91	90.15	93.27	**100.17**	70.54
Lakes/ponds	101.33	95.85	92.89	**102.52**	105.35

Optimal values are highlighted in bold font

Table 6 Error in predicting LULC classes for the year 2016 at pixel scale using the CA-Markov model developed based on the 2002–2012 LULC maps pair

LULC classes	Count of pixels		Area of each class (Km^2)		Error in prediction for pixels (%)
	Actual	Predicted	Actual	Predicted	
Agricultural cropland	69,254	67,539	62.3	60.7	−2.48
Agricultural plantation	65,408	65,713	58.9	59.1	0.47
Barren land	3960	4166	3.6	3.7	5.20
Built-up industrial	12,178	11,156	10.9	10.0	−8.39
Built-up rural	27,535	45,035	24.8	40.5	63.56
Built-up urban	190,727	174,862	171.6	157.3	−8.32
Forest	3048	3063	2.7	2.7	0.49
Lakes/ponds	15,282	15,687	13.7	14.1	2.65

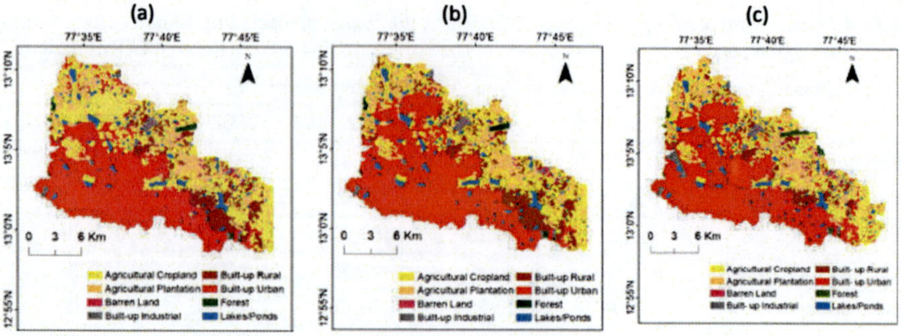

Fig. 5 Comparison of LULC maps projected for the year 2016 based on **a** Markov chain model and **b** CA-Markov models with the **c** actual LULC Map of 2016

Fig. 6 LULC map projected for the year 2026 based on CA-Markov model for 2002–2012 map pair

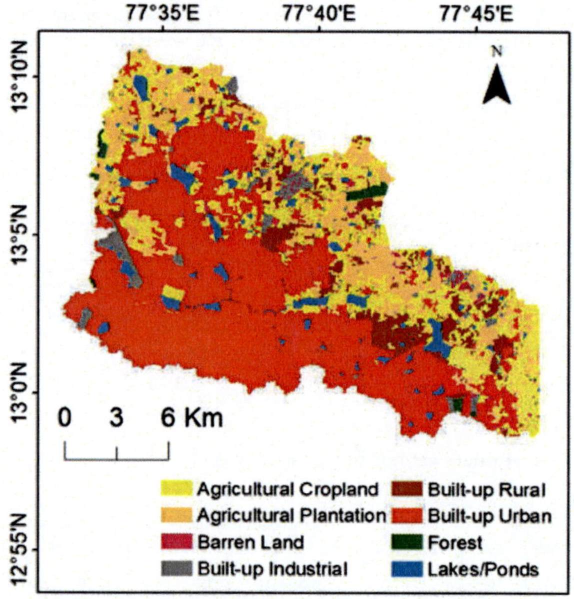

4 Summary and Concluding Remarks

The performance of Markov chain and CA-Markov models in simulating LULC changes witnessed in Yelahanka urban watershed, Bangalore, was investigated. Analysis of historical LULC maps of the watershed prepared for five typical years (1996, 2002, 2006, 2012, and 2016) in the last three decades indicated considerable changes in various LULC classes. The changes were found to be significant between the years 1996 and 2006, which were in the form of reduction in agricultural cropland and increase and in the built-up urban area. The increase in the latter could be primarily attributed to a reduction in (i) agricultural cropland during 1996–2006, (ii) built-up industrial area during the period 2002–2006, and (iii) built-up rural area during 2012–2016. Ten LULC models were developed, comprising five based on each model (Markov chain, CA-Markov) for a LULC map pair (e.g., 1996–2002, 2002–2006, 2006–2012, 2012–2016, and 2002–2012). The models utilized suitability maps corresponding to each LULC class and Markov transition probability matrix across the years 1996–2012 to arrive at LULC projections for the year 2016. Each model was validated by comparing the LULC map projected by it with the actual/true LULC map prepared for the year 2016. The error in projecting LULC was quantified at a 30 m resolution pixel scale in terms of correlation coefficient R and predicted LULC class in terms of percentage area predicted correctly. It is observed that among the models developed based on each of the LULC map pairs, the CA-Markov model exhibited better performance than the Markov chain model in yielding reliable projections of the land use classes. Furthermore, among the ten models, the performance was best for the CA-Markov model developed considering the 2002–2012 LULC map pair. The model was deemed suitable for use in projecting future LULC maps. Changes in the LULC map projected for the future year 2026 were quantified with respect to the true image for the year 2016. It indicated an increase in the built-up urban and industrial areas by about 6% and 1%, respectively, and a decrease/reduction in (i) built-up rural area by about 4%, and (ii) combination of agricultural cropland and plantation by about 3%. There is not much change in projections for barren land, forest, and lakes/ponds.

Acknowledgements Authors express their gratitude to the Department of Science and Technology (DST) for grants provided through Project No. DST/CCP/NUC/146/2018 (G), and to Ministry of Earth Sciences for the support provided to ICWaR of IISc through Project No. MOES/PAMC/H&C/41/2013-PC-II.

References

1. Ahmad, F., Goparaju, L., & Qayum, A. (2017). LULC analysis of urban spaces using Markov chain predictive model at Ranchi in India. *Spatial Information Research, 25*(3), 351–359.
2. Anderson, J. R. (1976). *A land use and land cover classification system for use with remote sensor data* (Vol. 964). US Government Printing Office.
3. Babbar, D., Areendran, G., Sahana, M., Sarma, K., Raj, K., & Sivadas, A. (2021). Assessment and prediction of carbon sequestration using Markov chain and InVEST model in Sariska Tiger Reserve, India. *Journal of Cleaner Production, 278*, 123333.
4. Bose, A., & Chowdhury, I. R. (2020). Monitoring and modeling of spatio-temporal urban expansion and land-use/land-cover change using Markov chain model: A case study in Siliguri Metropolitan area, West Bengal India. *Modeling Earth Systems and Environment, 6*(4), 2235–2249.
5. Dai, F. C., Lee, C. F., & Zhang, X. H. (2001). GIS-based geo-environmental evaluation for urban land-use planning: A case study. *Engineering geology, 61*(4), 257–271.
6. Eastman, J. R. (2006). *Idrisi Andes*. Clark Labs, Clark University, Worcester, MA.
7. Eastman, J. R. (2001). Guide to GIS and image processing volume. *Clark Labs, 2*, 1–144.
8. Eastman, J. R. (2012). *IDRISI Selva Tutorial, Manual Version 17.0*. Clark University.

9. Gidey, E., Dikinya, O., Sebego, R., Segosebe, E., & Zenebe, A. (2017). Cellular automata and Markov Chain (CA_Markov) model-based predictions of future land use and land cover scenarios (2015–2033) in Raya, northern Ethiopia. *Modeling Earth Systems and Environment, 3*(4), 1245–1262.
10. Hamad, R., Balzter, H., & Kolo, K. (2018). Predicting land use/land cover changes using a CA-Markov model under two different scenarios. *Sustainability, 10*(10), 3421.
11. Houet, T., & Hubert-Moy, L. (2006). Modeling and projecting land-use and land-cover changes with cellular automaton in considering landscape trajectories. *EARSeL eProceedings, 5*(1), 63–76.
12. Kafy, A. A., Rahman, M. S., Hasan, M. M., & Islam, M. (2020). Modelling future land use land cover changes and their impacts on land surface temperatures in Rajshahi, Bangladesh. *Remote Sensing Applications: Society and Environment, 18*, 100314.
13. Kantakumar, L. N., Kumar, S., & Schneider, K. (2019). SUSM: A scenario-based urban growth simulation model using remote sensing data. *Eur J Remote Sens, 52*(sup2), 26–41.
14. Lambin, E. F. (1997). Modelling and monitoring land-cover change processes in tropical regions. *Progress in Physical Geography, 21*(3), 375–393.
15. López, E., Bocco, G., Mendoza, M., & Duhau, E. (2001). Predicting land-cover and land-use change in the urban fringe: A case in Morelia city. *Mexico. Landscape and urban planning, 55*(4), 271–285.
16. Lu, Q., Chang, N. B., Joyce, J., Chen, A. S., Savic, D. A., Djordjevic, S., & Fu, G. (2018). Exploring the potential climate change impact on urban growth in London by a cellular automata-based Markov chain model. *Computers, Environment and Urban Systems, 68*, 121–132.
17. Mansour, S., Al-Belushi, M., & Al-Awadhi, T. (2020). Monitoring land use and land cover changes in the mountainous cities of Oman using GIS and CA-Markov modelling techniques. *Land Use Policy, 91*, 104414.
18. Mitsova, D., Shuster, W., & Wang, X. (2011). A cellular automata model of land cover change to integrate urban growth with open space conservation. *Landscape and urban planning, 99*(2), 141–153.
19. Mondal, M. S., Sharma, N., Kappas, M., & Garg, P. K. (2020). Cellular automata (CA) contiguity filters impacts on ca Markov modeling of land use land cover change predictions results. *The International Archives of the Photogrammetry, Remote Sensing and Spatial Information Sciences, XLIII-B3-2020*, 1585–1591. https://doi.org/10.5194/isprsarchives-XLIII-B3-2020-1585-2020
20. Mustafa, A., Ebaid, A., Omrani, H., & McPhearson, T. (2021). A multi-objective Markov Chain Monte Carlo cellular automata model: Simulating multi-density urban expansion in NYC. *Computers, Environment and Urban Systems, 87*, 101602.
21. Nasiri, V., Darvishsefat, A. A., Rafiee, R., Shirvany, A., & Hemat, M. A. (2019). Land use change modeling through an integrated multi-layer perceptron neural network and Markov chain analysis (case study: Arasbaran region, Iran). *Journal of Forestry Research, 30*(3), 943–957.
22. Niehoff, D., Fritsch, U., & Bronstert, A. (2002). Land-use impacts on storm-runoff generation: Scenarios of land-use change and simulation of hydrological response in a meso-scale catchment in SW-Germany. *Journal of Hydrology, 267*(1), 80–93.
23. Paul, S. S., Li, J., Wheate, R., & Li, Y. (2018). Application of object oriented image classification and Markov chain modeling for land use and land cover change analysis. *Journal of Environmental Informatics, 31*(1), 30–40.
24. Rahnama, M. R. (2020). Forecasting land-use changes in Mashhad Metropolitan area using cellular automata and Markov chain model for 2016–2030. *Sustainable Cities and Society*, 102548.
25. Ramachandra, T. V., & Mujumdar, P. P. (2009). Urban floods: Case study of Bangalore. *Disaster Development, 3*(2), 1–98.
26. Rimal, B., Zhang, L., Keshtkar, H., Wang, N., & Lin, Y. (2017). Monitoring and modeling of spatiotemporal urban expansion and land-use/land-cover change using integrated Markov chain cellular automata model. *ISPRS International Journal of Geo-Information, 6*(9), 288.

27. Schiff, J. L. (2011). *Cellular automata: a discrete view of the world* (Vol. 45). John Wiley & Sons.
28. Shen, L., Li, J. B., Wheate, R., Yin, J., & Paul, S. S. (2020). Multi-layer perceptron neural network and Markov chain based geospatial analysis of land use and land cover change. *Journal of Environmental Informatics Letters, 3*, 29–39.
29. Surabuddin Mondal, M., Sharma, N., Kappas, M., & Garg, P. K. (2013). Modeling of spatio-temporal dynamics of land use and land cover in a part of Brahmaputra River basin using Geoinformatic techniques. *Geocarto International, 28*(7), 632–656.
30. Talib, A., & Randhir, T. O. (2017). Climate change and land use impacts on hydrologic processes of watershed systems. *Journal of Water and Climate Change, 8*(3), 363–374. https://doi.org/10.2166/wcc.2017.064
31. Thakur, T. K., Patel, D. K., Bijalwan, A., Dobriyal, M. J., Kumar, A., Thakur, A., Bohra, A., Bhat, J. A. (2020) Land use land cover change detection through geospatial analysis in an Indian Biosphere Reserve. *Trees, Forests and People* (Vol. 2), 100018, ISSN 2666–7193.
32. Tong, S. T., Sun, Y., Ranatunga, T., & Yang, Y. J. (2012). Predicting plausible impacts of sets of climate and land use change scenarios on water resources. *Applied Geography, 32*(2), 477–489.
33. Wan, R., & Yang, G. (2007). Influence of land use/cover change on storm runoff—A case study of Xitiaoxi River Basin in upstream of Taihu Lake Watershed. *Chinese Geographical Science, 17*(4), 349–356.
34. Wang, S. W., Munkhnasan, L., & Lee, W. K. (2021). Land use and land cover change detection and prediction in Bhutan's high-altitude city of Thimphu, using cellular automata and Markov chain. *Environmental Challenges*, 2, 100017.
35. Wu, Q., Li, H. Q., Wang, R. S., Paulussen, J., He, Y., Wang, M., Wang, B. H., & Wang, Z. (2006). Monitoring and predicting land use change in Beijing using remote sensing and GIS. *Landscape and urban planning, 78*(4), 322–333.
36. Xiao, J., Shen, Y., Ge, J., Tateishi, R., Tang, C., Liang, Y., & Huang, Z. (2006). Evaluating urban expansion and land use change in Shijiazhuang, China, by using GIS and remote sensing. *Landscape and urban planning, 75*(1–2), 69–80.

Identification of Suitable Sites for Urban Solid Waste Disposal of GVMC Area, Visakhapatnam by Multicriteria Spatial Analysis

Pendyala Stephen, Kuna Bhindu Bhargavi, Gummapu Jai Sankar, and G. Viswanadha Kumar

Abstract Solid waste management (SWM) is becoming a most significant environmental issue. Improper solid waste management results in difficulty for the habitats. There is an increase in population rate in the urban areas in recent years because of migration and this in returns increases the commercial, residential and infrastructural improvements and these have many negative impacts on environment. Locating solid waste dumping sites in unsuitable areas have become one major impact. To overcome such problems proper site, need to be selected for solid waste dumping. Geographic Information System (GIS) will allow users to understand spatial and non-spatial data in many ways that disclose patterns and trends in the form of maps, reports and charts. This paper deals with the determination of suitable site for disposal of urban solid waste generated from Greater Visakhapatnam Municipal Corporation (GVMC) area and surroundings using geological information system (GIS) techniques. Several parameters considered while selecting suitable site for urban solid waste disposal are land use and landcover map, soil map, geomorphology map, geology map, drainage map, and ground water depths, by considering weighted overlay method the weightages were given to the considered parameters depending upon their relative importance. The results show the suitability map for urban solid waste disposal.

Keywords Environment · GIS · Municipal solid waste · Solid waste management · Spatial and attribute · Urban areas · Weighted overlay method

P. Stephen (✉) · G. Jai Sankar
Department of Geo-Engineering, Andhra University, Visakhapatnam, India
e-mail: stephen@andhrauniversity.edu.in

K. Bhindu Bhargavi · G. Viswanadha Kumar
Department of Architecture, Andhra University, Visakhapatnam, India

A. K. Dikshit et al. (eds.), *Innovative Trends in Hydrological and Environmental Systems*, Lecture Notes in Civil Engineering 234,
https://doi.org/10.1007/978-981-19-0304-5_23

1 Introduction

Due to unsustainable waste disposal practices, the environment is facing risk. Because of unexpected developments within the populated area, huge waste is being generated everywhere in the country. Solid waste predicts to extend to 677 million tonnes by 2025 [1]. At that point, low-income countries expect to contribute 70.9% of overall solid waste worldwide [1]. Because of this improper disposal of solid waste, some major environmental issues result in major problems like transmission of diseases, odour nuisance, air and water pollution, economic losses, etc., [2] [3] stated that the elementary goals of solid waste management methods are to handle the aesthetic, land use, economic considerations, health, and environmental aspects connected with the inappropriate disposal of waste. Geographical information system (GIS) is a using tool to identify suitable sites for solid waste disposal and the best method for selecting environmentally sustainable and socially acceptable disposal site/sites with precision in cost and time. In a perfect world, a well-designed landfill uses a designed strategy of transfer of solid waste on land in a way that minimizes environmental hazards by spreading the solid squander in thin layers, compacting the solid squander to the littlest viable volume, and applying a cover after each working day [4].

2 Study Area

The study area spans about 581 km^2, i.e. Greater Visakhapatnam Municipal Corporation (GVMC) in Visakhapatnam, Andhra Pradesh. As Visakhapatnam is a popular tourist destination, the towns are more attractive, including beaches, recreation parks, and various other attractions (Fig. 1).

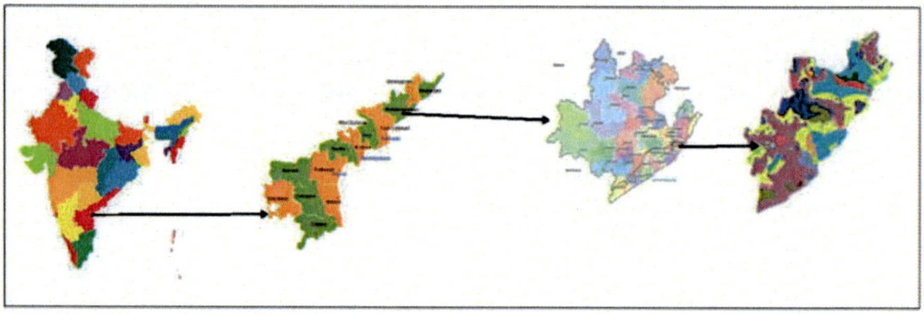

Fig. 1 Study area

3 Materials and Method

(A) **Data Used**
(B) Survey of India (SOI) Toposheets—65KO/1,65O/2 & O/3, 65O/5, 65O/6 of scale 1:50,000
(C) Sentinel-2A satellite image of GVMC study area with 10 m spatial resolution (April 24, 2020)
(D) ASTER-DEM (30 m) data
(E) ARC GIS 10.4
(F) Methodology

For the present study area, various data sources like Toposheets, remote sensing data, geological data, soil data, groundwater level data, surface water resources data are collected. Sentinel-2A satellite imagery data acquired on Mar 29, 2017 were used to generate various thematic maps. When choosing a solid waste disposal site, thematic maps are considered: (land use/land cover, geology, geomorphology, soil, slope, drainage network, groundwater level) (Fig. 2).

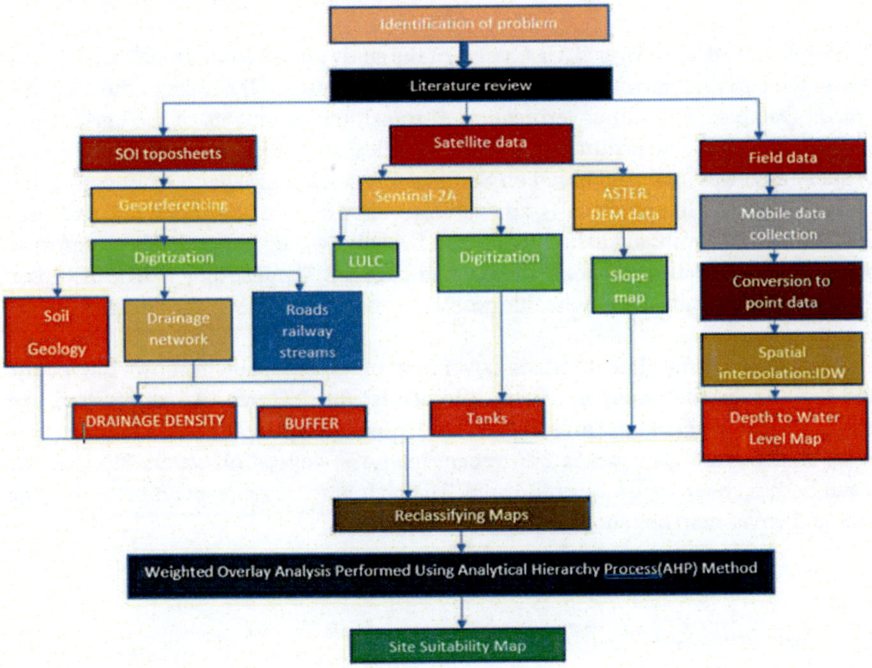

Fig. 2 Methodology flowchart

4 Results

Thematic maps

4.1 Slope

The slope map produced by the ASTER 30-m resolution digital elevation model (DEM) was obtained from the USGS portal; the slope map was prepared and divided into four categories based on the high value. Pollutants can travel farther from the control zone [5]. So, slopes with a lower degree are highly suitable than the higher degree to avoid the risk of water contamination, conveyance etc., the details of the slope map shown in Fig. 3.

4.2 Land Use and Land Cover

The land use and land cover (LULC) map of the study area provided data to help users obtain the current characteristics of the area. The annual LULC data in the National Spatial Database will allow verification of transitional elements of the agricultural ecosystem, forest conversion, surface water bodies, etc. The land use/land cover map of study area was analyzed based on Sentinel-2A satellite imagery. After consulting various types of literature, it is advisable to choose land that does not live near the habitat and shrubland called wasteland for solid waste disposal. The comfort of people when choosing a disposal site is very important, because they will be disturbed by flies and will cause many health problems and so on. The LULC map shown in Fig. 4.

The study area has land use/land cover map of Visakhapatnam shows the feature and their respective areas in square kilometres; the features like agriculture and beach cover 1% of the total area, barren land 5%, built-up areas 23%, and forest area 30%, respectively. Salt panels 2%, vacant land 4%, vegetation covers 3%, and the water bodies cover 4% of the entire area. The pictographic representation of the land use/land cover map has shown in Fig. 5.

4.3 Geomorphology Maps

Geomorphology takes high importance whilst selecting land for solid waste disposal. It consists of pediplain, denudation hill, inselberg, marine clay, marshy land, residual hill, structural hill, valley fills. Moreover, water bodies, dissected hills, and inter-mountain valleys are the dominant landforms found in the study boundary shown in

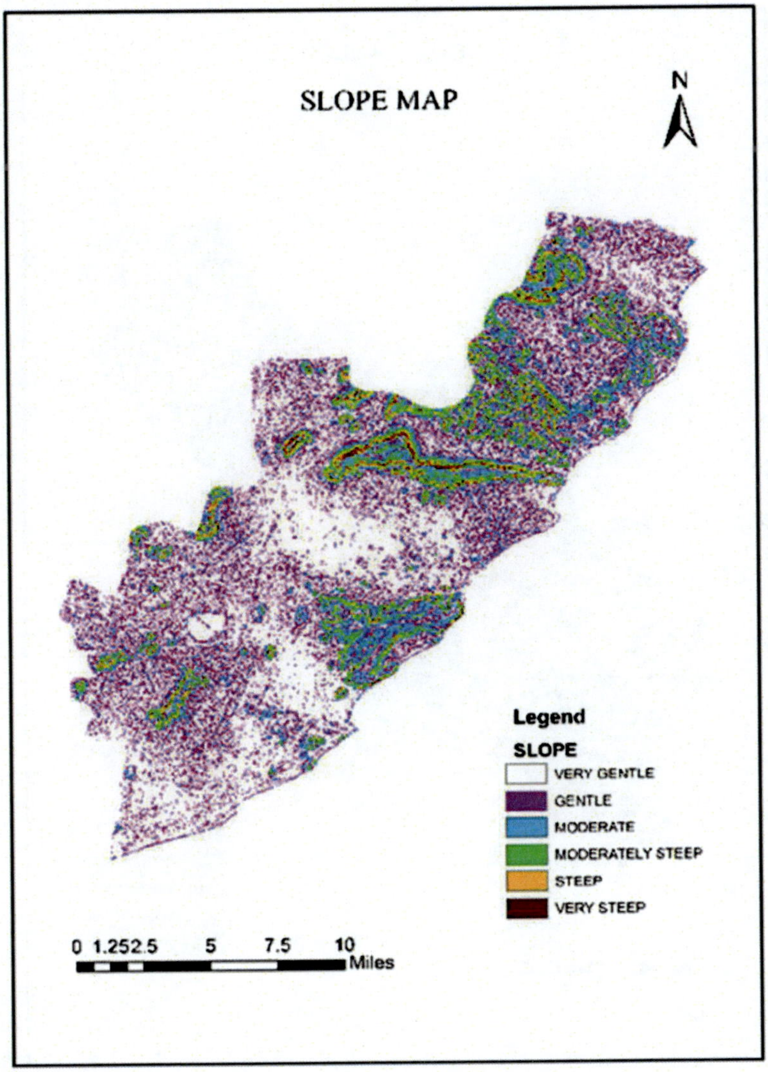

Fig. 3 Slope map

Fig. 6. The thickness and intensity of these landforms vary depending upon the region's slope and essential structural impact. The geomorphic features provide the best suitable dumpsite locations based on their weathering nature, plain nature, unsuitable hilly terrain features, and other coastal zone features.

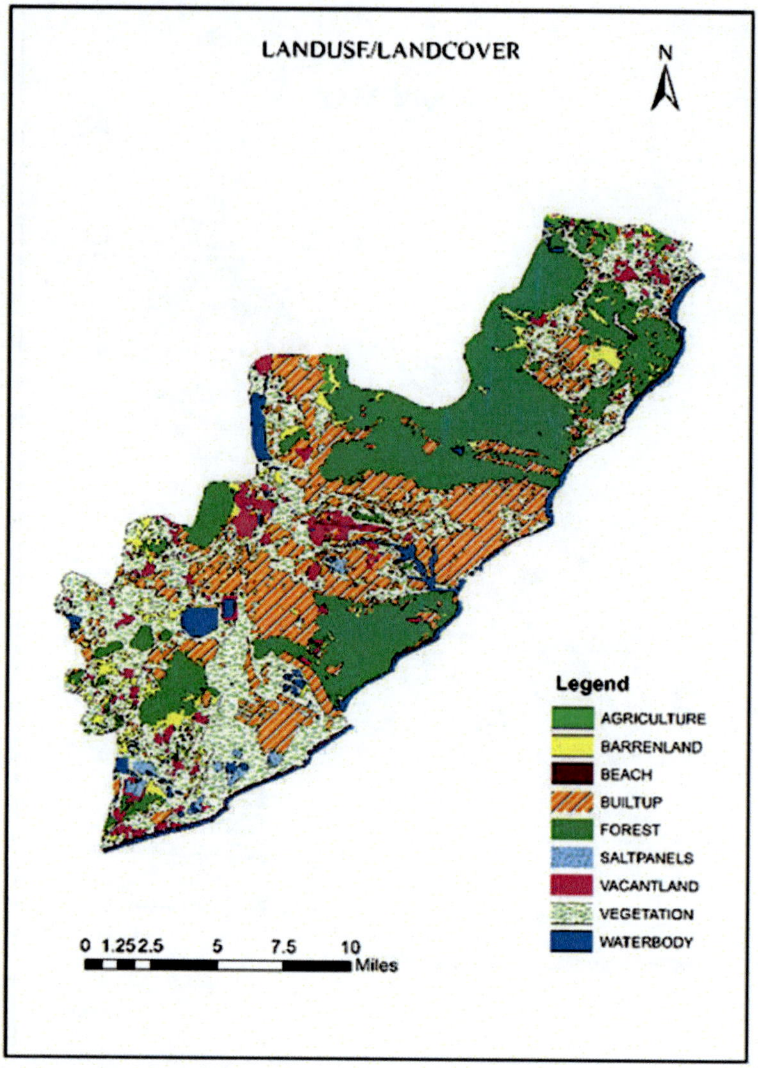

Fig. 4 Land use/land cover map

4.4 Soil Map

Generally, a soil map is a geographical representation that shows the variety of soils and soil properties in the study area. For land assessment, conventional soil maps are regularly utilized to understand the distribution of soils. However, they were required for more specific purposes, such as determining the suitability of the soil for various crops, construction of harvesting structures, and land drainage capabilities of an area.

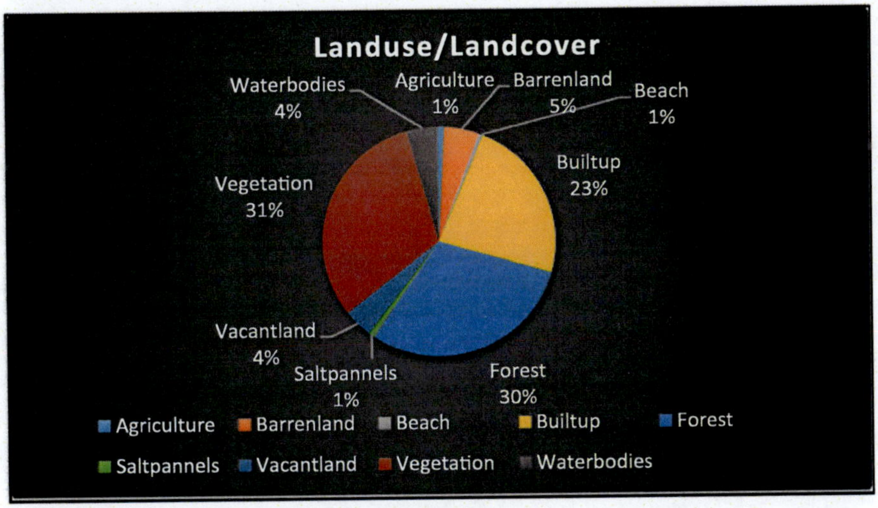

Fig. 5 Percentage representation of land use/land cover map

Whilst studying the soil map of the GVMC area, it reveals that brown clay soil, brown gravel clay soil, brown gravel loamy soil, clay soil, red clay soil, red coastal clay soil, red loamy soil, and sandy soil were well-known. Brown clay soil is mainly in Visakhapatnam rural mandal areas like Madhurwada, Kommadhi, and some parts of Peddha Gantyada Mandals.

Similarly, brown gravel clay soils cover areas like NAD, Narva, steel plant, and parts of Peddha Gantyada Mandals. Whereas, brown gravel loamy soil is covered only beside the Meghadrigedda area. The properties of clay soil are traced in the regions of Gopalapatnam, NAD, Sheelanagar. Red loamy soil is been scattered in industrial areas and sandy soil near Meghadrigedda and some places in Peddha gantyada and Gajuwaka Mandals. Fig. 7 shows the soil map. The less permeable soil type like clay that is less porous to the leachate to enter into the groundwater is best suitable than the loamy and sandy soils.

4.5 Geology Map

A geology map has been made using GSI data for Greater Visakhapatnam Municipal Corporation (GVMC), Visakhapatnam District. This area consists primary geological features like khondalite, charnockite, laterite, granite gneiss, tidal flat, and quartzite. Charnockite and quartzite ensue as intrusive bodies into the country-rock. Khondalite covers a significant study area portion. Khondalite and charnockite rocks are mostly occurring in the study area compared to other rocks. Charnockite is highly impermeable and highly suitable for disposal sites. Fig. 8 shows a geology map.

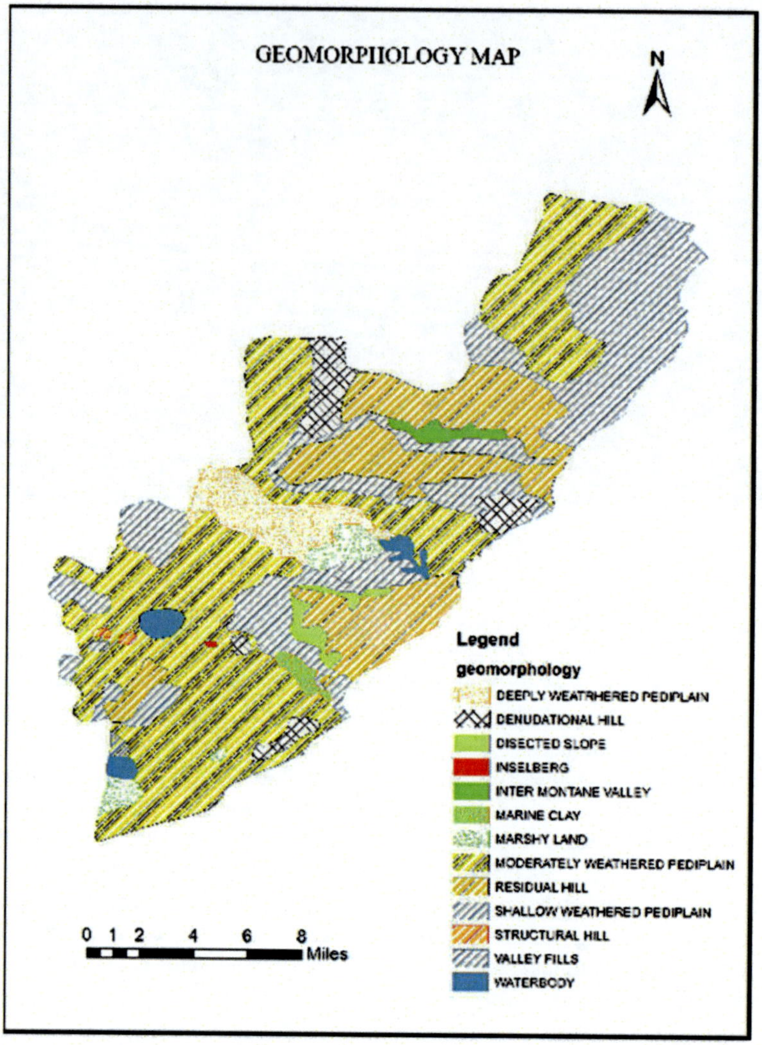

Fig. 6 Geomorphology map

4.6 Drainage Map

A drainage map has been generated from the survey of India (SOI) Toposheets for the Greater Visakhapatnam Municipal Corporation (GVMC) Visakhapatnam District. Drainage Area. The term '*drainage area*' is defined by the acreage where precipitation falls off into creeks, streams, rivers, lakes, and reservoirs. A land feature can be recognized by the most significant rise between two zones on a map, often a ridge.1 to 4 orders streams. Fig. 9 shows the drainage map.

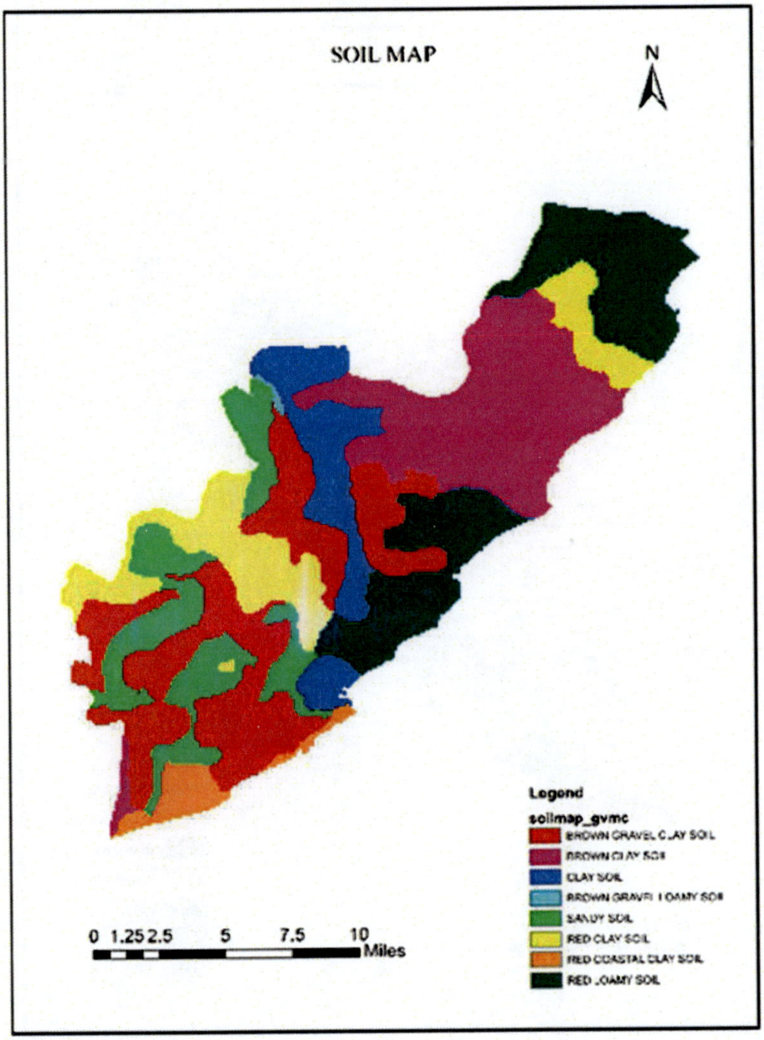

Fig. 7 Soil map

4.7 Transportation Map:

The road network comprises the national highway, state interstate, and other nearby roads that nearly interface towns and villages. The landfill site should not be placed too far away from existed road networks to avoid the expensive cost of constructing connecting roads. As a condiction, the waste disposal areas should not be too close to the road networks [6, 7]. The transportation map is shown in Fig. 10.

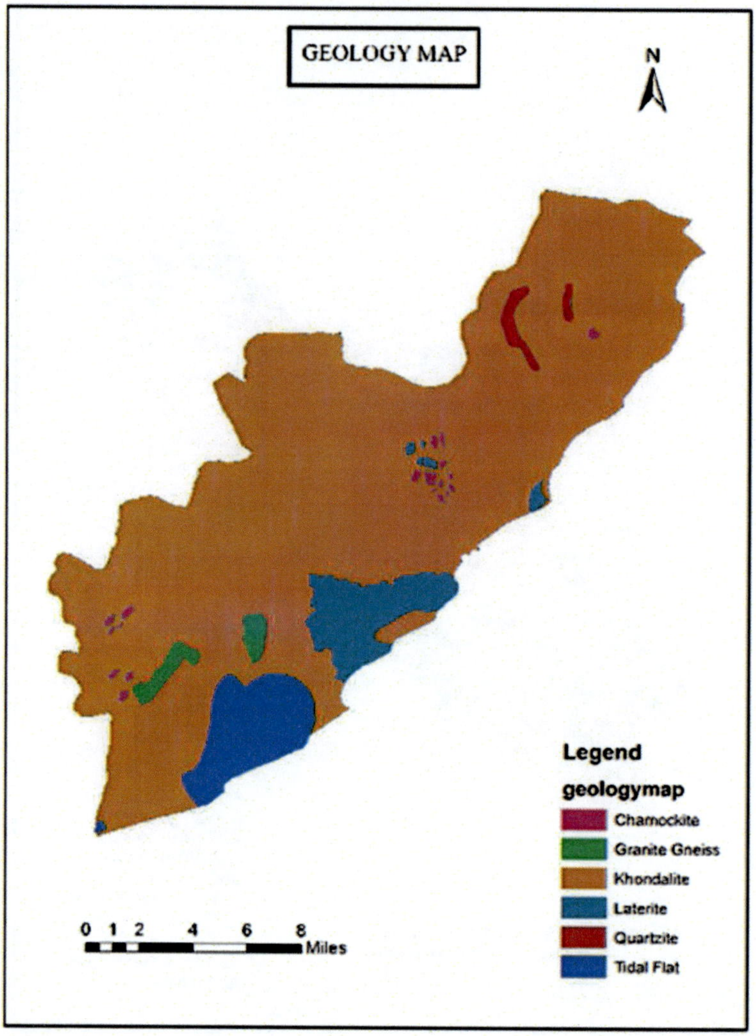

Fig. 8 Geology map

4.8 Tank Map

From the tanks map, we can notice the waterbodies density. Region having high waterbody density is not appropriate to consider that place for solid waste dumping as it exploits the surface and underground water bodies in the form of leachate. Hence, a region with low waterbody density is preferable. The tank map is made known in Fig. 11. The surface freshwater resources like tanks/water bodies and streams and reservoirs are most preferable to be located from the solid waste disposal

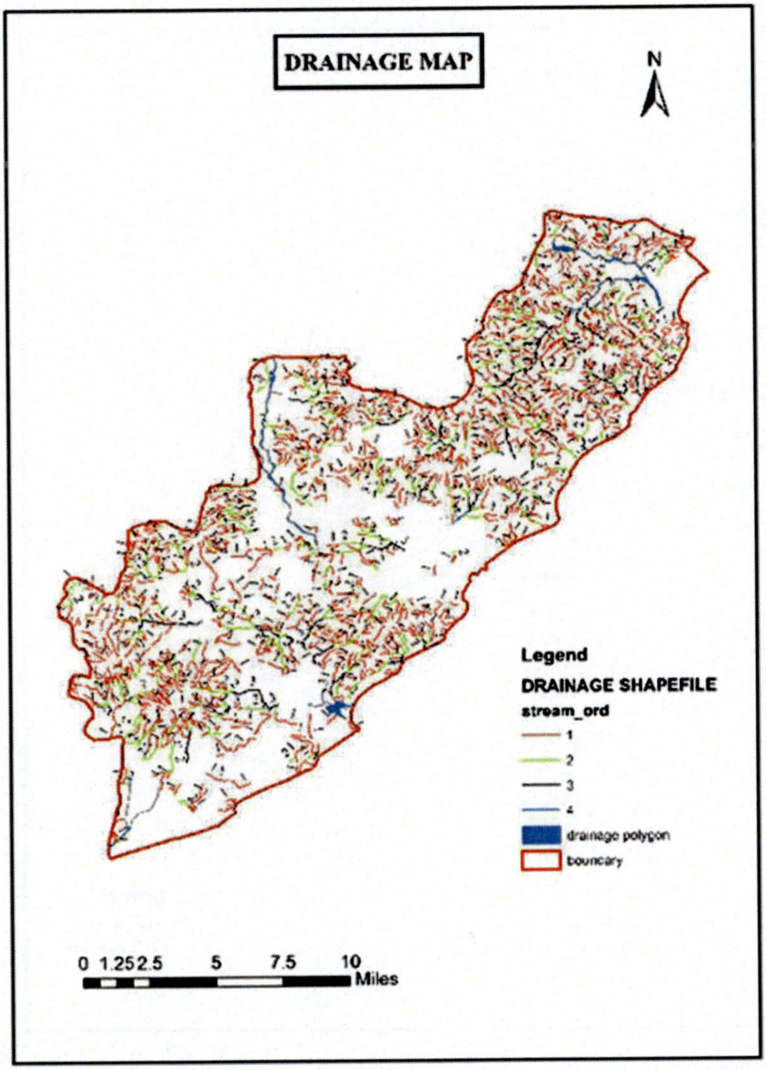

Fig. 9 Drainage map

sites according to the rules of the pollution control board as they are prone to contamination.

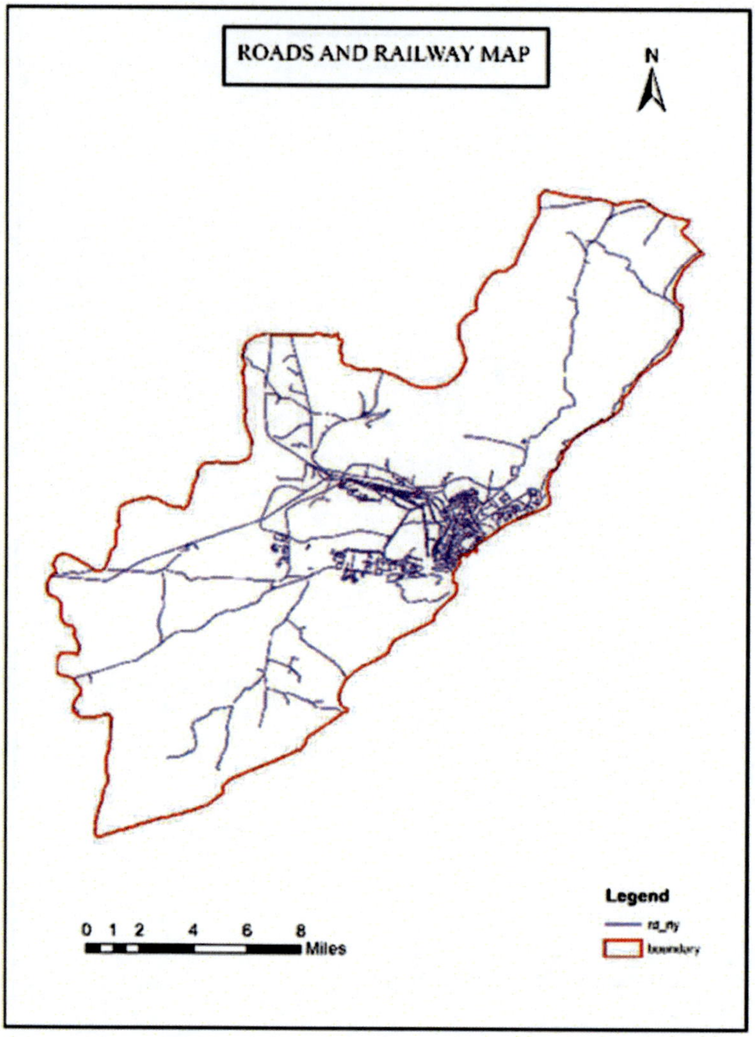

Fig. 10 Transportation map

4.9 Groundwater Table Map

The map was generated by the data found from good open wells and a questionnaire from people about bore wells. After generating the grown water depths map, it is reclassified and given weights as per the importance. High depth to groundwater level areas is more suitable than low depth to groundwater level as the probability of leachate contamination is more elevated. The groundwater table map is made known in Fig. 12. High depth to groundwater level areas is more suitable than the

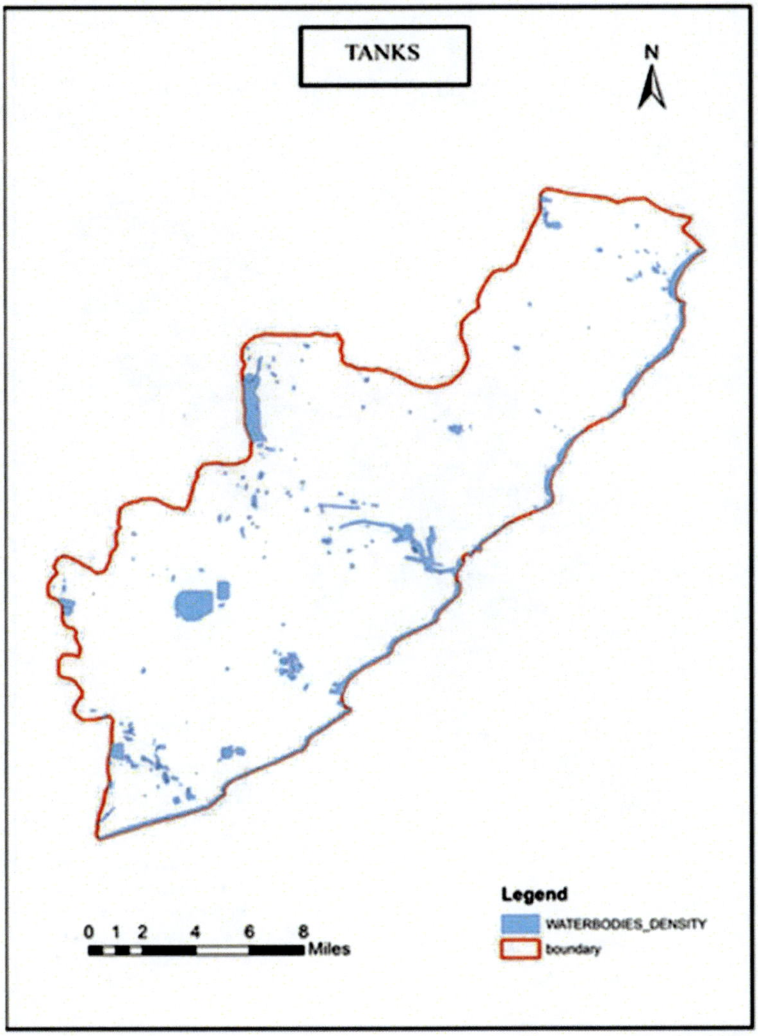

Fig. 11 Tanks map

low depth to groundwater level as the probability for leachate contamination is more diminutive.

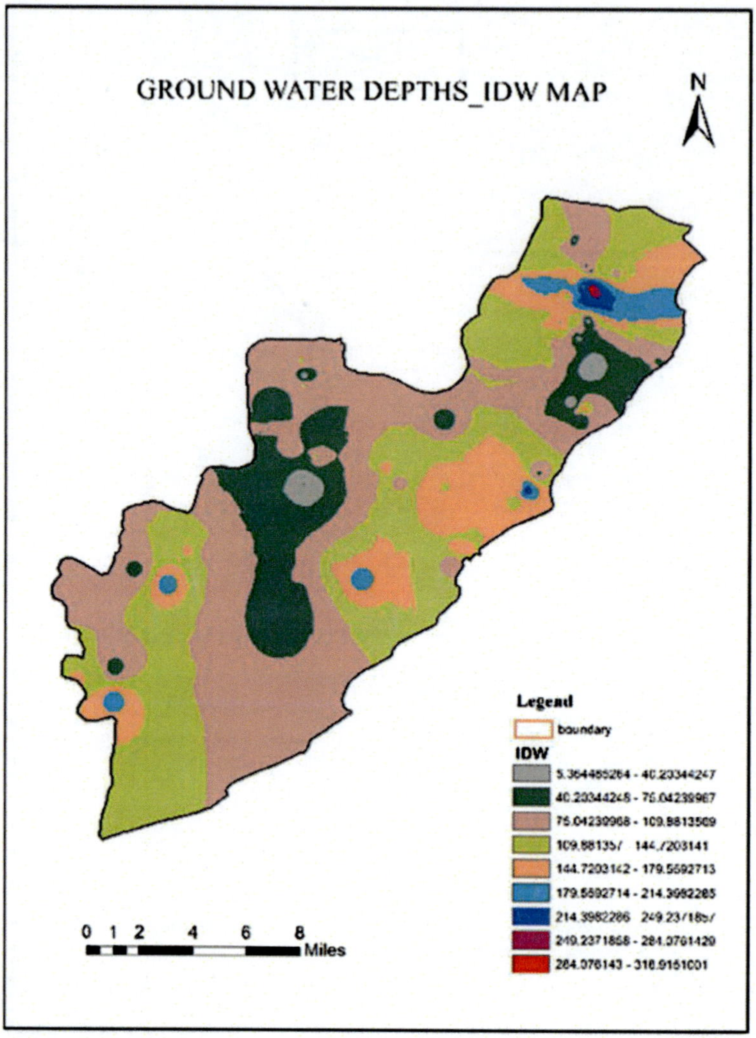

Fig. 12 Groundwater depths

4.10 Drainage Density Map

The drainage density map is generated based on the interpretation data of the topography and stratigraphic drainage network of the study area. Generally, solid waste landfills are not suitable for consideration in areas with high drainage density because it pollutes groundwater and surface water through the leaching process. Areas with higher drainage density are not suitable for landfill because they use soil and surface water through the leaching process. Therefore, the area where the drainage density is insufficient is preferable. The drainage density map is made known in Fig. 13.

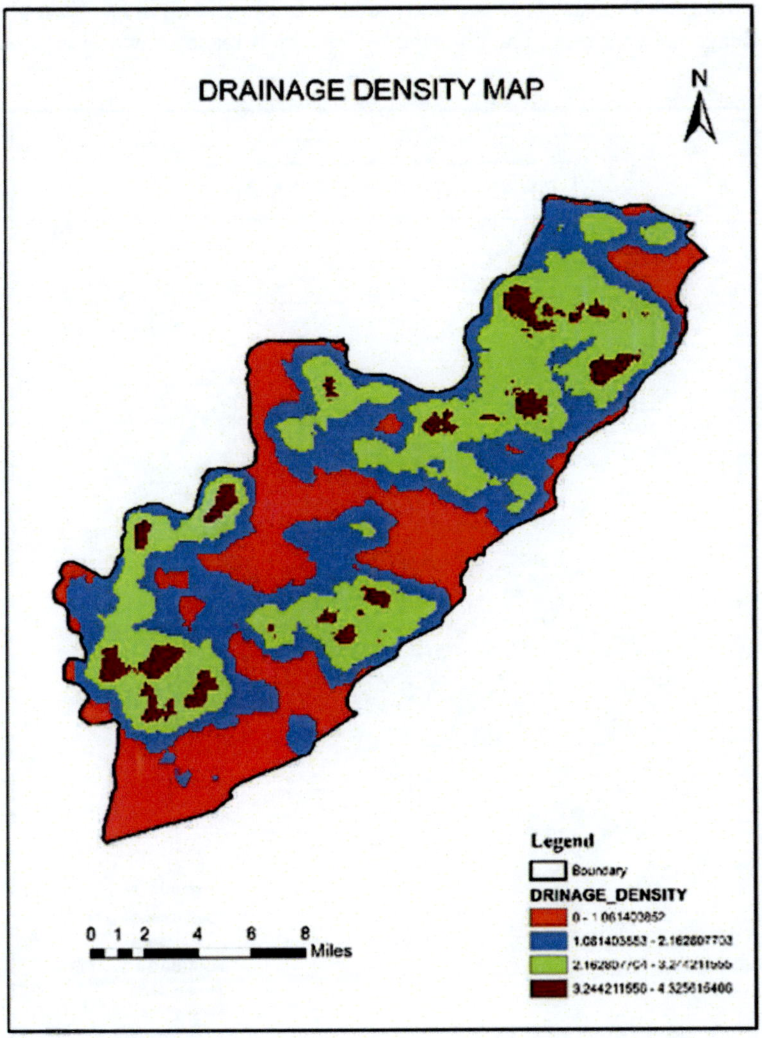

Fig. 13 Drainage density map

4.11 Drainage Buffer Map

The drainage buffer map has been made from the interpreted drainage network information from the Toposheets of the study area. A buffer of 100 m has been given to the drainage network map. The drainage buffer map is made known in Fig. 14.

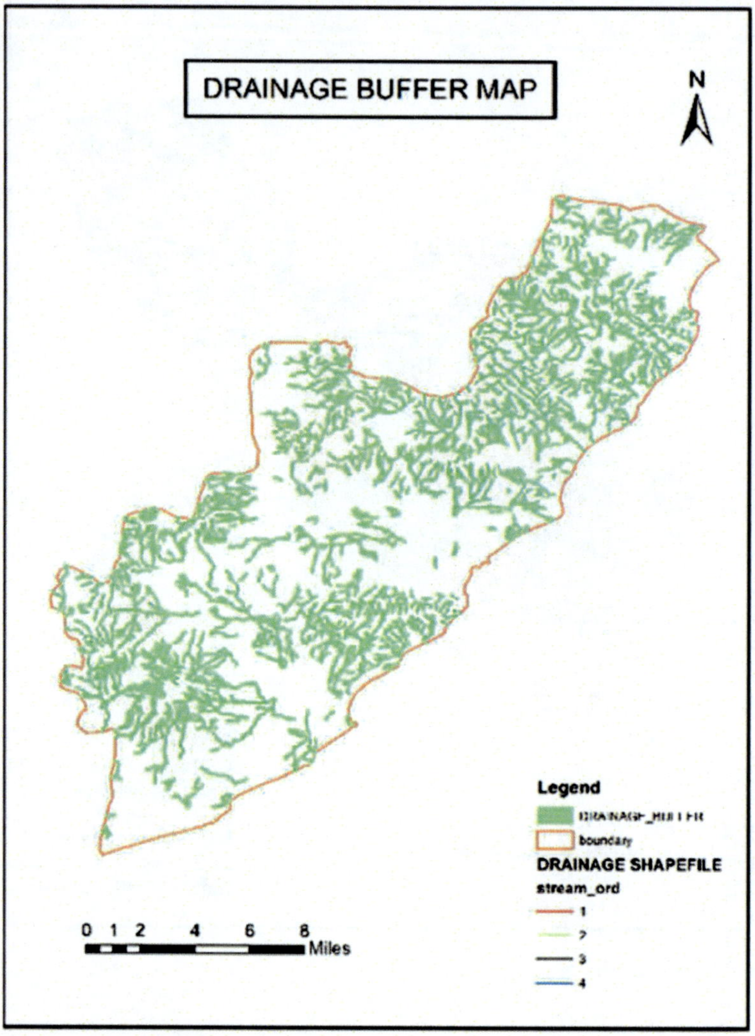

Fig. 14 Drainage buffer map

5 Discussions

Multicriteria Analysis: Weighted Overlay Analysis

This tool applies one of the utmost used approaches for overlay analysis to resolve multicriteria problems like suitability models and site selection. Weighted overlay was a procedure for applying a common estimation scale of values to diverse and different inputs to make an integrated investigation [8]. For each criterion in the weighted overlay analysis, its importance may be different. According to different thematic layers, the possible locations of waste disposal sites have been determined. Weights are assigned to each type of complete thematic map analyzed by GIS. The weight ranges from 1 to 5, where 1 means poor and 5 means excellent. The selected level is from 1 to 5, with a satisfactory applicability rating, as known in Table 1.

Analytical Hierarchy Process (AHP)

This method provides a tool for pair-wise comparison of various criteria and sub-criteria considered for landfill site selection. The relative importance of criteria/sub-criteria is expressed in numerical values from 1 to 9. This method is most appropriate when quantitative criteria such as distance from roads and habitation are considered. The technique tends to become complex when many parameters are required to be considered for any decision analysis. The sensitivity analysis results obtained for AHP gave a consistency ratio less than 0.1 justifying the results are acceptable [4, 7], and [2] is shown in Table 2.

Table 1 Suitability rating scale

Scale	Suitability rating
1	Poor
2	Fair
3	Good
4	Very good
5	Excellent

Table 2 Intensity of importance

Intensity of importance (weights)	Definition
1	Equivalent importance
5	Somewhat more important
6	Much more important
8	Very way more important
9	More important
2, 3, 4, 7	In-between values

Site Suitability Map

The ultimate site suitability map for urban solid waste disposal is made known in the below Fig. 15. It was separated into four categories such as not suitable, moderately suitable, suitable, and highly suitable. From the pictographic representation of the site suitability map, we have observed that not suitable area covers 3%, moderately suitable area as 50%, suitable area as 47%, and highly suitable area as 0.3%. The

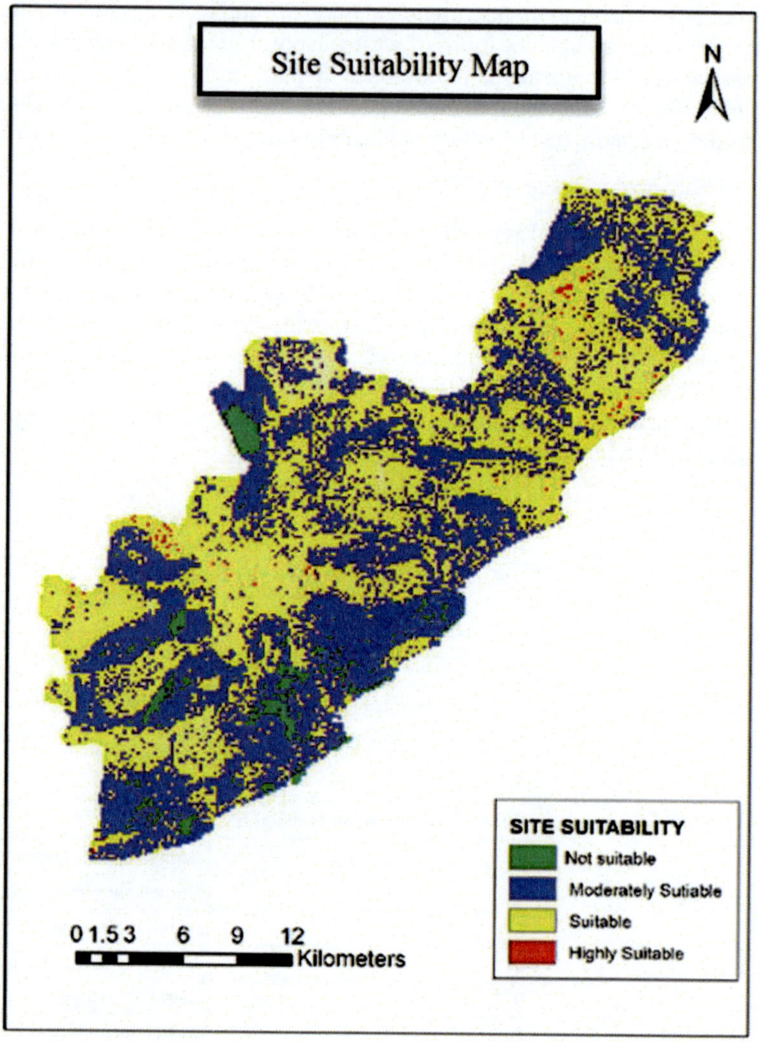

Fig. 15 Site suitability map

Table 3 Details of the area and percentage of site suitability map

Type	Area (km^2)	Percentage (%)
Not suitable	17.459548	2.94
Moderately suitable	283.069065	47.7
Suitable	267.409229	45.06
Highly suitable	1.869831	0.31

highly suitable area is near Marikavalasa junction near Gollallapalem and Madhurvada. The highly suitable areas are satisfying the existing criteria to dispose of the urban solid waste. The table below shows the details of site suitability (Table 3).

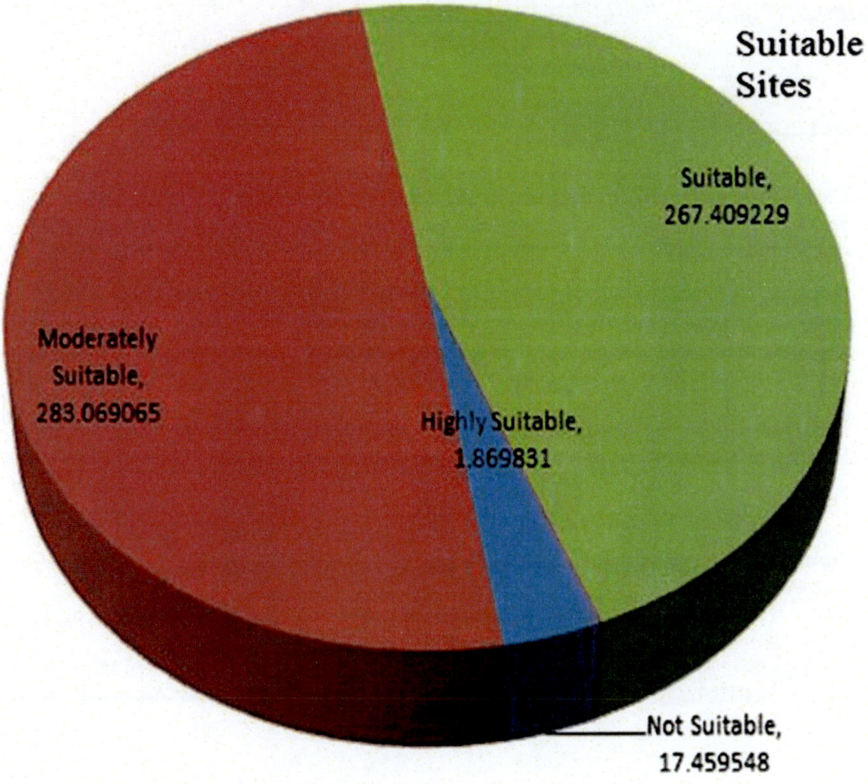

Fig. 16 Site suitability map in the area

6 Conclusion

The socio-economic and environmental problems related to the selection of the solid waste disposal site are complicated by the diversity of the quantity and nature of the waste. Each layer of data generated by will be judged based on environmental, social and community impacts, and appropriate weighting needs will be assigned before any decisions are made. Remote sensing and GIS technology are the most suitable tools for solving problems related to spatial decision-making. Highly suitable areas in the given study area constitute 1.87 km^2, suitable areas of 267 km^2, moderately suitable areas of 283 km^2, and not suitable areas of 17.5 km^2. The integration of different datasets in GIS has resulted in the multi-layer display of the data, which would have otherwise been difficult to do manually using the conventional site suitability methods (Fig. 16).

References

1. Chabuk, A., Al-Ansari, N., Hussain, H. M., Knutsson, S., & Pusch, R. (2016). *Landfill site selection using geographic information system and analytical hierarchy process: A case study Al-Hillah.*
2. Henry, R. K., Yongsheng, Z., & Jun, D. (2006). Municipal solid waste management challenges in developing countries–Kenyan case study. *Waste Management, 26*, 92–100.
3. Nemerow, N.L., Agardy, F.J., Sullivan, P. & Salvato, J.A. (2009). *Environmental Engineering: Environmental health and safety for municipal infrastructure, land use and planning, and industry* (6th ed.). https://doi.org/10.1002/9780470432822
4. Sumathi, V. R., Natesan, U., & Sarkar, C. (2008). GIS-based approach for optimized siting of municipal solid waste landfill. *Waste Management, 28*, 2146–2160.
5. Higgs, G. (2006). Integrating multi-criteria with technical geographical information systems in waste facility location to enhance public participation. *Waste Management Research, 24*, 105–117.
6. Nas, B., Cay, T., Iscan, F., & Berkta, A. (2010). Selection of MSW landfill Site for Konya, Turkey, using GIS and multicriteria evaluation. *Environmental Monitoring and Assessment, 160*, 491–500.
7. Sener, B., Suzen, L., & Doyuran, V. (2006). Landfill site selection by using geographic information systems. *Environmental Geology, 49*, 376–388.
8. Babu, S., & Sivasankar, S. (2015). GIS and remote sensing in urban waste disposal and management: A case study of Usilampatti municipality India. *International Journal of Applied Research, 1*(9), 1047–1051.

Effect of Identification of Extremes on Regional Flood Frequency Analysis

K. G. Kiran and V. V. Srinivas

Abstract Frequency analysis procedures are widely used to quantify the risk associated with floods that have devastating consequences worldwide. Conventionally, the frequency analysis is performed based on the annual maximum series (AMS) of peak flows extracted from the available streamflow records. Peaks over threshold or partial duration series (PDS) is deemed more efficient than AMS in depicting information on extremes. Despite its advantages, the use of PDS is less prevalent than AMS. It is due to the lack of a universally established systematic approach to select an appropriate threshold for PDS extraction. Another issue in flood risk assessment at target locations is sparsity or lack of data. In such situations, practitioners opt for regional flood frequency analysis (RFFA) approaches that involve regionalization (locating groups/regions comprising resembling watersheds) and pooling of flood-related information from outlets of the watersheds to estimate desired flood quantile(s) at the target sites. Most RFA approaches are focused on using AMS rather than PDS. This chapter investigates the effect of using AMS and PDS on the error in flood quantile estimation at ungauged sites in RFFA. For this purpose, fuzzy meta-regions (FMRs) delineated in peninsular India using fuzzy ensemble clustering and existing sub-zones in the area used by Central Water Commission (CWC) are considered. Errors were consistently lower in analysis with FMRs than CWC sub-zones, irrespective of the type of extremes (AMS or PDS) considered. Furthermore, PDS yielded lower errors for higher return periods.

Keywords Regional flood frequency analysis · Fuzzy ensemble clustering · Peaks over threshold · Annual maximum series · Automatic threshold selection · Meta-clustering

K. G. Kiran (✉) · V. V. Srinivas
Department of Civil Engineering, Indian Institute of Science, Bangalore 560012, India
e-mail: kirankg@iisc.ac.in

Interdisciplinary Center for Water Research (ICWaR), Indian Institute of Science, Bangalore 560012, India

A. K. Dikshit et al. (eds.), *Innovative Trends in Hydrological and Environmental Systems*, Lecture Notes in Civil Engineering 234,
https://doi.org/10.1007/978-981-19-0304-5_24

1 Introduction

Hydrologists conventionally use frequency analysis procedures to quantify the risk associated with extreme events such as floods. The risk estimates find use in studies related to the design of water control (e.g. barrages, dams, levees) and conveyance structures (e.g. culverts, storm sewers, spillways), economic and risk evaluation of flood protection projects, land use planning, and management, flood insurance assessment, and other hydrological applications. The procedures involve identifying extreme events from the time series of streamflow observations at the target site and performing frequency analysis on those events to estimate flood quantile function that relates the magnitude of a flood with its non-exceedance probability or risk. Approaches that are commonly used for statistical analysis of extremes could be broadly classified as those based on (i) annual maximum series (AMS) or block maxima (BM) and (ii) partial duration series (PDS) or peaks over threshold (POT) series. The AMS approach involves extracting annual maxima (i.e. BM) values from the available records, whereas the PDS approach considers values exceeding a predetermined high threshold. Several studies [1–4] have shown through Monte Carlo simulations and real-world applications that the PDS approach is more effective in using information from the available records, and variance of quantile estimates based on the PDS approach could be smaller than that based on the AMS approach.

Despite its advantages, the use of PDS is less prevalent than AMS due to the lack of a universally established systematic approach for extracting a PDS. A major issue with the PDS approach concerns the selection of an appropriate threshold value. The threshold should ensure the extraction of maximum information from the time series without violating the PDS approach's statistical assumptions. When the threshold is too high, the sample size of PDS tends to be small. Consequently, estimates of distribution parameters have high variance and small bias [5]. On the other hand, a low value for the threshold yields a large sample that could include values that may not necessarily be extremes. Therefore, the distribution of the PDS deviates from the generalized Pareto distribution (GPD), which results in significant bias in the quantile estimates [6]. Threshold selection is thus a trade-off between bias and variance. In recent decades, several methods have been proposed for the identification of optimal thresholds. The methods generally differ in their fundamental theoretical assumptions and performance with data exhibiting different extreme behaviour (e.g. fat or heavy-tailed, thin or light-tailed). There are various issues affecting the performance of different methods available for threshold selection. Recently, a Mahalanobis distance (MD)-based automatic threshold selection method [7] was proposed, which addresses those issues and is computationally efficient.

It is often necessary to estimate design flood quantiles at locations with limited or no data in a real-world scenario. Several regional frequency analysis (RFA) approaches are available for addressing this challenge. In the context of floods, the approaches generally involve (i) regionalization, which refers to the identification of a group of watersheds (i.e. region) resembling the target site's watershed in terms of attributes/characteristics influencing flood and (ii) RFA which involves the pooling of flood-related information from sites in the region and using it to estimate desired flood quantile(s) at the target site. Approaches to watershed regionalization for regional flood frequency analysis (RFFA) include those based on regression analysis (method of residuals), canonical correlation analysis, region of influence and its extensions, hierarchical approach, and cluster analysis. From the literature, one can note that each regionalization approach differs in its underlying assumptions and strategy. Therefore, different approaches often delineate a given set of watersheds into regions that differ in their composition. There is no prior knowledge of true regions in the real world. Hence, none of the available approaches could be established to be universally superior for determining true regions. Consequently, several approaches are in use, and there is ambiguity in their choice. Recently, a fuzzy ensemble clustering approach was proposed [8], which addresses these uncertainties by generating several sets of regions using different approaches and ensemble clustering of those sets to arrive at fuzzy meta-regions (FMRs).

After the regionalization step, the index-flood approach [9] and its variants [10–14] are widely used for RFA. However, most RFA approaches are focused on using AMS rather than PDS. There is a dearth of studies focusing on assessing the effect of different approaches for identifying extremes on the predicted quantiles through RFFA. The main objective of this study is to investigate the effect of using AMS and PDS on the error in flood quantile estimation at ungauged sites in peninsular India by RFFA. For this purpose, PDS is extracted using the MD-based automatic threshold selection method, which is ideal for use in such a large-scale application. In a previous study [15], the Indian subcontinent was delineated into seven zones containing 26 contiguous sub-zones. The sub-zones are widely used by practitioners in India for flood estimation by considering them to be homogeneous. This study's second objective is to assess the effectiveness of using FMRs instead of the Central Water Commission (CWC) sub-zones for RFFA with AMS and PDS. For use in this analysis, FMRs of watersheds delineated in peninsular India using fuzzy ensemble clustering in a previous study [8] are adopted.

The remainder of this paper is structured as follows. The methodology considered for the investigation is presented in Sect. 2. Following this, a case study is presented in Sect. 3. Finally, a summary of the study and its conclusions are provided in Sect. 4. The abbreviations used in the chapter are expanded in Appendix.

2 Methodology

The methodology considered for the proposed investigation involves (i) at-site analysis for identification and extraction of extreme events at each site, (ii) regionalization to delineate watershed regions, and (iii) RFFA with the identified extremes using the delineated regions.

(i) At-site analysis

PDS is extracted at each site by identifying the optimal threshold corresponding to the site using the recently proposed MD-based threshold selection method [7]. Let there be N_i-year-long sample/record of streamflow at site i. The steps involved are as follows.

(1) Identify independent peaks in the time series using a moving window. The length of the window (L_w) could be considered as $2\,s + 1$, where s is the minimum separation time between events in the time series. One could consider $s \geq 5 + \ln(A)$ with A as catchment's drainage area in square miles (e.g. [16]). As the window moves through the time series one time step at a time, the maximum value within the window is chosen as an independent peak if it coincides with the window's centre. An additional criterion is applied to ensure the independence of events causing the peaks. The criterion suggests that if the flows observed between two consecutive peaks do not drop below 0.75 times the magnitude of the lower peak among the two, then the lower peak value is to be discarded [16].

(2) Let n_p be the number of independent peaks chosen. Arrange them in ascending order of their magnitude so that the ordered sample appears as $X_{1:n_p} \leq X_{2:n_p} \leq \cdots \leq X_{n_p:n_p}$, where $X_{j:n_p}$ is j-th smallest independent peak. Following this, define a set of possible thresholds $\Xi = \{X_{1:n_s}, X_{2:n_s}, \ldots, X_{n_s:n_s}\}$ from the identified n_p peaks by excluding repeated values, such that $n_s \leq n_p$. It is to be noted that considering all the n_p peaks as candidates in Ξ is neither efficient nor necessary. The choice of candidates in Ξ depends on the available data, and it is left to the modeller's discretion.

(3) Choose one possible threshold at a time from the set Ξ and perform the following analysis using it.
   (a) Determine the set of excesses $\mathbf{Y}$ by deducting the threshold from independent peaks exceeding it. Fit GPD to $\mathbf{Y}$ and obtain sample estimates of the distribution's scale parameter α and shape parameter k using the method of L-moments. One could choose any other effective method for parameter estimation.
   (b) Transform or map $\mathbf{Y}$ onto $\mathbf{Z}$ as,

$$\mathbf{Z} = -\frac{1}{k}\log\left(1 - k\frac{\mathbf{Y}}{\alpha}\right) \tag{1}$$

The true population of the random variable $\mathbf{Z}$ (depicting the transformed values) is standard exponential distribution (with scale parameter equal to 1) if $\mathbf{Y}$ follows GPD. An advantage of the transformation is that it enables representing the true population of $\mathbf{Z}$ as a point in the L-moment space (considering any number of moments), which is not the case with the true population of **Y**. It facilitates the computation of Mahalanobis distance in the *L*-moment space.

(c) Estimate Mahalanobis distance between sample *L*-statistics (first *L*-moment l_1 and *L*-skew t_3) of **Z** and *L*-moments (first *L*-moment λ_1^E and *L*-skew τ_3^E) of exponential (1) distribution in the transformed space using Eq. (2).

$$\mathrm{MD} = \left[\frac{1}{1-\rho^2}\left\{\frac{\left(l_1 - \mu_{l_1^E}\right)^2}{\sigma_{l_1^E}^2} - \frac{2\rho\left(l_1 - \mu_{l_1^E}\right)\left(t_3 - \mu_{t_3^E}\right)}{\sigma_{l_1^E}\sigma_{t_3^E}} + \frac{\left(t_3 - \mu_{t_3^E}\right)^2}{\sigma_{t_3^E}^2}\right\}\right]^{\frac{1}{2}} \tag{2}$$

where $\mu_{l_1^E}$ and $\mu_{t_3^E}$ denote the mean, $\sigma_{l_1^E}$ and $\sigma_{t_3^E}$ represent the standard deviation of the sampling distribution of l_1^E and t_3^E respectively, and ρ is the coefficient of correlation between l_1^E and t_3^E. For Exp(1) distribution, $\mu_{l_1^E} = 1$ and $\sigma_{l_1^E} = 1/\sqrt{n}$ following the central limit theorem. Furthermore, $\rho = 0$ based on Basu's theorem [17], which states that when a sufficient statistic (in this case l_1^E) is boundedly complete, then all ancillary statistics are independent of the sufficient statistic. Estimate $\mu_{t_3^E}$ and $\sigma_{t_3^E}$ using approximate expressions relating them to sample size n (Eqs. 3 and 4).

$$\mu_{t_3^E} = \frac{1}{3} - 0.307n^{-1.076} \tag{3}$$

$$\sigma_{t_3^E} = \frac{166.507n + 102.384}{347.652n^{1.498} - 43.041n - 575.098} \tag{4}$$

(4) From the set Ξ, identify the optimal threshold for which estimate of Mahalanobis distance MD is minimum. The excesses corresponding to the identified threshold denote the required PDS or POT series.

The at-site estimates of quantiles are obtained using both AMS and PDS corresponding to different return periods. For AMS, appropriate frequency distributions can be identified from various three-parameter distributions (e.g. generalized logistic, generalized extreme value, Pearson type-III, generalized Pareto, and generalized normal), using L-moment-based goodness-of-fit test [18] and considering appropriate (say 90%) confidence level.

(ii) Regionalization of watersheds

Fuzzy meta-regions of watersheds delineated over India in a previous study [8] using a novel fuzzy ensemble clustering method (which addresses regionalization uncertainties), and existing CWC sub-zones are adopted for use in the present study. The fuzzy ensemble clustering approach involves the following two steps:

(a) Generation of several sets of regions using watershed attributes/features by considering different regionalization methods, such as region of influence [19] approach, clustering approach based on Gaussian mixture model (GMM; [20–22]), and a hybrid approach which combines canonical correlation analysis [23] with GMM-based clustering.
(b) Deriving an entropy-based watershed co-association matrix based on the generated ensemble of regions and, subsequently, partitioning of the matrix using the relational fuzzy c-means algorithm to arrive at the FMRs. Readers are referred to [8] for a detailed description of the methodology.

(iii) RFFA with identified extremes using the delineated regions

The index-flood method was considered for RFFA to predict flood quantiles at ungauged sites. In the case of analysis with n_r FMRs, an ungauged site s is assigned to n'_r regions in which its membership $u_{s,k}$ is greater than or equal to a threshold value chosen as $1/n_r$ [24]. To facilitate computing fuzzy memberships for ungauged sites, attributes for regionalization are chosen such that their information could be determined for watersheds of both gauged and ungauged sites. Flow chart of index-flood method [9], which is used to estimate regional flood quantiles for the ungauged site, involves the following steps:

(1) Normalize peak flows of gauged watersheds/sites in the region(s) to which the ungauged watershed/site is assigned. It requires dividing the peak flow values at each site by their respective mean value (i.e. index-flood value).
(2) In the case of analysis with AMS, use *L*-moment-based regional goodness-of-fit test [18] to identify an appropriate regional frequency distribution to fit peak flow information pooled from gauged sites in each region to which the ungauged site is assigned. In the case of analysis with PDS, GPD is used to fit peak flows. Refer to the cumulative distribution function of the frequency distribution fitted to region k as the regional flood quantile function $\hat{q}_k^R(\cdot)$.
(3) For the ungauged site, estimate regional flood quantile corresponding to T-year return period. In analysis with fuzzy regions, the estimation can be based on the following equation.

$$\hat{Q}_s^R(T) = \frac{\sum_{\forall k \text{ for which } u_{s,k} \geq n_r^{-1}} u_{s,k}\left(\overline{Q}_{s,k} \times \hat{q}_k^R(T)\right)}{\sum_{\forall k \text{ for which } u_{s,k} \geq n_r^{-1}} u_{s,k}} \quad (5)$$

where $\overline{Q}_{s,k}$ is the index-flood value for the ungauged site s. It could be estimated using multiple linear regression relationship developed between attributes (identified for regionalization) and index-flood values of gauged sites in region k to which the ungauged site is assigned. In analysis with AMS (PDS), the index-flood value is considered as the mean of AMS (PDS). Equation (5) expresses the regional flood quantile estimate as the weighted average of scaled quantile functions of the regions to which the ungauged watershed is assigned. Here, $\overline{Q}_{s,k}$ is considered as a scaling factor, and weights are in proportion to fuzzy memberships of the ungauged watershed in those regions. On the other hand, in analysis with CWC sub-zones which are hard regions, an ungauged site is assigned to the sub-zone (i.e. region k), which contains it geographically. For this case, the previous equation reduces to the following form:

$$\hat{Q}_s^R(T) = \overline{Q}_{s,k} \times \hat{q}_k^R(T) \quad (6)$$

3 Case Study

3.1 Study Area and Data

The proposed investigation was performed on stream gauges located at the outlets of eighty-nine watersheds in peninsular India (Fig. 1). The watersheds cover 12 major river basins (i.e. Godavari, Krishna, Cauvery, Mahanadi, Narmada, Tapi, Sabarmati, Subernarekha, Mahi, Tadri, Brahmani, and Pennar) and the area drained by a few west-flowing small rivers between Kanyakumari and Tadri. Records of daily flow at mouths/outlets of the watersheds were available from stream gauges monitored by the CWC, India. The records were at least 10-years long, and the flows were unregulated or minimally regulated. Figure 1 shows the FMRs and CWC sub-zones in the study area.

3.2 Results and Discussion

The at-site analysis involved extraction of AMS and PDS corresponding to each site and computation of quantiles for various return periods at those sites. Figure 2a–d shows the scatterplots between at-site quantiles estimated based on analysis with

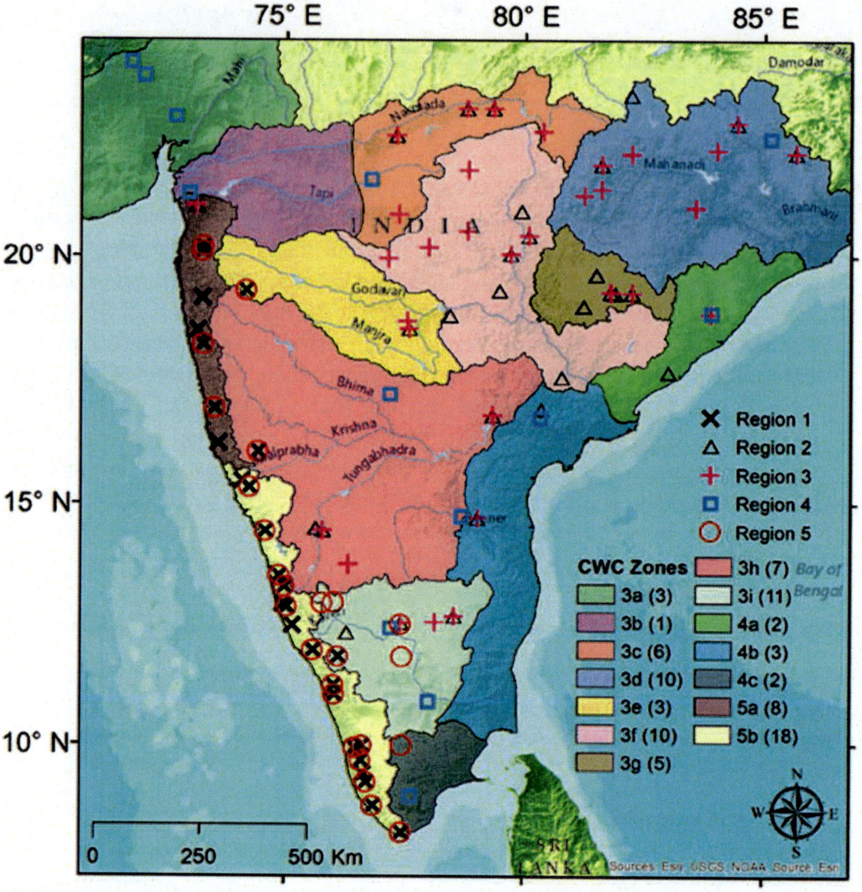

Fig. 1 Study area showing fuzzy meta-regions, CWC sub-zones, and the 89 stream gauges considered in the study

AMS and PDS and box plot showing variation in the ratios of those at-site quantiles (Fig. 2e–h) corresponding to different return periods (25, 50, 100, and 200 years). It can be inferred from these plots that the at-site quantile estimates based on PDS are (i) similar to those based on AMS for lower return periods (<100 years) and (ii) marginally lower for higher return periods (≥100 years).

The utility of FMRs and CWC sub-zones in arriving at flood quantile estimates at ungauged sites in RFFA based on AMS and PDS was evaluated through a leave-one-out cross-validation (LOOCV). It involved considering one watershed at a time to be ungauged and assigning it to the region(s) to which it belongs, following the procedure given in Sect. 2(iii). Peak flow information from gauged watersheds in the assigned region(s) was pooled to perform RFA using the index-flood approach and

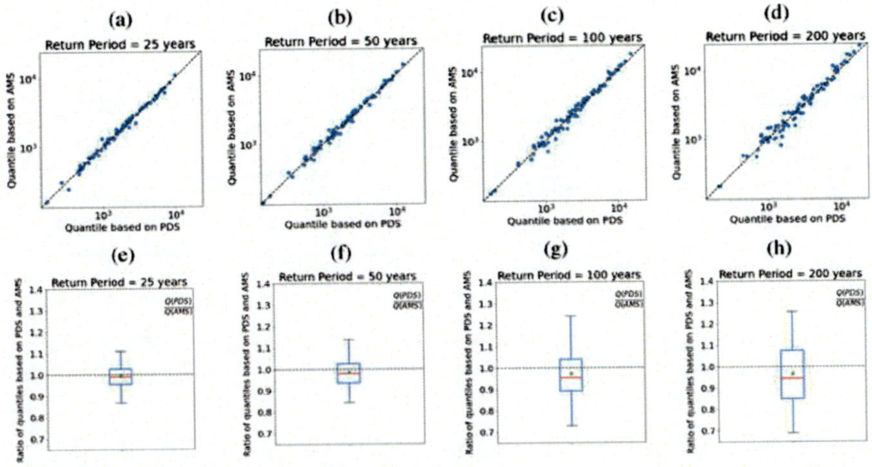

Fig. 2 **(a–d)** Scatter plot of at-site quantiles obtained based on AMS and PDS corresponding to different return periods. Plots **(e–h)** show the ratio of quantiles based on AMS and PDS corresponding to return periods in **(a–d)**, respectively

estimate regional flood quantile for the ungauged site corresponding to the aforementioned return periods. In the LOOCV, the index-flood value $\overline{Q}_{s,k}$ required to estimate flood quantile for the pseudo-ungauged site was considered equal to the mean of peak flows (AMS/PDS) observed at the site to avoid its effect on the performance evaluation.

Errors in the predicted flood quantiles at ungauged sites were quantified in analysis with each data set of extremes (AMS & PDS) and each set of regions (FMRs & CWC) in terms of three error measures: (i) Average relative bias (average R-Bias $\in (-\infty, \infty)$), (ii) Absolute relative bias (AR-Bias $\in (0, \infty)$), and (iii) Relative root mean-squared error (R-RMSE $\in (0, \infty)$).

$$\text{Average R-Bias }(T) = \frac{1}{N}\sum_{i=1}^{N} \text{R - Bias}_i(T) = \frac{1}{N}\sum_{i=1}^{N} \frac{\hat{Q}_i^R(T) - \hat{Q}_i^P(T)}{\hat{Q}_i^P(T)} \times 100 \tag{7}$$

$$\text{AR-Bias}(T) = \frac{1}{N}\sum_{i=1}^{N} \left| \frac{\hat{Q}_i^R(T) - \hat{Q}_i^P(T)}{\hat{Q}_i^P(T)} \right| \times 100 \tag{8}$$

$$\text{R-RMSE}(T) = \left[\sqrt{\frac{1}{N}\sum_{i=1}^{N} \left(\frac{\hat{Q}_i^R(T) - \hat{Q}_i^P(T)}{\hat{Q}_i^P(T)} \right)^2} \right] \times 100 \tag{9}$$

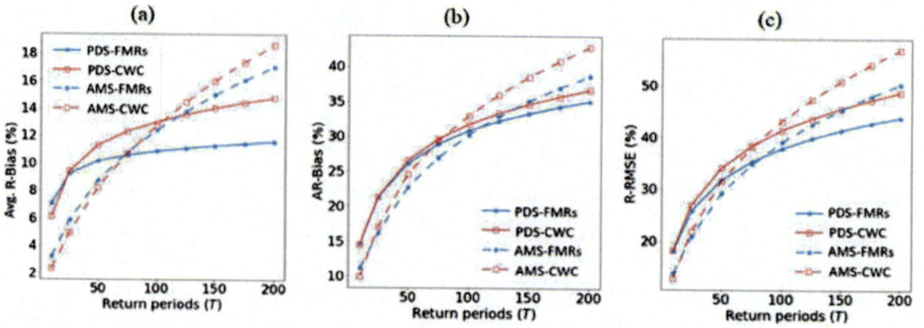

Fig. 3 Errors (**a**) average R-Bias, (**b**) AR-Bias, and (**c**) R-RMSE obtained in LOOCV with all the methods considered in the study

where N is the total number of watersheds considered in the study and $\hat{Q}_i^P(T)$ and $\hat{Q}_i^R(T)$ are, respectively, the true/population value and the regional estimate of flood quantile for i-th watershed corresponding to T-years return period. Since true/population value is unknown in reality, the at-site estimate of flood quantile corresponding to T-years return period was considered as $\hat{Q}_i^P(T)$.

The errors (average R-Bias, AR-Bias, and R-RMSE) obtained in LOOCV performed with both sets of regions (i.e. FMRs and CWC) using AMS and PDS are shown in Fig. 3a–c. In general, errors increase with an increase in the return period, as expected. Errors in quantile estimates for ungauged sites obtained in RFFA on FMRs are lower than those obtained based on CWC sub-zones, irrespective of the type of extremes (AMS or PDS) considered. This indicates that FMRs are more effective than CWC sub-zones in providing pooled information. Furthermore, in RFFA with PDS, the errors are lower for higher return periods ($\geq$100 yrs) and marginally higher for smaller return periods (<75 yrs) when compared to those obtained in analysis with AMS. This could be partly attributed to the fact that the at-site quantiles considered for quantifying errors in the LOOCV experiment with AMS and PDS are different. The rate of increase in error with the return period is lower for PDS than AMS. It indicates that the shapes of the flood quantile functions are better approximated with RFA using PDS.

4 Summary and Conclusions

The effect of identification of extremes in the form of AMS and PDS on the error in flood quantile estimation at ungauged sites in RFFA is investigated by considering (i) fuzzy meta-regions (FMRs) of watersheds delineated in India using fuzzy ensemble clustering and (ii) existing sub-zones in use by Central Water Commission, India. The RFFA with PDS does not require specifying the form of distribution for an ungauged site since it is known to be GPD, provided the threshold for PDS extraction is identified appropriately. It is concluded from the study that the use of PDS in RFFA provides significant advantages over AMS in quantile estimation corresponding to higher return periods. Furthermore, fuzzy meta-regions are more effective than CWC sub-zones in providing pooled information for reliable quantile estimation by RFA.

Acknowledgements Authors acknowledge all the organizations whose data were considered in the present study. They thank American Geophysical Union and John Wiley & Sons Ltd. for the permission to use parts of two articles published in Water Resources Research Journal. The second author acknowledges the Ministry of Jal Shakti (Department of Water Resources), Government of India, for grants provided through Project No. 16/22/2016-R&D/4015-4032, and the Ministry of Earth Sciences for the support provided through Project No. MoES/PAMC/H&C/41/2013-PC-II.

Appendix

Abbreviations

AMS	Annual maximum series
AR-Bias	Absolute relative bias
Average R-Bias	Average relative bias
BM	Block maxima
CWC	Central Water Commission
FMRs	Fuzzy meta-regions
GMM	Gaussian mixture model
GPD	Generalized Pareto distribution
LOOCV	Leave-one-out cross-validation
MD	Mahalanobis distance
PDS	Partial duration series
POT	Peaks over threshold
RFA	Regional frequency analysis
RFFA	Regional flood frequency analysis
R-RMSE	Relative root mean-squared error

References

1. Cunnane, C. (1973). A particular comparison of annual maxima and partial duration series methods of flood frequency prediction. *Journal of Hydrology, 18*(3–4), 257–271.
2. Wang, Q. J. (1991). The POT model described by the generalized Pareto distribution with Poisson arrival rate. *Journal of Hydrology, 129*(1–4), 263–280.
3. Madsen, H., Rasmussen, P. F., & Rosbjerg, D. (1997). Comparison of annual maximum series and partial duration series methods for modeling extreme hydrologic events: 1 at-site modeling. *Water Resources Research, 33*(4), 747–757.
4. Caires, S. (2009). *A comparative simulation study of the annual maxima and the peaks-over-threshold methods*. Deltares report 1200264-002 for Rijkswaterstaat, Waterdienst.

5. Coles, S. G. (2001). *An introduction to statistical modeling of extreme values*. Springer.
6. Adamowski, K. (2000). Regional analysis of annual maximum and partial duration flood data by nonparametric and L-moment methods. *Journal of Hydrology, 229*(3–4), 219–231.
7. Kiran, K. G., & Srinivas, V. V. (2021). A Mahalanobis distance-based automatic threshold selection method for peaks over threshold model. *Water Resources Research, 57*(1), e2020WR027534.
8. Kiran, K. G., & Srinivas, V. V. (2021). Fuzzy ensemble clustering approach to address regionalization uncertainties in flood frequency analysis. *Water Resources Research, 57*(3), e2020WR028412.
9. Dalrymple, T. (1960). *Flood-frequency analyses*. U.S. Geological survey water supply-paper, 1543-A, 11–51.
10. Stedinger, J. R. (1983). Estimating a regional flood frequency distribution. *Water Resources Research, 19*(2), 503–510.
11. Sveinsson, O. G., Boes, D. C., & Salas, J. D. (2001). Population index flood method for regional frequency analysis. *Water Resources Research, 37*(11), 2733–2748.
12. Sveinsson, O. G. B., Salas, J. D., & Boes, D. C. (2003). Uncertainty of quantile estimators using the population index flood method. *Water Resources Research, 39*(8).
13. Basu, B., & Srinivas, V. V. (2013). Formulation of a mathematical approach to regional frequency analysis, Water Resources. *Research, 49*, 6810–6833.
14. Basu, B., & Srinivas, V. V. (2016). Evaluation of the index-flood approach related regional frequency analysis procedures. *Journal of Hydrologic Engineering, 21*(1), 04015052.
15. CWC. (1983). Flood Estimation Report. Directorate of Hydrology (Small catchments) CWC, New Delhi.
16. Lang, M., Ouarda, T. B. M. J., & Bobée, B. (1999). Towards operational guidelines for over-threshold modelling. *Journal of Hydrology, 225*(3–4), 103–117.
17. Basu, D. (1955). On statistics independent of a complete sufficient statistic. *Sankhyā: The Indian Journal of Statistics (1933–1960), 15*(4), 377–380.
18. Hoskins, J. R. M., & Wallis, J. R. (1997). *Regional frequency analysis: An approach based on L-moments*. Cambridge University.
19. Burn, D. H. (1990). Evaluation of regional flood frequency analysis with a region of influence approach. *Water Resources Research, 26*(10), 2257–2265.
20. Bishop, C. M. (2006). *Pattern recognition and machine learning*. Springer.
21. Ojha, R. (2019). Identification of homogeneous regions of near surface air temperature lapse rates across India. *International Journal of Climatology, 39*(11), 4288–4304.
22. Ahani, A., Nadoushani, S. S. M., & Moridi, A. (2020). Regionalization of watersheds by finite mixture models. *Journal of Hydrology, 583*, 124620.
23. Cavadias, G. S. (1990). The canonical correlation approach to regional flood estimation. *Regionalization in Hydrology, 191*, 171–178.
24. Rao, A. R., & Srinivas, V. V. (2006). Regionalization of watersheds by fuzzy cluster analysis. *Journal of Hydrology, 318*(1–4), 57–79.

Modelling of Air Pollution Dispersion in Thiruvananthapuram Corporation

V. K. Sumadevi, U. Anila, Kiran P. Roy, V. B. Vishnu, and S. Minu

Abstract Air is a crucial element of the earth's ecosystem, and even minor changes in its composition can have a wide range of effects on the survival of creatures on earth. Deterioration of air quality is an important issue faced by many cities in India. Modelling of air pollution is a numerical method for describing the causal relationship between emissions, meteorology, atmospheric concentrations and deposition. The current study prepared annual and monthly air pollution dispersion maps at sensitive areas of Thiruvananthapuram Municipal Corporation, which is the administrative spot in the city of Thiruvananthapuram, the capital of Kerala. ADMS-Urban model was used in conjunction with GIS to produce the dispersion maps. The study has demonstrated a methodology for the development of emission inventory, dispersion modelling and mapping. Dispersion modelling and trend analysis were used to investigate the concentration of the pollutants and their intensity of dispersion in relation to meteorological conditions in the study area such as wind speed, wind direction, temperature and humidity. The present study calculates emission concentration of nitrogen dioxide (NO_2), sulphur dioxide (SO_2), suspended particulate matter (SPM) and respirable suspended particulate matter (RSPM), from various monitoring stations and industries within the study area from the year 2016–2020. It was found that concentration of pollutants lie within the Central Pollution Control Board limits. Also, trend analysis of pollutant concentration was done separately for the year 2020 and there was a significant reduction (>50%) in pollution concentration due to the lockdown scenario created by COVID-19 pandemic.

Keywords Dispersion modelling · ADMS-Urban ArcGIS

V. K. Sumadevi (✉) · U. Anila · K. P. Roy · V. B. Vishnu · S. Minu
Department of Civil Engineering, Government Engineering College Bartonhill, APJ Abdul Kalam Technological University, Thiruvananthapuram, Kerala, India

A. K. Dikshit et al. (eds.), *Innovative Trends in Hydrological and Environmental Systems*, Lecture Notes in Civil Engineering 234,
https://doi.org/10.1007/978-981-19-0304-5_25

1 Introduction

The numerical method for defining the causal relationship between emissions, atmospheric pollution concentration, meteorology and deposition is known as air pollution modelling. According to studies, air quality modelling and valuation may be done using both ground and satellite data [1]. Different models such as AERMOD, ISCST3 and ADMS were used for the data modelling and their modelling efficiencies compared. In assistance with these modelling data, using the modern ArcGIS software assists in the formation of high-resolution downtown area air pollution maps [2].

In this study, the air pollution dispersion model is prepared for the Thiruvananthapuram Municipal Corporation area, which is the administrative spot in the city of Thiruvananthapuram, the capital of Kerala. The capital city's air quality is poor. According to a study by the state pollution control board (SPCB) and the National Transportation Planning and Research Centre (NATPAC), pollutant such as particulate matter is higher than the limit set by National Ambient Air Quality Standards. This study looked at the varying trends in ambient air quality based on data collected at multiple monitoring stations in the Thiruvananthapuram Corporation area.

These are the main objectives of this study:

- To estimate and prepare an air pollution dispersion map of monitoring stations in Thiruvananthapuram Corporation using ADMS-Urban and ArcGIS.
- To do a statistical trend analysis of air pollution from 2016 to 2020 in the study area
- To analyse changes in air quality in the study area during the COVID-19 lockdown.

1.1 Study Area

Thiruvananthapuram Corporation is bordered on the west by the Arabian Sea and on the east by the Western Ghats. It is the most populous and largest city corporation in Kerala, India. This Municipal Corporation administrates the city of Thiruvananthapuram, Kerala's capital.

The area is located at latitude 8′ 25″ North and longitude 76′ 55″ East. The corporation is in the low-land region. The corporation was divided into 24 wards covering an area of 30.66 km^2 in 1940. In 1966, its area increased to 74.86 km^2 and further grew to 141.74 km^2 in 1990. Now the city corporation has grown to 100 wards with an area of 215.86 km^2. Population of 9,57,730 people having population density of 4457 per km^2. Trivandrum Corporation's primary industries are Travancore Titanium Products, Filatex, Travancore Chemical Industries, Hantex, and Chettinad Cement LTD, among others. In addition to the industries, the study area includes the Trivandrum Corporation Plastic/Waste Collection Centre and Aerobic Bin of Thiruvananthapuram City Corporation.

2 Methodology

2.1 Characteristics of the Study Area

The city has a climate that is in between tropical savanna and tropical monsoon. The average maximum temperature is 34 °C, while the average minimum temperature is 21 °C. The humidity is considerable, reaching above 90% during the monsoon season. The city receives heavy rainfall of around 1827 mm per year. By December, the dry season has started. In the city centre, the lowest and maximum temperatures were 17.8 °C and 38.0 °C, respectively. The study area's topographical map, boundary map, land use map and road maps were gathered.

2.2 Data Collection

The study requires emission concentration of nitrogen dioxide (NO_2), sulphur dioxide (SO_2), suspended particulate matter (SPM) and respirable suspended particulate matter (RSPM), from various monitoring stations and industries within the study area from the year 2016–2020. The data includes information on daily and hourly average which is expressed in $\mu g/m^3$. The data were mainly obtained from four monitoring stations of Cosmo, SMV, Plamoodu and Veli in Trivandrum Corporation (Table 1).

The study requires meteorological data which has been processed and provides information on a daily and hourly average of temperature in degree Celcius (°C), wind speed in metre per second (m/sec) and wind direction in degrees (°C)—the critical parameters that govern the intensity of emission dispersion. These were obtained from the official website of the Central Pollution Control Board.

Table 1 Details of monitoring stations in the study area

Station name	Cosmopolitan Hospital, Murinjapalam (COSMO)	Plamoodu	Filatex, Veli (VELI)	SMV School, Over Bridge (SMV)
Type	Sensitive	Residential	Industrial	Sensitive
Latitude	8° 31′ 1.4″ N	8° 30′ 49.6″ N	8° 30′ 16.9″ N	8° 29′ 20″ N
Longitude	76° 56′ 8.4″ E	76° 66′ 45.2″ E	76° 53′ 47″ E	76° 56′ 53.3″ E
Elevation (m)	6	9	6	15

2.3 *Estimate Air Pollution Dispersion Using ADMS-Urban Software*

ADMS-Urban model [3] is an advanced three-dimensional quasi-Gaussian model nested within a trajectory model. The model was used to forecast the distribution of key pollutants' 24-h average concentrations in the area of study [4]. Wind speed and direction are required for ADMS models, as well as one of the following parameters: cloud cover, length, heat flow and upper air data. The production pollution inventory is used directly as the source data for ADMS-Urban after the emission concentrations are exported from EMIT software. These databases were input into ADMS-Urban 5 software. Dispersion modelling is performed in this software, and these dispersion values were integrated into ArcMap 10.7.1. For each project in ADMS, an output file was created automatically as the project was run each time.

EMIT is a database tool for storing, manipulating and assessing emission data from a variety of sources. EMIT has been specially developed to facilitate the rapid calculation and assessment of emissions that can be directly exported to ADMS-Urban.

2.4 *Preparation of Dispersion Map and Trend Analysis in ArcGIS*

ArcGIS is a geographic information system (GIS) that allows the user to interact with maps and geographic data. It is used to produce and use maps, gather geographic data, evaluate mapped information, and distribute and discover geographic data [5]. The system establishes a foundation for making maps and geographic data available throughout an organization, across a community and publicly on the Internet.

3 Results and Discussion

3.1 *Dispersion Maps*

The contours from ADMS-Urban's dispersion model are combined with ArcGIS to produce dispersion maps as the final product. Annual average values of both pollutant and meteorological data are utilized for each year from 2016 to 2020 to create annual average dispersion maps of NO_2, SO_2, SPM and RSPM spread over from the four monitoring stations. Figures 1, 2, 3, 4 and 5 shows the annual average dispersion maps from 2016 to 2020, respectively, of the pollutants measured (i.e. NO_2, SO_2, SPM and RSPM).

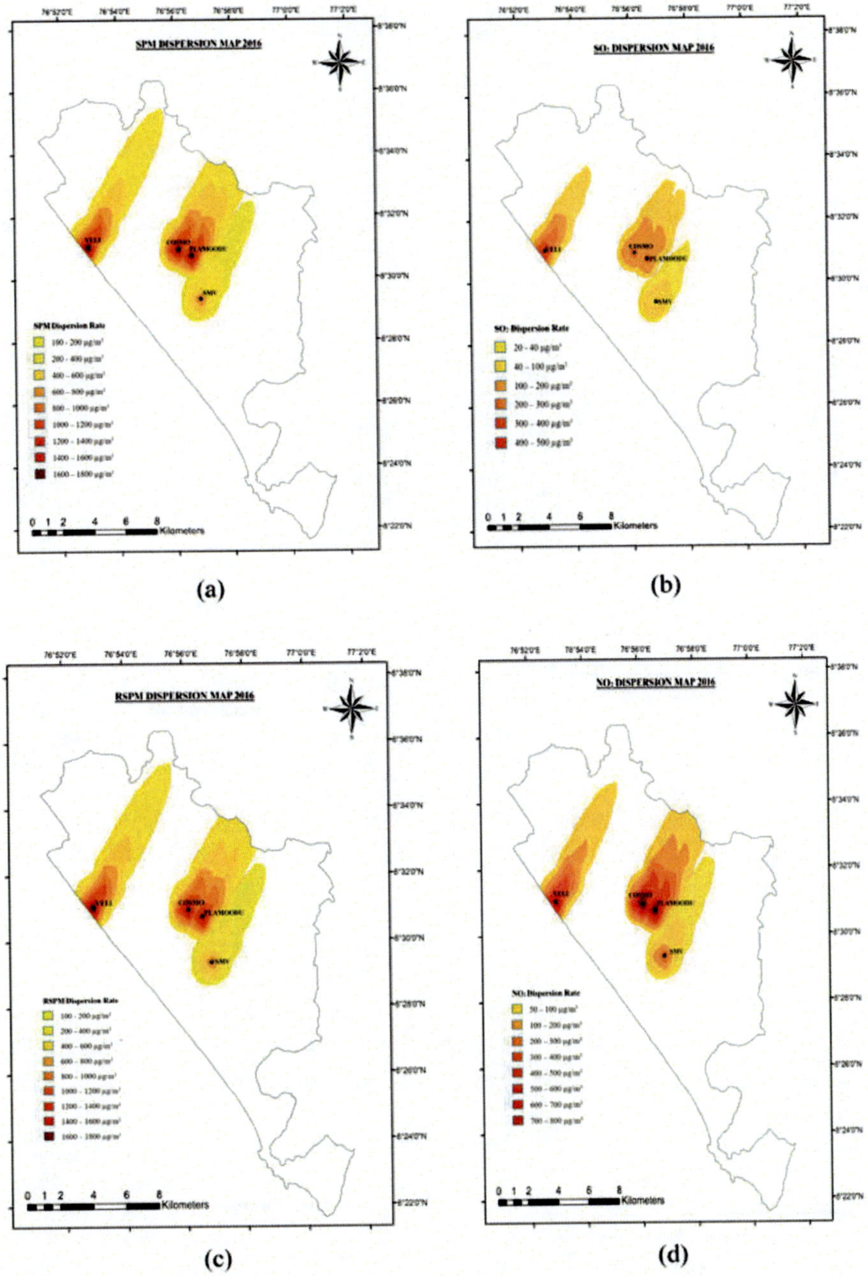

Fig. 1 Annual average dispersion map of 2016 for **a** SPM **b** SO_2 **c** RSPM **d** NO_2

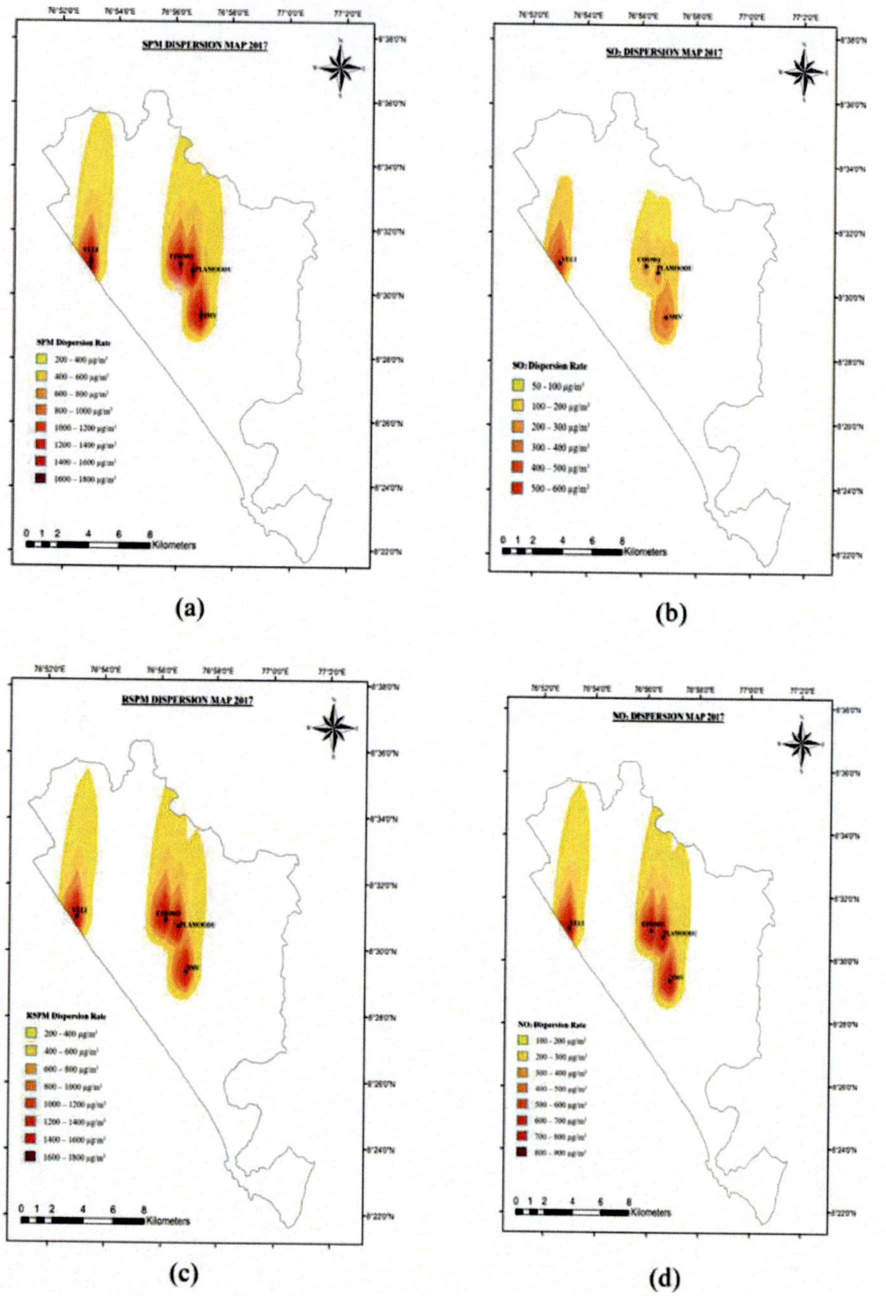

Fig. 2 Annual average dispersion map of 2017 for **a** SPM **b** SO_2 **c** RSPM **d** NO_2

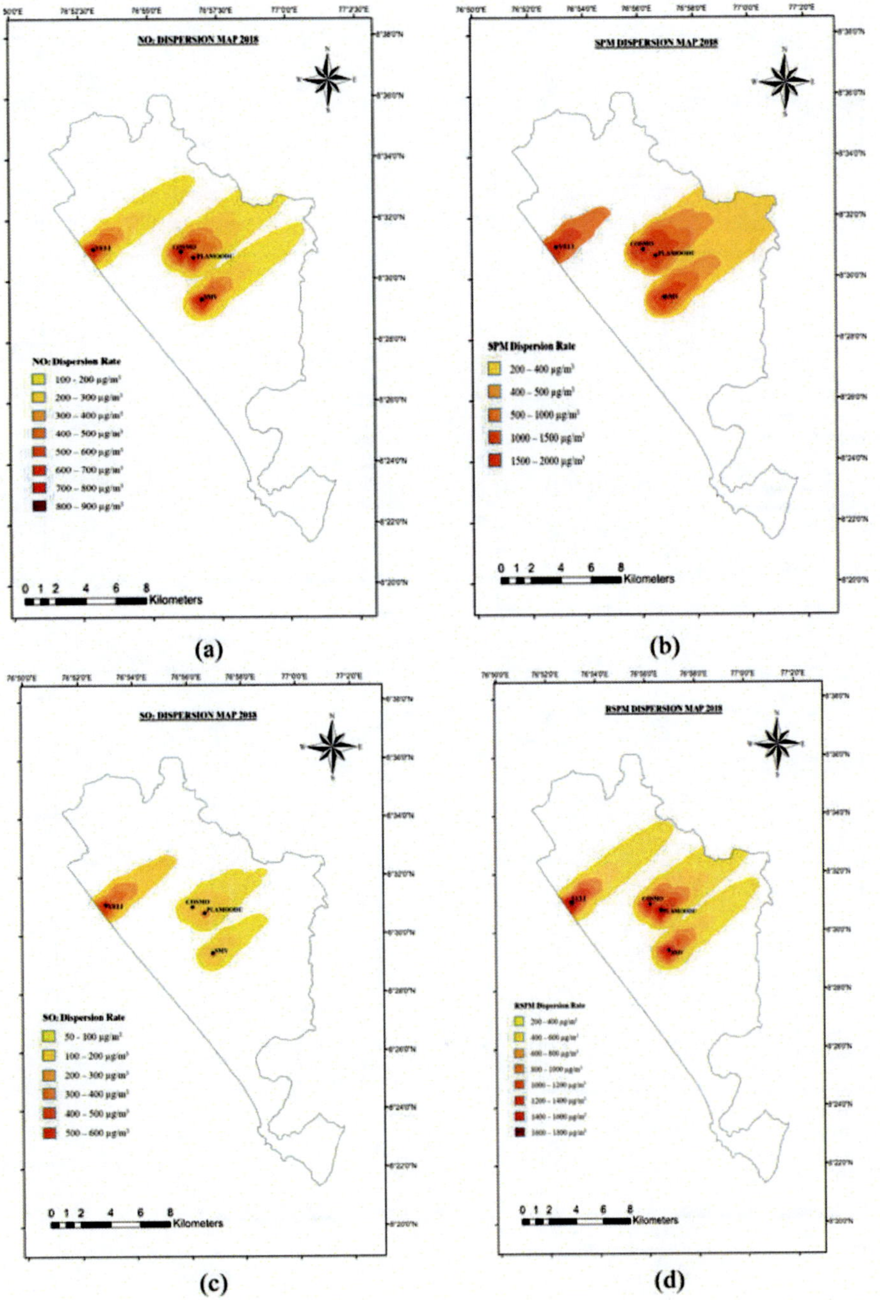

Fig. 3 Annual average dispersion map of 2018 for **a** NO_2 **b** SPM **c** SO_2 **d** RSPM

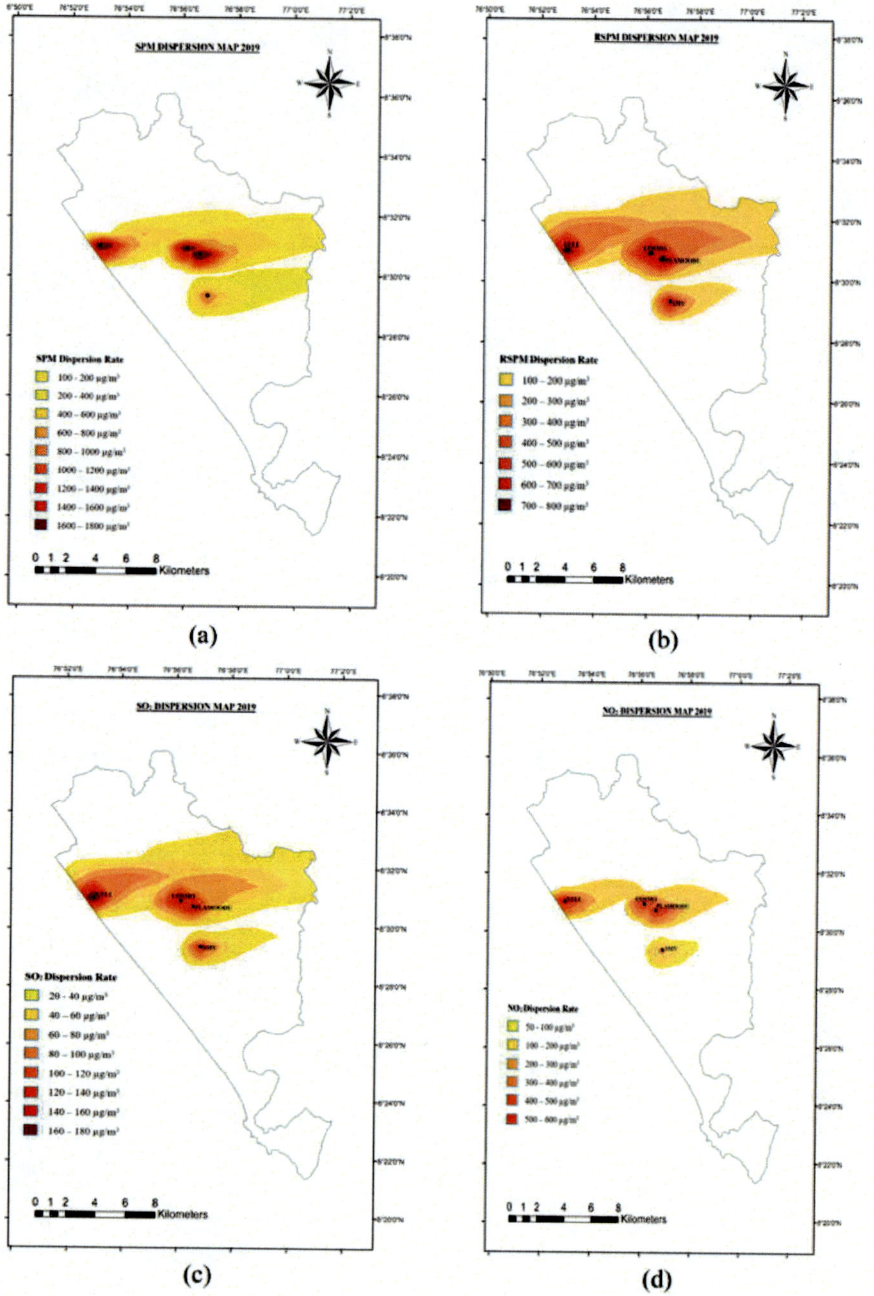

Fig. 4 Annual average dispersion map of 2019 for **a** SPM **b** RSPM **c** SO_2 **d** NO_2

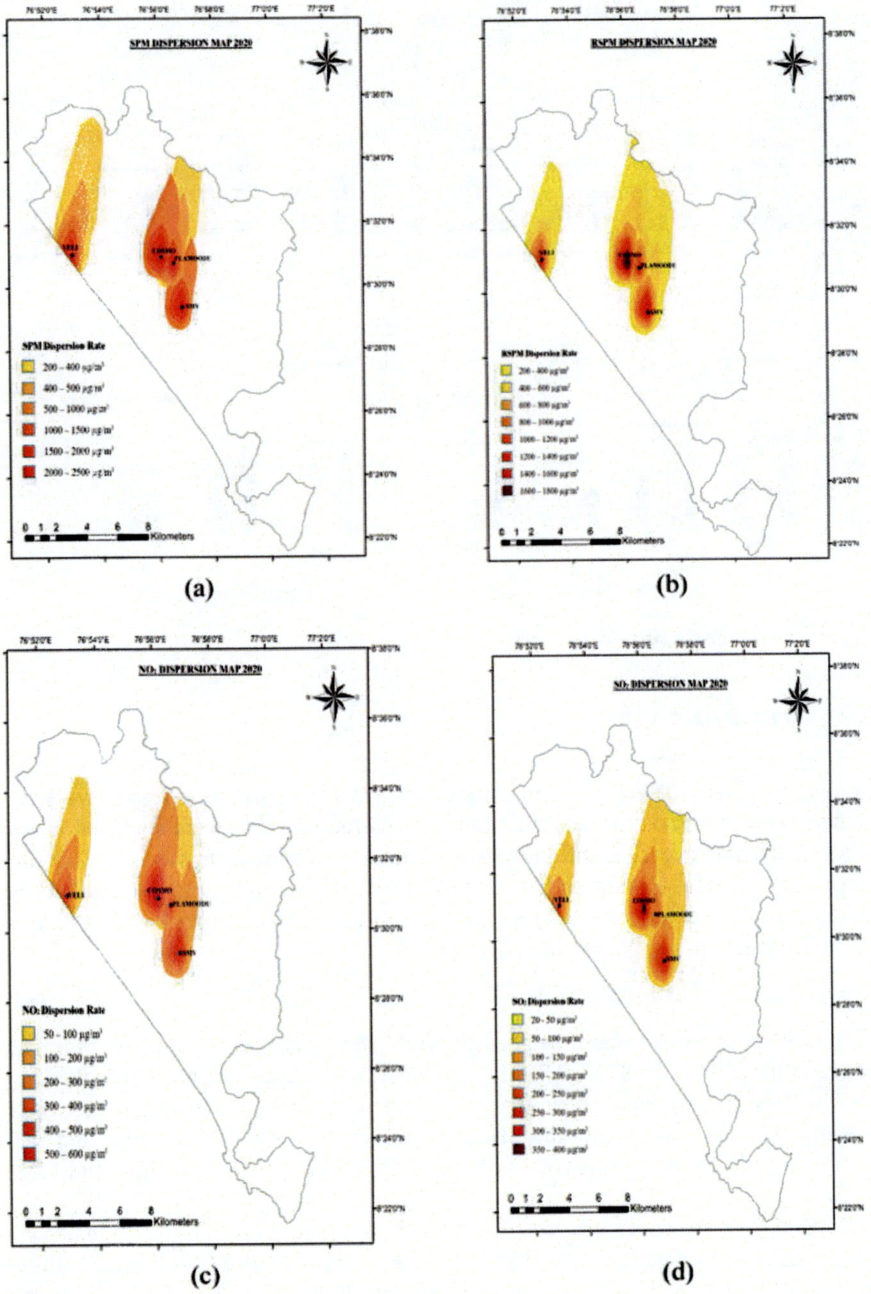

Fig. 5 Annual average dispersion map of 2020 for **a** SPM **b** RSPM **c** NO_2 **d** SO_2

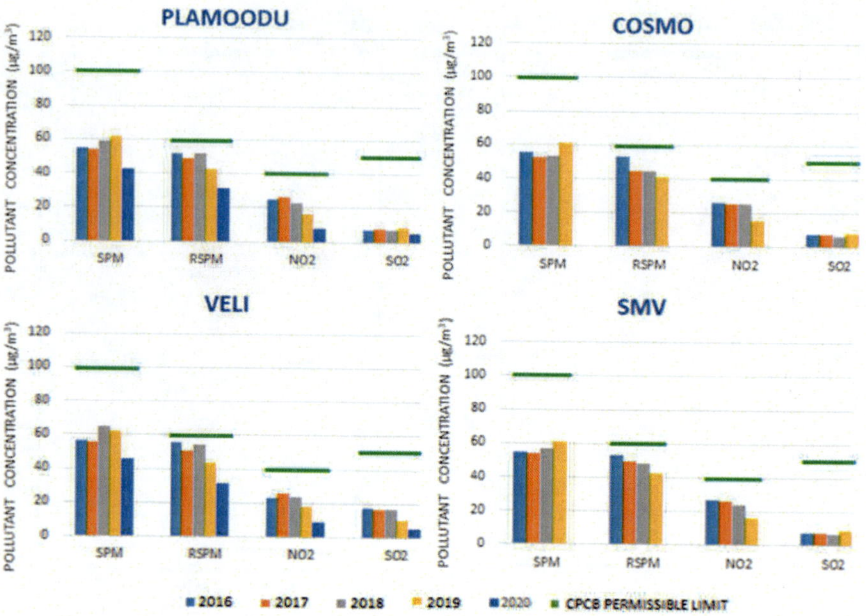

Fig. 6 Trend analysis 2016–2020

3.2 *Trend Analysis*

The pattern study of the pollutants concentration values for their respective years is carried out in trend analysis. Pollution concentration trends were analysed in two ways. The first is a yearly trend analysis of pollutant concentrations for the monitoring stations Plamoodu, Cosmo, Veli and SMV from 2016 to 2020 (Fig. 6). The second is a monthly trend study for the monitoring stations Plamoodu and Veli for the year 2020, from January to November (Fig. 7).

Trend Analysis 2016–2020

The Central Pollution Control Board established permissible annual pollutant concentration limits for SPM, RSPM, NO_2 and SO_2 of 100 $\mu g/m^3$, 60 $^{\mu g}/m^3$, 40 $^{\mu g}/m^3$ and 50 $^{\mu g}/m^3$, respectively. After analysing the four monitoring stations, it was found that all pollutants are within these permissible limits. SPM, RSPM, NO_2 and SO_2 had the highest concentrations of 64.65 $\mu g/m^3$, 55.74 $\mu g/m^3$, 26.84 $\mu g/m^3$ and 17.25 $\mu g/m^3$.

In COSMO, the concentration of SPM and SO_2 first reduced by 4.7%, and 14% (2016–18) and then increased by 15% and 36% (2018–19). But NO_2 remains almost the same during 2016–18 and then reduced by 40% (2018–19). RSPM continuously reduced by 22% (2016–19). Since it is a highly populated hospital region, the change in pollutant concentration may due to variation in emission from sources like combustion, automotive emissions and commercial sterilizers.

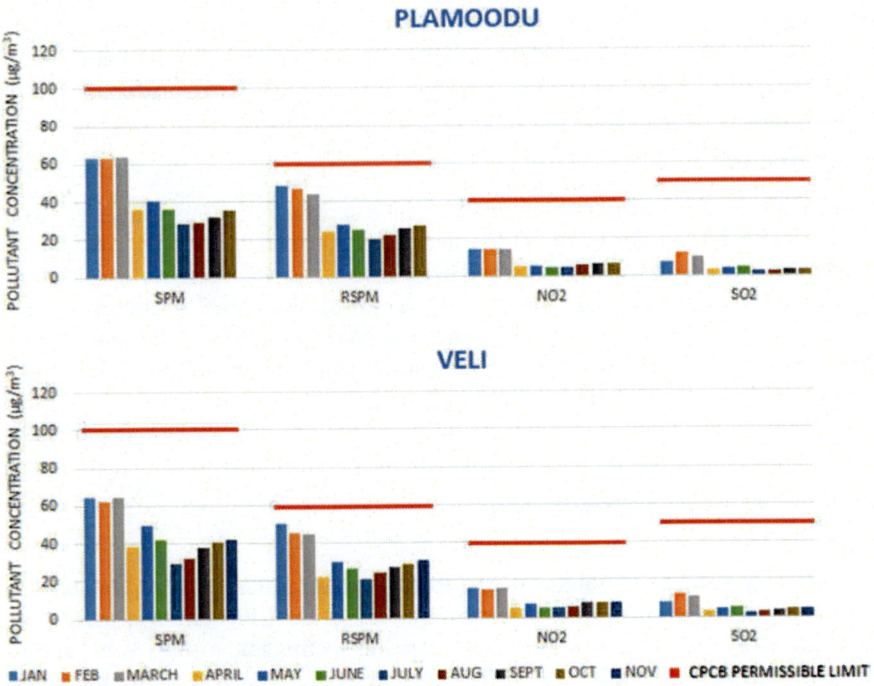

Fig. 7 Trend analysis 2020

In SMV, SO_2 and SPM first reduced by 12% (2016–17) and 6% (2016–18) then increased by 33% (2017–19) and 7% (2018–19). But NO_2, the concentrations were nearly identical from 2016 to 17, but decreased by 6% (2016–17) and 32% (2017–18).The concentration of RSPM decreased by 7% (2016–17), 3% (2017–18) and 11% (2018–19). SMV is heart of Thiruvananthapuram, near the Thampanoor railway station and bus station. The change in pollutant concentration by variation in emission from sources like vehicle emissions, engine combustion, garbage dumping and municipal wastes.

In Plamoodu, SPM concentrations are nearly identical for 2016–17, but increase by 10% (2017–18) and 4% (2018–19), then the value dropped by 30% (2019–20). RSPM decreased by 5% (2016–17), then climbed by 5.2% (2017–18) and cut by 18% (2018–19) and 26% (2019–20). The NO_2 concentration shows a 4% increase (2016–17) and decreased by 12% (2017–18), 28%(2018–19) and 51% (2019–20). SO_2 grew by 7% (2016–17), then decreased by 4% (2017–18). It then climbed by 16% (2018–19), before decreasing by 41% (2019–20). Traffic, minor industry, residential burning of fuels for cooking and natural sources such as dust may be the possible sources for the variations in this residential area.

In Veli, SO_2 and RSPM continuously reduce by 68 and 43% (2016 to 2020). But NO_2 first increased by 13% (2016–17) then reduced by 20% (2017–20). SPM first increased by 15% (2016–18) and then reduced by 30% (2018–20). It is an industrial area where the main source of pollution is industrial emission. Activities such as oil combustion, coal combustion or emissions from various industries cause main pollutant increase. Detrimental effect was due to the installation of emission analysers or filters in industries.

Trend Analysis 2020

Trend analysis of pollutant concentration is done separately for the year 2020 since it has significant variation due to the lockdown scenario created by COVID-19 pandemic. The concentration measurements reveal a sharp decrease in pollutant concentration from March to April, as illustrated by the graphical representation in Fig. 7. SPM, RSPM, NO_2 and SO_2 are reduced by 43%, 44%, 72% and 64% in Plamoodu, respectively. Similarly, the percentage reductions in Veli were 41%, 50%, 74% and 68%.

4 Summary and Conclusion

The concentration of the pollutants and their intensity of dispersion in relation to meteorological conditions were investigated. RSPM and SPM disperse more than NO_2 and SO_2 due to their higher concentration in the study area. Veli has the maximum dispersion, while SPM is the pollutant with the largest dispersion range (2000 $\mu g/m^3$).

The greatest concentrations of SPM, RSPM, NO_2 and SO_2 are found to be 64.65 $\mu g/m^3$, 55.74 $\mu g/m^3$, 26.84 $\mu g/m^3$ and 17.25 $\mu g/m^3$. Concentration limits for SPM, RSPM, NO_2 and SO_2 of 100 $\mu g/m^3$, 60 $\mu g/m^3$, 40 $\mu g/m^3$ and 50 $\mu g/m^3$, respectively. After analysing the four monitoring stations, it was found that all pollutants are within these permissible limits. The drop in pollutant levels up to 50% owing to the COVID-19 lockdown period is clarified by a 2020 trend study. The main cause is a rapid decline in vehicular traffic and the closure of industries.

To maintain the acceptable air quality in the study area, technology such as emission analysers, combustion and flue gas analysers, air quality monitoring and vehicle exhaust analysers can be used in conjunction with initiatives like AirNow and the Air Quality Flag Program.

References

1. El-Harbawi, M. (2013). Air quality modelling, simulation, and computational methods: A review. *Environmental Reviews, 21(3)*, 149–179. https://doi.org/10.1139/er-2012-0056
2. Gulliver, J., & Briggs, D. (2011) STEMS-air: A simple GIS-based air pollution dispersion model for city-wide exposure assessment. *Science of the Total Environment, 409*, 2419–2429.
3. Nagendra, S., Khare, M., Gulia, S., Vijay, P., Chithra1, V. S., Bell, M., & Namdeo, A. (2012). Application of ADMS and AERMOD models to study the dispersion of vehicular pollutants in

urban areas of India and the United Kingdom. *WIT Transactions on Ecology and the Environment* (pp. 3–12), ISSN, 157.
4. Behera, S. N., Sharma, M., Dikshit, O., & Shukla, S. P. (2011). GIS-based emission inventory, dispersion modeling, and assessment for source contributions of particulate matter in an urban environment. *Water Air Soil Pollution, 218*, 423–436.
5. Gupta, N. J. (2014). Fundamentals of geographical information system (GIS), map sources, and digital map preparation. Geospatial technologies and climate change (pp. 247–260), https://doi.org/10.1007/978-3-319-01689-4_15

Analysis of Unsteady Flow Using HEC-RAS and GIS Techniques

H. Pallavi and A. S. Ravikumar

Abstract Flooding is a global phenomenon that causes casualties and property loss on every inhabited continent. It is probably the most devastating, widespread and frequent natural disaster for human societies. To determine the flood in advance and to prevent the damages due to overflow of adjoining river banks, the Flood Inundation Modeling and Flood Inundation Mapping plays an important role in the field of Hydrology. Hence, in the present study an attempt has been made to analyze the flood inundation boundary by unsteady flow analysis for the downstream of T.Narsipura, discharge gauging station of upper Cauvery basin, Karnataka by using an integrated approach of HEC-RAS and ArcGIS techniques. Unsteady flow analysis is carried out by using HEC-RAS to detect the change in flooding pattern for 21 years and the results are mapped with ArcGIS. These flood inundation studies and the maps created will help the decision makers to assess the risk and to take precautions at the earliest.

Keywords Flood Inundation Modeling · Flood Inundation Mapping · HEC-RAS · ArcGIS

H. Pallavi (✉) · A. S. Ravikumar
Department of Civil Engineering, UVCE, Bangalore University, Bengaluru-56, India

A. K. Dikshit et al. (eds.), *Innovative Trends in Hydrological and Environmental Systems*, Lecture Notes in Civil Engineering 234,
https://doi.org/10.1007/978-981-19-0304-5_26

1 Introduction

India is one of the worst flood-affected countries in the world; the major flood-prone area covers 12.5% area of the country. About 40 Mha of land in India is prone to be flooded as per National Flood Commission [4]. In recent decades, flooding increased profoundly due to hydrological characteristics changes. When the river banks are overtopped, river water flows over the flood plain and thus causes damage to crops and property within the flood plain of river [7].

There is a constant endeavor to understand, assess and predict flood events and their impacts. Flood inundation models (FIMs) are therefore developed to serve this purpose [5]. FIM involves hydrologic modeling to estimate peak flows from storm events, hydraulic modeling to estimate water surface elevations and terrain analysis to estimate the inundation area [3].

Hydraulic modeling and flood inundation mapping are performed in order to predict important information's from a flood event including the extent of inundation and water surface elevations at specific locations [2].

There are many computer models exist to simulate this modeling analysis. From the review of literature, it is understood that the combined use of HEC-RAS with ArcGIS is very common in flood inundation modeling and mapping. In the present study an attempt has been made to analyze unsteady flow using HEC-RAS and ArcGIS.

2 Study Area

The study area lies below discharge station called T. Narsipura of Mysore district, which is considered as the input source of the flow of water into the defined river reach. The reason behind choosing this as the study area for the project is due to recent flood activities in this region. Geographically it lies between 76° 54′ 0″ E and 77° 8′ 0″ E, 12° 16′ 0″ N and 12° 9′ 30″ N, in the Upper Cauvery basin. The location map of the study area is as shown in Fig. 1.

3 Materials and Methodology

Various data are required in setting up of the HEC-RAS model, in case of 1 D unsteady flow analysis one of the vital information is the geometrical information of a specific river stretch, the geometrical data of a river created using SRTM-DEM in HEC-RAS tool called RAS Mapper which assists in preparing input file for the analysis, in case of 2D unsteady flow modeling 2D flow area is created using SRTM-DEM in geometric data tool of HEC-RAS software, then the required data of both are input into the software for performing the analysis. To prepare the map, the results

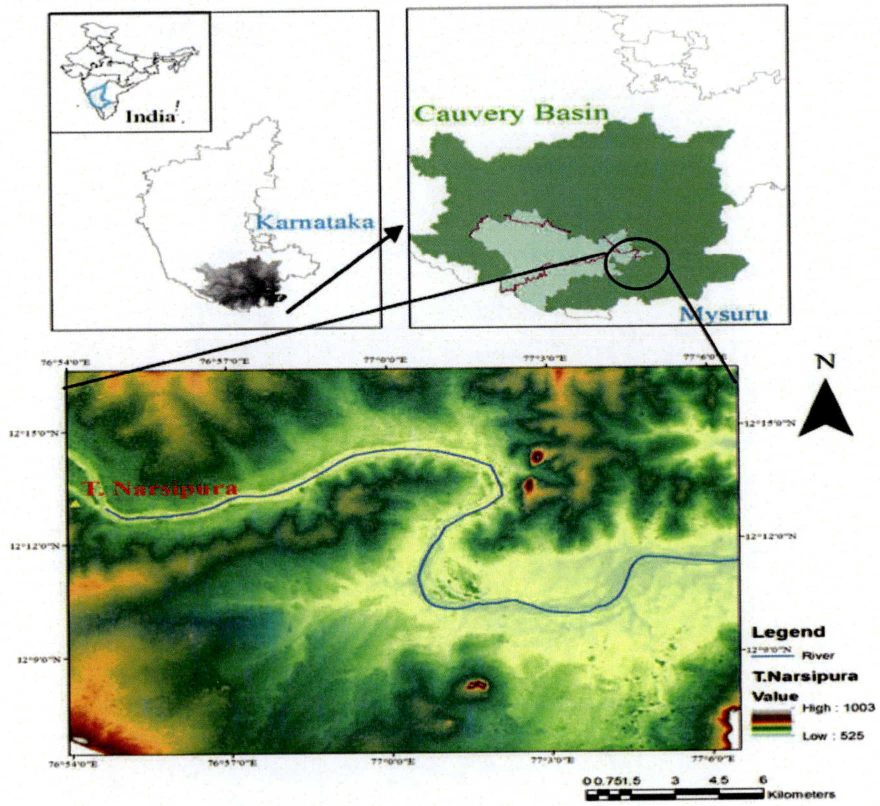

Fig. 1 Location map of the study area

thus obtained in HEC-RAS are then exported to the ArcGIS using RAS mapper. The data used for the analysis are summarized in Table 1 and flow chart depicting the methodology adopted in the study is shown in Fig. 2.

Steps involved in creating the models are as follows.

Table 1 Data Used

Sl. No.	Data	Purpose
01	SRTM 30 m DEM data	Extraction of study area
02	Discharge Data	1D, 2D modeling
03	Rainfall Data	2D modeling
Software used	Purpose	Source
05	ArcGIS (10.4)	Flood mapping
06	HEC-RAS (5.0.7)	Flow Analysis and Flood Plain Detection

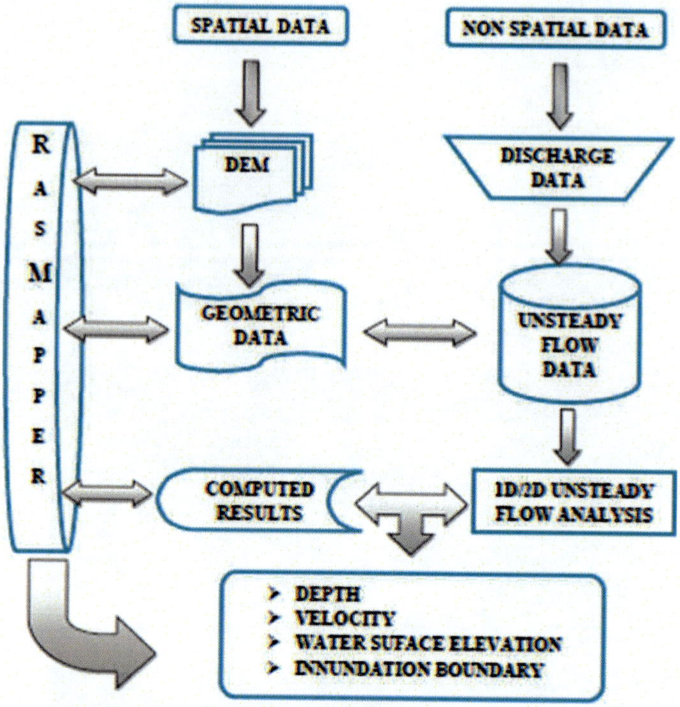

Fig. 2 Methodology adopted in the study

3.1 One-Dimensional Flow Modeling

Step 1: The SRTM-DEM image downloaded from USGS Earth Explorer is used for delineation of Upper Cauvery Basin on GIS platform.
Step 2: The delineated basin is imported to the HEC-RAS as it gives the reference for marking the geometric conditions of the river; for example, river centerline, bank lines, flow path and cross sections are created using RAS Mapper for further analysis. Creation of geometric section of the river is shown in Plate 1, and geometric data view in HEC-RAS is shown in Plate 2.

Plate 2Geometric data in HEC-RAS

Step 3: For the flow analysis data, the inputs like discharge data in the form of flow hydrograph for the upstream boundary condition and normal depth (0.001-slope of the river profile) for downstream boundary condition is fed to the software as shown in Plate 3.

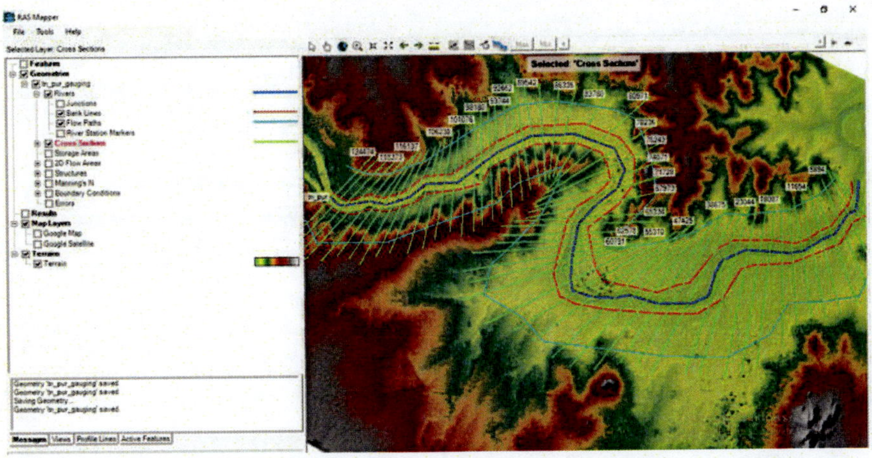

Plate 1 Creation of geometric features in RAS Mapper

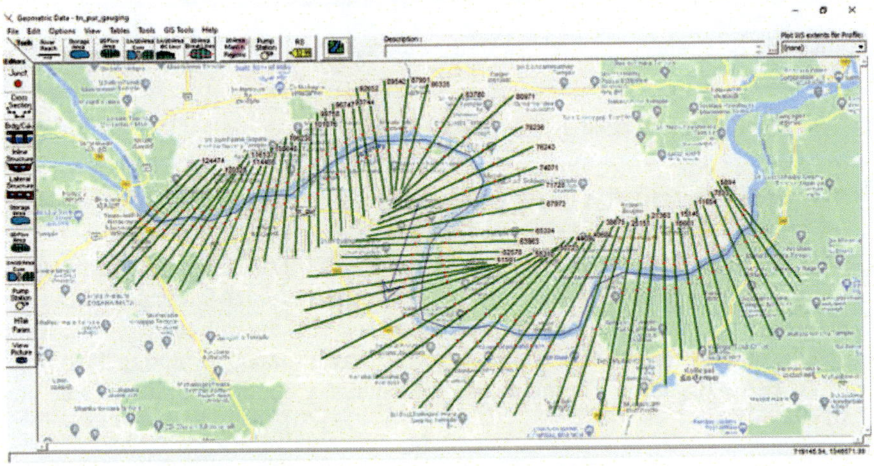

Plate 2 Geometric data in HEC-RAS

Step 4: The Manning's n value of 0.05 on LOB and ROB and 0.035 for the channel is provided for the river at all cross sections because it is a natural drain and also depends on the land use land cover, channel type, slope of the channel.

Step 5: Unsteady flow analysis is performed to get the final outputs namely inundation boundary, depth of the river, velocity, water surface elevation. The obtained results are ground truth verified with Google Earth explorer and mapping of the same is done in ArcGIS platform.

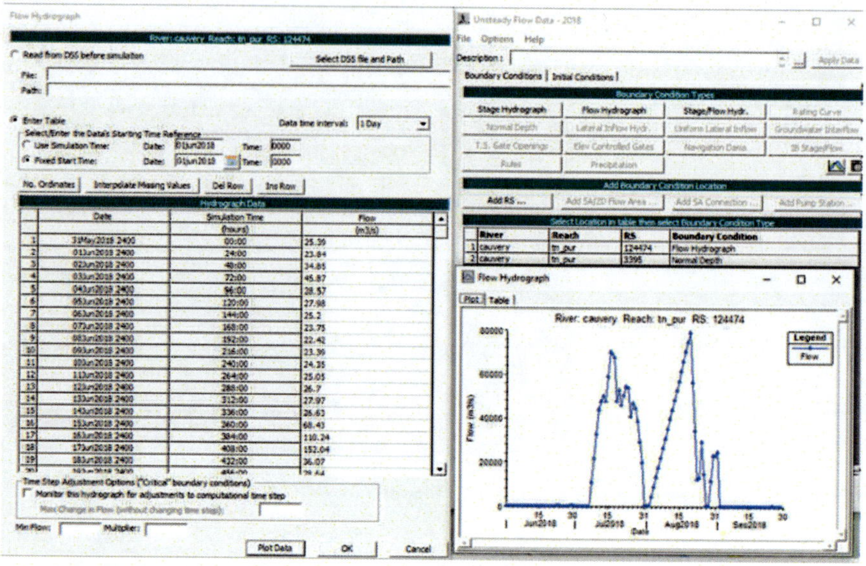

Plate 3 HEC-RAS windows of compiled 1D unsteady flow Data

3.2 *Two-Dimensional Flow Modeling*

Step 1: The SRTM-DEM image downloaded from USGS Earth Explorer is used for delineation of required study area on GIS platform.
Step 2: The delineated terrain is imported to the HEC-RAS as it gives the reference for marking the geometric conditions of the river. Under RAS Mapper a 2D flow perimeter is created and thus a mesh is formed. Upstream boundary condition and the downstream boundary conditions are drawn suitably by opting the option SA/2D area BC lines. The mesh and the boundary conditions thus created are shown in Plate 4.

Step 3: the required data for 2D analysis namely daily rainfall data of respective years are entered under the precipitation along with upstream and downstream data same as 1D unsteady flow analysis. The model proceeds to calculate the area affected throughout the interior points of the perimeter. The data for the 2D modeling is entered are shown in Plate 5.

Step 4: The unsteady flow analysis is performed to get the final outputs. The obtained results are ground truth verified with Google Earth explorer and mapping of the same is done in ArcGIS platform.

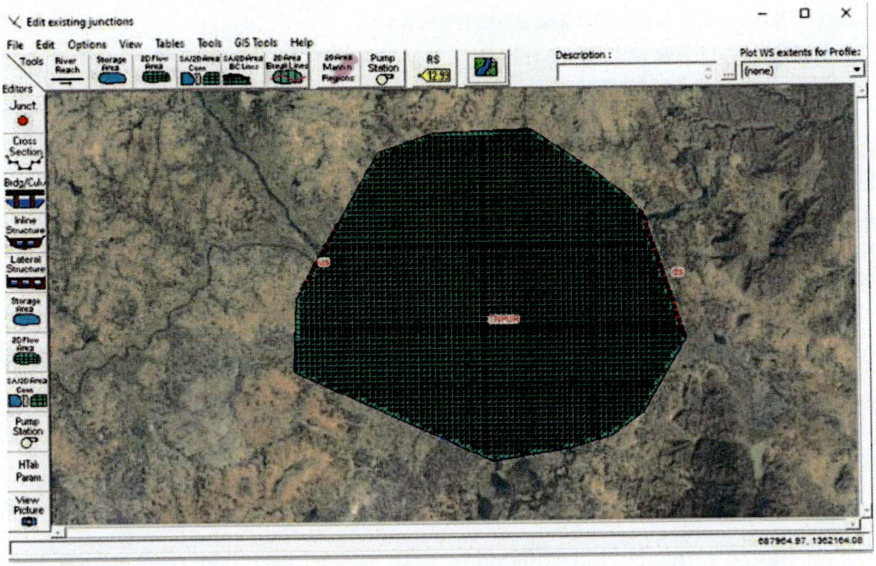

Plate 4 HEC-RAS Window of Mesh created for 2D modeling

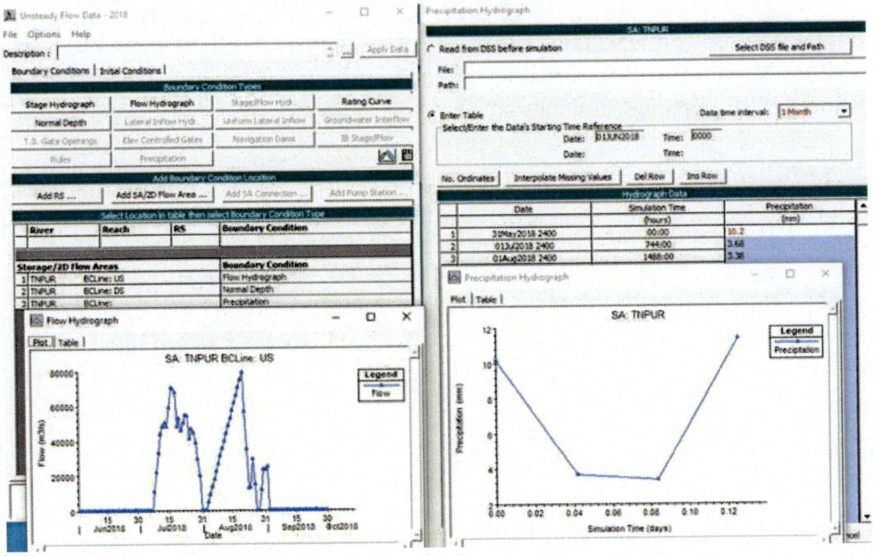

Plate 5 HEC-RAS windows of compiled 2D unsteady flow Data

4 Results and Concluding Remarks

Unsteady flow data analysis is performed for the period June 1 to September 30 (1998–2018) for both 1D and 2D modeling, and this time period of four peak monsoon months is considered because the discharge and water level are at peak during monsoon. Considering this criterion, the analysis is performed for the 21 different years (1998–2018), since in the unsteady flow analysis the discharge will be varying with time, i.e., for different days the particular discharge will be flowing through the stream. The inundation boundary from 1 and 2D modeling in terms of areal extent for 21 years is shown in Tables 2 and 3.

From Tables 2 and 3, the minimum and maximum inundation boundary of both 1D and 2D unsteady flow modeling is recorded in the year 2002 and 2018. The inundation mapping results are obtained from ArcGIS.

Inundation map for 1D unsteady flow analysis of the year 2018 and 2002 are shown in Plate 6 and Plate 7 respectively.

The ground truth has been verified on Google Map which is shown in Figs. 3, 4, 5 and 6.

As it was an extreme flood situation in 2018 due to heavy flow from Kabini and Krishnaraja Sagar (KRS) reservoir which have resulted in overflow of Cauvery cutoff over 30 villages at T.Narsipura and lower lying of Kabini river, since the study area is also a part of this flood zone, it reveals some of the inundated boundaries of flood-prone areas such as T.Narsipura, Madapura, Mudukuthore, Marahalli, Mavinahalli, Talakadu, Hosa Kukkur, Kaveripura, Sargur, Teramballi, and Kollegala.

Some paper clippings of floods are shown in Plate 8.

Table 2 1D modeling unsteady flow analysis inundation boundary of study area

Year	1998	1999	2000	2001	2002	2003	2004	2005	2006	2007
Area (km^2)	19.71	19.75	16.95	15.44	13.59	14.92	14.92	19.95	16.42	17.89
Year	2008	2009	2010	2011	2012	2013	2014	2015	2016	2017
Area (km^2)	22.24	16.20	18.59	15.64	13.97	26.28	26.39	18.42	15.64	17.27
Year	2018									
Area (km^2)	98.39									

Table 3 2D Modeling unsteady flow analysis inundation boundary of study area

Year	1998	1999	2000	2001	2002	2003	2004	2005
Area (km^2)	180.00	180.04	140.04	130.45	108.67	114.71	181.37	140.01
Year	2006	2007	2008	2009	2010	2011	2012	2013
Area (km^2)	210.45	145.11	192.82	139.85	150.05	144.21	109.89	193.27
Year	2014	2015	2016	2017	2018			
Area (km^2)	219.71	158.15	137.47	140.34	333.45			

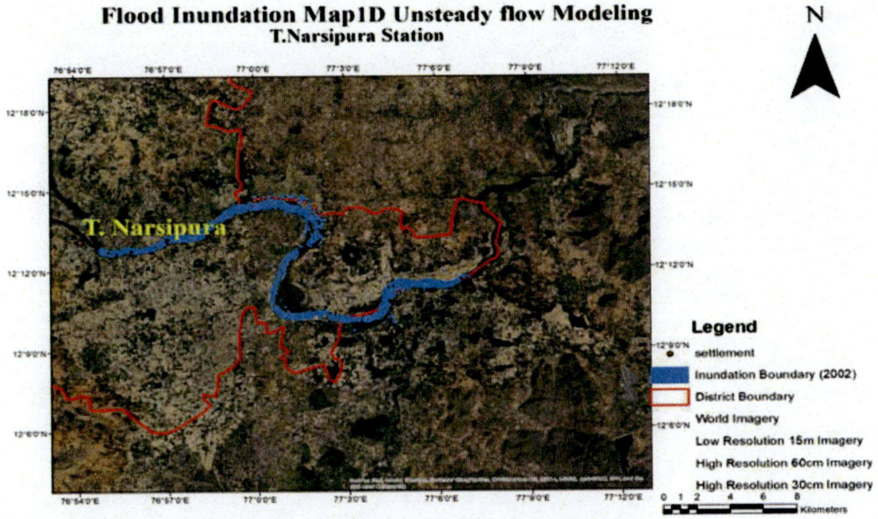

Plate 6 1D Unsteady flow analysis inundation boundary map-2002

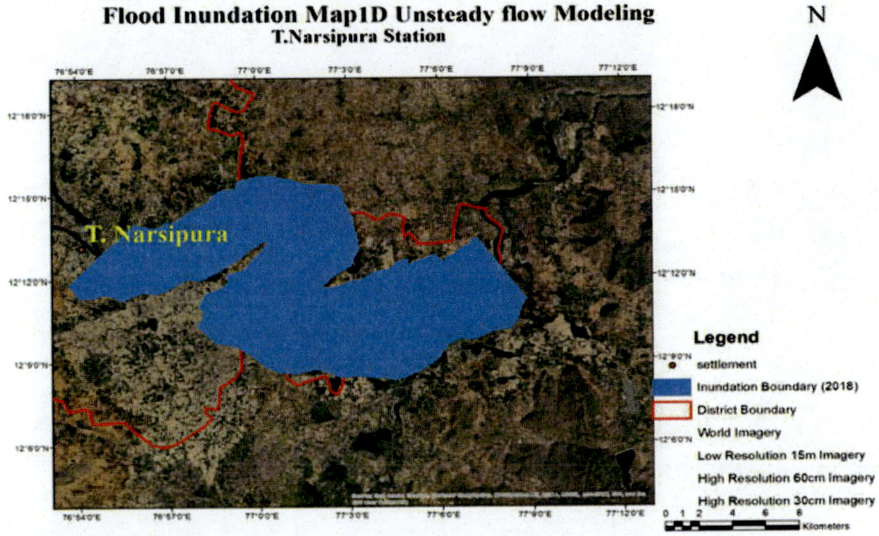

Plate 7 1D Unsteady flow analysis inundation boundary map-2018

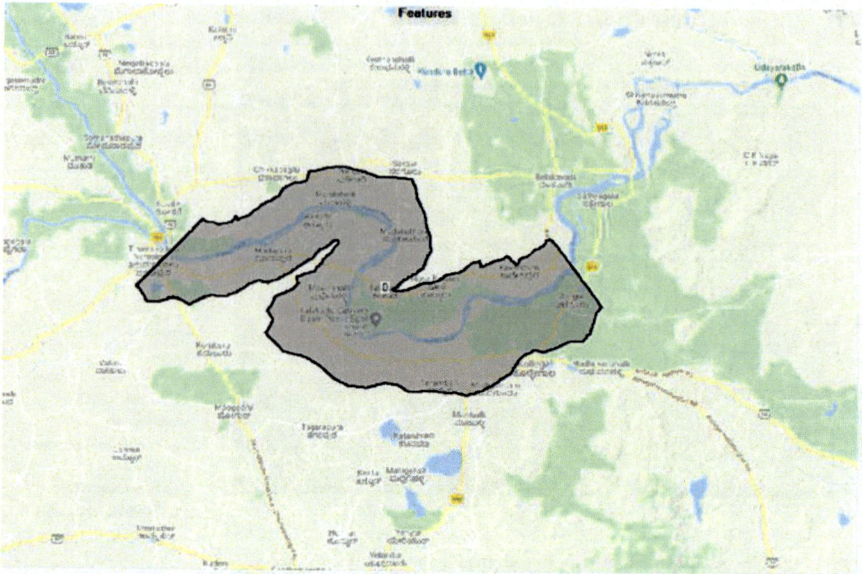

Fig. 3 Ground truth verified results of 1D unsteady flow analysis—2018 (RAS Mapper)

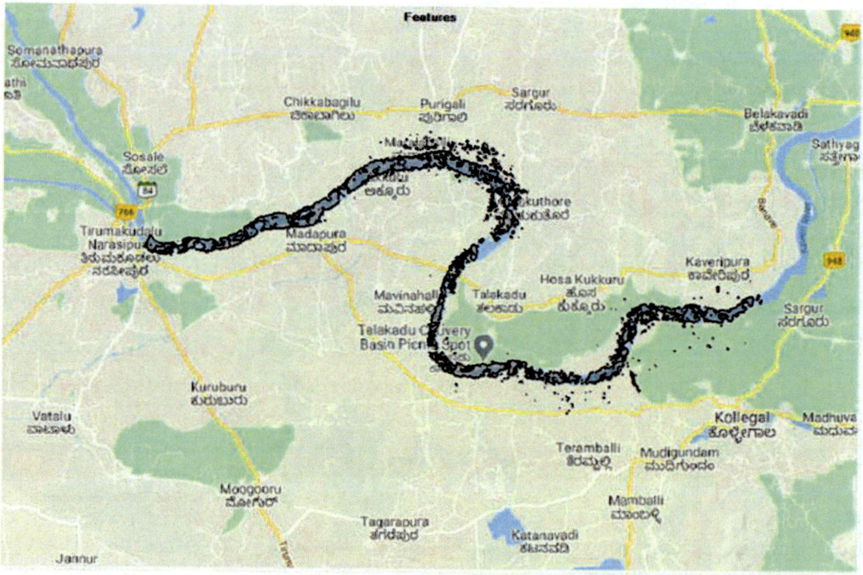

Fig. 4 Ground truth verified results of 1D unsteady flow analysis—2002 (RAS Mapper)

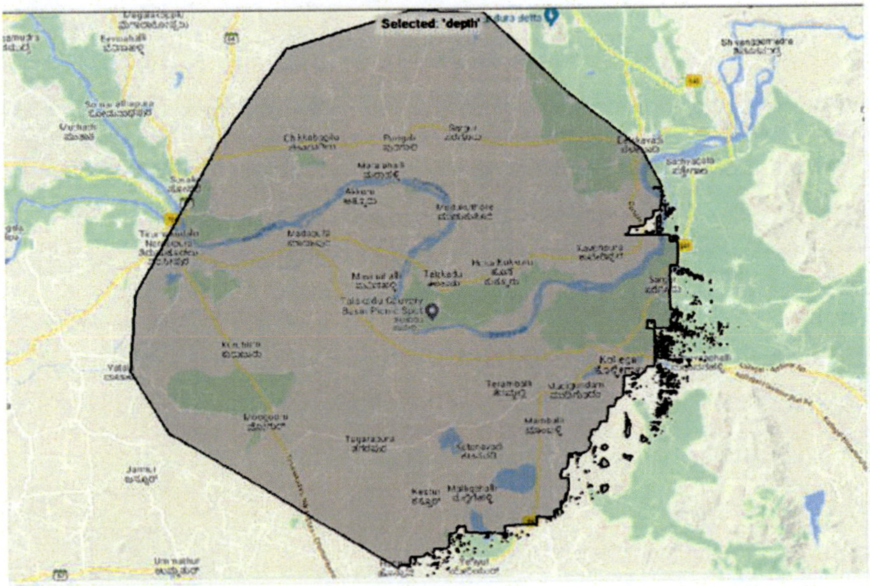

Fig. 5 Ground truth verified results of 2D unsteady flow analysis—2018 (RAS Mapper)

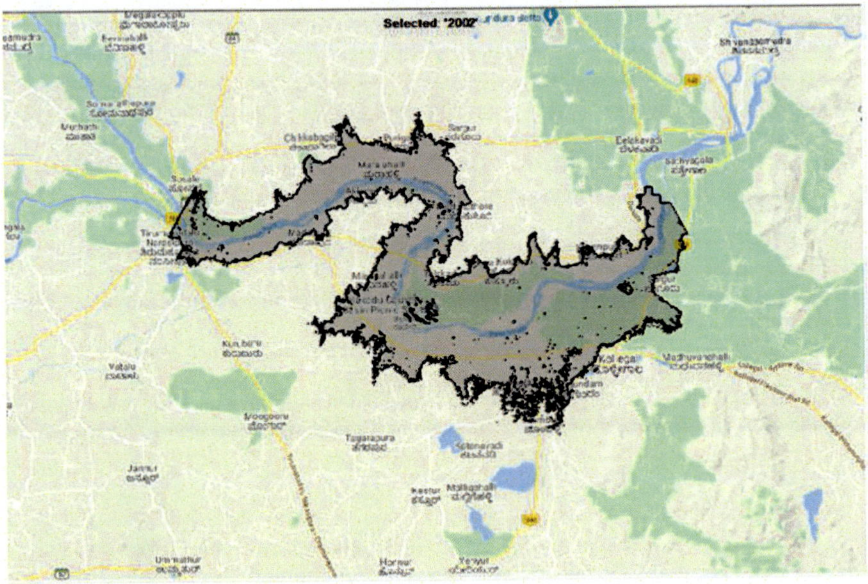

Fig. 6 Ground truth verified results of 2D unsteady flow analysis—2002 (RAS Mapper)

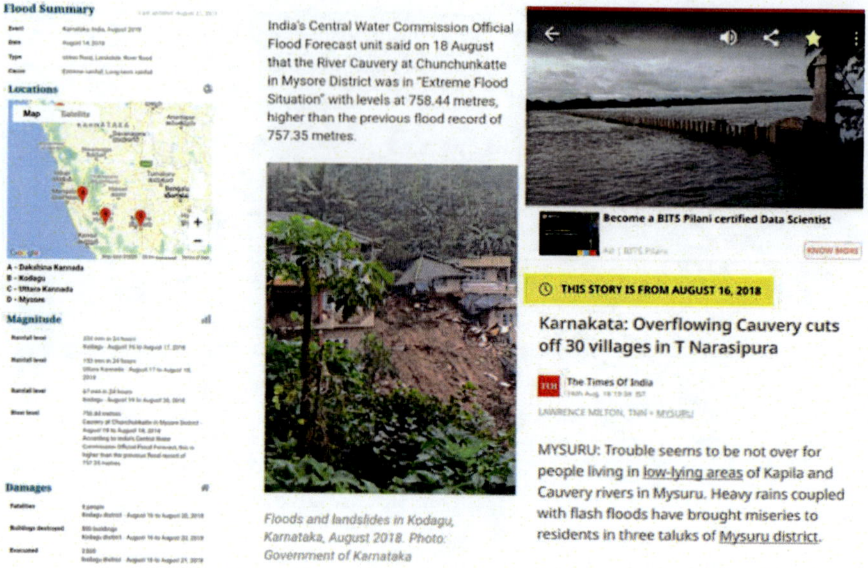

Plate 8 Flood 2018

References

1. Chow. (2009). *Open channel hydraulics* (pp. 675). New York: McGraw-HILL Book Company.
2. Horritt, & Bates. (2002). Evaluation of 1D and 2D numerical models for predicting river flood inundation. *Journal of Hydrology, 268*, 87–99.
3. Merwade, C., & Coonrod. (2008). GIS techniques for creating river terrain models for hydro dynamic modeling and flood inundation mapping. *Environmental Modeling & Software, 23*, 1300–1311.
4. Patel, C. G. (2016). Floodplain delineation using HEC-RAS model—A case study of Surat city. *Modern Journal of Hydrology, 6*, 34–42.
5. Teng, J., & Croke. (2017). *Flood inundation modelling: A review of methods, recent advances and uncertainty analysis.*
6. USACE. (2002). *HEC-RAS River Analysis System, Hydraulic Reference Manual, U.S Army Corps of Engineers (USACE).*
7. Younghun, D., & Munmo, & Lee. (2014). Simplified flood inundation mapping based on flood elevation-discharge rating curves using satellite images in gauged watersheds. *Water, 6*, 1280–1299.

Potential of Green Roofs in Heat Stress Mitigation—A Case of Nagpur City

Rakesh Kadaverugu, Suvha Lama, Vigna Purohit, Sravanthi Mopati, Chandrasekhar Matli, and Rajesh Biniwale

Abstract Urban areas are increasingly facing the challenges of extreme weather events (EWEs) mainly due to climate variability and rapid economic growth. Several international conventions have endorsed climate-smart urban development programs to improve the resilience of the cities. Nature-based solutions such as encouraging urban green spaces (UGSs) are mainly recognized as eco-friendly alternatives in combating the EWEs in cities. UGSs combat the effects of EWEs through heat mitigation and runoff retention and also provide an array of multiple ecosystem services, including but not limited to air purification, enhancing biodiversity, and cultural and aesthetic value. In the current chapter, the heat mitigation role of green roofs has been quantified for the present and two future land use scenarios of Nagpur City as a case study. InVEST (Integrated Valuation of Ecosystem Services and Tradeoffs) Urban Cooling Model has been utilized with the parameters tailored for the study area. Results indicate that the average ambient temperature has decreased by 0.23 °C and 0.44 °C in the scenarios of conversion of 10% and 20% built-up into green roofs, respectively. This scenario-based quantification of heat mitigation benefits of green spaces, including green roofs, will help in sound decision-making for a sustainable and resilient urban development.

Keywords Heat mitigation · InVEST model · Nature-based solutions · Urban green spaces · Urban microclimate · Urban planning

---

R. Kadaverugu (✉) · S. Lama · V. Purohit · S. Mopati · R. Biniwale
CSIR National Environmental Engineering Research Institute, Nagpur, Maharashtra 440020, India
e-mail: r_kadaverugu@neeri.res.in

C. Matli
National Institute of Technology Warangal, Warangal, Telangana 506004, India

A. K. Dikshit et al. (eds.), *Innovative Trends in Hydrological and Environmental Systems*, Lecture Notes in Civil Engineering 234,
https://doi.org/10.1007/978-981-19-0304-5_27

1 Introduction

In the twenty-first century, global warming due to certain anthropogenic activities is causing detrimental impacts worldwide, such as climate variability, sea level rise, and rise in the frequency and intensity of extreme weather events (EWEs) such as floods, droughts, and heatwaves [1]. These impacts are bound to affect the socio-economic and political fabric, especially in the developing and emerging economies [2]. With the endorsement of the Paris Agreement in 2015, ambitious climate mitigation strategies have been put in place to limit the temperature to 1.5 °C. These actions, if executed successfully, will reduce the severity of the hazards that the global society will have to face in the future. However, certain climate hazards will continue to escalate due to system feedback or climate inertia [1]. Hence, climate change adaptation is necessary to mitigate the impacts of EWEs, especially in urban areas, as more than 60% of the world's population resides in the cities.

Today, a significant portion of the global population lives in urban or peri-urban areas. With urbanization set to increase to 68% by 2050 (United Nations Department of Economic and Social Affairs, 2018), issues relating to climate variability are expected to accelerate in the coming decades. The changes in both population and land use and land cover dynamics due to urbanization result in the depletion of blue-green spaces, increased concretization, and increased use of public and private transportation [3]. These alterations in the urban ecosystem affect the energy and water fluxes, leading to changes in local ambient temperature, creation of urban heat islands, increase in greenhouse gas emissions, and depletion of the urban carbon storage and sinks [4, 5].

One of the significant ecological challenges modern cities faces is the changes in local ambient temperature leading to a regional climate phenomenon called urban heat island (UHI) effect [6, 7]. The degree of this effect is influenced by the population density, urban form (size and shape of its built-up area), the material used for construction, and infrastructure development and placement of various anthropogenic activities [6, 8]. Further, the space cooling energy demand will increase substantially due to high intense UHI [9]. The present refrigerant-based air conditioning is heavily dependent on fossil fuels, which further provides positive feedback to global warming and exerts more pressure on further space cooling, thus creates a vicious cycle. According to IEA [10], the present energy for space cooling contributes to nearly 10% of global CO_2 emissions and can increase the global temperature by 0.5 °C by 2100.

There is an urgent need to offset the energy needs for space cooling with more sustainable and nature-based solutions. Giguère [11] has listed down a comprehensive selection of UHI-related mitigation strategies and grouped them into four major categories: vegetation and cooling, sustainable urban structures, sustainable water management, and, lastly, reduction of ambient heat. However, these categories overlap in certain aspects. Looking at that, Aleksandrowicz [12] has proposed a classification system based on the type of intervention and listed four major groups: building envelope, urban landscaping, pavements, and street geometry.

1.1 Green Roofs for Urban Ecosystem Management

Green roofs and walls have been in use since ancient times. One famous example would be the Hanging Gardens of Babylon [13]. Green roofs were first used in modern construction methods as an alternative to black tar, which is a fire hazard, in the nineteenth century. Sand and gravel layers were used instead for protection against fire [14]. Today, green roofs can be seen as a paradigm shift in construction with the incorporation of the natural systems in modern architecture and construction. With the loss of urban green areas due to anthropogenic changes in land use patterns, it is critical to see urban green roofs and walls as the next biological active areas for improving the quality of the urban environment [15].

Structurally, the green roof comprises of different layers that include (i) a thermal insulation layer, (ii) a waterproofing layer, (iii) a drainage layer, and (iv) a soil layer [13, 15]. The most critical layer is the soil layer which is also called substrate layer, as this determines the type of green roof through the thickness of the soil substrate [13].

Under European guidelines, green roofs can be separated into two categories, (1) extensive green roofs (EGR) with soil substrate depth of 15 cm and (2) intensive green roofs (IGR) with soil layer depth more than 15 cm [12, 16]. This thickness also defines the type of vegetation that can be used in the green roof. The intensive green roofs consist of deeper substrate which supports a wide variety of plant species, whereas the extensive green roofs typically consist of much shallower substrate but covers most of the terrace. However, the difference between intensive and extensive green roofs is getting blurred with the overlap of multi-functional terrace gardens. The design aspects of green roofs consist of several layers of vegetation, substrate, filter media, and drainage.

Various studies have shown that the implementation of green roofs provides multi-functional benefits in improving urban environments 13. Green roofs can improve the urban micro-environment through: urban heat regulation and mitigation; runoff management; reduction of greenhouse gas emissions; reduction of gaseous and particulate pollutants; urban noise reduction; and enhancement of biodiversity [13, 15, 17, 18]. These are discussed in detail in subsequent sections.

1.1.1 Urban Heat Regulation and Mitigation

As mentioned earlier, the rapid urbanization and compaction of urban form is leading to the heating of urban areas. One of the most critical issues faced by urban areas is urban heat islands [15]. The most significant contributors for the formation of urban heat islands (UHI) are loss of vegetation and an increase in low-albedo surfaces [9, 17]. The interaction of temperature, wind, and solar radiation with natural and artificial structures plays an essential role in the heat exchange and the formation of urban heat islands [18–20].

The presence of vegetation in urban structures can reduce the wind speed, which is crucial in the propagation of heat islands [20, 21]. The vegetation height, type of soil substrate and depth, and type of vegetation, play a critical role in the reduction of ambient temperature [9]. In short, green roofs help mitigate heat through a combined effect of evapotranspiration and roof insulation, which reduces the transfer of heat flux [9, 12].

The application of green roofs has multiple benefits on the heat mitigation of a structure, both interior and exterior. Under green roof, the upper-most flats in a building show a decrease of cooling energy demand of 10–40% depending on climate and building parameters [15]. Further, during the winter season, the energy demand for heating the interior is much less when compared to conventional buildings. Another aspect that affects the UHI phenomenon is the discharge of exhaust heat by air conditioning (AC) units. According to Santamouris [22], electricity consumption on average rises by 4.6% to a degree rise in ambient temperature. Several studies have quantified that UGS reduces the ambient temperatures by 2–6.7 °C in the nearby localities [23]. The application of green roofs can mitigate the energy demand for cooling through a reduced need for space cooling [9]. Therefore, the mitigating effect of green roofs can be seen in urban areas with a hotter and drier climate like that of Nagpur City [18].

1.1.2 Runoff Management

Green roofs, in the urban setting, can act as a critical measure for control of urban flood events through the management of stormwater management (runoff reduction, delay, and attenuation) [24]. With the increase in urbanization and concretization, most spaces within a city have low permeability for rainwater [25]. With the rise in the intensity and frequency of EWEs, managing variable runoff creates issues for urban planners. Green roofs can offer a part solution as they can absorb and evaporate the rainfall and reduce the runoff volume from building roofs [16, 25]. Green roofs can significantly reduce runoff volume. Studies show a 63% attenuation of runoff with green roofs compared with traditional roofs [26]. Studies show that green roofs in case of light rain (precipitation rate is less than 2.5 mm/h) can reduce runoff by 100% and by 45% when rainfall is above 15 mm [27]. However, the runoff retention capacity of any green roof is controlled by various factors such as vegetation cover, soil type, and land use [25]. Different studies have argued that one of the crucial parameters is the substrate/soil depth [16].

1.1.3 Reduction of Greenhouse Gases

The implementation of green roofs can supplement the reduction of greenhouse gases, especially CO_2, through absorption by the vegetation [28]. However, the amount of CO_2 sequestered depends on the type and total areas of vegetation. Studies have shown that an extensive green roof system can sequester 375 g C/m^2 [18]. Several other studies have reported a high degree of variability in the carbon sequestered by the green roofs; the reported values varied between 73 and 276 g C/m^2 [29]. In a review, [30] has reported the carbon sequestered by green roofs in the range of 0.313–9.82 kg C/m^2/y. The variation could be influenced by the media depth, plants' age, type of irrigation, application of fertilizers, etc. Nevertheless, green roofs have the potential to sequester atmospheric carbon dioxide and contribute to reducing global warming. Studies also confirm that the ambient CO_2 concentration nearer to green roofs (4 m × 4 m area) is substantially lower than other areas by 4.3 mg/m^3 [31]. Depending upon the wind flow, the reduction of ambient CO_2 can reach up to 9.3% in the vicinity of green roofs [29].

1.1.4 Reduction of Gaseous and Particulate Pollutants

Major cities in the developing countries are facing the deterioration of air quality in the twenty-first century due to rapid development in the transportation sector, industrialization, and population growth. This leads to the emission of gaseous pollutants like CO_2, NO_x, SO_x, and particulate matter that are hazardous to human health [32]. Green roofs (GRs) have shown the capacity to reduce air pollutants [33]. Studies have shown that green roofs are apt in capturing CO_2, NO_x, and SO_x [32]. Since GRs control the ambient temperature, which affects the photochemical reactions and concentration of secondary pollutants. The subsurface layer of the GRs facilitates the removal of heavy metals like Cd, Cu, Pb, Ni, etc. [29]. A 4000 m^2 green roof surface can capture 37% of SO_2 present in the ambient air [34].

The European countries, especially Germany, have well-defined guidelines for the design of green roofs, followed by the USA. However, such guidelines or policy incentives are not explicitly available in the Indian context. Extensive studies on the benefits of green roofs in Indian context needs to be reported from multiple cities across different climatic zones of India. These studies will form basis for the defining guidelines which are specific to Indian cities considering the building design.

The greening of buildings with vegetation can change the energy balance and alter the cooling requirements [9, 20, 35]. This can have a compounding effect on the overall energy balance of the urban area by increasing more reflective and evaporative surfaces [9, 20]. The development of these heat mitigative zones in and around buildings can be termed heat mitigating envelopes. As listed by Aleksandrowicz [12], building envelopes can be categorized into (1) cool building envelopes, (2) green roofs, and (3) green shades. This chapter discusses the heat mitigation benefits of green roofs using the InVEST (Integrated Valuation of Ecosystem Services and Trade-offs) urban cooling model [36] as a case study of Nagpur City. The study also presents a scenario-based analysis to better understand the potential of green roofs in heat mitigation the city. Several challenges associated with the green roofs and its implementation are also briefly discussed in the chapter.

In the present chapter, we have quantified the heat mitigation services of the urban green spaces using the InVEST model. Here, the buildings with green roofs have been treated as green spaces for computing the heat mitigation. The study area's land use and land cover have been identified using the high-resolution Sentinel-2 satellite data collected during May 2020. Cooling properties of each pixel have been assigned in terms of shade, evapotranspiration, and albedo according to the underlying land use category. The model sums up the cooling spread by the pixels with the green cover and spreads its spatial profile in the study area. With this background of the model, we have attempted to identify the changes in the urban cooling capacity or heat mitigation in the study area according to the land use changes due to the implementation of green roofs in varying scenarios. The scenarios of green roof implementation include a 10% and a 20% conversion of built-up along the roadside in the city.

The rest of the chapter is organized into several sections. The details of InVEST model and data requirements are presented in Sect. 3. The results obtained in the study and its relevance are provided in Sect. 4 followed by the conclusions and way forward in the last section.

2 Materials and Methods

2.1 *Study Area*

Nagpur City is situated in Maharashtra State in the Deccan plateau of peninsular India (Fig. 1). The city extends over 222 km^2 area with a population of 2.41 million [37]. The city's terrain is almost flat with the average elevation of 310 m above mean sea level. The city experiences a tropical dry climate with the maximum temperature reaching up to 47 °C in summer (March to June), while the minimum temperature drops to 5 °C in winter months (November to February). The city receives around 1200 mm of rainfall annually, mainly during the months from July to October. The city's land use is predominantly covered with the built-up area, followed by vegetation and barren land. The city consists of several urban green spaces in the form of avenue plantations, institutional forest patches, home gardens, and public parks and forest lands. There are more than 100 public green spaces in the city as of 2020. The vegetation is characterized by a mix of deciduous trees, thorny shrubs, and climbers.

2.2 *InVEST Urban Cooling Model*

The InVEST is a suite of models which quantify the ecological benefits of various ecosystems, including urban systems. We have utilized the urban cooling model of the InVEST to quantify the decrease in the spatial average temperature over Nagpur City due to the multiple scenarios of urban built-up conversion into roof gardens. See Kadaverugu [9] for a detailed description of the InVEST urban cooling model and the data layers required.

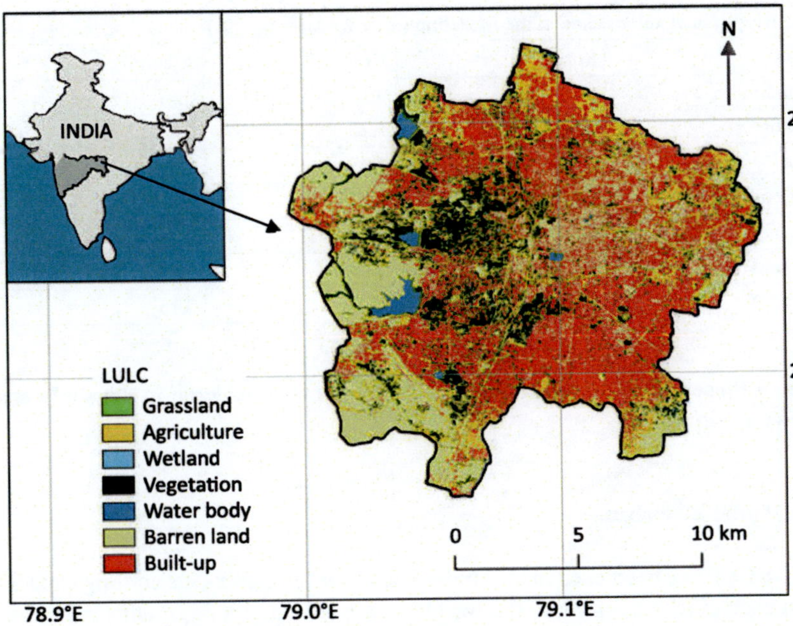

Fig. 1 Nagpur City, study area, is pointed out in the Maharashtra State of India. Land use and land cover is based on 2020

2.2.1 Data Preparation

The model requires input data in the form of the raster (land use land cover and evapotranspiration layer), vector (shape file of the city), and tabular data (biophysical variables) that define the structural and behavioral properties of the study area. The land use and land cover raster layer of 10-m spatial resolution has been prepared using the cloud-free Sentinel-2A image representing May 2021. The Google Earth Engine platform has been utilized to classify the land use categories using the support vector machine algorithm. The land use classes (code) defined in the study are grassland (1), agriculture (2), wetland (3), vegetation (4), water body (5), barren land (6), and built-up area (7). The accuracy of the land use analysis has been measured using the kappa coefficient.

The actual evapotranspiration (ET) for the study area has been prepared according to the reference evapotranspiration (ET_{ref}) and crop factor (K_c). The ratio between actual ET and ET_{max} has been calculated to obtain the ET_{factor} as an input layer into the model (Eq. 1). In the present chapter, the values of ET_{ref} and ET_{max} were considered to be 4.02 and 4.67 mm/day, respectively [38]. The crop factor values for varying land use categories are presented in Table 1.

The biophysical variables for each land use pixel include the weighted factors of albedo, shade, and evapotranspiration values. Depending upon the study area, the

Table 1 Biophysical values used in the modeling (after Kadaverugu [9])

Land use	Shade	Kc	Albedo
Grassland	0	0.1	0.27
Agriculture	0.3	1.22	0.2
Wetland	0	1.05	0.1
Vegetation	0.9	1.1	0.2
Water body	0	1.05	0.1
Barren land	0	0.1	0.27
Built-up	0.2	0.1	0.15

weights of these variables can be altered. The weights considered in the study area are shown in Eq. 2.

2.2.2 Model Description

InVEST-v3.8.0 has been used in the study. The model computes the cooling capacity (CC) on each pixel according to the weights assigned for albedo, shade, and evapotranspiration values (Eq. 2). Along with CC, the model also computes the cumulative cooling effect (CC_{UGS}) due to the green spaces (of area > 2 ha) on pixels within the search radius of d_{cool} (Eq. 3). The heat mitigation (HM) at a pixel is assigned a value equal to CC if the CC value is greater than the CC_{UGS} or the cumulative area of the UGS is less than 2 ha (Eq. 4). Otherwise, the HM value is assigned to be CC_{UGS}. This particular condition accounts for the spread of the cooling effect for a green space of considerable size in its vicinity (or search radius). The temperature blending distance of 2000 m is considered in the study, which is a model default value and used in other similar studies [9]. The heat mitigation value is considered responsible for reducing the ambient air temperature relative to the reference temperature (T_{ref}) by considering the UHI effect (Eq. 5). The values of the T_{ref} and UHI_{max} are considered to be 35 and 7 °C, respectively, in the study (according to Kadaverugu [9]). The values of the model parameters used in the study are presented in Table 2.

Table 2 Parameters used in the InVEST urban cooling model (after Kadaverugu [9])

Parameters	Value
Base line air temperature, T_{ref}	35 °C
Magnitude of UHI effect, UHI_{max}	7 °C
Air temperature maximum blending distance	2000 m
Green area maximum cooling distance	500 m
Shade	0.6
Abledo	0.2
Evapotranspiration	0.2

$$\mathrm{ET}_{\mathrm{factor}} = \frac{\mathrm{ET}_{\mathrm{ref}} \times K_c}{\mathrm{ET}_{\mathrm{max}}} \tag{1}$$

$$\mathrm{CC}_i = 0.6 \times \mathrm{Shade} + 0.2 \times \mathrm{Albedo} + 0.2 \times \mathrm{ET}_{\mathrm{factor}} \tag{2}$$

$$\mathrm{CC}_{\mathrm{UGS}} = \sum \mathrm{CC}_j \times \exp\left(\frac{d_{i,j}}{d_{\mathrm{cool}}}\right) \tag{3}$$

$$\mathrm{HM} = \begin{pmatrix} \mathrm{CC}_i & \text{if } \mathrm{CC}_i > \mathrm{CC}_{\mathrm{UGS}} \text{ or Area of UGS} < 2 \text{ ha within the search radius, } d_{\mathrm{cool}} \\ \mathrm{CC}_{\mathrm{UGS}} & \text{otherwise} \end{pmatrix} \tag{4}$$

$$T_{\mathrm{air}} = T_{\mathrm{ref}} + (1\ \mathrm{HM}) \times \mathrm{UHI}_{\mathrm{max}} \tag{5}$$

2.3 *Scenario-based Land Use Changes*

A scenario-based computational study has been performed on the study area, i.e., Nagpur City of central India using the present land use classification. Various land use classes like built-up, vegetation cover, agricultural land, open spaces, and water bodies perform differently in modifying the local climate and thereby affect heat mitigation.

The conversion of rooftops into greenery, as we know, alters the surface albedo and evapotranspiration, which modifies the cooling capacity and heat mitigation, hence, affects the urban microclimate and helps in heat stress reduction. In this study, we attempted to quantify the heat mitigation role of urban green roofs using the InVEST model. The pixels with built-up land use category along the city's arterial roads were converted into vegetation class (to account for green rooftops) in two future scenarios. In Scenario-10, only 10% of the total built-up area along the roads in the city would be converted into green rooftops, whereas in Scenario-20, 20% of the built-up area along the roads would be converted into green rooftops. As this analysis attempts to quantify the relative heat mitigation benefits due to the green roofs only, the rest of the land uses were kept unchanged. Built-up area along the roads has been selected, so as to reduce the effect of UHI, which is generally high in the core of the city and where population density is also high. The selection of built-up pixels for the conversion analysis has been carried out in QGIS-v3.16.

3 Results and Discussion

The results on LULC classification is discussed first, followed by the land use changes according to the future scenarios, and their corresponding heat mitigation or cooling effect by the green spaces (including vegetation and green roofs in the study area).

The supervised classification of the cloud-free Sentinel-2A image captured during May 2020 indicates a high degree of classification accuracy with the kappa coefficient of around 0.85. The city land area is categorized mainly into seven land use and land cover classes. Results show that majority of the area is occupied with built-up (39%), followed by barren land (26%), agriculture (20%), vegetation (14%), and others (Fig. 2). Here, barren land includes open spaces, grounds, and shrubs, which are mainly situated along the city's fringes. Kotharkar [39] has studied the city from an urban compactness point of view, and they mentioned that the population density in the city is high (250–1000 pph, people per hectare) in the mid-eastern parts of the city, in contrast to other parts of the city with the population density < 75 pph. This

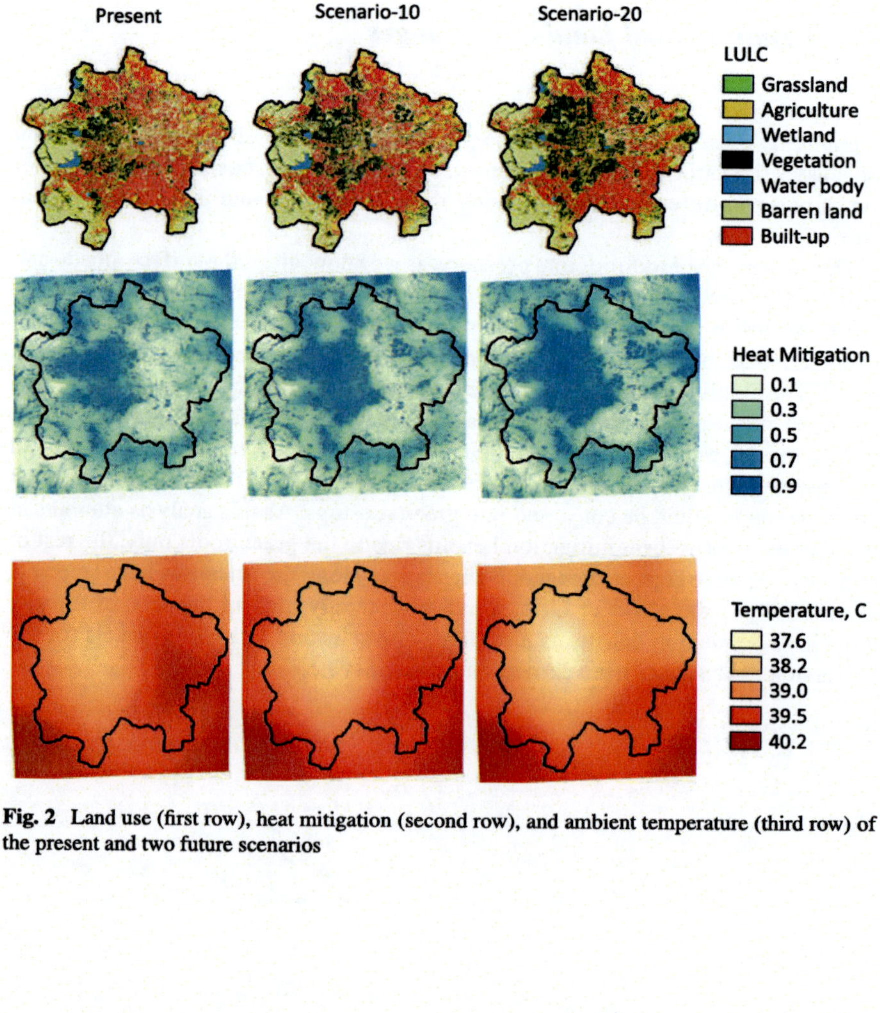

Fig. 2 Land use (first row), heat mitigation (second row), and ambient temperature (third row) of the present and two future scenarios

Table 3 Percentages of the land uses in the study area (222 km^2) for the present and two future scenarios

Land use	Present (%)	Scenario-10 (%)	Scenario-20 (%)
Grassland	0.04	0.04	0.04
Agriculture	19.98	19.98	19.98
Wetland	0.09	0.09	0.09
Vegetation	13.66	17.56	21.45
Water body	1.24	1.24	1.24
Barren land	26.02	26.02	26.02
Built-up	38.96	35.06	31.17

agrees with the findings of Kadaverugu [9] in terms of building density and urban heat.

The land use and land cover in scenario-10 have been allowed to change from built-up (along the city's major roads) to vegetation cover (as green rooftops) by 10% of the existing built-up pixels. The changes in land use and land cover percentages according to the scenarios are presented in Table 3. The newly formed land use and land cover include built-up as 35%, while vegetation cover around 18%. Similarly, a 20% conversion of built-up area along the roads, into green cover has resulted in a built-up of 31%, whereas vegetation cover of around 22%. The converted built-up pixels still represent buildings in reality, but with only difference in the green roofs that are treated as vegetation class for heat mitigation analysis for using in the InVEST model.

The analyses show that the ambient air temperature according to the present land use in the city varies in the range 38.23–40.18 °C, with an average of 39.17 °C. Results indicated that if 10% of the built-up area along the roads of Nagpur City is converted into green roofs, there will be a reduction of 0.23 °C in the average temperature of the city when compared with the current land use. Similarly, if 20% of the built-up is converted into green roofs, the reduction in ambient temperature is estimated to be 0.44 °C. The ambient temperature range is presented in Table 4, which shows that there will be a considerable decline in the ambient air temperature in both future scenarios. The results are also reflected in terms of the cooling capacity and heat mitigation values which are considerably lower in both future scenarios,

Table 4 Quantification of the heat mitigation by urban green spaces for the present and two future scenarios

Variables	Present	Scenario-10	Scenario-20
Average cooling capacity	0.28	0.31	0.34
Average heat mitigation	0.406	0.444	0.481
Average ambient temperature, °C	39.17	38.94 (−0.23)	38.73 (−0.44)
Temperature range, °C	38.23–40.18	37.89–40.18	37.60–40.18

in contrast with the present condition. Cooling capacity, as mentioned earlier, is a combination of the shade, albedo, and evapotranspiration factors according to the assigned weights. Due to the conversion of the land uses (built-up into green roofs/vegetation) the average cooling capacity has increased from 0.28 to 0.31 in scenario-10, while increased from 0.28 to 0.34 in scenario-20. The cooling capacity values for the three scenarios varied in the range of 0.074–0.799. Similarly, the average heat mitigation has increased from 0.406 to 0.444 in scenario-10, while increased from 0.406 to 0.481 in scenario-20. Heat mitigation in the region varied between 0.074 and 0.882. It can be interpreted that the more green roofs in the city, the better in the heat stress mitigation.

The aspect of spatial proximity plays a vital role if the heat mitigation is to be optimized. The effect of nearby surface vegetation on heat mitigation is predominant if the adjoining built-up is also equipped with green roofs. To obtain a more realistic quantification of heat mitigation, field-level experiments on temperature reduction due to green spaces is to be measured.

Results show that the cooling effect or heat mitigation is directly related to the proportion of the green spaces in the city (Fig. 2). High heat mitigation values are observed in scenarios-10 and 20, which have an increased green cover in terms of an increase in green roof cover. Kadaverugu [9] has estimated that in Nagpur City, there will be an additional space cooling energy demand of 25,310 units (kWh) for an hour for every degree of rise in ambient temperature. Further, the study quantified that for in a scenario with an increase in green spaces by 22.28%, there will be 17.4% more energy conserved for the space cooling needs of the city. Even for a marginal increase in the average temperature of a city, there will be compounded implications in the energy requirements for space cooling. Thermal comfort is one of the major factors in determining the working efficiency of people, which is related to the economic growth as well.

4 Challenges and Gaps

The challenges in the implementation of green roofs multifaceted comprising of technical, economic, social, and management aspects. The challenges can be categorized into technical, socio-cultural, and economic factors. Apart from these, many factors hinder the successful implementation of green roofs, especially in developing countries.

High initial Costs. One of the biggest challenges is the initial cost of the green roof during construction [15]. As the working of GRs is dependent on various parameters, they are sometimes considered long-term investments with short-term benefits [40]. The cost of an extensive green roof system in British Columbia, Canada, falls in the range of \$12–\$15 per feet [41]. However, the overall cost benefit analysis after factoring in all benefits show that GR systems are less expensive than traditional roofs [34].

High Maintenance Cost. Green roof systems require regular maintenance for a long life. Especially in countries like in India with a hot and dry summer, the plants must survive during this period to reap the benefits of the system all year around. Therefore, green roofs require watering, fertilization, and maintenance during period interval [41].

Material Usage and Disposal. The drainage and filtration layers on the green roof systems are made of up polymer materials. The manufacture of these polymers do cause some form of pollution; therefore, it is critical to expand the research on the different materials, which can extend the life and performance of green roof systems. Further, the disposal of green roof components are an issue of concern so the focus should shift toward using components that can be reused or recycled. When replacing the traditional roofs with the green roofs, studies have shown that GRs enhance the life of the roof [42]. However improper design and placement of GRs could lead to roof leaking and structural failure. Further, the local conditions also play an important part in the design of the GRs, so having a single optimal design that suits all conditions is a big challenge.

Socio-cultural. The green roofs have several benefits, however, still have issues in their implementations. Roggero [43] lists an interesting concept of "Free Rider" and "Collective Actions Problems" for small roof gardens and large roofs gardens, respectively. The author suggests that the free-rider problem is best solved by regulations, incentives, communication, awareness, and network-building. Public participation and commitment is the key to the implementation of the green roofs, for which incentives should be in place.

Management aspects. There are issues with scientific management and upkeep of green roofs, lack of a market space of green roofs, and lack of related research on market management [43, 44]. There are green-roof-related policies in western countries like the USA and EU; however, India still lacks a coherent policy or regulations. India can borrow experience from developed countries as they have identified four main policy instruments in terms of green roof implementation, such as: inclusion of green roofs in urban planning and regulations, ecological compensation policy, and direct and indirect financial policy for promotion of green roofs [44].

Application of high-resolution remote sensing images of urban infrastructure using Lidar imaging, digital terrain and surface modeling, and imaging of the urban 3D surface through unmanned aerial vehicles, etc., can be used for the identification of potential rooftops for developing greenery. However, the relative orientation of the buildings with respect to nearby structures and sunshine hours should also be considered in identifying the potential rooftops for roof gardens. For instance, Santos [45] showed that nearly 8.5% of rooftops in Lisbon City could be converted into green roofs by considering the slope of the terrace and sunshine hours. Studies in a similar fashion are required in India for fully exploring the potential of green rooftops. However, a minimum area of around 100 m^2 of the green area is required to be qualified as a green rooftop according to German standards. Such norms should be customized according to the building geometry. Further, these studies on potential green roofs should also evaluate the trade-offs between rooftop solar panels and greenery. A set of definitive guidelines should be in place to encourage the conversion of existing potential roofs into the greenery. Also, extensive field studies are required to provide accurate parameterization of the applied empirical models and to validate the seamless integration of field and computational modeling studies that can shape the future direction of research.

5 Conclusions

Nature-based solutions play an important role in modifying the local microclimate at the building-scale. Especially through promoting urban green spaces, heat stress can be mitigated at the city level. Additionally, the green spaces also reduce the runoff, help groundwater recharge, purify the air, and improve aesthetic value and biodiversity. There is a need to mainstream the benefits offered by green spaces in urban policy planning. India and other developing countries need to devise policies and guidelines for the large-scale implementation of green infrastructure. Despite the urban green infrastructure's benefits, very few case studies have quantified the heat stress mitigation, air purification, energy conserved, etc., due to the green roofs. City-specific empirical studies on the quantification of ecosystem services and other benefits of UGS are to be encouraged for getting valuable insights for decision-making.

The present study provides a methodology for quantifying the urban heat mitigation of green spaces or green roofs using the InVEST model. The analysis helps localize the sustainable development goal 11, which deals with sustainable cities and communities in light of rapid urbanization and economic development. Nagpur City, which is in a heatwave-affected zone of India, has been selected as the study area, and the results indicate that the average ambient temperature has decreased by 0.23 °C and 0.44 °C in the scenarios of conversion of 10% and 20% built-up into green roofs, respectively.

The utilization of geospatial tools provides a holistic view in the identification of potential hotspots for the implementation of urban green roofs for optimized benefits at the city level. Integration of future socio-economic scenarios with the empirical or process-based models will accommodate the future direction of the human systems and policy intervention, and the effect of such policies on the functioning of the socio-environmental systems can be quantified. The results provide valuable insights through various alternative scenarios that would not be possible through experimental observations. This approach can be replicated in other cities for effective policy action.

Acknowledgements Authors thank the manuscript processing services of Knowledge Resource Center of CSIR National Environmental Engineering Research Institute, Nagpur. The assigned number for this manuscript is CSIR-NEERI/KRC/2021/SEP/CTMD-EMD/1.

References

1. Masson-Delmotte, V., Zhai, P., Pörtner, H. -O., Roberts, D., Skea, J., Shukla, P. R., Pirani, A., Moufouma-Okia, W., Péan, C., Pidcock, R. (2018). Global warming of 1.5 C. *An IPCC Special Report on the Impacts of Global Warming of*, *1*.
2. WEF. (2020). Nature risk rising: Why the crisis engulfing nature matters for business and the economy.
3. Augusto, B., Roebeling, P., Rafael, S., Ferreira, J., Ascenso, A., Bodilis, C. (2020). Short and medium- to long-term impacts of nature-based solutions on urban heat. *Sustainable Cities and Society, 57*, 102122 (2020).https://doi.org/10.1016/j.scs.2020.102122
4. Flagg, D. D., & Taylor, P. A. (2011). Sensitivity of mesoscale model urban boundary layer meteorology to the scale of urban representation. *Atmospheric Chemistry and Physics, 11*, 2951–2972. https://doi.org/10.5194/acp-11-2951-2011
5. Whitford, V. (2001). "City form and natural process"—indicators for the ecological performance of urban areas and their application to Merseyside UK. *Landscape and Urban Planning.*, *57*, 91–103. https://doi.org/10.1016/S0169-2046(01)00192-X
6. Imran, H. M., Kala, J., Ng, A. W. M., & Muthukumaran, S. (2018). Effectiveness of green and cool roofs in mitigating urban heat island effects during a heatwave event in the city of Melbourne in southeast Australia. *Journal of Cleaner Production.*, *197*, 393–405. https://doi.org/10.1016/j.jclepro.2018.06.179
7. Saaroni, H., Amorim, J. H., Hiemstra, J. A., & Pearlmutter, D. (2018). Urban Green Infrastructure as a tool for urban heat mitigation: Survey of research methodologies and findings across different climatic regions. *Urban Climate.*, *24*, 94–110. https://doi.org/10.1016/j.uclim.2018.02.001
8. Arnfield, A. J. (2003). Two decades of urban climate research: A review of turbulence, exchanges of energy and water, and the urban heat island. *International Journal of Climatology, 23*, 1–26. https://doi.org/10.1002/joc.859
9. Kadaverugu, R., Gurav, C., Rai, A., Sharma, A., Matli, C., & Biniwale, R. (2021). Quantification of heat mitigation by urban green spaces using InVEST model—a scenario analysis of Nagpur City India. *Arabian Journal of Geosciences, 14*, 82. https://doi.org/10.1007/s12517-020-06380-w
10. IEA. (2018). *The future of cooling*. IEA, Paris.
11. Giguère, M. (2009). *Literature review of urban heat island mitigation strategies*. Institut national de santé publique Québec.
12. Aleksandrowicz, O., Vuckovic, M., Kiesel, K., & Mahdavi, A. (2017). Current trends in urban heat island mitigation research: Observations based on a comprehensive research repository. *Urban Climate, 21*, 1–26. https://doi.org/10.1016/j.uclim.2017.04.002
13. Knaus, M., & Haase, D. (2020). Green roof effects on daytime heat in a prefabricated residential neighbourhood in Berlin, Germany. *Urban Forestry & Urban Greening, 53*, 126738 (2020).https://doi.org/10.1016/j.ufug.2020.126738
14. Getter, K. L., & Rowe, D. B. (2006). The role of extensive green roofs in sustainable development. *HortSci, 41*, 1276–1285. https://doi.org/10.21273/HORTSCI.41.5.1276
15. Suszanowicz, D., & Kolasa-Wiecek, A. (2019). The impact of green roofs on the parameters of the environment in Urban areas-review. *Atmosphere, 10*, 792. https://doi.org/10.3390/atmos10120792
16. Mentens, J., Raes, D., & Hermy, M. (2006). Green roofs as a tool for solving the rainwater runoff problem in the urbanized 21st century? *Landscape and Urban Planning, 77*, 217–226. https://doi.org/10.1016/j.landurbplan.2005.02.010
17. Lalosevic, M. D., Komatina, M. S., Milos, M. V., & Rudonja, N. R. (2018). Green roofs and cool materials as retrofitting strategies for urban heat island mitigation case study in belgrade Serbia. *Thermal Science, 22*, 2309–2324. https://doi.org/10.2298/TSCI171120086L
18. Barriuso, F., & Urbano, B. (2021). Green roofs and walls design intended to mitigate climate change in Urban areas across all continents. *Sustainability, 13*, 2245. https://doi.org/10.3390/su13042245

19. Kadaverugu, R., Matli, C., & Biniwale, R. (2021). Suitability of WRF model for simulating meteorological variables in rural, semi-urban and urban environments of Central India. *Meteorology and Atmospheric Physics*. https://doi.org/10.1007/s00703-021-00816-y
20. Kadaverugu, R., Purohit, V., Matli, C., & Biniwale, R. (2021). Improving accuracy in simulation of urban wind flows by dynamic downscaling WRF with OpenFOAM. *Urban Climate, 38*, 100912. https://doi.org/10.1016/j.uclim.2021.100912
21. Kolokotroni, M., & Giridharan, R. (2008). Urban heat island intensity in London: An investigation of the impact of physical characteristics on changes in outdoor air temperature during summer. *Solar Energy., 82*, 986–998. https://doi.org/10.1016/j.solener.2008.05.004
22. Santamouris, M., Cartalis, C., Synnefa, A., & Kolokotsa, D. (2015). On the impact of urban heat island and global warming on the power demand and electricity consumption of buildings—A review. *Energy and Buildings, 98*, 119–124. https://doi.org/10.1016/j.enbuild.2014.09.052
23. Davtalab, J., Deyhimi, S. P., Dessi, V., Hafezi, M. R., & Adib, M. (2020). The impact of green space structure on physiological equivalent temperature index in open space. *Urban Climate, 31*, 100574. https://doi.org/10.1016/j.uclim.2019.100574
24. Mobilia, M., & Longobardi, A. (2017). Smart Stormwater Management in Urban Areas by Roofs Greening. In O. Gervasi, B. Murgante, S. Misra, G. Borruso, C. M. Torre, A. Rocha, T. Taniar, B. O. Apduhan, E. Stankova, & A. Cuzzocrea (Eds.), *Computational Science and Its Applications—Iccsa 2017, Pt Iii* (pp. 455–463). Springer International Publishing Ag.
25. Kadaverugu, A., Nageshwar Rao, C., & Viswanadh, G. K. (2020). Quantification of flood mitigation services by urban green spaces using InVEST model: a case study of Hyderabad city, India. *Modeling Earth Systems Environment*.https://doi.org/10.1007/s40808-020-00937-0
26. Pęczkowski, G., Orzepowski, W., Pokładek, R., Kowalczyk, T., Żmuda, R., & Wójcik, R. (2016). Retention properties of the type of extensive green roofs as an example of model tests. *Acta Scientiarum Polonorum Formatio Circumiectus, 15*, 113–120. https://doi.org/10.15576/ASP.FC/2016.15.3.113
27. Chen, H. (2011). Hall green roof stormwater retention and runoff reduction performance. https://www.cmu.edu/environment/campus-greendesign/green-roofs/documents/hh-green-roof-water-monitoring.pdf
28. Susca, T., Gaffin, S. R., & Dell'Osso, G. R. (2011). Positive effects of vegetation: Urban heat island and green roofs. *Environmental Pollution, 159*, 2119–2126. https://doi.org/10.1016/j.envpol.2011.03.007
29. Li, Y., & Babcock, R. W. (2014). Green roofs against pollution and climate change A review. *Agronomy for Sustainable Development, 34*, 695–705. https://doi.org/10.1007/s13593-014-0230-9
30. Shafique, M., Xue, X., & Luo, X. (2020). An overview of carbon sequestration of green roofs in urban areas. *Urban Forestry & Urban Greening, 47*, 126515.https://doi.org/10.1016/j.ufug.2019.126515
31. Li, J., Wai, O. W. H., Li, Y. S., Zhan, J., Ho, Y. A., Li, J., & Lam, E. (2010). Effect of green roof on ambient CO_2 concentration. *Building and Environment, 45*, 2644–2651. https://doi.org/10.1016/j.buildenv.2010.05.025
32. McRae, A. M. (2016). Case study: A conservative approach to green roof benefit quantification and valuation for public buildings. *The Engineering Economist, 61*, 190–206. https://doi.org/10.1080/0013791X.2016.1186255
33. Suszanowicz, D. (2018). Model research on the influence of green roofs on environmental parameters in Urban agglomerations. *E3S Web Conference, 45*, 00094. https://doi.org/10.1051/e3sconf/20184500094
34. Carter, T., & Keeler, A. (2008). Life-cycle cost–benefit analysis of extensive vegetated roof systems. *Journal of Environmental Management., 87*, 350–363. https://doi.org/10.1016/j.jenvman.2007.01.024
35. Kadaverugu, R., Sharma, A., Matli, C., & Biniwale, R. (2019). High resolution urban air quality modeling by coupling CFD and mesoscale models: A review. *Asia-Pacific Journal of Atmospheric Sciences*. https://doi.org/10.1007/s13143-019-00110-3

36. Sharp, R., Tallis, H. T., Ricketts, T., Guerry, A. D., Wood, S.A., Chaplin-Kramer, R., Nelson, E., Ennaanay, D., Wolny, S., Olwero, N., Vigerstol, K., Pennington, D., Mendoza, G., Aukema, J., Foster, J., Forrest, J., Cameron, D., Arkema, K., Lonsdorf, E., Kennedy, C., Verutes, G., Kim, C.K., Guannel, G., Papenfus, M., Toft, J., Marsik, M., Bernhardt, J., Griffin, R., Glowinski, K., Chaumont, N., Perelman, A., Lacayo, M., Mandle, L., Hamel, P., Vogl, A. L., Rogers, L., Bierbower, W., Denu, D., & Douglass, J. (2020). InVEST 3.8.0 User's Guide. The Natural Capital Project, Stanford University, University of Minnesota, The Nature Conservancy, and World Wildlife Fund.
37. Census. Census of India Website: Office of the Registrar General & Census Commissioner, India. http://censusindia.gov.in/
38. Ghare, A. D., Porey, P. D. (2008). Estimation of reference evapotranspiration of nagpur region using simplified approach. In *2008 First International Conference on Emerging Trends in Engineering and Technology*. https://doi.org/10.1109/ICETET.2008.74
39. Kotharkar, R., Bahadure, P., & Sarda, N. (2014). Measuring compact urban form: A case of Nagpur City India. *Sustainability, 6*, 4246–4272. https://doi.org/10.3390/su6074246
40. Bianchini, F., & Hewage, K. (2012). How "green" are the green roofs? Lifecycle analysis of green roof materials. *Building and Environment, 48*, 57–65. https://doi.org/10.1016/j.buildenv.2011.08.019
41. Shafique, M., Kim, R., & Rafiq, M. (2018). Green roof benefits, opportunities and challenges—A review. *Renewable and Sustainable Energy Reviews, 90*, 757–773. https://doi.org/10.1016/j.rser.2018.04.006
42. Kosareo, L., & Ries, R. (2007). Comparative environmental life cycle assessment of green roofs. *Building and Environment, 42*, 2606–2613. https://doi.org/10.1016/j.buildenv.2006.06.019
43. Roggero, M. (2020). Social dilemmas, policy instruments, and climate adaptation measures: The case of green roofs. *Mitigation and Adaptation Strategies for Global Change, 25*, 625–642. https://doi.org/10.1007/s11027-019-09883-4
44. Xiao, M., Lin, Y., Han, J., & Zhang, G. (2014). A review of green roof research and development in China. *Renewable and Sustainable Energy Reviews, 40*, 633–648. https://doi.org/10.1016/j.rser.2014.07.147
45. Santos, T., Tenedório, J., & Gonçalves, J. (2016). Quantifying the city's green area potential gain using remote sensing data. *Sustainability., 8*, 1247. https://doi.org/10.3390/su8121247

Evaluation of Groundwater Quality Using GIS Techniques in Part of Udupi District, Karnataka, Southern India

Vignesh Akshitha, Keshava Balakrishna, and Harikripa Narayana Udayashankar

Abstract Groundwater is a major source for drinking and agricultural purposes to most of the population in Udupi district, southern India. The objective of the current study is to evaluate the spatial distribution of major ions and drinking water quality index in the study area. Groundwater samples ($N = 100$) were collected during pre-monsoon season. The collected samples were analyzed for major cations and anions. Based on electrical conductivity, 90 open wells were classified as fresh (<1500 μS/cm), four open wells as brackish (1500–3000 μS/cm) and six open wells as saline (>3000 μS/cm). The evaluation of drinking water quality index identified that seven groundwater samples are good, four are poor, two are very poor, three are considered to be unsuitable for drinking and 84 groundwater samples are excellent for drinking purposes. Human health risk assessment revealed that infants were more vulnerable to health risk than other age groups. The concentrations of nitrate in the two open wells were above the prescribed limit (>45) which could cause human health risk for all the age groups. Water management plan in the study area should prioritize to reduce the nitrate contamination in the study area.

Keywords Saltwater intrusion · Human health risk assessment · Nitrate contamination · Water quality index

1 Introduction

The demand for groundwater is increasing in recent years because of increase in the global population, urbanization, industrialization and irrigation. The quality of groundwater relies on various factors such as the capacity of recharge, quality of recharged water, subsurface geochemical process overlying land use/land cover pattern and anthropogenic process [1]. The critical issue with groundwater is that quality is difficult to restore once it is contaminated. Thus, it is essential to preserve

V. Akshitha · K. Balakrishna (✉) · H. N. Udayashankar
Department of Civil Engineering, Manipal Institute of Technology, Manipal Academy of Higher Education, Manipal 576104, India
e-mail: k.balakrishna@manipal.edu

A. K. Dikshit et al. (eds.), *Innovative Trends in Hydrological and Environmental Systems*, Lecture Notes in Civil Engineering 234,
https://doi.org/10.1007/978-981-19-0304-5_28

the quality of groundwater for its sustainable development and management. Groundwater quality is evaluated by calculating drinking water quality indices from physical and chemical water quality parameters [2]. Geographical Information System (GIS) is an effective tool for assessing the spatial variability of physico-chemical parameters and groundwater quality [3, 4]. The spatial variability maps enable to identify the polluted areas and extent of pollution. Water quality index (WQI) is a useful tool for decision makers in order to develop new strategies for water resource management [5, 6]. Agricultural activities and discharge of domestic sewage lead to increased concentration of nitrate in the groundwater environment [7, 8]. The human health risk evaluation is essential to understand the impact of nitrate contamination on human health. Nitrates easily dissolve in water and commonly present in groundwater through leaching process [8, 9]. Increased concentration of nitrate in the groundwater can be hazardous to human health. Continuous monitoring of drinking water quality is necessary for determining the areas requiring water management plans. The main objective of the present study is to evaluate the ground water quality and human health risk of nitrate contamination in parts of Udupi district. There is no in-depth study that is carried out in the study area to evaluate drinking water quality and human health risk associated with consumption.

2 Materials and Methods

2.1 Study Area

The selected study area is a part of Udyavara river basin. The Udyavara river basin has 25 micro-watersheds from which nine micro-watersheds are selected for the present study. Udyavara River originates in the foothills of the Western Ghats, Karkala Taluk. It flows in Udupi and Karkala taluk. It flows parallel to the sea shore for about 10 km before it meets the Arabian Sea near Malpe. The climate is tropical, hot and humid with 4500 mm annual rainfall. The average atmospheric temperatures ranged from 22 to 33 °C. Geologically the river basin is composed of granite gneiss, laterite, migmatite gneiss and fine sand. Groundwater is present in weathered zone of granite and coastal alluvial sediments. The study area lies between longitudes 74°41′ and 74°48′ E and latitudes 13°12′ and 13°22′ N (Fig. 1) and covers an area of ~130 km^2 [10].

2.2 Sampling and Analytical Techniques

In the present study, 100 monitoring wells within the study area were chosen for ground water sampling during pre-monsoon season, 2019 (Fig. 2). The geographic

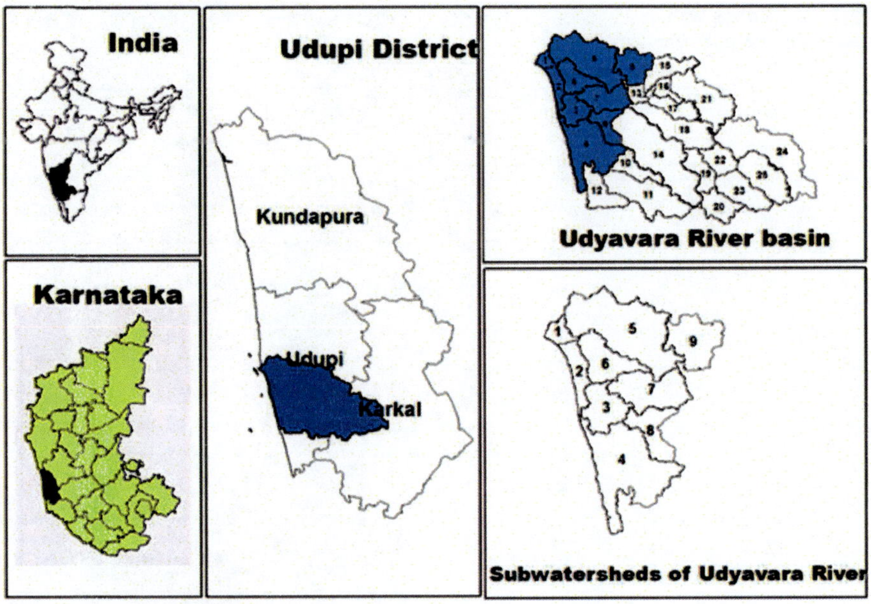

Fig. 1 Location map of the study area

coordinates of the well was recorded using Global Positioning System (GPS—Garmin make) and loaded to the GIS platform. Monitoring well location map was created using ArcGIS version 10.1. The collected samples were filled to 1000 ml pre-cleaned polypropylene bottles. The sample collected bottles were transported and stored at 4 °C cold room to avoid any significant chemical alteration [11]. The parameters total dissolved solids (TDS), electrical conductivity (EC) and pH were determined onsite using HACH multiparameter probes. Ion chromatography (IC, Dionex™ ICS1100) was used to analyze major cations (Na^+, Ca^{2+}, Mg^{2+}, K^+) and anions (Cl^-, SO_4^{2-}, NO_3^-, PO_4^{3-}). Bicarbonate was measured using Metrohm auto-titrator. The accuracy of the analysis was assessed by calculating the charge balance error (CBE). CBE values of the analyzed groundwater data were within 10%.

2.3 Water Quality Index Calculation

WQI is calculated using weighted arithmetic formula [12]. Many researchers across the globe calculated WQI and identified polluted and unpolluted groundwater [13–16]. The overall results of WQI are interpolated and spatial variability map of groundwater quality is prepared. WQI calculation has four steps. The steps include various water quality parameter measurements and its conversion into dimensionless number. Next, weight assignment is done depending on their significance and aggregation of

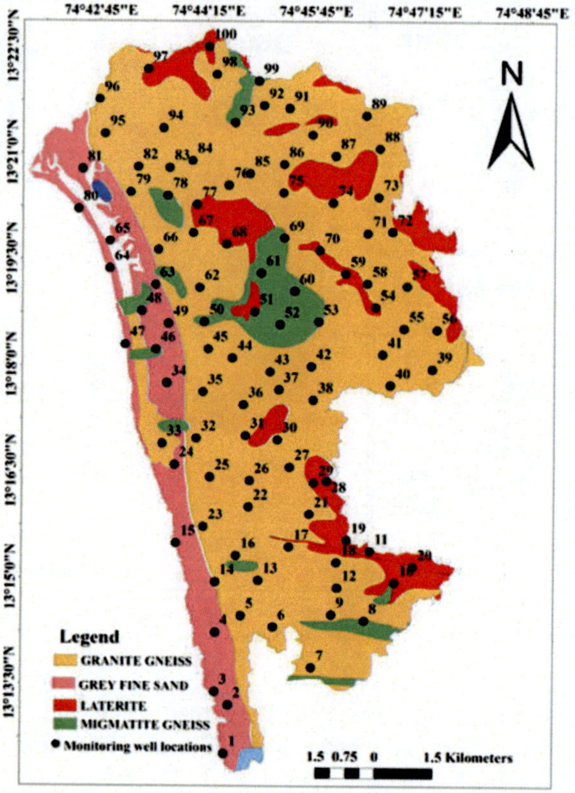

Fig. 2 Monitoring well locations in the study area

quality rating based on the final WQI values. The WQI is calculated using Eq. 1,

$$\text{WQI} = \sum_{i=1}^{n} W_i \times Q_i \tag{1}$$

where W_i is the relative weight of each water quality parameters and Q_i is the quality rating. The W_i is calculated using the Eq. 2,

$$W_i = \frac{w_i}{\sum_{i=1}^{n} w_i} \tag{2}$$

where w_i is the weight of each water quality parameter and n is the number of parameters. The Qi is calculated using the Eq. 3.

$$Q_i = 100 \times \frac{C_i}{S_i} \tag{3}$$

Table 1 Weight and relative weight of groundwater quality parameters used in the computation of WQI

Parameters	Weight (w_i)	Prescribed standard (S_i) as per WHO	Relative weight (W_i)
pH	4	6.5–8.5	0.11
TDS	5	500	0.13
Na^+ (mg/l)	4	200	0.11
K^+ (mg/l)	3	20	0.08
Mg^{2+} (mg/l)	3	30	0.08
Ca^{2+} (mg/l)	3	75	0.08
Cl^- (mg/l)	4	250	0.11
NO_3^- (mg/l)	5	45	0.13
SO_4^{2-} (mg/l)	2	200	0.05
HCO_3^- (mg/l)	3	200	0.08
	36		1

where C_i is the measured concentrations and S_i is the standard value of water quality parameters as per World Health Organization [17] drinking water quality standard. Weight for each parameter is assigned between 1 and 5 (Table 1).

2.4 *Assessment of Human Health Risk*

The present study used USEPA (United States Environmental Protection Agency) method to assess human health risk associated with consumption of groundwater contaminated with . Increased concentration of NO_3^- in drinking water lead to non-carcinogenic human health risk [18]. Human health risk is calculated for different age groups such as infants (0–1 years), kids (1–7 years), children (8–12 years), teens (13–20 years), adults (21–60 years) and aged people (61–69 years). The hazard quotient (HQ) due to ingestion through drinking water is calculated using the Eq. 4 [19].

$$\mathrm{HQ} = \frac{\mathrm{ADI}}{\mathrm{RfD}} \tag{4}$$

where ADI is average daily ingestion (mg/kg/day). RfD is reference dosage of the nitrate, which is 1.6 mg/kg/day based on USEPA [18] guideline. The ADI is calculated using the Eq. 5.

$$\mathrm{ADI} = \frac{\mathrm{CPW} \times \mathrm{IR} \times \mathrm{ED} \times \mathrm{EF}}{\mathrm{Bw} \times \mathrm{AT}} \tag{5}$$

CPW is measured concentration of nitrate in groundwater sample (mg/l). IR is average ingestion rate (in liters) which were considered as 0.8 (infants), 1.35 (kids), 1.9 (children), 2.2 (teens), 2.94 (adults) and 2.6 (aged people). ED is exposure duration (in years) which were taken as 0.75 (infants), 8 (kids), 13 (children), 40 (adults) and 64 (aged people). EF is exposure frequency which is 365 days. BW is body weight (in kg) which were noted as 11 (infants), 25 (kids), 35 (children), 55 (teens), 76 (adults) and 65 (aged people). AT is average life expectancy (in days) 274 (infants), 2920 (kids), 4745 (children), 6570 (teens), 14,600 (adults) and 23,360 (aged people).

2.5 Data Analysis in GIS

Geo-statistical interpolation method in ArcGIS is useful to identify the suitable and unsuitable zones for drinking water. In this study, Inverse Distance Weighted (IDW) analysis is used to examine the spatial variation of analyzed groundwater quality parameters, drinking water quality and human health risk in different age groups. IDW interpolates unknown values using weighted average of two nearest known points. The weight is determined based on the distance between known and unknown points.

3 Results and Discussions

3.1 General Hydrogeochemistry

The mean pH value in the pre-monsoon was 7.4 indicating slightly alkaline pH in the study area. pH of 32 groundwater samples were not within the prescribed drinking water standard limit (Fig. 3a). The spatial distribution of electrical conductivity (EC) during pre-monsoon season in collected groundwater samples ranged from 44.5 to 44000 μS/cm. Based on the EC values during pre-monsoon season, groundwater samples varied between fresh, brackish and saline water in nature. High EC was observed in the north-western part of the coastline indicating saline water intrusion. TDS ranged from 30 to 29,828 mg/l with mean value of 630 mg/l (Fig. 3b). Based on the TDS values, groundwater in the study area is classified into four groups [20]. Table 2 shows groundwater classification for the suitability of drinking purpose. The dominance of major anions and cations in the pre-monsoon season was $Na^+ > Ca^{2+} > Mg^{2+} > K^+ = Cl^- > HCO_3^- > SO_4^{2-} > NO_3^-$.

Cations: Sodium and calcium were dominant cations in pre-monsoon season (Fig. 4a, d). This is due to the presence of plagioclase feldspar minerals in the granitic rock. The maximum potassium values were 208 mg/l during pre-monsoon season (Fig. 4b). Potassium concentration was high in groundwater in coastal wells due to salinity

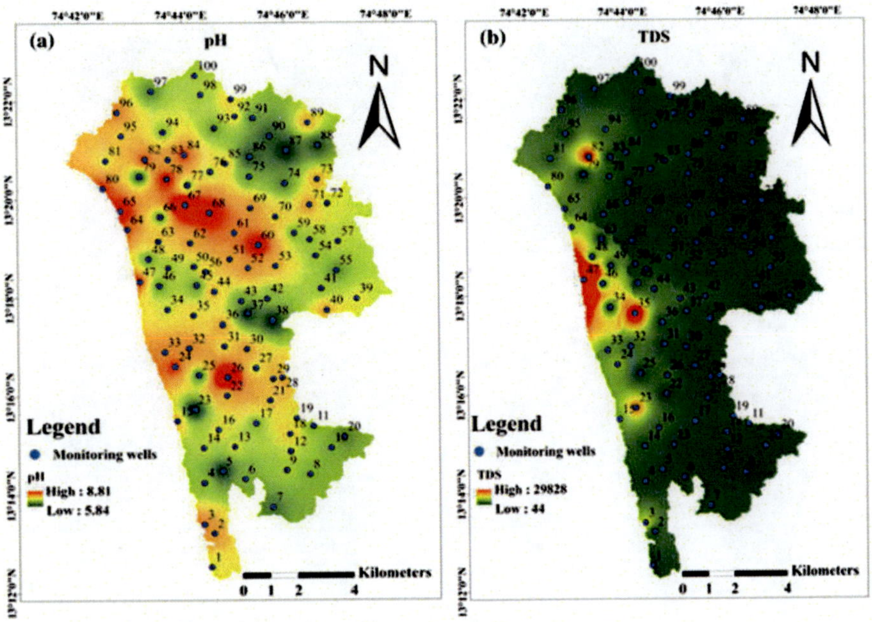

Fig. 3 Spatial distribution of **a** pH, **b** TDS in pre-monsoon

Table 2 Groundwater classification in the study area based on TDS values

TDS (mg/l)	Groundwater classification	Percentage of samples
<500	Desirable for drinking purposes	88
500–1000	Permissible for drinking purposes	3
1000–3000	Desirable for irrigation purposes	6
>3000	Unfit for drinking and irrigation	3

content [21]. The magnesium concentration ranged from 0.44 to 957 mg/l during pre-monsoon monsoon (Fig. 4c). The magnesium concentration was high in coastal wells and low in inland wells. The low concentration was because of the lack of carbonate rocks in the study area. Saltwater intrusion process resulted in high concentration of magnesium and potassium in groundwater present in coastal region.

Anions: Chloride was the dominant anion and ranged from 4.3 to 17,402 mg/l (Fig. 4e). Saltwater mixing led to the high chloride concentrations in the coastal wells [22]. Chloride may also contribute from wet atmospheric deposition through precipitation. The nitrate concentration ranged between 0.31 and 171 mg/l (Fig. 4f). The nitrate source in the groundwater is considered to be non-lithological [23]. Anthropogenic activities such as agricultural run-off and percolation from the soak pit are identified as causes of nitrate in the study area. Phosphate was below the detection limit (BDL) in the study area. Sulfate concentration varied between 0.43 and

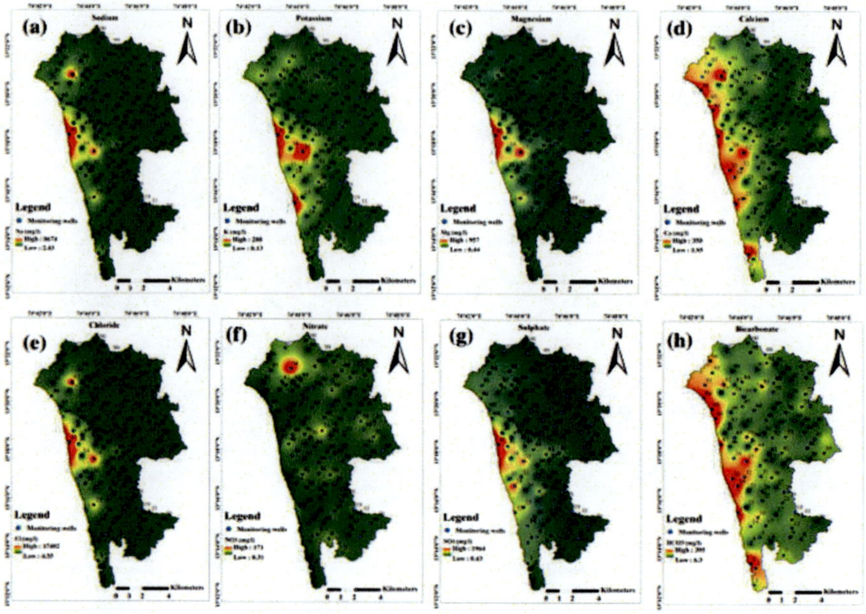

Fig. 4 Spatial distribution of **a** Na, **b** K, **c** Mg ,**d** Ca, **e** Cl, **f** NO_3, **g** SO_4 and **h** HCO_3 in pre-monsoon

1963 mg/l (Fig. 4g). Sulfate concentrations were high in the coastal wells were due to saltwater mixing process. Most of the inland wells showed very low concentrations of sulfate due to the absence of sulfate minerals bearing host rock. The maximum bicarbonate concentration was 395 mg/l. Bicarbonate concentration in the open wells near the coastal part was higher than the inland wells because of the impact of marine environment (Fig. 4h) [24].

3.2 *Groundwater Quality Index*

The five categories of water quality index are (i) water unsuitable for drinking (>300), (ii) very poor water (200–300), (iii) poor water (100–200), (iv) good water (50–100), (v) excellent water (0–50). The results of WQI estimated ranged from 13.9 to 2554. Based on the classification, 84 groundwater samples were excellent for drinking purposes. Whereas, seven were good, four were poor, two were very poor and three were considered to be unsuitable for drinking. The western part of study area near the coast is considered to be contaminated due the influence of saltwater. The increased concentration of major anions and cations lead to change in the quality of groundwater. The inland water samples posed excellent water quality and suitable for drinking. Figure 5 shows spatial distribution of water quality index in the study area.

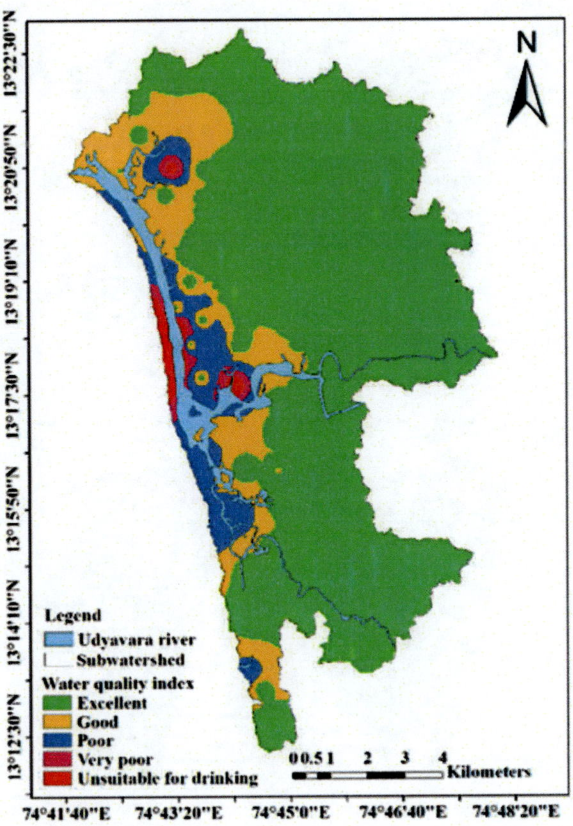

Fig. 5 Map showing spatial variability of WQI

3.3 Human Health Risk of Nitrate

Hazard quotient value >1 in the human health risk evaluation indicates potential non-carcinogenic health risk for humans and <1 indicates no potential health risk. Nitrate concentration >45 mg/l in groundwater can lead to health hazards [25]. The hazard quotient for infants, kids, children, teens, adults and aged people ranged from 0.01–7.78, 0.01–5.78, 0.01–5.81, 0.01–4.28, 0.01–4.14 and 0.01–4.28 respectively. The assessment result showed that 11%, 4%, 4%, 2%, 2% and 2% of the samples pose potential health risk (HQ > 1) for infants, kids, children, teens, adults and aged people respectively (Fig. 6). The infants were more vulnerable to health risk among other age groups. The open wells with high potential risk are present in residential area and nitrate contaminated due to leakage of soak pit in the study area. Nitrate concentration in two open wells (S94 and S51) were exceeding the prescribed drinking water standard. These two open wells pose high potential health risk for all the age groups. Therefore two open wells require attention in order to control the nitrate contamination.

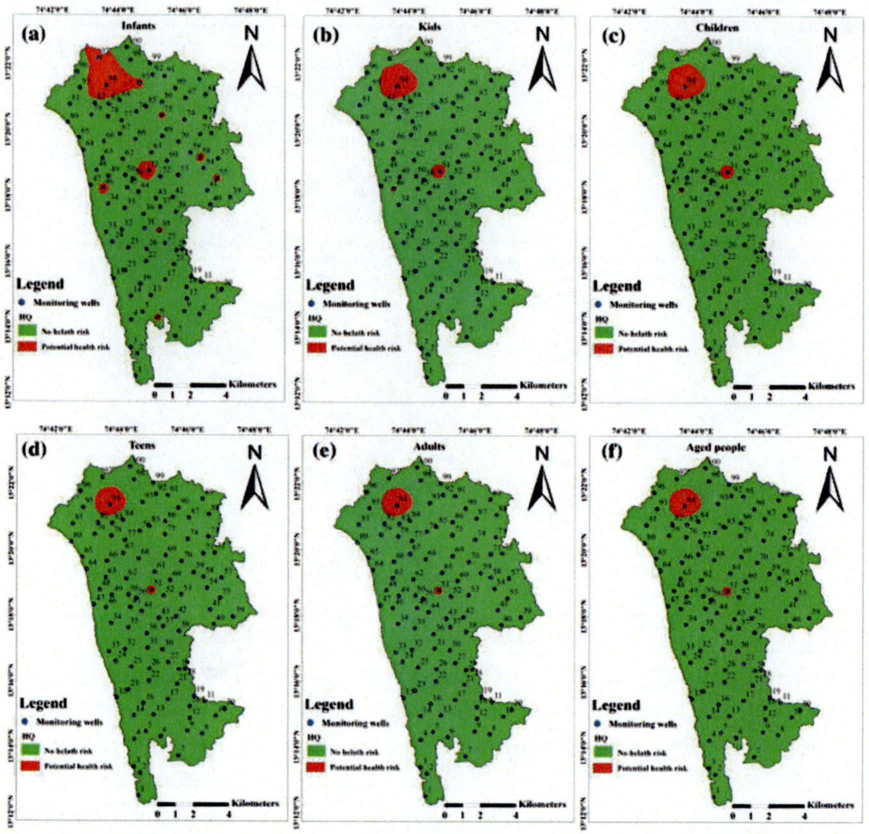

Fig. 6 Spatial distribution of HQ **a** infants, **b** kids, **c** children, **d** teens, **e** adults and **f** aged people in pre-monsoon

4 Conclusions

The present study evaluates the drinking water quality and nitrate contamination in the study area. WQI estimation revealed that 84 groundwater samples are excellent for drinking purposes, seven are good, four are poor, two are very poor and three are considered to be unsuitable for drinking. The western part of study area is considered to be contaminated due the influence of saltwater. Nitrate concentrations in two open wells exceeded the prescribed drinking water standard and pose high potential health risk for all the human age groups. Evaluation of groundwater quality using spatial interpolation of WQI delineates coastal and north-west region of study area need remedial measures to improve the drinking water quality. The findings of the study will be useful for the policy makers to make appropriate water quality management measures in the study area.

Acknowledgements The first author would like to acknowledge the support of Manipal Academy of Higher Education for providing Dr. T.M.A. Pai scholarship.

References

1. Vasanthavigar, M., Srinivasamoorthy, K., & Vijayaragavan, K. (2010). Application of water quality index for groundwater quality assessment: Thirumanimuttar sub-basin, Tamilnadu, India. *Environmental Monitoring and Assessment, 171*, 595–609. https://doi.org/10.1007/s10661-009-1302-1
2. Saeedi, M., Abessi, O., Sharifi, F., & Meraji, H. (2010). Development of groundwater quality index. *Environmental Monitoring and Assessment, 163*(1–4), 327–335. https://doi.org/10.1007/s10661-009-0837-5
3. Klimas, A. A. (1996). Methodology for mapping shallow groundwater quality in urbanized areas: A case study from Lithuania. *Environmental Geology, 27*(4), 320–328. https://doi.org/10.1007/BF00766701
4. Vaiphei, S. P., Kurakalva, R. M., & Sahadevan, D. K. (2020). Water quality index and GIS-based technique for assessment of groundwater quality in Wanaparthy watershed, Telangana India. *Environmental Science and Pollution Research, 27*(36), 45041–45062. https://doi.org/10.1007/s11356-020-10345-7
5. Ketata, M., Gueddari, M., & Bouhlila, R. (2012). Use of geographical information system and water quality index to assess groundwater quality in El Khairat deep aquifer (Enfidha, Central East Tunisia). *Arabian Journal of Geosciences, 5*, 1379–1390. https://doi.org/10.1007/s12517-011-0292-9
6. Rawat, K., Pradhan, S., Tripathi, V., Jeyakumar, L., & Singh, S. K. (2019) Statistical approach to evaluate groundwater contamination for drinking and irrigation suitability. *Groundwater for Sustainable Development, 9*, 100251.https://doi.org/10.1016/j.gsd.2019.100251
7. Kim, H., Yu, S., Oh, J., Kim, K., Lee, J., Moniruzzaman, M., Kim, H.K., & Yun, S. (2019). Nitrate contamination and subsequent hydrogeochemical processes of shallow groundwater in agro-livestock farming districts in South Korea. *Agriculture, Ecosystems & Environment, 273*, 50e61. https://doi.org/10.1016/j.agee.2018.12.010
8. Liu, J., Peng, Y., Li, C., Gao, Z., & Chen, S. (2021). Characterization of the hydrochemistry of water resources of the Weibei Plain, Northern China, as well as an assessment of the risk of high groundwater nitrate levels to human health. *Environmental Pollution, 268*, 115947.https://doi.org/10.1016/j.envpol.2020.115947
9. Chen, J., Wu, H., & Qian, H. (2016). Groundwater nitrate contamination and associated health risk for the rural communities in an agricultural area of ningxia, northwest China. *Exposure Health, 8*(3), 349–359. https://doi.org/10.1007/s12403-016-0208-8
10. Akshitha, V., Balakrishna, K., & Udayashankar, H. N. (2021). Assessment of hydrogeochemical characteristics and saltwater intrusion in selected coastal aquifers of southwestern India. *Marine Pollution Bulletin, 173*, 112989.https://doi.org/10.1016/j.marpolbul.2021
11. APHA. (2012). *Standard methods for the examination of water and wastewater*, (twenty-second ed.). American Public Health Association, American Water Works Association, Water Environment Federation.
12. Horton, R. K. (1965). An index number system for rating water quality. *Journal of the Water Pollution Control Federation, 37*, 300–306. https://doi.org/10.12691/ajwr-1-3-3
13. Loh, Y. S. A., Akurugu, B. A., Manu, E., & Aliou, A. S. (2020). Assessment of groundwater quality and the main controls on its hydrochemistry in some Voltaian and basement aquifers, northern Ghana. *Groundwater for Sustainable Development, 10*, 100296.https://doi.org/10.1016/j.gsd.2019.100296

14. Mallick, J., Kumar, A., Almesfer, M. K., Alsubih, M., Singh, C. K., Ahmed, M., & Khan, R. A. (2021). An index-based approach to assess groundwater quality for drinking and irrigation in Asir region of Saudi Arabia. *Arabian Journal of Geosciences, 14*(3), 1–17. https://doi.org/10.1007/s12517-021-06506-8
15. Singaraja, C. (2017). Relevance of water quality index for groundwater quality evaluation: Thoothukudi District, Tamil Nadu India. *Appl Water Sci, 7*(5), 2157–2173. https://doi.org/10.1007/s13201-017-0594-5
16. Singh, C. K., Shashtri, S., Mukherjee, S., Kumari, R., Avatar, R., Singh, A., & Singh, R. P. (2011). Application of GWQI to assess effect of land use change on groundwater quality in lower Shiwaliks of Punjab: Remote sensing and GIS based approach. *Water Resources Management, 25*(7), 1881–1898. https://doi.org/10.1007/s11269-011-9779-0
17. World Health Organization. (2011). *Guidelines for drinking-water quality* (4th ed.). World Health Organization. https://apps.who.int/iris/handle/10665/4458
18. USEPA (United States Environmental Protection Agency). (1989). Superfund Public Health Evaluation Manual (Washington, DC).
19. USEPA (US Environmental Protection Agency). (2001). Baseline Human Health Risk Assessment Vasquez Boulevard and I-70 Superfund Site. Denver CO. http://www.epa.gov/region8/superfund/sites/VB-170-Risk.pdf
20. Davis, S. N., & DeWiest, R. J. (1966). *Hydrogeology*. Wiley.
21. El Ghandour, M. F. M., Khalil, J. B., & Atta, S. A. (1983). Distribution of sodium and potassium in the groundwater of the Nile Delta region (Egypt). *CATENA, 10*(1–2), 175–187. https://doi.org/10.1016/S0341-8162(83)80015-0
22. Howard, K. W. F., & Lloyd, J. W. (1983). Major ion characterization of coastal saline ground waters. *Groundwater, 21*(4), 429–437. https://doi.org/10.1111/j.1745-6584.1983.tb00744.x
23. Ritzi, R. W., Wright, S. L., Mann, B., & Chen, M. (1993). Analysis of temporal variability in hydrogeochemical data used for multivariate analyses. *Ground Water, 31*, 221–222. https://doi.org/10.1111/j.1745-6584.1993.tb01814.x
24. Subba Rao, N., Saroja Nirmala, I., & Suryanarayana, K. (2005). Groundwater Quality in a Coastal Area: A Case Study from Andhra Pradesh India. *Environmental Geology, 48*(4–5), 543–550. https://doi.org/10.1007/s00254-005-1306-2
25. World Health Organization. (2017). Guidelines for drinking-water quality: Fourth edition incorporating the first addendum. ISBN978-92-4-154995-0.

MHD Flow Past Semi-Infinite Inclined Vertical Cylinder Under the Influence of Inclined Magnetic Field

H. P. Rani and Koragoni Naresh

Abstract Numerical investigation of natural convective time-dependent flow past an isothermally heated inclined semi-infinite vertical cylinder is carried out in the present study. The effects of a uniform constant inclined magnetic field that makes an angle with the horizontal axis is analysed. The fluid is assumed to be viscous incompressible and electrically conducting. The transient coupled equations that governing the considered problem are non-dimensionalized. These nonlinear equations are discretized by the finite difference method of implicit type, and the resulting system of equations is solved using the Thomas algorithm. The present study aims to analyse the impact of control parameters, such as the magnetic parameter, inclination angle of cylinder and of magnetic field on the field variables, velocity and temperature arising in the system. Profiles of velocity and temperature get deteriorated as the cylinder inclined more towards the horizontal position. Similar observations are made with respect to the inclined magnetic field. Therefore, rate of heat transfer gives better results, when the cylinder is kept in the vertical position in comparison with other positions. While these variables get accelerated with the increased values of inclined magnetic field. Response surface methodology is applied for the existing data sequence of Nusselt number so as to find its optimal response. The regression coefficients, in the equation of fitted line response data, are considered until third order for achieving a more accurate solution. The significance of values of regression coefficients is analysed based on the analysis of variance (ANOVA) method. Also, the optimum (maximum) value of Nusselt number is obtained by using the optimization technique.

Keywords Natural convection · Vertical cylinder · Nusselt number · RSM · ANOVA

H. P. Rani (✉) · K. Naresh
Department of Mathematics, National Institute of Technology, Warangal 506004, India
e-mail: hprani@nitw.ac.in

A. K. Dikshit et al. (eds.), *Innovative Trends in Hydrological and Environmental Systems*, Lecture Notes in Civil Engineering 234,
https://doi.org/10.1007/978-981-19-0304-5_29

Abbreviations

u	Axial velocity
v	Transverse (radial) velocity
T'	Temperature of the fluid
r	Radial coordinate
x	Axial coordinate
r_0	Radius of the cylinder
t'	Time
U	Dimensionless axial velocity
V	Dimensionless transverse velocity
T	Dimensionless temperature
R	Dimensionless radial coordinate
X	Dimensionless axial coordinate
t	Dimensionless time
g	Acceleration due to gravity
B_0	Constant magnetic field
M	Magnetic parameter
C_p	Specific heat at constant pressure
T'_∞	Ambient temperature
T'_w	Wall temperature
Pr	Prandtl number
Gr	Grashof number
Nu	Nusselt number

Greek Letters

ν	Kinematic viscosity
β	Coefficient of volume expansion
θ	Inclined angle of magnetic field
α	Inclined angle of cylinder
κ	Magnetic diffusivity
η	Thermal conductivity
ρ	Density
σ	Electrical conductivity
τ	Skin-friction coefficient

1 Introduction

Magnetohydrodynamics (MHD) offers the interplay among the electrically conducting fluids and electromagnetic forces. Such problems arise in conditions starting from the cause of the origin of Earth's field and therefore the prediction of space climate to damping of turbulent fluctuations in semiconductor melts during crystal boom and even in the size of flow rates of liquids within the food enterprise. The magnetic field plays an important role in controlling heat transfer; thus, the problems related to MHD have applications in a number of factual and experimental investigations, for example in metallurgical processes, electromagnetic stirring, strand casting approaches, metallic spinning, non-stop stretching of plastic films, glass-fibre production and so on.

The boundary layer flow past solid body plays a critical role due to its applications in aerodynamics, hydrodynamics, transportation, wind engineering and ocean engineering (e.g. airplanes, rockets, projectiles, ships, torpedoes, submarines, vans, cycles, automobiles, bridges, water towers, buildings, breakwaters, cables and buoys). In problems related to boundary layer, controlling of heat transfer performs a primary role Schlichting [1]. The understanding of air flow moving around the engine is vital in cooling the engine. As an instance, the heat transfer problems associated with the air-cooled motorbike based on numerical simulations are paid attention in the literature Takahashi and Gokan [2] and Gokan et al. [3]. Consequently, the present study is aimed to achieve the understanding of flow field around the inclined isothermally heated cylinder under the impact of magnetic field.

Laminar free convective boundary layer flow emanating from the surface of cylinder is analysed in the literature extensively. For one such problem, Sparrow and Gregg [4] first obtained the analytical solutions by using the Laplace transform method. For the similar problem related to thin needles and vertical cylinders, using the power-law wall temperature variation, Lee et al. [5] obtained the analytical solution. Transient solutions of such problems are given by Goldstein and Briggs [6] and Velusamy and Garg [7]. It is found that the increasing radius of the cylinder decreases the thermal capacity. When the above flow field is kept under the magnetic field, the fluid velocity gets decelerated and Nusselt and Sherwood characteristics increase with the magnetic control parameter Ganesan and Rani [8]. Further, under similar conditions the flow past the oscillating infinite cylinder Agarwal et al. [9], vibrating coaxial infinite cylinder Raptis and Agarwal [10] and moving vertical cylinder Ganesan and Loganathan [11] are studied. They noticed that both the presence and an increase in the strength of magnetic field lead to a decelerated flow.

For the case of MHD flow past a permeable cylinder with absorption and chemical reaction, the heat transfer rate proportionally increases with the rate of suction or injection Chamkha [12]. For the flow in a rotating cylinder containing the liquid metal pushed by a rotating magnetic field, the amplifying magnetic force stabilizes the flow in counter-rotating field while it destabilizes the flow in the field without rotation Pedchenko and Grants [13]. Initially when the cylinder is packed with cold air, the flow of cold air sustained due to the magnetizing force by overcoming the buoyancy force, but it gets reversed in the case of hot air Tagawa et al. [14]. For natural convective laminar fluid flow between two conducting concentric cylinders, the velocity of the fluid and induced magnetic field are determined to be rapidly reducing Kumar and Singh [15] with increased magnetic parameter.

The inclined magnetic field effect on symmetric or asymmetric channel with the peristaltic flow of a Williamson fluid revealed that by increasing the inclination angle of magnetic field flow get decelerated [Nadeem and Akram [16]. Similar problem for the case of free and forced convection together with the stagnation point stratified flow over an inclined cylinder confirmed, the velocity profiles get improved with the increasing inclination angle of cylinder Hayat et al. [17]. The natural convective flow over vertical cone under the effect of growing strength of the induced magnetic field slows down the convection system Vanita and Anand [18].

From the literature survey, it can be seen that when the effect the magnetic field is considered, most of the existing studies related to the free convection flow past the cylinder are in the steady state. The present study deals with the time-dependent natural convective flow over the semi-infinite cylinder kept in the inclined magnetic field with the electrically conducting fluid. Also, the cylinder is assumed to be inclined to the horizontal. Such type of flow field dynamics can be used in geophysics and electric technology problems. The boundary layer nonlinear governing equations with Boussinesq's approximation are considered and are made dimensionless. The equations are discretized by using the Crank–Nicolson method, and resulting system is solved by using tri-diagonal matrix algorithm. The solutions of velocity and temperature are presented graphically for different values of magnetic parameter, inclination angles of cylinder and magnetic field with respect to radial coordinate and time. The mean wall shear stress and heat transfer rate are also discussed. In addition, the numerical solutions are used to predict the optimum value of rate of heat transfer using the RSM and ANOVA statistical tools.

2 Mathematical Formulation

A semi-infinite vertical cylinder with radius r_0 is taken into consideration in the present study. The flow around the cylinder is assumed to be of two-dimensional because of axisymmetric nature of the flow. The cylindrical coordinates are assumed with x-axis measured along the axis of cylinder while the radial (r) coordinate measured normal to the axis. The velocity (u, v) is considered along (x, r)-coordinates, respectively. The cylinder is supposed be kept in the applied magnetic field of constant strength B_0, and the induced magnetic field is supposed to negligible. Both the cylinder and magnetic field are assumed to be inclined at angles α and θ, respectively, to the horizontal. The physical model is schematic in Fig. 1.

The considered laminar flow is assumed to be unsteady. The fluid is assumed to be viscous, incompressible and electrically conducting. Initially, when $t' = 0$, the cylinder and its surrounding fluid are assumed to be kept in the ambient temperature,

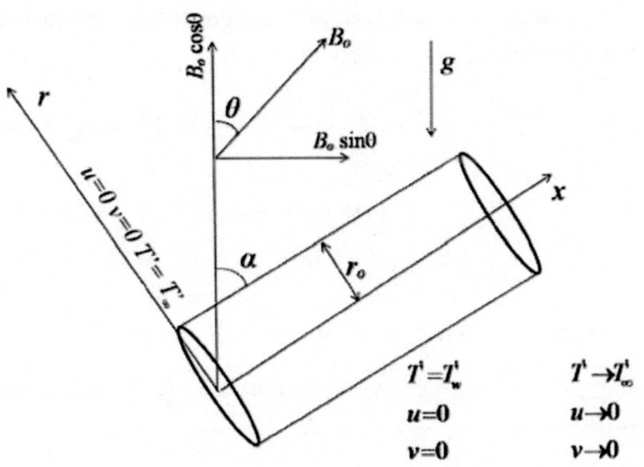

Fig. 1 Schematic of the physical model

T'_∞. As time increases, i.e. $t' > 0$, the surface temperature of cylinder amplified to T'_w and continued to be at the same level for all $t' > 0$. Due to this temperature difference between the cylindrical wall and the adjacent fluid, there is a density variation between the fluid, which is adjacent to the cylindrical wall ($r = r_0$) and far from the cylindrical wall ($r \to \infty$). This density variation interacts with the gravitational force, g, and, thus, there exists a free convective flow. With these assumptions and along with the buoyancy-driven convection, the boundary layer governing equations are as follows [1]:

Continuity equation

$$\frac{\partial u}{\partial x} + \frac{\partial v}{\partial r} + \frac{v}{r} = 0 \tag{1}$$

Momentum equation along axial direction

$$\frac{\partial u}{\partial t'} + u\frac{\partial u}{\partial x} + v\frac{\partial u}{\partial r} = \nu\frac{1}{r}\frac{\partial}{\partial r}\left(r\frac{\partial u}{\partial r}\right) + g\beta\left(T' - T'_\infty\right)\cos\alpha + \frac{\sigma B_0^2}{\rho}\left(v\sin\theta\cos\theta - u\cos^2\theta\right) \tag{2}$$

Momentum equation along radial direction

$$\frac{\partial v}{\partial t'} + \left(u\frac{\partial v}{\partial x} + v\frac{\partial v}{\partial r}\right) = \nu\frac{1}{r}\frac{\partial}{\partial r}\left(r\frac{\partial v}{\partial r}\right) + \frac{\sigma B_0^2}{\rho}\left(u\sin\theta\cos\theta - v\sin^2\theta\right) \tag{3}$$

Energy equation

$$\frac{\partial T'}{\partial t'} + v\frac{\partial T'}{\partial r} + u\frac{\partial T'}{\partial x} = \frac{\kappa}{\rho c_p}\nabla' T' \tag{4}$$

The notations and symbols are described in the Nomenclature. The conditions that govern the flow are given by

$$\begin{aligned} t' = 0 &: u = 0, v = 0, T' = T'_\infty \text{ for all } x \text{ and } r \\ t^2 > 0 &: u = 0, v = 0, T' = T'_w \text{ at } r = r_0 \\ &u = 0, v = 0, T' = T'_\infty \text{ at } x = 0 \\ &u \to 0, v \to 0, T' \to T'_\infty \text{ as } r \to \infty \end{aligned} \quad (5)$$

With the scaling $X = \frac{x}{r_0}$, $R = \frac{r}{r_0}$, $t = \frac{\nu t'}{r_0^2}$, $U = \frac{u r_0}{\upsilon}$, $V = \frac{\upsilon r_0}{\nu}$, $T = \frac{T' - T'_\infty}{T'_w - T'_\infty}$, $M = \frac{\sigma B_0^2 r_0^2}{\rho \nu}$, $\text{Gr} = \frac{g\beta r_0^3 (T'_w - T'_\infty)}{\nu'}$ and $\text{Pr} = \frac{\nu}{\alpha}$, the non-dimensional form of Eqs. (1)–(4) as follows:

$$\frac{\partial U}{\partial X} + \frac{\partial V}{\partial R} + \frac{V}{R} = 0 \quad (6)$$

$$\begin{aligned} \frac{\partial U}{\partial t} + U\frac{\partial U}{\partial X} + V\frac{\partial U}{\partial R} &= \frac{1}{R}\frac{\partial}{\partial R}\left(R\frac{\partial U}{\partial R}\right) \\ &+ \text{Gr}\, T \cos\alpha + M\big(V \sin\theta \cos\theta - U \cos^2\theta\big) \end{aligned} \quad (7)$$

$$\frac{\partial V}{\partial t} + U\frac{\partial V}{\partial X} + V\frac{\partial V}{\partial R} = \frac{1}{R}\frac{\partial}{\partial R}\left(R\frac{\partial V}{\partial R}\right) + M\big(U \sin\theta \cos\theta - V \sin^2\theta\big) \quad (8)$$

$$\frac{\partial T}{\partial t} + U\frac{\partial T}{\partial X} + V\frac{\partial T}{\partial R} = \frac{1}{\text{Pr}}\frac{1}{R}\frac{\partial}{\partial R}\left(R\frac{\partial T}{\partial R}\right) \quad (9)$$

Also, the non-dimensional conditions are given by

$$\begin{aligned} t = 0 &: U = 0, V = 0, T = 0 \text{ for all } X \text{ and } R \\ t > 0 &: U = 0, V = 0, T = 1 \text{ at } R = 1 \\ &U = 0, V = 0, T = 0 \text{ at } X = 0 \\ &U \to 0, V \to 0, T \to 0 \text{ as } R \to \infty \end{aligned} \quad (10)$$

To solve the transient coupled nonlinear governing Eqs. (6)–(9) subject to (10), the Crank–Nicolson finite difference method has been applied. The resultant discretized tri-diagonal system of equations is solved by applying the TDMA algorithm. The details about the solution methodology are detailed in Ganesan and Rani [11]. Region of solution is considered with boundaries 0 to 1 along the X-axis and 1–15 ($R_{\max}$) along the R-axis. The boundaries are chosen such that the boundary conditions are satisfied. The initial and boundary values of the field variables U, V and T are known from the initial and boundary conditions, respectively, given in Eq. (10). Numerical results are analysed for tolerance rate of 10^{-6} in order to obtain more realistic form of the output results with reasonable convergence. To choose the optimal gird which represents the real flow physics, a grid independence test was conducted. It is noted that the steady-state velocity and temperature values acquired with the 100 × 500 grid system vary in the second decimal place from the results obtained with the 50 × 250 grid system and in the third decimal place from those with the 200 × 1000. Further, in Table 1, the L^2—relative error between different grid systems is shown. The grid system of 100 × 500 and 200 × 1000 shows less error in comparison to that of grid system of 50 × 250 and 100 × 500. Thus, the mesh size of 100 × 500 is considered to be appropriate for further calculations. Thus, for all the subsequent analyses, the grid system of 100 × 500 was chosen, with mesh size of 0.01 and 0.03 in X and R directions, respectively.

To validate the employed solution methodology, the simulated results are corroborated with those available in the literature. As there are no experimental results and analytical solutions available in the literature for the present model, the simulated results are compared with the numerical steady-state results of Lee et al. [5] without magnetic field ($M = 0$) and inclination ($\alpha = 0$) for Pr = 0.7, and the comparison results are depicted in Fig. 2. As expected, the agreement between the two results of field variables, such as velocity and temperature profiles, is observed to be good.

Table 1 L^2—relative errors for different grid sizes

Field variable	L^2—relative error of grid system of 50 × 250 and 100 × 500	L^2—relative error of grid system of 100 × 500 and 200 × 1000
Velocity	0.001329	0.000472
Temperature	0.000485	0.000441

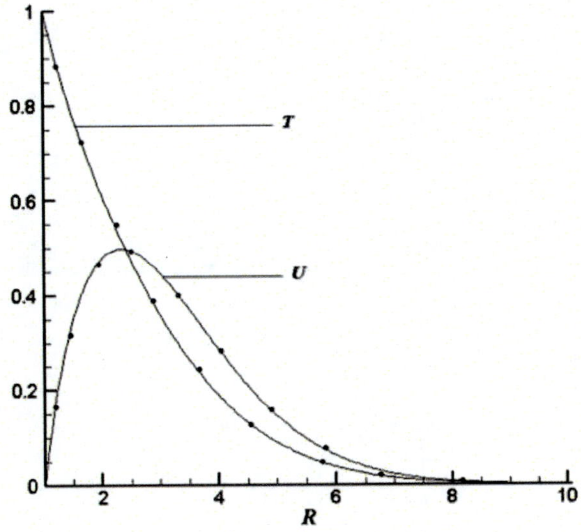

Fig. 2 Profiles of velocity and temperature (.) Lee et al. [5] and (_____) present result

3 Results

In this section, the influence of control parameters on the flow variables, namely the velocity field, U and temperature, T, by means of graphical representations is explored. The expressions for the wall shear stress and heat transfer coefficient of the fluidic system are also detailed in this section. The effect of inclined magnetic field arising due to the motion of an electrically conducting fluid flow is taken into consideration. The effect of magnetic parameter, M and the inclination angles of cylinder ($0 \leq \alpha \leq \pi/2$) and magnetic field ($0 \leq \theta \leq \pi/2$) on velocity and temperature variations has been discussed for fixed Gr = 5 and Pr = 0.71 and is shown in Figs. 3, 4, 5, 6, 7, 8, 9, 10, 11, 12, 13 and 14.

Figure 3 depicts the variation in velocity filed for varying values of α with respect to R. It is noted that for increasing R, i.e. starting from the cylindrical wall ($R = 1$) to far away from the cylinder ($R = 15$), the velocity, U, starts from zero, reaches its maximum then decreases to zero, thus satisfying the boundary conditions. The velocity boundary layer thickness gets reduced for larger values of α, since with the amplifying values of α, the frictional force increases. For the varying values of M, the velocity profiles with respect to R are shown in Fig. 4. The fluid flow slows down as the magnetic field parameter increases. This can be attributed to the existence of Lorentz force serves as a resisting force to the fluid flow.

The velocity, U, boundary layer profiles with respect to R for varying θ, are plotted in Fig. 5. The values of U observed to increase along the radial direction and get its

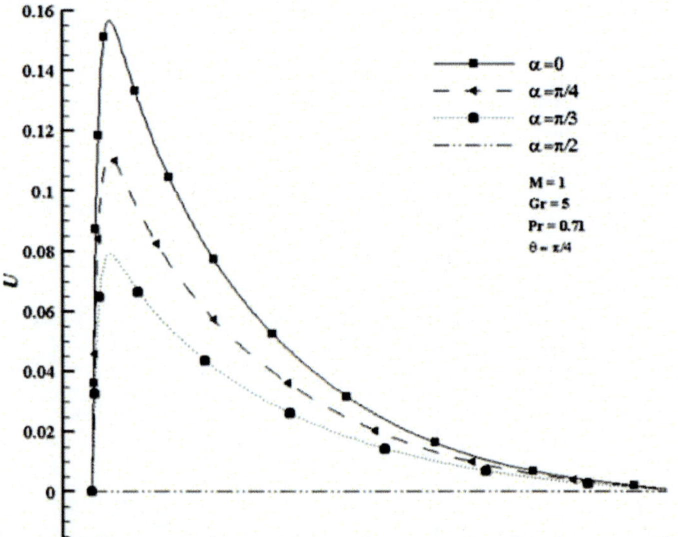

Fig. 3 Velocity, U, profiles for different values of α against R

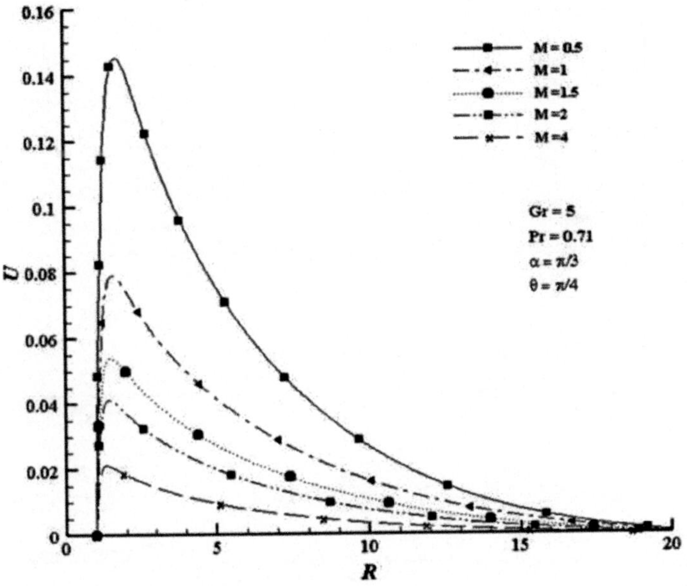

Fig. 4 Velocity profile, U, for different values of M against R

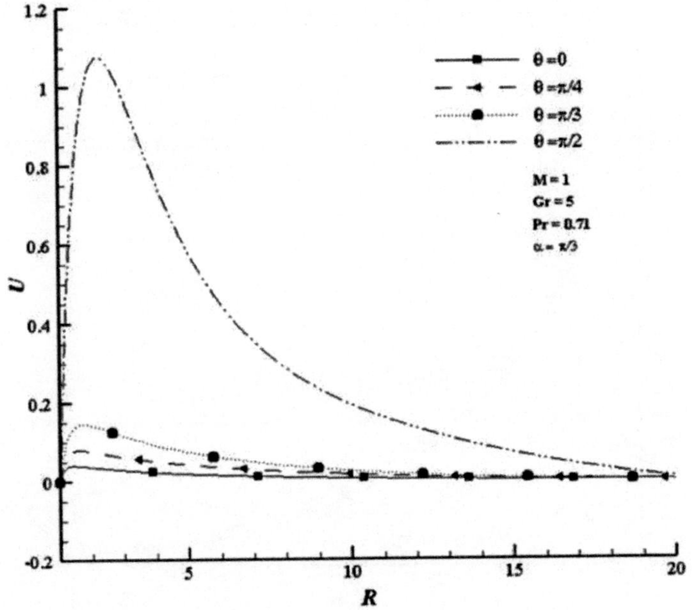

Fig. 5 Velocity profiles, U, for different values of θ against R

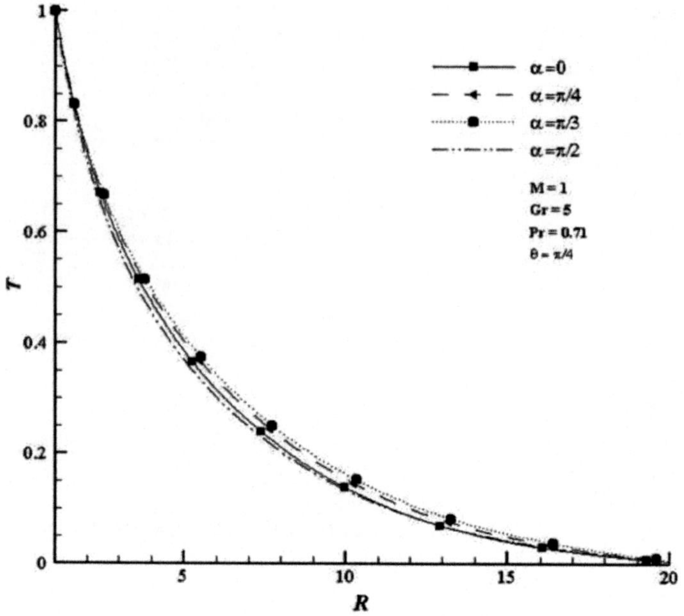

Fig. 6 Temperature profiles for different values of α with respect to R

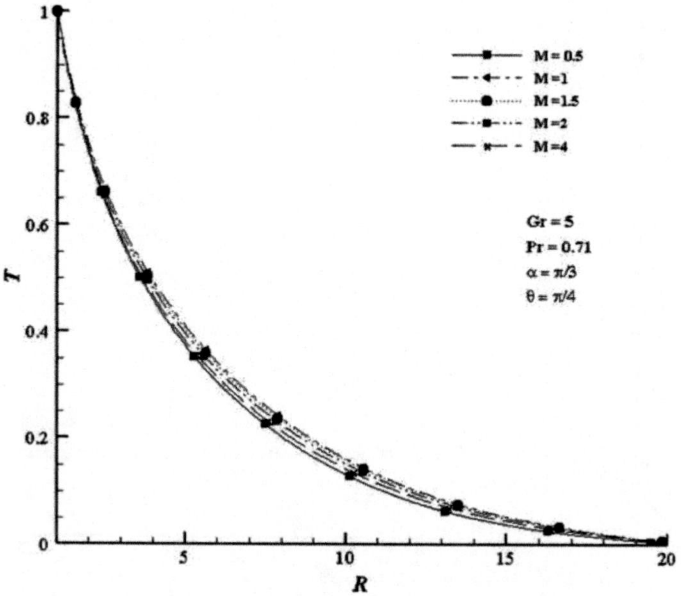

Fig. 7 Temperature profiles for different values of M with respect to R

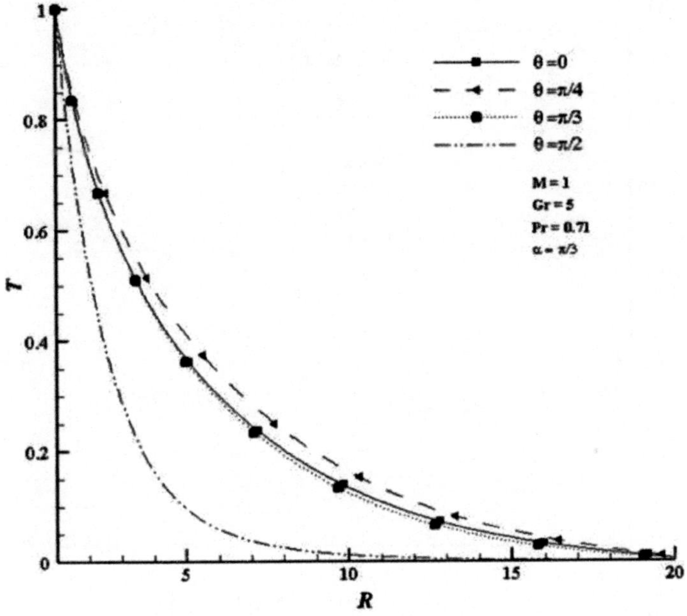

Fig. 8 Temperature profiles for different values of θ with respect to R

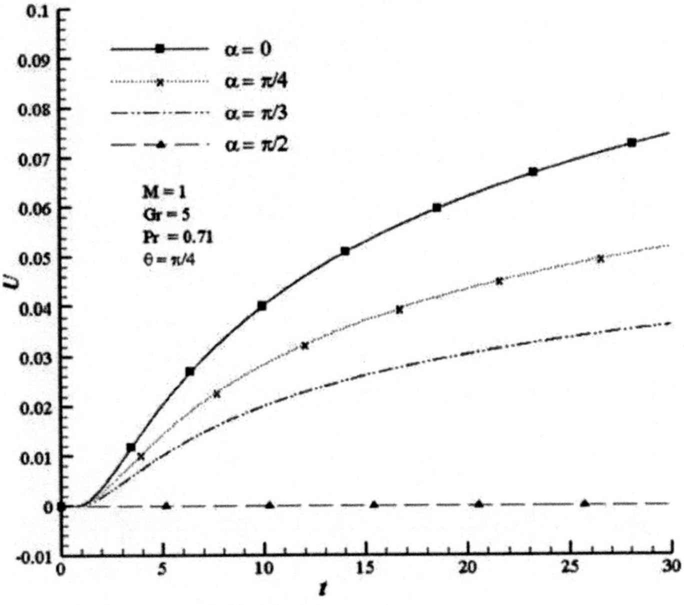

Fig. 9 Velocity profiles for different values of α against t

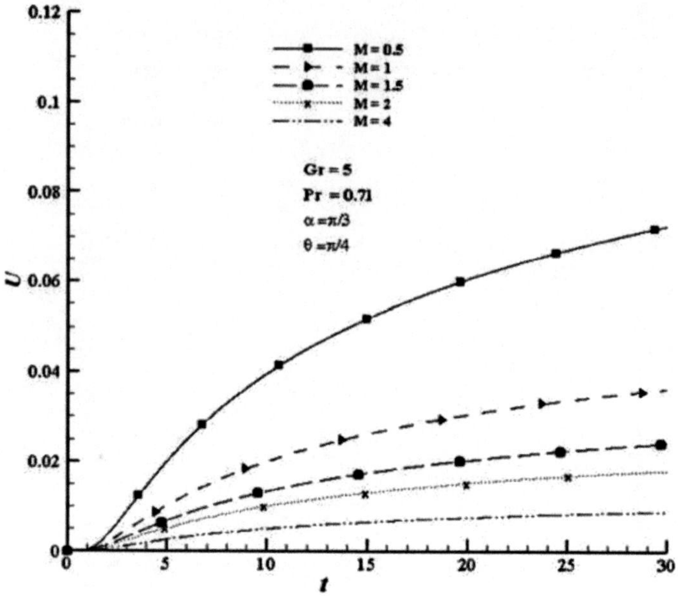

Fig. 10 Velocity for various values of M with respect to t

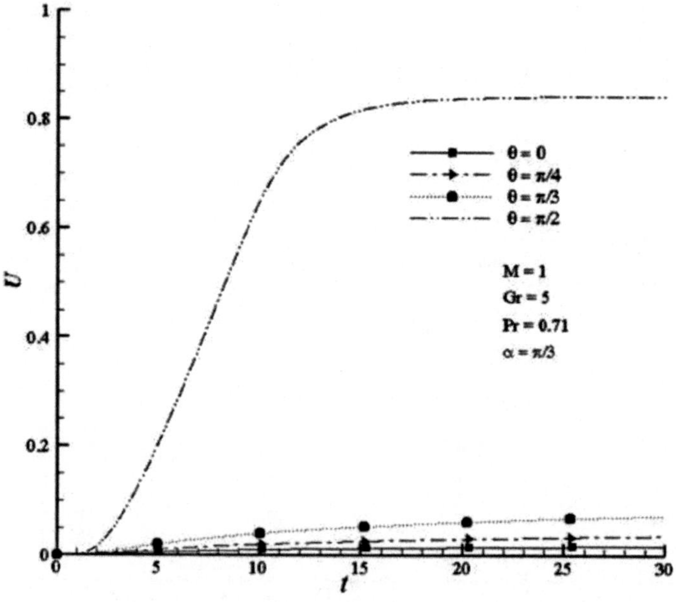

Fig. 11 Velocity profiles for different values of θ with respect to t

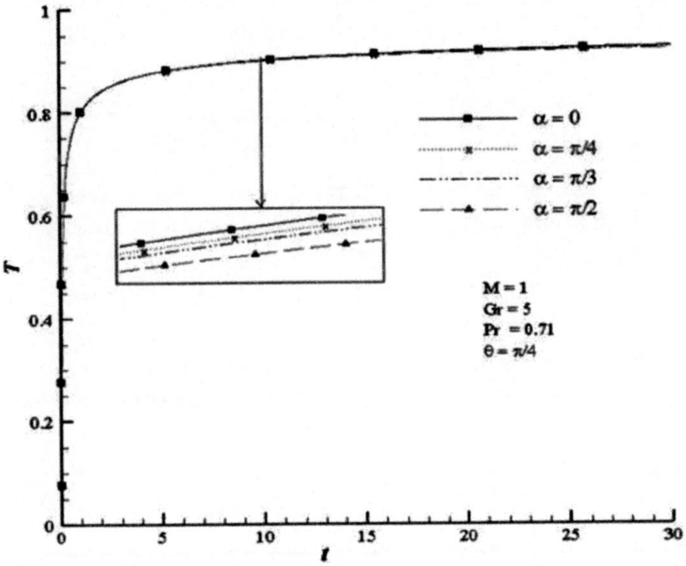

Fig. 12 Temperature profiles for different values of α with respect to t

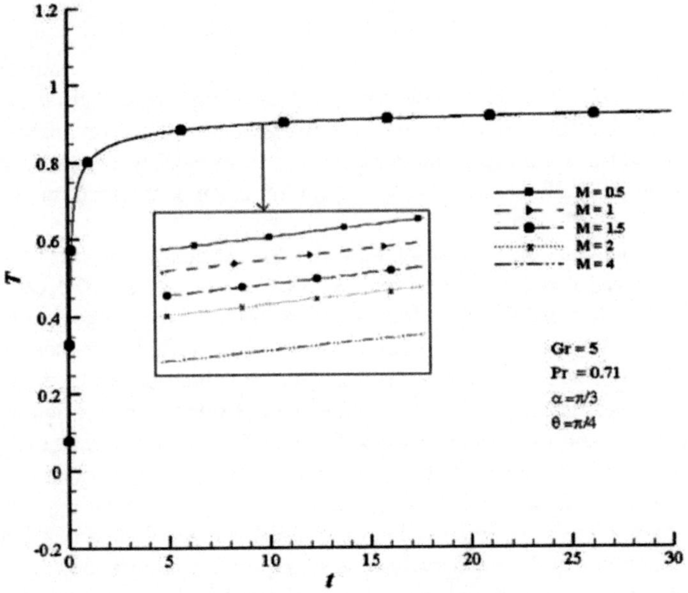

Fig. 13 Temperature, T, profiles for different values of M against t

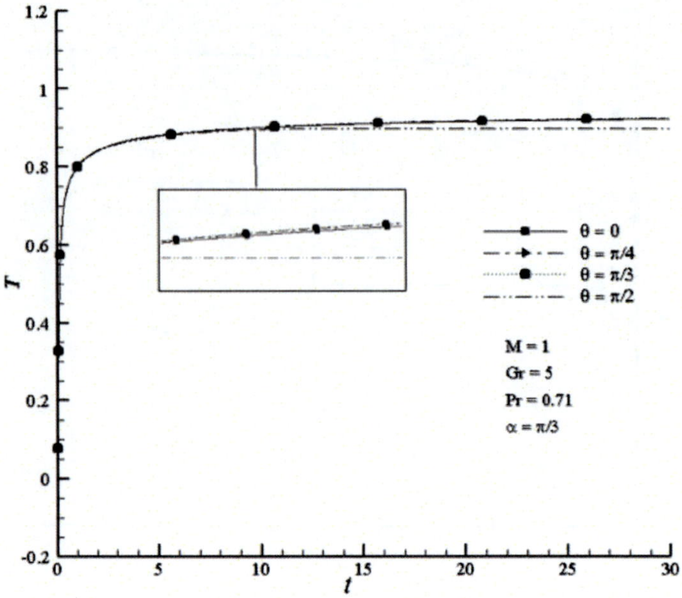

Fig. 14 Temperature, T, profiles for different values of θ against t

maximum and asymptotically decreases to zero. Also, the velocity boundary layer thickness increases with θ, due to the fact that increasing θ influences the magnetic field on the fluid particles. When the inclination angle increases, it increases the fluid flow since the normal striking of magnetic field towards the flow produces minimum resistance than the inclined striking. Thus, the influence of inclined magnetic field increases the fluid velocity.

Figure 6 depicts that the temperature profiles for various values of α with respect to R. It is observed that T increases with α, while for $\alpha = \pi/2$, T decreases as R increases (far away from the cylindrical wall) because of the improving buoyancy effect. Figure 7 shows that the temperature profiles for different values of M with respect R. It is observed that the temperature profiles increase when M increases. When M increases the magnetic field produces the Lorentz force; this force slows down the velocity and increases the temperature in flow field. It can be seen that the impact of increasing magnetic field parameter is noticeably accelerated by the fluid temperature within the solution region. Figure 8 shows the temperature profiles for different values of θ with respect to R. It is observed that T decreases when θ increases. This is due the fact that the electrically conducting fluid generates the resistive force with increasing θ and due to which the thermal boundary layer decreases.

Figure 9 depicts the unsteady velocity profiles for various values of α. It is noticed that U starts from its minimum value (zero), reaches its maximum and then attains the steady state as time t increases. When α increases, the frictional forces reduce the velocity in axial and radial directions; thus, U decreases when α increases as shown in Fig. 9. It can be observed from the figure there are no fluctuations with respect to time in the velocity profile in the case of free convective flow when $\alpha = \pi/2$.

Figure 10 shows that the velocity profiles for different values of magnetic parameter, M against t. For all values of M, it is observed that the flow starts with zero, increases to maximum and gets steady state for amplifying t values. Also observed that U decelerates when M increases. When M increases, the electrically conducting fluid produces the resistive force, due to which the velocity boundary layer thickness decreases.

Momentum boundary layer profiles for varying values of θ against t are shown in Fig. 11. The figure shows that when the angle θ increases, the velocity boundary layer increases with the time and attains the steady state. For increasing θ, the magnetic field produces the Lorentz force, so that the velocity variable reaches the steady state when time increases.

Figure 12 depicts the thermal profiles for increasing values of α against time t. It is noted that thermal boundary layer profiles decrease when α increases with increasing values of t and the time to attain the steady is almost same for all values of α. When α increases, the buoyancy force increases and, thus, T decreases with time t.

Temperature profiles for amplifying M against t are showed in Fig. 13. The observation is that T decreases when M increases with increasing time. When M increases, the electrically conducting fluid produces resistive force, and this force reduces the heat transfer with increasing time. The rate of heat transfer, Nu, is studied in detail in the next section. Figure 14 shows the temperature profiles for varying θ values with respect to t. It is observed that T decreases when θ and time increases, because the magnetic field produces Lorentz force due to increasing θ and with increasing t. This force increases the heat transfer.

Average skin-friction coefficient or shear stress ($\overline{\tau}$) and Nusselt number ($\overline{\text{Nu}}$): The coefficients of skin-friction (wall shear stress) and Nusselt number play a key role in the heat transfer studies and are calculated from the following equations:

$$\overline{\tau} = \int_0^1 \left(\frac{\partial U}{\partial R}\right)_{R=1} dX \tag{11}$$

$$\overline{\text{Nu}} = -\int_0^1 \left(\frac{\partial T}{\partial R}\right)_{R=1} dX \tag{12}$$

In Eqs. (11) and (12), the five-point finite difference approximation formula is used to calculate the derivatives employed and the Newton–Cotes closed integration formula to calculate the integrals.

Table 2 shows the $\overline{\text{Nu}}$ and $\overline{\tau}$ profiles for different values of M, α and θ. The rate of heat transfer observed to decrease with M. Also, the $\overline{\text{Nu}}$ increases when θ increases, since in the early time intervals there is the occurrence of conduction only and it dominates the convective heat transfer. When α increases, the $\overline{\text{Nu}}$ decreases due to the increased buoyancy force. The $\overline{\tau}$ decreases when M increases, since when M increases the Lorentz force reduces the velocity so skin friction decreases. The $\overline{\tau}$

Table 2 Nusselt number and skin-friction profiles for different parameter values

Pr	Gr	M	α	θ	$\overline{\text{Nu}}$	$\overline{\tau}$
0.71	5	0.5	$\pi/3$	$\pi/4$	2.75037	0.52599
		1			1.97961	0.46089
		1.5			1.62000	0.43660
		2			1.40185	0.42625
		4			0.98471	0.41190
0.71	5	1	0	$\pi/4$	3.90899	0.52471
			$\pi/4$		2.78357	0.48995
			$\pi/3$		1.97961	0.46089
			$\pi/2$		0.39556	0.00000
0.71	5	1.5	$\pi/3$	0	1.39400	0.44722
				$\pi/4$	1.97953	0.46108
				$\pi/3$	2.75056	0.53187
				$\pi/2$	6.59772	0.89816

increases when θ gets amplified since the magnetic force generates the frictional force with increasing θ, and due to this, the skin friction gets increased. While the reversing trend is observed when α increases.

4 Optimum Nusselt Number Analysis

The numerical simulation results of Nusselt number, obtained for the different combinations of control parameters M, Pr, Gr,α and θ arising in the system, are used for further analysis to find the optimum value. Regression analysis is used to derive the following Eq. (13) based on the response surface method (RSM):

$$y = \beta_c + \sum_{i=1}^{n} \beta_i x_i + \sum_{i=1}^{n} \beta_{ii} x_i x_i + + \sum_{i,j=1}^{n} \beta_{ij} x_i x_j + + \sum_{i,j,k=1}^{n} \beta_{ijk} x_i x_j x_k + \in \quad (13)$$

where y denotes the dependent variable, i.e. $\overline{\text{Nu}}$ and $\in$ is the value of the error. This equation includes β_c as the intercept, β_i as the i-th factor linear regression coefficient, β_{ii} as the i-th factor quadratic regression coefficients, β_{ij} as the i-th and j-th factor interaction, and β_{ijk} as the i-th, j-th and k-th factor cubic regression coefficients. Tahari et al. [19] used the quadratic regression coefficients for Eq. (13), while the cubic regression coefficients are used in the present study. The major analyses include the normalization of errors and analysis of errors of variance. The remaining and predicted values for all tested instances are shown in Table 3 through the regression equations. The analysis of variance (ANOVA) methodology is employed to determine the interaction between these coefficients and the impact of each coefficient on the response function. For each of the nonlinear regression coefficient, shown in Eq. (13),

Table 3 Significance of coefficients using the ANOVA

Term	Coefficient	*p*-value	Significant
Const	0.39536	0.00000	Yes
M	−0.00001	0.99942	No
Pr	0.18785	0.00000	Yes
Gr	0.00306	0.79051	No
α	−0.67440	0.00000	Yes
θ	0.38779	0.00000	Yes
M^2	0.00000	0.99193	No
Pr^2	0.03636	0.01152	Yes
Gr^2	0.00000	0.99141	No
α^2	0.67428	0.00000	Yes
θ^2	−0.98395	0.00000	Yes
M Pr	−0.00405	0.00000	Yes
M Gr	−0.00006	0.00000	Yes
$M\,\alpha$	0.00015	0.78157	No
$M\,\theta$	−0.00110	0.04754	Yes
Pr Gr	0.00872	0.00000	Yes
Pr α	0.24852	0.00000	Yes
Pr θ	−0.13795	0.00000	Yes
Gr α	0.00089	0.42594	No
$Gr\,\theta$	−0.00080	0.47406	No
$\alpha\,\theta$	0.88221	0.00000	Yes
M^3	0.00000	0.98064	No
M^2Pr	0.00001	0.00037	Yes
M^2 Gr	0.00000	0.00000	Yes
$M^2\,\alpha$	−0.00001	0.00009	Yes
$M^2\,\theta$	0.00000	0.01345	Yes
$M\,\mathrm{Pr}^2$	0.00075	0.20283	No
M Pr Gr	−0.00002	0.00000	Yes
$M\,Pr\,\alpha$	0.00084	0.00127	Yes
M Pr θ	0.00056	0.02948	Yes
$M\,\mathrm{Gr}^2$	0.00000	0.01084	Yes
M Gr α	0.00001	0.00000	Yes
M Gr θ	0.00001	0.00045	Yes
$M\,\alpha^2$	0.00089	0.00002	Yes
$M\,\alpha\,\theta$	−0.00043	0.02052	Yes
$M\,\theta^2$	0.00139	0.00000	Yes
Pr^3	−0.09149	0.00000	Yes

(continued)

Table 3 (continued)

Term	Coefficient	*p*-value	Significant
Pr^2 Gr	−0.00151	0.19776	No
$Pr^2\ \alpha$	0.13071	0.00000	Yes
$Pr^2\ \theta$	−0.17037	0.00000	Yes
Pr Gr^2	−0.00003	0.01292	Yes
Pr Gr α	−0.00231	0.00001	Yes
Pr Gr θ	0.00188	0.00032	Yes
Pr α^2	−0.22882	0.00000	Yes
Pr $\alpha\ \theta$	−0.21449	0.00000	Yes
Pr θ^2	0.38068	0.00000	Yes
Gr^3	0.00000	0.95737	No
$Gr^2\ \alpha$	0.00002	0.00614	Yes
$Gr^2\ \theta$	−0.00003	0.00061	Yes
Gr α^2	−0.00233	0.00000	Yes
Gr $\alpha\ \theta$	−0.00160	0.00002	Yes
Gr θ^2	0.00325	0.00000	Yes
α^3	−0.19648	0.00000	Yes
$\alpha^2\ \theta$	−0.25110	0.00000	Yes
$\alpha\ \theta^2$	−0.28623	0.00000	Yes
θ^3	0.42769	0.00000	Yes

the following hypothetical test is performed in ANOVA:

$$H_0 : \beta_1 = \beta_2 = \cdots = \beta_k = 0$$
$$H_1 : \beta_j = 0 \text{ for at least one } j.$$

Primarily based on the *p*-values, the impact of every parameter is determined based on ANOVA methodology and the results are shown in Table 3. It can be noted that the parameters *M*, Pr, Gr, α and θ have significant impacts on $\overline{\text{Nu}}$.

Based on the ANOVA, the average heat transfer rate is calculated from

$$\begin{aligned}\overline{Nu} = {} & 0.39536 + 0.18785\ \text{Pr} - 0.6744\alpha + 0.387790\theta + 0.03636\ \overset{2}{\text{Pr}} \\ & + 0.67428\ \alpha^2 - 0.98395\ \theta^2 - 0.00405\ M\ \text{Pr} - 0.00006\ M\,\text{Gr} \\ & - 0.0011\ M\theta + 0.00872\ \text{Pr Gr} + 0.24852\ \text{Pr}\,\alpha - 0.13795\ \text{Pr}\,\theta \\ & + 0.88221\ \alpha\theta + 0.00001\ M^2\,\text{Pr} - 0.00001\ M^2\alpha \\ & - 0.00002\ M\,\text{Pr Gr} + 0.00084\ M\ \text{Pr}\,\alpha + 0.00056\ M\ \text{Pr}\,\theta \\ & + 0.00001\ M\,\text{Gr}\alpha + 0.00001\ M\,\text{Gr}\theta + 0.00089\ M\alpha^2 \\ & - 0.00043\ M\alpha\theta + 0.00139\ M\theta^2 - 0.09149\,r^3 + 0.13071\ \overset{2}{\text{Pr}}\,\alpha\end{aligned}$$

$$
\begin{aligned}
&- 0.17037\ \mathrm{Pr}^2\theta - 0.00003\ \mathrm{Pr}\ \mathrm{Gr}^2 - 0.00231\ \mathrm{Pr}\ \mathrm{Gr}\alpha \\
&+ 0.00188\,\mathrm{Pr}\ \mathrm{Gr}\theta - 0.22882\ \mathrm{Pr}\,\alpha^2 - 0.21449\ \mathrm{Pr}\,\alpha\theta + 0.38068\ \mathrm{Pr}\,\theta^2 \\
&+ 0.00002\ \mathrm{Gr}^2\alpha - 0.00003\,\mathrm{Gr}^2\theta - 0.00233\,\mathrm{Gr}\alpha^2 \\
&- 0.0016\ \mathrm{Gr}\alpha\theta + 0.00325\ \mathrm{Gr}\,\theta^2 - 0.19648\ \alpha^3 - 0.2511\ \alpha^2\theta \\
&- 0.28623\ \alpha\theta^2 + 0.42769\ \theta^3.
\end{aligned}
$$

Mathematica software is used to determine the best instances among all the tested and untested studies, and it is found that the maximum of $\overline{\mathrm{Nu}}$ (= 1.64558) occurs for the values of control parameters, $M = 5.5387$, $\mathrm{Pr} = 1$, $\mathrm{Gr} = 55.1096$, $\alpha = 0$ and $\theta = \frac{\pi}{2}$.

Thus, from the numerically simulated results the optimum value of mean rate of heat transfer for the considered problem is obtained using the RSM, and further validation is carried out with the ANOVA methodology. This study can be extended to boundary layer flow past different types of solid bodies kept under different conditions. Since the solution of the governing equations consume more memory and time, using RSM and ANOVA methodology, the general equation to calculate the rate of heat transfer can be proposed.

5 Conclusion

The natural convective flow of a viscous incompressible electrically conducting fluid is important in boundary layer flow related to heat transfer studies. Such fluid flow along with transient effects is given very less importance in the literature. Hence in the present study, the unsteady flow past semi-infinite isothermally heated inclined cylinder has been considered under the effect of inclined magnetic field. The non-dimensional governing equations are solved numerically. The simulated velocity and temperature profiles for different values of dimensionless parameters such as M, θ and α for fixed values of Pr and Gr are analysed. First an attempt is made to derive the regression equations for calculating rate of heat transfer as a function of other parameters. The regression coefficients involved in the equation are tested using the ANOVA statistical tool. The optimum value of rate of heat transfer is obtained without numerical calculations. The following observations are noted: Velocity and temperature distributions decreased as the direction of inclined cylinder changes from vertical to horizontal. Velocity and temperature distributions get increased with the angle of inclined magnetic field. Similar observations are made with respect to heat transfer rate. Skin-friction coefficients are decreased with the angle of inclined cylinder. Therefore, the heat transfer rate in the horizontal position is better than that in other positions of the inclined magnetic field. For the present mathematical model, the maximum of $\overline{Nu} = 1.64558$, which occurs for the values of control parameters, $M = 5.5387$, $\mathrm{Pr} = 1$, $\mathrm{Gr} = 55.1096$, $\alpha = 0$ and $\theta = \frac{\pi}{2}$.

References

1. Schlichting, H. (1955). *Boundary-layer theory* (pp. 28–32). McGraw-Hill.
2. Takahashi, Y. & Gokan, Y. (2006). *CFD analysis of air flow of air-cooled motorcycle engines.* SAE paper 2006-32-0005. https://doi.org/10.4271/2006-32-0005
3. Gokan, Y., Takahashi, Y., Inayoshi, M., Ishima, T., & Obokata, T. (2007). *Development of air/oil-cooled motorcycle engine using thermal and fluid analyses.* SAE paper 2007-01-0538. https://doi.org/10.4271/2007-01-0538
4. Sparrow, E. M., & Gregg, J. L. (1956). Laminar free convection heat transfer from the outer surface of a vertical circular cylinder. *Transactions of ASME, 78*, 1823–1829.
5. Lee, H. R., Chen, T. S., & Armaly, B. F. (1988). Natural convection along slender vertical cylinders with variable surface temperature. *Journal of Heat Transfer, 110*, 103–108. https://doi.org/10.1115/1.3250439
6. Goldstein,R. J., & Briggs, D. G. (1964). Transient free convection about a vertical plates and circular cylinders. *Transactions of ASME C: Journal Heat Transfer,* 86, 490–500. https://doi.org/10.1115/1.3688728
7. Velusamy, K., & Garg, V. K. (1992). Transient natural convection over a heat generating vertical cylinder. *International Journal Heat Mass Transfer*, 35, 1293–1306. https://doi.org/10.1016/0017-9310(92)90185-U
8. Ganesan P., & Rani, H. P. (1998). Transient natural convection along vertical cylinder with heat and mass transfer. *Heat Mass Transf*er, *33*, 449–55. https://doi.org/10.1007/s002310050214
9. Agarwal, A. K., Kishor, B., & Raptis, A. (1989). Effect of MHD free convection and mass transfer on the flow past a vibrating infinite vertical circular cylinder. *Warme-und Stoffubetragung, 24*, 243–250. https://doi.org/10.1007/BF01589995
10. Raptis, A., & Agarwal, A. K. (1991). Effect of MHD free convection and mass transfer on the flow past an oscillating infinite coaxial vertical circular cylinder. *Warme-und Stoffubetragung, 26*, 253–256. https://doi.org/10.1007/BF01589995
11. Ganesan, P., & Loganathan, P. (2003). Magnetic field effect on a moving vertical cylinder with constant heat flux. *Heat and Mass Transfer, 39*, 381–386. https://doi.org/10.1007/s00231-002-0383-y
12. Chamkha, A. J. (2011). Heat and mass transfer from MHD flow over a moving permeable cylinder with heat generation or absorption and chemical reaction. *Communications in Numerical Analysis, ID cna-00109*. https://doi.org/10.5899/2011/cna-00109
13. Pedchenko, A., Grants, I. (2005). Instability of rotating magnetic field driven flow in a counter-rotating cylinder. *Physics of Fluids, 17*(104102), 1–7. https://doi.org/10.1063/1.2083967
14. Tagawa, T., Ozoe, H., Sassa, K., & Asai, S. (2010). Convective and diffusive phenomena of air in a vertical cylinder under a strong magnetic field. *An International Journal of Computation and Methodology, 41*, 383–395. https://doi.org/10.1080/104077902753541078
15. Kumar, A., & Singh, A. K. (2013). Effect of induced magnetic field on natural convection in vertical concentric annuli heated/cooled asymmetrically. *Journal of Applied Fluid Mechanics, 6*(1), 15–26
16. Nadeem, S., & Akram, S. (2010). Influence of inclined magnetic field on peristaltic flow of a Williamson fluid model in an inclined symmetric or asymmetric channel. *Mathematical and Computer Modelling, 52*, 107–119. https://doi.org/10.1016/j.mcm.2010.02.001
17. Hayat, T., Farooq, M, & Alsaedi, A. (2015). Inclined magnetic field effect in stratified stagnation point flow over an inclined cylinder. *Zeitschrift für Naturforschung A, 70*(5)a, 317–324. https://doi.org/10.1515/zna-2014-0342
18. Vanita, A. K. (2016). Numerical study of effect of induced magnetic field on transient natural convection over a vertical cone. *Alexandria Engineering Journal, 55*, 1211–1223 (2016). https://doi.org/10.1016/j.aej.2016.04.007
19. Tahari, M., Ghorbanian, A., Hatami, M., & Jing, D. (2017). Physical effect of a variable magnetic field on the heat transfer of a nanofluid-based concentrating parabolic solar collector. *The European Physical Journal Plus, 132*(549). https://doi.org/10.1140/epjp/i2017-11813-1

Selection of Suitable General Circulation Model Outputs of Precipitation for a Humid Tropical Basin

Alka Abraham and Subrahmanya Kundapura

Abstract Climate change has an observed effect on the environment and ecosystem. Climate change study aids in visualisation of changes in the environmental processes as well as development of planning and adopting the strategies. General circulation models (GCMs) are widely used in understanding present and projecting future climate change. The uncertainties associated with the climate projections is a major risk in impact assessment studies and could be reduced with considering GCMs suitable for the region. The study attempts to select a suitable subset of GCMs for precipitation simulation in the humid tropical basin, Achencoil, Kerala, India. Three statistical indicators, correlation coefficient (CC), normalised root mean square deviation (NRMSD) and absolute normalised mean bias deviation (ANMBD), are considered to evaluate the GCMs with the historical observations. The entropy technique is assigned to determine the weight of each performance indicator. Multi-criterion decision-making approaches, compromise programming (CP) and preference ranking organisation method of enrichment evaluation (PROMETHEE-2) are applied individually to rank the GCMs. The ranking is then integrated with group decision-making approach. The GCMs, MIROC-ESM-CHEM, MIROC-ESM, BCC-CSM1-1 and NorESM1-M occupied the first four positions in replicating the historical rainfall in the basin.

Keywords General circulation models · Multi-criterion decision-making · Compromise programming · PROMETHEE-2

A. Abraham (✉) · S. Kundapura
Department of Water Resources and Ocean Engineering, National Institute of Technology Karnataka, Surathkal, Mangaluru 575025, India

S. Kundapura
e-mail: subrahmanyakundapura@nitk.edu.in

A. Abraham
Department of Civil Engineering, Mar Athanasius College of Engineering, Kothamangalam, Kerala, India

A. K. Dikshit et al. (eds.), *Innovative Trends in Hydrological and Environmental Systems*, Lecture Notes in Civil Engineering 234,
https://doi.org/10.1007/978-981-19-0304-5_30

1 Introduction

Climate change is a major threat that the world is facing [1, 2]. This could decline the socio-economic conditions and sustainability of the ecosystem at local, regional and national levels [3, 4]. The effect of climate change on the environment has to be examined, and the most utilised resource for this is the general circulation models (GCMs) [5]. GCMs consist of various models which represent the earth's climate system [6]. These models generate the past climate and project future climate [7]. Now GCMs became the common tools utilised for obtaining climate data [8], in climate change impact studies [9] in various sectors and helps in developing suitable adaptation and mitigation measures for the simulated climate change impacts [10].

The GCMs used for impact studies should be able to represent temporal variability and spatial pattern of the climate at a region [1]. Considering climate scenario, data from all GCMs may not be acceptable and sometimes could lead to increased uncertainties. The uncertainties arise due to various factors and increases when scaled down to local levels [6]. In order to reduce uncertainty in assessing climate change impacts, it is better to select models performing well regionally [5, 11]. The uncertainties in climate projections can be minimised by selecting appropriate set of GCMs for a region. This could further reduce the uncertainties associated with impact assessment and associated adaptation measures [11, 12]. Also, ranking GCMs increases the confidence in using the outputs of GCMs in planning adaptation strategies [6]. Selecting a suitable subset of GCM will play a key role in climate studies and effective water management [13]. The performance assessment of a GCM in simulating historic climate has been considered widely for the selection of GCMs [10]. The assessment helps to obtain better projections by considering an ensemble of selected GCMs from a pool of GCMs [13]. For selecting the GCMs, multi-criteria decision-making approach can be considered as an effective method [14]. In this context, the aim of the present study is to prioritise the suitable set of GCMs based on their ability to simulate precipitation in the Achencoil basin, Kerala. Three statistical performance metrics, correlation coefficient (CC), normalised root mean square deviation (NRMSD) and absolute normalised mean bias deviation (ANMBD) are considered to understand the relation between GCM and IMD gridded dataset. Further entropy method is utilised to determine the weights of each criteria. Two multi-criteria decision-making approaches, compromise programming and preference ranking organisation method of enrichment evaluation (PROMETHEE-2), are utilised followed by group decision-making to rank the GCMs.

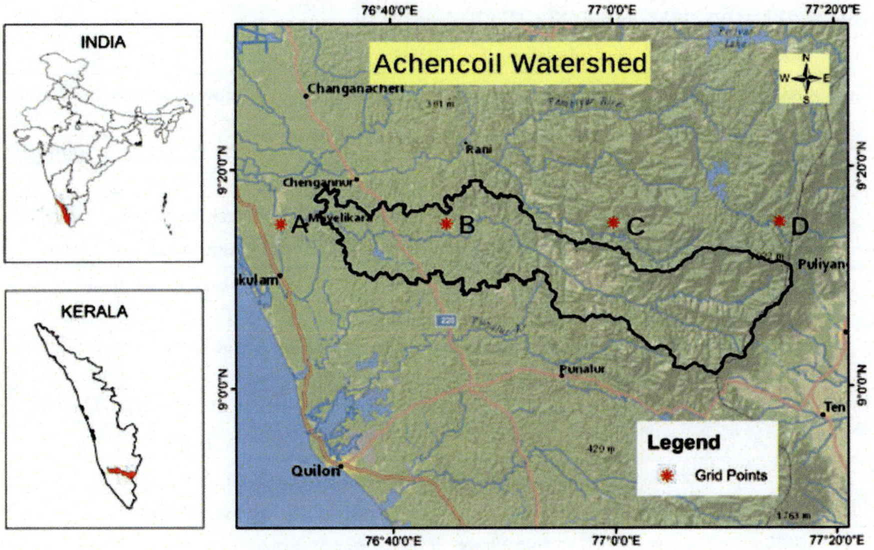

Fig. 1 Location of Achencoil river basin

2 Materials and Methods

2.1 Study Area

Achencoil river basin (Fig. 1), a humid tropical basin located in Kerala, India, is considered for the study. The river is 128 km long, originating from Western Ghats flowing westward and drains into the Vembanad Lake. The basin is located between 76° 24′–77° 18′ E longitude and 9° 2′–9°19' N latitude and having a basin area of 1484 km^2. The area receives an average annual rainfall of above 3000 mm, and temperature varies between 28 °C and 33 °C. The river basin witnessed severe floods in the years 2018 and 2019 due to extreme rainfall events.

2.2 Datasets

The National Aeronautics and Space Administration (NASA) Earth Exchange Global Daily Downscaled Projections (NEX-GDDP) dataset is evaluated in replicating the spatial pattern of precipitation during the time frame 1951–2005 in the Achencoil river basin, Kerala. The NEX-GDDP dataset consists of 21 GCM based on Coupled Model Intercomparison Project 5 (CMIP5) simulations, statistically downscaled and bias corrected with bias correction spatial disaggregation method [15]. Two greenhouse gas emission scenarios, RCP 4.5 and RCP 8.5, are available with a spatial

Table 1 21 CMIP5 GCMs in the NEX-GDDP dataset

S. No.	Model	Modelling Institution
1	ACCESS1-0	Commonwealth Scientific and Industrial Research Organisation and Bureau of Meteorology, Australia
2	BNU-ESM	Institute of Global Change and Earth System Sciences, Beijing Normal University, China
3	CCSM4	National Centre for Atmospheric Research, America
4	CESM1-BGC	National Centre for Atmospheric Research, America
5	CNRM-CM5	Centre National de Recherches Météorologiques, France
6	CSIRO-Mk3-6-0	Commonwealth Scientific and Industrial Research Organisation/Queensland Climate Change Centre of Excellence, Australia
7	CANESM2	Canadian Centre for Climate Modelling and Analysis, Canada
8	GFDL-CM3	Geophysical Fluid Dynamics Laboratory, America
9	GFDL-ESM2G	Geophysical Fluid Dynamics Laboratory, America
10	GFDL-ESM2M	Geophysical Fluid Dynamics Laboratory, America
11	IPSL-CM5A-LR	Institute Pierre-Simon Laplace, France
12	IPSL-CM5A-MR	Institute Pierre-Simon Laplace, France
13	MIROC-ESM	Atmosphere and Ocean Research Institute, Japan
14	MIROC-ESM-CHEM	Atmosphere and Ocean Research Institute, Japan
15	MIROC5	Atmosphere and Ocean Research Institute, Japan
16	MPI-ESM-LR	Max Planck Institute for Meteorology, Germany
17	MPI-ESM-MR	Max Planck Institute for Meteorology, Germany
18	MRI-CGCM3	Max Planck Institute for Meteorology, Germany
19	NorESM1-M	Norway Consumer Council, Norway
20	BCC-CSM1-1	Beijing Climate Centre, China
21	INMCM4	Institute of Numerical Calculation, Russia

resolution of 0.25° × 0.25° in the NEX-GDDP dataset [16]. The GCMs and the respective institutes developed it are shown in Table 1. This dataset helps in understanding the behaviour of climate at finer spatial scales such as urban areas and catchments [17]. This is one of the most accurate datasets based on CMIP5 scenarios available in India for climate change studies [18]. The reference dataset considered for evaluating the GCM is the Indian Meteorological Department (IMD) gridded daily precipitation data of 0.25° × 0.25° resolution [19].

2.3 *Performance Indicators*

The performance of GCMs in simulating the considered climate variables can be evaluated with various metrics [20]. It is considered as a measure of performance of a GCM in simulating the data [6] and helps to determine the confidence of the GCM simulated climate parameters [21]. Various performance indicators are chosen by researchers in evaluating the performance of GCMs [1, 6, 8, 12, 20]. The present study considered three performance indicators; correlation coefficient (CC), normalised root mean square deviation (NRMSD) and absolute normalised mean bias deviation (ANMBD). The evaluations are performed on monthly scale.

Correlation Coefficient (CC)

The indicator, correlation coefficient (CC), estimates the relationship between observed and simulated values [11], and its value varies between −1 and 1 [4]. The indicator, CC, is expressed as,

$$\mathrm{CC} = \frac{\sum_{j=1}^{m} (x_i - \overline{x})(y_i - \overline{y})}{(m-1)(\sigma_{\mathrm{obs}} - \sigma_{\mathrm{sim}})} \tag{1}$$

where x_i is the observed values and y_i is the simulated values, $\overline{x}$ and $\overline{y}$ are the average of the observed and simulated values respectively, m is the number of observations and σ_{obs} and σ_{sim} are the standard deviations of the observed and simulated values.

Normalised Root Mean Square Deviation (NRMSD)

The indicator, NRMSD, considers the mean and standard deviation of the simulated values [8]. The value of NRMSD varies between 0 and positive infinity, and lower values represent best simulation. A smaller value of this indicator represents that the simulated values are closer to the observed values [3]. NRMSD is calculated as,

$$\mathrm{NRMSD} = \frac{\sqrt{\frac{1}{m}\sum_{i=1}^{m} (x_i - y_i)^2}}{\overline{x}} \tag{2}$$

Absolute Normalised Mean Bias Deviation (ANMBD)

The ratio of the mean of the differences between observed and simulated values is given by the indicator ANMBD, and a good agreement of simulated values is represented by smaller ANMBD [14].

$$\mathrm{ANMBD} = \left| \frac{\frac{1}{m}\sum_{i=1}^{m} (y_i - x_i)}{\overline{x}} \right| \tag{3}$$

2.4 Weight Computing Techniques

The performance indicator matrix has to normalise for multi-criterion evaluation. This helps in the transformation of inconsistent indicators to the same space [14]. The normalisation (f_{aj}) is done using the expression,

$$f_{aj} = \frac{F_j(a)}{\sum_{a=1}^{N} F_j(a)} \tag{4}$$

where *Fj*(*a*) is the indicator *j* for the GCM 'a' and *N* is the number of GCMs. For computing weights, entropy method [22] is applied in the study. In this method, the relative importance of the selected indicators [20] and the weights are determined without the interference of a decision maker [6]. The decision can be considered with the influence of three different indicators optimised to the weight of each one [21]. After determining the performance indicators, the payoff matrix is evaluated and then the entropy of each indicator (*Ej*) is calculated as,

$$E_j = -\frac{1}{\ln(N)} \sum_{a=1}^{N} f_{aj} \ln(f_{aj}) \tag{5}$$

$$D_j = 1 - E_j \tag{6}$$

where j is the number of indicator and f_{aj} is the payoff matrix. Then the degree of diversification (D_j) of each indicator is calculated, and the weight of each selected indicator (W_j) can be determined as

$$W_j = \frac{D_j}{\sum_{j=1}^{J} D_j} \tag{7}$$

If the entropy value is high, the diversification is low, so the corresponding indicator is having less importance[6].

2.5 Multi-criterion Decision-Making (MCDM)

MCDM techniques help to arrive at a better alternative for further study from a set of different alternative scenarios [11, 22]. Selecting suitable GCMs with certain statistical metrics and timeframe is challenging as the degree of accuracy varies with the metrics and timeframe selected [1]. So, two MCDM techniques, compromise programming (CP) and preference ranking organisation method of enrichment evaluation (PROMETHEE-2), are considered with the historical evaluation metrics to select a suitable subset of GCMs. The suitable GCMs obtained at various grids are used in the MCDM technique to consider the ability of GCMs in simulating the spatial and temporal variability of rainfall [10].

Compromise Programming (CP)

The method is based on the least distance taken from an ideal value [20] for each alternative. The alternative with the smallest distance obtained is the best for the ideal point [23]. This method identifies a set of GCMs which might satisfy the selected objectives in a compromise manner [12]. The least distance, *Lp* metric is expressed as,

$$Lp = [\sum_{j=1}^{J} Wj|f_j^* - f_j(a)|^p]^{1/p} \tag{8}$$

where f_j^* is the normalised ideal value of the indicator, f_j is the normalised value of the indicator and p is the parameter. The value of p is taken as 1 in this study, as linear measure was considered for determining *Lp* metric. The *Lp* metric value is considered in ranking the GCMs with CP, GCM with lowest *Lp* metric is given the first rank and vice versa.

Preference Ranking Organisation Method of Enrichment Evaluation (PROMETHEE-2)

PROMETHEE-2 is a MCDM approach of outranking nature [24, 25]. This approach considers a preference function, $P_j(a, b)$, which rely upon the pairwise difference $d_j(a, b)$ between $fj(a)$ and $fj(b)$ of GCMs a and b for the chosen indicator j. The study considered usual criterion function, and the preference function can be calculated as

$$P_j(a,b) = \begin{bmatrix} 0 \text{ if } d_j(a,b) \leq 0 \\ 1 \text{ if } d_j(a,b) > 0 \end{bmatrix} \tag{9}$$

Once the preference function is determined, the multi-indicator preference index $\pi(a, b)$ has to be determined, which is defined as the weighted average of preference functions for all the chosen indicators and is expressed as

$$\pi(a,b) = \frac{\sum_{j=1}^{J} W_j P_j(a,b)}{\sum_{j=1}^{J} W_j} \tag{10}$$

$$\Phi^+(a,b) = \frac{\sum_N \pi(a,b)}{(N-1)} \tag{11}$$

$$\Phi^-(a,b) = \frac{\sum_N \pi(a,b)}{(N-1)} \tag{12}$$

$$\Phi(a, b) = \Phi^{+}(a, b) - \Phi^{-}(a, b) \quad (13)$$

The outranking index Φ^+ (a, b), outranked index Φ^- (a, b) and net index $\Phi(a, b)$ have to be determined, and the GCM obtaining higher net ranking indicates the most suitable one.

2.6 Spearman Rank Correlation

The correlation between ranks achieved by a GCM by the two scenarios can be evaluated with Spearman rank correlation coefficient (R) [26], and its value varies between −1 and 1 [24]. R can be determined as,

$$R = 1 - \frac{6\sum_{a=1}^{N} Da^2}{N\left(N^2 - 1\right)} \quad (14)$$

where Da is the difference between the ranks obtained by GCM with different approaches.

2.7 Group Decision-Making

The ranking obtained from various techniques was integrated by group decision-making technique to arrive at a final ranking of GCMs [14]. The ranking obtained from the various techniques is considered to evaluate strength (STa), weakness (WEa) and net strength (NSa) of each GCM considered in the study. The GCM with higher net strength indicates the most suitable GCM.

$$\text{STa} = \sum_{k=1}^{n} \sum_{z}^{X} (X - z + 1) q_{az}^{k} \quad (15)$$

$$\text{WEa} = \sum_{k=1}^{n} \sum_{z=y}^{N} (z - Y + 1) q_{az}^{k} \quad (16)$$

$$\text{NSa} = \text{STa} - \text{WEa} \quad (17)$$

where $X = N/2$ for even number of GCMs; $X = N/2 + 1$ for odd number of GCMs; $Y = X + 1$; z is the position in the upper portion from the first position to the Xth position; k denotes the ranking technique; $q_{az}k = 1$ if the GCM, 'a' is in the position z for the considered ranking technique or 0, otherwise.

3 Results and Discussion

The grid-wise monthly precipitation is considered to rank GCMs using CP and PROMETHEE2 with three statistical performance indicators. The methodology adopted for the study is explained for one sample grid 9.25°N × 76.5°E. Three performance indicators, CC, NRMSD and ANMBD, are evaluated for each of the grids. At the sample grid for the indicator CC, MIROC-ESM-CHEM is the most preferred GCM; with CC value of 0.7117 and least preferred is MRI-CGCM3. For NRMSD, MIROC-ESM-CHEM is the most preferred with a value of 0.7066 and MRI-CGCM3 is the least preferred with a value of 1.0604. The GCM, GFDL-ESM2G was the preferred one with the indicator ANMBD, and MRI-CGCM3 is the least preferred one with ANMBD values of 0.1609 and 0.2058, respectively. After finding the three selected indicators for all the grids, normalisation of indicators is done and weights are assigned to each of the indicators based on the entropy method. The entropy method enables to differentiate the weights of the performance metrics which eventually influence the ranking of GCMs [9]. Among the three statistical indicators, CC is given more prominence with 55% weight assigned, followed by NRMSD with 32.5% and ANMBD with 12.5% for the sample grid. The effect of CC in ranking the GCMs is prominent for the sample grid. The indicators and weights obtained for each GCM vary with grids. The weights obtained for all the grids varied between 6 and 76% for CC, 3.8% to 32% for NRMSD and 6% to 64% for ANMBD. Further, the procedure for applying CP and PROMETHEE2 is considered for all the grids, and group decision technique is applied to arrive at a final ranking of the GCMs suitable for the study area.

3.1 Compromise Programming for Selection of GCMs

The GCMs were ranked initially with the MCDM technique CP for all the grids. Considering the weights assigned with the entropy method and applying the CP, the Lp metric is evaluated for all the GCMs and rankings are assigned at all grid points. The Lp metric is used to rank the GCMs in estimating the monthly rainfall, the GCM which obtain the lowest Lp metric is ranked first and so on. The result of the sample grid point 9.25°N × 76.5°E for CP is shown in Table 2. At the sample grid point minimum and maximum Lp metric are found for the GCMs, MIROC-ESM-CHEM and MRI-CGCM3, respectively. The minimum value obtained is 0.0049, and maximum obtained is 0.2057. The Lp metric for the first four ranked GCMs is 0.0049, 0.0114, 0.0208 and 0.0468, respectively. The least preferred models are ACCESS1-0 and MRI-CGCM3, with *Lp* metric of 0.1982 and 0.2057, respectively. From the analysis, it is noticed that the GCMs, MIROC-ESM-CHEM and MIROC-ESM from Atmosphere and Ocean Research Institute, Japan, BCC-CSM1-1 from Beijing Climate Centre, China and NorESM1-M from Norway Consumer Council, Norway, are ranked as the first four GCMs from the CP method.

Table 2 Indicator values for 21 GCMs (inputs), *Lp* metric value and rank with compromise programming

GCM	CC	NRMSE	ANMBD	*Lp*	Rank
ACCESS1-0	0.4158	1.0501	0.1759	0.1982	20
BNU-ESM	0.6065	0.8115	0.1684	0.0676	6
CCSM4	0.5730	0.8530	0.1812	0.0903	10
CESM1-BGC	0.5549	0.8680	0.1709	0.1014	11
CNRM-CM5	0.5787	0.8433	0.1705	0.0860	8
CSIRO-Mk3-6-0	0.5151	0.9799	0.1971	0.1405	19
CanESM2	0.6102	0.8037	0.1663	0.0646	5
GFDL-CM3	0.5419	0.8647	0.1641	0.1071	13
GFDL-ESM2G	0.5278	0.9226	0.1609	0.1237	17
GFDL-ESM2M	0.5587	0.9464	0.1659	0.1153	16
IPSL-CM5A-LR	0.5498	0.8554	0.1751	0.1018	12
IPSL-CM5A-MR	0.5498	0.8894	0.1666	0.1076	14
MIROC-ESM	0.6952	0.7211	0.1657	0.0114	2
MIROC-ESM-CHEM	0.7117	0.7066	0.1654	0.0049	1
MIROC5	0.5429	0.8840	0.1694	0.1098	15
MPI-ESM-LR	0.6051	0.8223	0.1648	0.0701	7
MPI-ESM-MR	0.5810	0.8515	0.1622	0.0865	9
MRI-CGCM3	0.4034	1.0604	0.2058	0.2057	21
NorESM1-M	0.6387	0.7789	0.1767	0.0468	4
BCC-CSM1-1	0.6795	0.7365	0.1694	0.0208	3
INMCM4	0.5285	0.9457	0.1643	0.1279	18

3.2 PROMETHEE-2 for Selection of GCMs

In the PROMETHEE-2 method, the pairwise difference function for each indicator is calculated. Then the preference function matrix is determined for each indicator. The usual criteria function is considered, so the preference function is taken as 0 or 1 based on the value of the pairwise difference function ≤ 0 and > 0, respectively. Then multi-indicator preference index matrix is determined based on the weights obtained for each indicator by entropy technique. The outranking index, outranked index and net index for each GCM are evaluated. The outranking index is calculated as sum of all elements in a row of multi-indicator preference index matrix divided by the number of elements subtracted 1. The sum of all elements in a column of the multi-indicator preference index matrix divided by the number of elements minus 1 is taken as the outranked index. The difference between the outranking index and the outranked index is taken as the net index. Based on the net index, ranks are assigned to GCMs, higher the net index, more priority is given to that particular model. Table 3 shows the results with PROMETHEE-2 method for the sample grid.

Table 3 Outranking index, outranked index, net index and rank with PROMETHEE-2

GCM	Outranking index	Outranked index	Net index	Rank
ACCESS1-0	0.1407	0.8588	−0.7181	20
BNU-ESM	0.7194	0.2801	0.4393	5
CCSM4	0.5921	0.4074	0.1846	9
CESM1-BGC	0.4637	0.5358	−0.0721	12
CNRM-CM5	0.6221	0.3774	0.2446	7
CSIRO-Mk3-6-0	0.2028	0.7415	−0.5387	17
CanESM2	0.7236	0.4967	0.2269	8
GFDL-CM3	0.2142	0.6749	−0.4607	16
GFDL-ESM2G	0.1644	0.8075	−0.6431	19
GFDL-ESM2M	0.3837	0.6322	−0.2485	14
IPSL-CM5A-LR	0.4472	0.5247	−0.0775	13
IPSL-CM5A-MR	0.3456	0.6263	−0.2807	15
MIROC-ESM	0.8433	0.1286	0.7147	3
MIROC-ESM-CHEM	0.8812	0.0907	0.7904	1
MIROC5	0.6224	0.6694	−0.0470	11
MPI-ESM-LR	0.6392	0.3603	0.2788	6
MPI-ESM-MR	0.5608	0.4387	0.1220	10
MRI-CGCM3	0.1210	0.8785	−0.7575	21
NorESM1-M	0.8496	0.1499	0.6996	4
BCC-CSM1-1	0.8572	0.1362	0.7210	2
INMCM4	0.1999	0.7780	−0.5781	18

The net index obtained for the GCMs varied between −0.7575 and 0.7904 in the sample grid. The highest value is obtained for MIROC-ESM-CHEM and lowest for MRI-CGCM3. The procedure is done for all the grids considered for the study area and follows a similar pattern. The GCMs MIROC-ESM-CHEM, BCC-CSM1-1, MIROC-ESM and NorESM1-M occupied the first four positions by this approach with net index as 0.7904, 0.7210, 0.7147 and 0.6996, respectively.

The correlation between the ranks obtained by CP and PROMETHEE-2 is assessed with Spearman rank correlation. The average value of R is 0.92 with minimum and maximum values being 0.86 and 0.963 for the grids, respectively. The sample grid is found to have the highest Spearman rank correlation coefficient of 0.963. The value obtained shows a good agreement in ranking the GCMs with the selected MCDM approaches.

3.3 Group Decision-Making

MIROC-ESM-CHEM obtained the first position in CP and PROMETHEE-2, MIROC-ESM obtained second position in CP and third position with PROMETHEE-2. The position obtained for the GCM BCC-CSM1-1 was third in CP and second in PROMETHEE-2, and NorESM1-M is placed in the fourth position with both methods. MRI-CGCM3 and ACCESS1-0 are observed as the lowest ranked GCMs with both the MCDM techniques. Considering the results from the MCDM approach, the group decision-making process is applied at each of the grid that covers the study. The strength and weakness of each GCM are determined. From the group decision-making method, it is observed that the net strength obtained for first occupied GCM is 20, followed by 17, 16 and 14 for the second, third and fourth occupied GCMs. The least net strength obtained is −20 for ACCESS1-0 and -19 for GFDL-ESM2G. The net strength as well as the final ranking of GCMs at the sample grid considered is shown in Table 4. Integrating the two MCDM techniques with group decision-making, MIROC-ESM-CHEM, MIROC-ESM, BCC-CSM1-1 and NorESM1-M are

Table 4 Net strength and rank of GCM rank with group decision-making

GCMs	Nsa	Rank
MIROC-ESM-CHEM	20	1
MIROC-ESM	17	2
BCC-CSM1-1	16	3
NorESM1-M	14	4
BNU-ESM	11	5
CanESM2	9	6
MPI-ESM-LR	9	7
CNRM-CM5	7	8
CCSM4	3	9
MPI-ESM-MR	3	10
INMCM4	0	11
CESM1-BGC	−3	12
IPSL-CM5A-LR	−5	13
MIROC5	−6	14
GFDL-CM3	−9	15
IPSL-CM5A-MR	−9	16
GFDL-ESM2M	−10	17
MRI-CGCM3	−12	18
CSIRO-Mk3-6-0	−16	19
GFDL-ESM2G	−19	20
ACCESS1-0	−20	21

the GCMs occupied in the first four positions. The ensemble of these models can be used for climate change and hydrological studies in this basin.

4 Conclusions

The 21 CMIP5 GCMs in the NEX-GDDP data are analysed to rank the GCMs for precipitation in the Achencoil river basin. The three performance indicators adopted, CC, NRMSE and ANMBD helped in improving the evaluation of GCMs, representing the historical precipitation during 1951–2005. The weights for the indicators are assigned with the entropy technique. The CC obtained the higher preference among the indicators in the selected basin. The ranks obtained by CP and PROMETHEE2 are integrated with group decision-making to arrive at final ranking of GCM. The results revealed the better performance of MIROC-ESM-CHEM, MIROC-ESM, BCC-CSM1-1 and NorESM1-M in replicating the historic rainfall. Consideration of more statistical metrics may change the ranking obtained which can be the future scope of the study. Considering multiple GCMs is better for climate change studies, the results of the study help in determining the suitable multi-model ensemble for the climate studies in the region. The multi-model ensemble has to be bias corrected before using in climate impact studies. The study can be considered as an attempt to reduce uncertainty in the projection of precipitation and climate change studies over the Achencoil basin.

Declarations

Funding: No significant financial support for this work that could have influenced its outcome.

Availability of data: Rainfall data downloaded from Indian Meteorological Department (IMD) website and NEX-GDDP data downloaded using Earth Engine Code Editor.

References

1. Nashwan, M. S., & Shahid, S. (2020). A novel framework for selecting general circulation models based on the spatial patterns of climate. *International Journal of Climatology, 40*, 4422–4443. https://doi.org/10.1002/joc.6465
2. Bokhari, S. A. A., Ahmad, B., Ali, J., Ahmad, S., Mushtaq, H., & Rasul, G. (2018). Future Climate Change Projections of the Kabul River Basin Using a Multi-model Ensemble of High-Resolution Statistically Downscaled Data. *Earth System Environ.*, 2, 477–497. https://doi.org/10.1007/s41748-018-0061-y
3. Yoo, C., Cho, E.: Comparison of GCM precipitation predictions with their RMSEs and pattern correlation coefficients. Water. 10, (2018). https://doi.org/10.3390/w10010028
4. Miao, C., Duan, Q., Yang, L., & Borthwick, A. G. L. (2012). On the Applicability of Temperature and Precipitation Data from CMIP3 for China. *PLoS ONE, 7*, 1–10. https://doi.org/10.1371/journal.pone.0044659

5. Ahmadalipour, A., Rana, A., Moradkhani, H., & Sharma, A. (2017). Multi-criteria evaluation of CMIP5 GCMs for climate change impact analysis. *Theoretical and Applied Climatology, 128*, 71–87. https://doi.org/10.1007/s00704-015-1695-4
6. Raju, K. S., & Kumar, D. N. (2014). Ranking of global climate models for India using multicriterion analysis. *Clim. Res., 60*, 103–117. https://doi.org/10.3354/cr01222
7. Srinivasa Raju, K., & Nagesh Kumar, D. (2020). Review of approaches for selection and ensembling of GCMS. *J. Water Clim. Chang., 11*, 577–599. https://doi.org/10.2166/wcc.2020.128
8. Shi, F., Wang, Z., Qi, L., Chen, R.: An assessment of GCM performance at a regional scale using a score-based method. Adv. Meteorol. 2018, (2018). https://doi.org/10.1155/2018/7641019
9. Raju, S. K., Sonali, P., & Nagesh Kumar, D. (2017). Ranking of CMIP5-based global climate models for India using compromise programming. *Theoretical and Applied Climatology, 128*, 563–574. https://doi.org/10.1007/s00704-015-1721-6
10. Shiru, M. S., Shahid, S., Chung, E. S., Alias, N., & Scherer, L. (2019). A MCDM-based framework for selection of general circulation models and projection of spatio-temporal rainfall changes: A case study of Nigeria. *Atmospheric Research, 225*, 1–16. https://doi.org/10.1016/j.atmosres.2019.03.033
11. Panjwani, S., Naresh Kumar, S., Ahuja, L., & Islam, A. (2019). Prioritization of global climate models using fuzzy analytic hierarchy process and reliability index. *Theoretical and Applied Climatology, 137*, 2381–2392. https://doi.org/10.1007/s00704-018-2707-y
12. Salman, S. A., Nashwan, M. S., Ismail, T., & Shahid, S. (2020). Selection of CMIP5 general circulation model outputs of precipitation for peninsular Malaysia. *Hydrology Research, 51*, 781–798. https://doi.org/10.2166/NH.2020.154
13. Hassan, I., Kalin, R.M., White, C.J., Aladejana, J.A.: Selection of CMIP5 GCM ensemble for the projection of spatio-temporal changes in precipitation and temperature over the Niger Delta, Nigeria. Water. 12, (2020). https://doi.org/10.3390/w12020385
14. Raju, K.S., Kumar, N.: Impact of climate change on water resources. (2018)
15. Thrasher, B., Maurer, E. P., McKellar, C., & Duffy, P. B. (2012). Technical Note: Bias correcting climate model simulated daily temperature extremes with quantile mapping. *Hydrology and Earth System Sciences, 16*, 3309–3314. https://doi.org/10.5194/hess-16-3309-2012
16. Sajjad, H., & Ghaffar, A. (2019). Observed, simulated and projected extreme climate indices over Pakistan in changing climate. *Theoretical and Applied Climatology, 137*, 255–281. https://doi.org/10.1007/s00704-018-2573-7
17. Yu, R., Zhai, P., Lu, Y.: Implications of differential effects between 1.5 and 2 °C global warming on temperature and precipitation extremes in China's urban agglomerations. Int. J. Climatol. 38, 2374–2385 (2018). https://doi.org/10.1002/joc.5340
18. Singh, V., Jain, S.K., Singh, P.K.: Inter-comparisons and applicability of CMIP5 GCMs, RCMs and statistically downscaled NEX-GDDP based precipitation in India. Sci. Total Environ. 697, 134163 (2019). https://doi.org/10.1016/j.scitotenv.2019.134163
19. Pai, D.S., Sridhar, L., Rajeevan, M., Sreejith, O.P., Satbhai, N.S., Mukhopadhyay, B.: Development of a new high spatial resolution (0.25° × 0.25°) long period (1901–2010) daily gridded rainfall data set over India and its comparison with existing data sets over the region. Mausam. 65, 1–18 (2014)
20. Sreelatha, K., & Anand Raj, P. (2019). Ranking of CMIP5-based global climate models using standard performance metrics for Telangana region in the southern part of India. ISH J. Hydraul. Eng. 1–10 (2019). https://doi.org/10.1080/09715010.2019.1634648
21. Kheireldin, K., Roushdi, M., & Aboelkhear, M. (2020). Selection of the optimum global circulation model that mimics the circumstances of Egypt. *International Journal of Recent Technology an Engineering, 9*, 784–793 (2020). https://doi.org/10.35940/ijrte.f1229.059120
22. Pomerol, J. C., & Romero, S. B. (2000). *Multicriterion decision in management: Principles and practice.* Kluwer Academic.
23. Zeleny, M. (2011). Multiple criteria decision making (MCDM): From paradigm lost to paradigm regained? *Journal of Multi-Criteria Decision Analysis, 18*, 77–89. https://doi.org/10.1002/mcda.473

24. Raju, K. S., & Kumar, D. N. (1999). Multicriterion decision making in irrigation planning. *Agricultural Systems, 62*, 117–129. https://doi.org/10.1016/S0308-521X(99)00060-8
25. Brans, J. P., Vincke, P., & Mareschal, B. (1986). How to select and how to rank projects : The PROMETHEE method. *European Journal of Operational Research, 24*, 228–238.
26. Gibbons, J. D. (1971). *Nonparametric statistical inference*. McGraw-Hill.

Extraction of Water Surface Bodies for Pakhal Lake, India Using GEE

Ashok Amgoth, H. P. Rani, and K. V. Jayakumar

Abstract Water is an essential natural resource that creates the foundation for all life on earth. It is therefore crucial that our water supplies are maintained and managed. To address these challenges, an attempt is made in this paper to study the changes in water bodies using the Google Earth Engine (GEE) algorithms. The chosen study area is the Pakhal lake situated in the Warangal district of Telangana State, India. The NDVI and NDWI indices are used in extracting water bodies over the lake. The harmonic time series model is used to calculate the missing data and to reduce the noise in the data. The calculated suitability index, based on the area and perimeter of the lake, is below 0.18, and hence, this lake can be strongly categorized as a wetland. The high-resolution multispectral satellite imagery (Sentinel-2) is used for change detection of the lake using the spectral water indices. The image classification of Pakhal lake is carried out using the Landsat images. Random forest image classification is performed to estimate the training and the validation accuracy based on the Kappa coefficient. The water change intensity and seasonality water change over the wetland for the period 1984 to 2020 are studied using the Global Water Occurrence Explorer (GWOE). The study found significant changes in pixel range with the presence of water across the study region, as well as changes in location in terms of seasonality and persistence.

Keywords Water bodies · GEE · Suitability index · NDVI · NDWI · Harmonic model · Image classification

A. Amgoth (✉) · K. V. Jayakumar
Department of Civil Engineering, National Institute of Technology, Warangal, India
e-mail: ashoka6@student.nitw.ac.in

H. P. Rani
Department of Mathematics, National Institute of Technology, Warangal, India

A. K. Dikshit et al. (eds.), *Innovative Trends in Hydrological and Environmental Systems*, Lecture Notes in Civil Engineering 234,
https://doi.org/10.1007/978-981-19-0304-5_31

1 Introduction

Water, the vital natural resource, forms the basis for life, and hence it is essential to manage and maintain the water resources around the globe. Remote sensing (RS) techniques are widely employed in estimating the fluxes such as evapotranspiration, the meteorological variables like precipitation and temperature, and the hydrological state variables like land surface characteristics and soil moisture and in delineating the surface water bodies. For the past three decades, the availability of RS data with better spectral and temporal resolutions has led to many regional-scale studies. This vast amount of data is used in extracting and detecting surface water areas and analyzing their dynamics.

The land use/ land cover (LULC) change detection in the study related to upper Citarum Watershed in the West Java Province of Indonesia, showed that water body in this region got decreased by 12% during the period starting from 1997 to 2005, and 10% during the period from 2005 to 2014 [1]. A simple geospatial model was developed by Fang et al. [2] using GIS. As an alternative method for spatial analysis, this study focused on the effect of groundwater hydrodynamics on water quality. Multitemporal satellite images are processed using the post-classification comparison methods and object-based image analysis to track the changes in trajectories and used for landscape metrics analysis [3]. These methods showed the overall accuracies of 76.64% in the year 1972, 83.52% in 1987, and 88.84% in 2013 with Kappa statistics of 69%, 78%, and 86%, respectively. Using the multispectral satellite images from Landsat 2, 5, and 8, the land use/land cover changes of the Halda watershed over 40 years were studied. This data facilitated the implementation of the best land management option for the Halda watershed by planners, policy makers and associated development workers [4]. The land use in the Yangtze River delta between 2000 and 2017 has been classified by visual interpretation of Landsat's satellite images. [5]. In this region, it was noted there was a rapid expansion, especially in the core area with a significant increase in the built-up extents. The evolution of urbanization between 1963 and 2005 in major Lebanese cities was analyzed by processing and interpreting the satellite images and topographical maps which were acquired from the different space platforms [6].

The studies carried out by Du et al. [7] related to the spatial impact of urban expansion on surfaces of water bodies aim to raise the issues of surface water management in urban development and spatial planning processes. The Yangtze river basin's water bodies were identified and extracted using RS techniques. [8]. These techniques were used to classify the river basin into three major categories, such as rivers, natural lakes, and artificial reservoirs. It was identified that the reservoir construction led to the shrinkage of natural lakes. The time series data of the land cover assessed using the confusion matrices exhibited the land cover errors and they were spatially auto-correlated [9]. Hence, the spatio-temporal errors influence environmental monitoring, change detection, and modeling activities. The principal component analysis was employed in extracting the details about multi-temporal spatial accuracy. This spatial approach provides more accurate and informative measures, hence, allowing users and map producers to evaluate the time series thematic maps more comprehensively.

Tracking the spatio-temporal changes in the inundation patterns of water bodies remains a challenge for RS studies. In general, processing of downloaded RS images from the enormous long time series information is time-consuming since it may involve thousands of gigabytes of data. Due to the free availability of MODIES / Sentinel/Landsat series data, this load of information has been chronicled by Google and is connected with the Google cloud. The Google Earth Engine (GEE) is an online distributed computing programming apparatus for planetary-scale environmental information examination. In GEE online algorithm, the information on big data investigation in RS is used with high effect empowered data-driven science, and considerable advancement will be accomplished dependent on the global challenges including huge geospatial datasets. GEE is intended for processing and storing the informational index images at the petabyte scale.

Vegetation indices obtained from the earth observation satellites play a vital role in applications ranging from vegetation monitoring, agricultural activities, drought studies, hydrologic and climate modeling. The difference between the reflectance radiated in the visible red and near-infrared wavebands will explore the stress or health of plants and the density of plant growth. The Enhanced Vegetation Index (EVI) and the Normalized Difference Vegetation Index (NDVI) are two commonly used vegetation indices for monitoring vegetation on a local and global scale [10, 11]. At relatively high spatial resolution, EVI provides near-real-time data over large areas and it has been widely used in monitoring the vegetation condition.

Landsat satellite images cover the multispectral and thermal data, about once in every two weeks over the Earth's surface at a 30-m resolution. Remotely sensed images are in the form of raw data. The image classification technique is used to analyze the image. In this technique, the information is extracted within the pixel of the image raster and then it is transformed as a thematic map. The general practice in classifying the image is to understand the land cover within the study region. This classification is further subdivided as unsupervised and supervised classification. In the supervised classification, with the assistance of the analysts, the training samples are processed, while in the unsupervised classification, the spectral classes or clusters are obtained without the analyst's intervention. Classification of an image can be executed in different GIS platforms, like, QGIS, ArcGIS, ENVI, etc. From the high-resolution Landsat imagery, extraction and detection of the surface water body area using the spectral water indexing methods, like, NDWI, NDVI, NDMI, and MNDWI using the unsupervised classification were carried out [12].

To summarize, the considered classification model along with the errors can be analyzed using the confusion matrix. For characterizing quantitatively the image classification accuracy, the confusion/error matrix is employed. It gives the relationship between the reference image and the characterization result. The present study employs the GEE platform for extracting the RS data with the focus on the spatial and temporal dynamics of Pakhal lake situated in India with the following objectives:

i. Identify the presence of vegetation and water over the lake using GEE based on NDVI and NDWI;
ii. Estimate area, perimeter and identify the permanent water bodies present in this region;
iii. Calculate the SI using the area and perimeter of the lake over the years and perform harmonic time series analysis for missing data.
iv. Analyze the temporal distribution of water surface between 1984 and 2020.
v. Carry out change detection using the Canny edge detection and Otsu thresholding.
vi. Perform the supervised image classification with overall accuracy and validate the accuracy by utilizing the Kappa coefficient.

2 Study Area

Pakhal lake is an artificial lake constructed during the Kakatiya dynasty in the year 1213 A.D., across the Munneruvagu stream situated near Ashoknagar village of Khanapur Mandal of Warangal district, Telangana, India. The project, having a catchment area of 271.95 km^2, was commissioned as a medium irrigation project during the year 1967. This lake, situated among afforested valleys and undulating hills, is a popular tourist retreat with Pakhal Wild Life Sanctuary spread over an area of 839 km^2. The main problem identified over the lake is its continuous shrinking in the size over the years. Figure 1a, b show the Landsat image of Pakhal lake taken in the years 2014 and 2020.

2.1 Collection of Data

The NDWI and NDVI time series between 2013 and 2020 of the Landsat 7 and 8 image Collection 1 Tier 1 calibrated Top of Atmosphere (TOA) reflectance is used. Table 1 provides details of the Landsat data. Bands 1 to 7 are used in the study with an original spatial resolution. The GEE contains Landsat specific processing methods for calculating the radiance at the sensor, TOA, surface reflection (SR), cloud score and composites without any clouds. GEE also includes a basic cloud scoring algorithm, ee.Algorithms. Landsat. simpleCloudScore can be used to identify Landsat pixels based on their relative cloudiness. The processing of Landsat data is done on this platform using cloud computing technology. Big data processing and parallel computing are now possible in the research area.

(a)

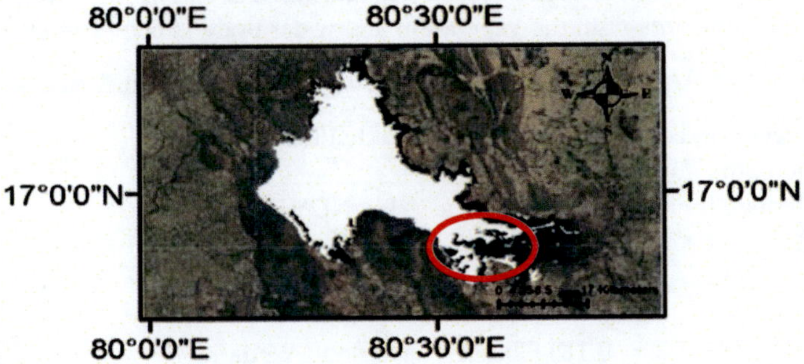

(b)

Fig. 1 Landsat image of Pakhal lake in the years **a** 2014 and **b** 2020

Table 1 Description of landsat data

ID	Description
LANDSAT /LC08 /C01 /T1_RT	Landsat 8, collection 1, tier 1 + real time
LANDSAT /LE07 /C01 /T1_TOA	Landsat 7, collection 1, tier 1

3 Methods

3.1 Methodology

The study of the temporal variations of NDWI and NDVI values employs an index-based classification methodology. The Landsat images of Pakhal lake are used in the GEE platform with JavaScript. The multi-decadal patterns are recorded along with the temporal changes in the lake region. Based on the literature [13], an algorithm is created to map the Pakhal lake based on its unique features such as the annual phonological land cover change of the uncovered soils, plants, and shallow flooding water. Due to the general flooding of shallow water bodies during the ante-tillering stage, the NDVI values are greater than zero but less than the NDWI values. But in the conceptive and maturing stages, the NDVI values are positive and greater than NDWI value. These NDVI values are greater than the NDWI with negative NDWI values, in the post-harvest wetland planting stage.

To mask the wetland utilizing an automatic cloud masking strategy, the cloud-free Landsat 8 TM images of late May and early June are captured. The output information depends on the contrast between the NDVI and NDWI values. Any pixel covered by the vegetation has a connection between NDVI and NDWI. The chosen images of wetland plants are in the vegetative development, conceptive and aging stages (sprouting period), and they are described as $0 < (\text{NDVI–NDWI}) < 1$. The permanent water bodies have a relationship $-1 < (\text{NDVI–NDWI}) < 0$. Based on the algorithm proposed in [13], for the summer season i.e., late April to early May during the years 2014 to 2020, the area with the vegetation and water body is determined.

GEE Workflow Algorithm Proposed For Mapping Of Wetland Areas

(a) Extract Landsat 7 and 8 Images of Pakhal lake
(b) Calculate NDVI and NDWI
(c) Calculate the water covered area: $-1 < \text{NDVI-NDWI} < 0$
(d) Extract $-1 < \text{NDVI} < 0$
(e) Calculate $\sqrt{\text{Area/ Perimeter}}$
(f) If (SI > 0.18) Classify the rivers and ponds
Else If (SI < = 0.18) Classify Paddy and Wetlands
(g) Display the result

3.2 NDVI and NDWI

The NDVI is a key metric for quantifying green vegetation because it normalizes green leaf scattering in near-infrared wavelengths. It accounts for chlorophyll absorption in red wavelengths and the diffraction of green leaves in infrared wavelengths. The NDVI is calculated from $\text{NDVI} = \frac{\text{NIR}-\text{Red}}{\text{NIR}+\text{Red}}$, where NIR and red denote the spectral reflectance measurements obtained in the near-infrared and red visible regions, respectively. The values of NDVI range from -1 to 1 and for green vegetation, this value is between 0.2 and 0.8. Negative NDVI corresponds to water. For barren areas of sand, rock or snow regions, NDVI values range approximately between -0.1 and 0.1. Low positive values (~0.2–0.4) represent grassland and shrub but the high values (approaching to 1) indicate the tropical and temperate rainforests.

The NDWI is used in monitoring the changes concerning water content present in water bodies [14]. Because water absorbs the infrared electromagnetic spectrum, the NDWI uses the green and near-infrared bands to highlight water bodies.. This index is less sensitive to atmospheric scattering effects than the NDVI and is an indicator of vegetation liquid water content, rather than atmosphere reflectivity. NDWI is calculated using $\text{NDWI} = \frac{\text{NIR}-\text{SWIR}}{\text{NIR}+\text{SWIR}}$, where NIR and SWIR represent the reflectance of Bands 5 and 6 used in Landsat 8 and Bands 4 and 5 used in Landsat 7, respectively.

NDWI can be employed in exploring the water content at a single leaf level also at the canopy/satellite level.

3.3 Image Classification

The image classification is done using the training samples of a polygon or shapefile class. The organization of the feature class characteristic must be coordinated while preparing the sample attributes. To confirm, the reference dataset using the 'Training Samples Manager' is created to read and write the dataset. The overall assessment of the accuracy of the classification is computed based on the following Kappa's statistics, $\hat{\kappa}$

$$\hat{\kappa} = \frac{N \sum_{i=1}^{n} m_{i,i} - \sum_{i=1}^{n} (G_i C_i)}{N^2 - \sum_{i=1}^{n} (G_i C_i)} \quad (1)$$

where i and N denote the class number and the total number of classified values, respectively; $m_{i,i}$ denotes values found along the diagonal of the matrix; C_i denotes the sum of each row values belonging to class i; and G_i denotes the sum of each column values belonging to class i.

Supervised image classification is the classifier package that is handled by the traditional machine learning algorithms running in GEE. These classifiers include random forest, CART, naïve Bayes, and support vector machine (SVM). The following procedure was performed over the Pakhal lake through random forest image classification:

- Collecting the training data and assembling the features, which store the realized class name and numeric qualities for predictors.
- Instantiate classifier and set its boundaries.
- Train classifier utilizing the above training data. Classify an image or feature collection with a property for storing the class label with consecutive integers starting from zero and also properties of predictor variables, which are of numeric type.
- Using independent validation data, calculate the classification error.

3.4 Global Surface Water Explorer

The Global Surface Water Explorer (GSWE) deals with the water dataset and the temporal distribution of water surfaces at a global scale. It also deals with statistics on the degree and change of water surface, seasonal water occurrence, and the varieties in ingenuity between change intensity and transition among permanent water bodies. GEE collaborates and quantifies the changes in GSWE by interacting maps that highlight the changes in earth surface water over 36 years. Pakhal lake water dataset is gathered by utilizing the Landsat 8 image collection during the period beginning from 1984 to 2020. The Global Surface Water Occurrence (GSWO) map gives the data on water elements by capturing both intra and inter-annual changes and their variability. As a single product, the frequency of water on the surface between March 1984 and October 2020 was captured.

3.5 *Water Occurrence Change Intensity (WOCI)*

Based on WOCI calculations, the temporal changes in surface water occurrence are shown. During the years 1984 to 2020, the directions of change, such as decrease, increase, or no change, as well as their intensity, were determined. This variation is calculated using data collected over time, as well as seasonal variations and valid observation frequency. This intensity is calculated using a homologous pair of months, i.e., the same months with valid perceptions in both time frames, and the difference is expected due to the presence of surface water. The average of these distinctions prompts the surface water occurrence change intensity. This information additionally gives a steady estimation of the event of change.

Using the above methods, the extraction and change detection over the Pakhal lake is carried out.

4 Results

For the period between 2013 and 2020, the temporal characteristics of NDVI using the Landsat 7 and 8 images over the Pakhal lake are shown in Fig. 2. These NDVI values lie between the average of 0.2 and 0.7 and they depict the presence of thick

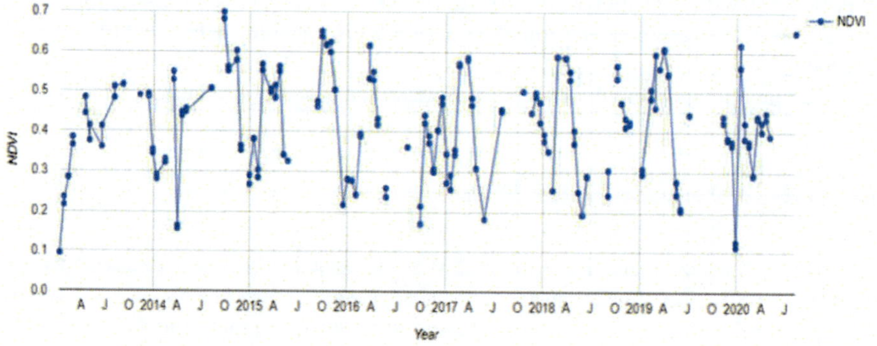

Fig. 2 NDVI data using landsat 8 images from 2013 to 2020

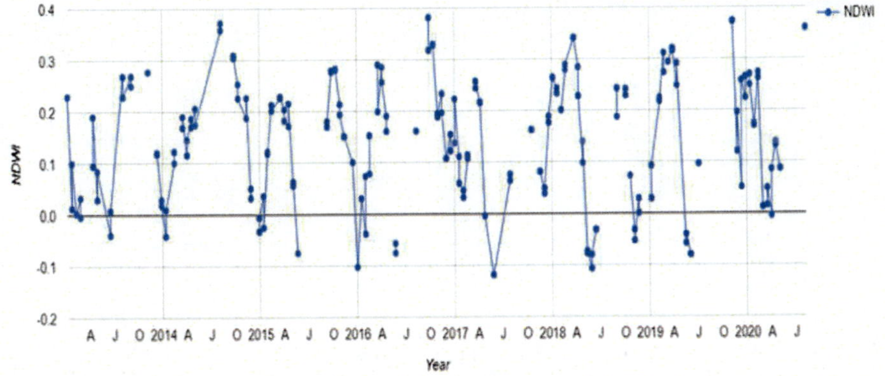

Fig. 3 NDWI data using Landsat 8 images from 2013 to 2020

vegetation shade. This normalized data is also continuous. The density of green on a patch of land is determined by the distinct colors of visible and near-infrared sunlight reflected by the plants. The NDVI values, around the range of 0.2 and 0.5, address the shrubs and senescing crops or grasslands [15]. The higher values of NDVI refer to healthy and dense vegetation, while the lower values show sparse vegetation. The increase in leaf widths and vegetation growth has resulted in high NDVI values over the Pakhal lake.

The surface water body in terms of NDWI for the period beginning from 2013 to 2020 is shown in Fig. 3. The NDWI values typically range from −0.1 to + 0.35. High and low NDWI values correspond to high and low plant water content and vegetation content, respectively. Low NDWI values indicate low vegetation water content and portion cover, both of which indicate water stress. It can be noted that in the years 2016 to 2019, the NDWI values are negative (−0.1) and this standardized NDWI data is continuous.

The continuous NDVI data over the lake using Landsat 8 image collection, with a smooth curve is shown in Fig. 4. The harmonic time series data with decreased noise along with the fitted data are displayed in this figure. As far as land cover classes are concerned, positive and negative patterns are observed in barren and sparsely vegetated, grassland and forest areas. Negative patterns are also present in open shrublands. The presence of permanent water bodies and dense vegetation is indicated by NDVI values ranging from 0 to 0.57. The average fitted NDVI values are observed to be within the range of 0.1 to 0.18 with a uniform smooth curve. This shows that NDVI values during the study period are significant with a continuous sequence of observations and the quantitative estimation of vegetation growth. The size of fitted information has a consecutive uniform pattern and is seen to be more deterministic.

Based on the proposed algorithm, Table 2 shows the vegetation areas of Pakhal lake. It can be observed that these areas got increased during the years 2014, 2015, and 2019. According to the characterization of wetlands, the calculated SI values

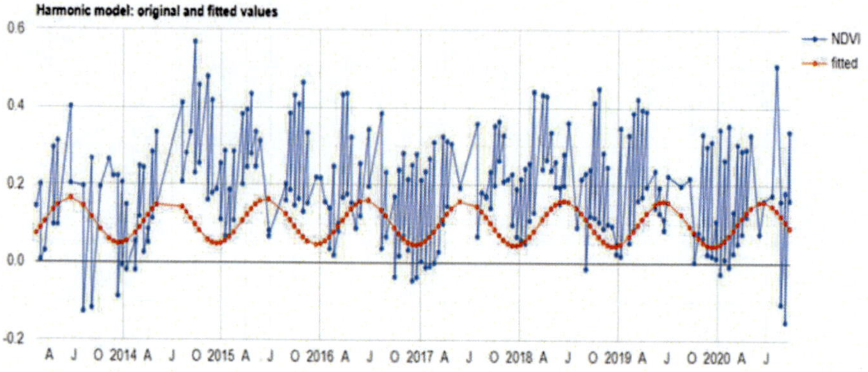

Fig. 4 Harmonic time series using landsat 8 image collection and observed data of NDVI

Table 2 Water spread area of Pakhal lake

Year	Vegetation Area (m^2) 0 < NDVI-NDWI < 1	Waterbody −1 < NDVI < 0.3		
		Area (m^2)	Perimeter (m)	SI = $\sqrt{\text{Area}}/\text{Perimeter}$
2013	217,057.07	95,269.4	2277.0	0.135
2014	272,763.28	98,175.3	3320.8	0.094
2015	272,993.72	91,918.2	1616.04	0.187
2016	271,881.71	89,087.6	13,317.4	0.022
2017	271,961.77	101,172.0	4641.3	0.068
2018	272,726.54	90,787.9	5579.6	0.054
2019	272,566.22	82,240.4	1209.4	0.237
2020	267,780.94	95,037.4	7502.8	0.041

have to be below 0.18 as suggested in [13], and so, the Pakhal lake can be considered under the category of wetlands. In the year 2015, the SI value slightly increases to 0.187 and this represents the lake as the paddy field and wetland. While in 2019 this index drastically increased to 0.237, classifying the lake as the river and pond. This is because there is an increase in rainfall in the year 2019. Further to analyze the wetland characteristics, image classification is performed using the supervised classification.

Supervised image classification is performed over the Pakhal lake and the results are shown in Fig. 5. A classifier's accuracy is determined using the confusion matrix. It depicts how the classification result and the reference image are similar. This matrix necessitates the use of ground truth data such as cartographic data and the manual digitization of an image result. To ensure this, a reference dataset is created, which is read and written using the 'training samples manager'. The Kappa's statistical of agreement provides an overall assessment of the classification's accuracy. Table 3 shows the statistical values, with accuracy ranging from 0 to 1, with 1 representing 100 percent accuracy.

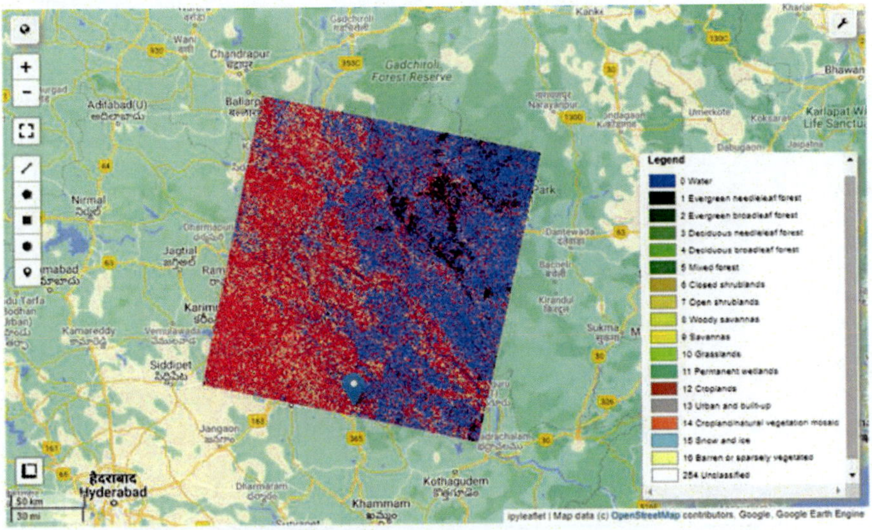

Fig. 5 Supervised image classification

Table 3 Image classification using landsat 7 and 8 image collections over the Pakhal lake for the year 2014–2020

	Landsat 7	Landsat 8
Overall accuracy	0.92	0.95
Validation accuracy	0.80	0.82
Kappa coefficient	0.85	0.87

Figure 6 shows the raw image of Pakhal lake using a high-resolution multispectral image of Sentinel-2. This imagery layer can be used for multiple purposes including but not limited to the applications like forest monitoring, precision agriculture, and vegetation stress detection. The change in water volume can be detected by the change in the shape of the lake and water surface. The edges of Pakhal lake are detected using Canny's method.

Figure 7 shows the edge change detection of Pakhal lake using the Canny edge detection method and the threshold is computed using the Otsu method by clipping the image edges. The pixel that detected the edge is shown in white color and the false pixel edge is shown in red color. The large number of edges getting detected does not imply that a good result is obtained because some of these pixels may be falsely detected edges. Otherwise, getting detected a too-small number of edges is also not good, since it implies that many edge pixels are not rightly detected as the edge. Some false edges are derived from the texture of vegetation around the Pakhal lake. Most of the vegetation and their shadows may have the same color. Some lake edges can be detected well, but some objects or vegetation which have the same color as the lake give the negative impression. False detection gets minimized by adjusting the kernel size and sigma values. These results show the continuous observation of

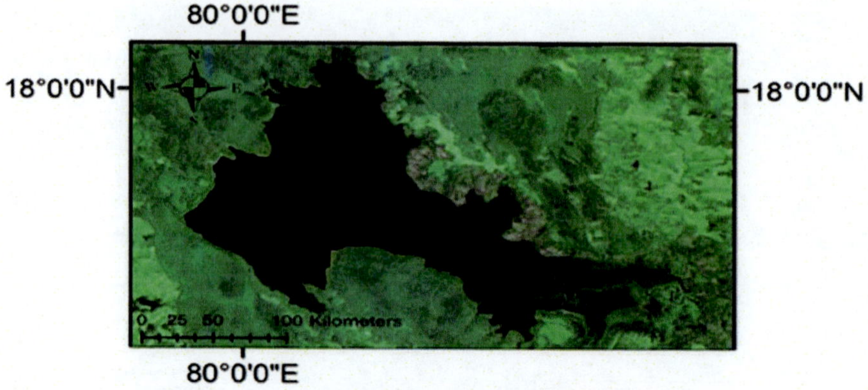

Fig. 6 Raw image of Pakhal lake using Sentinel-2 multispectral imagery from 2015–2020

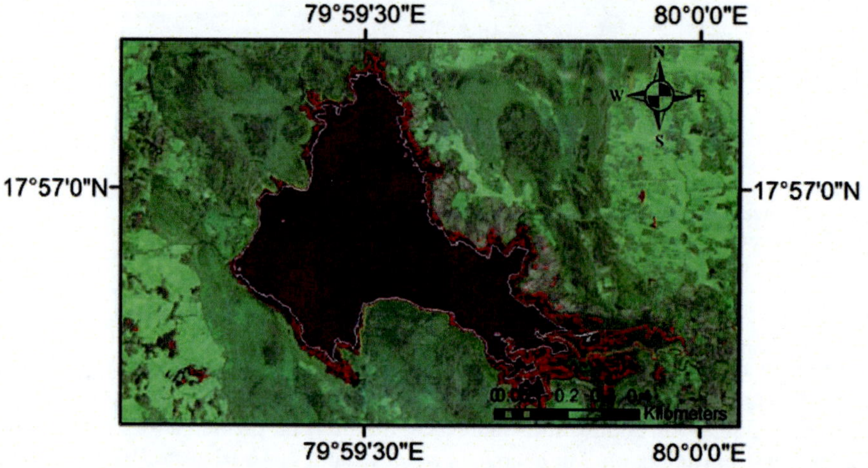

Fig. 7 Change detection over the Pakhal lake using Sentinel-2 multispectral imagery from 2015–2020

the Pakhal region which show a clear distinction between the land and water over the study period.

Waterbody extraction is a vital part of water resource management and has been performed over the Pakhal lake for the period from 2015 to 2020 and is shown in Fig. 8. The NDWI analysis shows that the method successfully extracts the water body areas over the lake using a Sentinel-2 image. The Pakhal region captures the spatial and temporal variations of surface water. Figure 9 depicts the occurrence of surface water between 1984 and 2020. This offers information on the overall water dynamics by taking into account water frequency and occurrence in monthly stages on the surface of the land. The method captures changes in both inter and

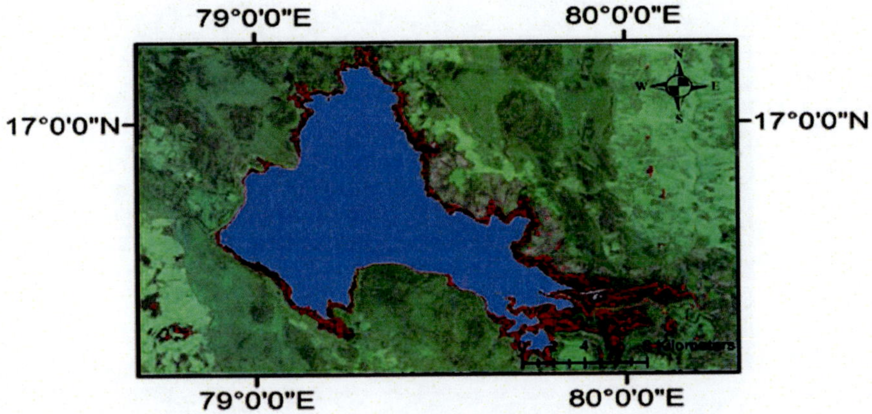

Fig. 8 Extraction of surface water bodies over the Pakhal lake using Sentinel-2 imagery from 2015–2020

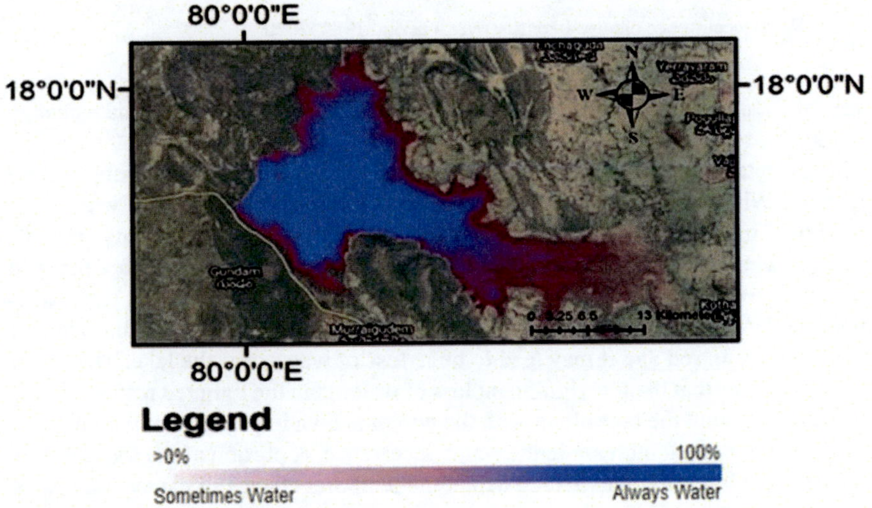

Fig. 9 Water occurrence over the period 1984–2020

intra-annual variability. Some locations are underwater: (i) for the duration of the observation period, (ii) for a few months each year, and (iii) on an episodic basis. Some areas have never been submerged, with a water content of 0%. The water accessibility on the east side and all edges of the Pakhal lake are exceptionally less in examination with different regions. This decrease in recurrence with shifting tones from pink to purple, 0 < occurrence < 100% shows there is an adjustment of the water body region concerning space and time. The outcomes show the water events in the Pakhal lake are essentially diminishing both spatially and temporally. The lowering

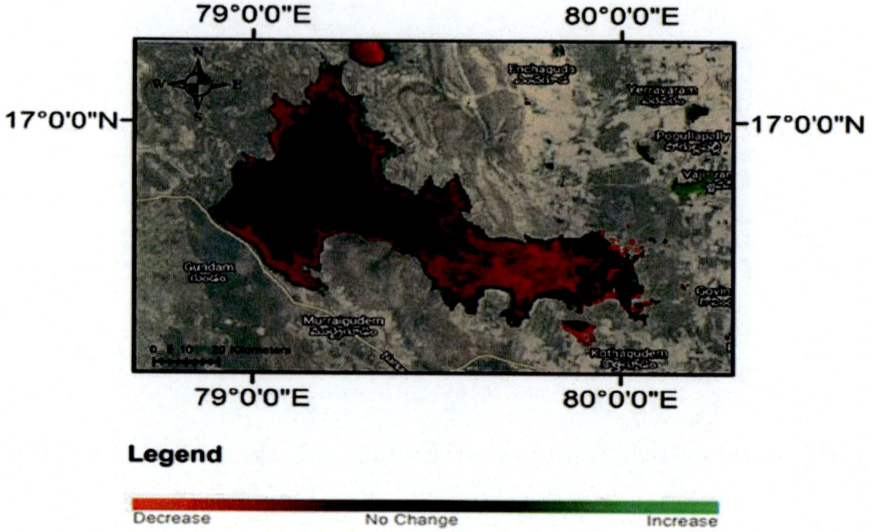

Fig. 10 Water occurrence change intensity over a period 1984–2020

levels of water in lakes and other water bodies are good indicators of climate change (Fig. 9).

In percentage terms, Fig. 10 depicts the degree of water occurrence change intensity map. With a tolerance of ± 15 percent, the detected water areas with equal occurrence in both periods are depicted in black on the map. The study period's locations where surface water occurrence has decreased and increased are mapped in red and green, respectively.. In these two cases, the brighter tone shows substantial changes over the lake. The water body changes concerning the time in the east side of the lake are in red and it may lead to huge loss of water over the lake. The bright red tone implies that there is significant loss of water than the light red regions. Some of the areas around the lake along with the border shown in green color which imply that there is a gradual increase in the water occurrence. A portion of the space is dark because there is insufficient data to depict the temporal change in the availability of water in these areas. However, the outcomes show that there is a substantial increase in water levels all around the Pakhal lake.

Figure 11 shows the differences in surface water classes in the form of a transition water map. There are three types of water bodies: no water, seasonal water bodies, and permanent water bodies. The water spread areas over the Pakhal lake are computed with the transition changes and illustrated in Fig. 11a. The long-term temporal profiles of the study region are recorded in each pixel to note the presence or absence of water during the period. The water transition increased during the rainfall seasons and very little variation is noticed in the summer season.

The transitions in the study region were mapped and are depicted as a pie chart in Fig. 11b. Seasonal variation of water surface area frequently occurs in the lake

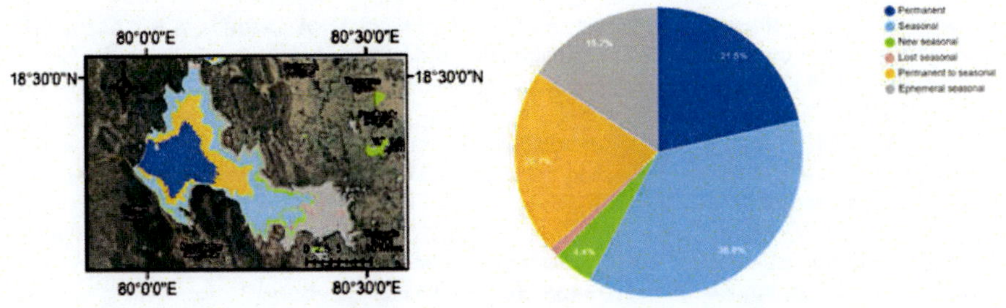

Fig. 11 **a** Transition class over a Pakhal lake and **b** Pie chart of transition class

and lost seasonal area is abruptly increased. This quantified study of water transition over the Pakhal lake suggests that loss of water body should be prevented and lake should be preserved and improved.

5 Conclusion

The NDVI and NDWI indexes of Pakhal lake were determined from the Landsat 8 images utilizing the Google Earth Engine (GEE) over period 2014 to 2020. To obtain a smooth curve, missing data were filled using the harmonic time series by decreasing the noise in the Landsat images. The computed results of fitted NDVI information using Landsat images showed a uniform smooth curve over the investigated time frame. Supervised image classification utilizing random forest showed better precision with a reasonable Kappa coefficient of 0.85. In this case study, the surface water mask is computed using the Canny Edge detector, Otsu thresholding and high-resolution multispectral satellite imagery from 2015 to 2020. Using the Global Surface Water Explorer map water events, change intensity and water transitions were studied. The water changes over the study area as far as the pixel range alongside of the seasonality and persistence were determined. The results show the long-term changes in the surface water occurrence and lowering the water levels over the study period.

References

1. Agaton, M., Setiawan, Y., & Effendi, H. (2016). Land use/land cover change detection in an urban watershed: A case study of upper Citarum Watershed, West Java Province, Indonesia. *Procedia Environmental Sciences, 33*, 654–660. https://doi.org/10.1016/j.proenv.2016.03.120.
2. Fang, Y., Zheng, T., Zheng, X., Peng, H., Wang, H., Xin, J., Zhang, B. (2020). Assessment of the hydrodynamics role for groundwater quality using an integration of GIS, water quality index, and multivariate statistical techniques. *Journal of Environmental Management, 273*, 111185. https://doi.org/10.1016/j.jenvman.2020.111185.

3. Badjana, H. M., Helmschrot, J., Selsam, P., Wala, K., Flügel, W. A., Afouda, A., & Akpagana, K. (2015). Land cover changes assessment using object-based image analysis in the Binah River watershed (Togo and Benin). *Earth and Space Science, 2*(10), 403–416. https://doi.org/10.1002/2014EA000083
4. Chowdhury, M., Hasan, M. E., & Abdullah-Al-Mamun, M. M. (2020). Land use/land cover change assessment of Halda watershed using RS and GIS. *The Egyptian Journal of Remote Sensing and Space Science, 23*(1), 63–75. https://doi.org/10.1016/j.ejrs.2018.11.003
5. Sun, W., Shan, J., Wang, Z., Wang, L., Lu, D., Jin, Z., Yu, K. (2020). Geospatial analysis of urban expansion using remote sensing methods and data: A case study of Yangtze River Delta, China. *Complexity*.https://doi.org/10.1155/2020/3239471.
6. Faour, G. (2015). Evaluating urban expansion using remotely-sensed data in Lebanon. *Lebanese Science Journal, 16*(1), 23–24.
7. Du, N., Ottens, H., Sliuzas, R. (2010). Spatial impact of urban expansion on surface water bodies—A case study ofWuhan, China. *Landscape and Urban Planning, 94*(3–4), 175–185. https://doi.org/10.1016/j.landurbplan.2009.10.002.
8. Yang, X., & Lu, X. X. (2013). Delineation of lakes and reservoirs in large river basins: An example of the Yangtze River Basin, China. *Geomorphology, 190*, 92–102. https://doi.org/10.1016/j.geomorph.2013.02.018
9. Tsutsumida, N., & Comber, A. J. (2015). Measures of Spatio-temporal accuracy for time series land cover data. *International Journal of Applied Earth Observation and Geoinformation, 41*, 46–55. https://doi.org/10.1016/j.jag.2015.04.018
10. Wardlow, B. D., & Egbert, S. L. (2010). A comparison of MODIS 250-m EVI and NDVI data for crop mapping: A case study for southwest Kansas. *International Journal of Remote Sensing, 31*(3), 805–830. https://doi.org/10.1080/01431160902897858
11. Garroutte, E., Hansen, A., & Lawrence, R. (2016). Using NDVI and EVI to map spatiotemporal variation in the biomass and quality of forage for migratory elk in the Greater Yellowstone Ecosystem. *Remote Sensing., 8*(5), 404. https://doi.org/10.3390/rs8050404
12. Naik, B. C., & Anuradha, B. (2018). Extraction of water-body area from high-resolution Landsat imagery. *International Journal of Electrical and Computer Engineering, 8*(6), 4111. https://doi.org/10.11591/ijece.v8i6.pp4111-4119
13. Dong, Z., Wang, Z., Liu, D., Song, K., Li, L., Jia, M., & Ding, Z. (2014). Mapping wetland areas using Landsat-derived NDVI and LSWI: A case study of West Songnen plain, Northeast China. *Journal of the Indian Society of Remote Sensing, 42*(3), 569–576. https://doi.org/10.1007/s12524-013-0357-1
14. Gao, B. C. (1996). NDWI—A normalized difference water index for remote sensing of vegetation liquid water from space. *Remote Sensing of Environment, 58*(3), 257–266. https://doi.org/10.1016/s0034-4257(96)00067-3
15. Pettorelli, N. (2013). *Climate and the NDVI: A complex story*. Oxford University Press. https://doi.org/10.1093/acprof:osobl/9780199693160.003.0004

Evaluation of Soil Water Crop Compatibility Using RS and GIS Techniques

D. Sindhu, A. S. Ravikumar, and B. L. Shivakumar

Abstract A basic knowledge of soil water interaction for a particular crop is very much essential to decide the frequency of irrigation, amount of water to be supplied and other factors which affect the planning of irrigation system. Studies show that due to inadequate knowledge of soil water and plant relationship, farmers have a tendency of growing crops which are not suited for the cultivation in a particular area resulting in crop stress and decreasing yield of crops. Choosing suitable crops for good yield is very much essential in the development of agriculture. A farmer should have a proper knowledge of choosing suitable crops based on water quality, soil quality and fertility for optimum growth of plant and to get optimum yield with minimum risk. In the present study, an attempt has been made on Siddapura catchment, Harangi Command area, Karnataka to evaluate the soil water crop compatibility using Perur's method. Soil water crop compatibility has been evaluated based on the water sample analysis from different sources (Bore well, open well, surface waters), soil texture and crops grown within study area. The results shows that the salinity hazard in the water is very high, sodium hazard and bicarbonate hazard is low, hence crops with good salt tolerance can be grown in the study area. Soil permeability in the study area is moderate to rapid, crops which are tolerant, semi tolerant and sensitive for the salts are grown in the study area. Overall, the soil water crop compatibility in the study area is found to be good to moderate.

Keywords Soil water crop compatibility · Salt tolerance · Soil permeability

D. Sindhu (✉) · B. L. Shivakumar
Department of Civil Engineering, RV College of Engineering, Bengaluru, Karnataka 560059, India
e-mail: sindhud@rvce.edu.in

A. S. Ravikumar
Department of Civil Engineering, UVCE, Bangalore University, Jnanabharathi, Bengaluru, Karnataka 560056, India

A. K. Dikshit et al. (eds.), *Innovative Trends in Hydrological and Environmental Systems*, Lecture Notes in Civil Engineering 234,
https://doi.org/10.1007/978-981-19-0304-5_32

1 Introduction

Water is one of the most essential natural sources required for the survival of people. The demand for food crops will double during the next 50 years. With limited land and water resources, farmers need to increase the output from existing cultivated areas to satisfy the food demand of increasing population [1]. Studies shows that due to inadequate knowledge of soil water and plant relationship, farmers have a tendency of growing crops which are not suited for the cultivation in a particular area resulting in crop stress and decreasing yield of crops. Evaluation of soil water crop compatibility is very much essential to achieve a good crop yield. The farmer should have a proper knowledge of choosing suitable crops based on water quality, soil quality and fertility for optimum growth of plant and to get optimum yield with minimum risk [2]. The farmers of the Harangi command area are growing number of crops but the good yield of the crops are not achieved. Hence in the present study an attempt has been made to evaluate the soil water crop compatibility using Perurs's method.

2 Study Area

Study area chosen is Siddapura catchment part of Harangi command area which falls in Hassan, Kodagu and Mysore districts. Geographically the study area lies between 75° 38' E and 76° 15′ E longitude and 12° 24' N and 12° 46' N latitude, having an area of 1522.4 km^2 and is covered in Survey of India (SOI) toposheet numbers 48 P/10, 48 P/11, 48 P/14, 48 P/15, 57 D/1, 57 D/2 and 57 D/3 of 1:50,000 scale. The maximum length and width of the study area is approximately equal to 37.53 km and 68.88 km, respectively. Figure 1 shows the location map of the study area.

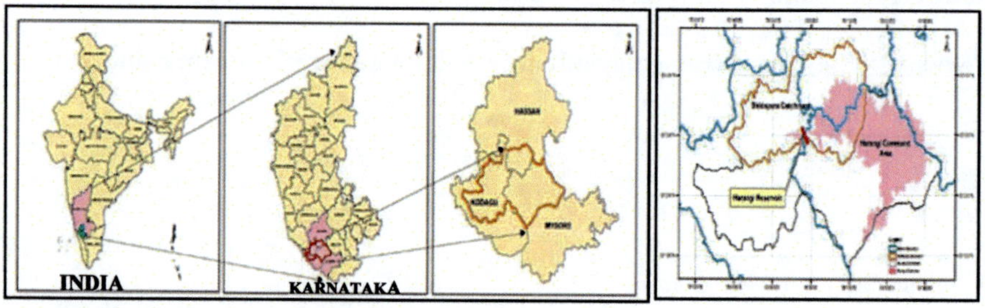

Fig. 1 Location map of the study area

3 Materials and Methodology

Perur et al. [3] recommended an empirical rule for computing soil water crop compatibility by using soil, water and plant factors.

Soil: The soil properties such as soil texture and soil permeability will have a direct influence on irrigation. Therefore, these two soil properties are used in the evaluation of soil water crop compatibility. The different textural and permeability classes and their class numbers are presented in Tables 1 and 2, respectively.

Water: The properties such as salinity, SAR and RSC are considered in deciding the suitability of water for irrigation with the crop and soil. Irrigation water is grouped into different classes based on the values of salinity, SAR and RSC and are presented in Table 3, 4 and 5, respectively.

Crops: The salt present in the irrigation water and soils, influence the type of crop to be grown on certain locality with particular soil. Hence, while choosing a crop to be irrigated in particular soil on a certain locality the information on the salt tolerance

Table 1 Texture based on soil type

Soil texture	Class
Sand, loamy Sand	T2
Sandy loam	T2.5
Loam	T3
Clay Loam	T3.5
Clay	T4

Perur [3]

Table 2 Soil permeability

Permeability	Class
Rapid	P1
Moderate	P2
Slow	P3

Perur [3]

Table 3 Salinity Hazard in water

Conductivity	Class	Interpretation
Less than 250	C1	Excellent
250–750	C2	Good
750–2250	C3	Doubtful
Above 2250	C4	Unsuitable

Table 4 Sodium Hazard in water

SAR	Class	Interpretation
Less than 10	S1	Excellent
10–18	S2	Good
18–26	S3	Fair
Above 26	S4	Poor

Table 5 Bicarbonate Hazard in water

RSC (meq/l)	Class	Interpretation
Less than 1.25	RSC1	Can be safely used
1.25–2.5	RSC2	Marginal
Above 2.5	RSC3	Unsuitable

Table 6 Salt tolerance of crop

Salt tolerance	Class
Salt tolerant	CP1
Semi-tolerant	CP2
Sensitive	CP3

Perur [3]

of crops is very much required. On the basis of salt tolerance, the crops are grouped into salt tolerant crops, semi tolerant crops and sensitive crops and are assigned with class numbers for each individual group which are shown in Table 6.

In order to determine the salinity, SAR and RSC representative water samples were collected from different sources in the study area during June 2018 and November 2019. The water sampling locations obtained using GPS were then converted into spatial data using ArcGIS software. Table 7a, b shows the location details of the water sample locations in study area. Figure 2 shows the water sampling locations in study area.

4 Results and Discussion

In compatibility assessment using Perur's method, the information's provided in Tables 1, 2, 3, 4, 5 and 6 are used in classifying the soil, water and crops into distinct groups. Individual class numbers are assigned to soil, water and crops system, the sum of which provides an index. The sum of the class numbers pertaining to the soil texture, permeability of the soil, salt tolerance of the crops grown and basic characteristics of irrigation water such as SAR, RSC and salinity at a particular

Table 7 Details of water sample locations

(a)

Sample No.	Sample location	Longitude €	Latitude (N)	Water source
W1	T. Mayagaudanahalli	76° 14′ 16.8″	12° 41′ 16′	Bore Well
W2	Tejuru Lake	76° 14′ 16.8″	12° 40′ 58.8″	Surface Water
W3	Badakyatanahalli	76° 13′ 12″	12° 39′ 25.2″	Bore Well
W4	Bidaraka	76° 11′ 24″	12° 38′ 27.6″	Bore Well
W5	Nerole	76° 11′ 20.4″	12° 37′ 12″	Bore Well
W6	Dumme Mallapura	76° 7′ 19.2″	12° 43′ 4.8″	Bore Well
W7	Machagondanahalli	76° 0′ 21.6″	12° 42′ 50.4″	Bore Well
W8	Mokli	76° 4′ 44.4″	12° 43′ 12″	Bore Well
W9	Dodda Magge	76° 5′ 6″	12° 41′ 2.4″	Bore Well
W10	Konanuru	76° 3′ 7.2″	12° 37′ 51.6″	Bore Well
W11	Ramanathapura	76° 4′ 55.2″	12° 36′ 36.9″	Bore Well
W12	Doddahalli	76° 6′ 18′	12° 39′ 0″	Bore Well
W13	Kote Kapparavalli	76° 6′ 46.8″	12° 39′ 57.6″	Bore Well
W14	Koratigere	76° 6′ 25.2″	12° 40′ 51.6″	Bore Well
W15	Ankanatapur	76° 2′ 16.8″	12° 40′ 44.4″	Surface Water
W16	Keregodu	76° 10′ 58.8″	12° 43′4.8″	Bore Well
W17	Belavadi	76° 8′ 38.4″	12° 39′54″	Bore Well
W18	Karahalli	76° 9′ 32.4″	12° 40′ 30″	Bore Well
W19	Koratikere Kaval	76° 7′ 51.6″	12° 38′ 49.2″	Bore Well
W20	Vaddarahalli	76° 6′ 3.6″	12° 34′ 58.8″	Bore Well
W21	Mallapura	76° 6′ 0″	12° 32′ 6″	Surface Water
W22	Bettadpur	76° 6′ 7.2″	12° 27′ 46.8″	Bore Well

(b)

Sample No.	Sample location	Longitude €	Latitude (N)	Water source
W23	Barse	76° 7′ 30″	12° 25′ 58.8″	Bore Well
W24	Adaguru	76° 3′ 18″	12° 30′ 50.4″	Bore Well
W25	Bayalakuppe	76° 3′ 18″	12° 30′ 50.4″	Bore Well
W26	Kanagalu	76° 1′ 37.2″	12° 33′ 25.2″	Bore Well
W27	Tamadahalli	76° 10′ 51.6″	12° 28′33.6″	Bore Well
W28	Maluganahalli	76° 13′ 12″	12° 36′ 32.4″	Bore Well
W29	Maduranahalli	76° 9′ 3.6″	12° 33′ 39.6″	Bore Well
W30	Mavanur	76° 13′ 15.6″	12° 32′ 9.6″	Bore Well
W31	Ullenahalli	76° 1′ 48.8″	12° 36′ 49.8″	Bore Well
W32	Konanur	76° 02′ 33.0″	12° 37′ 42.4″	Surface Water
W33	Keshavatur	76° 00′ 20.6″	12° 39′ 04.9″	Bore Well

(continued)

Table 7 (continued)

(b)				
W34	Handrangi	75° 59′ 59.8″	12° 38′ 51.5″	Surface Water
W35	Yelakanur	75° 54′ 39.1″	12° 32′ 32.6″	Bore Well
W36	Madalapura	75° 56′ 15.1″	12° 30′ 14.1″	Surface Water
W37	Kudige	75° 57′ 24″	12° 29′ 39.8″	Bore Well
W38	Hulse	75° 58′ 8.4″	12° 30′ 56.8″	Bore Well
W39	Hebbale	75° 58′ 45.2″	12° 32′ 3.1″	Bore Well
W40	Kanive	75° 57′ 47.1″	12° 31′ 25.3″	Surface Water
W41	Kudumangalore	75° 55′ 4.8″	12° 29′ 38.4″	Bore Well
W42	Anekad	75° 55′ 12″	12° 28′ 33.6″	Bore Well
W43	Guddahosuru	75° 52′ 44.4″	12° 30′ 46.8″	Bore Well
W44	Angagove	75° 51′ 39.6″	12° 28′ 22.8″	Bore Well
W45	Madikeri	75° 44′ 2.4″	12° 25′ 8.4″	Bore Well
W46	Karanageri	75° 44′ 16.8″	12° 25′ 51.6″	Bore Well
W47	Kalakeri	75° 42′ 23.7″	12° 26′ 18.4″	Open Well
W48	Monnageri (SW)	75° 41′ 42″	12° 27′ 3.6″	Surface Water
W49	Monnageri (OW)	75° 41′ 38.4″	12° 26′ 56.4″	Open Well
W50	Garagandur	75° 48′ 38.2″	12° 30′ 36″	Bore Well
W51	Madapura Hole	75° 48′ 39.6″	12° 30′ 32.4″	Surface Water
W52	Garavale	75° 46′ 1.2″	12° 34′ 22.8″	Bore Well
W53	Kerehosahalli	75° 62′ 32.2″	12° 38′ 17.9″	Bore Well

region are determined. Soil water crop compatibility is considered as excellent if the total class number is equal to 9, good if the total class number varies from 9 to 11, moderate for total class number 11 to 13 and if the total class number is more than 13 the compatibility is considered poor.

In the current study, soil water crop relationship has been evaluated based on the water sample analysis from different sources falling within study area and is summarized in Table 8.

The values of electrical conductivity, SAR and RSC for the representative water samples indicates that the salinity hazard in the water is very high, sodium hazard and bicarbonate hazard is low, hence crops with good salt tolerance can be grown in the study area. Soil permeability in the study area is moderate to rapid. Crops which are tolerant, semi tolerant and sensitive for the salts can only be grown in the study area. The results obtained by adding the class numbers with respect to EC, SAR, RSC, soil texture and salt tolerance of the crops varies from 9 to 12.5. Therefore, soil water crop compatibility in the study area is found to be good to moderate. But for water sample number 26, compatibility is poor, which implies that tobacco crop grown in that particular soils with the available water is not suitable. For coconut plant which is a salt tolerant crop, the total class becomes less than 13. Therefore, the

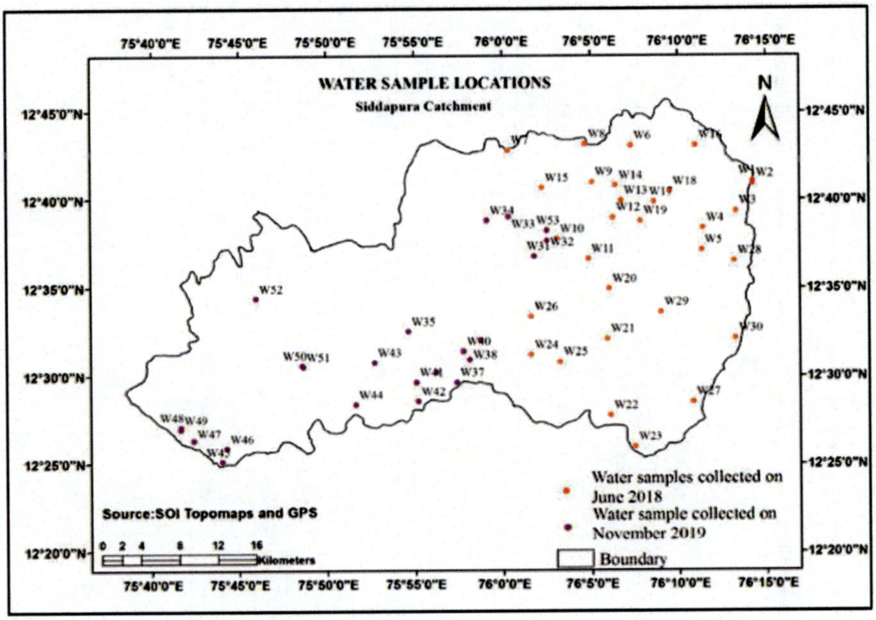

Fig. 2 Water sample locations in the study area

existing system of water and soil unsuitable for tobacco will be suitable moderately for growing coconut using water sample 26.

5 Conclusions

In the Perur's method, evaluation of compatibility is done by assigning respective class numbers to the soil texture, permeability, salt tolerance of the crops grown, salinity, RSC, SAR for water. The sum of these classes is used in determining the suitability of crops. The study reveals that semi tolerant crops such as paddy, Bengal gram, ginger, banana, ragi, potato, maize, chili, coffee, pepper and tolerant crops such as coconut, arecanut can be suitably grown in the study area as the soil water crop compatibility is good to moderate.

Table 8 Soil water crop compatibility assessment in the study area (Pre monsoon, June 2018), soil water crop compatibility assessment in the study area (Post monsoon, November 2019)

(a)

Water Sample No./ Source	EC (micro mhos/cm)	EC class	SAR	SAR Class	RSC	RSC Class	Soil Texture	*Soil Texture Class*	*Soil Permeability Class*	*Crops grown*	*Salt tolerant class*	*Total class No.*	Compatibility
*W*1-BW	1425	C3	0.93	*S*1	−6.04	*RSC*1	*Sandy Loam*	*T*2.5	*P*2	Tobacco, Coconut	CP3 *CP*1	12.5	Moderate
W2-SW	780	C3	0.64	S1	−0.96	RSC1	Sandy	T2	P1	Tobacco, Coconut	CP3 CP1	11.0	Good
W3-BW	3940	C4	1.49	S1	−4.44	RSC1	Sandy Loam	T2.5	P2	Bengal Gram	CP1	11.5	Moderate
W5-BW	2385	C4	1.41	S1	−1.92	RSC1	Sandy	T2	P1	Ginger	CP2	11.0	Good
W7-BW	1470	C3	0.47	S1	−3.37	RSC1	Loam	T3.5	P2	Banana	CP2	12.5	Moderate
W8-BW	1150	C3	0.98	S1	−0.93	RSC1	Sandy Loam	T2.5	P2	Ragi, Tobacco	CP2 CP3	12.5	Moderate
W9-BW	1880	C3	0.84	S1	−2.86	RSC1	Sandy	T2	P1	Arecanut	CP1	9.0	Good
W12-BW	2870	C4	0.96	S1	−2.19	RSC1	Sandy	T2	P1	Tobacco, Coconut, Jowar	CP3 CP1 CP2	12.0	Moderate
W15-SW	510	C2	0.65	S1	−1.29	RSC1	Loam	T3.5	P2	Paddy	CP2	11.5	Moderate
W17-BW	380	C2	0.81	S1	−0.74	RSC1	Sandy	T2	P1	Potato	CP2	9.0	Good
W18-BW	3120	C4	1.41	S1	−5.38	RSC1	Sandy	T2	P1	Maize, Potato	CP2	11.0	Good
W19-BW	1020	C3	0.62	S1	−1.10	RSC1	Sandy	T2	P1	Maize, Potato	CP2	10.0	Good

(continued)

Table 8 (continued)

(a)

Water Sample No./ Source	EC (micro mhos/cm)	EC class	SAR	SAR Class	RSC	RSC Class	Soil Texture	*Soil Texture Class*	*Soil Permeability Class*	*Crops grown*	*Salt tolerant class*	*Total class No.*	Compatibility
W24-BW	5010	C4	1.79	S1	-6.73	RSC1	Clay Loam	T3.5	P2	Arecanut	CP1	12.5	Moderate
W26-BW	3800	C4	1.65	S1	−4.80	RSC1	Loam	T3.5	P2	Tobacco	CP3	14.5	Poor
W27-BW	4800	C4	1.55	S1	−7.36	RSC1	Sandy	T2	P1	Chili	CP2	11.0	*Good*

(b)

Water Sample No./ Source	EC (micro mhos/cm)	EC class	SAR	SAR Class	RSC	RSC Class	Soil Texture	*Soil Texture Class*	*Soil Permeability Class*	*Crops grown*	*Salt tolerant class*	*Total class No.*	Compatibility
W1-BW	920	C3	0.44	S1	−1.82	RSC1	Loamy Sand	T3.5	P2	Cassava	CP2	12.5	Moderate
W4-SW	1120	C3	0.59	S1	−2.45	RSC1	Loamy Sand	T2.5	P2	Jowar, Maize	CP2	11.5	Moderate
W5-BW	1720	C3	0.89	S1	−2.90	RSC1	Loamy Sand	T2.5	P2	Jowar, Maize	CP2	11.5	Moderate
W6-SW	240	C1	0.47	S1	−1.01	RSC1	Sandy Loam	T2.5	P2	Jowar, Maize	CP2	9.5	Good
W7-BW	2320	C4	0.74	S1	−3.60	RSC1	Loam	T2	P2	Jowar, Maize	CP2	12	Moderate
W8-BW	1720	C3	0.75	S1	−2.85	RSC1	Sandy Loam	T2.5	P2	Ragi	CP2	11.5	Moderate
W10-SW	740	C2	0.31	S1	−1.29	RSC1	Sandy Loam	T2.5	P2	Ragi, Maize	CP2	10.5	Good

(continued)

Table 8 (continued)

(b)

W17-OW	520	C2	0.33	S1	−0.55	RSC1	Loam	T3.5	P2	Coffee	CP2	11.5	Moderate
W19-OW	390	C2	0.19	S1	−0.42	RSC1	Loam	T3	P2	Coffee, Pepper	CP2	11	Good
W22-BW	1280	C3	0.48	S1	-1.62	RSC1	Loam	T3.5	P2	Maize	CP2	12.5	Moderate
W23-BW	1800	C3	0.70	S1	−3.55	RSC1	Loam	T3.5	P2	Coffee, Arecanut, Pepper	CP2,CP1, CP2	12.5	Moderate

References

1. Gaurav, P., Gupta., P., & Nale., J. (2010). Crop and Irrigation Water Requirement Estimation by Remote Sensing and GIS: A Case Study of Karnal district, Haryana, India. *International Journal of Engineering and Technology, 2*(4), 207–211.
2. Shivakumar, B. L. (2008). Integrated study on watershed for sustainable management of water, Soil and Crop. Ph.D. Thesis, Department of Civil Engineering, UVCE, Bangalore University, Bangalore.
3. Perur, N. G., Subramanian, C. K., Muhr, G. R., & Bay, H. E. (1973). *Soil Fertility Evaluation to Serve Indian Farmers*. U. S. Agency for International Development and University of Agricultural Sciences.

Diagnosing the Important Morphometric Parameters for Water Resource Management of Geba River Basin, Tigray, Ethiopia

Gebremedhin Godif, B. R. Manjunatha, and H. Gangadhara Bhat

Abstract Characterizing of morphometric parameters in the watershed level is important for water resources management. Geba drainage basin is a source of water for many lives. However, its geomorphology is not yet been investigated. This study is an attempt to diagnose the important morphometric parameters that have high nexus with the hydrologic condition of the drainage basin. For this purpose, the Digital Elevation Model (DEM) was used as input data and GIS was used as data processing platform. From the processed DEM data, a total of 16 parameters expected to have high nexus with the basin hydrology including, Bifurcation ratio, Stream frequency, Drainage density, Drainage texture, Infiltration number, Length of overland flow, Elongation ratio, Circulatory ratio, Form factor, Compactness coefficient, Constant of channel maintenance, Basin relief, Relief ratio, Ruggedness number, Channel gradient, and Basin slope were quantified, and their values are 4.3, 1.1 km^{-2}, 1.3 km^{-1}, 8 km^{-1}, 1.4 km^{-3}, 0.4 km, 0.2, 0.1, 0.2, 0.8, 2.9 km, 2.4 km, 0.01, 3, 0.01, 0.02, respectively. Based on morphometric characterization, values of Stream frequency, Infiltration number, Elongation ratio, Form factor, Constant of channel maintenance, Relief ratio, Channel gradient, and Basin slope are low. Drainage density is coarse, and the remained parameters fall under moderate category. All these results indicate that the Geba river basin is characterized by elongated shape, relatively moderate infiltration capacity, and hence with moderate groundwater potential.

Keywords Geomorphology · GIS · Remote sensing · DEM

G. Godif (✉)
Department of Physics, Raya University, Tigray, Ethiopia

B. R. Manjunatha · H. Gangadhara Bhat
Department of Marine Geology, Mangalore University, Mangalagangothri, Karnataka 574199, India

A. K. Dikshit et al. (eds.), *Innovative Trends in Hydrological and Environmental Systems*, Lecture Notes in Civil Engineering 234,
https://doi.org/10.1007/978-981-19-0304-5_33

1 Introduction

In arid and semi-arid regions, like Ethiopia, where water resources are limited and groundwater recharge depends mainly on rainwater, investigating, modelling, and managing water resources at the watershed scale are important. Understanding of groundwater recharge and water resources budget allocation highly depends on the characterization of drainage basin parameters. The term morphometric was derived from two Greek words 'Morpho' and 'Metric' in which Morpho represents earth and metric represents a measurement [2]. Therefore, morphometry as a whole is measuring or quantifying some landmass of the earth [5].

Quantifying and characterizing drainage basin parameters have an immense utility in the practical world, for example, it relates to the different aspects of the drainage pattern of the basin [18]. Above all, the morphometric analysis provides valuable information about the hydrologic characteristics of drainage basins in cases when the absence of hydrologic data is observed [28]. Before the development of geospatial technologies, information about drainage systems has been extracted from topographic maps on a manual basis [14, 22, 32]. This method is labour, resource, and time-intensive. Further advancements in remote sensing and computer technologies with DEM as a data source and GIS as an information extraction platform greatly improved the process. DEM data are faster, precise, updated, and inexpensive compared to that of topographic maps [19, 25] while GIS is preferable over the manual technique in visualizing, processing, and quantifying the morphometric information [1]. Over the years, many pieces of research have been carried out using the integrated approach of remote sensing and GIS technologies [6, 10, 13, 26, 27, 30]. The outputs of these researches have been utilized for different purposes, such as rainwater harvesting [26], soil and water conservation measures [8], hydrogeological characterization [17], and flash flood hazards evaluation [36].

Geba river basin, found in the northern part of Ethiopia, is the major tributary of the Tekeze river basin which is the upper tributary of the Nile River. It is mainly characterized by a semi-arid climate with limited and erratic rainfall which makes it one of the most water-stressed and food-insecure areas of the region [11]. Despite the importance of characterizing basin morphometry in mitigating the existing problems, researches related to this issue have not yet been documented, except a few that mainly focused on the eastern and south-eastern parts of the river basin [9, 10]. Therefore, in this study, an attempt has been made to address the identified gaps concerning water and suggest alternative solutions.

2 Methods and Materials

2.1 Description of the Study Area

Geba river basin, with a drainage area of 5137 km^2, is found in the Tigray region, the northern part of Ethiopia. It is one of the tributaries of the Tekeze river basin, the basin which starts from the northern Ethiopian high lands and drains to the Nile river basin at Atbara, Sudan. The drainage basin is situated between latitudes 12° 55′ 00″ to 14° 10′ 00″ north and longitudes 38° 45′ 00″ to 39° 35′ 00″ east (Fig. 1). The topographic elevation of the river basin above means sea level ranges from 926 to 3300 m, in which its lowest topographic elevation is on the western part and the highest point of elevation is projected towards the northern part. The maximum length of the basin from its source to the outlet is about 240 km. The longitudinal profile of distance versus elevation (Fig. 2) shows that the elevation declines as the distance increases towards the outlet. Both temporal and spatial variations in precipitation and temperature are observed in the drainage basin. A high amount of rainfall is observed in the summer season which lasts from mid-June to mid-September. Compared to its eastern part with mean annual precipitation of about 400 mm, its western and northern parts get high mean annual precipitation, which is about 950 mm. Similarly, the northern high land part of the basin shows a minimum average temperature of 6.5 °C, while the western low land parts of the basin experience a maximum average

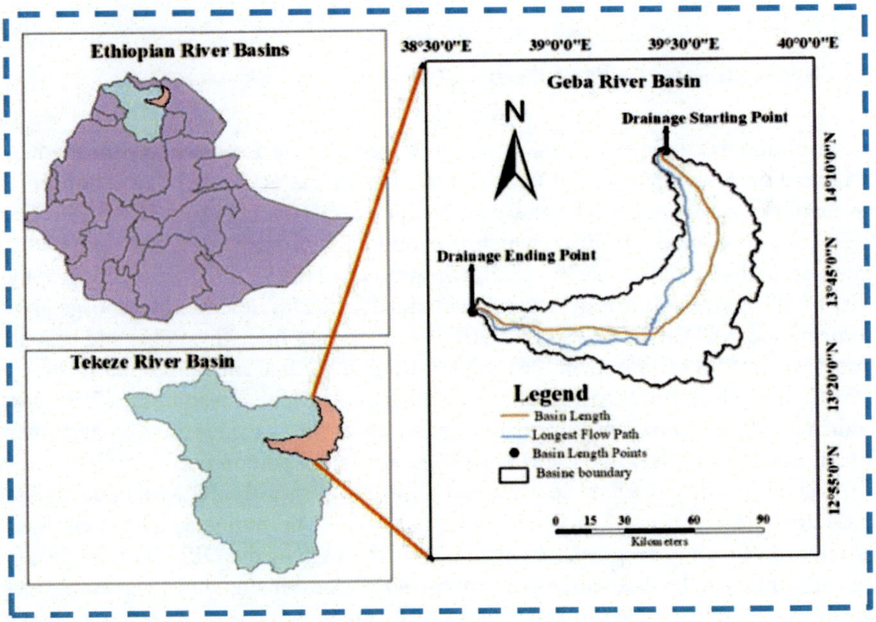

Fig. 1 Location map of Geba drainage basin

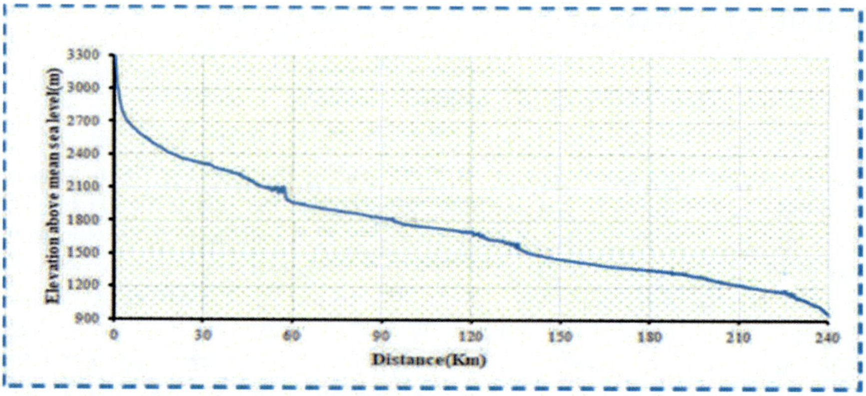

Fig. 2 Horizontal distance versus elevation profile of Geba drainage basin

temperature of 32 °C. The southwest parts of the area are covered with Precambrian, metamorphosed volcanic, and sediments, the central part is comprised of Palaeozoic and Mesozoic sediments, Dolerites and the northern part of the basin are structured by localized quaternary sediments. Most land cover of the area is cultivated due to the extensive agricultural practices. The hydrogeological condition of the river basin is not yet known in detail. Field observations show that with exception of some areas with shallow groundwater, groundwater levels in most of the areas are deep.

2.2 *Data and Data Processing*

In conventional techniques, drainage system information is obtained from the manual extraction of topographic maps. Recently, the manual extraction of information from the topographic maps is replaced by automatic extraction of information from the digital elevation model (DEM), which is economical, manageable, and with better accuracy. In this study, the DEM which is the product of the Shuttle Radar Topography Mission (STRM) satellite sensor was downloaded from http://earthexplorer.usgs.gov. Compared to ASTER-DEM, SRTM-DEM databases are freely available and easy to process in the GIS environment [16] and have higher vertical and horizontal accuracy [30]. Commonly, DEMs are available with 30 and 90 m spatial resolutions. It is quite evident that the accuracy of information increases as the spatial resolution gets finer [4]. Hence, in this study, the DEM with 30 m spatial resolution was adopted.

After identifying the data sources, the different tiles of the DEM that cover the study area were projected to WGS–1984, zone 37°. The projected tiles were then mosaicked to a single representative data with the help of the GIS platform. In the present study, Arc Hydro, which is one of the ArcGIS extension tools, was employed due to its capability for automatic processing of DEM [9]. Using the different functions in the menu of this tool, stepwise procedures such as DEM depression filling, flow direction processing, flow accumulation grids identification, raster to polygon processing, and parameter calculations [3] were performed. Filling the depression was aimed to remove the local depressions from the DEM so that the flow continuity could be maintained. To initiate and predict the nature of the drainage network, 0.06% of the total pixels were taken as threshold values.

2.3 *Morphometric Analysis*

After delineating the watershed and extracting the drainage network, quantifying the morphometric parameters was the next procedure followed. In this respect, the basic basin parameters, like the basin area, basin perimeter, basin length, maximum elevation, minimum elevation, stream numbers, and stream length were directly calculated in the GIS platform using hydrology tool (Tables 1 and 2). Using the basic parameters as inputs, a total of 16 parameters expected to have high nexus with the basin hydrology, including Bifurcation ratio (R_b), Stream frequency (F_s), Drainage density (D_d), Drainage texture (D_t), Infiltration number (I_f), Length of overland flow (L_o), Elongation ratio (R_e), Circulatory ratio (R_c), Form factor (F_f), Compactness coefficient (C_c), Constant channel maintenance (C), Basin relief (R), Relief ratio (R_r), Ruggedness number (R_n), Channel gradient (C_g), and Basin slope (S_b) were computed based on standard mathematical equations (Table 3). Furthermore, the drainage density, slope, and aspect maps were extracted from the DEM. The hydrologic condition of the drainage basin was then inferred from the details of the maps.

Table 1 Stream numbers and stream length of Geba drainage basin

S. No	Parameters	Stream orders							Total
		1	2	3	4	5	6	7	
1	Stream numbers (N_u)	4311	943	230	48	7	2	1	5542
2	Stream length (L_u)(km)	3473	1771	715	402	200	12	137	6710

Table 2 Basic geometry parameters of Geba drainage basin

Parameters	Area (A) (km)	Perimeter (P) (km)	Basin length (L_b) (km)	Minimum elevation (h) (m)	Maximum elevation (H) (m)
Calculated values	5137	747	168	926	3300

Table 3 Morphometric parameters and their methods of calculation

S. No	Morphometric parameters	Representing symbol	Mathematical methods	Unit	References
1	Bifurcation ratio	R_b	$R_b = N_u/N_{(u+1)}$		[14]
2	Stream frequency	F_s	$F_s = N_u/A$	km^{-2}	[14]
3	Drainage density	D_d	$D_d = L_u/A$	km^{-1}	[15]
4	Drainage texture	D_t	$D_t = N_u/P$	km^{-1}	[14]
5	Infiltration number	I_f	$I_f = D_d \times F_s$	km^{-3}	[7]
6	Length of overland flow	L_o	$L_o = 1/2D_d$	km	[14]
7	Elongation ratio	R_e	$R_e = (2/L_b) \times (A/\pi)^{1/2}$		[29]
8	Circulatory ratio	R_c	$R_c = 4\pi A/P^2$		[22]
9	Form factor	F_f	$F_f = A/L_b^2$		[15]
10	Constant of channel maintenance	C	$C = 1/D_d$	km	[29]
11	Compactness coefficient	C_c	$C_c = P/2(\pi A)^2$		[12]
12	Basin relief	R	$R = H - h$	km	[34]
13	Relief ratio	R_r	$R_r = R/L_b$		[29]
14	Ruggedness number	R_n	$R_n = D_d \times R$		[29]
15	Channel gradient	C_g	$C_g = R/(\pi/2 \times L_b)$		[30]
16	Basin slope	S_b	$S_b = H/L_b$		[22]

3 Results and Discussion

In this section, a qualitative description of different parameters and their implications on the hydrology of the drainage system is presented. Some parameters, like Bifurcation ratio (R_b), are related to the one-dimensional view (linear aspect), others like Stream frequency (F_s), Drainage density (D_d), Drainage texture (D_t), Infiltration number (I_f), Length of overland flow (L_o), Elongation ratio (R_e), Circulatory ratio (R_c), Form factor (F_f), Compactness coefficient (C_c), and Constant of channel maintenance (C) are related to two-dimensional view (areal aspects), and the remained, notably Basin relief (R), Relief ratio (R_r), Ruggedness number (R_n), Channel gradient (C_g), and Basin slope (S_b) are indicators of three-dimensional view (relief aspect) of a drainage basin. The calculated values of all the parameters is Table 4.

Table 4 Calculated values of morphometric parameters in Geba drainage basin

Morphometric parameters	R_b	F_s	D_d	D_t	I_f	L_o	R_e	R_c	F_f	C	C_c	R	R_r	R_n	C_g	S_b
Calculated values	4.3	1.1	1.3	8	1.4	0.4	0.2	0.1	0.2	0.8	2.9	2.4	0.01	3	0.01	0.02

3.1 Bifurcation Ratio

The bifurcation ratio is one-dimensional property of the drainage basin. Its value is estimated by dividing the stream number of the lowest order by the stream number of the next higher order. It is indicative of the degree to which the geological structure controls the drainage pattern [30]. The drainage pattern, in turn, gives a clue on the hydrological potentiality of the drainage basin, such as the recharge and discharge rates. An increase in the number of streams results in a low degree of infiltration due to the high surface runoff. The calculated mean bifurcation ratio of the Geba river basin is 4.3 (Table 4). For drainage basins with less distorted stream channels and structural uniformity, the bifurcation ratio value ranges from 3 to 5 [14]. The calculated bifurcation ratio of the Geba river basin is in the specified range, which indicates that the Geba river basin is characterized by a less distorted drainage network and uniform geological structure. The drainage structure of the Geba river basin is a dominantly dendritic pattern arranged in seven different orders (Fig. 3). Among the different forms of drainage patterns, the dendritic drainage pattern is the most stable. The more the value of the bifurcation ratio gets lower, the less the structural disturbance [24]. Furthermore, the bifurcation ratio is related to the groundwater recharge

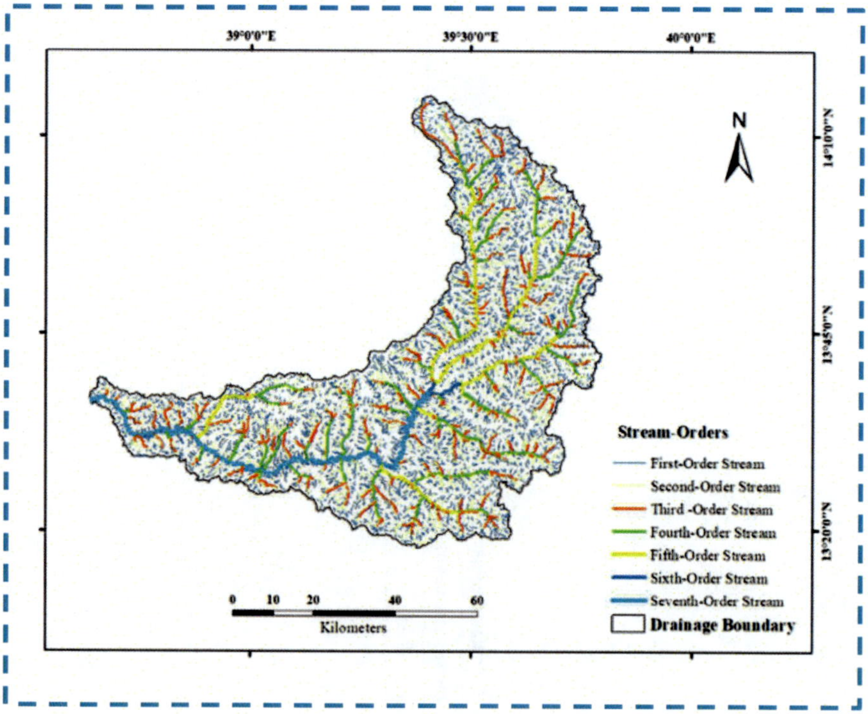

Fig. 3 Drainage network pattern and stream orders of Geba drainage basin

in such a way that for a high bifurcation ratio, land flow increases and get sufficient time for infiltration to increase groundwater recharge.

3.2 Form Factor

The form factor is a dimensionless property of the drainage basin which is used to indicate whether the basin is elongated or circular. It gives information on the rate of flow of water. The area and the length of the drainage basin are the determinant factors for its magnitude. The calculated value of the form factor for the Geba river basin is 0.2 (Table 4). According to Strahler [33], the form factor of any given drainage basin is within a range of 0 and 1. Drainage basins with a form factor approaching 1 are elongated, and these with the form factor values approaching 0 are circular. Under normal circumstances, river basins with less than 0.7854 form factor value are elongated in shape, and with greater than 0.7854 form factor value are circular [31]. Some researchers (e.g. [13, 30]) also argued that the value of the form factor of river basins can go up to 0.7854. The form factor of the Geba river basin (i.e., 0.2) is by far smaller than 0.7854 and more approaches to 0, which confirms that its shape is narrowly elongated. River basins, like the Geba river basin, having low form factor experience low peak flows of longer duration compared to river basins with high form factor that experience high peak flows of shorter duration [30].

3.3 Elongation Ratio

The elongation ratio is a two-dimensional property that describes the shape of the drainage basin. Its value depends on the diameter of a circle of the same area as the drainage basin and the length of the basin. The increase in the length of the basin results in the decrease of the elongation ratio, whereas the increase in the area of the drainage basin favours the increase in the elongation ratio value [29]. The computed value of the elongation ratio of the Geba river basin is 0.2 (Table 4). Based on their elongation ratio values, Schumm [29] classified basin shapes into five classes, i.e., river basins with elongation ratio values 0.9–1.0, 0.8–0.9, 0.7–0.8, 0.5–0.7, and $<$ 0.5 are categorized as circular, oval, less elongated, elongated, and more elongated shapes, respectively. Based on this scheme, the Geba river basin with an elongation ratio value of 0.2 is categorized under an elongated shape. Unlike circular-shaped drainage basins, elongated shape basins are characterized by smooth ground slopes [33]. Hence, as the drainage basins get more elongated, they are less likely sensitive to high run-off, soil erosion, and sediment load.

3.4 Circulatory Ratio

The circularity ratio is another important two-dimensional parameter to describe the shape of the drainage basin. Its value depends on the magnitude of the area and the perimeter of the drainage basin. As the area of the river basin increases, the circulatory ratio follows an increasing trend, and as the perimeter of the drainage basin gets higher, the circulatory ratio reflects a decreasing tendency. The value of the circulatory ratio of the Geba river basin is 0.1 (Table 4). According to Miller [22], the circularity ratio value approaching 0 is indicative of elongated shape, and the value approaching 1 is indicative of circular shape. Hence, the Geba river basin with a circularity ratio value of 0.12 is more approached to the elongation shape.

3.5 Compactness Coefficient

Like the form factor, elongation ratio, and circularity ratio, the Compactness coefficient, is a two-dimensional property that determines the shape of the drainage basin. The area and perimeter of the drainage basin determine its magnitude [12]. The calculated value of the compactness coefficient of the Geba river basin is 2.9 (Table 4), which is relatively high. Drainage basins with high values of compactness coefficient are elongated, while those with low values are more circular [27]. Unlike drainage basins with a high value of compactness coefficient, these with low values have low infiltration capacity and are at high risk of soil erosion and high surface run-off.

3.6 Stream Frequency

As its name indicates, stream frequency is an expression of the degree of distribution of streams with the given area. It is an indirect measure of the permeability and infiltration capacity of drainage basins [21]. For any increase in the number of stream segments of the drainage basin, an increase in stream frequency is observed [14]. For the Geba river basin, the computed value of stream frequency is 1.1 km^{-2} (Table 4), which is relatively low. The low value of stream frequency is indicative of permeable surface material.

3.7 Drainage Density

Drainage density is an indicator of the degree to which the drainage networks are close to each other. The quantitative average length of all channels in the drainage basin is inferred from the drainage density [23]. The magnitude of drainage density was quantified by Horton [15], in which its magnitude is directly related to the number of streams and inversely related to the area. According to the classification scheme suggested by Smith [32], the drainage density can be labelled as very coarse, coarse, moderately coarse, fine coarse, and very fine coarse for the drainage density values < 2 km^{-1}, 2 km^{-1}–4 km^{-1}, 4 km^{-1}–6 km^{-1}, 6 km^{-1}–8 km^{-1} and > 8 km^{-1}, respectively. The calculated drainage density of the Geba river basin is 1.3 km^{-1} (Table 4), which falls under the very coarse category. This indicates that the Geba River is less dissected with less concentration of the stream channels. Compared to the drainage basins with a high concentration of stream channels, these with less concentration of stream channels are more permeable with low surface run-off and high infiltration capacity. The calculated drainage density (i.e., 1.3 km^{-1}) only indicates the cumulative value. To better understand the spatial variation of the drainage density across the drainage basin, the distribution of the density was mapped (Fig. 4) and numerically tabulated (Table 5). As shown in the map, the

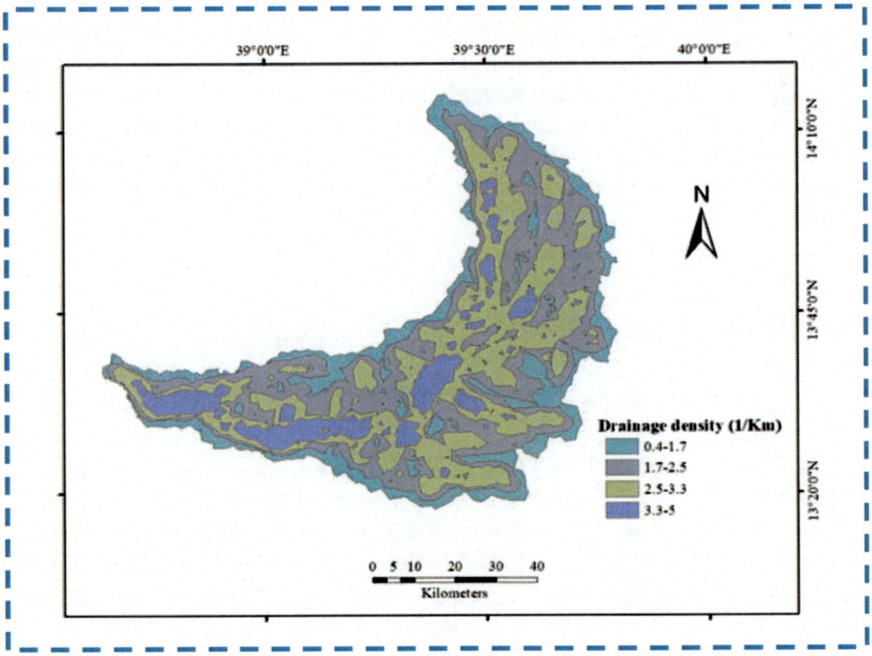

Fig. 4 Map showing the drainage density distribution of Geba drainage basin

Table 5 Areal distribution of drainage density in Geba drainage basin

Drainage density (km^{-1})	0.4–1.7	1.7–2.5	2.5–3.3	3.3–5
Area covered (km^{-1})	1000	1810	1729	598
% of Total area	19	35	34	12

drainage density ranges from 0.7 to 5 km^{-1}. The smallest value belongs to the lower-order channels found at the side points, while the highest value is related to the higher-order channels found on the mainstream. Most area (i.e., 35% of the total area) of the drainage basin is found in the density range 1.7–2.5 km^{-1}.

3.8 Drainage Texture

Drainage texture is the property of a drainage basin to describe the spacing of stream channels. The infiltration capacity and the nature of the lithology of the terrain are some of the attributes that can be inferred from the analysis of the drainage texture. The total number of streams and the perimeter of the terrain are important properties to determine the magnitude of the drainage texture [14]. Based on the drainage texture values, Smith [32] categorized drainage texture of drainage basins into four classes, coarse (< 4 km^{-1}), moderate (4–10 km^{-1}), fine (> 10 km^{-1}), and ultra-fine (> 15 km^{-1}). The drainage texture for the Geba river basin is 8 km^{-1} (Table 4), which is under the moderate category. It is worthy to note that fine drainage texture is indicative of the high concentration of stream channels. River basins with such drainage textures are prone to soil erosion [20]. On the other hand, drainage basins as in the case of the Geba river basin with a sparse relative spacing of the streams are characterized by high infiltration capacity and high base flow which lead to the low peak flow.

3.9 Infiltration Number

The infiltration capacity of a drainage basin is described by the infiltration number. It is the product of stream frequency and drainage density [7]. The calculated value of the infiltration number of the Geba river basin is 1.4 km^{-3} (Table 4). Drainage basins with a high value of infiltration number exhibit low infiltration capability, and those with low infiltration number are characterized by high infiltration capacity [33]. Hence, the relatively low infiltration number (i.e., 1.4 km^{-2}) of the Geba river basin suggests that the infiltration capacity of the basin is relatively high. The low infiltration number with relatively high infiltration capability further confirms that the river basin has a lower run-off potential.

3.10 Length of Overland Flow

Water in drainage basins flows from the channel of the lowest order to the next higher order channel. The total path that the water flows over the land surface before it gets drained into the next stream channel is the length of overland flow. Its qualitative expression was suggested by Horton [14], in which its value is half of the given drainage density. Therefore, for the Geba river basin, with a drainage density of 1.3 km^{-1}, the length of overland flow is 0.4 km (Table 4). Length of overland flow is also an indirect description of the relief of landscapes. Lower relief landscapes are related to the high length of overland flow, and high relief landscapes are characterized by lower values of length of overland flow [8]. As a result, these drainage basins with low values of length of overland flow are at high risk of high surface run-off due to their steeper slopes and fine textures [35]. The steeper slopes and fine textures, in turn, reduce the infiltration capability of drainage systems.

3.11 Constant of Channel Maintenance

Constant of channel maintenance stands for the capability of an area measured in Kilometres in maintaining a unit kilometre of stream length [29]. Since the drainage density is directly related to the channel length and inversely related to the area of the drainage basin, the constant of channel maintenance is a reciprocal form of drainage density. For the Geba river basin, with a drainage density of 1.3 km^{-3}, the constant of channel maintenance is approximately 0.8 km (Table 4). Based on the morphometric classification, this value is relatively high. Such drainage basins with a relatively high constant channel maintenance have smooth slopes, which enabled them to have a better infiltration rate.

3.12 Basin Relief

Naturally, drainage basins have high elevation at their source and lower towards their outlet. The difference between the highest and lowest points on the valley floor of a basin is expressed in terms of basin relief. It is a determining factor for many landscape characteristics, notably permeability, drainage development, erosional status, and surface run-off conditions. It also gives a clue on the stream gradient, flood pattern, and sediment volume that can be transported. The calculated value of the basin relief for the Geba drainage basin is 2374 m (i.e., 2.3747 km) (Table 4) above mean sea level, which is relatively medium. River basins with relatively medium basin relief experience less probability for surface run-off and have good infiltration capacity [20].

3.13 Relief Ratio

The relief ratio interconnects the basin relief and basin length. Its value is larger in river basins whose basin length is shorter and their relief is larger [29]. For the

Geba river basin, with a basin relief of 2.3747 km and basin length of 170 km, the relief ratio is 0.01 (Table 4). High values of relief ratio are indicative of steep slope, while small values of relief ratio are indicative of mild slopes [10]. In drainage basins, like the Geba river basin whose relief ratio is low, resistant basement rocks are observed. Furthermore, the relatively small value of the relief ratio is indicative of a comparatively gentle slope. In gentle slopes, the sediment load transport is low compared to steep slopes [35].

3.14 Ruggedness Number

Ruggedness number provides information on the slope steepness indicating whether the land surface is stable. For the value of the ruggedness number to be large, both the drainage density and the basin relief have to increase, i.e., the increase in both the drainage density and the basin relief results in a considerable increase in the ruggedness number [29]. In the Geba river basin, whose drainage density is 1.3 km^{-1} and basin relief of 2.3747 km, the ruggedness number is 3 (Table 4). In contrast to drainage basins with a low value of ruggedness number, drainage basins with high values of ruggedness experience low infiltration capacity and are prone to high surface run-off.

3.15 Channel Gradient

The channel gradient is one of the basic parameters to describe the geomorphologic characteristics of drainage basins. It is the height difference of points in the river basin when the water flows through streams. This difference in height makes water flow from the higher potential point to the lowest potential point, enhanced by the force of gravity. The increase in basin relief and the decrease in basin length favour the value of channel gradient to increase [30]. In the Geba river basin, where the basin length is 170 km and the basin relief with 2.3747 km, the value of the channel gradient is estimated to be 0.01 (Table 4), which is a low value, indicating that the drainage basin is characterized by moderate slope, and hence better infiltration capacity.

3.16 Basin Slope

The basin slope is the measure of the degree of steepness or smoothness of a terrain. The study of the slope is very important to characterize drainage basins in terms of their erosional status, infiltration capability, and surface run-off conditions. Besides, it is useful to indicate the balance between run-off response and soil infiltration rates [10]. According to Miller [22], the magnitude of the basin slope is determined by dividing the basin relief by the basin length. In the case of the Geba river basin, where the maximum elevation is 3.300 km and the basin length is 170 km, the basin slope is 0.02 (Table 4), which is very small indicating that the drainage basin ranges from gentle to moderate slope. For detailed information, the slope of the drainage basin was mapped (Fig. 5) and its areal extent at all corners were displayed (Table 6). As shown in the map, the slope ranges from 0° to > 27°. Almost 50% of the area is found at the slope range 0° and 7°. This slope range is indicative of a relatively smooth area with better infiltration capacity.

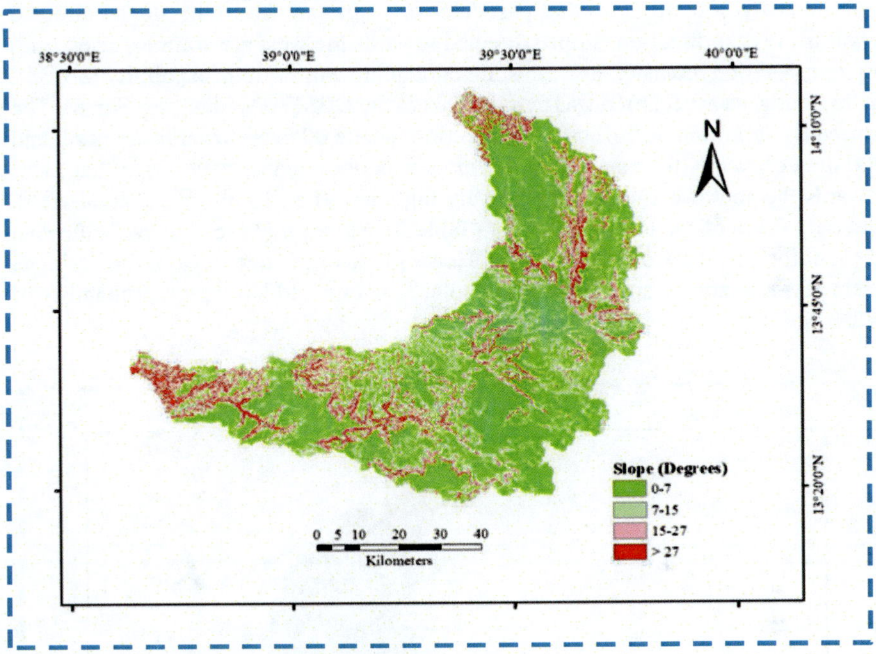

Fig. 5 Slope map of Geba drainage basin

Table 6 Areal slope distribution of Geba drainage basin

S. No	Slope in degrees	Class	Area covered (km^2)	% of total area
1	0.7	Low	2570	50
2	7–15	Moderate	1513	29
3	15–27	High	760	15
4	> 27	Very high	294	6

3.17 Aspect Map

The aspect map is an important parameter that indicates the direction to which the slopes of the landscape faces [20, 30]. The direction of the slope away or towards sunlight is important to determine the hydrologic condition of drainage basins. It is evident that areas exposed to high intensity of sunlight exhibit high evaporation. The high evaporation makes them drier with less moisture content. The less moisture content, in turn, declines the infiltration capacity. Slopes of a terrain faced towards the western direction are more sensitive to high sunlight in the afternoon due to the high intensity of sunlight at that particular time. In contrast to the slopes directed towards the east direction, slopes directed towards the west are warmer, drier, with high evaporation intensity, low moisture content, low infiltration capability [30, 35]. In this study, an aspect map was extracted from the DEM raster data set (Fig. 6). The aspect, by definition, is the compass direction, where 00 faces towards the north and 900 direct towards the east. From the map, it is observed that the slope that faces towards the southwest covers a relatively high area (i.e., 13.8% of the total area), and only 0.1% of the total area is flat (Table 7). Slopes that faced towards the east account for only 12% of the total area. Hence, 12% of the total area of the drainage basin gets a relatively small amount of solar radiation and has a high probability of infiltration rate.

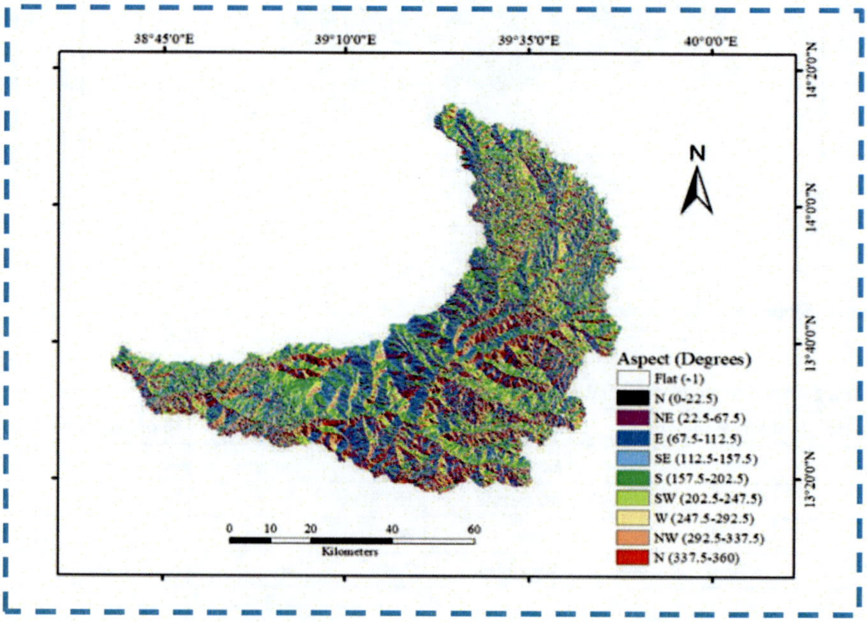

Fig. 6 Aspect map of Geba drainage basin

Table 7 Areal aspect distribution of Geba drainage basin

S. No	Aspect	Degree covered	Area coverage(km^2)	% of total area
1	Flat	− 1	5.9	0.1
2	North	0–22.5	314.7	6.1
3	North east	22.5–67.5	597.8	11.6
4	East	67.5–112.5	617.8	12
5	South east	112.5–157.5	646.9	12.6
6	South	157.5–202.5	695.1	13.5
7	South west	202.5–247.5	710	13.8
8	West	247.5–292.5	668	13
9	North west	292.5–337.5	610.3	11.9
10	North	337.5–360	271.1	5.3
Total			5137	100

4 Conclusion

In this study, attempts have been made to assess the morphometric parameters in relation to the hydrologic condition of the Geba drainage basin. For this purpose, the remote sensing and GIS technologies were used as a data source and data processing tool, respectively. The study was mainly focused on characterization of the morphometric parameters that are mainly related to hydrology, notably, Bifurcation ratio, Stream frequency, Drainage density, Drainage texture, Infiltration number, Length of overland flow, Elongation ratio, Circulatory ratio, Form factor, Compactness coefficient, Constant of channel maintenance, Basin relief, Relief ratio, Ruggedness number, Channel gradient, and Basin slope. After processing the DEM data, the numerical values of these parameters were tabulated using different mathematical equations suggested by different experts. The calculated values were then classified into high, moderate, and low based on the morphometric characterization scheme. It was found that values of Stream frequency, Infiltration number, Elongation ratio, Form factor, Constant of channel maintenance, Relief ratio, Channel gradient, and Basin slope are low. Drainage density is coarse, and the remained parameters fall under moderate category. All these results indicate that the Geba river basin is characterized by elongated shape, relatively moderate infiltration capacity, and hence with moderate groundwater potential. The output of this study can help planners as a precondition for water and soil conservation and assessment of the groundwater potential.

Acknowledgements This research was carried out by the support of Raya University and Ethiopian Ministry of Science and Higher Education.

Bibliography

1. Altaf, F., Meraj, G., & Romshoo, S. A. (2013). *Morphometric analysis to infer hydrological behaviour of Lidder watershed*. Western Himalaya.
2. Aravinda, P. T., & Balakrishna, H. B. (2013). Morphometric analysis of Vrishabhavathi watershed using remote sensing and GIS. *Int J Res Eng Technol, 2*(8), 514–522.
3. Bai, R., Li, T., Huang, Y., Li, J., & Wang, G. (2015). An efficient and comprehensive method for drainage network extraction from DEM with billions of pixels using a size-balanced binary search tree. *Geomorphology, 238*, 56–67. https://doi.org/10.1016/j.geomorph.2015.02.028
4. Chen, L., Yuan, Y., Yuan, X., Yang, X., Huang, J., Yu, Y. (2018). Threshold selection of river network extraction based on different DEM scales using ATRIC algorithm. In: *IOP Conference Series: Materials Science and Engineering* (p. 052047).
5. Clarke, J. (1966). Morphometry from maps. *Essays in Geomorphology* 235–274.
6. Dubey, S. K., Sharma, D., & Mundetia, N. (2015). Morphometric analysis of the Banas River Basin using the geographical morphometric analysis of the Banas River Basin using the geographical information system, Rajasthan, India. *Hydrology, 3*(5), 47–57. https://doi.org/10.11648/j.hyd.20150305.11
7. Faniran, A. (1968). The index of drainage intensity: a provisional new drainage factor. *Australian Journal of Science, 31*(9), 326–330.
8. Farhan, Y. (2017). Morphometric assessment of Wadi Wala Watershed, Southern Jordan Using ASTER (DEM) and GIS. *Journal of Geographic Information System, 9*(2), 158–190. https://doi.org/10.4236/jgis.2017.92011
9. Fenta, A. A., Yasuda, H., Shimizu, K., Haregeweyn, N., & Woldearegay, K. (2017). Quantitative analysis and implications of drainage morphometry of the Angulo watershed in the semi-arid northern Ethiopia. *Applied Water Science, 7*(7), 3825–3840. https://doi.org/10.1007/s13201-017-0534-4
10. Gebre, T., Kibru, T., Tesfaye, S., & Taye, G. (2015). Analysis of watershed attributes for water resources management using GIS: The case of Chelekot micro-watershed. *Journal of Geographic Information System, 7*(2), 177–190. https://doi.org/10.4236/jgis.2015.72015
11. Gebreyohannes, T., De Smedt, F., Walraevens, K., Gebresilassie, S., Hussien, A., Hagos, M., Amare, K., Deckers, J., & Gebrehiwot, K. (2017). Modèle d'écoulement régional d'eaux souterraines du bassin de Geba Nord de l'Ethiopie. *Hydrogeology Journal, 25*(3), 639–655. https://doi.org/10.1007/s10040-016-1522-8
12. Gravelius, H. (1941). *Flusskunde*. Goschen'sche Verlagshandlung.
13. Hindersah, R., Handyman, Z., Indriani, F. N., Suryatmana, P., & Nurlaeny, N. (2018). Azotobacter population, soil nitrogen and groundnut growth in mercury-contaminated tailing inoculated with Azotobacter. *Journal of Degraded and Mining Lands Management, 5*(53), 2502–2458. https://doi.org/10.15243/jdmlm
14. Horton, R. E. (1945). Erosional development of streams and their drainage basins; hydrophysical approach to quantitative morphology. *Geological Society of America Bulletin, 56*(3), 275–370.
15. Horton, R. E. (1932). Drainage-basin characteristics. *EOS, Transactions of the American Geophysical Union, 13*(1), 350–361.
16. Huggel, C., Schneider, D., Miranda, P. J., Granados, H. D., & Kääb, A. (2008). Evaluation of ASTER and SRTM DEM data for lahar modeling: A case study on lahars from Popocatépetl Volcano, Mexico. *Journal of Volcanology and Geothermal Research, 170*(1–2), 99–110. https://doi.org/10.1016/j.jvolgeores.2007.09.005
17. Kabite, G., & Gessesse, B. (2018). International soil and water conservation research hydrogeomorphological characterization of Dhidhessa River basin. *International Soil and Water Conservation Research, 6*(2), 175–183. https://doi.org/10.1016/j.iswcr.2018.02.003
18. Kumar, P., & Kshitij, R. (2017). A GIS-based approach in drainage morphometric analysis of Kanhar River Basin, India. *Applied Water Science, 7*(1), 217–232. https://doi.org/10.1007/s13201-014-0238

19. Maathuis, B. H. P., & Wang, L. (2008). Digital elevation model based hydro-processing digital elevation model based hydro-processing. *Geocarto International, 21*(1), 21–26. https://doi.org/10.1080/10106040608542370
20. Magesh, N. S., Chandrasekar, N., & Soundranayagam, J. P. (2011). Morphometric evaluation of Papanasam and Manimuthar watersheds, parts of Western Ghats, Tirunelveli district, Tamil Nadu, India: A GIS approach. *Environment and Earth Science, 64*(2), 373–381. https://doi.org/10.1007/s12665-010-0860-4
21. Meshram, S. G., & Sharma, S. K. (2017). Prioritization of watershed through morphometric parameters: A PCA-based approach. *Applied Water Science, 7*(3), 1505–1519. https://doi.org/10.1007/s13201-015-0332-9
22. Miller, V. C. (1953). A quantitative geomorphologic study of drainage basin characteristics in the Clinch Mountain area, Virginia and Tennessee. Technical report 3, Columbia University, New York
23. Moges, G., & Bhole, V. (2015). Morphometric Characteristics and the Relation of Stream Orders to Hydraulic Parameters of River Goro: An Ephemeral River in Dire-Dawa, Ethiopia. Univers J Geosci 3(1):13–27. https://doi.org/10.13189/ujg.2015.030102
24. Nag, S. (1998). Morphometric analysis using remote sensing techniques in the Chaka sub-basin, Purulia district, West Bengal. *Journal of the Indian Society of Remote Sensing, 26*(1):69–76
25. Panhalkar, S. S. (2014). RETRACTED: Hydrological modeling using SWAT model and geoinformatics techniques. *The Egyptian Journal of Remote Sensing and Space Science, 17*(2), 197–207. https://doi.org/10.1016/j.ejrs.2014.03.001
26. Pareta, K., Pareta, U., & Decisions, S. (2011). Quantitative morphometric analysis of a watershed of Yamuna Basin, India using ASTER (DEM) data and GIS. *International Journal of Geomatics and Geosciences, 2*(1), 248–269.
27. Patel, D. P., Gajjar, C. A., & Srivastava, P. K. (2013). Prioritization of Malesari mini-watersheds through morphometric analysis: Remote sensing and GIS perspective. *Environment and Earth Science, 69*(8), 2643–2656. https://doi.org/10.1007/s12665-012-2086-0
28. Perucca, L. P., & Angilieri, Y. E. (2011). Morphometric characterization of Del Molle Basin applied to the evaluation of flash floods hazard, Iglesia Department, San Juan, Argentina. *Quaternary International, 233*(1), 81–86. https://doi.org/10.1016/j.quaint.2010.08.007
29. Schumm, S. A. (1956). Evolution of drainage systems and slopes in bad lands at Perth Amboy, New Jersey. *Geological Society of America Bulletin, 67*(5), 597–646.
30. Singh, P., Gupta, A., & Singh, M. (2014). Hydrological inferences from watershed analysis for water resource management using remote sensing and GIS techniques. *The Egyptian Journal of Remote Sensing and Space Science, 17*(2), 111–121. https://doi.org/10.1016/j.ejrs.2014.09.003
31. Singh, V., & Singh, U. C. (2011). Basin Morphometry of Maingra River, district Gwalior, Madhya Pradesh. *International journal of Geomatics and Geosciences, 1*(4), 891–902.
32. Smith, K. G. (1950). Standards for grading texture of erosional topography. *American Journal of Science, 248*, 655–668.
33. Strahler, A. N. (1964). Part II. Quantitative geomorphology of drainage basins and channel networks. In *Handbook of Applied Hydrology*. McGraw-Hill, New York, pp. 4–39
34. Strahler, A. N. (1952). Hypsometric (area-altitude) analysis of erosional topography. *Geological Society of America Bulletin, 63*(11), 1117–1142.
35. Thomas, J., Joseph, S., Thrivikramji, K. P., Abe, G., & Kannan, N. (2012). Morphometrical analysis of two tropical mountain river basins of contrasting environmental settings, the southern Western Ghats, India. *Environmental Earth Sciences, 66*(8), 2353–2366. https://doi.org/10.1007/s12665-011-1457-2
36. Wilford, D. J., Sakals, M. E., Innes, J. L., Sidle, R. C., & Bergerud, W. (2004). Recognition of debris flow, debris flood and flood hazard through watershed morphometrics. *Landslides, 1*(1), 61–66. https://doi.org/10.1007/s10346-003-0002-0

Validation and Correction of GSMaP_Gauge Product Over the Nagavali Basin in Eastern India

G. Venkata Rao and Keesara Venkata Reddy

Abstract Accurate estimate of precipitation is important for many hydro-climatological applications. In recent years, the use of satellite-based precipitation estimates has been increased in hydrological simulations due to the availability of high spatial and temporal resolutions. However, with the implementation of additional data and continuous improvements in data processing algorithms, it is necessary to understand and document satellite-based precipitation estimates. In the present study, the error characteristics of GSMaP_Gauge (GG) rainfall estimates have been assessed using gridded data provided by India Meteorological Department (IMD). The error characteristics were assessed using various validation matrices such as statistical indices, contingency statistics, and error decomposition. A simple multiplication scheme was used to apply bias correction to GG rainfall estimates. After applying the bias correction, significant improvement was observed in the GG rainfall estimates. The findings from the present analysis will be further used to establish the better links between satellite-based rainfall estimates and observed data.

Keywords Bias correction · Contingency statistics · Error decomposition · GSMaP · Statistical indices

1 Introduction

Climate monitoring, weather forecasting, water resource management, reservoir operation, agricultural modeling, food security, and emergency response planning all rely on precipitation, which is a fundamental element of the global hydrological cycle [1–4]. The estimation and accurate representation of precipitation are required planning and execution of water-related projects [5]. However, the accurate

G. Venkata Rao (✉) · K. Venkata Reddy
Department of Civil Engineering, National Institute of Technology Warangal, Telangana 506004, India
e-mail: gvenkat@sudent.nitw.ac.in

K. Venkata Reddy
e-mail: kvreddy@nitw.ac.in

A. K. Dikshit et al. (eds.), *Innovative Trends in Hydrological and Environmental Systems*, Lecture Notes in Civil Engineering 234,
https://doi.org/10.1007/978-981-19-0304-5_34

measurements of precipitation over large spatial extent are still unknown because of the spatiotemporal variability [6]. The availability of water resources in a river or catchment area is directly influenced by the spatiotemporal distribution of rainfall [7].

Rain gauges, weather radars, and sensors on-board satellites are the most commonly used instruments for measuring precipitation [8, 9]. Rain gauge provides point observations of precipitation [10] and is assumed to be most accurate [1, 11]. Rain gauge measures the precipitation for the very small area and is often taken to be representative of a large area [12]. However, the data is constrained on spatiotemporal scales [13].

To overcome the drawbacks of gauge-based measurements, weather radar is a useful tool for monitoring rainfall at both the temporal and spatial scales [14, 15]. The data obtained from the radar is subjected to a large number of errors and is not available all over the world [16]. However, due to poor technological infrastructure, inaccessible topography, or excessive cost, both rain gauge and radar are not feasible in some areas. Therefore, the use of satellite data is an alternative for quantitative estimation of precipitation [17, 18].

A number of high-resolution multi-satellite rainfall products have been developed since the launch of the Tropical Rainfall Measuring Mission (TRMM) in late 1997, such as Climate Prediction Center (CPC), CPC MORPHing technique (CMORPH), TRMM Multi-satellite Precipitation Analysis (TMPA), Global Precipitation Climatology Project (GPCP), Global Satellite Mapping of Precipitation (GSMaP), Precipitation Observation (TAMSAT) [19]. The passive microwave radiometers are frequently combined with geostationary infrared data to provide these products [20].

The GSMaP project began in 2002 with the purpose of providing high-precision and high-resolution global precipitation estimates and was supported by the Japan Science and Technology Agency and the Japanese Aerospace Exploration Agency (JAXA) [19]. The GSMaP product uses almost all passive microwave radiometer data and provides a one hourly worldwide gridded rain rate with a horizontal precision of 0.1° for the 60°N–60°S latitude range.

At various spatial scales, several researchers have attempted to evaluate and validate the GSMaP product with other precipitation products. Tian et al. [21] validated the GSMaP rainfall estimates and found an overestimation in summer rainfall and underestimation in winter rainfall. Similarly, in the arid and semi-arid regions of northwest India, GSMaP products underestimated rainfall occurrences in the wetter region than in drier regions [22]. Shrestha et al. [23] compared GSMaP rainfall products over the Nepal Himalayas using gauge observation and reported that the performance of GSMaP rainfall estimates showed inaccuracies with the increase in altitude. Prakash et al. [24] evaluated the performance of GSMaP_MVK and GSMaP_NRT products against the India Meteorological Department (IMD) for the southwest monsoon and found GSMaP products underestimated the rainfall over most parts of the India. With the implementation of additional data and updation of data processing algorithms, GSMaP version 6 products showed better accuracy in the estimation of low to moderate rainfall events across India [25]. In another study by Parida et al. [20] for a rainfall extreme event over Uttarakhand state India, GSMaP_Gauge underestimated rainfall amount in the southern and eastern parts of the state and overestimated amount and intensity of rainfall in the northern parts.

From previous studies, it is evident that GSMaP is able to capture the spatial variations of rainfall at regional and spatial scales. However, the data suffer from underestimation or overestimation compared with gauge observations. Additionally, rainfall products are updated continuously in the input datasets as well as data processing algorithms.

Hence, it is quite important to have region or basin-specific bias-corrected rainfall be- fore using them for hydrological applications. As a result, the aim of this research is to validate the performance of the GSMaP_Gauge product in order to gain a better understanding of the Nagavali river basin in India. The study area, data used, methodology, and findings are provided in the following sections.

2 Study Area

The river Nagavali, also known as Langulya (Fig. 1), is an independent and interstate river. The river originates at Bijipur hills of the Eastern ghat near Lakhabahal village in Kalahandi district of Odisha [7]. The catchment area of the river is about 9510 km^2 with a total length of the river being about 256 km. The average annual

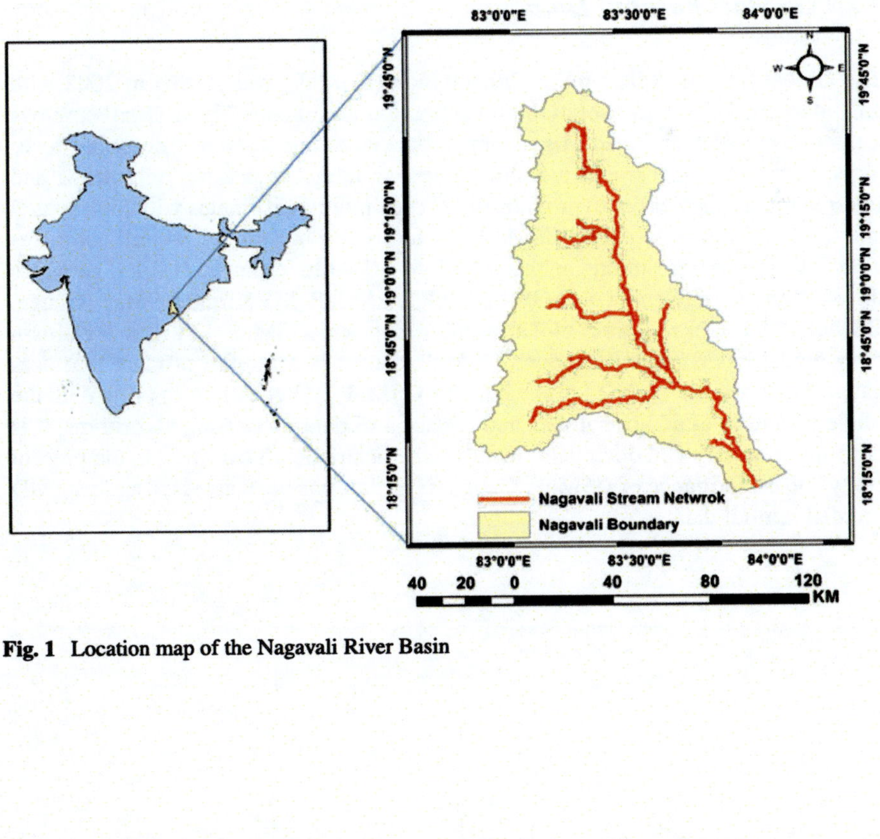

Fig. 1 Location map of the Nagavali River Basin

rainfall over the basin is 1230 mm. The river flows through three districts of Odisha and two districts of Andhra Pradesh and drains into the Bay of Bengal (BoB). The people in the catchment area are mainly dependent on agriculture. The basin has both Kharif and Rabi types of crops such as paddy, sugarcane, cotton, red gram, pulses, sunflower, maize, and sesame.

3 Data

3.1 Observed Rainfall Data

For the evaluation of GSMaP_Gauge rainfall estimates, daily rainfall records for a period of four years (i.e., 2015–2019) available at the India Meteorological Department (IMD) are used. The data has horizontal resolution of 0.25° and more information about the data may be obtained at Pai et al. [26]. In the daily time series, there are no missing data. The GSMaP_Gauge was verified against the observed rainfall data provided by Mahanadi & Eastern Rivers Organization (M&ERO), Central Water Commission (CWC), Bhubaneswar as a quality control measure, and a good correlation (0.79) was found.

3.2 GSMaP Rainfall Data

The Global Satellite Mapping of Precipitation (GSMaP) was started in 2002 with an aim to produce high-resolution global precipitation maps. The GSMaP employs a microwave-IR Combined Algorithm, a backward and forward morphing technique using IR images, and a Kalman filter to integrate passive microwave and infrared retrievals in order to provide global precipitation estimates with high spatial and temporal resolution. The GSMaP provides one-hour global rainfall estimates with 0.1° spatial resolution for the 60° N–60° S latitude range. GSMaP provides three different rainfall data, i.e., GSMaP_NRT, GSMaP_MVK, and GSMaP_Gauge. GSMaP_NRT is a real-time global rainfall map and GSMaP_MVK is a Kalman filter model-based rainfall product. GSMaP_Gauge is a rainfall product that uses gauge data provided by NOAA to adjust the GSMaP_MVK estimates [19]. With the implementation of additional data and updation of data processing algorithms, it is crucial to quantify and document the error characteristics. Therefore, in the present study, the performance of GSMaP_Gauge (GG) data has been assessed against IMD gridded rainfall data.

4 Methods

4.1 Validation Metrics

In order to understand the error characteristics of GG rainfall estimates, several statistical and categorical indices such as statistical indices, contingency statistics, and error decomposition techniques have been used in the present analysis. Statistical indices namely, correlation coefficient (CC), root mean square error (RMSE), relative error (RBIAS), and mean error (ME) are used to quantify the difference between GG rainfall estimates and IMD gridded rainfall dataset. Contingency statistics including probability of detection (POD), false alarm ration (FAR), critical success index (CSI), and true skill statistics (TSS) are used to examine the correspondence between GG rainfall estimates and reference datasets. For a better understanding of ME, it is further divided into three independent error components such as hit bias (HB), missed bias (MB), and false bias (FB). This method of partitioning the overall error has demonstrated that it can help people better understand and identify the error characteristics of GG rainfall estimates. The mathematical expressions for the statistical indices, contingency statistics, and error decomposition are presented in Table 1 (source: https://www.cawcr.gov.au/projects/verification/).

Where, O is the IMD gridded rainfall data, S is the satellite rainfall data, H, M, and F are the number of rain events correctly detected, not detected, and wrongly detected. Detailed information about H, M, and F is presented in Table 2.

4.2 Bias Correction of GSMaP_Gauge Rainfall

The use of accurate satellite-based precipitation estimates has the potential to reduce uncertainties in hydrological simulations. Bias correction of simple GG rainfall estimates is conducted by using a simple multiplication bias correction scheme. The ratio between the gridded and satellite precipitation estimates is calculated and that is multiplied by the satellite estimates to arrive at the bias-corrected rainfall estimates. The mathematical expression for calculating the bias factors (BF) is as follows.

$$\mathrm{BF} = \frac{\sum_{i=1}^{n} O_i}{\sum_{i=1}^{n} S_i}$$

where, i is the grid location and n is the total number of grids analyzed. To calculate the bias factors, The intensity of precipitation is divided into seven bins: (1) 1 mm < R < 2 mm, (2) 2 mm < R < 4 mm, (3) 4 mm < R < 8 mm (4) 8 mm < R < 16 mm, (5) 16 mm < R < 32 mm, (6) 32 mm < R < 64 mm, (7) 64 mm < R < 128 mm. The bias factor for each bin was then calculated.

Table 1 Validation matrices

Statistics	Formula	Range	Unit	Best value
Statistical indices				
Correlation coefficient	$CC = \frac{\sum_{i=1}^{n}(O_i = \overline{O})(S_i - \overline{S})}{\sqrt{\sum_{i=1}^{n}(O_i = \overline{O})^2}\sqrt{\sum_{i=1}^{n}(S_i - \overline{S})^2}}$	0–1	None	1
Relative error	$RBIAS = \frac{\sum_{i=1}^{n}(S_i - O_i)}{\sum_{i=1}^{n} O_i}$	0–∞	None	1
Mean error	$ME = \frac{1}{n}\sum_{i=1}^{n}(S_i - O_i)$	− ∞ to ∞	mm	0
Root mean square error	$RMSE = \sqrt{\frac{1}{n}\sum_{i=1}^{n}(S_i - O_i)^2}$	0–∞	mm	0
Contingency statistics				
Probability of detection	$POD = \frac{H}{H+M}$	0–1	None	1
False alarm ratio	$FAR = \frac{F}{H+F}$	0–1	None	0
Critical success index	$CSI = \frac{H}{H+M+F}$	0–1	None	1
True skill statistics	$TSS = \frac{(H \times CS) - (M \times F)}{(H+M) - (F+CS)}$	0–1	None	1
Error decomposition				
Hit bias	$HB = \frac{1}{n}\sum_{h=1}^{H}(S_h - O_h)$	− ∞ to ∞	mm	0
Missed bias	$MB = \frac{1}{n}\sum_{m=1}^{M} O_m$	− ∞ to ∞	mm	0
False bias	$FB = \frac{1}{n}\sum_{f=1}^{F} S_f$	− ∞ to ∞	mm	0

Table 2 Contingency table

GG rainfall	IMD rainfall	
	Rain (> 1 mm/d)	No rain (≤ 1 mm/d)
Rain	Hit (H)	False (F)
No rain	Miss (M)	Correct negative (CS)

5 Results and Discussions

The error characteristics of GG data were analyzed on a daily time scale against the IMD gridded rainfall product. The results of the present analysis: spatial analysis of the error components, error decomposition analysis, and bias correction are presented in the following sections. This analysis will provide an overview of the error characteristics of GSMaP_Gauge product over the Nagavali river basin.

5.1 *Spatial Analysis of the Statistical Indices*

In the present analysis, CC, ME, RMSE, and RBIAS were calculated on a daily time scale to analyze the GG rainfall product over the Nagavali river basin. The CC for the daily data ranged from 0.567 to 0.702. The ME values ranged from -0.824 to 1.146. From the ME values, it is observed that the GG data showed both underestimation and overestimation of rainfall over the Nagavali basin. RMSE showed larger error values which ranged from 10.097 to 16.716. The results shown by RBIAS are similar to ME. The spatial distribution of CC, ME, RMSE, and RBIAS are shown in Fig. 2.

From the figure, it is evident that the GG data showed highest correlation with observed data in the middle of the basin and the lowest in the lower portion of the basin. In the upper portion of the Nagavali basin, the correlation coefficient was moderate. From the ME, is observed that GG data underestimated the precipitation

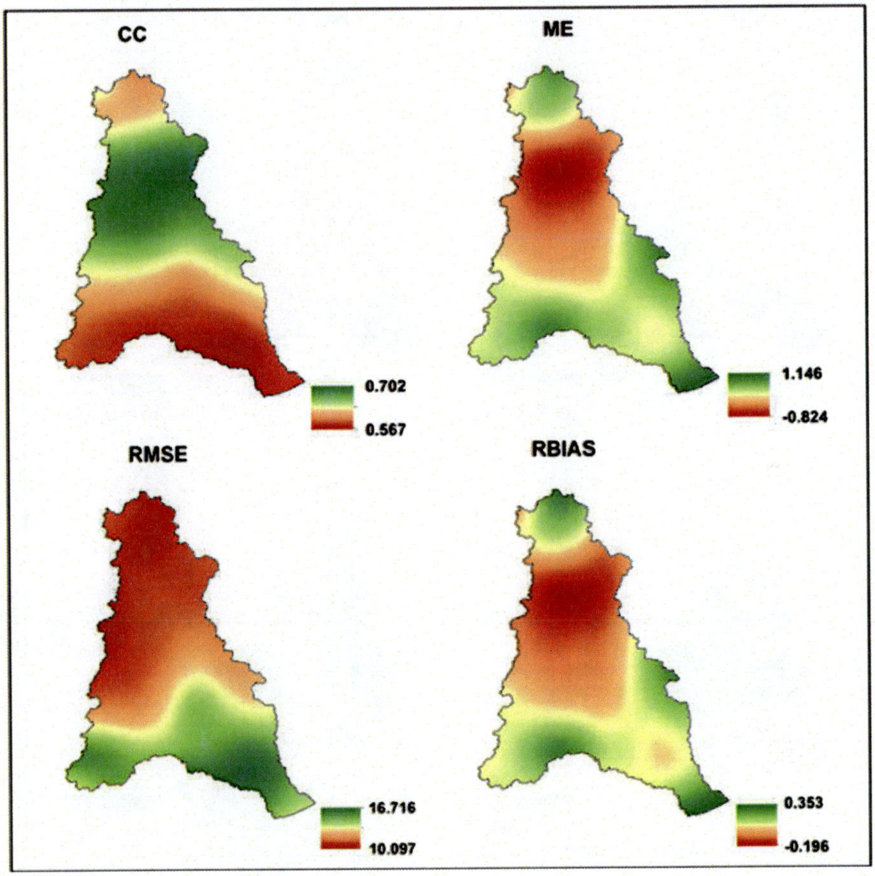

Fig. 2 Spatial plots of statistical indices (CC, ME, RMSE, and RBIAS)

in the middle portion and overestimated in both the lower and upper portions of the basin. The spatial patterns of RBIAS are similar to ME. The GG data showed lower RMSE values in the middle and upper portions of the basin and higher values in the lower portion.

5.2 *Spatial Analysis of the Contingency Statistics*

Contingency statistics are calculated with a rain/no-rain threshold of 1 mm/day. These statistics can reveal the performance of the GG product. The spatial distribution of contingency statistics is presented in Fig. 3. The POD values for GG data ranged from 0.511 to 0.653. The higher POD values are located all over Nagavali basin except in

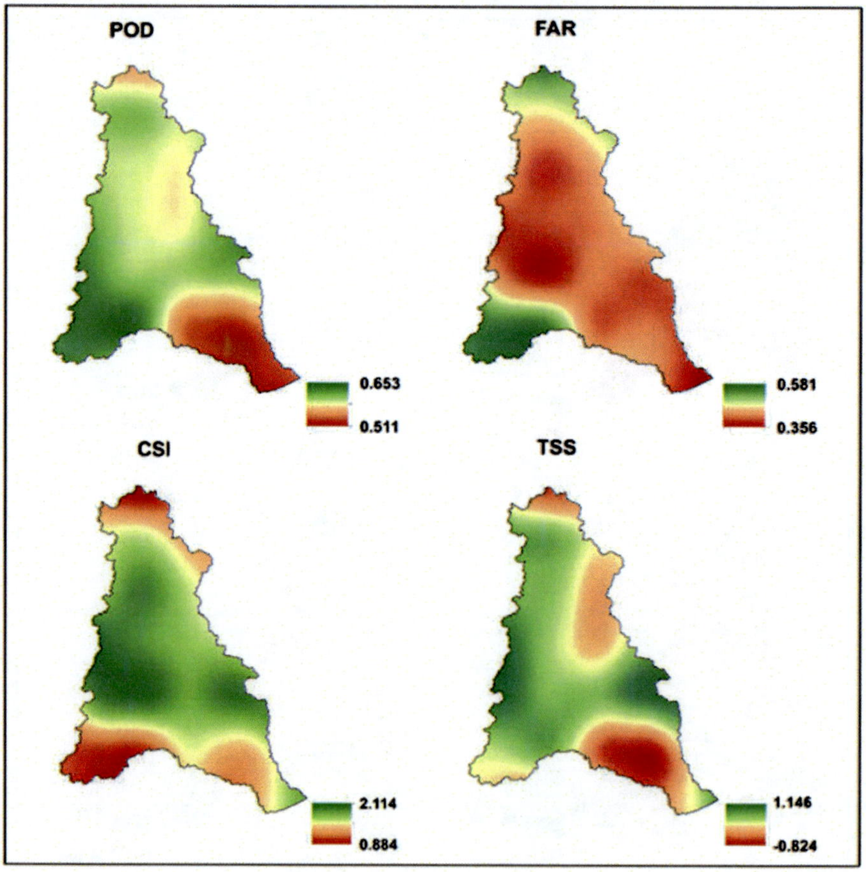

Fig. 3 Spatial plot of contingency statistics

the lower delta region. The FAR values ranged from 0.356 to 0.581. The lower values of FAR were found all over the basin except in some small areas. The values of CSI and TSS ranged between 0.330 to 0.446 and 0.355 to 0.459, respectively. From the spatial patterns, it is observed that the higher values of CSI and TSS are located in the middle and upper portions of the basin.

5.3 Error Decomposition

The ME of GG rainfall product is further divided into three independent components: Hit Bias, Missed Bias, and False Bias. The box plot of these biases is presented in Fig. 4. When compared to HB, the GG rainfall product showed higher MB and FB values. The higher values of MB and FB indicated that the GG rainfall product has more missed and false rainfall. The spatial distribution of HB, MB, FB, and total bias is presented in Fig. 5. From the spatial plots, it is evident that the GG data underestimated the hit precipitation events in the upper and upper-middle portions of the basin and overestimated in the lower middle and lower portions of the basin. The GG data similar smaller missed bias values in the inner middle and lower portions

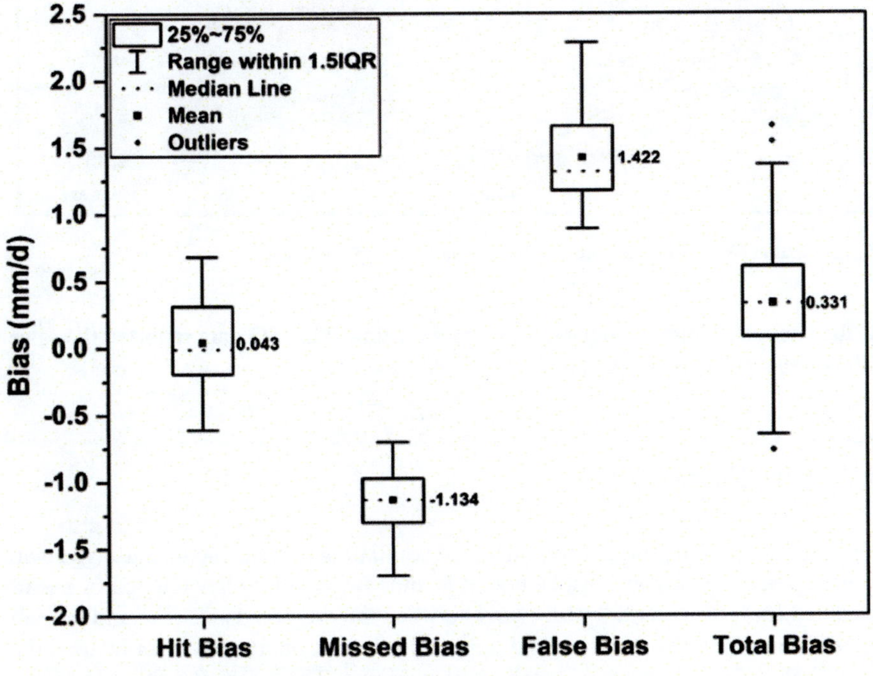

Fig. 4 Box plot of error decomposition of GG rainfall product

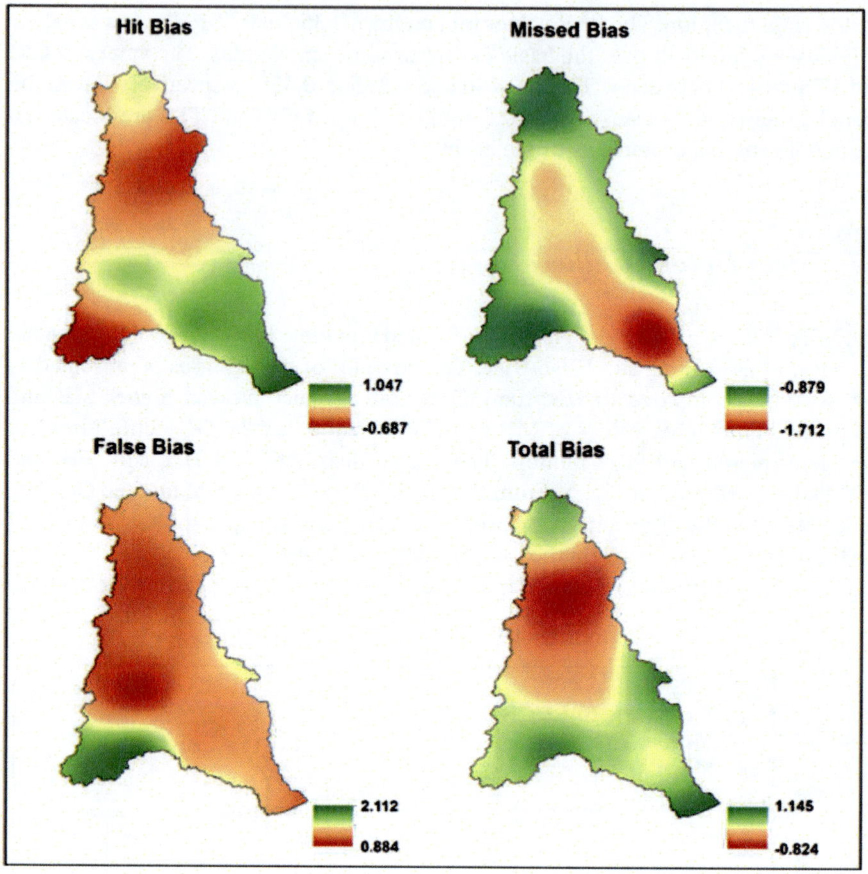

Fig. 5 Spatial plots of hit bias, missed bias, false bias, and total bias

of the basin and higher values in the upper portion. The GG data reported the false rainfall events across the basin.

5.4 Bias Correction

The satellite rainfall products were updated continuously in the input datasets as well as data processing algorithms. Hence, it is quite important to have region or basin-specific bias factors to minimize errors between the estimated and observed rainfall before using them for hydrological applications. To calculate the bias factors, the precipitation intensities are grouped into seven bins. The bias factor was calculated as the ratio of sum of precipitation amount in each bin to the sum of observed

Table 3 Bias factor for GSMaP rainfall product at different intensities

Bin	Bias factor	Bia	Bias factor
1 mm < *R* < 2 mm	0.926	16 mm < *R* < 32 mm	0.721
2 mm < *R* < 4 mm	1.188	32 mm < *R* < 64 mm	0.802
4 mm < *R* < 8 mm	1.321	64 mm < *R* < 128 mm	0.685
8 mm < *R* < 16 mm	1.333		

precipitation amount in that particular bin. The bias factors for the rainfall intensities in each bin are given in Table 3.

In order to assess if there was an improvement in the GG rainfall products after applying the bias factors to the respective bins, the bias-corrected GG rainfall products were evaluated by recalculating statistical indices and the spatial plots are presented in Fig. 6. After bias correction, significant improvement was observed in the statistical indices. The range of CC increased from 0.609 to 0.735. The range of ME and RMSE decreased from − 0.824 to 1.146 and 10.097 to 16.716 to − 0.565 to 0.656 and 7.191 to 12.893, respectively. The results shown by RBIAS are similar to that of ME.

6 Conclusions

The GSMaP global rainfall estimates have been widely used in various hydro-meteorological applications due to the availability of high spatial and temporal resolutions. In the present analysis, the error characteristics of GSMaP_Gauge have been assessed using various validation matrices such as statistical indices, contingency statistics, error decomposition, and bias correction. From the results, it is observed that the missed and false rainfall events in GSMaP_Gauge data are responsible for higher values of ME and RMSE between the GSMaP rainfall estimates and observed datasets. To calculate the bias factors, the entire data was divided into seven bins and the bias factors for each bin were calculated. After applying the bias correction to GG rainfall estimates, significant improvement was observed in the statistical indices. From the analysis, it is observed that, before applying the GSMaP rainfall estimates for hydro-meteorological application, it is recommended to perform bias correction with region-specific bias factors.

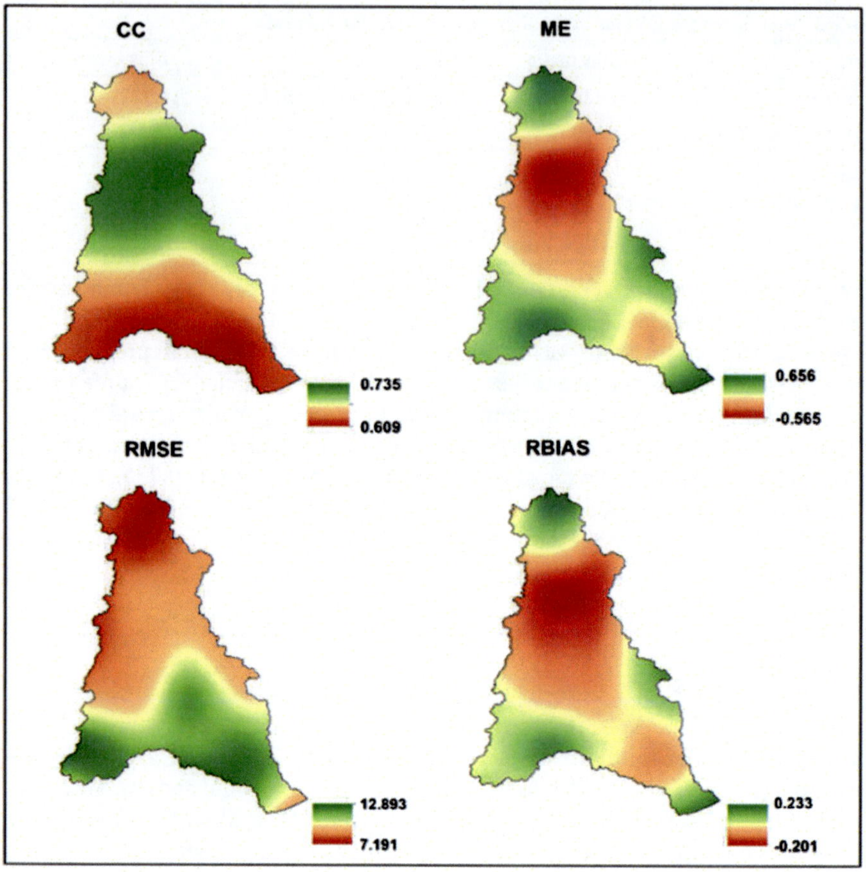

Fig. 6 Spatial plots of statistical errors after the bias correction of GG rain-fall product

Acknowledgements The research described in this paper was carried out with funds made available by Ministry of Human Resource Development (MHRD), Government of India under the Scheme for Promotion of Academic and Research Collaboration (SPARC) through project number P270.

References

1. Brocca, L., Pellarin, T., Crow, W.T., Ciabatta, L., Massari, C., Ryu, D., Su, C.- H., Rüdiger, C., Kerr, Y. (2016). Rainfall estimation by inverting SMOS soil moisture estimates: A comparison of different methods over Australia. *Journal of Geophysical Research: Atmospheres* 79–103. https://doi.org/10.1002/2016JD025382
2. Sehad, M., Lazri, M., & Ameur, S. (2017). Novel SVM-based technique to improve rainfall estimation over the Mediterranean region (north of Algeria) using the multispectral MSG

SEVIRI imagery. *Advances in Space Research, 59*, 1381–1394. https://doi.org/10.1016/j.asr.2016.11.042
3. Tarpanelli, A., Massari, C., Ciabatta, L., Filippucci, P., Amarnath, G., & Brocca, L. (2017). Exploiting a constellation of satellite soil moisture sensors for accurate rainfall estimation. *Advances in Water Resources, 108*, 249–255. https://doi.org/10.1016/j.advwatres.2017.08.010
4. Cecinati, F., Rico-Ramirez, M. A., Heuvelink, G. B. M., & Han, D. (2017). Representing radar rainfall uncertainty with ensembles based on a time-variant geostatistical error modelling approach. *Journal of Hydrology, 548*, 391–405. https://doi.org/10.1016/j.jhydrol.2017.02.053
5. Rawat, K. S., Mishra, A. K., & Ahmad, N. (2017). Surface runoff estimation over heterogeneous foothills of Aravalli mountain using medium resolution remote sensing rainfall data with soil conservation system-curve number method: A case of semi-arid ungauged Manesar Nala watershed. *Water Environment Journal, 31*, 262–276. https://doi.org/10.1111/wej.12243
6. Derin, Y., Anagnostou, E., Anagnostou, M. N., Kalogiros, J., Casella, D., Marra, A. C., Panegrossi, G., & Sano, P. (2018). Passive microwave rainfall error analysis using high-resolution X-band dual-polarization radar observations in complex Terrain. http://ieeexplore.ieee.org/document/8264698/ https://doi.org/10.1109/TGRS.2017.2763622
7. Rao, G. V., Venkata Reddy, K., Srinivasan, R., Sridhar, V., Umamahesh, N. V., & Pratap, D. (2020). Spatio-temporal analysis of rainfall extremes in the flood-prone Nagavali and Vamsadhara Basins in eastern India. *Weather and Climate Extremes, 29*, 100265. https://doi.org/10.1016/j.wace.2020.100265
8. Gires, A., Tchiguirinskaia, I., Schertzer, D., Schellart, A., Berne, A., & Lovejoy, S. (2014). Influence of small scale rainfall variability on standard comparison tools between radar and rain gauge data. *Atmospheric Research, 138*, 125–138. https://doi.org/10.1016/j.atmosres.2013.11.008
9. Waken, R. J., Song, J. J., Kwon, S., Min, K. H., & Lee, G. W. (2017). A flexible and efficient spatial interpolator for radar rainfall estimation. *Journal of Applied Statistics, 0*, 1–16. https://doi.org/10.1080/02664763.2017.1317723
10. Upadhyaya, S., & Ramsankaran, R. (2013) Review of satellite remote sensing DATA based rainfall estimation methods. In *Proceedings of Hydro 2013 International*. 4–6 December 2013, IIT Madras, India.
11. Hasan, M. M., Sharma, A., Mariethoz, G., Johnson, F., & Seed, A. (2016). Improving radar rainfall estimation by merging point rainfall measurements within a model combination framework. *Advances in Water Resources, 97*, 205–218. https://doi.org/10.1016/j.advwatres.2016.09.011
12. Kidd, C. (2001). Satellite rainfall climatology: A review. *International Journal of Climatology, 21*, 1041–1066. https://doi.org/10.1002/joc.635
13. Aksoy, H., Gedikli, A., Unal, N. E., Yilmaz, M., Eris, E., Yoon, J., & Tayfur, G. (2016). Rainfall-runoff model considering microtopography simulated in a laboratory erosion flume. *Water Resources Management, 30*, 5609–5624. https://doi.org/10.1007/s11269-016-1439-y
14. Van De Beek, C. Z., Leijnse, H., Hazenberg, P., & Uijlenhoet, R. (2016). Close-range radar rainfall estimation and error analysis. *Atmospheric Measurement Techniques, 9*, 3837–3850. https://doi.org/10.5194/amt-9-3837-2016
15. Kim, H. L., Suk, M. K., Park, H. S., Lee, G. W., & Ko, J. S. (2016). Dual-polarization radar rainfall estimation in Korea according to raindrop shapes obtained by using a 2-D video disdrometer. *Atmospheric Measurement Techniques, 9*, 3863–3878. https://doi.org/10.5194/amt-9-3863-2016
16. Rabiei, E., Haberlandt, U., Sester, M., Fitzner, D., & Wallner, M. (2016). Areal rainfall estimation using moving cars—computer experiments including hydrological modeling. *Hydrology and Earth System Sciences, 20*, 3907–3922. https://doi.org/10.5194/hess-20-3907-2016
17. Veerakachen, W., & Raksapatcharawong, M. (2015). Rainfall estimation for real time flood monitoring using geostationary meteorological satellite data. *Advance in Space Research, 56*, 1139–1145. https://doi.org/10.1016/j.asr.2015.06.016
18. Gado, T. A., Hsu, K., & Sorooshian, S. (2017). Rainfall frequency analysis for ungauged sites using satellite precipitation products. *Journal of Hydrology, 554*, 646–655. https://doi.org/10.1016/j.jhydrol.2017.09.043

19. Deng, P., Zhang, M., Guo, H., Xu, C., Bing, J., & Jia, J. (2018). Error analysis and correction of the daily GSMaP products over Hanjiang River Basin of China. *Atmospheric Research, 214*, 121–134. https://doi.org/10.1016/j.atmosres.2018.07.022
20. Parida, B. R., Behera, S. N., Bakimchandra, O., Pandey, A. C., & Singh, N. (2017). Evaluation of satellite-derived rainfall estimates for an extreme rainfall event over Uttarakhand, Western Himalayas. *Hydrology, 4*, 22. https://doi.org/10.3390/hydrology4020022
21. Tian, Y., Peters-Lidard, C. D., Adler, R. F., Kubota, T., & Ushio, T. (2010). Evaluation of GSMaP precipitation estimates over the contiguous United States. *Journal of Hydrometeorology, 11*, 566–574. https://doi.org/10.1175/2009JHM1190.1
22. Dinku, T., Ruiz, F., Connor, S. J., & Ceccato, P. (2010). Validation and intercomparison of satellite rainfall estimates over Colombia. *Journal of Applied Meteorology and Climatology, 49*, 1004–1014. https://doi.org/10.1175/2009JAMC2260.1
23. Shrestha, M. S., Takara, K., Kubota, T., & Bajracharya, S. R. (2011) Verification of GSMaP rainfall estimates over the Central Himalayas. *Journal of Japan Society of Civil Engineers, Ser. B1 (Hydraulic Engineering), 67*, I_37–I_42. https://doi.org/10.2208/jscejhe.67.I_37
24. Prakash, S., Mitra, A. K., Rajagopal, E. N., & Pai, D. S. (2016). Assessment of TRMM- based TMPA-3B42 and GSMaP precipitation products over India for the peak southwest monsoon season. *International Journal of Climatology, 36*, 1614–1631. https://doi.org/10.1002/joc.4446
25. Prakash, S., Mitra, A. K., AghaKouchak, A., Liu, Z., Norouzi, H., & Pai, D. S. (2018). A preliminary assessment of GPM-based multi-satellite precipitation estimates over a monsoon dominated region. *Journal of Hydrology, 556*, 865–876. https://doi.org/10.1016/j.jhydrol.2016.01.029
26. Pai, D. S., Sridhar, L., Rajeevan, M., Sreejith, O. P., Satbhai, N. S., Mukhopadhyay, B. (2014). Development of a new high spatial resolution (0.25° × 0.25°) long period (1901–2010) daily gridded rainfall data set over India and its comparison with existing data sets over the region. *MAUSAM, 65*, 1–18.

Estimation of Soil Loss Using RS and GIS Techniques

S. Supriya and A. S. Ravikumar

Abstract Flooding is a global phenomenon that causes casualties and property Soil loss is a most severe environmental problem. To take measures in controlling soil erosion it is required to estimate soil loss occurring in the area. In the present study, watersheds of Kollegal taluk, Chamarajanagar district, Karnataka, are chosen. The 18 watersheds having more than 20 km^2 are considered for analysis and they are represented by watershed-1 (WS1) to watershed-18 (WS18). The Universal Soil Loss Equation (USLE) is applied to predict soil loss magnitude and Geographic Information System (GIS) software ArcGIS 10.1 is used to simulate the soil loss in spatial distribution. USLE has proved to be very popular and it estimates long-term average annual gross soil erosion. This equation multiplies six parameters: rainfall erosivity factor, soil erodibility factor, slope length factor, slope gradient factor, crop cover or crop management factor and conservation practices factor. For this study, rainfall data, soil maps, land use/and cover Map and a Digital Elevation Model are used. The annual average soil erosion for the watersheds of Kollegal taluk varies from 0.88 to 10.58 t/ha/yr. WS11 shows the moderate soil loss, and WS15, WS16 and WS17 show slight to moderate erosion (ISRO-NNRMS-TR-103-2002).

Keywords GIS · Remote sensing · Soil loss · USLE · Kollegal taluk

1 Introduction

Soil is a precious gift of nature to mankind. Ironically, soil is the most neglected commodity on the earth. Soil erosion is one of the most serious problems for environmentalists, which must be taken into consideration to prevent ecological imbalance in nature, basically among natural resource like soil, water and plants. Soil erosion is the detachment and transportation of soil materials from one place to another, resulting in the removal of the uppermost fertile soil layer, thus decreasing its productivity [7].

S. Supriya (✉) · A. S. Ravikumar
Department of Civil Engineering, UVCE, Bangalore University, Bengaluru-56, India

A. K. Dikshit et al. (eds.), *Innovative Trends in Hydrological and Environmental Systems*, Lecture Notes in Civil Engineering 234,
https://doi.org/10.1007/978-981-19-0304-5_35

Overgrazing, deforestation, faulty cultivation and carelessly build roads in catchment areas have led to enormous soil erosion.

The soil erosion process is modified by biophysical environment comprising soil, climate, terrain, ground cover and interactions between them. Important terrain characteristics influencing the mechanism of soil erosion are slope, length, aspect and shape [2]. The accuracy of estimating soil erosion risk depends on prediction model and its factors. Researchers have developed many predictive models that estimate soil loss and identify areas where conservation measures will have the greatest impact on reducing soil loss for soil erosion assessments. United States Department of Agriculture (USDA), Wischmeier and Smith [11] developed Universal Soil Loss Equation (USLE) is the most widely accepted empirical model to estimate soil erosion.

2 Study Area

The study area chosen is Kollegal taluk which is the largest taluk of Chamarajanagar district in Karnataka. It lies to the East of Chamarajanagar, covering an area of 2796.47 km^2. the area is bounded by the latitude 11° 45′ 0″ N and 12° 45′ 20″ N and longitude of 77° 00′ 40″E and 77° 46′ 30″ E. The area comprises Survey of India (SOI) Toposheets bearing numbers 57D/16, 57H/3, 57H/4, 57H/7, 57H/8, 57H/12, 57H/16, 58E/1, 58E/5 and 58E/9 (Scale 1:50,000). Figure 1 shows the location map of the study area. The river Cauvery and its tributaries, viz. Palar, Suvarnavathi, Uduthoraihalla, Minnattuhalla and Doddahalla drains the study area.

3 Methodology

The topographical information of the watersheds on I:50,000 scale acquired from SOI was digitized and geo-referenced using ArcGIS 10.1 tool. Plate 1 shows the topographical map of Kollegal taluk. The mosaicking of these toposheets has been done, which is then used to delineate taluk boundary. The Cartosat-3.1 DEM data which is downloaded from Bhuvan Geoportal of ISRO is clipped from delineated taluk boundary. Plate 2 shows the Digital Elevation Model of Kollegal taluk. Digital Elevation Model is used to delineate watersheds by spatial analyst tool of ArcGIS to understand the hydrological features. 18 watersheds having area more than 20 km^2 are considered for present study. Figure 2 shows the watersheds delineated within Kollegal taluk. The spatial data Land use/land cover (LU/LC) map of Kollegal taluk is obtained from Karnataka State Remote Sensing and Application Centre (KSRSAC) Bengaluru and National Bureau of Soil Survey and Land Use Planning (NBSS and LUP). Daily rainfall data of 19 years from 1998 to 2016 for Kollegal taluk is collected from the Department of Statistics and Economics, Bengaluru. Soil data was downloaded from Food and Agriculture Organization of the United Nations (FAO). All these datasets are used in Universal Soil Loss Equation (USLE) to obtain

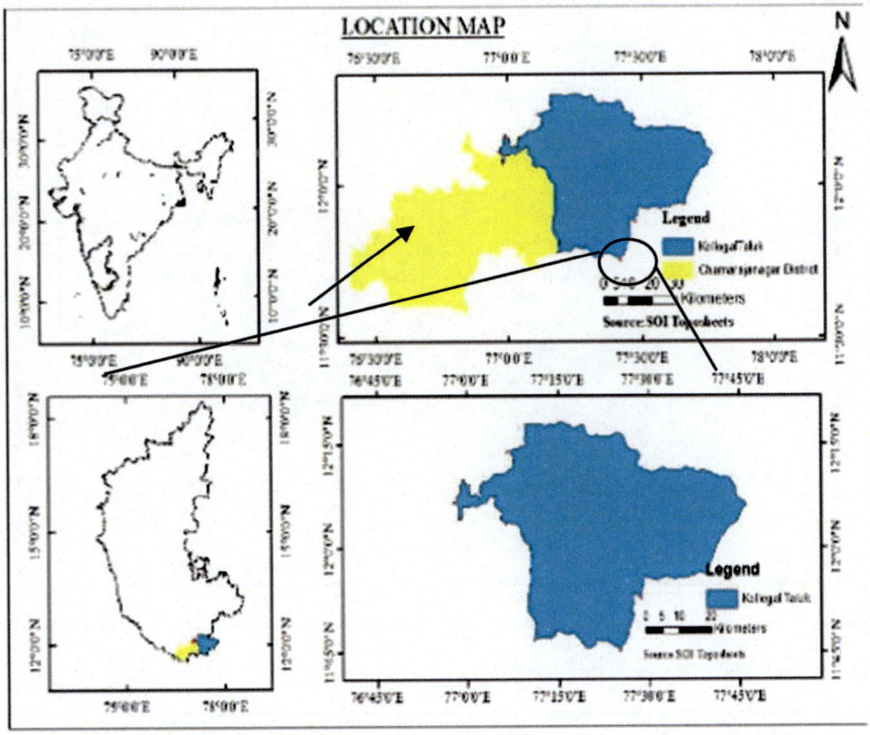

Fig. 1 Location map of the study area

annual average soil loss for the watersheds of Kollegal taluk. Methodology adopted in study is shown in Fig. 3.

3.1 Universal Soil Loss Equation (USLE)

Wischmeier and Smith, 1965 suggested USLE model, it is an erosion model designed to compute longtime average soil losses from sheet and rill erosion under specified conditions. It is also useful for construction sites and other non-agricultural conditions, but it does not predict deposition and does not compute sediment yields from gully, stream bank, and streambed erosion.

The equation is as follows:

$$A = RKLSCP \tag{1}$$

where

A Computed soil loss (t/ha/year).

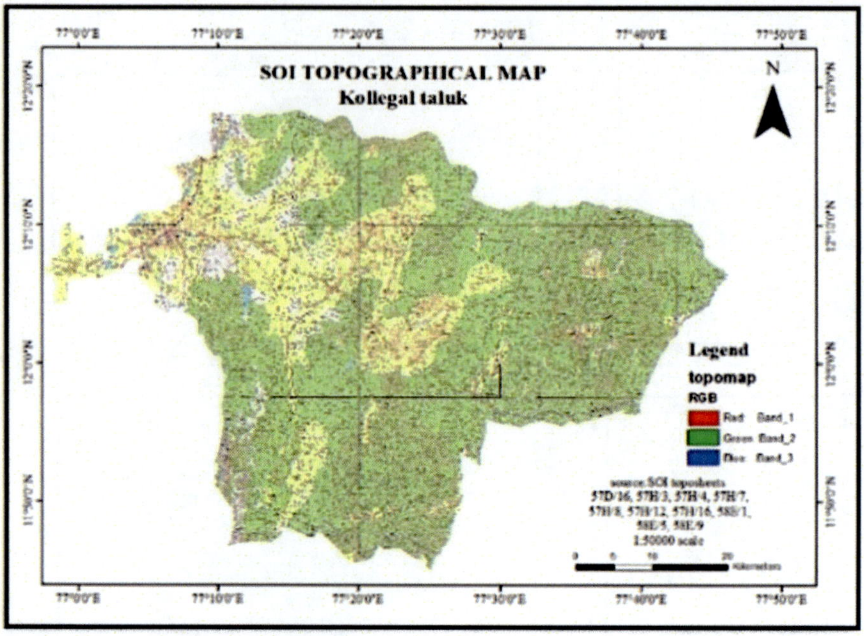

Plate 1 Topographical map of Kollegal taluk

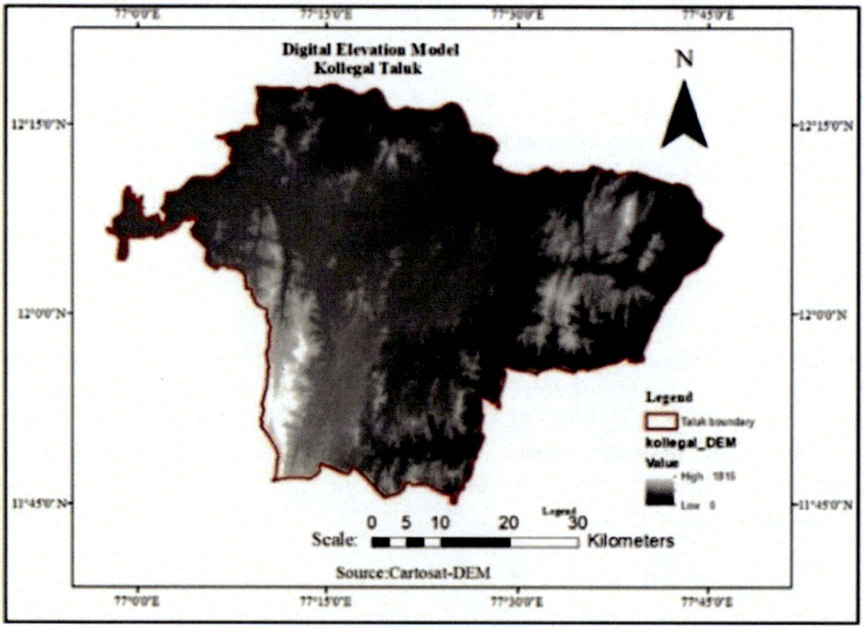

Plate 2 Digital elevation model of Kollegal taluk

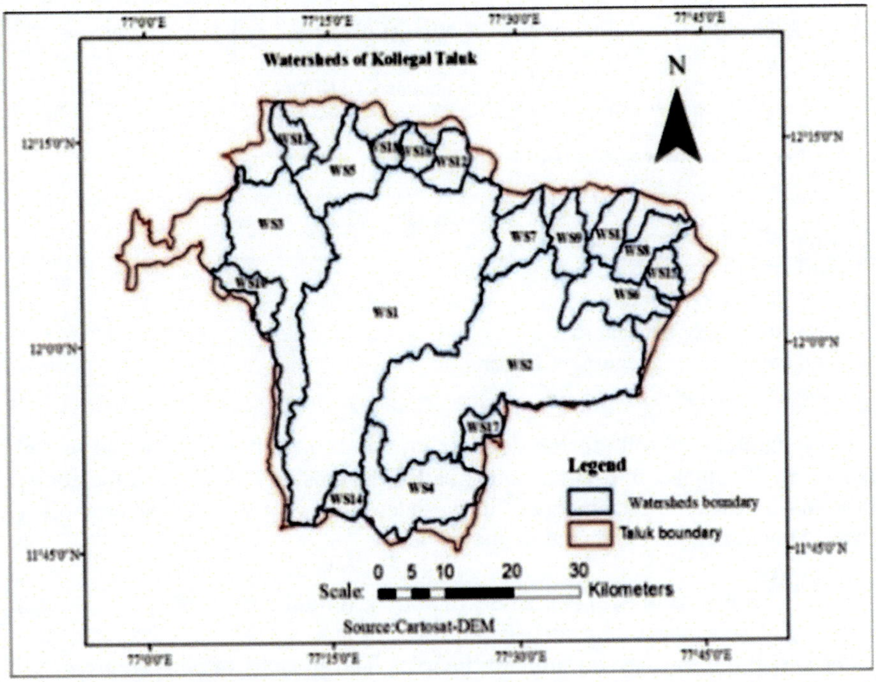

Fig. 2 Watersheds of Kollegal taluk

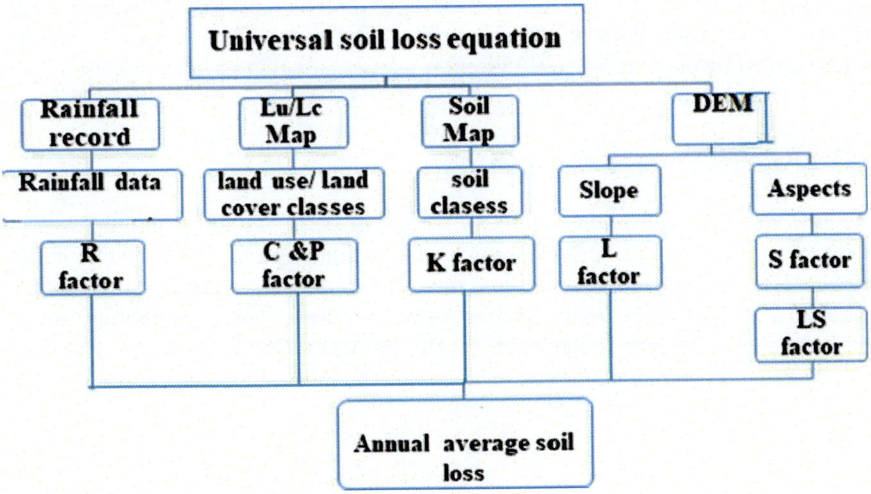

Fig. 3 Methodology adopted for estimation of soil loss

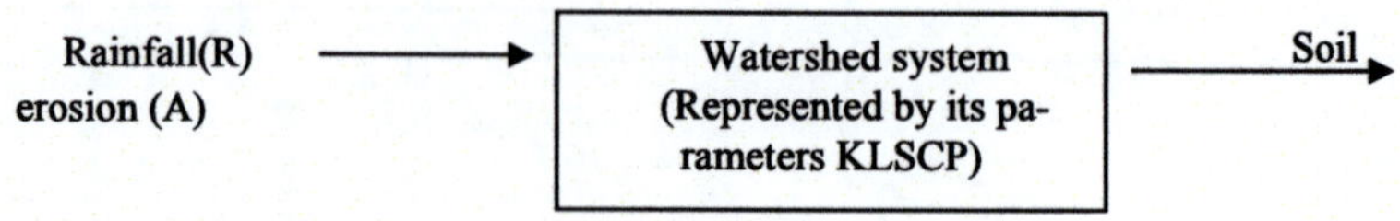

Fig. 4 Universal soil loss equation (USLE) model

R Rainfall erosivity factor.
K Soil erodibility factor.
L The slope length factor.
S The slope steepness factor.
C The cover and management factor.
P Conservation practice factor.

The magnitude of soil erosion depends on two forces—the detachment of soil particles by the impact of rainfall energy called the erosivity of rain, and the ability of the soil to resist the detachment of its particles by this force called the erodibility of soil. This relation is expressed as shown below

$$\text{Soil erosion} = f\left[(\text{erosivity of rain}) \times (\text{erodibility of soil})\right] \tag{2}$$

The USLE is also based on similar principles. The erosivity of rain is represented by the factor R and the erodibility of soil surface system by the multiples of the factors KLSCP. In systems terminology considering the watershed as a system represented by the multiples of the factors KLSCP, the input force is represented by the rainfall erosivity factor R and the output (the response to the input), which is the soil erosion is represented by the letter A. The system model of USLE is shown in Fig. 4.

3.1.1 Rainfall Erosivity Factor (R)

The erosivity factor of rainfall (R) is a function of falling raindrop and the rainfall intensity, and is the product of kinetic energy of the raindrop and the 30 min maximum rainfall intensity. But in Indian context that kind of detailed meteorological data is less available. Therefore, empirical equation [6] has been used for estimating annual and seasonal R factors in Indian context. The annual erosion index was as follows:

Table 1 The mean annual rainfall of rain gauge stations

S. No	Station code	Station name	Latitude (*N*)	Longitude (*E*)	Altitude (*M*)	MAP (mm)
1	80301	Bailur	11° 49′ 00″	77° 15′ 00″	745.16	88.54
2	80303	Gajanur	12° 4′ 00″	77° 5′ 00″	726.42	59.52
3	80304	Gundapur	12° 1′ 00″	77° 21′ 00″	755.87	56.49
4	80306	Kollegal	12° 7′ 00″	77° 7′ 00″	748.15	78.34
5	80309	Lokkanahally	12° 1′ 00″	77° 15′ 00″	744.97	57.84
6	80310	Mudigundam	12° 5′ 00″	77° 5′ 00″	726.41	74.69
7	80312	M. M. Hills	12° 3′ 00″	77° 36′ 00″	783.07	96.75
8	80313	Saggiam	12° 54′ 00″	77° 11′ 00″	761.14	67.66
9	80401	B. R. Hills	12° 0′ 00″	77° 8′ 00″	731.92	124.39
10	210301	Halagur	12° 26′ 00″	77° 14′ 00″	742.42	68.80

$$R = 79 + 0.363 * \text{MAP} \tag{3}$$

where

R is the average annual rainfall erosivity factor.
MAP is the mean annual precipitation (mm).

In the present study, R is computed by analyzing the rainfall data available from ten rain gauge stations, namely Bailur, Gundapura, Gajanur, Saggiam, Kollegal, Halagur, M. M. Hills, B. R. Hills, Lokkanahally and Mudigundam located in the Kollegal taluk and its adjoining area. The Spatial distribution of R factor for the study area is estimated using inverse distance weighting (IDW) method of interpolation using ArcGIS 10.1. For IDW interpolation method, mean annual rainfall is calculated by taking daily rainfall data of 10 rain gauge stations for a period of 19 years (1998–2016). Table 1 shows the mean annual rainfall of rain gauge stations in mm. The R factor has been determined for all the watersheds of Kollegal taluk by IDW method using ArcGIS 10.1. The R factor estimated for study area is shown in Fig. 5.

3.1.2 Soil Erodibility Factor (K)

The soil erodibility factor *(K)* relates the rate at which different soils erode under the conditions of equal slope, rainfall. Some soils erode more easily than others due to inherent soil characteristics such as texture, structure, permeability and organic matter content.

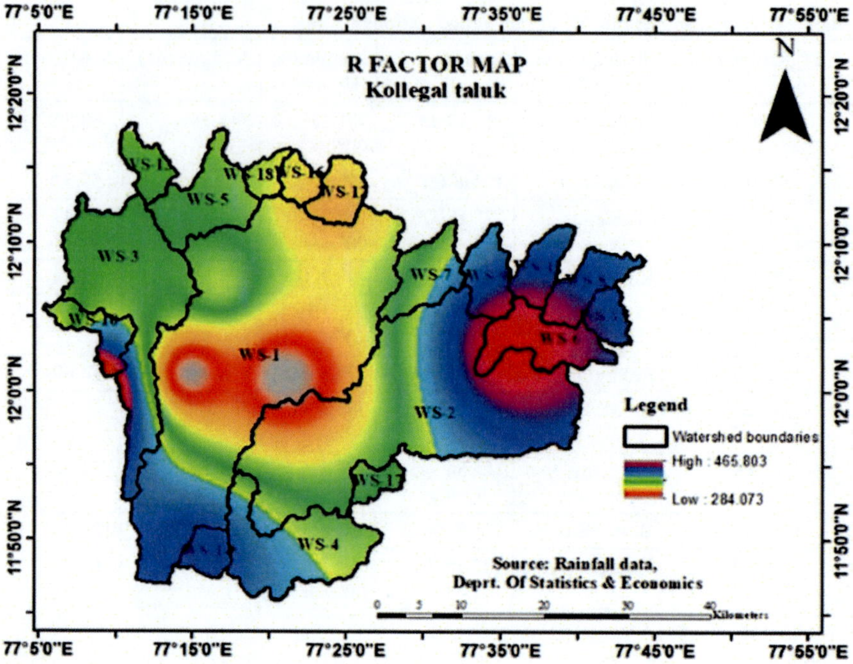

Fig. 5 R factor map of Kollegal taluk

In the present study K factor has been determined using Williams [9] equation and world soil data obtained from Food and Agriculture Organization of the United Nations (FAO). The world soil data has been downloaded from FAO portal and then it is clipped for the study area using ArcGIS 10.1. From the attribute table of the clipped file, soil unit symbol for the particular soil is noted. Table 2 shows soil distribution for Kollegal taluk. World soil data provides the grain size distribution and the percentage of soils present. Using this data and using the following equations K factor value for the watersheds of Kollegal taluk has been determined. Figure 6 shows the K factor value for the study area.

$$K_{\text{USLE}} = f_{\text{csand}} * f_{\text{cl-si}} * f_{\text{orgc}} * f_{\text{hisand}} \quad (4)$$

Table 2 Shows soil distribution of Kollegal taluk

Soil mapping unit number	Soil unit symbol	Sand % topsoil	Silt % topsoil	Clay % topsoil	OC % topsoil
3727	I	58.9	16.2	24.9	0.97
3781	LC	64.3	12.2	23.5	0.63
3714	I	65	15	20	1

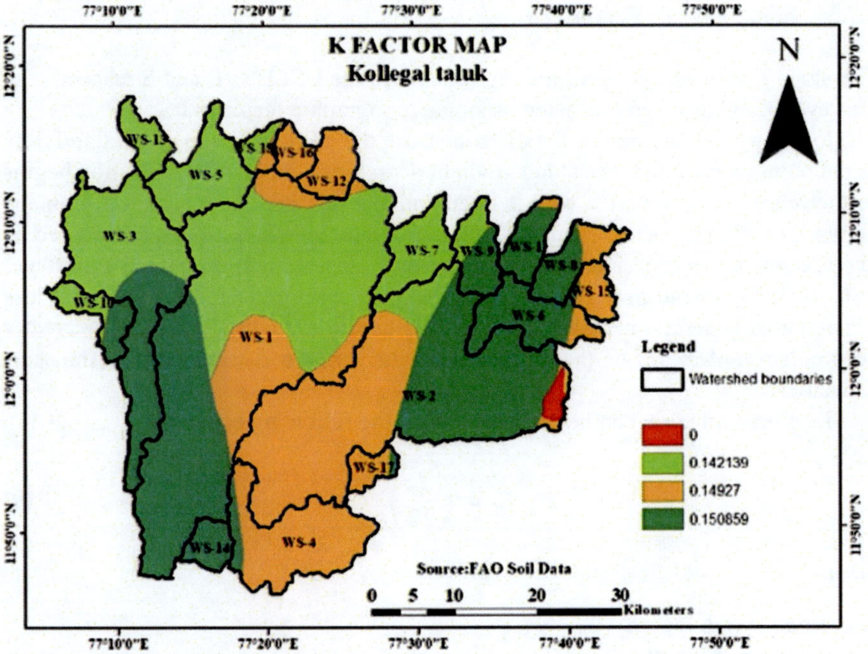

Fig. 6 K factor map of Kollegal taluk

$$f_{\text{csand}} = [0.2 + 0.3 * \exp(-0.256 * m_s * (1 - (m_{\text{silt}}/100))] \tag{5}$$

$$f_{\text{cl-si}} = [m_{\text{silt}}/(m_c + m_{\text{silt}})]^{0.3} \tag{6}$$

$$f_{\text{orgc}} = \left[1 - (0.0256 * \text{orgc})/(\text{orgc} + \exp\left[3.72 - 2.95 * \text{orgc}\right)\right] \tag{7}$$

$$f_{\text{hisand}} = \left[1 - (0.7 * (1 - \text{ms}/100))/((1 - \text{ms}/100) + \exp\left[-5.51 + 22.9(1 - \text{ms}/100)\right)\right] \tag{8}$$

where

ms	the sand fraction content (0.05–2.00 mm diameter) [%].
m_{silt}	the silt fraction content (0.002–0.05 mm diameter) [%].
m_c	the clay fraction content (< 0.002 mm diameter) [%].
orgC	the organic carbon (SOC) content [%].

3.1.3 Slope Length Factor (L)

The slope length and gradient are represented in the USLE as L and S respectively. However, they are often evaluated as single topographic factor as LS.

Slope length is defined as the distance from the point of origin of overland flow to the point where either the slope gradient decreases enough that deposition begins or the runoff water enters a well-defined channel that may be a part of a drainage network or a constructed channel. However, slope length has been considered as average length of overland flow. The effect of slope length on annual runoff per unit area of cropland may generally be assumed negligible. However, the soil loss per unit area generally increases substantially as slope length increases. The greater accumulation of runoff on the longer slopes increases its detachment and transport capacities.

Slope length factor, can be computed from the following equation

$$L = \left(\frac{l}{22}\right)^m \tag{9}$$

where

L slope length factor.
l slope length in m.
m dimensionless exponent.

= 0.5 for slopes > 4%; 0.4 for 4% slope; 0.3 for slopes < 3%

3.1.4 Slope Steepness Factor (S)

The slope gradient factor is expressed as the ratio of soil loss from a plot of known slope to soil loss from a unit plot under identical conditions. On steep slopes the flow velocity is more, which leads to scouring and cutting of soil. As per Wischmeier and Smith, slope gradient factor is determined by the formula,

$$S = \frac{0.43 + 0.3(\theta) + 0.043(\theta)^2}{6.574} \tag{10}$$

where

S slope steepness factor.
θ field slope in percent.

3.1.5 Crop Management Factor (C)

Factor C in the soil loss equation is the ratio of soil loss from land cropped under specified conditions to the corresponding loss from clean tilled, continuous fallow.

Table 3 The C factor values given for different land cover classes based on literature review

S. No	Land cover classes	C factor	Researchers/author/source
1	Agriculture plantation	0.63	Karaburun [4]
2	Barren Rocky	0.95	Zante et al. [13]
3	Cropland	0.4	Devatha et al. [1]
4	Deciduous (Dry/moist/thorn)	0.3	Karaburun [4]
5	Evergreen/semi evergreen	0.3	
6	Fallow land	0.5	Renard et al. [5]
7	Forest plantation	0.4	Karaburun [4]
8	Gullied/Ravenous land	0.95	Zante et al. [13]
9	Lake/pond	0	Karaburun [4]
10	Reservoir/tank	0	
11	River	0	
12	Rural	0.2	Devatha et al. [1]
13	Salt affected land	0.95	Zante et al. [13]
14	Scrub forest	0.25	Jebari [3]
15	Scrub land	0.25	Jebari [3]
16	Temperate/sub tropical	0.3	Karaburun [4]
17	Tree clad area	0.95	Zante et al. [13]
18	Urban	0.2	Devatha et al. [1]

This factor measures the combined effect of all the interrelated cover and management variables.

According to the USLE crop management factor determines the amount of erosion process [8]. C factor map was prepared on the basis of land use–land cover map of the study area. The land use–land cover of watersheds of Kollegal taluk obtained from Karnataka State Remote Sensing and Application Centre (KSRSAC) which is then classified into eighteen major types of land use–land cover classes using ArcGIS software. From the literature survey the value for crop management factor for different land cover is given. Table 3 shows the C factor values given for different land cover classes based on literature review. The area covered under each land cover classes is determined, and weighted average C factor for each watershed is determined.

3.1.6 Conservation Practice Factor (P)

Conservation practice factor is the ratio of soil loss with a specific supporting practice to the corresponding to soil loss with up and down cultivation. In general, whenever sloping land is to be cultivated and exposed to erosive rain, the protection offered by soil or close growing crops in the system needs to be supported by practices that will slow runoff and thus reduce the amount of soil it carries. The most important support practices are contour cultivation; strip cropping, terrace system and waterways for the disposal of excess rainfall. The values are selected based on the recommendations of Wischmeier and Smith (1959). Since the study area comprised of only field bund conservation P factor was taken as unity. The soil loss estimated from the USLE for all the watersheds is shown in Table 4. Figure 7 shows the soil erosion in the study area. Table 5 shows the limits of the soil loss.

4 Results and Discussion

The study has been carried out to quantify the soil loss in the watersheds of Kollegal taluk, Chamarajanagar district. The use of USLE model integrated with RS and GIS is found to be effective for assessing the soil loss vulnerability within the taluk. The USLE parameters, i.e., R, K, LS, C and P factors, are combined to find out annual average soil loss in the study area. The result shows that the Rainfall erosivity factor (*R*) varies from 32.87 to 42.08. The soil erodibility factor (*K*) is evaluated using Williams equation and FAO soil data, found to be 0.14, 0.15 and 0.151 for three different soil series delineated within the watersheds. The slope length and slope steepness are calculated using the Wischmeier and Smith equation (LS), and the obtained values vary from 0.53 to 4.28. The crop management factor (*C*) was taken considering the land cover in the entire area and varies between 0.3 and 0.39. Since the watershed comprises only bunds around the agricultural lands, no conservation was followed, and the conservation practice factor P is taken as unity. The average annual soil loss for the watersheds of Kollegal taluk varies from 0.88 to 10.58 t/ha/yr. WS11 shows the moderate soil loss. WS15, WS16, WS17 show slight to moderate erosion (ISRO-NNRMS-TR-103-2002).

5 Conclusions

Soil loss estimation has been carried out to identify the annual soil loss using USLE model. The quantitative soil loss ranges are estimated and classified the watersheds into different levels of soil erosion severity and also soil erosion index map. The watershed is classified according to Indian condition into different erosion classes such as slight, moderate, high, very high, severe, and very severe. Soil loss in the WS11 is found to be 10.58 t/ha/yr which is having moderate soil erosion, WS15, WS16, WS17 show slight to moderate erosion (ISRO-NNRMS-TR-103-2002). Soil and water conservation measures should be provided to prevent further soil erosion from these watersheds. The result obtained from the study helps in prioritization of soil and water conservation measures for sustainable development and management of natural resources within the watersheds of Kollegal taluk.

Table 4 Soil loss estimated for watersheds

Watershed name	Area (km^2)	K factor	R factor	LS	C factor	P factor	Erosion (A) t/ha/yr	Density of soil taken (kg/m^3)	Soil loss (mm/yr)
WS1	772.26	0.15	34.20	0.54	0.35	1	0.95	1450	0.65
WS2	550.80	0.14	36.24	0.53	0.32	1	0.88	1450	0.61
WS3	253.08	0.15	38.73	0.57	0.36	1	1.17	1450	0.81
WS4	147.23	0.15	37.21	0.59	0.31	1	1.01	1450	0.70
WS5	92.26	0.14	34.71	0.53	0.34	1	0.89	1450	0.61
WS6	77.91	0.15	42.09	1.71	0.31	1	3.36	1450	2.32
WS7	63.00	0.14	36.35	0.65	0.33	1	1.10	1450	0.76
WS8	51.43	0.15	40.33	2.13	0.30	1	3.88	1450	2.67
WS9	48.00	0.15	39.68	2.45	0.31	1	4.44	1450	3.07
WS10	41.04	0.15	39.85	1.46	0.31	1	2.63	1450	1.82
WS11	40.95	0.15	39.87	4.52	0.39	1	10.58	1450	7.30
WS12	40.58	0.15	32.88	0.98	0.33	1	1.59	1450	1.10
WS13	33.92	0.14	34.80	0.79	0.32	1	1.25	1450	0.86
WS14	28.59	0.15	39.53	1.17	0.34	1	2.37	1450	1.63
WS15	27.36	0.15	40.22	4.12	0.30	1	7.42	1450	1.63
WS16	23.37	0.15	33.15	3.51	0.31	1	5.37	1450	3.71
WS17	22.52	0.15	34.94	4.28	0.30	1	6.71	1450	4.63
WS18	20.42	0.14	33.81	2.80	0.30	1	4.11	1450	2.83

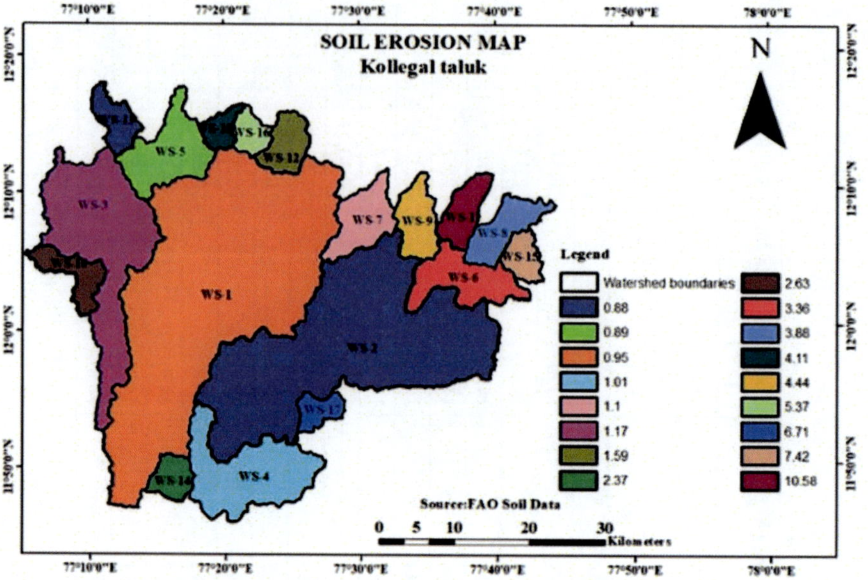

Fig. 7 Soil erosion map of watersheds of Kollegal taluk

Table 5 Soil loss limits

S. No	Particulars	Soil loss (t/ha/yr)
1	Nil to slight	Less than 5
2	Slight to moderate	5–10
3	Moderate	11–25
4	Severe	25–50
5	Very severe	Greater than 50

Source ISRO-NNRMS-TR-103-2002

References

1. Devatha, C. P., Deshpandeb, V., & Renukaprasad, M. S. (2015). Estimation of soil loss using USLE model for kulhan watershed, Chattisgarh- A case study. *ScienceDirect Aquatic Procedia, 4*, 1429–1436.
2. Ganasri, B. P., & Ramesh, H. (2016). Assessment of soil erosion by RUSLE model using remote sensing and GIS—A case study of Nethravathi Basin. *Geoscience Frontriers, 7*(6), 953–961.
3. Jebari, S. (2009). *Water erosion modeling using fractal rainfall disaggregation – A study in semiarid Tunisia.* Water resources engineering, Lund University, Sweden.
4. Karaburun, A. (2010). Estimation of C factor for soil erosion modeling using NDVI in Buyukcekmece watershed. *Ozean Journal of Applied Sciences, 3*(1), 77–85.

5. Renard, K., Foster, G., Weesies, G., McCool, D., & Yoder, D. (1997). *Predicting soil erosion by water: a guide to conservation planning with the revised universal soil loss equation (RUSLE).* US Government Printing Office, Washington, DC.
6. Singh, G., Rambabu, V. V., & Chandra, S. (1981). Soil loss prediction research in India. Bulletin No. T-12/D9, Central Soil & Water Conservation Research & Training Institute, Dehradun, India, 70 p Stoddart.
7. Tideman, E. M. (1996). *Watershed management: Guidelines for Indian conditions* (p. 101). Omega Scientific Publisher.
8. Vinay, M., Ramu, & Mahalingam, B., (2015). Quantification of soil erosion by water using GIS and Remote Sensing techniques: A study of Pandavapura Taluk, Mandya District, Karnataka, India. *ARPN Journal of Earth Sciences, 4*(2), 103–110.
9. Williams, J. R. (1995). Chapter 25: The EPIC model. In V. P. Singh (Ed.). (1995). *Computer models of watershed hydrology.* Water Resources Publications (pp. 909–1000).
10. Wischmeier, W. H. (1959). A rainfall erosion index for a universal soil-loss equation. *Soil Science Society of America Proceedings, 23,* 246–249.
11. Wischmeier, W. H., & Smith, D. D. (1965). Predicting rainfall erosion losses from Cropland East of the Rocky Mountains. *Handbook no. 282, United States Department of Agriculture,* Washington DC, USA.
12. Wischmeier, W. H., & Smith, D. D. (1978). Predicting rainfall erosion losses: A guide to conservation planning. *Agriculture Handbook, Vol. 537.* US Department of Agriculture, US Government Printing Office, Washington DC, USA.
13. Zante, P., Collinet, J., & Leclerc, G. (2001). Cartographie des risques érosifs sur le basin versant de la retenue collinaire d'Abdessadok (nord dorsale tunisienne), Institut de Recherche pour dèveloppement, Tunisia.

Investigating the Drought Teleconnections of Peninsular India Using Partial and Multiple Wavelet Coherence

S. Shamna, S. Adarsh, and V. Sreedevi

Abstract This study investigated the relationships of four large-scale climate oscillations with meteorological drought index of Peninsular India (PI) employing different forms of wavelet coherence methods. The 6-month Standardized Precipitation Index (SPI) is used as the representative drought index and its teleconnections with El Nino Southern Oscillations (ENSO), North Atlantic Oscillation (NAO), Pacific Decadal Oscillation (PDO) and Indian Ocean Dipole (IOD) are studied using wavelet-based approaches. The continuous wavelet transform, wavelet coherence, partial wavelet and multiple wavelet coherence analysis were conducted. It was observed that PDO had the strongest influence on 6-month SPI. It is also observed that ENSO influences the most in the PDO-SPI teleconnection of PI region. The multiple wavelet coherence analysis showed that the PDO-IOD and NAO-PDO-IOD combinations have the highest influence on SPI of PI region for the two and three variable cases, respectively.

Keywords Teleconnection · Peninsular India · Standardized precipitation index · Wavelet analysis

1 Introduction

Drought is a natural phenomenon that arising from considerable deficiency in precipitation. It is known to cause extensive damage and affects a significant number of people. To characterize and quantity droughts, drought indices are developed [1]. The Standardized Precipitation Index (SPI) is a meteorological drought index that helps to quantify droughts that occur mainly due to the lack of precipitation. Many statistical procedures ranging from simple correlation analysis to advanced multi-scale spectral analysis methods were recognized as potential tools for performing teleconnection problems in hydrology [2].

S. Shamna (✉) · S. Adarsh · V. Sreedevi
TKM College of Engineering, APJ Abdul Kalam Technological University, Kollam, Kerala 601005, India

A. K. Dikshit et al. (eds.), *Innovative Trends in Hydrological and Environmental Systems*, Lecture Notes in Civil Engineering 234,
https://doi.org/10.1007/978-981-19-0304-5_36

Continuous wavelet transform and its extensions like wavelet coherence (WC) were identified to be one of the most popular approaches for investigating hydroclimatic teleconnections by considering different hydroclimatic pairs [3–5]. It is very well proven that multiple climate oscillations modulate the Indian summer monsoon rainfall [6]. Therefore, the interdependency between teleconnection variables is to be accounted while performing multiscale teleconnection studies in Indian context. The multiple wavelet coherence (MWC) analysis and partial wavelet coherence (PWC) analysis are two popular extension algorithms of WC which helps to investigate the role of multiple climatic signals on hydrologic variables [7–9]. Recently, Rathinasamy et al. [10] investigated the teleconnections of extreme precipitation of thirty different locations in India. Their study analyzed the partial and joint influences of three prominent climate indices, namely Nino 3.4, Pacific Decadal Oscillation (PDO) and Indian Ocean Dipole (IOD) on Indian precipitation extremes by performing a graphical comparison of different wavelet coherence plots. This study considers the Standardized Precipitation Index (SPI) time series at 6-month timescale computed for Peninsular India (PI) as the representative index to investigate the teleconnections of drought with El Nino Southern Oscillation (ENSO), PDO, IOD and the North Atlantic Oscillation (NAO) variability. In this study, the results of bivariate WC, PWC and MWC are quantified by two statistical coherence measures: average power of wavelet coherence (AWC) and percentage of significant coherence (PoSC).

2 Methods

A brief description on the SPI and wavelet-based analysis is provided in the following section.

2.1 Standardized Precipitation Index (SPI)

The Standardized Precipitation Index (SPI), a meteorological drought index, is commonly used to measure the standard deviation of precipitation. It can be derived from parametric and nonparametric methods, while nonparametric SPI is widely recommended for drought assessment over India [11]. The nonparametric method of computing SPI is obtained from the marginal probabilities of precipitation using the empirical Gringorten plotting position formula:

$$p(Xi) = \frac{i - 0.44}{n + 0.12} \tag{1}$$

where n is the total number of years in the time series, i denotes the rank of nonzero precipitation data from the smallest, and $p(Xi)$ is the corresponding empirical probability. The SPI can be derived as:

$$SPI = \emptyset^{-1}(p) \tag{2}$$

where $\emptyset$ is the standard normal distribution function and p is probability derived from Eq. 1.

2.2 Continuous Wavelet Transform

The wavelet transforms are capable to decompose the time series into different timescales and it is capable to capture the high and low frequencies in the series. Continuous and discrete variants of wavelets are defined in literature [12], and the latter one is more appropriate for time–frequency characterization and teleconnection studies. The Morlet wavelet is one of the popular complex wavelet functions and provides a good balance between time and frequency localization. It allows separating the phase and amplitude of any time series and is localized in scale, which helps in achieving high resolution in frequency [12]. The wavelet power spectrum developed by continuous wavelet transform (CWT) provides such information on time–frequency localization of the time series. The power spectrum can be averaged either over the time domain or over the scale, and the time average of power at a given frequency is known as global wavelet spectrum (GWS). GWS was used to test for the presence of significant periodic components in the time series, and the peak of the GWS indicates the dominant periodicity.

2.3 Bivariate Wavelet Coherence (BWC)

The wavelet coherence evaluates the linkage between two time series within the time–frequency space by measuring the correlation between the time series, whose values vary between 0 and 1. This technique enables us to understand the dynamic behavior of hydro-climatological variables through analyzing the variability structure of hydrological components. Following Torrence and Compo [12], the wavelet coherence of two time series is defined as:

$$R_n^2(s) = \frac{\left|S\left(s^{-1}W_n^{XY}(s)\right)\right|^2}{S\left(s^{-1}\left|W_n^X(s)\right|^2\right).S\left(s^{-1}\left|W_n^Y(s)\right|^2\right)} \tag{3}$$

where $R_n^2(s)$ denotes the coherence coefficient with maximum coherence at 1 and no coherence at 0, $W_n^{XY}(s)$ represents cross-wavelet transform of two time series, $W_n^X(s)$ and $W_n^Y(s)$ are the continuous wavelet transform of each time series, s^{-1} is used to convert to energy density and S denotes the smoothing operator.

2.4 Partial Wavelet Coherence (PWC)

The partial wavelet coherence (PWC) is a technique similar to partial correlation that helps find the resulting BWC between two time series y and x_1 after eliminating the influence of the time series x_2. The PWC can be squared (after the removal of the effect of x_2) and can be defined by an equation similar to the partial correlation squared [8] as:

$$\mathrm{RP}^2(y, x_1, x_2) = \frac{|R(y, x_1) - R(y, x_2).R(y, x_1)^*|^2}{[1 - R(y, x_2)]^2[1 - R(x_{2,} x_1)]^2} \tag{4}$$

This squared value, like the simple WTC, ranges from 0 to 1 and $R(y, x_1)$, $R(y, x_2)$, $R(x_{2,}, x_{1,})$ is given by Eqs. 5, 6 and 7, respectively, as:

$$R(y, x_1) = \frac{s[W(y, x_1)]}{\sqrt{s[W(y)].s\{\ W(x_1)]}} \tag{5}$$

$$R(y, x_2) = \frac{s[W(y, x_2)]}{\sqrt{s[W(y)].s\{\ W(x_2)]}} \tag{6}$$

$$R(x_1, x_2) = \frac{s[W(x_1, x_2)]}{\sqrt{s[W(x_1)].s\{\ W(x_2)]}} \tag{7}$$

2.5 Multiple Wavelet Coherence (MWC)

The multiple wavelet coherence (MWC) analysis is based on auto- and cross-wavelet power spectra among the analyzed variables. The response variable (y_t) is the SPI. The MWC $\rho_m^2(s, t)$ at scale s and time t is expressed as [13]:

$$\rho_m^2(s, t) = \frac{W_t^{y,x}(s, t) W_t^{y,x^-}(s, t)}{W_t^{x,x}(s, t) W_t^{y,y}(s, t)} \tag{8}$$

$W_t^{y,x}(s, t)$ is the matrix of the smoothed cross-wavelet power between the response and predictor variables; $W_t^{x,x}(s, t)$ is the matrix of the smoothed auto- and cross-wavelet power among the multiple predictor variables; and $W_t^{y,y}(s, t)$ is the matrix of the smoothed wavelet power of the response variable. The $W_t^{y,x^-}(s, t)$ represents the complex conjugate of $W_t^{y,x}(s, t)$.

2.6 Computation of Significant Coherence

The average wavelet coherence (AWC) and percentage of significant coherence (PoSC) can be used as measurements to examine the relative dominance of individual teleconnections or combinations of teleconnections on precipitation variability [14]. The AWC can be calculated by averaging the wavelet coherence produced over all scales according to the coherence values produced. The PoSC can be obtained by calculating the ratio of the number of significant values of power over the total number of values of the power produced in the MWC computation. Significant power occurs when the ratio of the power over the significance level is greater than 1. Higher overall AWC and PoSC values indicate more dominance. Although an additional teleconnection may increase the coherence with precipitation, the increase in PoSC by at least 5% should be observed before concluding the additional teleconnection has a practical significance [7].

2.7 Methodology

The following steps are used in this study to investigate the multiscale investigations of SPI-6 with different climatic oscillations.

1. Perform bivariate WC (BWC) analysis of SPI with climatic oscillations independently and generate BWC plots. Quantify the influence of each oscillation by estimating AWC and PoSC.
2. Examine the influence of multiple oscillations with SPI-6 using MWC varying the number of climate oscillations (independent variables, IVs) from 2 to 4. Quantify the coherences by estimating AWC and PoSC from the developed MWC plots.
3. Examine the partial influence of oscillations with SPI-6 using PWC analysis by eliminating the influence of single climate oscillations on the considered bivariate pair. Quantify the coherences by estimating AWC and PoSC by developing PWC plots.
4. Analyze the BWC, MWC, PWC plots along with AWC and PoSC statistics to draw useful conclusions on teleconnections of multiple climatic influences on SPI-6 of PI.

3 Study Area and Data

On the basis of similarity in rainfall characteristics, India is divided into five rainfall homogeneous regions like North West, North East, Central North East, West Central and Peninsular India (PI). The rainfall of Peninsular India gets influenced significantly by the large-scale climatic oscillations, as the region is bounded by oceans in its three sides. Therefore, PI is selected as study area in this research work, and the location of the region is shown in Fig. 1.

In this study, the monthly rainfall time series of Peninsular India for the period of 1950–2015 is collected from Indian Institute of Tropical Meteorology (IITM), Pune. Subsequently, the SPI at 6-month timescale (SPI-6) is computed following nonparametric distribution. The Nino 3.4 index, PDO, NAO and IOD data is collected from National Oceanic and Atmospheric Administration (NOAA). The Standardized Drought Analysis Toolbox (SDAT) by Farahmand and AghaKouchak [11] is used to compute the 6-month SPI of the region. The BWC, MWC and PWC analyses are performed by using the respective tool boxes by Grinstead et al. [15], Hu and Si [13] and Ng and Chan [8] by incorporating user-defined modifications for statistical computations of correlation measures and graphical representations.

Fig. 1 Peninsular India

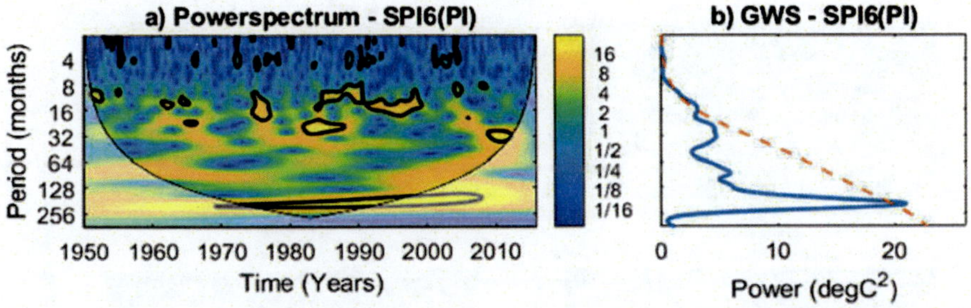

Fig. 2 Wavelet power spectrum and global wavelet spectrum of SPI-6 of Peninsular India

4 Results and Discussions

The results of the wavelet analysis and the corresponding discussions are provided in this section.

4.1 Continuous Wavelet Transform

The wavelet transform analysis on climate oscillations revealed the periodicities as interannual for ENSO (16–64 months), PDO (8–16 months) and NAO (8–32 months). A decadal periodicity of 128 months is also observed for NAO, while no significant periodicity was observed for the case of IOD. From the analysis of SPI, periodicities at 16–32 months (interannual) and at 128–256 months (interdecadal) are observed. The wavelet power spectrum and the global wavelet spectrum of wavelet transform analysis of SPI are shown in Fig. 2.

4.2 Wavelet Coherence Analysis

The wavelet coherence analysis between SPI and individual oscillations is performed, and the wavelet power spectrum obtained from the analysis is shown in Fig. 3. The contours where significant coherences are present around 16 months, between 16–32 months and 64–128 months for the case of ENSO, while for NAO, it is present around 16, 32 and 64 months. For the case of PDO, the significant coherence is found at the scale of 16–128 months, and for the case of IOD, it is observed as small and scattered around 16 months and between 32 and 64 months. It is further noticed that the influence of different climate oscillations on SPI is mostly anti-phase in character. But one cannot ignore that such relationships can be in-phase at some localized time spells over the time domain (e.g., in 1970s in SPI-NAO link, 1960s and 1990s in SPI-IOD link, etc.). Also, coexistence of more than one climatic drivers is

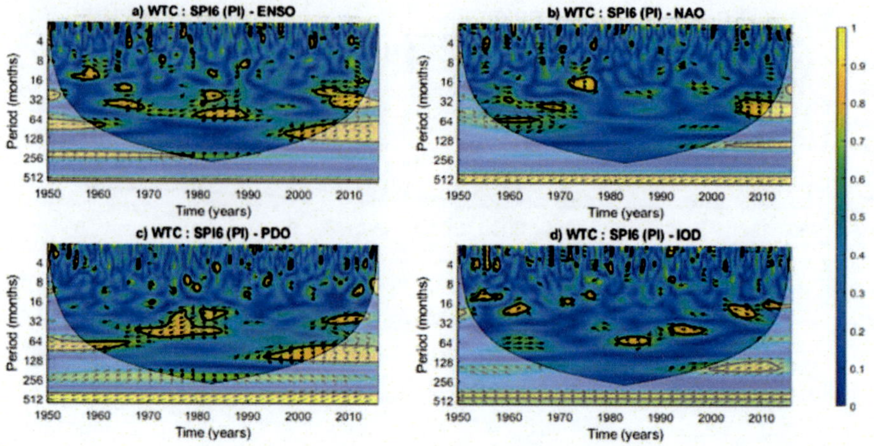

Fig. 3 Power spectrum of wavelet coherence analysis of SPI-6 of Peninsular India with climate oscillations

also possible (e.g., the occurrences of IOD, ENSO and PDO in 1980s at annual-inter annual scales).

4.3 Partial Wavelet Coherence Analysis

As the concurrent existence of multiple drivers is possible in a climate system, one cannot ignore its role on the rainfall and drought conditions, and its effect may be either to magnify or to minify the rainfall magnitude. Therefore, the roles of other oscillations are to be excluded from a typical SPI-climate oscillation link using the PWC analysis. The power spectrum obtained from partial wavelet coherence analysis is shown in Fig. 4. On comparing Figs. 4 with 3, it is noted that the regions of significant contours get disappeared or reduced in all the cases of BWC analysis. For example, on considering SPI-PDO link and on removing the influence of ENSO, the regions of significant contours are practically absent (Fig. 4g), which means that influence of ENSO on SPI-6 is very much strong.

4.4 Multiple Wavelet Coherence Analysis

To examine the influence of coexistence of multiple climate oscillations, the MWC analysis is performed by considering the concurrent role of multiple drivers. Here, two factor, three factor and all factor combinations are considered, and eleven such possibilities are considered. Figure 5 presents the power spectrum obtained for the

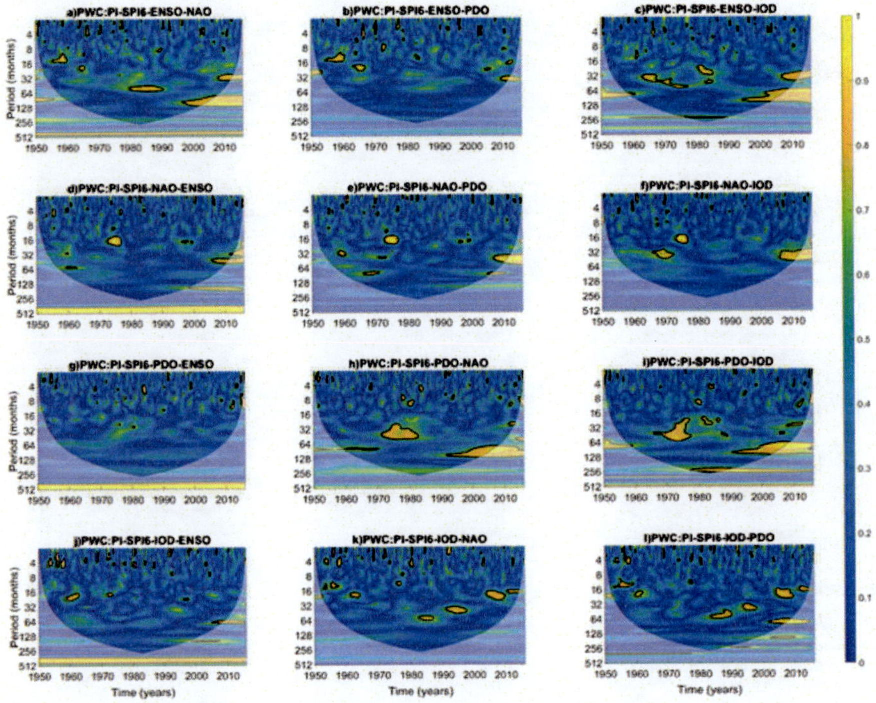

Fig. 4 Power spectrum of partial wavelet coherence analysis of SPI-6 of Peninsular India with climate oscillations. In each panel, the last variable is the excluding variable

analysis. Figure shows that the addition of another oscillation to the two variable combinations caused a significant increase in the regions of significant power except for the cases when ENSO was added to PDO-IOD and when NAO is added to PDO-IOD combination of SPI. Also, a significant increase in power is not observed when NAO is added to the ENSO-PDO combination. When a fourth oscillation is added to different triplet oscillation combinations, no significant change in AWC is observed.

4.5 Analysis of Significant Coherence

Along with the power spectral representations of coherence, two statistical measures AWC and PoSC as defined in section are adopted for quantifying the coherence strength. All the AWC and PoSC values obtained from the wavelet coherence analysis are shown in Table 1.

The bivariate coherence analysis shows that highest AWC is observed for PDO for the region. The AWC values observed for wavelet coherence analysis of SPI with ENSO, NAO, PDO and IOD, respectively, are 0.3862, 0.3598, 0.4245 and

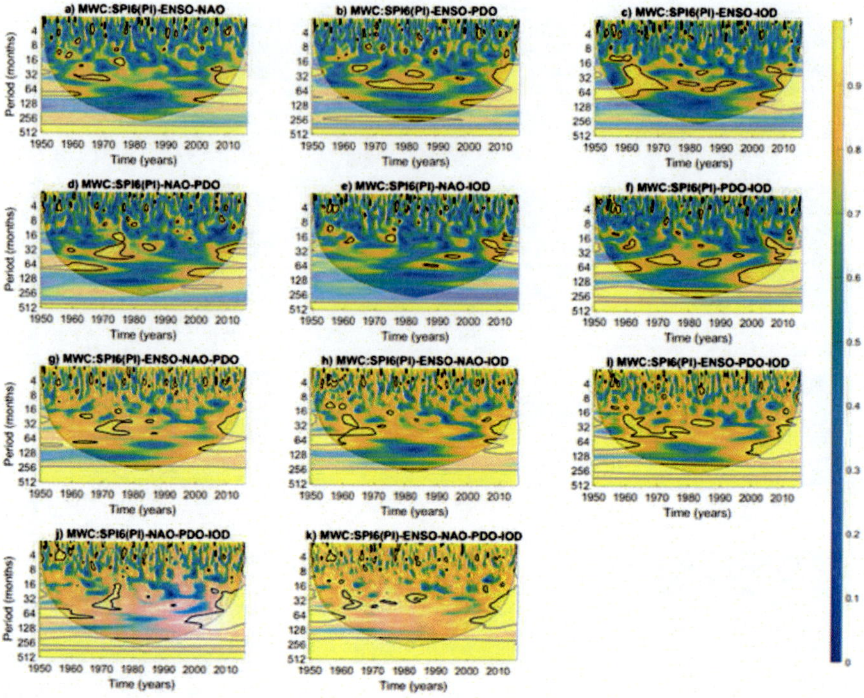

Fig. 5 Power spectrum of multiple wavelet coherence analysis of SPI-6 of Peninsular India with climate oscillations

0.3610. Table 1 further shows that, all the AWC and PoSC values of partial wavelet coherence are observed to be less than that of the wavelet coherence. The max AWC of the analysis is observed for the case of PDO with IOD as an excluding variable with the value 0.3125. The maximum decrease in AWC of 0.1734 occurred upon eliminating the possible influence of PDO on ENSO, and a decrease of 0.17 was observed for the case of ENSO with PDO as the excluding variable. Removing any of the oscillations from SPI-PDO link, there is significant reduction in PoSC value, indicating a strong association of other oscillations with PDO. On the contrary, for IOD-SPI link, the reduction in PoSC was not significant (less than 5% reduction) on the removal of ENSO and PDO. This indicates that these two oscillations have no strong and direct link on IOD at least for the selected region and specified period of study. Also, the removal of ENSO from SPI-NAO link and its converse does not display significant reduction in coherence value. The AWC values for MWC analysis are much higher (0.57–0.9) than the values for bivariate wavelet coherence (BWC) analysis and PoSC ranges from 10.36 to 26.79%. The max AWC for two oscillation combinations is observed for PDO-IOD combinations with values being 0.6847, while for three oscillation combinations, the max AWC is observed for the NAO-PDO-IOD combination with value being 0.8188.

Table 1 AWC and PoSC values of wavelet analysis of SPI-6 of Peninsular India

Analysis	Combination	AWC	PoSC (%)
BWC	SPI-6-ENSO	0.39	11.95
	SPI-6-NAO	0.36	9.52
	SPI-6-PDO	0.42	15.78
	SPI-6-IOD	0.36	8.53
PWC	*SPI-6-ENSO-NAO*	0.31	7.93
	SPI-6-ENSO-PDO	0.21	2.07
	SPI-6-ENSO-IOD	0.28	6.49
	SPI-6-NAO-ENSO	*0.26*	*9.24*
	SPI-6-NAO-PDO	0.21	2.57
	SPI-6-NAO-IOD	0.20	2.70
	SPI-6-PDO-ENSO	0.25	7.09
	SPI-6-PDO-NAO	0.31	5.07
	SPI-6-PDO-IOD	0.31	9.57
	SPI-6-IOD-ENSO	*0.25*	*5.95*
	SPI-6-IOD-NAO	0.22	2.66
	SPI-6-IOD-PDO	*0.25*	*4.13*
MWC	SPI-6-ENSO-NAO	0.64	14.49
	SPI-6-ENSO-PDO	0.64	16.82
	SPI-6-ENSO-IOD	0.63	18.02
	SPI-6-NAO-PDO	0.63	13.31
	SPI-6-NAO-IOD	0.57	10.36
	SPI-6-PDO-IOD	0.68	25.64
	SPI-6-ENSO-NAO-PDO	0.81	22.07
	SPI-6-ENSO-NAO-IOD	0.79	22.09
	SPI-6-ENSO-PDO-IOD	0.81	25.71
	SPI-6-NAO-PDO-IOD	0.82	23.40
	SPI-6-ENSO-NAO-PDO-IOD	0.90	26.79

The values in italics show that such combinations are not resulting in significant reduction in coherence

5 Conclusions

The present study investigated the teleconnections between climatic oscillations (ENSO, NAO, PDO and IOD) with 6-month SPI of Peninsular India. The specific conclusions of the study are:

- The oscillations with most influence on SPI-6 are ENSO and PDO.
- The influence of ENSO on SPI-6 is affected by PDO and vice versa.

- The best two oscillation combination influencing SPI-6 of the region is PDO-IOD combination
- NAO-PDO-IOD combination is observed as the best three-variable combination influencing SPI-6 of the region.

The novel insights gained from the study can be further extended to identify most influential climate driver for short-term and long-term drought in India to improve drought characterization, management and drought risk analysis.

Acknowledgements Authors would like to the thank the support provided by Amina B., Arunkrishna P. J. and Surya Suresh of TKM college of Engineering in this study.

References

1. Zargar, A., Sadiq, R., Naser, B., & Khan, F. I. (2011). A review of drought indices. *Environmental Reviews, 19*, 333–349. https://doi.org/10.1139/a11-013
2. Adarsh, S., & Janga Reddy, M. (2021). Multi-scale spectral analysis in hydrology: From theory to practice. ISBN 9780367622015, CRC Press, New York.
3. Araghi, A., Mousavi-Baygi, M., Adamowski, J., & Martinez, C. (2017). Association between three prominent climatic teleconnections and precipitation in Iran using wavelet coherence. *International Journal of Climatology, 37*(6), 2809–2830. https://doi.org/10.1002/joc.4881
4. Das, J., Goyal, M. K., & Jha, S. (2020). On the relationship of climatic and monsoon teleconnections with monthly precipitation over meteorologically homogenous regions in India: Wavelet and global coherence approaches. *Atmospheric Research, 23*. https://doi.org/10.1016/j.atmosres.2020.104889
5. Massei, N., Durand, A., Deloffre, J., Dupont, J. P., Valdes, D., & Laignel, B. (2007). Investigating possible links between the North Atlantic oscillation and rainfall variability in northwestern France over the past 35 years. *Journal of Geophysical Research, 112*, D09121. https://doi.org/10.1029/2005JD007000
6. Adarsh, S., & Janga Reddy, M. (2019). Links between global climate teleconnections and Indian Monsoon Rainfall. In C. Venkataraman, T. Mishra, S. Ghosh, S. Karmakar (Eds.) *Climate change signals and response*. Springer (pp. 61–72). https://doi.org/10.1007/978-981-13-0280-0_4s
7. Nalley, D. P., Adamowski, J., Biswas, A., Gharabaghi, B., & Hu, W. (2019). A multiscale and multivariate analysis of precipitation and streamflow variability in relation to ENSO, NAO and PDO. *Journal of Hydrology, 574*, 288–307. https://doi.org/10.1016/j.jhydrol.2019.04.024
8. Ng, E. K., & Chan, J. C. (2012). Geophysical applications of partial wavelet coherence and multiple wavelet coherence. *Journal of Atmospheric and Oceanic Technology, 29*, 1845–1853. https://doi.org/10.1175/JTECH-D-12-00056.1
9. Su, L., Miao, C., Duan, Q., Lei, X., & Li, H. (2019). Multiple-wavelet coherence of world's large rivers with meteorological factors and ocean signals. *Journal of Geophysical Research: Atmospheres, 124*. https://doi.org/10.1029/2018JD029842
10. Rathinasamy, M., Agarwal, A., Sivakumar, B., Marwan, N., & Kurths, J. (2019). Wavelet analysis of precipitation extremes over India and teleconnections to climate indices. *Stochastic Environmental Research and Risk Assessment, 33*, 2053–2069. https://doi.org/10.1007/s00477-019-01738-3
11. Farahmand, A., & Aghakouchak, A., (2015). A generalized framework for deriving nonparametric standardized drought indicators. *Advances in Water Resources, 76*, 140–145. https://doi.org/10.1016/j.advwatres.2014.11.012

12. Torrence, C., & Compo, G. P. (1998). A practical guide to wavelet analysis. *Bulletin of the American Meteorological Society, 79*(1), 61–78.
13. Hu, W., & Si, B. C. (2016). Technical note: Multiple wavelet coherence for untangling scale-specific and localized multivariate relationships in geosciences. *Hydrology of Earth System Sciences, 20*, 3183–3191.
14. Nalley, D. P. (2020). *The use of wavelet transform-based methods to analyze variability in hydrological data, multiscale linkages to large-scale climate oscillations, and for hydrological record extension*. Ph.D. thesis submitted to McGill University, Motreol, Canada.
15. Grinsted, A., Moore, J. C., & Jevrejeva, S. (2004). Application of the cross wavelet transform and wavelet coherence to geophysical time series. *Nonlinear Process Geophysics, 11*(5), 561–566.
16. Joshi, N., Gupta, D., Suryavanshi, S., Adamowski, J., & Madramootoo, C. A. (2016). Analysis of trends and dominant periodicities in drought variables in India: A wavelet transform based approach. *Atmospheric Research*. https://doi.org/10.1016/j.atmosres.2016.07.030
17. Mallenahalli, N. K. (2020). Comparison of parametric and nonparametric standardized precipitation index for detecting meteorological drought over the Indian region. *Theoretical and Applied Climatology, 142*, 219–236. https://doi.org/10.1007/s00704-020-03296-z
18. Mishra, A. K., & Singh, V. P. (2010). A review of drought concepts. *Journal of Hydrology, 391*(1–2), 202–216. https://doi.org/10.1016/j.jhydrol.2010.07.012
19. Song, X., Zhang, C., Zhang, J., Mo, Y., & Tian, Y. (2020). Potential linkages of precipitation extremes in Beijing-Tianjin-Hebei region, China, with large-scale climate patterns using wavelet-based approaches. *Theoretical and Applied Climatology, 141*(3–4), 1251–1269. https://doi.org/10.1007/s00704-020-03247-8

Development of Hybrid Wavelet Artificial Neural Network Model for Downscaling Precipitation and Temperature

S. R. Saipriya and N. R. Chithra

Abstract Climate change is likely to pose huge challenges on different sectors and is a topic of intense discussion in the twenty-first century. General Circulation Models (GCMs) are important tools for the assessment of climate change impact. As they are available in coarser resolution, those projections are needed to be downscaled to station scale. As there is a wide-spread use of Artificial Intelligence (AI) in the domain of hydrology and water resources, Artificial Neural Network (ANN) models-based transfer function is developed for downscaling GCM projections to station scale in this study. Being non-stationary data, wavelet which has various application in pre-processing of data was adopted in the study for the development of the model. Based on the values of correlation coefficient between the large-scale atmospheric variables (predictors) and the observed predictand (precipitation or temperature) data, separate sets of potential predictors were selected at the station points. Hybrid Wavelet ANN (WANN) model was developed and the performance of the trained WANN models was evaluated using RMSE and coefficient of correlation and was found to be satisfactory. The developed model can be used to downscale the grid-based coarse resolution GCMs data to station scale.

Keywords General circulation models · Wavelet · Hybrid wavelet ANN

1 Introduction

Climate change is emerging as one of the main challenges in the global scenario. Projections from the General Circulation Models (GCMs) play a key role for understanding future climate changes. They provide us with the information of future climatic variables up to coming 100 years considering the anthropogenic activities but this information is available at such a coarse resolution (about 300 km grid size) that it cannot be used directly for any analysis at local or regional level [1]. Inferring general conclusions on the regional climate change impacts on water resources is not

S. R. Saipriya (✉) · N. R. Chithra
Department of Civil Engineering, National Institute of Technology, Calicut, Kerala, India

A. K. Dikshit et al. (eds.), *Innovative Trends in Hydrological and Environmental Systems*, Lecture Notes in Civil Engineering 234,
https://doi.org/10.1007/978-981-19-0304-5_37

exceptionally simple. Many studies have been conducted to assess climate change impact on water availability [2, 3]. Studies of regional impacts of climate change in water resources around the globe have shown that the implications are not similar as they vary across regions. It has been widely recognized that the impact of climate change varies regionally, thereby necessitating impact assessment studies at regional scale. For the study of regional impact of climate change on water resources, most of the hydrological models require local scale input data. The spatial resolution in which GCMs run is too coarse to obtain reliable projections at regional and local scales. So, to convert this coarser scale output data of GCM to finer scale, downscaling techniques are used. Dynamical downscaling and statistical downscaling are the two most commonly adopted methods of downscaling; statistical downscaling has been widely used to generate future climate scenarios at local scale [4]. Statistical downscaling tools are most commonly used by the researchers and scientists worldwide due to its simplicity and low computational cost. Methods like Artificial Neural Networks (ANNs), linear and nonlinear regression, canonical correlation, Relevance Vector Machine (RVM) and Support Vector Machine (SVM), etc., have been used for statistical downscaling. Out of these, the Artificial Intelligence (AI) methods like SVM and ANN are most commonly used for statistical downscaling owing to the superior capabilities to handle nonlinear input–output relationship.

The data-driven methods of Artificial Intelligence (AI) have shown significant promise in modeling and forecasting nonlinear hydrological processes and in handling large amounts of dynamicity and the noise concealed in the datasets [5]. ANN-based transfer functions have been widely used for the downscaling of large-scale climatic variables. Over the years, the use of wavelet–AI models in hydrology has steadily increased and attracted interest because of the robustness and accuracy of the approach [5]. Sang [6] highlighted the multifaceted information that can be drawn from such analysis: characterization and understanding of hydrologic series' multi-temporal scales, identification of seasonality and trends, and data denoising. Therefore, the ability of the wavelet transforms to decompose non-stationary signals into sub-signals at different temporal scales (levels) is helpful in better interpreting hydrological processes [7–10]. The wavelet transform is applicable in extracting nontrivial and potentially useful information, or knowledge, from the large data sets available in experimental sciences (historical records, reanalysis, global climate model simulations, etc.) [5].

Depending on the individual capacities of wavelet and AI methods, a hybrid model comprising of both would have advantages of both the techniques. The objective of this study is to develop hybrid Wavelet Artificial Neural Networks (WANN)-based downscaling model for downscaling daily precipitation, daily maximum temperature, and daily minimum temperature.

2 Study Area and Data

2.1 Study Area

The study area is the Chaliyar River Basin in Kerala, India. The fourth longest river in Kerala is the Chaliyar River with a length of 169 km. It originates from Elambalari hills, Nilgiris district of Tamil Nadu. The river basin lies between latitudes 11° 06′ N and 11° 36′ N and longitudes 75° 48′ E and 76° 33′ E. Total area of the basin is 2933 km^2 out of which 2535 km^2 area is in Kerala. The average annual precipitation in the river basin is 3012 mm and the annual maximum and minimum temperatures in the basin are 34 °C and 24 °C, respectively.

2.2 Data

The Indian Meteorological Department (IMD) gridded precipitation data, IMD gridded maximum and minimum temperature data, and the reanalysis data of daily mean atmospheric variables prepared by the National Centre for Environmental Prediction-National Centre for Atmospheric Research (NCEP-NCAR) are required for the study. Table 1 lists the source of data and the spatial resolution of the data.

The IMD gridded daily precipitation data, daily maximum and minimum temperature data, and reanalysis data of the atmospheric variables such as air temperature, relative humidity, specific humidity, zonal velocity, meridional velocity, surface pressure, and sea-level pressure at different pressure levels are collected for a period of 35 years (1980–2014). Figure 1 shows the map of Chaliyar Basin with NCEP grid points and IMD precipitation and temperature grid points. The downscaling for the precipitation is performed at five locations (Table 2) coinciding IMD grid points that

Table 1 Data description and source of data

Datasets	Spatial resolution	Source
IMD (gridded) daily precipitation	0.25° × 0.25°	India Meteorological Department (IMD) (https://imdpune.gov.in/Clim_Pred_LRF_New/Grided_Data_Download.html)
IMD (gridded) daily maximum and minimum temperature	1° × 1°	India Meteorological Department (IMD) (https://imdpune.gov.in/Clim_Pred_LRF_New/Grided_Data_Download.html)
NCEP/NCAR reanalysis data	2.5° × 2.5°	National Centre for Environmental Prediction-National Centre for Atmospheric Research (https://psl.noaa.gov/data/gridded/data.ncep.reanalysis.html)

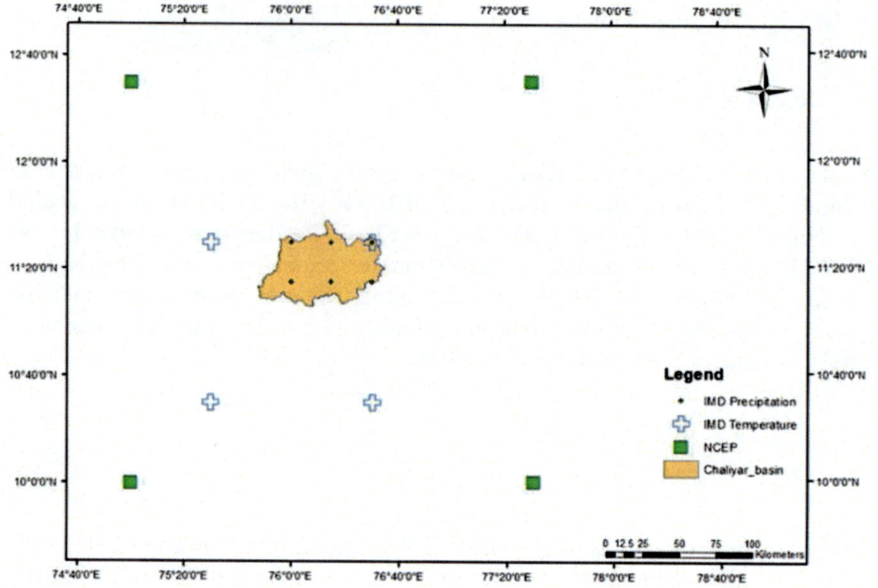

Fig. 1 Map of the Chaliyar Basin with NCEP and IMD grid points

Table 2 Location of station points

Station	Latitude	Longitude
A	11.25° N	76° E
B	11.25° N	76.25° E
C	11.5° N	76° E
D	11.5° N	76.25° E
E	11.5° N	76.5° E

fall in the basin (named A, B, C, D, and E for convenience). For temperature, only one grid point falls in the basin and that station point (E) is considered. Figure 2 shows the map of Chaliyar Basin with the station points. All the NCEP predictors are re-gridded to the IMD grid locations using Inverse Distance Weightage (IDW) method, which assumes that the influence of grid value on a required location depends on the distance—lesser the distance higher the influence and vice versa.

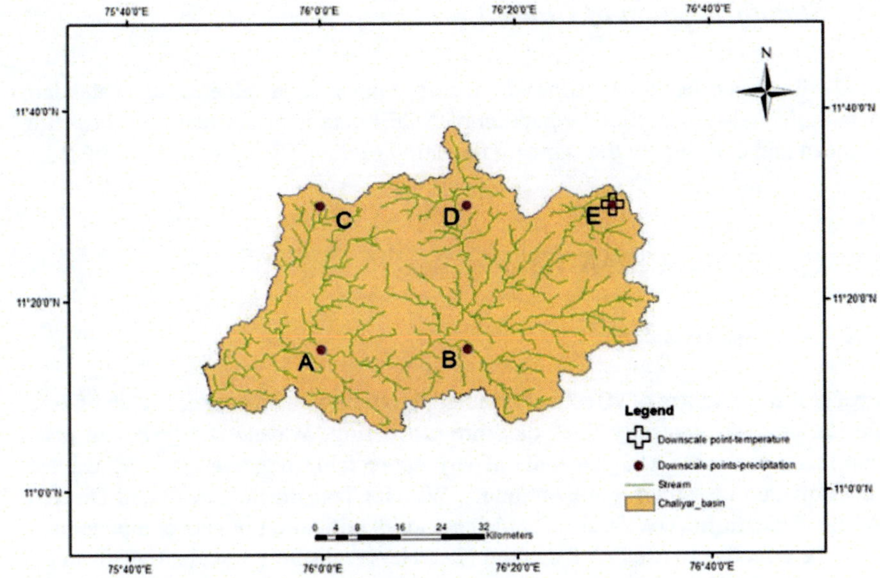

Fig. 2 Map of the Chaliyar Basin with the station points

3 Methodology

3.1 Selection of Potential Predictors

Selecting the potential predictors is important in statistical downscaling models in order to establish useful relation between the predictors and predictand. According to Maraun et al. [11], selected predictors should be informative and can be identified through statistical analyses, such as correlation between predictors and predictands. In this study, the values of coefficient of correlation (R) are used to determine the appropriate predictors.

$$R = \frac{\sum_{i=1}^{n}(x_i - \overline{x})(y_i - \overline{y})}{n\sigma_x\sigma_y} \quad (1)$$

where x is the predictand, y is the predictor, n is the number of observations, σ_x and σ_y are the standard deviations of the predictor and predictand, respectively.

3.2 *Standardization of Data*

Standardization refers to rescaling of the data to have mean value zero and standard deviation equal to one. Standardization of NCEP data is performed by subtracting the mean and dividing by the standard deviation of the NCEP predictor variable.

3.3 *Hybrid Wavelet ANN Model*

3.3.1 Wavelet Transform

Wavelet is a waveform of effectively limited duration with an average value of zero. Wavelet analysis basically does data pre-processing. Wavelet analysis can often denoise or compress a signal without any appreciable degradation. The wavelet transforms are classified into Continuous Wavelet Transform (CWT) and Discrete Wavelet Transform (DWT). For denoising and identification of true components, it would be effective to use DWT method [12]. Hence, discrete wavelet transform has been adopted in this study.

Figure 3 presents the flowchart of the WANN structure. Wavelet decomposition is applied to the input data for the development of model. The main steps involved in wavelet decomposition are selection of mother wavelet and determination of decomposition level for wavelet decomposition. Wavelet analysis performs pre-processing of the input data by decomposing the original input data into approximate and detail components using the selected mother wavelet and reconstructing the effective components chosen by correlation analysis back into a de-noised single sum series.

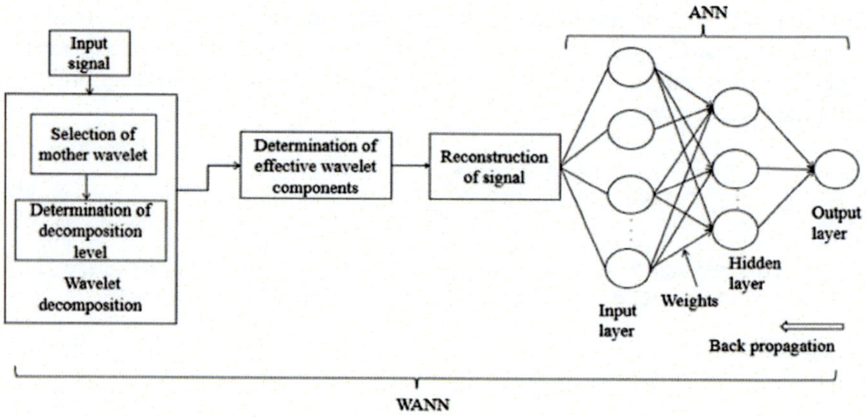

Fig. 3 Flowchart of the WANN structure

3.3.2 Selection of Mother Wavelet

Generally, mother wavelets are of two types, orthogonal and non-orthogonal wavelets. Torrence and Compo [13] suggested the use of DWT for orthogonal wavelets. Orthogonal wavelets are found to be ideal for hydrological variables because they are efficient in wavelet decomposition, denoising, multiresolution analyses, and many other wavelet analyses. According to Maheswaran and Khosa [14], wavelets coming under Daubechies family often yield reliable results in hydrological time series problems. These is a family of wavelets with orthogonal properties and is compactly supported with extreme phase.

3.3.3 Wavelet Decomposition

The accuracy of the features identified in a time series depends on the decomposition level selected. So, the choice of an appropriate level of decomposition or temporal scale is very important. In earlier studies, trial and error procedure was used for identifying appropriate level of decomposition. Later, a new formula was introduced by Nourani et al. [15] and Wang and Ding [16] to determine the minimum level of decomposition as,

$$L = \text{int}\,[\log\ N] \tag{2}$$

where N is number of data points.

According to Lei et al. [17], the formula for maximum decomposition level for a DWT is given by:

$$L = \text{int}\,[\log_2\ N] \tag{3}$$

The decomposition level was chosen based on the correlation coefficient analysis between the observed data and the wavelet coefficients at various decomposition levels within the minimum and maximum levels. The decomposition level yielding highest correlation coefficient value was selected as the decomposition level for this study.

3.3.4 Determination of Effective Wavelet Components and Reconstruction

Correlation analysis was performed to determine the correlation between the observed time series data (output) and the decomposed components at various levels so as to choose the effective components. The components found effective (the one which gave the higher correlation coefficient) were added up into a single sum series and were reconstructed back. Thus, the reconstructed signal will give a higher correlation coefficient compared to the original input time series data. This decomposition and reconstruction were carried out for all the input variables. The reconstructed data was then input to the ANN model.

3.3.5 Artificial Neural Network Model

ANN has been developed as a generalization of mathematical models of human cognition or neurobiology. A neural network is characterized by its architecture that represents the pattern of connection between nodes, its method of determining the connection weights, and the activation function. Wavelet analysis was applied to the large-scale climate variables extracted at the station locations and was fed as input to the input layers which further gets transformed to the hidden layers with the help of weights and the bias present at each node. The process is carried out from one hidden layer to the next till it reaches output layer. The predictand fed at the output layer measures the error and arranges the weights and bias at each node in the backward direction and this is termed as back-propagation technique. The Levenberg–Marquardt feed-forward back-propagation algorithm is used in the present study. Training and validation of the network are done using the neural network toolbox in MATLAB. Calibration of the model is performed using the data for the period from January 1, 1980, through December 31, 2007. Validation of the model is performed using the data for the period January 1, 2008, through December 31, 2014. The validation results are then compared with the observed IMD data.

3.4 Performance Indices

The Root Mean Square Error (RMSE) and Coefficient of Correlation (R) are used to evaluate the performance of the developed model.

- Root Mean Square Error (RMSE)

  The RMSE is commonly used to measure the difference between the values predicted by the model and the values observed. The individual differences are known as the residuals and the RMSE aggregates these residuals to measure the predictive power. RMSE is expressed as,

$$\text{RMSE} = \sqrt{\frac{1}{n}\sum_{i=1}^{n}(O_i - P_i)^2} \tag{4}$$

where O_i and P_i are the observed and estimated precipitation or temperature and n is the number of data points in the series. The closer the value to 0, the better is the model performance.

- Coefficient of Correlation (R)

Equation (1) is used to find the Coefficient of Correlation. R varies from [− 1, 1], with higher values indicating higher responsiveness of the predicted time series to be observed.

4 Results and Discussion

Based on the values of the correlation coefficient between the observed predictand at the station points and the NCEP predictors, the potential predictors were identified and selected. Table 3 presents the list of selected predictors for downscaling daily precipitation, daily maximum, and minimum temperature.

The Daubechies family of wavelets was selected as the mother wavelet for the present study following the guidelines by Maheswaran and Khosa [14]. The decomposition level 6 is chosen as the appropriate decomposition level as it yielded higher correlation coefficient for the observed data and the wavelet coefficients at various levels. The correlation coefficient for the model developed for downscaling precipitation for the station points was found to be 0.63, 0.64, 0.66, 0.61, and 0.56, respectively, and that for downscaling daily maximum and minimum temperature for the station point was found to be 0.85 and 0.84, respectively. The RMSE for the models for precipitation was found to be in between 10 and 12 mm and that for maximum and minimum temperature was found to be 0.8 and 1.2 °C. Figure 4 shows the comparison of simulated vs. observed precipitation. Figures 5 and 6 show the comparison of simulated vs. observed maximum and minimum temperatures, respectively.

Table 3 Selected predictors at various pressure levels (mb) for precipitation, maximum temperature, and minimum temperature

S. No	Predictor variables	Precipitation	Maximum temperature	Minimum temperature
1	Air temperature	250 mb	850 mb,100 mb, surface	1000 mb, 850 mb
2	Geopotential height	850 mb	700 mb, 500 mb	1000 mb, 100 mb
3	Surface pressure	Surface	–	Surface
4	Relative humidity	1000 mb, 500 mb	1000 mb, 850 mb, 500 mb	–
5	Specific humidity	–	–	1000 mb, 700 mb
6	Mean sea-level pressure	Surface	–	Surface
7	Eastward wind	1000 mb, 850 mb, 700 mb, 100 mb	700 mb	1000 mb, surface

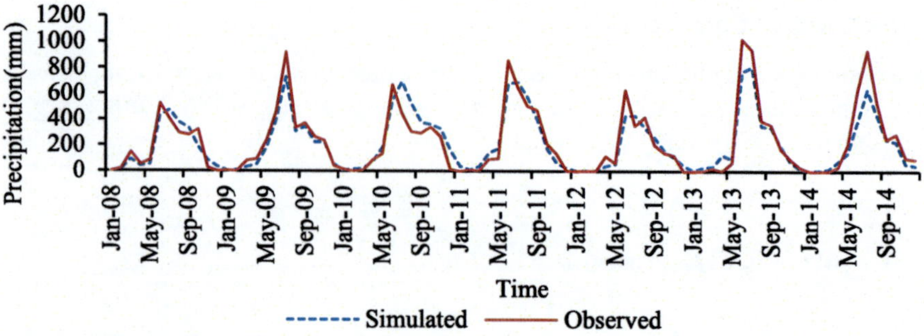

Fig. 4 Simulated versus observed precipitation for the validation period for station A

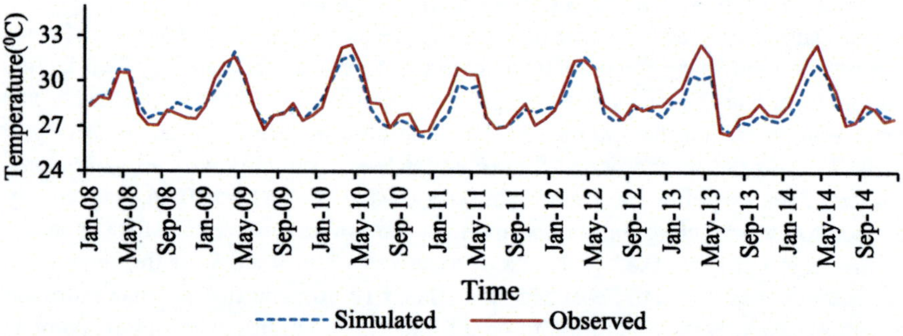

Fig. 5 Simulated versus observed maximum temperature for the validation period for station E

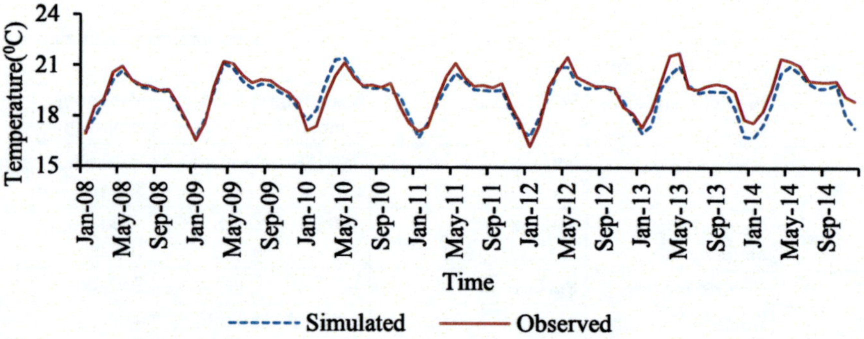

Fig. 6 Simulated versus observed minimum temperature for the validation period for station E

5 Summary and Conclusions

Hybrid Wavelet ANN-based models were developed to downscale daily precipitation and daily maximum and minimum temperatures for the station points in the Chaliyar river basin in Kerala, India. Results of the correlation coefficient analysis performed to identify the potential predictors show that air temperature, geopotential height, surface pressure, relative humidity, eastward wind at various pressure levels and surface levels were the influencing predictors for downscaling precipitation. In the case of temperature, the influencing predictors were air temperature, geopotential height, relative humidity, specific humidity, mean sea-level pressure and eastward wind at various pressure levels and surface. The Daubechies family of wavelets was selected as the suitable mother wavelet. Based on the correlation coefficient analysis carried out, the decomposition level 6 was found to be appropriate. The components found effective were reconstructed and were used as input to ANN. The performance of the trained wavelet ANN models was evaluated using RMSE and coefficient of correlation and was found to be satisfactory. The developed hybrid WANN model can be used to downscale grid-based coarse resolution General Circulation Models (GCMs) data to the station scale.

References

1. Wigley, T. M. L., Jones, P. D., Briffa, K. R., & Smith, G. (1990). Obtaining sub-grid-scale information from coarse-resolution general circulation model output Ttx. *Journal of Geophysical Research, 95*, 1943–1953. https://doi.org/10.1029/JD095iD02p01943
2. Chithra, N. R., & Thampi, S. G. (2017). Downscaling future projections of monthly precipitation in a catchment with varying physiography. *ISH Journal of Hydraulic Engineering, 23*, 144–156. https://doi.org/10.1080/09715010.2016.1264895
3. Hoanh, C. T., Jirayoot, K., Lacombe, G., & Srinetr, V. (2010). *Impacts of climate change and development on Mekong flow regime, First assessment—2009*. Mekong River Comm Vientiane, Lao PDR
4. Tripathi, S., Srinivas, V. V., & Nanjundiah, R. S. (2006). Downscaling of precipitation for climate change scenarios: A support vector machine approach. *Journal of Hydrology, 330*, 621–640. https://doi.org/10.1016/j.jhydrol.2006.04.030
5. Nourani, V., Hosseini, A., Adamowski, J., & Kisi, O. (2014). Applications of hybrid wavelet—Artificial Intelligence models in hydrology: A review. *Journal of Hydrology, 514*, 358–377. https://doi.org/10.1016/j.jhydrol.2014.03.057
6. Sang, Y. (2013). A review on the applications of wavelet transform in hydrology time series analysis. *Atmospheric Research, 122*, 8–15. https://doi.org/10.1016/j.atmosres.2012.11.003
7. Adamowski, J. F. (2008). River flow forecasting using wavelet and cross-wavelet transform models. *Wiley Interscience, 4891*, 4877–4891. https://doi.org/10.1002/hyp
8. Adamowski, K., Prokoph, A., & Adamowski, J. (2009). Development of a new method of wavelet aided trend detection and estimation. *Wiley Interscience, 2696*, 2686–2696. https://doi.org/10.1002/hyp

9. Mirbagheri, S. A., Nourani, V., Rajaee, T., & Alikhani, A. (2010). Neuro-fuzzy models employing wavelet analysis for suspended sediment concentration prediction in rivers Neuro-fuzzy models employing wavelet analysis for suspended sediment concentration prediction in rivers. *Hydrological Sciences Journal, 55*, 1175–1189. https://doi.org/10.1080/02626667.2010.508871
10. Sang, Y. (2012). A practical guide to discrete wavelet decomposition of hydrologic time series. *Water Resource Management* 3345–3365. https://doi.org/10.1007/s11269-012-0075-4
11. Maraun, D., Wetterhall, F., Ireson, A. M., Chandler, R. E., Kendon, E. J., Widmann, M., Brienen, S., Rust, H. W., Sauter, T., et al. (2010). Precipitation downscaling under climate change: Recent developments to bridge the gap between dynamical models and the end user. *Reviews of Geophysics, 48*, 1–34. https://doi.org/10.1029/2009RG000314
12. Sang, Y., Singh, V. P., & Sun, F., et al. (2016). Wavelet-based hydrological time series forecasting. Journal of Hydrologic Engineering 21. https://doi.org/10.1061/(ASCE)HE.1943-5584.0001347
13. Torrence, C., & Compo, G. P. (1998). A practical guide to wavelet analysis. *Bulletin of the American Meteorological Society, 79*, 61–78. https://doi.org/10.1175/1520-0477(1998)079%3c0061:APGTWA%3e2.0.CO;2
14. Maheswaran, R., & Khosa, R. (2012). Computers & geosciences comparative study of different wavelets for hydrologic forecasting. *Computers & Geosciences, 46*, 284–295. https://doi.org/10.1016/j.cageo.2011.12.015
15. Nourani, V., Komasi, M., & Mano, A. (2009). A multivariate ANN-wavelet approach for rainfall—Runoff modeling. *Water Resources Management, 23*, 2877–2894. https://doi.org/10.1007/s11269-009-9414-5
16. Wang, W., & Ding, J. (2003). Wavelet network model and its application to the prediction of hydrology. *Natural Science, 1*, 67–71.
17. Lei, L., Wang, C., & Liu, X. (2013). Discrete wavelet transform decomposition level determination exploiting sparseness measurement. *International Science Index, Electrical and Computer Engineering, 7*, 1182–1185. https://doi.org/10.5281/zenodo.1087700

Computational Fluid Dynamics (CFD) Modelling and Design for Computing the Internal Hydrodynamic Behaviour of Final Clarifier

R. Ravi Varma

Abstract To obtain effluent low in suspended solids is critical for operation of the final clarifier tank in waste water treatment facilities. Flocculants solids tend to form sludge at the bottom of the tank and at peak flow rates; this sludge may fill the entire tank and overflow the weirs. Inadequate design of the tank may cause the suspended solids to be discharged into the effluent weir. CFD model was used to analyse the flow of a laboratory-based clarifier tank. This CFD model was further extended to analyse flow features in a full-scale clarifier tank. Design modification to the clarifier tank was analysed to optimise its performance in terms of limiting the carryover of suspended solids to the effluent weir.

Keywords Clarifier tank · CFD · Effluent

1 Introduction

Secondary sedimentation or final clarifier tanks are used to separate suspended solids from water. Design of final clarifier tank is based on the function of separation of solids in the activated sludge process in order to provide well-clarified effluent [1]. Separation of solids in the activated sludge treatment plant is governed by two functions—sufficient time needed to provide gravity settling and thickening of settled solids. Design of physical facilities of the final clarifier is a key step to ensure solids separation for production of a stable effluent that is low in suspended solids [2, 3]. Large volumes of flocculent solids are present in the mixed liquor and therefore requires careful design of the clarifier. The flocculent solids tend to form a sludge blanket in the bottom of the tank. This sludge blanket may fill the entire depth of the clarifier tank at peak flow rates. If the return sludge pumping capacity or the clarifier tank capacity is inadequate, sludge may overflow the weirs, and solids may be discharged into the effluent. In addition, the mixed liquor entering the tank has a tendency to flow as a density current. This phenomenon interferes with the separation

R. Ravi Varma (✉)
Centre for Environment, Institute of Science and Technology, JNTUH, Hyderabad, India

A. K. Dikshit et al. (eds.), *Innovative Trends in Hydrological and Environmental Systems*, Lecture Notes in Civil Engineering 234,
https://doi.org/10.1007/978-981-19-0304-5_38

of solids and thickening of sludge. Weir placement and baffle provision are one of the strategies to improve the performance of the clarifier.

CFD modelling technique has become an essential tool for predicting and gathering information for design and optimization of wastewater engineering systems [4–8]. Hydrodynamic modelling of secondary settling tanks has been developed and validated with experimental works in the field [9, 10]. Different mechanical structures were evaluated based on prediction of flow, sludge concentrations over a range of conditions for wastewater treatment plants [11, 12].

2 Methodology

2.1 CFD Model

Flows observed in a clarifier tank have nature of mixture of phases known as multiphase. In multiphase flow, different phases can be assumed to have different phase responses such as settling velocity, slip velocity, and phase interactions, respectively. Also, particles with different material size can be treated as different phases.

Mixture model available in fluent CFD software code can be used to model multiphase flows where different phases can be specified with particular settling velocities and slip velocities. Further, the model can simulate multiphase flows by solving momentum and continuity equations for the mixture. Equations for the volume fractions of secondary phases and their relative velocities are also solved. Mixture model allows for secondary phase to move at different velocities thus applying the concept of slip velocities.

In the sedimentation process, what we intend to model involves a relative velocity of the secondary phase (sludge) with respect to the primary phase (water). As previously known, the activated sludge is a flocculant which means that as concentration of flocculants increases the particle diameter increases. Therefore, with an increase in concentration, we can also observe increase in settling velocity. In mixture model, it was assumed that the two phases act as fluids with varying density to simulate this phenomenon. To solve for momentum, a source term is included which represents the change in forces due to variable density of the secondary phase. Variation in density of the secondary phase depends on the variation in concentration of the fluid in secondary phase (sludge). Concentration of the sludge particles is obtained from the solution of particulate mass conservation equation. Also, this equation solves for the change in relative velocities between the phases due to variation in density. This change in density with respect to the change in secondary phase concentration is incorporated into the code as follows:

$$\rho = \rho_r + C(1 - \frac{\rho_r}{\rho_s}) \tag{1}$$

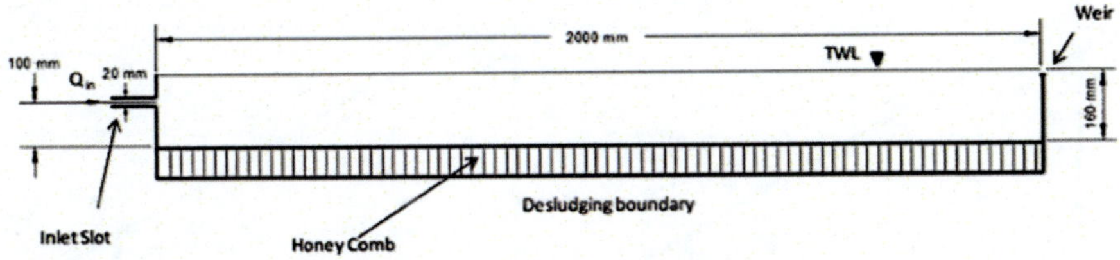

Fig. 1 Schematic diagram of the laboratory clarifier of Krebs (1998)

where ρ_r is the reference density (water), ρ_s is the density of the dry particle's solids, and C is the solids concentration.

2.2 *CFD Domain and Boundary Conditions*

Benchmark studies based on laboratory clarifier of [13] was selected to analyse the performance of the model. Figure 1 shows the physical dimensions of the laboratory clarifier. Sludge flow is represented by flow carrying glass bead suspensions which enter at the inlet slot on the left. These solids will settle at the bottom and exit through the desludging boundary, whilst the clear water will flow at the exit through the weir on the right.

Computational mesh was generated using hexahedral cells for analysis in two dimensions. Standard k-epsilon turbulence model with standard wall functions was selected to resolve the viscous flow. Mixture model was selected to model multiphase flows where phases move at different velocities. In the mixture model, granular phase was selected which applies for liquid–solid flows. The outlet weir was modelled as pressure outlet boundary. The top surface was modelled using zero-shear slip walls to simulate the free surface.

2.3 *Full-Scale Model of a Final Clarifier Tank*

This section describes the CFD study performed for a final clarifier tank in operation in Hyderabad area. Flow features and internal hydrodynamics of clarifier tank were analysed applying the CFD methodology described in the previous section. Influent design modification including a baffle was modelled, and CFD analysis performed. The CFD study allowed for comparison of flow features including velocity profiles, concentration distribution in the downstream and outlet locations. The tank was of circular diameter, $D = 5.5$ m and side wall depth of $h = 1$ m and with floor inclined at approximately 7° to the horizontal. Models were created for two tank arrangements without baffle and with baffle near the outlet weir as shown in Fig. 2.

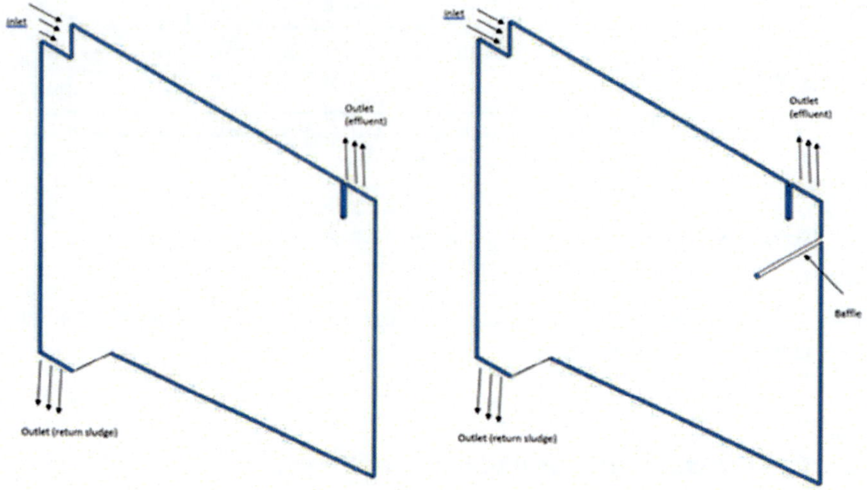

Fig. 2 Model showing internal tank modifications without baffle and with baffle near outlet weir

3 Results and Discussion

Momentum equations were solved using second-order differencing. Steady-state solutions were achieved with a maximum of 10,000 iterations. The plot in Fig. 3 shows the concentration distribution of solids volume fractions and streamlines of velocity. Results were also extracted at three axial locations shown in the Fig. 3. These axial positions correspond to ratio of axial location to length of the tank, $x/L = 0.15$, 0.3, and 0.65. The contour plot indicates difference in concentration distribution compared to the inlet and downstream locations. Concentration of solids is greater at the bottom of the tank.

Grid sensitivity studies were performed for coarse mesh (8,000 cells), medium mesh (12,000 cells), and fine mesh (23,000 cells). Radial velocity profiles were plotted against the depth of the tank for each location. Shape of the flow is captured with the model, whereby peak velocity occurs at the shear layer separation just above the solids settlement (Fig. 4). Negative values of the radial flow velocity indicate flow reversal due to recirculation. Velocity profile shapes indicate little difference for the three grids considered.

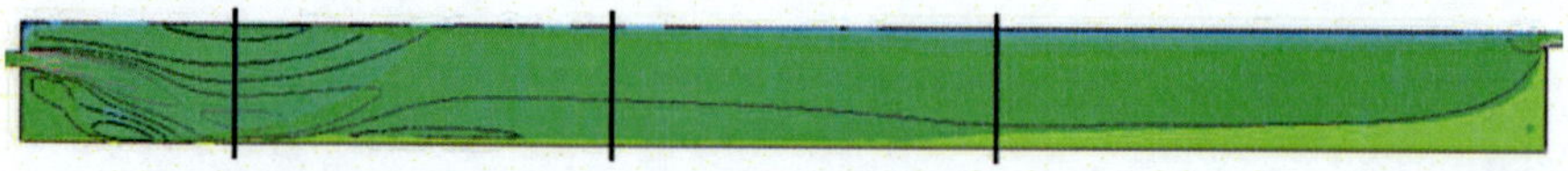

Fig. 3 Contour plot of solids concentration distribution and streamline plot of the velocity magnitude with axial locations used for recording velocity profiles along the tank

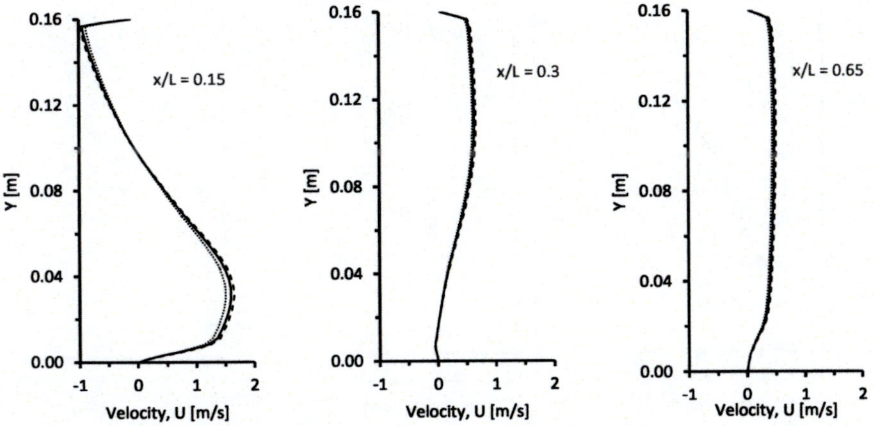

Fig. 4 Grid sensitivity study comparing the velocity profiles for (—) coarse, (___) medium, and (....) fine mesh grid at the three axial locations

Further, this CFD model was applied to study internal design of a clarifier tank with and without baffle arrangement near the outlet weir. Velocity contour plots are shown in the clarifier tank with and without baffle along with axial locations (Fig. 5). These axial positions correspond to ratio of axial location to length of the tank, x/L = 0.65, 0.8, and 0.95.

Results in Fig. 6 show the velocity profiles which indicates the shear layer separation between the settled solids and the suspended solids. Shear layer separation maybe indicated by the flow reversal as shown by the velocity profiles. Flow reversal is indicated by change in velocity from positive to negative values with depth of the tank. In the presence of side wall baffle, there is significant influence on the

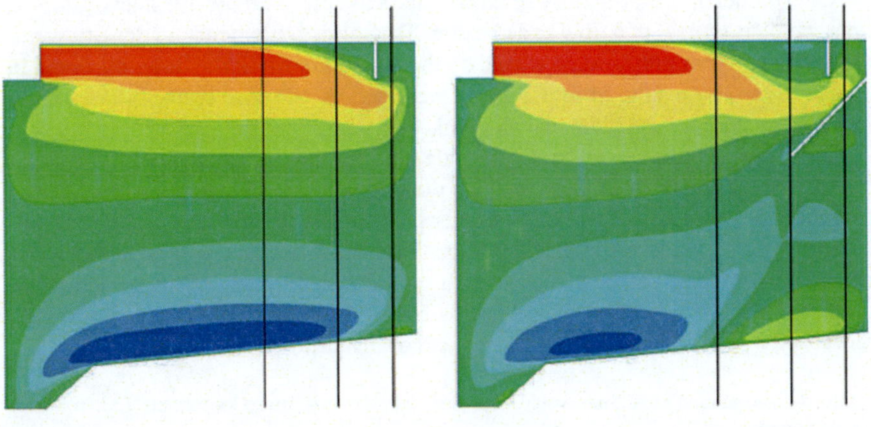

Fig. 5 Velocity contours plots of internal flow dynamics without baffle and with side wall baffle plate with three axial locations used for recording velocity profiles along the tank

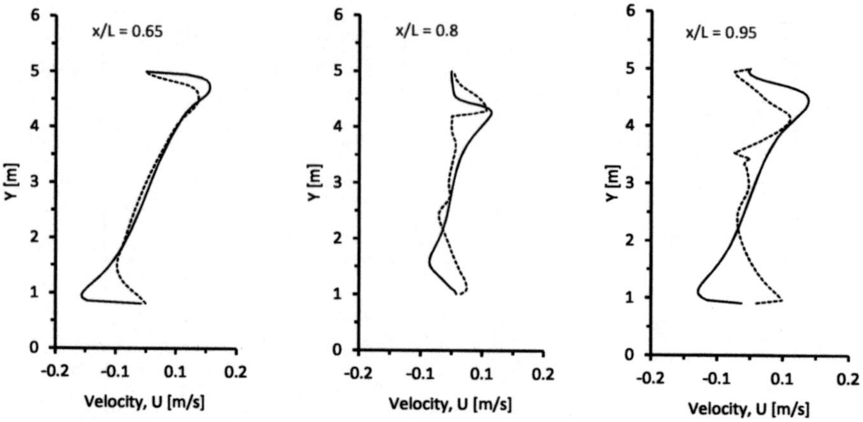

Fig. 6 Velocity profiles with baffle (____) and without baffle (—)

shear layer separation and flow dynamics. Without the baffle flow profiles indicate a non-uniform shear layer separation. Near the effluent weir, flow velocity profiles indicate uniform separation with a baffle and non-uniform flow pattern without a baffle. These results suggest that including a baffle could perform better in terms of having a uniform flow separation. A uniform flow separation could be an indicator of separation of suspended solids. Thus, including a baffle could promote lowering the suspended solids discharged in to the effluent weir.

4 Conclusions

The CFD model of the laboratory scale clarifier was able to predict the density-driven flows which is typical of a flow in a clarifier. The solids settlement and density variation and the momentum field predicted by the model are similar to flows observed in the literature. The results from this study were satisfactory and convincing to apply for a design study of a typical clarifier tank.

Application of this model to a full-scale clarifier tank associated with a sewage treatment works enabled us to investigate the internal flow distributions. With aid of this model, it was possible to examine internal flow distribution due to design modifications such as a side-wall baffling option. With these design changes, an improvement in performance can be achieved such as limiting carryover of suspended solids to effluent weir. Application of the results from this study is relevant to improving the capacity of existing clarifier tank instead of installing new equipment.

Acknowledgements The author would like to thank ID Structural Consultants Ltd. for their valuable inputs.

References

1. Tchobanoglous, G., Burton, F. L., Stensel, H. D., Eddy, M. (2003). Wastewater engineering: Treatment and reuse.
2. Burt, D. J., & Ganeshalingam, J. (2005). Design and optimisation of final clarifier performance with CFD modelling. In *Design and Operation of Activated Sludge Plants*, UK.
3. Das, S., Bai, H., Wu, C., Kao, J.-H., Barney, B., Kidd, M., & Kuettel, M. (2016). Improving the performance of industrial clarifiers using three-dimensional computational fluid dynamics. *Engineering Applications of Computational Fluid Mechanics, 10*, 130–144. https://doi.org/10.1080/19942060.2015.1121518
4. Ghawi, A., & Kris, J. (2008). Design and optimisation of final clarifier performance with CFD modelling. In *Proceedings of the International Scientific Conference*, Bratislava, Slovakia.
5. Mohanarangam, K., & Stephens, D. W. (2009). CFD modelling of floating and settling phases in settling tanks. https://doi.org/10.13140/RG.2.1.5078.7686
6. Korobeinichev, O. P., Shmakov, A. G., Chernov, A. A., Bol'shova, T. A., Shvartsberg, V. M., Kutsenogii, K. P., & Makarov, V. I. (2010). Fire suppression by aerosols of aqueous solutions of salts. *Combustion Explosion and Shock Waves, 46*, 16–20. https://doi.org/10.1007/s10573-010-0003-y
7. Wicklein, E., Batstone, D. J., Ducoste, J., Laurent, J., Griborio, A., Wicks, J., Saunders, S., Samstag, R., Potier, O., & Nopens, I. (2016). Good modelling practice in applying computational fluid dynamics for WWTP modelling. *Water Science and Technology, 73*, 969–982. https://doi.org/10.2166/wst.2015.565
8. Xanthos, S., Gong, M., Ramalingam, K., Fillos, J., Deur, A., Beckmann, K., & McCorquodale, J. A. (2011). Performance assessment of secondary settling tanks using CFD modeling. *Water Resources Management, 25*, 1169–1182. https://doi.org/10.1007/s11269-010-9620-1
9. Ekama, G.A., & Marais, P. (1997). Hydrodynamic modelling of secondary settling tanks.
10. Tarpagkou, R., & Pantokratoras, A. (2013). CFD methodology for sedimentation tanks: The effect of secondary phase on fluid phase using DPM coupled calculations. *Applied Mathematical Modelling., 37*, 3478–3494. https://doi.org/10.1016/j.apm.2012.08.011
11. Gao, H., & Stenstrom, M. K. (2020). Computational fluid dynamics analysis for improving secondary settling tank performance. In *World Environmental and Water Resources Congress 2020* (pp. 212–224). American Society of Civil Engineers, Henderson, Nevada (Conference Cancelled)
12. Karpinska, A. M., & Bridgeman, J. (2016). CFD-aided modelling of activated sludge systems—A critical review. *Water Research, 88*, 861–879. https://doi.org/10.1016/j.watres.2015.11.008
13. Krebs, P., Armbruster, M., & Rodi, W. (1998). Laboratory experiments of buoyancy-influenced flow in clarifiers. *Journal of Hydraulic Research., 36*, 831–851. https://doi.org/10.1080/00221689809498606

Surface Soil Moisture Retrieval Over Partially Vegetated Areas from the Remote Sensing Data Using a Modified Water Cloud Model

Athira Krishnankutty and D. Sathish Kumar

Abstract Surface soil moisture is an essential parameter for the hydrological modelling of the watershed. Field techniques adopted for soil moisture measurements are laborious and time-consuming. In a watershed, the soil moisture varies both spatially and temporally. Making continuous on-field observations simultaneously at several locations is practically impossible for large watersheds. Soil moisture estimation using remote sensing techniques is considered a viable alternative. In this study, it is proposed to apply a modified water cloud model for surface soil moisture (SSM) retrieval over partially vegetated regions. This method combines the microwave data obtained from Sentinel-I and optical data obtained from Landsat Operational Land Image (OLI) for SSM estimation. It is necessary to remove the effect of vegetation moisture content for better SSM estimation. Therefore, the index derived from the Landsat OLI spectral bands is applied to build a model for the vegetation water content estimation. A modified water cloud model (MWCM) is developed by integrating the vegetation index with the original water cloud model. In this study, the backscatter coefficients measured using the C-band (5.405 GHz) synthetic aperture radar (SAR) sensor onboard the Sentinel-IA satellite has been used. The developed MWCM has been used to prepare multi-temporal SSM maps.

Keywords Surface soil moisture · Synthetic aperture radar · Backscatter coefficients · Modified water cloud model

1 Introduction

Surface soil moisture is an important parameter in explaining how water and energy exchanges occur on the earth surface [1]. The determination of spatially varying surface soil moisture plays a vital role in water management processes and agricultural applications. Specifically, estimation of surface soil moisture is critical for developing the drought indices to quantify the spatial extent and severity of the drought

A. Krishnankutty (✉) · D. Sathish Kumar
Department of Civil Engineering, National Institute of Technology Calicut, Calicut, Kerala, India

A. K. Dikshit et al. (eds.), *Innovative Trends in Hydrological and Environmental Systems*, Lecture Notes in Civil Engineering 234,
https://doi.org/10.1007/978-981-19-0304-5_39

[2, 3]. It is not possible to use conventional techniques to monitor the spatially and temporally varying surface soil moisture. Advances in remote sensing techniques, specifically in microwave remote sensing, come in handy in solving this issue [4, 5]. The microwave remote sensing has unique abilities like day/night data acquisition capability, scanning in all weather conditions, high penetration power, etc. The SAR or microwave remote sensing has been used to surface soil moisture estimations because the dielectric property of the material influences the radar backscattering (σ^0). The soil dielectric property (ε) can be more related to the changes in surface soil moisture than other soil properties.

There are several remote sensing techniques available for retrieving information on surface soil moisture. However, there are several limitations in estimating the SSM due to the vegetation cover, cloud cover, etc. The moisture content present in the vegetation influences the radar backscattering values and also blocks the backscattered signals from the ground. As a result, backscattering values obtained in the SAR images are due to the mixed response of the SSM and vegetation. Therefore, it is essential to eliminate the effect of vegetation in the combined backscattering values (σ^0_{total}) obtained. Several studies indicate that the effect of vegetation can be eliminated by combining SAR data with optical remote sensing data [6]. This process improves the accuracy of the SSM estimation.

2 Study Area and Data Collection

2.1 Study Area

A part of the Thrissur district in Kerala is selected as the study area to retrieve soil moisture using the proposed methodology. The geographical extent of the study area is between 10°32′ 06″ N–10° 33′ 17″ N latitudes and 76° 13′ 41″ E–76° 15′ 23″ E longitudes. It covers a total area of about 101 km^2. The district experiences a humid tropical climate with hot summer and receives fairly good rainfall in the monsoon season. Fourteen random in situ sampling points were identified in the study area to collect soil samples. Initially (two sampling days), the samples were collected from nine points and later, the sampling points were increased to fourteen. The soil sampling was carried out exactly on the days of Sentinel-IA data acquisition used in the study. The study area map and soil sampling locations are shown in Fig. 1.

2.2 Satellite Data Collection and Processing

In the present study, the Sentinel-1A C-band (5.405 GHz) satellite images acquired on 02/11/2020, 26/11/2020, 08/12/2020, and 01/01/2021 were used. The image preprocessing of SAR data was carried out using Sentinel application platform (SNAP)

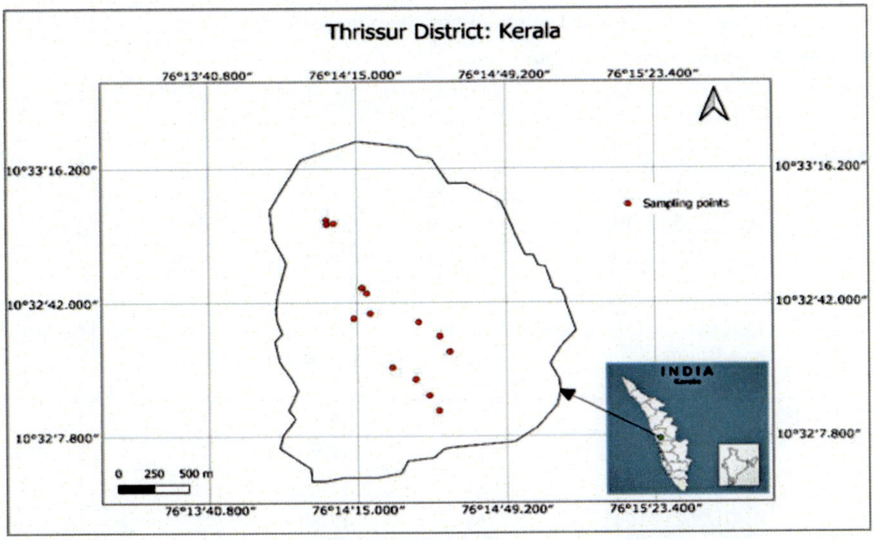

Fig. 1 Study area map showing the field sampling points

tools. Landsat 8 OLI datasets were downloaded for the same dates or the nearest possible dates to the date of acquisition of SAR images. These images were used for computing different vegetation indices like normalized difference vegetation index (NDVI), enhanced vegetation index (EVI), and normalized difference water index (NDWI) [7], which can be used to describe vegetation in WCM and MWCM. The radiometric calibration and atmospheric correction were carried out using ENVI-5.3. Finally, the images were co-registered as per Sentinel-1A. Table 1 lists the details of the satellite data used in this study.

3 Methodology

3.1 Field Measurement of Surface Soil Moisture (SSM)

The surface soil moisture is influenced by the physical and chemical properties of the soil on the earth's surface, such as dielectric properties and surface roughness. The soil sampling was conducted in conjunction with the SAR image acquisition dates (02/11/2020, 26/11/2020, 08/12/2020, and 01/01/2021). The locational details of soil sampling points are given in Table 2.

The soil moisture from the field samples collected is determined by applying the gravimetric technique. The soil samples were weighed and then oven-dried at 110 °C for 18–20 h. The average water content of the soil sample was used to calculate the surface soil moisture percentage [8]. The soil moisture can be calculated using the

Table 1 The details of the SAR and optical datasets used in the study

(a) Synthetic aperture radar (SAR) data

Satellite	Date of acquisition	Mode	Polarization	Product type	Resolution (m*m)	Level
Sentinel-1A	02/11/2020	IW	VV/VH	GRD	5*20	L1
	26/11/2020	IW	VV/VH	GRD	5*20	L1
	08/12/2020	IW	VV/VH	GRD	5*20	L1
	01/01/2021	IW	VV/VH	GRD	5*20	L1

(b) Optical data

Satellite	Date of acquisition	No. of bands	Spatial resolution (m)	Radiometric resolution	Swath (km)	Level
Landsat 8 (OLI)	10/11/2020	11	30	12-bit	185	L1
	26/11/2020	11	30	12-bit	185	L1
	12/12/2020	11	30	12-bit	185	L1
	01/01/2021	11	30	12-bit	185	L1

Table 2 Location of sampling points for SSM measurements in the field

Study area	Sampling point	Latitude	Longitude	Vegetation cover type
Thrissur (Kerala)	1	10°32′ 33″ N	76°14′ 36″ E	Natural vegetation
	2	10° 32′ 29″ N	76° 14′ 36″ E	Natural vegetation
	3	10° 32′ 25″ N	76° 14′ 23″ E	Agriculture (paddy)
	4	10° 32′ 14″ N	76° 14′ 33″ E	Natural vegetation
	5	10° 32′ 44″ N	76° 14′ 17″ E	Natural vegetation
	6	10° 32′ 38″ N	76° 14′ 14″ E	Natural vegetation
	7	10° 32′ 39″ N	76° 14′ 18″ E	Natural vegetation
	8	10°33′ 03″ N	76° 14′ 08″ E	Natural vegetation
	9	10°33′ 02″ N	76° 14′ 08″ E	Agriculture
	10	10°32′ 22″ N	76° 14′ 28″ E	Natural vegetation
	11	10° 32′ 18″ N	76° 14′ 31″ E	Bare soil
	12	10° 32′ 37″ N	76° 14′ 29″ E	Paddy
	13	10° 32′ 45″ N	76° 14′ 16″ E	Natural vegetation
	14	10° 32′ 55″ N	76° 14′ 10″ E	Natural vegetation

following relationship.

$$\text{Soil Moisture}(\%) = \frac{W_{\text{wet}} - W_{\text{dry}}}{W_{\text{dry}}} * 100 \quad (1)$$

Table 3 The details of the field soil samples and volumetric soil moisture (%)

Sampling date	Nature of the field	Volumetric soil moisture (%) range	No. of samples
02/11/2020	Partially vegetated	2.65 to 29.83	9
26/11/2020	Partially vegetated	4.28 to 19.97	9
08/12/2020	Partially vegetated	6.12 to 26.82	14
01/01/2021	Partially vegetated	17.04 to 27.58	14

where W_{wet} and W_{dry} are the weights of soil samples collected before and after oven drying. The volumetric soil moisture percentage can also be obtained using the bulk density of the soil. It is calculated using the following relationship.

$$\text{Vol. Soil Moisture}(\%) = \frac{W_{\text{wet}} - W_{\text{dry}}}{W_{\text{dry}}} * 100 * \text{Bulk density} \tag{2}$$

Table 3 shows the details of the soil moisture content in the samples collected from the field. However, during model calibration and validation phases, only 32 samples were utilized due to the limitations in the acquired SAR datasets.

3.2 Water Cloud Model (WCM)

The total backscatter coefficient (σ^0) is related to soil moisture and vegetation in the water cloud model (WCM). The total backscatter coefficient is the combined response of the vegetation backscatter (σ^0_{veg}) and soil backscatter (σ^0_{soil}). The σ^0_{veg} in equation three depends upon the vegetation characteristics. The water cloud model is based on the radiation transport model. In this model, the vegetative canopy is considered as the homogeneous horizontal clouds, and thus, multiple scattering is not taken into account. The model is relatively simple and practical in explaining radar scattering mechanisms in crop-covered regions. As a result, this model is frequently used to extract relevant information in the areas covered with vegetation.

The water cloud model is expressed as

$$\sigma^0 = \sigma^0_{\text{veg}} + T^2\sigma^0_{\text{soil}} \tag{3}$$

$$\sigma^0_{\text{veg}} = AV_1\cos(\theta)\left(1 - T^2\right) \tag{4}$$

$$T^2 = \exp(-2BV_2\sec(\theta)) \tag{5}$$

where σ^0 is the total backscattering coefficient (dB); σ^0_{veg} is the vegetation backscattering coefficient (dB); σ^0_{soil} is the soil backscattering coefficient (dB). Direct canopy backscattering and vegetation attenuation are indicated using the vegetation descriptors V_1 and V_2. In this study, normalized difference vegetation index (NDVI), enhanced vegetation index (EVI), and normalized difference water index (NDWI1, NDWI2) are the vegetation descriptors applied. A and B are the estimated model coefficients that depend on the vegetation descriptor and radar configuration, respectively. T^2 is the double attenuation factor for the radar waves passing through vegetation; θ is the incident angle (°) [9].

The previous studies have shown that there is a linear relationship exist between the soil backscattering coefficient and volumetric soil moisture (SSM). Therefore, the following relationship is used to relate the soil backscattering coefficient σ^0_{soil} with SSM.

$$\sigma^0_{\text{soil}} = (C * SSM) + D) \tag{6}$$

where C and D are the calibrated model coefficients that depend on the characteristics of the soil.

3.3 Estimation of Vegetation Indices Using Landsat 8 OLI Data

One of the most significant inputs in the water cloud model is the vegetation water content (M_V) expressed in the form of vegetation descriptors V_1 and V_2. These parameters are included to eliminate the influence of the plant canopy on the surface soil moisture estimation. According to the published research, vegetation indices derived from satellite data might be used to estimate vegetation water content [10]. NDVI, EVI, and NDWI are the vegetation indices that are frequently used to estimate vegetation water content.

The vegetation water content is estimated using vegetation indices derived using Landsat 8 OLI datasets. The Landsat 8 satellite comprises onboard sensors that acquire data in two shortwave infrared bands: SWIR1 (band range: 1.57–1.65 m) and SWIR2 (band range: 2.11–2.29 m). These two shortwave infrared bands were utilized in this study to create the NDWI indices, namely NDWI1 and NDWI2. These indices were used to estimate the vegetation water content for WCM and the modified water cloud model (MWCM).

3.4 Estimation of Vegetation Water Content for WCM and MWCM

In WCM, the vegetation moisture content is derived by assuming that it has a linear relationship with the vegetation index. Therefore, the vegetation water content (M_V)

for WCM is given as follows.

$$M_v = bVI \tag{7a}$$

In MWCM, the vegetation water content (M_V) is assumed to have a nonlinear relationship with the vegetation index. Therefore, the vegetation water content (M_V) for MWCM is expressed using a single variable quadratic model as shown in Eq. 7b.

$$M_v = aVI^2 + bVI + c \tag{7b}$$

Where *VI* denotes vegetation index; *a*, *b*, and *c* are model coefficients. The vegetation descriptors V_1 and V_2 in Eqs. 4 and 5 are replaced with vegetation water content (M_V) yielding the Eqs. 8 and 9. The M_V values estimated using Eq. 7a are applied for the WCM, and values estimated using Eq. 7b are applied for the MWCM.

$$\sigma^0_{veg} = AM_V \cos(\theta)\left(1 - T^2\right) \tag{8}$$

$$T^2 = \exp(-2BM_V \sec(\theta)) \tag{9}$$

3.5 Development of the SSM Model

The final equation for SSM estimation is derived by substituting Eqs. 6, 8, and 9 in Eq. 3. By applying Taylor series expansion, the term $\exp(-2BM_v \sec(\theta))$ is expanded. Finally, the following semi-empirical model for estimating surface soil moisture is developed [11].

$$\begin{aligned} SSM = {} & k_1 + k_2\sigma^0 + k_3VI + k_4VI^2 + k_5VI^3 + k_6VI^4 \\ & + k_7\sigma^0 \sec\theta + k_8\sigma^0 VI \sec\theta + k_9\sigma^0 VI^2 \sec\theta \end{aligned} \tag{10}$$

where $k_1, k_2, \ldots .k_9$ are best-fit model coefficients obtained using regression analysis.

4 Results and Discussion

4.1 WCM and MWCM Calibration

The coefficients for WCM and MWCM were obtained through model calibration by applying the regression technique in R studio. The surface soil moisture values obtained from the field measurements (14 samples) were used for the calibration of the models. Statistical measures, such as the correlation coefficient (R) and the root mean square error (RMSE), were used to assess the agreement between the modelled SSM and the observed volumetric soil moisture.

The Sentinel-1 SAR datasets that are available in VV and VH polarizations were studied for their effectiveness in surface soil moisture estimation. The results indicated that the backscattering coefficients obtained in VV polarization had a better correlation with the SSM. As far as vegetation indices are considered, EVI had a good correlation compared to other vegetation indices. Table 4 shows the correlation coefficients (R) and RMSE values for WCM and MWCM models combined with different vegetation indices. The scatter plot between measured (samples used for calibration) and MWCM estimated soil moisture using EVI and VV polarization is shown in Fig. 2. The model produced good results with a correlation coefficient of 0.783 and an RMSE value of 2.43%. The calibrated MWCM model (EVI, VV polarization) with best-fit model coefficients ($k_1, k_2, \ldots .k_9$) for SSM estimation is given in Eq. 11.

Table 4 Correlation coefficients and RMSE values between measured and estimated SSM

Model	Polarization	Statistical measures	NDVI	EVI	NDWI1	NDWI2
WCM	VV	R	0.555	0.627	0.284	0.278
		RMSE (%)	4.992	4.678	5.756	5.766
	VH	R	0.353	0.524	0.398	0.264
		RMSE (%)	5.617	5.114	5.507	5.791
MWCM	VV	R	0.690	0.783	0.347	0.553
		RMSE (%)	4.346	2.430	5.635	5.004
	VH	R	0.629	0.668	0.703	0.447
		RMSE (%)	4.670	5.166	4.269	5.371

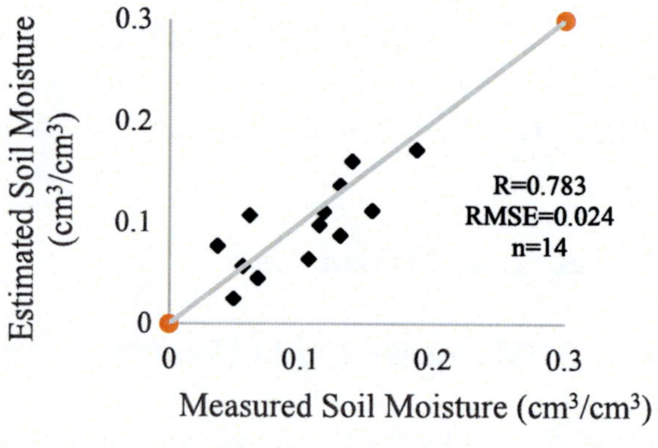

Fig. 2 Scatter plot between the observed (calibration samples) and estimated soil moisture by the MWCM (EVI and VV polarization)

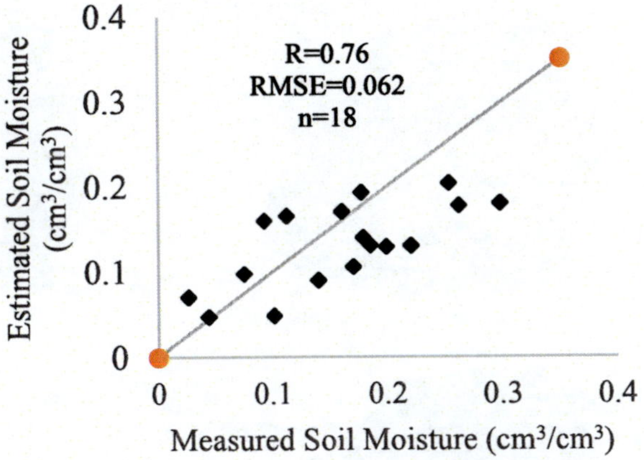

Fig. 3 Scatter plot between the observed (validation samples) and estimated soil moisture by the MWCM model

$$SSM = 19.45 + 828.472\sigma^0 - 126.501\,VI + 311.64\,VI^2 + -363.893\,VI^3 + 181.524\,VI^4 - 525.773\sigma^0 \sec\theta - 2.69223\sigma^0\,VI\,\sec\theta + 2.85233\sigma^0 VI^2 \quad (11)$$

4.2 Validation of MWCM

The calibration results indicate that MWCM is better for SSM retrieval compared to the WCM. Overall calibration results also indicate that EVI is the best index to estimate vegetation fraction that can be used as vegetation descriptors in the WCM and MWCM models. The best MWCM model for SSM estimation, which is presented in Eq. 11, was further validated with the observed soil moisture datasets. Eighteen field observations were used for validation. The results obtained from the MWCM model had good agreement with the SSM estimated from field samples. The validation of the developed model resulted in a correlation coefficient value of 0.76 and an RMSE value of 6.2% between the field observations and MWCM model results. The scatter plot between the estimated SSM (from the MWCM model) and measured soil moisture (for the validation samples) is shown in Fig. 3.

4.3 Multi-temporal Surface Soil Moisture Maps

The statistical measures of the MWCM model in the calibration and validation stages indicated that it could effectively estimate the SSM using the SAR datasets. Therefore, multi-temporal SSM maps for the study area have been prepared using this model. The SSM maps were prepared using the MWCM model for three days

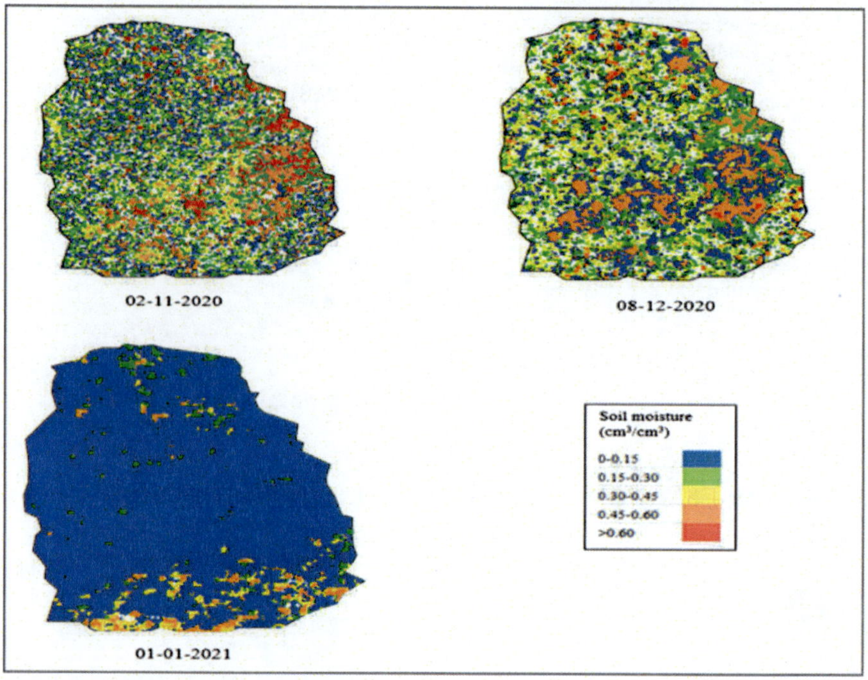

Fig. 4 Multi-temporal SSM maps estimated using the MWCM model

(02/11/2020, 08/12/2020, and 01/01/2021). The maps are shown in Fig. 4. The maps prepared for November and December show that SSM was between 15 and 30% for most regions. However, there was a substantial reduction in SSM at the beginning of January month.

5 Conclusions

The surface soil moisture estimation is essential for several hydrological studies. The backscattering values obtained using SAR have proved to have a reasonable correlation with the SSM. The vegetation moisture content often influences the backscattering values. This study indicates that EVI obtained using Landsat 8 optical datasets better correlate with the vegetation water content. Similarly, SAR backscattering values obtained using VV polarization had shown a better relationship with SSM. The results also indicate that the vegetation moisture content estimated using a nonlinear relationship with the vegetation index holds good. Finally, the semi-empirical MWCM model developed in this study using regression analysis has a good capability to estimate SSM. It is possible to prepare multi-temporal SSM maps using the developed model. These maps provide a convenient means for studying the temporal changes in the soil moisture drought.

References

1. Ambrosone, M., Matese, A., Di Gennaro, S. F., Gioli, B., Tudoroiu, M., Genesio, L., Miglietta, F., Baronti, S., Maienza, A., Ungaro, F., & Toscano, P. (2020). Retrieving soil moisture in rainfed and irrigated fields using Sentinel-2 observations and a modified OPTRAM approach. *International Journal of Applied Earth Observation and Geoinformation, 89*, 102113. https://doi.org/10.1016/j.jag.2020.102113
2. Drisya, J., Kumar, D. S., & Roshni, T. (2021). Hydrological drought assessment through streamflow forecasting using wavelet enabled artificial neural networks. *Environment, Development and Sustainability, 23*, 3653–3672. https://doi.org/10.1007/s10668-020-00737-7
3. Drisya, J., Kumar, D.S., & Roshni, T. (2018). Spatiotemporal variability of soil moisture and drought estimation using a distributed hydrological model. *Integrating Disaster Science Management,* 451–460. https://doi.org/10.1016/B978-0-12-812056-9.00027-0
4. Zhao, T., Shi, J., Lv, L., Xu, H., Chen, D., Cui, Q., Jackson, T. J., Yan, G., Jia, L., Chen, L., Zhao, K., Zheng, X., Zhao, L., Zheng, C., Ji, D., Xiong, C., Wang, T., Li, R., Pan, J., Wen, J., Yu, C., Zheng, Y., Jiang, L., Chai, L., Lu, H., Yao, P., Ma, J., Lv, H., Wu, J., Zhao, W., Yang, N., Guo, P., Li, Y., Hu, L., Geng, D., & Zhang, Z. (2020). Soil moisture experiment in the Luan River supporting new satellite mission opportunities. *Remote Sensing of Environment, 240*. https://doi.org/10.1016/j.rse.2020.111680
5. Ramsankaran, R., Kumar, D. S., & Eldho, T. I. (2017). Remote sensing and geographical information systems in watershed management: An overview. *Sustainable Water Resources Management*, 51–79
6. Alexakis, D. D., Mexis, F. D. K., Vozinaki, A. E. K., Daliakopoulos, I. N., & Tsanis, I. K. (2017). Soil moisture content estimation based on Sentinel-1 and auxiliary earth observation products. A hydrological approach. *Sensors (Switzerland), 17*, 1–16. https://doi.org/10.3390/s17061455
7. Zhang, J., Yao, F., Wang, P., Guo, W., Li, L., & Yang, L. (2010). Advances in the estimation methods of vegetation water content based on optical remote sensing techniques. *Science China Technology Science*
8. Yadav, V. P., Prasad, R., Bala, R., & Vishwakarma, A. K. (2020). An improved inversion algorithm for spatio-temporal retrieval of soil moisture through modified water cloud model using C- band Sentinel-1A SAR data. *Computers and Electronics in Agriculture, 173* 105447.https://doi.org/10.1016/j.compag.2020.105447
9. Qiu, J., Crow, W. T., Wagner, W., & Zhao, T. (2019). Effect of vegetation index choice on soil moisture retrievals via the synergistic use of synthetic aperture radar and optical remote sensing. *International Journal of Applied Earth Observation and Geoinformation, 80*, 47–57. https://doi.org/10.1016/j.jag.2019.03.015
10. Singh, K., Kumar, S., Kumar, R. (2019). Remote sensing applications : Society and environment synergetic methodology for estimation of soil moisture over agricultural area using Landsat-8 and Sentinel-1 satellite data.*Remote Sensing Applications Society Environment, 15*, 100250.https://doi.org/10.1016/j.rsase.2019.100250
11. Bao, Y., Lin, L., Wu, S., Kwal Deng, K. A., & Petropoulos, G. P. (2018). Surface soil moisture retrievals over partially vegetated areas from the synergy of Sentinel-1 and Landsat 8 data using a modified water-cloud model. *International Journal of Applied Earth Observation and Geoinformation, 72*, 76–85. https://doi.org/10.1016/j.jag.2018.05.026

Bias Correction of Regional Climate Models for Streamflow Forecasting

N. Sinsha and D. Sathish Kumar

Abstract Climate change is recognised as a serious phenomenon affecting socio-economic agricultural development activities around the globe. Regional climate models (RCMs) are found to be more reliable to study the impact of climate change at a regional scale. However, the RCMs are not free from bias errors. Therefore, it is necessary to apply a bias correction before the application of models for water resources research. Linear scaling (LS), power transformation (PT), local intensity scaling (LOCI), delta change correction (DC) and distribution mapping (DM) are the methods commonly applied for the bias correction of the precipitation data. On the other hand, temperature datasets are corrected by using linear scaling (LS), variance scaling (VS), delta change correction (DC) and distribution mapping (DM) techniques. In the proposed study, various bias correction techniques are evaluated to determine a suitable method for correcting the meteorological datasets. The delta change method showed better performance for both climate variables with good statistical metrics (frequency and time-series-based). The corrected datasets were used as a forcing input to the SWAT hydrological model to project streamflow for three future time periods in the twenty-first century. Simulations were carried out for the initial period (2021–2040), mid-period (2041–2070) and end-period (2071–2100) under RCP 4.5 and 8.5 scenarios. Predicted streamflow shows significant changes in all time intervals, with the highest increase during the end of this century. The result indicates that mean annual streamflow is likely to increase by about 7.4% and 23.1% under RCP 4.5 and RCP 8.5 scenarios, respectively.

Keywords Regional climate models · Bias correction · SWAT · Streamflow projections

N. Sinsha (✉) · D. Sathish Kumar
Department of Civil Engineering, National Institute of Technology, Calicut, India

A. K. Dikshit et al. (eds.), *Innovative Trends in Hydrological and Environmental Systems*, Lecture Notes in Civil Engineering 234,
https://doi.org/10.1007/978-981-19-0304-5_40

1 Introduction

Water is one of the most important natural resources and has a significant role in the survival of human beings and other living things. It is an integral part of agricultural, economic and industrial development activities. However, it has become a scarce resource due to urban development, land mining and industrialisation. Understanding the possible climate change impact on water resources is important for ensuring their proper utilisation and management. Prediction of hydro-climatic processes is not an easy task, as the various processes are interrelated and are highly variable with respect to time and space. The variation in the pattern of climate variables (precipitation, humidity, temperature, etc.) in a given region over an extended period of time is referred to as climate change. It has been considered as a critical environmental issue in the twenty-first century and has a significant impact on the redistribution of water through the hydrological cycle, ecology and environment [1]. Understanding changes in concentrations of various atmospheric components is essential for predicting future climate change. The industrial revolution and human activities resulted in a significant increase in greenhouse gas concentrations in the atmosphere. It is now widely accepted as a major cause of global climate change.

The Intergovernmental Panel on Climate Change (IPCC) forecasts a temperature rise of 0.4–1.7 °C under the RCP 2.6 scenario, 1.1–2.6 °C under the RCP 4.5 scenario, 1.4–3.1 °C under the RCP 6.0 scenario, and 2.6–4.8 °C under the RCP 8.5 scenario by the year 2100. It results in variation in the intensity, timing and frequency of precipitation events. These changes have a significant impact on the hydrological processes of the basin, such as subsurface flow, surface runoff, evapotranspiration, interception, infiltration and flood regime. As exemplified by changes in precipitation patterns and an increase in temperature, climate change is expected to impact the hydrological cycle, altering the quantity and timing of streamflow [2]. Recently, many studies have been carried out on climate change and its impact on hydrological issues and water resources. The combined influence of climate and land use land cover changes on the hydrological regime of the watershed is well established [3]. General circulation models (GCMs) are a powerful tool for analysing the global warming effect on climate and hydrologic cycles at different time scales [4]. It is well known that these GCMs can represent gradually varying variables such as surface pressure, wind speed or temperature in the spatial domain. However, the coarse resolution GCMs (100–200 km) has limitations in simulating local precipitation patterns. Hence, it is not possible to use precipitation related variables obtained from these GCMs directly for mesoscale climate change impact studies [5, 6]. Therefore, regional studies on the meteorological variables are essential to understand the impact of climate change at a watershed level [7].

The RCMs are of finer spatial resolution found to be more reliable to study the impact of climate change at a regional scale. The accuracy of RCMs in simulating climate is highly dependent on seasonal and regional characteristics. Recently, many researchers have focused on performance evaluation of the CORDEX- South Asia experiments [8]. Applying bias correction to the RCM is essential to match the

RCM output with the observed datasets. Different bias correction techniques were developed, ranging from simple linear scaling to complex distribution mapping [9]. In recent years, several studies are being carried out to assess the impact of climate change on basin hydrology by integrating the hydrological model with RCM output [10, 11]. The reliability of these studies depends on the selection of climate models and the performance of bias correction methods. As reliable streamflow forecasts are necessary for managing extreme events such as droughts and floods [12], applying an appropriate RCM input is essential for forecasting such events.

2 Study Area and Data

2.1 *Study Area*

The study area is the Chaliyar river basin in Kerala, India. The river basin is located between latitudes 11° 06' N–11° 36' N and 75° 48' E–76° 33' E longitudes. The origin of this river is at the Elambalari hills at an elevation of 2067 m above mean sea level. The total drainage area of the basin is 2933 $km^{2,}$ of which 2530 km^2 lie in Kerala and the rest in Tamil Nadu. The average annual precipitation in the basin is about 3000 mm. The principal rainy seasons are southwest monsoon (June to August) and northeast monsoon (September to November), and pre-monsoon (April to May), which contributes about 60%, 25% and 15% of the annual rainfall, respectively. December to March is considered to be the driest period. The average annual minimum and maximum temperatures are 24 °C and 34 °C, respectively. The average annual relative humidity ranges between 60–90% in summer and 65–85% during winter. The location map of the study area is shown in Fig. 1.

2.2 *Data*

Daily precipitation data for five stations in the Chaliyar river basin, namely, Manjeri, Nilambur, Kottamparamba, Ambalavayal and Gudalur, were collected for the period 1991–2008. The rainfall data at the Kottamparamba station was obtained from the Centre for Water Resources Development Management (CWRDM), Kunnamangalam, Kerala and the precipitation datasets for Nilambur, Manjeri, Ambalavayal and Gudalur were collected from the Indian Meteorological Department (IMD). Other meteorological data such as maximum and minimum temperature, relative humidity, wind speed, and duration of the sunshine hour at the Kottamparamba station was also collected for the same period. Climate projections of RCM, RCA4 (including precipitation, maximum temperature, minimum temperature, wind speed

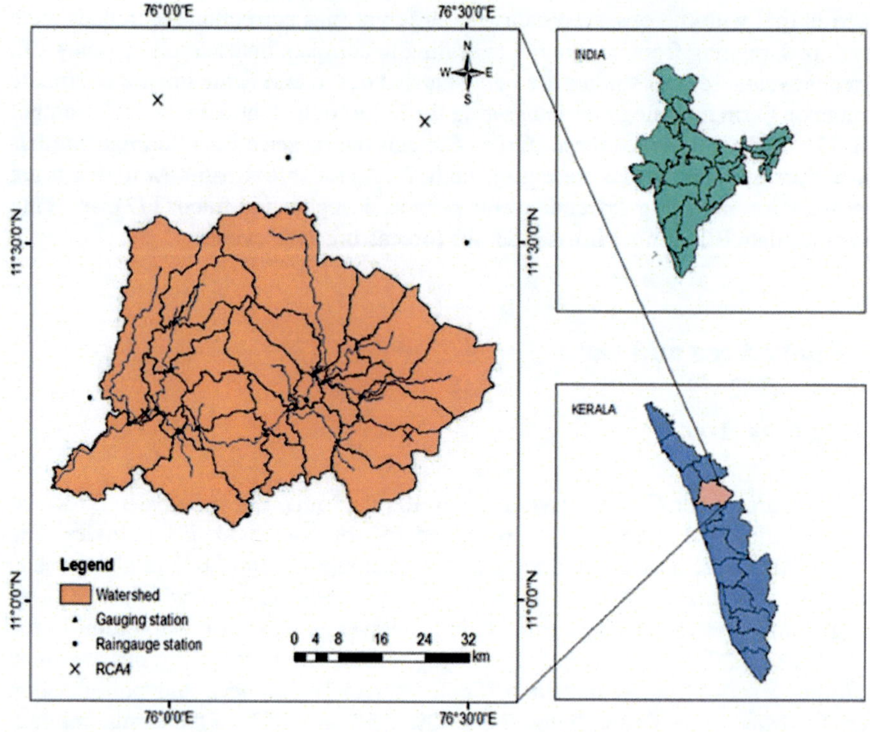

Fig. 1 Study area showing stations and RCM grid point locations

and relative humidity) under RCP 4.5 and 8.5 scenarios are available on the WCRP-CORDEX-South Asia website. The RCM data for the years 1991–2005 under historical scenarios and for the years 2006–2100 under RCP scenarios were downloaded. Driving GCM is ICHEC -EC-EARTH, and the resolution of the RCM data is 0.44° × 0.44°.

3 Methodology

3.1 Bias Correction

Climate models exhibit systematic or random model errors in their output. A range of factors in the climate models can cause errors or biases. They are imperfect parameterisation, inappropriate boundary conditions, incomplete knowledge of climate processes, simplified physics, thermodynamic processes and limited spatial resolution. Thus, the use of uncorrected climate variables in hydro-climatic studies leads to unrealistic results. The elimination of these errors is carried out by comparing the model output with unbiased observations. Before performing the bias correction, the preprocessing of the data is essential. Various bias correction methods developed by different researchers viz., LS, LOCI, PT, DM and DC [9, 13] are used for precipitation and temperature, whereas multiplicative correction is applied for relative humidity.

Linear scaling is the simplest bias correction technique in which a monthly correction factor is applied to correct climate variables. It does not account for the frequency distribution of the precipitation. In the LOCI method, wet day frequencies and intensities were corrected by applying a monthly correction factor to the daily precipitation. The linear and local intensity scaling method adjusts the bias in the mean alone; it does not consider biases in the variance. Therefore, the PT method uses a nonlinear correction algorithm in an exponential form ($a.P^b$), to adjust the variance statistics of the precipitation dataset. The power transformation is an effective method to adjust both the mean and variance statistics; however, it is limited to precipitation datasets because of the use of power functions [14]. A corresponding method to correct both the mean and the variance of temperature datasets is referred to as variance scaling. In the distribution mapping method, biases in the mean, standard deviation and quantiles were adjusted by matching the distribution function of the raw RCM datasets to that of observations. Moreover, it preserves the extreme values in the precipitation. However, this method also has its limitation due to the stationarity assumption, i.e. both observed and raw meteorological variables follow the same distribution function during the present and future scenarios. It may introduce new biases in the climate output. The delta change correction is a simple, stable and robust method that uses observed data as a basis to produce future time-series with dynamics similar to current conditions [9]. So this method does not consider potential changes in climate dynamics for the future scenario, which implies that the number of wet and dry days is constant. The best bias correction method can be identified by calculating time-series (correlation, Nash Sutcliffe efficiency, etc.) and frequency-based statistics (mean, median, variance, percentile values, etc.).

3.2 Description of the SWAT Model

The Soil and Water Assessment Tool (SWAT) is a long-term yield physically based conceptual model developed by the Agricultural Research Service of the United States Department of Agriculture-Agricultural Research Service (USDA-ARS) [15]. This model has wide application in watershed modelling. It is useful to study the impact of land management practices on water, agriculture, chemical and sediment yield in large basins with varying land use and soil conditions. During the modelling process, the catchment is partitioned into sub-basins or sub-watersheds based on topography. Further, it is divided into a series of Hydrological Response Units (HRUs) having unique land use soil and slope combinations. The SWAT model is applied to simulate various hydrological processes like surface runoff, infiltration, interception lateral flow, evapotranspiration, percolation to shallow and deep aquifers. The hydrologic water balance equation is considered as governing equation for the simulation of the SWAT model. Using daily weather data, SWAT simulates runoff for each HRU separately. SWAT uses the soil conservation service curve number method (USDA, SCS-CN, 1972) for the estimation of surface runoff volume.

3.3 SWAT Model Setup

The first step in modelling is the delineation of the watershed. In this process, the watershed is divided into a number of sub-watersheds. With that the user can analyse the spatial variability of the hydrological parameters of different regions within the watershed. DEM is the basic topographical data input given to the ArcSWAT model. From DEM data, flow accumulation and direction are computed, and stream networks along with outlets were generated. In this study, the entire watershed area is divided into 32 sub-watersheds. The land use/ land cover map and soil map of the Chaliyar basin were obtained from Kerala State Land Use Board (KSLUB) for the year 2008. After reclassification, it was observed that agricultural land covers the major portion of the river basin, around 50.16%, followed by the forested area around 44.17%, mainly deciduous type (29.84%). Urban areas, wetlands and the rocky area covers less than 1% of the total area. With regard to the soil texture, the basin is dominated by loam texture (41.91%), followed by clay (29.31%), clay loam (24.21%) and sandy loam (4.57%). A unique combination of land use/ land cover, soil and slope class is known as Hydrologic Response Units (HRU). The analysis resulted in 229 HRUs for the study watershed. Input tables for various spatial inputs have been created before the simulation. The weather generator (WXGEN), a built-in tool, is available in the SWAT model to generate missing variables (if any) for specific geographic locations. The model's calibration, validation and uncertainty analysis have been carried out using the Sequential Uncertainty Fitting version 2 (SUFI-2) algorithm of SWAT-CUP [16].

4 Results and Discussion

4.1 Bias Correction of RCM Data

The frequency-based statistics of daily precipitation (raw RCM simulated, bias-corrected and observed) at the Kottamparamba station are listed in Table 1. The daily mean observed and raw RCM precipitations for this station are 8.3 mm and 4.3 mm, respectively. The raw RCM simulated precipitation shows significant deviation from observations, with underestimation of frequency-based statistics. It can be observed that all bias correction techniques improves the RCM simulation resulting

Table 1 Frequency-based statistics of daily precipitation

Dataset	Mean (mm)	Median (mm)	Standard deviation (mm)	99th percentile (mm)
Obs	8.3	0.0	19.5	90.0
Raw	4.3	1.7	6.7	29.4
LS	8.3	1.5	15.6	79.6
LOCI	8.3	0.0	16.0	79.7
PT	8.3	0.3	19.5	99.5
DM	8.5	0.0	21.1	104.7
DC	8.3	0.0	19.5	90.0

in a different range of corrected statistics. In all the methods applied, the mean of the RCM datasets got adjusted closely to the mean of the observed datasets. A slight overestimation of the mean value has been observed for the DM method. The LS and LOCI methods are good in correcting the mean values, but there is underestimation in standard deviation, and therefore, the 99th percentile. The PT method provides a good estimation of the standard deviation but poorly estimates the other statistical measures. Compared to other methods, the performance of DC method is good in adjusting all the statistical measures of interest.

The time series-based performance evaluation measures of the bias correction methods are listed in Table 2. The delta change method was found to be very accurate across all stations, with R^2 values greater than 0.80 and NSE values greater than 0.74.

Table 3 lists the frequency-based statistics of daily maximum temperature (raw RCM simulated, bias-corrected and observed) at the Kottamparamba station. The mean, median and 90th percentiles of the observed temperature are 31.9 °C, 32 °C and 34.5 °C, respectively. The raw temperature shows a slight deviation from observations, underestimating the mean, median, 90th percentile values and overestimating the standard deviation. Improvement in frequency statistics is attained by applying all correction methods. In all the methods applied, the mean and standard deviation of the RCM datasets got adjusted closely to that of the observed datasets. The mean and standard deviation are adjusted by all methods. However, there is a slight overestimation in Standard deviation with the LS method. The DM and VS have good estimations of mean, Standard deviation and 90th percentile, while it shows biases in the 10th percentile. The DC method has good estimations of all the frequency-based metrics.

4.2 *SWAT Model Calibration and Validation*

The SWAT model simulation was carried out on a daily as well as on a monthly basis for a period of 18 years (1988–2005) with initial three years as a warm-up period. Daily observed streamflow data for 15 years was used for calibration (1991–2000) and validation (2001–2005). Figure 2 presents the results of the model simulation. The

Table 2 Time series-based metrics of bias-corrected climate variables

Station	RCM variable	Bias correction method	R^2	NSE
Kottamparamba	Precipitation	LS	0.54	0.37
		LOCI	0.53	0.36
		PT	0.46	0.17
		DM	0.45	0.13
		DC	0.89	0.87
	Maximum temperature	LS	0.68	0.62
		VS	0.78	0.76
		DM	0.75	0.72
		DC	0.92	0.89
	Minimum temperature	LS	0.64	0.60
		VS	0.62	0.56
		DM	0.64	0.60
		DC	0.90	0.86
	Relative humidity	DC	0.96	0.94
Nilambur	Precipitation	LS	0.66	0.64
		LOCI	0.56	0.47
		PT	0.44	0.17
		DM	0.37	-0.16
		DC	0.88	0.86
Manjeri	Precipitation	LS	0.49	0.32
		LOCI	0.47	0.36
		PT	0.40	0.10
		DM	0.33	-0.44
		DC	0.84	0.78
Ambalavayal	Precipitation	LS	0.63	0.60
		LOCI	0.58	0.51
		PT	0.40	0.02
		DM	0.33	-0.44
		DC	0.80	0.74
Gudalur	Precipitation	LS	0.48	0.46
		LOCI	0.42	0.37
		PT	0.32	0.12
		DM	0.28	-0.12
		DC	0.82	0.80

Table 3 Frequency-based statistics of daily maximum temperature

Dataset	Mean (°C)	Median (°C)	Standard deviation (°C)	10th percentile (°C)	90th percentile (°C)
Obs	31.9	32.0	2.3	29.0	34.5
Raw	28.8	28.9	2.9	24.9	32.8
LS	31.9	32.0	2.6	28.3	35.1
VS	31.9	32.1	2.3	28.9	34.7
DM	31.9	32.2	2.3	28.5	34.7
DC	31.9	32.0	2.3	29.0	34.5

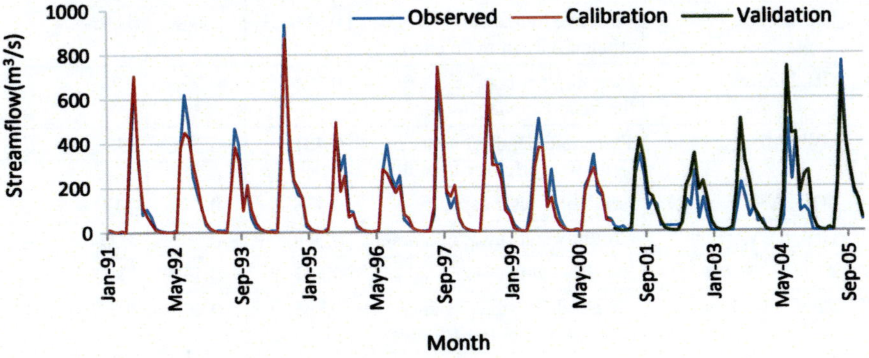

Fig. 2 Comparison of observed and simulated streamflow

performance evaluation of the hydrological model is generally carried using statistical –metrics [17]. The model performance was evaluated using the coefficient of determination (R^2), Nash Sutcliffe efficiency (NSE), percentage bias (PBIAS), r-factor and p-factor. The pre-calibration results indicated that the performance evaluation metrics were slightly below the acceptable range, and the model needs calibration. In this study parent model (SWAT) was coupled with SWAT- CUP for calibration and sensitivity analysis. Global sensitivity analysis for 15 selected parameters was carried out at the monthly time step to identify the most sensitive parameters. The initial ranges of parameters were identified by referring to absolute values (maximum and minimum) available in the SWAT-CUP. The sensitive parameters were further adjusted for better simulation of runoff. Here, sensitivity analysis was performed using the Latin hypercube sampling technique. The parameters which have a smaller p-value and larger absolute value of t-stat are the most sensitive parameters. It is observed that CN2 is the most sensitive parameter, followed by SLSUBBSN, SURLAG and ESCO (Table 4).

The NSE and R^2 values are found to be 0.94 and 0.95, respectively, which indicates a strong correlation between the observed and calibrated results. The NSE and R^2 values for the validation phase are 0.78 and 0.83, respectively, indicating the good

Table 4 Parameters and their best fitted values used for calibration

Parameter	Parameter description	P- value	t-Stat	Fitted Value
V__CN2.mgt	SCS curve number	0.000	5.603	78.7
V__SLSUBBSN.hru	Average slope length	0.000	−4.807	117.1
V__SURLAG.bsn	Surface runoff lag coefficient	0.001	3.609	7.24
V__ESCO.hru	Soil evaporation compensation factor	0.115	−1.592	0.835
V__SOL_ALB(…).sol	Moisture soil albedo	0.115	−1.591	0.015
V__RCHRG_DP.gw	Deep aquifer percolation fraction	0.189	−1.324	0.032
V__GW_DELAY.gw	Groundwater delay time	0.194	−1.310	31.55
V__CH_K2.rte	Effective hydraulic conductivity in main channel	0.316	−1.008	0.097
V__CH_N2.rte	Manning's n-coefficient for the main channel	0.323	−0.993	0.085
V__GWQMN.gw	Threshold depth of water in shallow aquifer for base flow	0.431	−0.791	196
R__SOL_AWC(…).sol	Available water capacity	0.438	0.779	0.17
V__GW_REVAP.gw	Groundwater revap coefficient	0.549	−0.602	0.003
R__ALPHA_BF.gw	Baseflow alpha factor	0.631	0.482	13.75
V__REVAPMN.gw	Threshold depth of water in the shallow aquifer for revap	0.835	−0.208	541.7
R__SOL_K (…).sol	Effective hydraulic conductivity of the soil layer	0.997	0.003	0.13

performance of the model for the monthly flow simulation. The obtained PBIAS value of −17.7 indicates an overestimation of streamflow for some months during the validation period. The calibration result shows a higher value for p-factor (0.86) which indicates the higher percentage (86%) of bracketing of observed data in the 95ppu plot. The relative width of the 95% probability band can be obtained by using the r-factor (0.42). The larger p-factor and lower r-factor indicate less uncertainty range in model simulation and good calibration results. The r-factor and p-factor for the validation phase are 0.65 and 0.42, respectively. Relatively smaller p-factor indicates less percentage (42%) bracketing observed data in 95ppu plot.

4.3 Future Projections of Streamflow

The impact of climate change on streamflow on monthly and annual timescales was assessed. The comparison of mean annual streamflow for the future period and base period is represented in the form of box plots (Fig. 3) which helps to understand the variation of the distribution of the data around the median. The remaining years

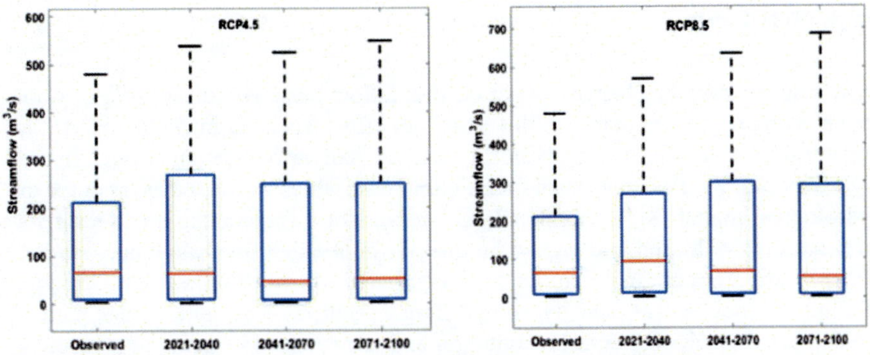

Fig. 3 Boxplots for forecasted streamflow for different future climate projections

in the twenty-first century was divided into three different periods for streamflow forecasting, i.e. initial period (2021–2040), mid-period (2041–2070) and end-period (2071–2100). The forecasted stream flows indicated a change of −4.9 to −30.1% in the median values and 9.3 to 14.1% in the upper extremities for the RCP 4.5 scenario. For the RCP 8.5 scenario, these changes were between −17.7 and 43.2% and 18.8–42.9%, respectively.

The percentage change in monthly average streamflow under RCP scenarios are shown in Fig. 4. During the monsoon season (June to August), the streamflow discharges are expected to increase by 12.3% under the RCP 4.5 and 26.1% under the RCP 8.5 scenario. During post-monsoon (September to November), these changes are reduced to −2.2% and 17.8%, respectively. The mean annual streamflow showed an increase of 7.4% and 23.1% under RCP 4.5 and RCP 8.5 scenarios, respectively.

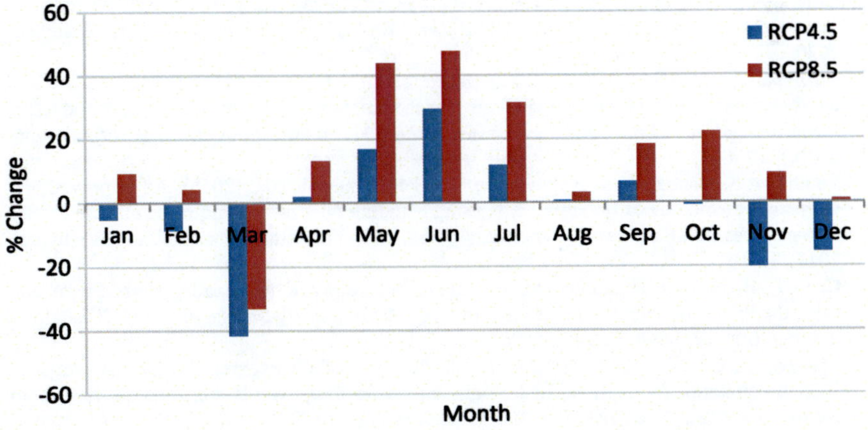

Fig. 4 Percentage change in monthly average streamflow for different climate projections

5 Conclusion

Assessing the climate change impact on streamflow has a significant role in water resources management plans. In this study, an effort has been made to analyse the impacts of climate change on streamflow in the Chaliyar river basin through hydrological modelling. Bias-corrected future climate variables were used as input for the hydrological model. RCM simulated climate variables (precipitation and temperature) are always biased and cannot be directly applied for hydro-climatic studies. These biases are more in precipitation as compared to temperature. Bias correction of climate variables was done by using different methods. The delta change method yielded good results for both precipitation and temperature in terms of frequency-based and time-series-based indices. SWAT model simulation was carried out for daily and monthly streamflow for the period 1991–2005. The model yielded satisfactory and reliable results for both time steps. Sensitivity analysis indicates that parameters CN2 followed by SLSUBBSN, SURLAG, ESCO and SOL_ALB are most sensitive and greatly impact the streamflow. The projected streamflow shows a significant increasing trend as compared to the baseline conditions, with the biggest change in the end-period for both RCPs. This variation is more in the RCP 8.5 scenario than in the RCP 4.5 scenario. The percentage increase in the streamflow will be higher during the monsoon season compared to other seasons.

References

1. Aawar, T., & Khare, D. (2020). Assessment of climate change impacts on streamflow through hydrological model using SWAT model: A case study of Afghanistan. *Modeling Earth Systems and Environment, 6*, 1427–1437. https://doi.org/10.1007/s40808-020-00759-0
2. Sathya, A., G Thampi, S. (2020). Impact of projected climate change on streamflow and sediment yield—A case study of the Chaliyar River Basin, Kerala. In *Roorkee Water Conclave*, 2020.
3. Chanapathi, T., & Thatikonda, S. (2020). Investigating the impact of climate and land-use land cover changes on hydrological predictions over the Krishna river basin under present and future scenarios. *Science of the Total Environment, 721*, 137736.https://doi.org/10.1016/j.scitotenv.2020.137736
4. Smitha, P. S., Narasimhan, B., Sudheer, K. P., & Annamalai, H. (2018). An improved bias correction method of daily rainfall data using a sliding window technique for climate change impact assessment. *Journal of Hydrology, 556*, 100–118. https://doi.org/10.1016/j.jhydrol.2017.11.010
5. Chen, J., Brissette, F. P., & Leconte, R. (2011). Uncertainty of downscaling method in quantifying the impact of climate change on hydrology. *Journal of Hydrology, 401*, 190–202. https://doi.org/10.1016/j.jhydrol.2011.02.020
6. Sharma, D., Gupta, A. Das, & Babel, M. S. (2007). Spatial disaggregation of bias-corrected GCM precipitation for improved hydrologic simulation: Ping River Basin, Thailand. *Hydrology Earth System Sciences, 11*, 1373–1390. https://doi.org/10.5194/hess-11-1373-2007
7. Kabbilawsh, P., Sathish Kumar, D., & Chithra, N. R. (2020). Trend analysis and SARIMA forecasting of mean maximum and mean minimum monthly temperature for the state of Kerala, India. *Acta Geophysica, 68*, 1161–1174. https://doi.org/10.1007/s11600-020-00462-9

8. Sannan, M. C., Nageswararao, M. M., & Mohanty, U. C. (2020). Performance evaluation of CORDEX-South Asia simulations and future projections of northeast monsoon rainfall over south peninsular India. *Meteorology and Atmospheric Physics, 132*, 743–770. https://doi.org/10.1007/s00703-019-00716-2
9. Teutschbein, C., & Seibert, J. (2012). Bias correction of regional climate model simulations for hydrological climate-change impact studies: Review and evaluation of different methods. *Journal of Hydrology, 456–457*, 12–29. https://doi.org/10.1016/j.jhydrol.2012.05.052
10. Nilawar, A. P., & Waikar, M. L. (2019). Impacts of climate change on streamflow and sediment concentration under RCP 4.5 and 8.5: A case study in Purna river basin, India. *Science Total Environmental, 650*, 2685–2696. https://doi.org/10.1016/j.scitotenv.2018.09.334
11. Fekadu, G., Zelalem, B., Gonfa, B., & Mamo, G. (2019). Impacts of climate change on stream flow and water availability in Anger sub—basin, Nile Basin of Ethiopia. *Sustainable Water Resources Management, 5*, 1755–1764. https://doi.org/10.1007/s40899-019-00327-0
12. Drisya, J., Kumar, D. S., & Roshni, T. (2021). Hydrological drought assessment through streamflow forecasting using wavelet enabled artificial neural networks. *Environment Development Sustainability, 23*, 3653–3672. https://doi.org/10.1007/s10668-020-00737-7
13. Chen, J., Brissette, F. P., Chaumont, D., & Braun, M. (2013). Finding appropriate bias correction methods in downscaling precipitation for hydrologic impact studies over North America. *Water Resources Research, 49*, 4187–4205. https://doi.org/10.1002/wrcr.20331
14. Leander, R., & Buishand, T. A. (2007). Resampling of regional climate model output for the simulation of extreme river flows. *Journal of Hydrology, 332*, 487–496. https://doi.org/10.1016/j.jhydrol.2006.08.006
15. Arnold, J. G., Srinivasan, R., Muttiah, R. S., Williams, J. R. (1998). Large area hydrologic modeling and assessment part I: Model development.
16. Abbaspour, K. C., Yang, J., Maximov, I., Siber, R., Bogner, K., Mieleitner, J., Zobrist, J., & Srinivasan, R. (2007). Modelling hydrology and water quality in the pre-alpine/alpine Thur watershed using SWAT. *Journal of Hydrology, 333*, 413–430. https://doi.org/10.1016/j.jhydrol.2006.09.014
17. Drisya, J., & Sathish Kumar, D. (2018). Automated calibration of a two-dimensional overland flow model by estimating Manning ' s roughness coef ficient using genetic algorithm. *Journal of Hydroinformatics, 20*(2), 440–456.https://doi.org/10.2166/hydro.2017.110

Performance Evaluation of Hydraid Zeolite Filter

Bonala Sneha and Perla Harish

Abstract In the prevailing conditions of increased population, people demands huge quantity of water for their needs, and it should have good quality. The present study evaluates the performance of a new filtering technology with easily available materials named as hydraid zeolite filter (HZF). A synthetic zeolite called clinoptiolite being the largest part of filter media, specially selected sand, and gravel as other materials, the HZF has been set up. Water samples were collected from three areas (RGUKT sump water, Borgaon (P) (V) pond water, Kulaspur canal water). For each of the sample water, quality parameters such as P^H, TS, TDS, hardness, alkalinity, BOD and COD were measured before and after passing through the filter bed. The results showed that there is a significant reduction in all the parameters under consideration, and especially the hardness has been reduced by 25–45% due to the presence of clinoptiolite in the filter media. Particularly, the filtered water is free from pathogens and bacteria. Therefore, the applicability of the HZF to different water sources with different water quality is important due to seasonal fluctuations.

Keywords Hydraid zeolite filter (HZF) · Clinoptiolite · Performance evaluation · Water quality comparison with WHO standards

1 Introduction

Water covers 71% of surface of planet, but potable water had very small quantity which used to fulfil the domestic needs [1]. In the present scenario, available water resources being the same, the increased population needs a very huge quantity of water for their daily needs. But the quality of water is degrading due to anthropogenic activities. Use of contaminated water can cause infections and diseases that kill

B. Sneha (✉)
Department of Civil Engineering, Rajiv Gandhi University of Knowledge Technologies—IIIT, Basar, Telangana, India

P. Harish
Department of Civil Engineering, Government College of Engineering Kalahanadi, Bhawanipatna, Odisha, India

A. K. Dikshit et al. (eds.), *Innovative Trends in Hydrological and Environmental Systems*, Lecture Notes in Civil Engineering 234,
https://doi.org/10.1007/978-981-19-0304-5_41

millions and sicken more than a billion people every year [2]. At rural areas, the facilities are inadequate for treating the raw water and to supply to the consumer, which is leading to the usage of treacherous water that turns the human's life into risk. The awareness developed about the consequences of contaminants is demanding for superior quality water, which requires a sustainable, reliable and simple water treatment such that it can be maintained and repaired or cleaned easily. Water soluble salts (bicarbonates and carbonates, chlorides, sulphates of calcium and magnesium, nitrites, nitrates), solids and pathogens need to be removed to improve water quality [3].

The locally developed sand filters are effectively removing solids from raw water [4] by surface straining and interparticle attraction mechanisms [5]. The distinguished design of bio-sand filters from slow sand filters contributes to the sustainment of biofilm/Schmutzdecke formation during the intermittent flow [6]. The microbiology in the bio-sand filters is helpful for the removal of particulate matter and pathogen and also removes the chemicals like iron, manganese, sulphur, low concentrations of gases by the operations of filtration, oxidation and natural dye [7]. The trivial modifications in the design set-up of these bio-sand filters are by introducing the locally available materials like natural zeolites and modified zeolites for effective removal of hardness, alkalinity and different salts present in water [8].

The application of natural zeolite (clinoptilolite) is an exceptional and cheapest solution for providing clean water [9]. In comparison with other chemical and biological processes, the natural zeolites are having the advantage of removing impurities by the properties of ion exchange and adsorption [8]. In household consumption, the hardness of water can be removed by the zeolite filter because of its cationic exchange nature. The water which is free from suspended solids, inorganic solids and pathogens represents the wholesomeness of water as per World Health Organization—WHO (1971). Since a mere 2% drop in our body's water supply can trigger signs of dehydration, it would make sense that quality of water should be just as important as the quantity [10].

The present study was aimed to evaluate the performance of a new filtering technology with easily available materials named as hydraid zeolite filter (HZF) to reduce the quantities of pH, total solids(TS), total dissolved solids(TDS), hardness, alkalinity, biochemical oxygen demand (BOD) and chemical oxygen demand (COD) for the raw water.

2 Materials and Methods

The hydraid zeolite filter was set up as shown in Figs. 1 and 2, by using the locally available materials like sand, gravel and zeolite. The bottom layer of HZF is filled with *gravel* (depth 5, 50–70 mm), second layer from the bottom is *coarse sand* (3, 2.6–4.75 mm), and the next layer is *zeolites*. Mainly in synthetic zeolite group, clinoptilolite (10, 5–7 *mm*) is used for filtration process, followed by *sand* (3, 1.8–2.36 *mm*). A *mesh* is used to separate the various layers, and it is also used for

Fig. 1 .

Fig. 2 .

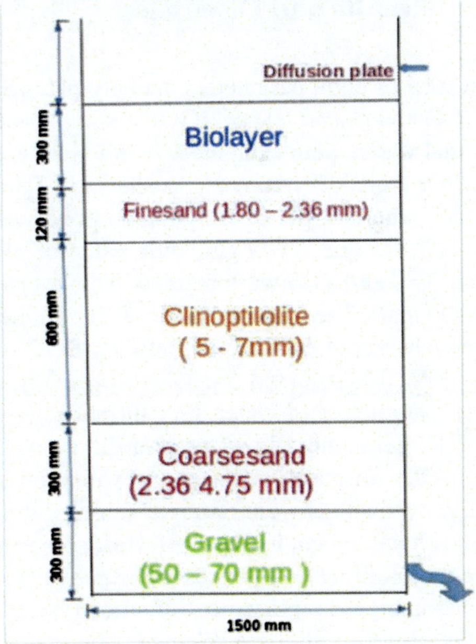
Diffusion plate
Biolayer
Finesand (1.80 – 2.36 mm)
Clinoptilolite
(5 - 7mm)
Coarsesand
(2.36 4.75 mm)
Gravel
(50 – 70 mm)
300 mm
120 mm
600 mm
300 mm
300 mm
1500 mm

clogging of materials. A metal *diffusion plate* (holes diameter—4 mm) is used to equally distribute the water to the layers. It is like a shallow basin and rests on the inner edge of the concrete body. It is also helpful to avoid the disturbance on the biolayer by the charge of water.

The hydraid zeolite filter has five distinct zones: inlet reservoir zone—where water is poured into the filter. Standing water zone—this water keeps the sand wet while letting oxygen pass to the biolayer. Biological zone develops at the top 5–10 cm of the sand surface. The filtration sand removes pathogens, suspended particles and other contaminants. As in slow sand filters, a biological layer of microorganisms develops at the top 1–2 cm of the sand surface. Non-biological zone contains virtually no living microorganisms due to the lack of nutrients and oxygen. Gravel zone holds the sand in place and protects the outlet tube from clogging.

Water samples were collected from three areas (RGUKT sump water, Borgaon pond water and Kulaspur canal water). The quality of water was tested for pH, TS, TDS, hardness, alkalinity, BOD and COD before and after passing through the filter set-up by using the standard procedures. The flow rate observed during the filtration is 4.167 ml/sec. The flow rate through the filter will slow down over time as the biolayer develops and suspended solids are trapped in the upper layer, then the cleaning of the filter by swirl and dump method is required, even though the filter is still effectively treating the water, but it takes too long time to be inconvenient for the users.

3 Results and Discussion

A total of eight parameters, namely pH, TS, TDS, TSS, alkalinity, hardness, BOD, COD and *E. coli* test of RGUKT sump water, Borgaon(P) pond water and Kulaspur canal water, were examined. Aesthetically, the three water samples' colour and bad odour are 100% removed. As shown in Table 1, pH of the RGUKT sump water was 7.52, whereas, pH of Borgaon(P) pond water was 7.96 and Kulaspur canal water was 7.82. The WHO guideline value of pH is 6.5–8.5. The removal efficiency of TS, TSS and TDS was obtained in the range of 50–88%. WHO guideline for TDS is 1000 mg/l. The hardness of water for all the three different samples is considerably reduced about 60–80%. Alkalinity, BOD and COD values are also reduced about 5–35%, 10–70% and 5–15%, respectively, and filtered water free from pathogens. All parameters of water after filtration were in the range of BIS 10500–2012 and WHO guidelines. So water after filtration can be used for domestic purpose.

After filtration, pH increased which makes water alkaline. Total solids and total dissolved solids were removed during filtration because of straining, impingement and entanglement in filter bed. Hardness was removed due to the presence of zeolite (clinopitolite) having excessive negative charge present on its surface, which results in isomorphic replacement of silicon by aluminium in the primary structural units that exhibits cationic exchange. The cage like structure, with nearly uniform size of pores and channels, allows to draw and trap positively charged heavy metal ions and also provides high internal surface area available for adsorption such that it acts

Table 1 Concentration and efficiency of water quality parameters after filtration on comparison BIS 10500-2012 & WHO guidelines

S. No.	Water quality parameter	RGUKT		Boragaon (P)		KULASPUR		Permissible limits as per BIS 10500-2012
		Output	Efficiency	Output	Efficiency	Output	Efficiency	
1	pH	7.52	–	7.96	–	7.82	–	6.5–8.5
2	TS (mg/l)	627	65.55	602	81.19	505	57.92	2000 mg/l
3	TDS (mg/l)	206	71.59	103	88.56	145	51.67	500 mg/l
4	Alkalinity (mg/l)	167.6	6.89	190	21.41	72	25.47	200 mg/l
5	Hardness(mg/l)	20	77.15	65	63.89	50	64.92	200 mg/l
6	BOD (ppm)	0.8	70.38	0.8	11.12	0.72	31.43	1 (WHO standards)
7	COD (ppm)	9.8	6.67	9.5	13.64	9.35	3.61	10 (WHO standard)
8	*E. coli*	NIL	100	NIL	100	NIL	100	Nil/100

like molecular chemical sieve [8]. In the spaces between the sand grains of fine sand, suspended solids and pathogens are physically trapped, and it is having a static charge which admits the pathogens to attach to the top surface of HZF. As the new water enters the filter, the bacterium present in the biolayer which is developed on the sand consumes the harmful bacteria and parasites. *E. coli* test results proved that other microorganisms present in the biological layer consume the pathogens present in water leading [11].

4 Conclusion

- Total solids and total dissolved solids were removed during filtration because of straining, impingement and entanglement in filter bed.
- Hardness was removed due to the presence of zeolite (clinopitolite) having excessive negative charge present on its surface, which results in isomorphic replacement of silicon by aluminium in the primary structural units that exhibits cationic exchange.
- The biolayer that forms on top of the sand contains bacteria that consume harmful bacteria and parasites as water enters the filter.
- In the spaces between the sand grains of fine sand, suspended solids and pathogens are physically trapped, and it is having a static charge which admits the pathogens to attach to the top surface of HZF.

5 Appendix

HZF	Hydraid zeolite filter
TS	Total solids
TDS	Total dissolved solids
BOD	Biochemical oxygen demand
COD	Chemical oxygen demand
WHO	World Health Organization

References

1. Haydar, S., Arshad, M., & Aziz, J. A. (2016). Evaluation of drinking water quality in urban areas of Pakistan: A case study of Southern Lahore. *Pakistan Journal of Engineering and Applied Sciences.*
2. Clark, P. A., Pinedo, C. A., Fadus, M., & Capuzzi, S. (2012). Slow-sand water filter: Design, implementation, accessibility and sustainability in developing countries. *Medical Science Monitor International Medicine Journal of Experimental Clinical Research, 18*, RA105.
3. Malakootian, M., Amirmahani, N., Yazdanpanah, G., Nasiri, A., Asadipour, A., Ebrahimi, A., & Moghaddam, S. D. (2017). Performance evaluation of household water treatment systems used in Kerman for removal of cations and anions from drinking water. *Applied Water Science, 7*, 4437–4447.
4. Jenkins, M. W., Tiwari, S. K., & Darby, J. (2011). Bacterial, viral and turbidity removal by intermittent slow sand filtration for household use in developing countries: Experimental investigation and modeling. *Water Research, 45*, 6227–6239.
5. Weber-Shirk, M. L., & Dick, R. I. (1997). Physical-chemical mechanisms in slow sand filters. *American Water Works Association on Journal, 89*, 87.
6. Lee, T. (2000). Biosand household filter project in Nepal.
7. Collin, C. (2009). Biosand filtration of high turbidity water: Modified filter design and safe filtrate storage.
8. Margeta, K., Logar, N. Z., Šiljeg, M., & Farkaš, A. (2013). Natural zeolites in water treatment–how effective is their use. *Water Treatment, 5*, 81–112.
9. Valentukeviciene, M., & Rimeika, M. (2007). Development of a fluidized batch process using natural powdered Zeolite (Clinoptilolite). *Polish Journal Environmental Studies, 16.*
10. Candau, M. G., & Organization, W. H. (1972). The work of WHO, 1971: Annual report of the director-general to the world health assembly and to the united nations.
11. Unger, M., & Collins, M. R. (2008). Assessing Escherichia coli removal in the schmutzdecke of slow-rate biofilters. *Journal of American Water Works Assocication, 100*, 60–73.

Flood Mapping Using Sentinel-1 SAR Data

Rijun T. Raju, Santosh G. Thampi, and D. Sathish Kumar

Abstract Satellite-based flood monitoring is a powerful tool to map inundated areas and distinguish water, vegetation, and urban settlements. In this study, Synthetic Aperture Radar (SAR) images from Sentinel-1 satellite were used for this purpose as images are available round the clock. Also, being a passive microwave sensing radar, the images obtained are not affected by cloud cover. The satellites Sentinel-1A and 1B combined have a 6-day revisit frequency which can be used to create a near-real-time flood inundation map. The study area is the areal extent of the Nilambur municipality in the Chaliyar river basin in Kerala, India. This area was very badly affected during the floods of 2018 and 2019. Google Earth Engine (GEE) was used to process the Sentinel-1A and 1B images. Manual selection of threshold backscattering coefficient values was used to determine open water (−16 dB) and submerged urban areas with water (−5 dB). The extent of flooding was analyzed by plotting the time series of the number of flooded pixels in the images of the area during the major flood events of 2018 and 2019. Flood maps were generated for pre and post-flood events.

Keywords Synthetic Aperture Radar · Sentinel-1A · Google Earth Engine · Backscatter coefficient

1 Introduction

Floods are one of the most frequent disasters and can occur due to excessive rainfall, cyclone, dam failure, etc. Kerala witnessed one of the worst floods in recorded history in the years 2018 and 2019 and this has been attributed to some extent to excessive rainfall. According to media reports about 683 people lost their lives and property worth about ₹400 billion was reported to have been damaged during the floods of 2018. The death toll during the 2019 floods was lower with about 121 reported deaths. Flood inundation mapping, probabilistic evaluation of flood damages and risks, and

---

R. T. Raju (✉) · S. G. Thampi · D. Sathish Kumar
Department of Civil Engineering, National Institute of Technology Calicut, Kozhikode, Kerala, India

A. K. Dikshit et al. (eds.), *Innovative Trends in Hydrological and Environmental Systems*, Lecture Notes in Civil Engineering 234,
https://doi.org/10.1007/978-981-19-0304-5_42

development of emergency action plans and master plans for flood mitigation have to be taken up as part of flood management strategies in order to reduce flood-related hazards. Nearly real-time flood inundation mapping can facilitate fast and effective decision making. Remote sensing is a very powerful tool in this regard as it can distinguish water bodies over large areas. Optical remote sensing has been used for dynamic flood monitoring. However, under high cloud cover, its effectiveness in flood monitoring is very low [1]. Synthetic Aperture Radar (SAR) images from Sentinel-1 satellite are not affected by cloud cover and these images are available round the clock. In general, radar signals can penetrate through vegetation (wavelength dependent) and hence flooded areas with vegetation can be identified. Groeve [2] used passive microwave remote sensing images for detecting flooded areas in Namibia during 2009 and demonstrated the capability of this data for identifying areas vulnerable to flooding in the future. Tiwari et al. [3] used the Google Earth Engine to delineate flood inundated areas in Kerala during the 2018 floods using SAR images. They applied the Otsu algorithm to differentiate submerged and non-flooded areas. Vishnu et al. [4] used Sentinel-1 data for inundation mapping in some highly flooded districts in Kerala during the 2018 floods and compared it with an average pre-flood, water-cover map obtained from Sentinel-2 using the Modified Normalized Difference Water Index (MNDWI). Vanama et al. [5] implemented two change detection techniques combined with semiautomatic thresholding technique on descending pass SAR images and statistical-based thresholding for ascending pass SAR images for analyzing the 2018 Kerala flood event. The study showed that combined use of ascending and descending pass SAR images yielded a better understanding of flood conditions.

The aim of this study is to demonstrate the potential of threshold detection techniques using multi-temporal earth observation images. The C-band Sentinel-1A and 1B Synthetic Aperture Radar (SAR) images were processed in Google Earth Engine. Time series of Sentinel-1 SAR data of pre and post-flood events were utilized for mapping flood inundated areas.

2 Study area

The study area is the areal extent of the Nilambur municipality in Malappuram district of Kerala, India, which experienced severe flooding during 2018 and 2019. The geographical extent of the study area was delineated from the website http://delimitation.lsgkerala.gov.in. It is spread over an area of approximately 28 km^2. The principal rainy seasons in the area are the southwest (June–September) and the northeast (October–November) monsoons.

Sentinel-1A and 1B VV (vertically transmitted and vertically received) polarized data with descending pass direction and relative orbit equal to 165 for the period 2018–2019 was used in this study. VV polarized data has the potential to detect partially submerged areas [6]. The study area is presented in Fig. 1.

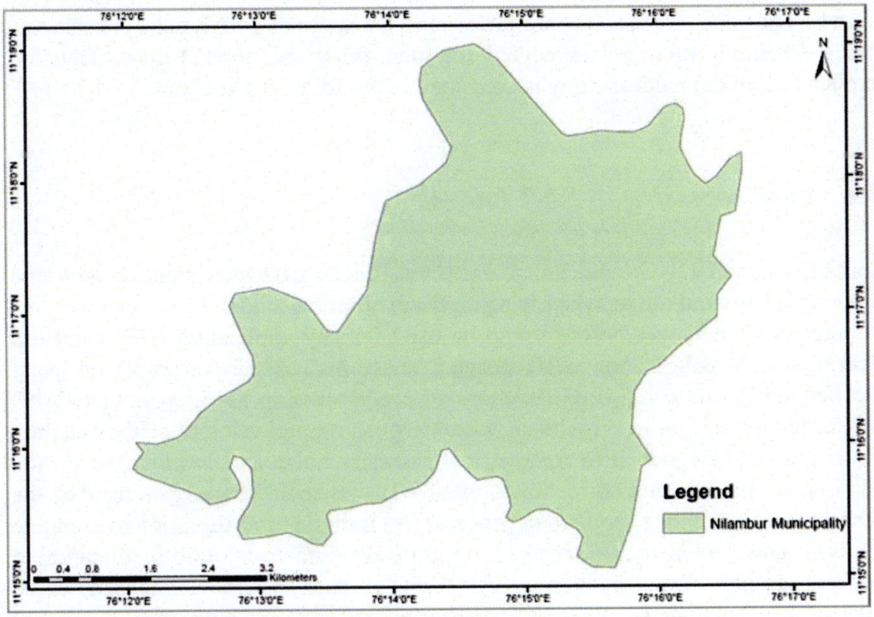

Fig. 1 Map of the areal extent of Nilambur Municipality

3 Methodology

In this study, flood mapping of the areal extent of the Nilambur municipality was performed using Sentinel-1 SAR data. A flow chart of the overall methodology adopted in this study is presented in Fig. 2. GEE has unrestricted access to satellite image collections, thereby facilitating analysis over large areas and for longer time

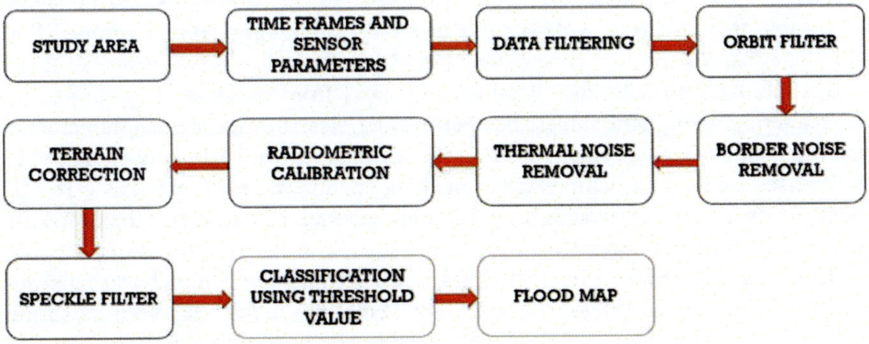

Fig. 2 Steps in pre-processing of SAR images

frames. The GEE was used for pre-processing the Sentinel-1 SAR data. A backscattering threshold value was specified for identifying the flooded areas. Detailed explanation of the methodology is presented in the following sections.

3.1 Pre-processing of SAR Images

Sentinel-1A and 1B Ground Range Detected (GRD) data were multi-looked and projected to ground range by applying an Earth ellipsoid model. Every satellite data requires pre-processing before it can be used for any application. GEE provides Sentinel-1 VV polarization GRD images. These images are pre-processed using the Sentinel-1 algorithm in GEE. Steps in pre-processing the images include (i) orbital file correction to provide an accurate position and velocity of the satellite, (ii) border noise removal to remove low intensity noise and invalid data on the edges, (iii) thermal noise correction to remove any noise in the data generated by the sensors during the data acquisition process, (iv) radiometric calibration to compute the backscatter intensity, (v) terrain correction, also known as ortho-rectification to convert Sentinel-1 images into a suitable form for maps by removing terrain related geometric distortions like layover, foreshortening, or shadowing effects, and (vi) speckle filtering to remove granular noise which is caused by the interference of waves reflected from many elementary scatterers. In this study, focal median filter was applied to remove the speckle noise with a circular kernel of 50 m in the GEE environment.

3.2 Classification of the Flooded Area

After pre-processing of the Sentinel-1 VV SAR images, these were classified into open water areas and partially submerged urban areas. In the pre-processed images, open water areas, such as water bodies and fully submerged areas, exhibited low values of backscattering coefficient, resulting in dark pixels. As the surface of these areas is relatively smooth, the signal scatters away from the satellite, yielding low backscattering coefficient values. For performing classification, threshold value of the back scattering coefficient for the open water areas was manually specified in GEE. Based on the observations mentioned above, this value was fixed as −16 dB. Those pixels in the images with backscattering coefficient values less than −16 dB were classified as open water.

The submerged urban areas in the image exhibited relatively high backscattering coefficient values. Flood water and walls of the buildings in the urban area constitute smooth surfaces which are approximately at right angles resulting in double bounce of the signal, thereby yielding bright pixels with relatively high backscattering coefficient values. This can also happen in the non-flooded urban regions, where roads or paved surfaces and the walls of buildings cause double bounce of the signal. Based on these observations, the threshold value for the submerged urban areas was fixed as −5 dB and those pixels with a backscattering coefficient value greater than −5 dB were classified as submerged urban area.

The time series of both these classes can provide information on the extent of flooding in terms of a sudden increase in the number of inundated pixels. This time series was used to determine the date on which maximum inundation occurred in the study area. The JavaScript used for pre-processing and classification is presented in Appendix.

4 Results and Discussion

The time series of number of pixels classified as open water and partially submerged during the years 2018 and 2019 are presented in Figs. 3 and 4 respectively. In both cases, there was a sudden increase in the number of pixels in these classes in the month of August. The number of pixels in these classes during August 2019 was higher than the corresponding value for August 2018 which indicated more severe flooding in 2019, matching with field observations. Landslides in the upper reaches of the Chaliyar river basin had resulted in very severe flooding in 2019. It is also observed from Fig. 3 that the flood waters receded relatively slowly, particularly in 2018. The peak of the time series of partially submerged areas was higher in August 2018 August when compared to that in August 2019 (Fig. 4). Faster recession of flood was observed in 2018 when compared to that in 2019 in this Figure as well.

The area under water in the Nilambur municipality before, during and after the floods of 2018 and 2019 are presented in Figs. 5 and 6 respectively. During the 2018 flood, the maximum inundation was observed on 9th August whereas during the 2019 flood, the maximum inundation occurred on 10th August. From the time series

Fig. 3 Time series of number of pixels classified as open water

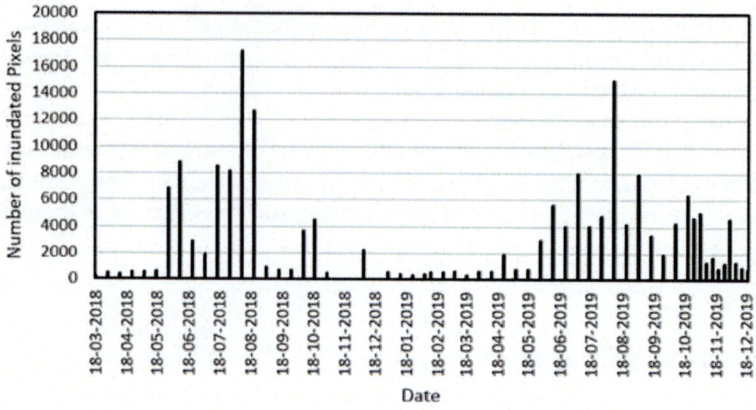

Fig. 4 Time series of number of pixels classified as partially submerged

plots (Figs. 3 and 4), it is also observed that floodwaters started receding by the end of August 2018 and August 2019.

5 Conclusions

In this study, a methodology to assess the extent of inundation during floods using Sentinel-1A and 1B SAR data is outlined and further demonstrated by applying it to the 2018 and 2019 floods that occurred in the areal extent of the Nilambur municipality in Malappuram District of Kerala, India. Results of the study indicated a sharp increase in the area under water during the month of August. The study can be further extended using SAR images with ascending pass direction and the results can be compared with that obtained in this study. Also, images with VH polarization can be used for the analysis. GEE facilitates application of different analysis techniques and faster identification of the regions impacted by floods, thereby aiding rescue and relief operations. As the data used need not be downloaded into the system, high end computing systems are not required for performing this analysis.

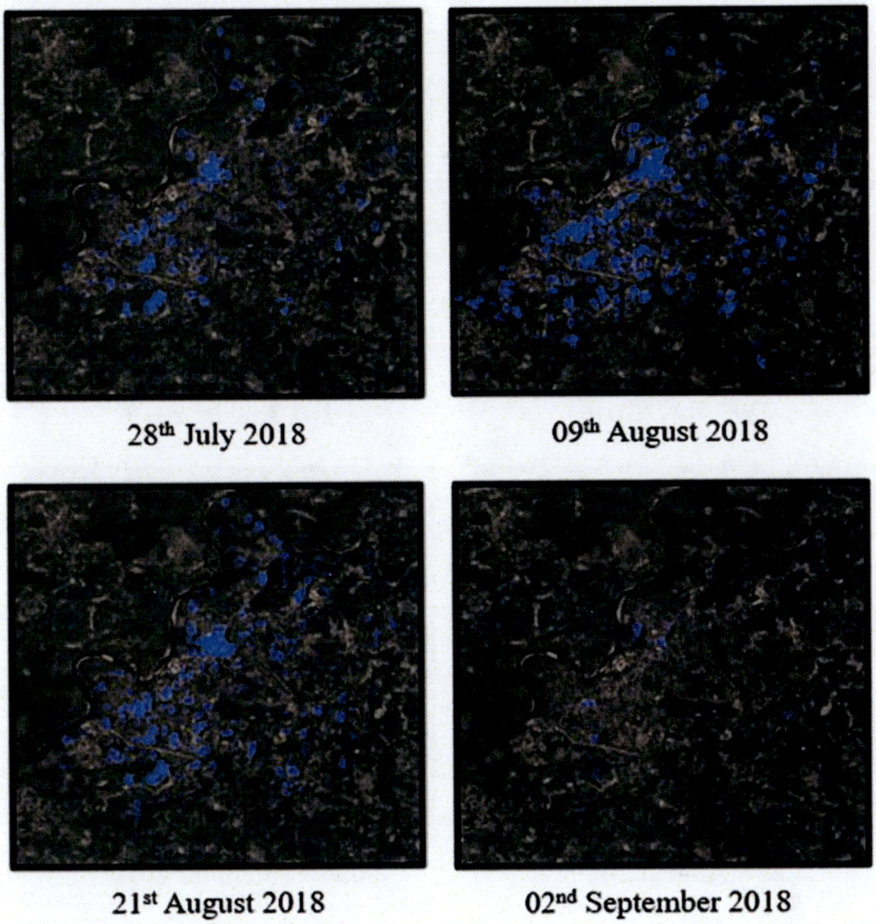

Fig. 5 Flood map of Nilambur municipality in 2018

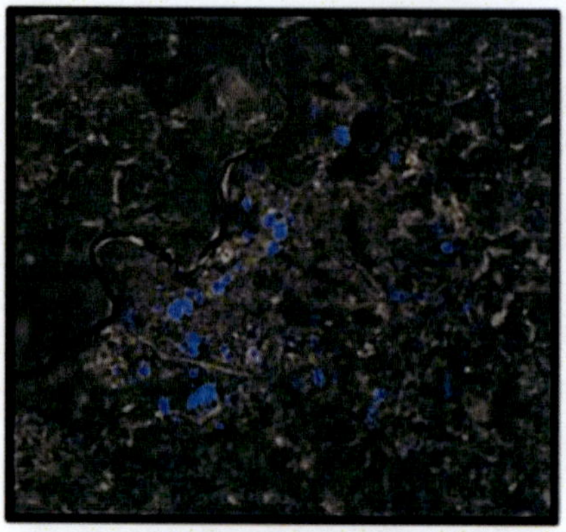

29th July 2019

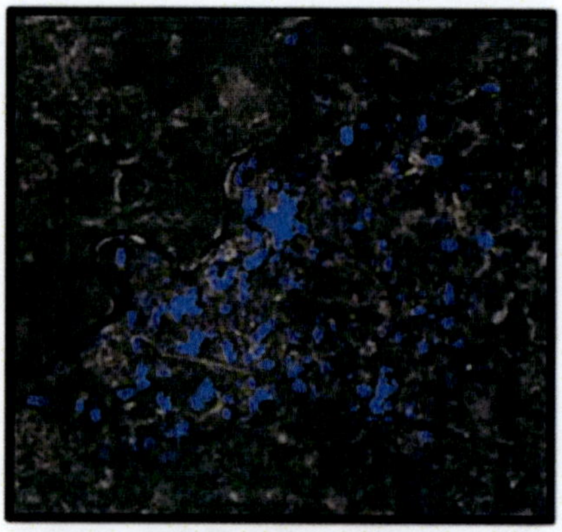

10th August 2019

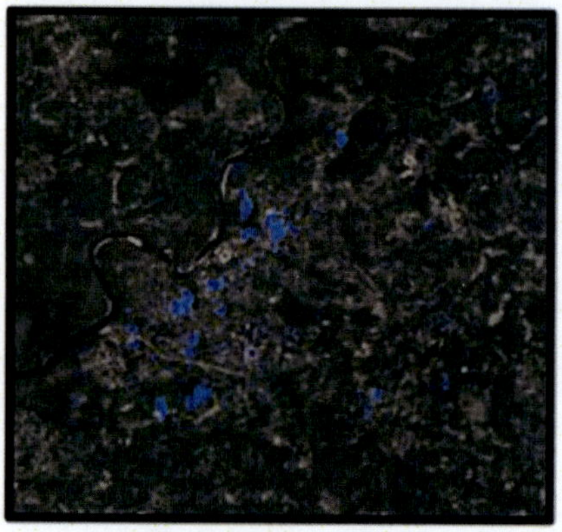

22nd August 2019

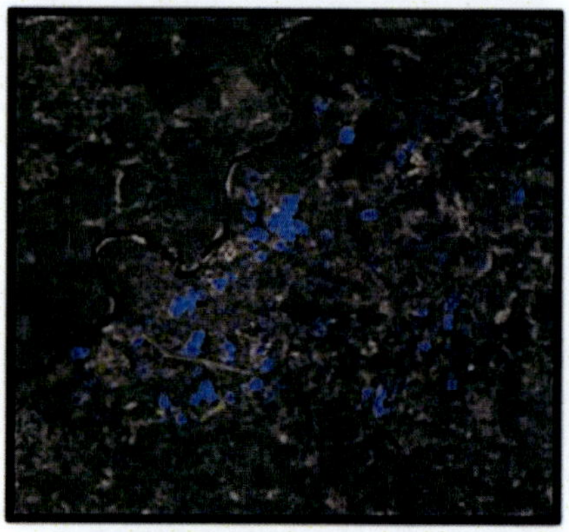

03rd September 2019

Fig. 6 Flood map of Nilambur municipality in 2019

Appendix

Program Code

```
// Load region defined by polygon and add it to the map
Map.addLayer(table2, {}, 'ROI')
Map.centerObject(table2)
var pass_direction='DESCENDING';
var relative_orbit= 165;
```

```
//Load Sentinel-1 SAR collection and filter according to
data collection type
var S1 = ee.ImageCollection('COPERNICUS/S1_GRD')
  .filterBounds(table2)
  .filter(ee.Filter.eq('instrumentMode','IW'))

.filter(ee.Filter.eq('orbitProperties_pass',pass_directio
n))



  .filter(ee.Filter.eq('resolution_meters',10))
  .filterDate('2018-03-11','2019-12-25')

.filter(ee.Filter.listContains('transmitterReceiverPolari
sation', 'VV'))

.filter(ee.Filter.eq('relativeOrbitNumber_start',relative
_orbit ))
  .select('VV');

//Add first image to map to get an idea of what a SAR im-
age looks like
Map.addLayer(S1.first(),{bands: 'VV',min: -18, max: 0},
'SAR image')

// Filter speckle noise
var filterSpeckles = function(img) {
  var vv = img.select('VV') //select the VV polarization
band
  var vv_smoothed =
vv.focal_median(50,'circle','meters').rename('VV_Filtered
```

```
') //Apply a focal median filter
  return img.addBands(vv_smoothed) // Add filtered VV
band to original image
}

// Map speckle noise filter across collection. Result is
same collection, with smoothed VV band added to each im-
age
S1 = S1.map(filterSpeckles)
//Add speckle filtered image to map to sompare with raw
SAR image
Map.addLayer(S1.first(),{bands: 'VV_Filtered',min: -18,
max: 0}, 'Filtered SAR image')

//Classify water pixels using a set threshhold
//Here we are using -16.
var classifyWater = function(img) {
  var vv = img.select('VV_Filtered')
  var water = vv.lt(-16).rename('Water')  //Identify all
pixels below threshold and set them equal to 1. All other
pixels set to 0
  water = water.updateMask(water) //Remove all pixels
equal to 0



  return img.addBands(water)  //Return image with added
classified water band
}

//Map classification across sentinel-1 collection and
print to console to inspect
S1 = S1.map(classifyWater)
print(S1)
```

```
//Make time series of water pixels within region
var ClassChart = ui.Chart.image.series({
  imageCollection: S1.select('Water'),
  region: table2,
  reducer: ee.Reducer.sum(),
  scale: 10,
})
  .setOptions({
      title: 'Inundated Pixels',
      hAxis: {'title': 'Date'},
      vAxis: {'title': 'Number of Inundated Pixels'},
      lineWidth: 2
    })

//Set the postion of the chart and add it to the map
ClassChart.style().set({
    position: 'bottom-right',
    width: '500px',
    height: '300px'
  });

Map.add(ClassChart)

// Create a label on the map.
var label = ui.Label('Click a point on the chart to show
the image for that date.');
Map.add(label);

//Create callbakc function that adds image to the map
coresponding with clicked data point on chart
ClassChart.onClick(function(xValue, yValue, seriesName) {
    if (!xValue) return;  // Selection was cleared.

    // Show the image for the clicked date.
    var equalDate = ee.Filter.equals('system:time_start',
xValue);
```

```
    //Find image coresponding with clicked data and clip
water classification to roi
    var classification =
ee.Image(S1.filter(equalDate).first()).clip(table2).selec
t('Water');
    var SARimage =
ee.Image(S1.filter(equalDate).first());
    //Make map layer based on SAR image, reset the map
layers, and add this new layer
    var S1Layer = ui.Map.Layer(SARimage, {
      bands: ['VV'],
      max: 0,
      min: -20
    });
    Map.layers().reset([S1Layer]);

    //Add water classification on top of SAR image
    Map.addLayer(classification,visParams,'Water')

    // Show a label with the date on the map.
    label.setValue((new Date(xValue)).toUTCString());
  });


  //Classify  submerged water pixels using a set thresh-
hold
//Here we are using -5.
var classifyWater1 = function(img) {
  var vv1 = img.select('VV_Filtered')
  var water1 = vv1.gt(-5).rename('S_Water')
  water1 = water1.updateMask(water1)
  return img.addBands(water1)
}

//Map classification across sentinel-1 collection and
print to console to inspect
S1 = S1.map(classifyWater1)
print(S1)

//Make time series of water pixels within region
var ClassChart = ui.Chart.image.series({
  imageCollection: S1.select('S_Water'),
  region: table2,
  reducer: ee.Reducer.sum(),
  scale: 10,
```

```
})
  .setOptions({
      title: 'Inundated Pixels',
      hAxis: {'title': 'Date'},
      vAxis: {'title': 'Number of Inundated Pixels'},
      lineWidth: 2
    })

//Set the postion of the chart and add it to the map
ClassChart.style().set({
    position: 'bottom-right',
    width: '500px',
    height: '300px'
  });

Map.add(ClassChart)

// Create a label on the map.
var label = ui.Label('Click a point on the chart to show
the image for that date.');
Map.add(label);

//Create callbakc function that adds image to the map
coresponding with clicked data point on chart
ClassChart.onClick(function(xValue, yValue, seriesName) {
    if (!xValue) return;  // Selection was cleared.

    // Show the image for the clicked date.
    var equalDate = ee.Filter.equals('system:time_start',
xValue);
    //Find image coresponding with clicked data and clip
water classification to roi
    var classification =
ee.Image(S1.filter(equalDate).first()).clip(table2).selec
t('S_Water');
    var SARimage =
ee.Image(S1.filter(equalDate).first());
    //Make map layer based on SAR image, reset the map
layers, and add this new layer
    var S1Layer = ui.Map.Layer(SARimage, {
      bands: ['VV'],
      max: 0,
   min: -20
    });
    Map.layers().reset([S1Layer]);
```

```
    //Add water classification on top of SAR image
    Map.addLayer(classification,visParams,'S_Water')

    // Show a label with the date on the map.
    label.setValue((new Date(xValue)).toUTCString());
  });
```

References

1. Notti, D., Giordan, D., Caló, F., Pepe, A., Zucca, F., & Galve, J. P. (2018). Potential and limitations of open satellite data for flood mapping. *Remote Sensing, 10*. https://doi.org/10.3390/rs10111673
2. de Groeve, T. (2010). Flood monitoring and mapping using passive microwave remote sensing in Namibia. *Geomatics, Natural Hazards and Risk, 1*, 19–35. https://doi.org/10.1080/19475701003648085
3. Tiwari, V., Kumar, V., Matin, M. A., Thapa, A., Ellenburg, W. L., Gupta, N., & Thapa, S. (2020). Flood inundation mapping-Kerala 2018; Harnessing the power of SAR, automatic threshold detection method and Google Earth Engine. *Plos One, 15*, 1–17. https://doi.org/10.1371/journal.pone.0237324
4. Vishnu, C. L., Sajinkumar, K. S., Oommen, T., Coffman, R. A., Thrivikramji, K. P., Rani, V. R., & Keerthy, S. (2019). Satellite-based assessment of the August 2018 flood in parts of Kerala, India. *Geomatics, Natural Hazards and Risk, 10*, 758–767. https://doi.org/10.1080/19475705.2018.1543212
5. Vanama, V. S. K., Rao, Y. S., & Bhatt, C. M. (2021). Change detection based flood mapping using multi-temporal earth observation satellite images: 2018 flood event of Kerala India. *European Journal of Remote Sensing, 54*, 42–58. https://doi.org/10.1080/22797254.2020.1867901
6. Manjusree, P., Prasanna Kumar, L., Bhatt, C. M., Rao, G. S., & Bhanumurthy, V. (2012). Optimization of threshold ranges for rapid flood inundation mapping by evaluating backscatter profiles of high incidence angle SAR images. *International Journal of Disaster Risk Science, 3*, 113–122. https://doi.org/10.1007/s13753-012-0011-5

Application of a SWAT Model for Estimating Runoff and Other Hydrological Parameters in Cauvery River Sub-basins

Pandurang and M. Inayathulla

Abstract SWAT Model has been used successfully for exploring hydrological characteristics of the Cauvery River Basin using HRU based approach. The automatic watershed delineation at HRU level clearly shows how the basic features like land use, soil, and slope influence the hydrology of the catchment. The pre-processing step includes the usage of spatial data such as DEMs, soil, and land use, which is then supplied into the SWAT model via the interface. Daily climate data were sourced and prepared according to the SWAT. SWAT model is used to estimate the runoff and other hydrological parameters like Water yield, Actual Evapotranspiration (AET), Potential Evapotranspiration (PET), and Sediment yield during period 2000–2014. The results obtained by this model are Rainfall-Runoff correlation is satisfactory in all sub-basins. Water yield is more in the upper Cauvery Sub-basin (C1) due to Western Ghats region gets rain and less loses. Actual Evapotranspiration (AET) and Potential Evapotranspiration (PET) is more in the Suvarnavati (C3) and Shimsha (C4) Sub-basin respectively. When compared to the other Cauvery River Sub-basins in Karnataka, the Suvarnavati Sub-basin (C3) and Shimsha (C4) Sub-basin has a higher and growing sediment production. The coefficient of determination (R^2) of all sub-basins are under Very Satisfactory.

Keywords SWAT · Hydrological modeling · Surface runoff and other parameters · Cauvery basin · Karnataka

1 Introduction

Water is a vital resource for long-term economic and social growth. Because of the human activities (e.g., increasing global population, land-use change, water pollution) and climate change, water shortages have become the major crises of sustainable development of communities all over the world. As a result, effective management of water resources is essential for human society's long-term development. Proper

Present Address:
Pandurang (✉) · M. Inayathulla
Department of Civil Engineering, UVCE, Bangalore University, Karnataka, India

A. K. Dikshit et al. (eds.), *Innovative Trends in Hydrological and Environmental Systems*, Lecture Notes in Civil Engineering 234,
https://doi.org/10.1007/978-981-19-0304-5_43

management of water, which is one of the only natural resources, is the ultimate challenge faced by the modern world. One of the main challenges for hydrologists is the rising population, modern urbanization, and meeting their demands. Developments in agricultural practices often add up to the rising demand of the water with these variables, hydrological models play a crucial role in determining the activity of water at the watershed scale. Watersheds are the resource centers of the developing India; they need to be managed in a proper way to sustain the environment. For those studies should be conducted to assess the process taking place in the watersheds and any disturbances such as major soil losses, sudden depletion of land cover, excessive land use manipulations etc.

In this present study, Cauvery River Sub-basins in Karnataka are selected for the estimation of surface runoff and other hydrological parameters using GIS based approach, SWAT (Soil and Water Assessment Tool). SWAT model will simulate amount and motion of water through the catchment/Sub-basin. Hydrological model is simulating the transport of water through the Cauvery River Sub-basins and for the present study 6 Sub-basins are considered. ArcGIS is a proprietary Geographic Information System (GIS) that is used to show geographical information on a map and provides a standard framework for working with different spatial data collected from various sources.

The main objective of this study is to test the performance and feasibility of the SWAT model for estimation flow in the Cauvery River basin of Karnataka, which will contribute to the water resources management in the Cauvery River basin and thereby is useful for the sustainable development of the country. SWAT provides an integrated platform for the analysis of these things in a system.

The study's main goal is to record and simulate the spatiotemporal changes of runoff and other hydrological parameters. Cauvery River Sub-basins are selected for the estimation of surface runoff and other hydrological parame-ters using a GIS-based approach, SWAT (Soil and Water Assessment Tool). In the examination, the SWAT model was used to simulate/to check the possible yearly changes in the Precipitation, Runoff, Water yield, Actual Evap-otranspiration (AET), Potential Evapotranspiration (PET), Sediment yield, Lateral Flow and Runoff in the period 2000–2014, under land-use land cover and Soil conditions [1]. SWAT weather generator was used to supplement the missing meteorological data and Hydrological model is simulating the amount and transport of water through the Cauvery River Sub-basins.

2 Study Area

The Cauvery River, also known as the Dakshin Ganga, is one of South India's most important interstate peninsular rivers. Before draining into the Bay of Bengal, the Cauvery River rises in the Western Ghats and flows eastward through the states of Karnataka, Tamil Nadu, Kerala, and the union territory of Pondicherry.

Table 1 Sub-basins of the Cauvery River Basin and drainage area

Sl. No.	Sub-basin name	Sub-basin code	Drainage area (km^2)
1	Upper Cauvery sub-basin	C1	10,875
2	Kabini sub-basin	C2	4882
3	Suvarnavati sub-basin	C3	1303
4	Shimsha sub-basin	C4	8735
5	Arkavati sub-basin	C5	4162
6	Middle Cauvery sub-basin	C6	2638

The study area is a part of the 6 sub-basins of the Cauvery River Basin in Karnataka. Study area covering the river basin area around 32,595 km^2 and area covers Sub-basins are Upper Cauvery Sub-basin (C1), Kabini Sub-basin (C2), Suvarnavati Sub-basin (C3), Shimsha Sub-basin (C4), Arkavati Sub-basin (C5) and Middle Cauvery Sub-basin (C6). Sub-basins of the Cauvery River Basin drainage area are mentioned (refer Table 1) and location map of the Cauvery River Sub-basins (refer Fig. 1).

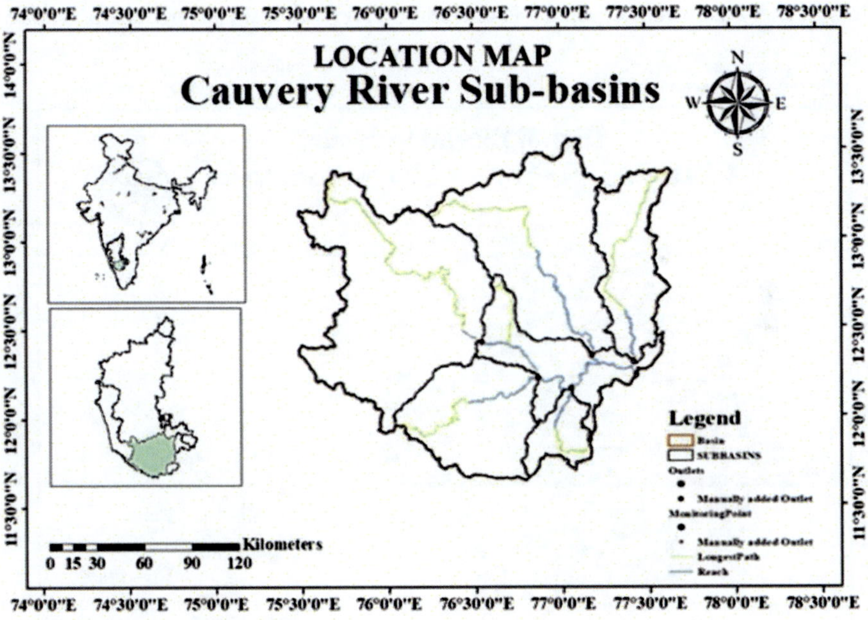

Fig. 1 Location map of study area

3 Input Data

The SWAT hydrological model requires basic input data such as topography, weather, rainfall, land use, and soil data. The SRTM (Shuttle Radar Topogra-phy Mission) provided topographic data in the form of DEM (Digital Elevation Model) at 30 m resolution (refer Fig. 2), which was used to delineate a basin into multiple sub-basins and calculate topographic related parameters such as slope class, stream length, and locate monitoring points. 1:50,000 scale land use data and daily precipitation data for 140 rain gauge stations across the river basin is collected from Advanced Centre for Integrated Water Resources Management (ACIWRM) (refer Fig. 3). Soil map physical prop-erties database was obtained from the SWAT-FAO data. Climatic data was gathered daily from climate stations operated by the Indian Meteorological Department (IMD).

3.1 *Land Use/Land Cover*

The major classes of Land Use and Land Cover are classified as per SWAT input classes. Sub-basins major portion is covered by Agricultural land-generic and Rice

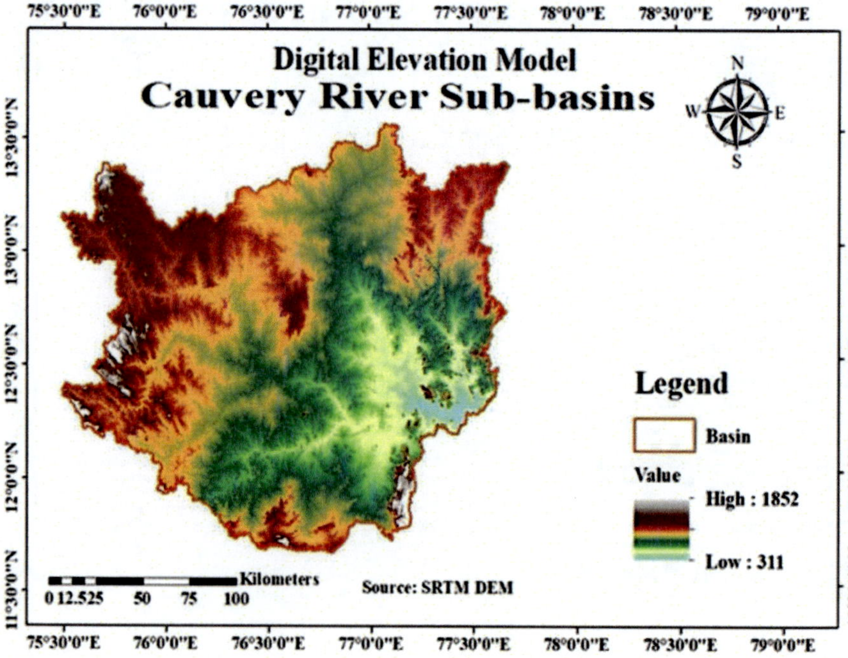

Fig. 2 Digital elevation model

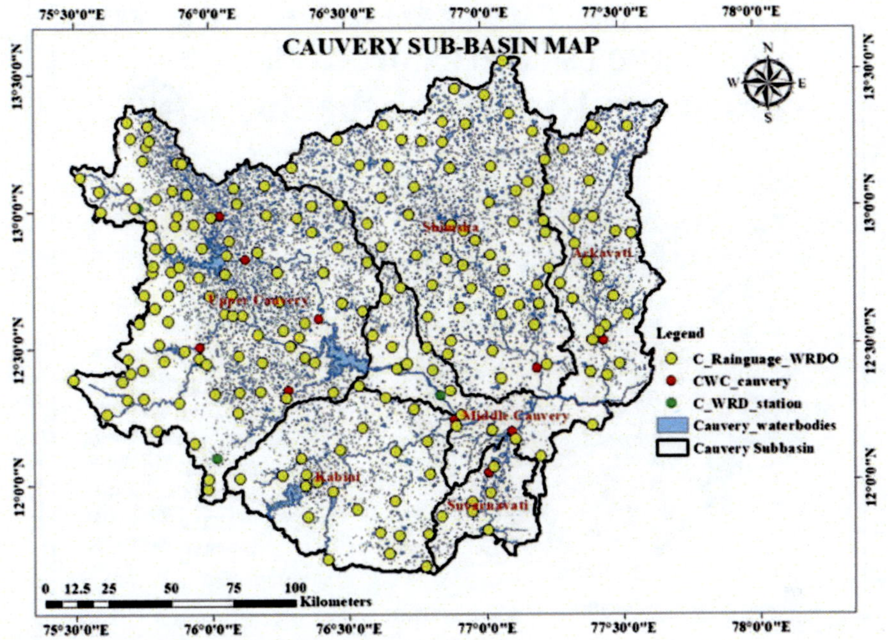

Fig. 3 Sub-basins and gauge station

crop (73.75%), Built-up (Residential) (10.2%), Forest and Wetlands (6.5%), Barren and Waste land (4.75%), Pasture and Brushes (4.5%), and Waterbodies (0.6%). Land Use and Land Cover class of Cauvery River Sub-basins (refer Fig. 4).

3.2 *Soil Classes*

The soil map is generated using SWAT-FAO-Soil under GIS environment. Soil map of the Sub-basins Chromic Luvisols Soil occupies maximum area of 17,606.32 Km^2 (54%) the Sub-basins. It is followed by Eutric Nitosols Soil with an area of 11,252.13 Km^2 (35%), Plinthic Acrisols Soil occupies area of 3419.50 Km^2 (11%) and further water occupies the minimum area is 200 Km^2 (1%). Soil class of Cauvery River Sub-basins (refer Fig. 5).

4 Methodology

The SWAT model is limited in that it does not explicitly allow for the inclusion of spatial data as model inputs. Data must be processed into a form that the model can

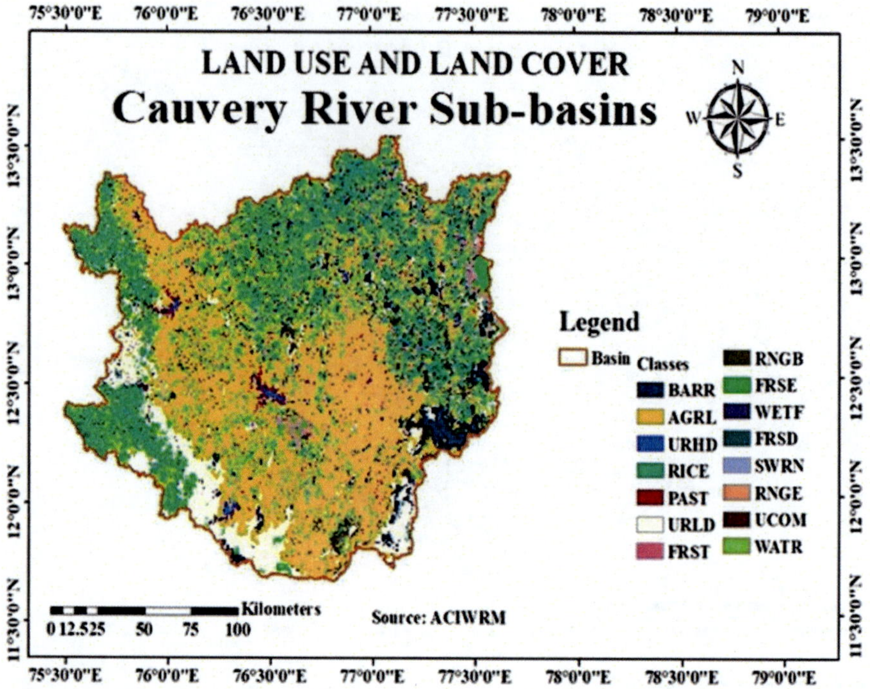

Fig. 4 Land use/land cover map

use. Processing these data, even with the use of a GIS is tiresome and time consuming due to the large number of model parameters required for executing SWAT.

4.1 SWAT Model

SWAT Model is a physically based semi distributed hydrologic model which can operate on different time steps. It is a comprehensive tool that enables the impacts of land management practices on water, sediment, and agricultural chemical yields for the watersheds with varying soils, land use, and management practices. Spatial data DEM, soil, and land use is used in the pre-processing phase and fed into the SWAT model through the interface [2]. Daily climate data were sourced and prepared according to the SWAT.

This model considers two phases to simulate hydrology of the Sub-basin. Namely, the land phase (accounts for water, sediment, nutrient, and pesti-cide loadings to the main channel in each sub-basin) and the routing phase (accounts for movement of water, sediments, nutrients, and organic chemicals through the channel network of the watershed to the outlet) [3]. SWAT Model has been used successfully for exploring

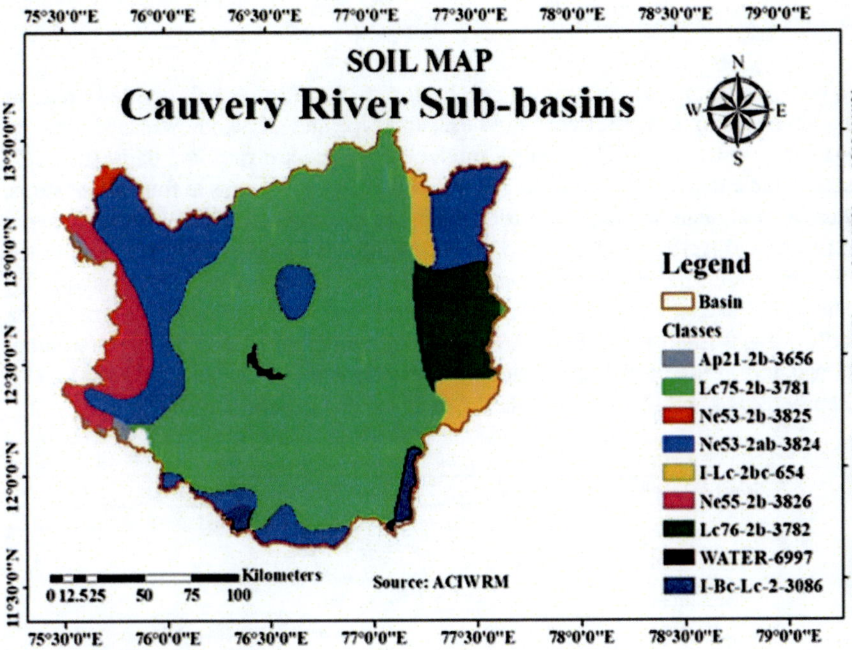

Fig. 5 Soil map

hydrological characteristics of the Cauvery River Basin using Hydrological Response Unit (HRU), which describes local diversity, which constitutes the land cover, soil characteristics, and land slope characteristics of unique land use.

SWAT model is based on the principle of water balance equation:

$$\mathrm{SW}_t = \mathrm{SW}_o + \sum_{i=1}^{p} (\mathrm{R}_{\mathrm{day}} - \mathrm{Q}_{\mathrm{surf}} - \mathrm{E}_a - \mathrm{W}_{\mathrm{seep}} - \mathrm{Q}_{\mathrm{gw}} \tag{1}$$

where, SW_t is the ultimate water content in (mm), SW_o is the amount of water content on the first soil of the day i (mm), t is time (days), $\mathrm{R}_{\mathrm{day}}$ is the amount of rainfall on day i (mm), $\mathrm{Q}_{\mathrm{surf}}$ is the amount of surface runoff on specific day i (mm), E_a is the amount of evapotranspiration on day i (mm), $\mathrm{W}_{\mathrm{seep}}$ is the amount of percolated water into the vadose zone from the soil profile on day i (mm) and Q_{gw} is the amount of return flow on day i (mm).

Methods used in SWAT calculate Surface runoff ($\mathrm{Q}_{\mathrm{surf}}$) for each HRU is calculated by using Soil Conservation Services (SCS) curve number method. MUSLE (Modifies Soil Loss Equation) for Sediment calculation [4]. Penman Monteith Method for evapotranspiration. The modified Universal soil loss equation for soil erosion. Green Ampt method for infiltration [5].

4.2 SWAT Model Setup

In SWAT, a Basin is delineated to sub-basins, which further sub divided areas of homogeneous similarity between land use, soil type and slope, known as Hydrological Response Units (HRUs). SWAT model operations are run on a daily time step based on the concept of hydrologic response units. SWAT works in four major steps, Watershed delineation, HRU definition, Weather data input, Edit SWAT input, setup and run model (refer Fig. 6). Based on DEM, stream network and by selecting the watershed outlet, the entire basin was divided into sub basins. In HRU analysis, by feeding land use, soil maps and by defining HRU'S with threshold percentage, an HRU's were formed [6]. After watershed delineation and HRU analysis, weather data was fed to the model and simulation was done from 1st January 2000 to 31st December 2014 and Model setup, User Interface and execution (refer Fig. 7).

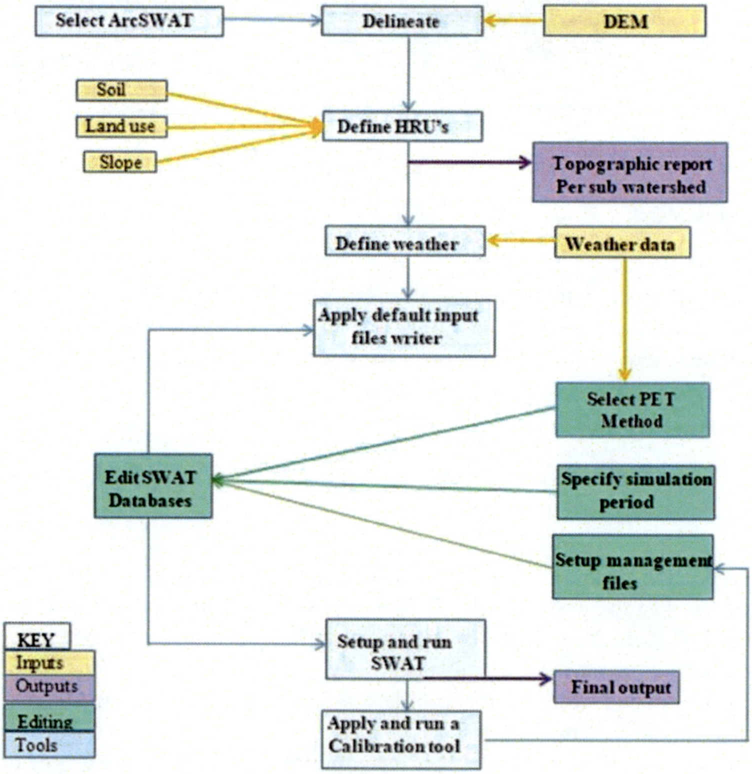

Fig. 6 Flowchart showing model setup and execution

Fig. 7 ArcSWAT user interface and model execution

5 Results and Discussions

During the study Agriculture Land and Paddy area was found that the 73.75% and Chromic Luvisols Soil occupies maximum area of 17,606.32 Km^2 (54%) the Cauvery Sub-basins in Karnataka. Also, Elevation of basin is varying 311–1852 m from mean sea level.

The SWAT output reports included details of area distribution in hydro-logic response units and their unit combinations of soil, land use, and slope. Also output of the Sub-basins includes the desired time step outputs of rainfall, runoff, sediment yield, water yield, groundwater flow, actual evapotranspiration, potential evapotranspiration, and other parameters. During the period 2000–2014, the SWAT model was used to estimate runoff, water yield, actual evapotranspiration (AET), potential evapotranspiration (PET), and sediment production for all sub-basins (refer Fig. 8, 9, 10, 11, 12, 13). Estimated hydrologi-cal parameters are varying periodically with increase/decrease trends. Also, extreme conditions and maximum values of runoff and other hydrological parameters of Sub-basins is tabulated (refer Table 2).

Rainfall in mm						
Years	Upper Cauvery Sub-basin (C1)	Kabini Sub-basin (C2)	Suvarnavati Sub-basin (C3)	Shimsha Sub-basin (C4)	Arkavati Sub-basin (C5)	Middle Cauvery Sub-basin (C6)
2000	1971.968	1693.344	1730.064	787.678	1243.697	1284.331
2001	2654.161	1134.566	1322.153	788.437	838.036	1304.971
2002	2453.513	1464.181	1515.185	891.872	1271.765	1164.696
2003	2645.008	1830.875	2074.3	928.011	871.134	891.558
2004	2705.217	1712.576	2016.702	867.287	739.607	1249.24
2005	2417.445	1685.709	2128.581	713.219	1006.736	1320.806
2006	2408.274	1548.082	1680.756	456.246	792.248	1138.105
2007	2296.419	1614.033	1798.672	596.309	1083.403	1135.97
2008	2853.118	1183.087	1528.577	655.545	901.537	1084.623
2009	2252.869	1608.831	1922.911	606.95	899.582	1238.558
2010	2877.081	1935.312	1795.108	910.052	1025.96	1220.792
2011	2542.106	1699.469	2232.645	801.659	1025.163	1240.263
2012	2479.369	1527.67	1734.003	748.848	941.763	1268.442
2013	2168.003	1842.383	2176.448	760.005	1056.215	1063.611
2014	3062.655	1761.428	1807.175	899.238	779.454	1308.147
Trend						

Fig. 8 Average annual rainfall in the sub-basins

Runoff in mm						
Years	Upper Cauvery Sub-basin (C1)	Kabini Sub-basin (C2)	Suvarnavati Sub-basin (C3)	Shimsha Sub-basin (C4)	Arkavati Sub-basin (C5)	Middle Cauvery Sub-basin (C6)
2000	1049.53	687.68	752.20	192.83	513.01	558.28
2001	1736.50	332.58	514.33	226.94	305.13	567.38
2002	1504.74	603.85	623.29	288.07	470.65	411.23
2003	1789.90	805.48	1085.06	351.71	161.78	177.07
2004	1870.18	704.42	1136.26	279.43	135.09	523.81
2005	1539.48	713.45	1223.87	139.28	350.08	514.20
2006	1527.91	600.79	840.88	26.72	168.61	392.38
2007	1445.28	708.13	983.20	98.74	315.00	333.60
2008	2017.05	394.59	699.34	216.03	217.85	407.59
2009	1404.08	645.45	1033.64	91.34	220.38	476.13
2010	2086.95	1018.67	923.02	242.78	269.22	445.64
2011	1700.86	719.96	1324.24	181.45	304.29	480.17
2012	1598.01	625.13	877.89	134.57	269.93	510.15
2013	1305.96	983.18	1328.25	198.13	354.12	369.67
2014	2160.14	801.43	892.69	285.43	141.12	513.60
Trend						

Fig. 9 Average annual runoff in the sub-basins

Actual Evapotranspiration (AET) in mm						
Years	Upper Cauvery Sub-basin (C1)	Kabini Sub-basin (C2)	Suvarnavati Sub-basin (C3)	Shimsha Sub-basin (C4)	Arkavati Sub-basin (C5)	Middle Cauvery Sub-basin (C6)
2000	632.929	824.42	861.354	590.53	712.981	702.895
2001	654.012	711.737	752.226	556.238	522.331	659.928
2002	656.632	702.67	740.516	620.007	741.192	706.499
2003	623.297	788.554	821.311	586.012	677.384	674.172
2004	621.142	767.936	791.135	576.908	573.333	650.178
2005	648.521	773.852	791.474	598.222	640.515	748.382
2006	666.925	806.032	797.961	473.629	633.501	702.256
2007	605.928	722.418	729.32	485.873	713.956	726.624
2008	614.828	710.126	737.705	404.283	641.361	614.867
2009	613.281	754.788	764.139	530.492	635.411	708.425
2010	603.152	746.631	726.156	653.803	725.906	742.151
2011	625.781	719.128	771.832	597.079	674.679	687.558
2012	616.731	780.866	787.061	628.453	668.184	720.826
2013	605.359	696.892	766.837	583.044	665.56	641.479
2014	635.938	757.81	771.931	602.739	617.106	722.455
Trend						

Fig. 10 Average annual AET in the sub-basins

Potential Evapotranspiration (PET) in mm						
Years	Upper Cauvery Sub-basin (C1)	Kabini Sub-basin (C2)	Suvarnavati Sub-basin (C3)	Shimsha Sub-basin (C4)	Arkavati Sub-basin (C5)	Middle Cauvery Sub-basin (C6)
2000	1176.91	1580.421	1592.612	1823.647	1562.422	1623.416
2001	1175.164	1664.788	1648.897	1880.212	1625.665	1694.557
2002	1179.632	1584.749	1574.372	1783.413	1511.04	1598.255
2003	1134.859	1531.138	1528.004	1783.232	1512.037	1581.508
2004	1111.94	1506.064	1499.044	1856.374	1543.639	1617.134
2005	1139.275	1524.299	1513.096	1789.375	1509.296	1568.144
2006	1198.747	1624.941	1624.73	1926.399	1620.335	1657.538
2007	1166.548	1568.061	1547.966	1796.49	1507.251	1555.672
2008	1141.007	1580.007	1575.305	1857.927	1538.759	1609.193
2009	1147.641	1591.27	1579.344	1813.102	1508.195	1578.912
2010	1155.478	1496.956	1483.384	1769.49	1487.676	1535.246
2011	1150.673	1539.274	1518.662	1807.963	1515.039	1557.491
2012	1113.35	1549.368	1536.577	1801.521	1533.355	1591.667
2013	1184.924	1547.996	1542.998	1771.633	1486.886	1540.244
2014	1139.398	1557.89	1525.207	1773.619	1529.778	1578.182
Trend						

Fig. 11 Average annual PET in the sub-basins

The Cauvery basin receives rainfall from both the monsoons of South India—the south-west (SW) monsoon (June–September), and the north-east (NE) monsoon (November–January). The SW monsoon generates heavy precipitation in the uplands of the Upper Cauvery Sub-basin (C1) and average annual rainfall is 2800 mm, and due to Western Ghats region sub-basin produce more water yield and less loses comparing to all other Sub-basins. Suvarnavati Sub-basin (C3) having Actual Evapotranspiration

Water Yield in mm						
Years	Upper Cauvery Sub-basin (C1)	Kabini Sub-basin (C2)	Suvarnavati Sub-basin (C3)	Shimsha Sub-basin (C4)	Arkavati Sub-basin (C5)	Middle Cauvery Sub-basin (C6)
2000	1299.259	828.721	828.947	164.721	489.896	577.624
2001	1993.185	404.341	559.689	194.294	302.471	618.099
2002	1787.518	742.031	736.844	244.833	484.484	472.52
2003	2016.535	981.734	1184.74	298.811	202.204	207.275
2004	2060.915	879.248	1200.967	244.716	139.592	538.532
2005	1791.273	913.484	1316.421	123.782	341.285	580.118
2006	1718.487	706.581	871.721	25.964	161.257	435.619
2007	1678.342	826.984	1067.437	84.57	326.308	397.44
2008	2229.792	487.23	739.352	180.286	228.436	436.606
2009	1628.355	789.48	1117.942	83.533	244.051	504.176
2010	2295.49	1158.582	1046.586	210.277	292.212	498.915
2011	1882.616	962.414	1411.982	159.143	323.579	529.613
2012	1857.744	706.191	921.909	117.336	246.022	518.961
2013	1568.954	1141.007	1408.57	170.333	385.195	422.965
2014	2368.545	931.371	972.659	242.497	127.854	555.294
Trend						

Fig. 12 Average annual water yield in the sub-basins

Sediment yield in ton/ha/yr						
Years	Upper Cauvery Sub-basin (C1)	Kabini Sub-basin (C2)	Suvarnavati Sub-basin (C3)	Shimsha Sub-basin (C4)	Arkavati Sub-basin (C5)	Middle Cauvery Sub-basin (C6)
2000	89.717	66.449	250.171	137.918	39.642	179.811
2001	147.761	36.156	140.942	171.743	53.808	228.604
2002	123.339	52.581	195.947	171.444	43.378	125.992
2003	161.824	68.412	301.001	245.354	17.673	67.391
2004	170.826	68.077	347.549	213.923	20.796	175.241
2005	134.755	69.466	372.712	112.975	36.254	146.118
2006	139.319	66.65	315.771	33.982	25.348	125.388
2007	119.056	69.931	297.276	68.632	25.575	117.711
2008	171.755	36.217	192.141	171.536	21.293	172.185
2009	117.325	66.756	306.059	77.555	23.353	162.845
2010	197.28	79.563	268.42	163.719	24.676	149.85
2011	150.221	68.831	369.749	116.595	30.998	170.38
2012	126.864	63.841	277.012	106.913	31.57	126.484
2013	106.585	88.247	389.862	155.634	28.964	116.318
2014	182.819	64.473	279.747	188.07	13.944	151.791
Trend						

Fig. 13 Average annual sediment yield in the sub-basins

(AET) and Potential Evap-otranspiration (PET) has increased trend due to climate change and Soil moisture condition.

The highest soil erosion was calculated more than 400t/ha/year in Kodagu, North/east and west of Mysuru, north of Chamarajanagar, Chikmangaluru, Hassan and South Parts of Bangalore Rural. Soil erosion and Hard rocky land is more in Suvarnavati Sub-basin (C3) has a higher and growing sediment production.

Table 2 Maximum hydrological parameters of Cauvery River Sub-basins

Parameters/Sub-basins code	C1	C2	C3	C4	C5	C6
Rainfall in mm	3062.66	1935.31	2232.65	928.01	1271.77	1320.81
Runoff in mm	2160.14	1018.67	1328.25	351.71	513.01	567.38
PET in mm	1198.75	1664.79	1648.90	1926.40	1625.67	1694.56
AET in mm	666.93	824.42	861.35	653.80	741.19	748.38
Water Yield in mm	2368.55	1158.58	1411.98	298.81	489.90	618.10
Sediment yield in ton/ha/yr	197.28	88.25	389.86	245.35	53.81	228.60
R2	0.98	0.92	0.96	0.83	0.86	0.90

Coefficient of determination or correlation (R^2) is calculated using the average annual rainfall along with the annual runoff scatter diagram is drawn. The rainfall-runoff relationship of all Sub-basins indicates a good correlation between both properties and the values of rainfall and runoff are largely correlated. Over the years, the crests and troughs in rainfall data are impacting the runoff data to act similarly. Coefficient of determination or correlation (R^2) for all the Sub-basins has the under Very Satisfactory condition.

6 Conclusions

The objective of this study was to test the robustness of the SWAT model and project goal was not to produce highly accurate results for immediate deci-sion making, but rather to evaluate the ability of SWAT to perform at higher spatial-temporal resolutions. While the DEM inputs featured high spatial res-olution, the soil and land use classification layers details represented correctly the in Sub-basins. The SWAT model is setup for 15 years that is from 2000 to 2014, estimated that average of 25% rainfall is occurring as runoff. The results suggested that the decrease in precipitation was the main cause of runoff reduction. The rainfall-runoff relationship of all Sub-basins indicates a good correlation between both properties and the values of rainfall and runoff are largely correlated. Over the years, the crests and troughs in rainfall data are impacting the runoff data to act similarly. Coefficient of determination or correlation (R^2) for all the Sub-basins has the under Very Satisfactory.

Water yield is more in the Upper Cauvery (C1) Sub-basin due West-ern Ghats region gets rain and less loses. Actual Evapotranspiration (AET) and Potential Evapotranspiration (PET) is more in the Suvarnavati (C3) and Shimsha (C4) Sub-basin respectively. AET and PET has the increased trend due to increase in climate change, Climate, and Soil Moisture condition. When compared to the other Cauvery River Sub-basins in Karnataka, the Suvarnavati (C3) and Shimsha (C4) Sub-basin has a higher and growing sediment production. Suvarnavati Sub-basin having Actual Evapotranspiration (AET) and Potential Evapotranspiration (PET) hass increased trend due to climate change and Soil Moisture condition.

The research findings can improve changes in hydrological variables and making decisions about regional agricultural water management. SWAT model development should be focused on the application in specific hydrologic sit-uations (such as the occurrence of flood waves and improvement of the base runoff component).

Further Research. Scope for further research is ongoing, collection of field records coupled with the initial watershed model will yield future opportuni-ties to test model estimates of groundwater movement, pesticide and bacteria transport, nutrient cycling, erosion, and non-point pollutant flows against recorded data.

Acknowledgements The authors are grateful for the assistance provided by the Department of Civil Engineering, UVCE, Bangalore University, Karnataka, India. We would also like to thank the Advanced Centre for Integrated Water Resources Management (ACIWRM) and Government Departments for providing required data.

References

1. Gull, S., Ahangar, M., & Dar, A. (2017). Prediction of stream flow and sediment yield of lolab watershed using swat model. *Hydrology: Current Research, 8*(01), 1–9.
2. Lewarne, M. (2009). *Setting up arcswat hydrological model for the verlorenvlei catchment.* Ph.D. thesis [University of Stellenbosch, Stellenbosch].
3. Arnold, J. G., Srinivasan, R., Muttiah, R. S., & Williams, J. R. (1998). Large area hydrologic modeling and assessment part i: Model development 1. *JAWRA Journal of the American Water Resources Association, 34*(1), 73–89.
4. Rokhsare, R., Aazam, J., Ma Jid, A., Seyed, F., Manouchehr, H., Ahmad, J., & Karim, C. (2008). Application of a swat model for estimating runoff and sed-iment in two mountainous basins in central Iran. *Hydrological Sciences Journal, 53*(5), 977–988.
5. Simi´c, Z., Milivojevi´c, N., Prodanovi´c, D., Milivojevi´c, V., & Perovi´c, N. (2009). Swat-based runoff modeling in complex catchment areas: Theoretical background and numerical procedures. *Journal of Serbian Society for Computational Mechanics, 3*(1), 38–63.
6. Ning, J., Gao, Z., & Lu, Q. (2015). Runoff simulation using a modified swat model with spatially continuous hrus. *Environmental Earth Sciences, 74*(7), 5895–5905.

Recent Advances in Pluvial Flash Flood Forecasting of India

A. Brahm Parkash Yadav, B. S. K. Asok Raja, C. Rahul Saxena, D. Hemlata Bharwani, E. Ashok Kumar Das, F. Ram Kumar Giri, G. S. K. Manik, and H. Deepak Yadav

Abstract Flash floods are typically associated with high-intensity rainstorms with short response time. They have the potential to severely impact and damage communities at different climatic settings especially in a densely populated region of South Asia. Recent years witnessed an increased effort to understand the dynamics of flash floods with the availability of high-resolution hydro-meteorological and topographical data. Despite their scientific significance and social impacts, the fundamental processes triggering a flash-flood response are yet not fully understood. Operational flash flood warnings over small watersheds with high spatial and temporal resolution have become feasible with (a) the development of new approaches for the extraction of useful information from radar and satellite remotely sensed data and rain gauge sensors, (b) the availability of high resolution digital spatial data and the ability to derive useful hydrologic information through geographic information systems applications, (c) advances in computer technology for fast processing of data over large areas with high resolution by inclusion of uncertainty propagation computations (Adams and Pagano in Flash flood guidance system (3.4.5) in flood forecasting in the United States NOAA/National Weather Service, Academic Press, pp 249–310, 2016). In this context, the operationalization of South Asia Flash Flood Guidance Services is one such attempt to integrate the hydrological mechanisms causing saturation runoff in response to intense rainfall causing flash floods over the watersheds of Indian region. The purpose and objective is to provide location specific flash flood guidance up to watershed level on pluvial flash floods in the form of risks with 24-h lead time based on numerical weather forecasts and threats with 6 h of lead time based on near real-time observations based on 00,06,12 and 18 UTC observations. Though still evolving, the efficacy of South Asia Flash Flood Guidance System at operational level is validated by many accurately forecasted Threat and Risk potentials of various flash flood events in the South Asia region. This article presents the novel concept and efforts adapted in India, in assessing the capability of this state-of-the-art Flash Flood Guidance System developed to provide effective operational guidance on flash floods at watershed level. This work

A. Brahm Parkash Yadav (✉) · B. S. K. Asok Raja · C. Rahul Saxena · D. Hemlata Bharwani · E. Ashok Kumar Das · F. Ram Kumar Giri · G. S. K. Manik · H. Deepak Yadav
India Meteorological Department, New Delhi 110003, India

A. K. Dikshit et al. (eds.), *Innovative Trends in Hydrological and Environmental Systems*, Lecture Notes in Civil Engineering 234,
https://doi.org/10.1007/978-981-19-0304-5_44

enhances the utility of this system along with NOWCAST by the operational meteorologists and hydrologists by bridging the gap for better predictability of hydro-meteorological events like flash floods at local level.

Keywords Flash flood · Watersheds · Threshold runoff · FFG · Soil moisture · Bankfull

1 Introduction

In recent times, extremes of hydrological and meteorological events are increasing significantly impacting the society and economy. During 1995–2015, around 3062 flood disasters have occurred in the world where 43% accounts to natural disasters. In South Asia, India is one of the worst floods affected country in the world after Bangladesh which accounts to one-fifth of the global death counts. In India, Nepal and Bangladesh "flash floods triggered by heavy monsoon rains have killed dozens of people and dispatched more than a million" in the middle of June 2018 [1].

Flash floods are natural hazards that are defined as fast surface flows with high peak discharge values, often limited in their spatial extent [2]. The most frequent cause of this type of flood is heavy rainfall events [3]; hence, the expression pluvial flash flood is used [4, 5]. Pluvial flooding occurs when rainfall with a high intensity (high amount of precipitation during very short period) exceeds the infiltration capacity of soil, or the discharge capacity of drainage systems, and water flows uncontrolled in urban areas [6]. The rainfall-induced runoff and flow processes are highly complex and vary in space and time with respect to terrain and climate conditions [7]. According to World Meteorological Organization (WMO), flash floods are natural hydro-meteorological hazards with highest mortality rate (defined as the number of deaths per number of people affected) and cause devastating economic loss every year [8]. In terms of the human loss, natural disasters result in approximately 1.7 million fatalities between 1980 and 2016. More than 49% of these fatalities were due to geophysical events (earthquake, tsunami, volcanic activity), 26% were due to meteorological events (tropical cyclone, convective storm, lightening), 14% were due to hydrological events (floods, landslides), and 11% were due to climatological events (extreme temperature, drought, forest fire). Slightly less than 80% of the 16,500 disaster events that caused fatalities were hydrological or meteorological (39%) [9].

Due to the physical characteristics of convective heavy rainfall cells, the forecasting time of pluvial flash floods is, unlike river (fluvial) floods, very short [10]. Urbanization, altering processes in the hydrological response, and continuous changing climatic conditions are the main factors for an increasing intensity of pluvial flash floods. The findings of Intergovernmental Panel on Climate Change (IPCC, 2014) are indicating that there is a possibility of more flood occurrences with short time period as both the frequency and intensity of extreme rainfall are expected in increase. However, the precise relationship between climate change and hydrological extreme events remains insufficiently researched [11].

Simultaneously, urban populations are increasing at an unprecedented rate, generating higher risks due to flood hazards [12]. According to The World Bank, it is expected that by 2050 more than 70% of the world's population will live in urban areas [13]. Current early warning systems (EWS) are based largely on measuring rainfall intensity or monitoring of rainfall over a spatial extent, water levels, whereby the real danger due to torrential flow is just as insufficiently considered as the vulnerability. In South Asia, there exists no formal platform, or EWS capable for flash floods, there is a lack of general capacity to develop effective warnings for these quick response events. Although there is a huge demand for understanding flash floods, and there are upcoming discussions about strategies to adapt to climate change and new concepts for water sensitive cities [9], there is a lack of effective information systems for risk assessment [4].

2 Background

Recognizing that flash floods have a particularly disastrous impact on lives and properties of the affected populations, the Fifteenth WMO Congress had approved the implementation of a Flash Flood Guidance System (FFGS) project with global coverage that had been developed by the WMO Commission for Hydrology (CHy) jointly with the WMO Commission for Basic Systems (CBS) and in collaboration with the US National Weather Service, the US Hydrologic Research Center (HRC) and USAID/OFDA [14].

The South Asia Flash Flood Guidance System (SASIAFFGS) is a part of Global Flash Flood Project Initiative by World Meteorological Organization (WMO), in partnership with the U.S. Agency for International Development/Office of the U.S. Foreign Disaster Assistance (USAID/OFDA), the National Weather Service (NWS) of the U.S. National Oceanic and Atmospheric Administration (NOAA), and the Hydrologic Research Center (HRC). The Flash Flood Guidance is a robust system designed to provide the necessary products in real time to support the development of warnings for flash floods about 6–24 h in advance at the watershed level with resolution of 4kmx4km for the Flash Flood-prone South Asian countries (see Fig. 1), viz. India, Nepal, Bhutan, Bangladesh and Sri Lanka.

The FFGS is a dynamic automatic information system used as an impact-based guidance tool, designed to provide the necessary products to support the development of warnings for flash floods from rainfall events through the use of remote sensed precipitation (e.g., radar and satellite-based rainfall estimates) and hydrological models. In order to assess the threat potential and likelihood of a local flash flood, the FFGS is designed to allow product adjustments based on forecaster experience with local conditions, incorporation of other information (e.g., Numerical Weather Prediction output) and any last-minute local observations (e.g., non-traditional rain gauge data) or local observer reports [2].

Fig. 1 Member nations covered under South Asia Flash Flood Guidance System

3 Materials and Methods

3.1 Hydrological Cycle

The hydrological cycle has many interconnected components, with runoff connecting precipitation to bodies of water. Surface runoff is precipitation that does not infiltrate into the soil and runs across the land surface into surface waters (streams, rivers, lakes or other reservoirs). Surface runoff varies by time and location, with about one-third of the precipitation that falls on land turning into runoff; the other two-thirds is evaporated, transpired or infiltrated into the soil. By returning excess precipitation to the oceans and controlling how much water flows into stream systems, runoff is important in balancing the hydrological cycle.

The water balance equation governs the hydrological cycle by describing the flow of water into and out of a system for a specific period of time (shown in Eq. 1)

$$\mathrm{Qss} = P - \mathrm{ET} - \Delta S - \Delta G \quad (1)$$

where P is precipitation, Qs is surface runoff, ET is evapotranspiration, ΔSM is change in soil moisture, and ΔGW is the change in groundwater storage).

The amount of surface runoff is influenced by soil properties, land cover, hill-slope, vegetation, and storm properties such as rainfall duration, amount, and intensity. Runoff is generated by a combination of two mechanisms; saturation excess and infiltration excess [15]. Saturation excess occurs when the soil becomes fully saturated with water, exceeding the water holding capacity of the soil; when the surplus rainfall can no longer be held in the soil, the water is directed to another location through overland flow [16].

Infiltration excess occurs when rainfall intensity exceeds the maximum rate that water can infiltrate into the soil, and water must flow over land to a different area [15].

Excess rainwater flowing over land picks up debris and chemicals along the flow path. The debris may include sediments, organic matter, nutrients, pesticides, and other materials which impact the quality of receiving surface water [17]. Surface runoff is therefore an important area of interest for monitoring water resources, as well as solving water quality and quantity problems such as flood forecasting and ecological and biological relationships in the water environment [18]. Runoff is also the main driver in contaminant transport due to excess nutrients and pesticides from agricultural lands being washed into waterways by rain events. High runoff rates, along with unmanaged drainage systems, cause flooding and erosion that damage vegetation and manmade structures [17].

Sediment transport by runoff can change stream morphology and alter stream biodiversity. When runoff reaches the stream, along with the transported sediment, it is added to the natural baseflow of the stream. Baseflow supplies stream channels with water in the absence of runoff or precipitation events, creating a persistent habitat for aquatic life that responds slowly to precipitation events.

3.2 Physical Modeling

Physical models, also called process-based or mechanistic models, are based on the understanding of the physics related to the hydrological processes. Physically based equations govern the model to represent multiple parts of real hydrologic responses in the catchment. The general physics laws and principles used include water balance equations, conservation of mass and energy, momentum, and kinematics. St. Venant, Boussinesq's, Darcy, and Richard's are some of the equations adopted by physical models.

The semi-discrete form of the St. Venant Equation below is used in the Penn State Integrated Hydrological Modeling System:

$$\left(\frac{\partial h}{\partial t} = P_o - I - E_o - Q_{oc} + \sum_{j=1}^{3} Q_s^{IJ}\right)_i$$

In Equation. Where $\partial h/\partial t$ is water depth at time t, Qsij is the surface flow from element i to j, P_o is the precipitation, I is infiltration, E_o is evaporation, and Q_{oc} is the interaction between overland flow and channel routing.

Spatial and temporal variations within the catchment are incorporated into physical models. A physical model has a logical structure similar to the real-world system. The greatest strength of a physical model is the connection between model parameters and physical catchment characteristics which make it more realistic. They are best used when precise data are available, physical properties of the hydrological processes are accurately understood, and applied on fine scales due to computational time. A large number of physical and process parameters are needed to calibrate the model.

The spatial processes in runoff models provide a means of representing the catchment for modeling. They are based on input data and how runoff is generated and routed over the catchment. Variability in geology, soils, vegetation, and topography affect the relationship between rainfall and runoff within a catchment and should be considered in modeling [19]. Lumped models do not consider spatial variability within the catchment; semi-distributed models reflect some spatial variability; and fully distributed models process spatial variability by grid cells.

Lumped models treat the catchment area as a single homogenous unit. Spatial variability of catchment parameters is disregarded in lumped models [20]. Averaged values over the catchment are used such as mean soil storage and uniform precipitation amounts [19]. The catchment characteristics are set as equal for the entire area and often cause over- or under-parameterization. A single runoff output value is calculated at the river outlet point of the catchment area in these models.

3.3 Lumped Model

A lumped model is designed to simulate total runoff and stream flow at the outlet point, not specific flows within a catchment. For this reason, lumped models adequately simulate average runoff conditions with fast computational times. Average and annual runoff conditions produced by lumped models are used for regulatory purposes that look at long-term conditions. Lumped models include a lot of assumptions about the hydrological processes. Because of these assumptions, lumped models have a tendency to over- or under-estimate runoff values. They do not consider changes within a watershed, or if the changes affect the runoff process.

Land use changes may alter the runoff process in specified areas, but a lumped model averages these over the entire catchment. Inputs in a lumped model are all lumped or "averaged" data. Such data are relatively easy to attain or create by averaging data across the study area. All data including input, output, and parameters are constant over space and time in a lumped model. By assuming homogeneity over the catchment, lumped models lose spatial resolution of the data. For example, rainfall and runoff patterns vary over space and time but, in lumped models they are considered stationary. "There is spatial variability in rainfall across a catchment which is not captured" when used in a lumped model. Empirical and conceptual models are usually run spatially as lumped. Due to the many assumptions and averaged conditions that lumped models incorporate, they do not represent large watersheds and catchments accurately [20].

3.4 SAC-SMA Hydrological Modeling

The hydrologic model consists of time-continuous components that simulate the evolution of the state variables that represent the natural system, e.g., snow mass and energy properties, and soil and channel water contents at a single time step. The snow model component is forced by mean areal temperature and precipitation. The mean areal temperature serves as an index for energy exchange across the snow–air interface and is used to compute snow melt heat exchange and to determine snowfall from rain. During periods with no rain, temperature is used as a melt index with a seasonally varied melt factor. For rain on snow events, a simplified energy balance approach that considers temperature and precipitation data are used. The model also accounts for snow–soil interface interaction, and the liquid storage of the pack, together with delay and attenuation due to liquid transmission through the snow pack.

The operational hydrologic model used to produce the operational short-term forecasts consists of the Sacramento soil moisture accounting model SAC-SMA Burnash et al. [21] to simulate runoff production, the SNOW-17 model Anderson to simulate snow melt and snow water equivalent, and a routing scheme that conveys runoff to the location of interest. Unit hydrograph procedures are used to route the flows for simulations of the operational model with historical data, while the cascade of linear conceptual reservoirs [22] is used to route the flows for the real-time short-term flow forecasts and for the ensemble stream flow forecasts.

The SAC-SMA is a continuous model of the wetting and drying processes in the soil. It simulates changes in the water content of upper and lower soil zones and in conceptual storages of tension water that can only be depleted by evapotranspiration and free gravitational water. The model is forced by snow melt and/or precipitation and reference potential evapotranspiration and produces as output runoff from impervious and pervious areas, soil water content, and actual evapotranspiration.

The previous area runoff consists of surface and subsurface runoff, with surface runoff generation depending on both the saturation levels of the soils and the soils ability to percolate water to deeper layers. A variable impervious area model function allows for the increase and decrease of the saturated soil area near the streams. The discrete-time form of the model is described in Burnash et al. [21], while a continuous-time form [22]. Model parameters consist of 13 SAC-SMA parameters, 7 SNOW-17 parameters, and the unit hydrograph coordinates or the

Table 1 State property name used in SACSMA model

Property name	Description
UZTWC	Upper zone tension water contents; units of MM
UZFWC	Upper zone free water contents; units of MM
LZTWC	Lower zone tension water contents; units of MM
LZFSC	Lower zone free supplemental contents; units of MM
LZFPC	Lower zone free primary contents; units of MM
ADIMC	Tension water contents of the ADIMP area; units of MM. If not known then use UZTWC + LZTWC
FGIX	Initial value of the frost index; units of DEGC. If starting with no frozen ground the value is zero
UNIT	This entry is optional. The presence of "UNIT = METRIC" in state file is only for visual information, because the state is always in METRIC units, never in ENGLISH units; An exception will be thrown if "UNIT = ENGLISH" is present in the state file

two parameters of the kinematic channel routing model. Estimation of the values of these parameters for specific drainage basins is performed by the RFCs following an interactive calibration methodology that recognizes explicitly the function of the physical/conceptual components of the models, e.g., see NWS, Koren et al., Smith et al., Shamir et al. [23], regarding SAC-SMA; and Anderson, regarding snow model parameter estimations.

SAC-SMA uses an XML representation of model parameters where each parameter is captured within a separate XML tag. Like all the other models, SAC-SMA model states are also defined in a property file format. An example is shown below. The sequence of property names in the state file or in the Table 1 has no any significance. The model state property names are shown in Table 1.

The SAC-SMA model has six Mods; all the Mods must be within the observation time period, not applied in the future period. All the new state values are checked against the maximal values specified in the parameter XML file.

The Flash Flood Guidance (FFG) value is the actual amount of rainfall needed over a prescribed area, for a given duration to bring small streams to bank full flood conditions. Based on these FFG values, the system automatically generates both Risk and Threat perception of a Flash Flood over a watershed at varied time intervals.

3.5 *Operational Concept of Flash Flood Guidance System*

The Flash Flood Guidance System (FFGS) ingests real-time satellite precipitation data, on-site gauge precipitation and temperature data, model-forecasted precipitation, and, on the basis of available spatial databases, produces flash-flood-occurrence diagnostic and prognostic indices over small flash flood-prone catchments. The indices are updated regularly, and they include: mean areal precipitation (1-, 3-, 6-, and 24-h accumulations); gauge-interpolated mean areal precipitation (6- and 24-h accumulations); gauge-interpolated mean areal temperature (6-h average), areal snow cover fraction, snow water equivalent, snow melt (24- and 96-h accumulations), upper soil moisture saturation fraction; flash flood guidance (for 1, 3 and 6 h in the future); forecast mean areal precipitation (1, 3, 6, 24 h); imminent flash flood threat (for the last 1, 3 and 6 h); persistence flash flood threat (for 1, 3, and 6 h into the future), forecast flash flood threat (for 1, 3 and 6 h into the future) and flash flood risk (for 12, 24 and 36 h into the future).

In this context, flash flood guidance of duration T for a small catchment is the volume of rainfall accumulated over a future period T that is just enough to cause bankfull flow at the outlet of the draining stream. Thus, if the volume of accumulated rainfall is greater than flash flood guidance during the future period, then overbank flow is expected (minor flash flooding). The diagnostic flash flood guidance index may then be used with nowcast or forecast rainfall volumes of the appropriate durations to identify the likelihood of flash flooding at the outlet of specific small catchments.

The physical basis of the data integration within the flash flood guidance system and theoretical performance analysis has been described in Carpenter et al. [24], and Georgakakos [2].

The primary mission of the South Asia Flash Flood Guidance (SAsiaFFG) System is to provide real-time informational guidance products pertaining to the imminence of potential small-scale flash flooding. The SAsiaFFG System is installed on specially configured equipment, hosted at the India Meteorological Department (IMD), and is responsible for the real-time data acquisition and ingest of all input products, flash flood guidance computations, and numerical product generation for various fields and various durations. The flash flood guidance systems (FFGS) provide the user with remote real-time access to the FFGS products for online review and/or download onto their local computer for potential adjustments and further application in forecasting activities. The schema of the guidance system is illustrated precisely in the Fig. 2.

3.6 SAsiaFFGS System Configuration and Interface

The SAsiaFFGS system resides on two cooperative and specially configured servers hosted at IMD: the FFGS Computational Server and the FFGS Dissemination Server. The FFGS Computational Server (or CS) is responsible for continuous real-time data acquisition, pre-processing, model processing and the generation output product exports. After these phases of processing have completed, the real-time and recent historical products are provided to authorized users through a secure web interface, the dissemination interface, which is hosted on the FFGS Dissemination Serve (or DS) and may be accessed by the forecaster's web browser

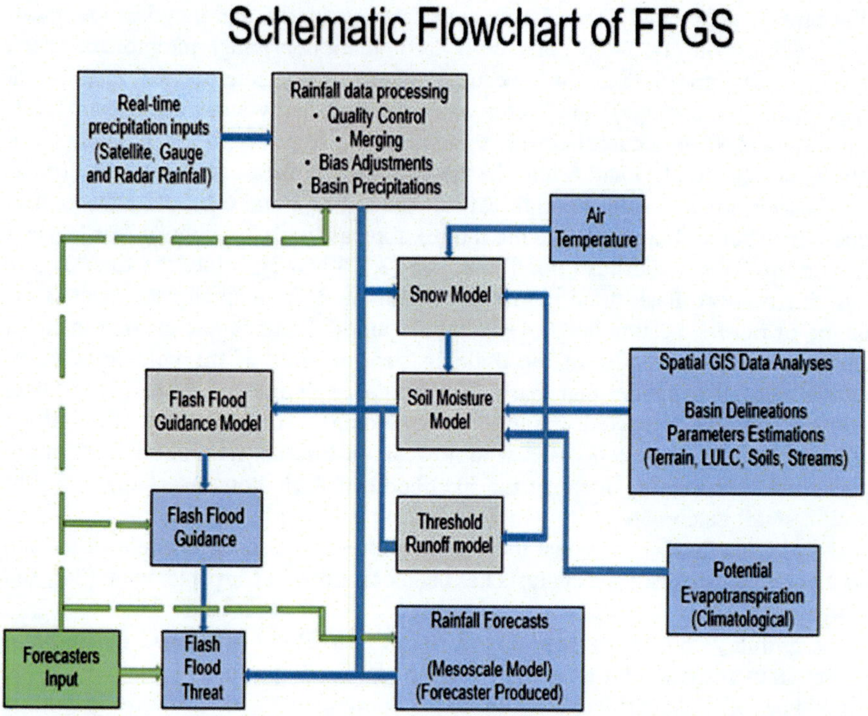

Fig. 2 Technical components of Flash Flood Guidance System Schema

using HTTPS. This user interface is accessible through a password protected website and is composed of two main parts: the Product Console and the Dashboard. The primary page for hydrologists and meteorologists on duty is the Product Console as in Fig. 3. The Product Console presents the complete system products in overview features access links to detailed output which forecasters are trained to use.

The SAsiaFFGS The primary software components of the SAsiaFFG System are its CentOS Linux operating system, a PostgreSQL relational Database server, an Apache web server and the FFGS software. (It should be noted that the PostgreSQL and Apache servers are both standard components of and provided by the Linux distribution content.)

The System Design and Processing flow of the FFGS, designed to carry out all operational functions toward the provision of flash flood guidance information in a completely automated fashion. This highly automated aspect of the FFGS design allows the users and system administrators to focus their attention on system output product application and quality control. Toward this end, the primary point of routine user interaction is through the web-based Dissemination Interface data product console.

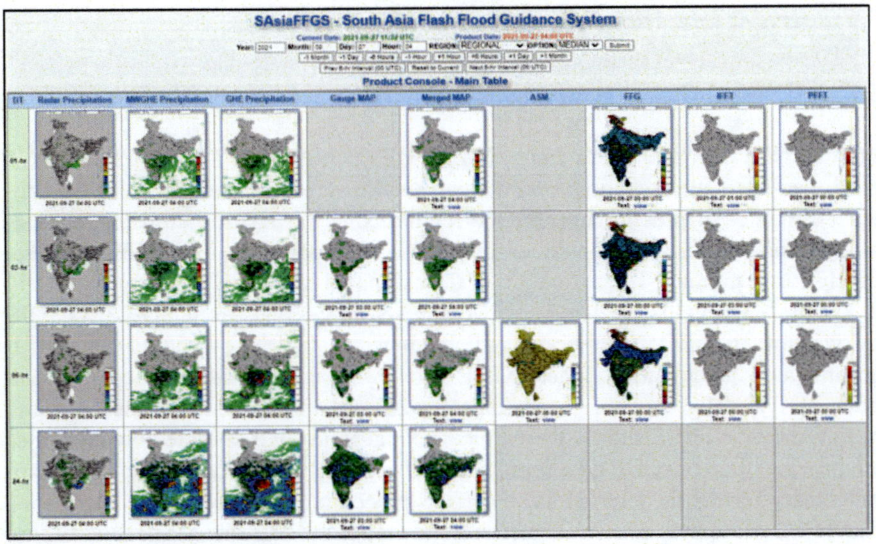

Fig. 3 SAsiaFFGS product console interface page

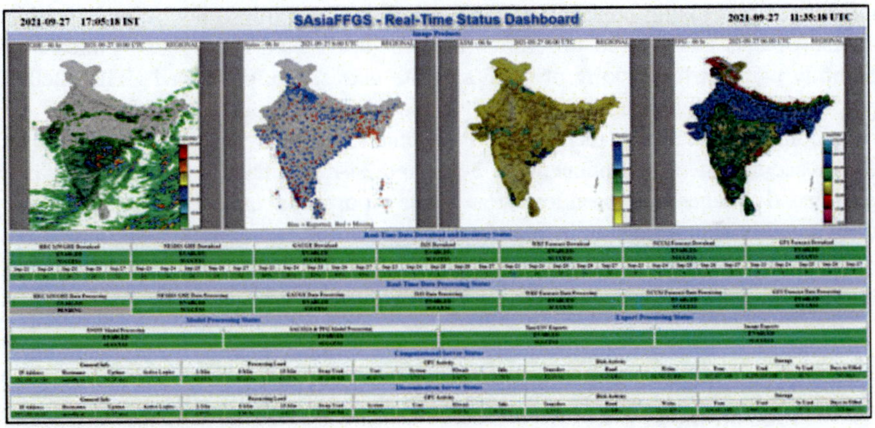

Fig. 4 SAsiaFFGS dashboard page

The Dashboard is designed primarily for system administrators and provides a summary of system processes. The Dashboard also helps forecasters to get a brief inspection of the system status (Fig. 4). Information is available for each 28,780 small watersheds delineated through 30 m DEM from 153 sub-basins of India. There are about 17 types of dynamic products and 19 types of base layers available in this system for the forecasters to visualize their area of interest and take quick decision accordingly.

3.7 Real-Time Input Data to SASIAFFGS

3.7.1 Gridded Global Hydro-Estimator (GHE) Satellite Precipitation

Infrared (IR) data combined from GOES, METEOSAT and MTSAT satellites is used together with operational numerical weather prediction model fields to produce estimates of precipitation for the region using the Hydro-Estimator methodology. This product is known as the Global Hydro-Estimator (GHE) Satellite Precipitation product. The resolution of the estimated precipitation is approximately $4 \times 4\ \text{km}^2$. The fields are being retrieved on hourly intervals from NOAA NESDIS by way of an HRC FTP server and are ingested hourly by the FFGS as 1-h accumulated precipitation. These GHE satellite-based estimates, processed each hour, are typically available approximately 20 min after the turn of the hour and retrieved by the FFGS shortly thereafter. The real-time GHE satellite download files are provided in a NetCDF data format and subsequently pre-processed to a standard raw binary format for internal use by FFGS processing and modeling components. The FFGS processing phases incorporate the real-time GHE satellite data according to parameterized strategies specified by the forecaster in order to produce the 1-h mean areal precipitation (MAP) estimates which serve as the precipitation forcing for the soil and FFG modeling components. Depending on the configuration within the system processing control, the GHE satellite data values may be processed to adjust for bias and filtered to remove areas of designated anomalies by means of monthly parametric masking of GHE satellite pixels. The processed GHE satellite precipitation is subsequently merged with additional input products during the development of real-time merged MAP estimates. The real-time GHE satellite data is also accumulated into independent 3-, 6- and 24-h grid data products from which unadjusted images are produced for display purposes only in addition to the unadjusted hourly image as in Fig. 5.

More information on the Global Hydro-Estimator (GHE) Satellite Precipitation product can be found at [25]: http://www.ospo.noaa.gov/Products/atmosphere/ghe.

3.7.2 Gridded Microwave Adjusted GHE (MWGHE) Satellite Precipitation

Following initial acquisition of the NOAA NESDIS Global Hydro-Estimator (GHE) Satellite Precipitation product, HRC performs additional post-processing bias-adjustment using the CPC MORPHing (CMORPH) technique data product to produce the Microwave-adjusted Global Hydro-Estimator (MWGHE) Satellite Precipitation input product to the FFGS. Analogous to the GHE input product, the resolution of the estimated precipitation is approximately $4 \times 4\ \text{km}^2$. The fields are being retrieved on hourly intervals from the HRC FTP server and are ingested hourly by the FFGS as 1-h accumulated precipitation. These MWGHE satellite-based estimates, processed each hour, are typically available approximately

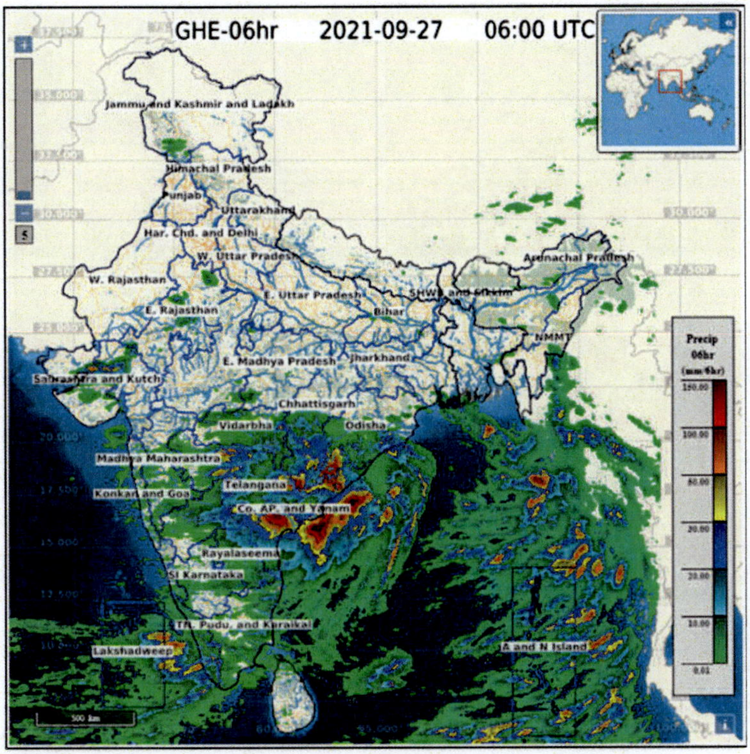

Fig. 5 Gridded global hydro-estimator (GHE) product

45 min after the turn of the hour and retrieved by the FFGS shortly thereafter. The real-time MWGHE download files are provided in a raw gridded binary data format and subsequently pre-processed to a standard raw binary format for internal use by FFGS processing and modeling components. The FFGS processing phases incorporate the real-time MWGHE satellite data according to parameterized strategies specified by the forecaster in order to produce the 1-h mean areal precipitation (MAP) estimates which serve as the precipitation forcing for the soil and FFG modeling components. Depending on the configuration within the system processing control, the MWGHE satellite data values may be processed to adjust for bias and filtered to remove areas of designated anomalies by means of monthly parametric masking of MWGHE satellite pixels. The processed MWGHE satellite precipitation is subsequently merged with additional input products during the development of real-time merged MAP estimates. The real-time MWGHE satellite data is also accumulated into independent 3-, 6- and 24-h grid data products from which unadjusted images are produced for display purposes only in addition to the unadjusted hourly image. The images and text provide gridded 1-, 3-, 6- and 24-h accumulations of satellite-based rainfall estimates (mm) ending on the current hour

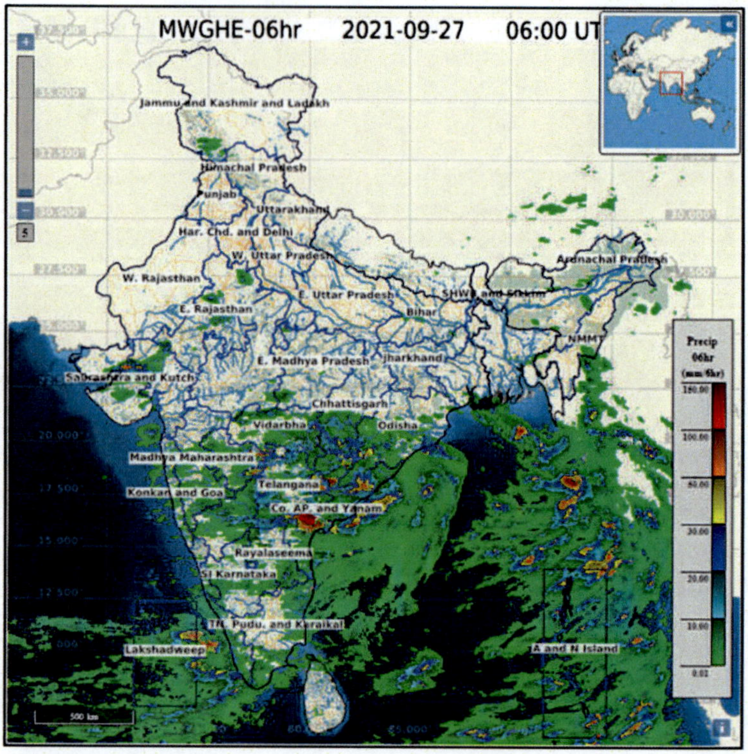

Fig. 6 Gridded microwave adjusted GHE (MWGHE) product

from the NOAA-NESDIS Global Hydro-Estimator (infrared-based) and adjusted by the NOAA-CPC CMORPH microwave-based satellite rainfall product. The satellite-based rainfall estimates are provided on a grid which is displayed over a background of the system sub-basin boundaries as in Fig. 6.

The MWGHE data products are updated every hour with a latency of approximately 45 min and are not bias-corrected (see Merged MAP product). This product is provided for visual quality control assessment of the adjusted satellite input.

- MWGHE 01-hr: Accumulations of rainfall estimates from the microwave-adjusted Global Hydro-Estimator over the last hour ending on the current navigation hour. (mm/1 h)
- MWGHE 03-hr: Accumulations of rainfall estimates from the microwave-adjusted Global Hydro-Estimator over the last 3 h ending on the current navigation hour. (mm/3 h)
- MWGHE 06-hr: Accumulations of rainfall estimates from the microwave-adjusted Global Hydro-Estimator over the last 6 h ending on the current navigation hour. (mm/6 h)

MWGHE 24-h: Accumulations of rainfall estimates from the microwave-adjusted Global Hydro-Estimator over the last 24 h ending on the current navigation hour. (mm/24 h).

3.7.3 Regional/ National Gauge Surface Meteorological Data

Reported gauge precipitation and temperature data from the various stations within the region of system coverage are collected by IMD and provided on the designated external data exchange service for retrieval by the FFGS as in Fig. 7. These stations report provide the FFGS with accumulations of precipitation and temperature observations which are then extracted, quality-controlled and processed for model application.

The gauge precipitation observations are used with corresponding satellite pixel comparisons as part of the bias adjustment processing for these products. Additionally, the precipitation values, when enabled for a given gauge, are included in the gauge interpolation processing to produce the gauge-based mean areal precipitation product (referred to as GMAP). The GMAP product is produced at the 6- and 24-hourly processing intervals from which text and images are produced and is included in the generation of the Merged MAP and, subsequently, as input to the system models.

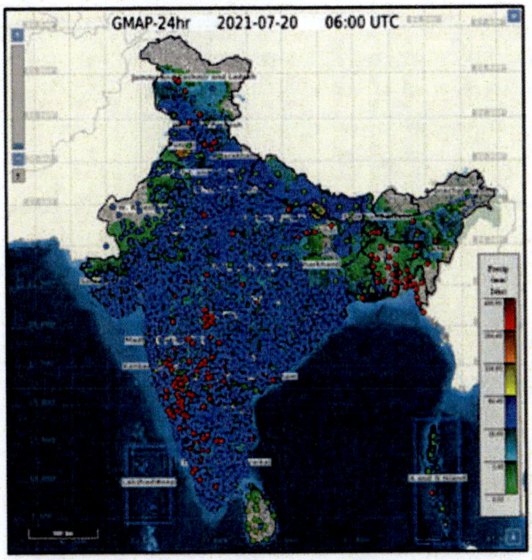

Legend		Description
Red Dot	●	Missing Data
Blue Dot	●	Precipitation Data
Green Dot	●	Precipitation & Temperature Data
Orange Dot	●	Temperature Data

Fig. 7 Regional Gauge in situ data

3.7.4 Gridded Numerical Weather Prediction Data

IMD WRF

Gridded WRF model numerical precipitation and temperature forecasts of 3-h lead time intervals out 48 h are generated by IMD twice per day at 00 and 12 UTC and provided for application within the FFGS. These WRF model forecasts are retrieved by the FFGS and are processed for each update cycle to provide 3-, 6- and 24-hourly total precipitation on the original model computational grid. The forecasts are each retrieved by the FFGS and subsequently pre-processed into a standard raw binary format for application within system processing and modeling components. Each processed model forecast precipitation domain is used to derive a distinct Forecast Mean Areal Precipitation (FMAP) data product over the sub-basins. Additionally, the gridded precipitation and temperature forecasts lead time are applied in the flash flood risk outlook module processing to generate the associated FFR data product index.

IMD GFS

Gridded IMD model numerical precipitation and temperature forecasts of 3-h lead time intervals out 48 h are generated by IMD twice per day at 00 and 12 UTC and provided for application within the FFGS. These GFS model forecasts are retrieved by the FFGS and are processed for each update cycle to provide 3-, 6- and 24-hourly total precipitation on the original model computational grid. The forecasts are each retrieved by the FFGS and subsequently pre-processed into a standard raw binary format for application within system processing and modeling components. Each processed model forecast precipitation domain is used to derive a distinct Forecast Mean Areal Precipitation (FMAP) data product over the sub-basins. Additionally, the gridded precipitation and temperature forecasts lead time are applied in the flash flood risk outlook module processing to generate the associated FFR data product index.

NCMRWF NCUM

Gridded NCUM model numerical precipitation and temperature forecasts of 3-h lead time intervals out 48 h are generated by IMD twice per day at 00 and 12 UTC and provided for application within the FFGS. These NCUM model forecasts are retrieved by the FFGS and are processed for each update cycle to provide 3-, 6- and 24-hourly total precipitation on the original model computational grid. The forecasts are each retrieved by the FFGS and subsequently pre-processed into a standard raw binary format for application within system processing and modeling components. Each processed model forecast precipitation domain is used to derive a distinct Forecast Mean Areal Precipitation (FMAP) data product over the sub-basins. Additionally, the gridded precipitation and temperature forecasts lead time are applied in the flash flood risk outlook module processing to generate the associated FFR data product index.

3.8 Satellite Bias Adjustment

The bias adjustment component performs the estimation of rainfall bias for both the MWGHE and GHE input products. The bias factor for each input product is estimated using adaptive state estimation (a combination of Kalman filtering with a stochastic approximation algorithm) and a window of 120 6-hourly intervals (30 days) to develop stable estimator parameters prior to evaluating the current bias factor. The bias factor is based on the ratio of the real-time gauge reports averaged over the region to the average of the corresponding embedding grid boxes of the corresponding input product for a given hour. A number of conditional processing tests ensure that significant and widespread rainfall is present during the selected hours for which bias adjustment is performed. The real-time bias adjustment processing can be disabled in the processing control configuration and substituted solely with static monthly climatological bias factors. The default configuration of the FFGS enables the processing and application of dynamic real-time bias adjustment factors.

4 Deliverables

4.1 Diagnostic Products

4.1.1 Merged Mean Areal Precipitation (MAP)

The Merged MAP processing component derives mean areal precipitation estimates by selecting the best-available mean areal precipitation input product for each sub-basin from the bias-adjusted microwave-adjusted Global Hydro-Estimator satellite precipitation, bias-adjusted Global Hydro-Estimator satellite precipitation or the interpolated gauges with preference for selection in that order. The Merged MAP data product serves as the precipitation input to the Snow and Sacramento Soil Moisture Accounting Models. The merging process considers available data and parametric indicators and controls for the incorporation of each data source on a sub-basin level. The merging process selects as the Merged MAP value the bias-adjusted MWGHE precipitation. For any sub-basin masked for the MWGHE precipitation input products, or in the event that the MWGHE input products is unavailable, the merging process selects the GHE satellite precipitation as the Merged MAP value and so forth. During various scenarios of data

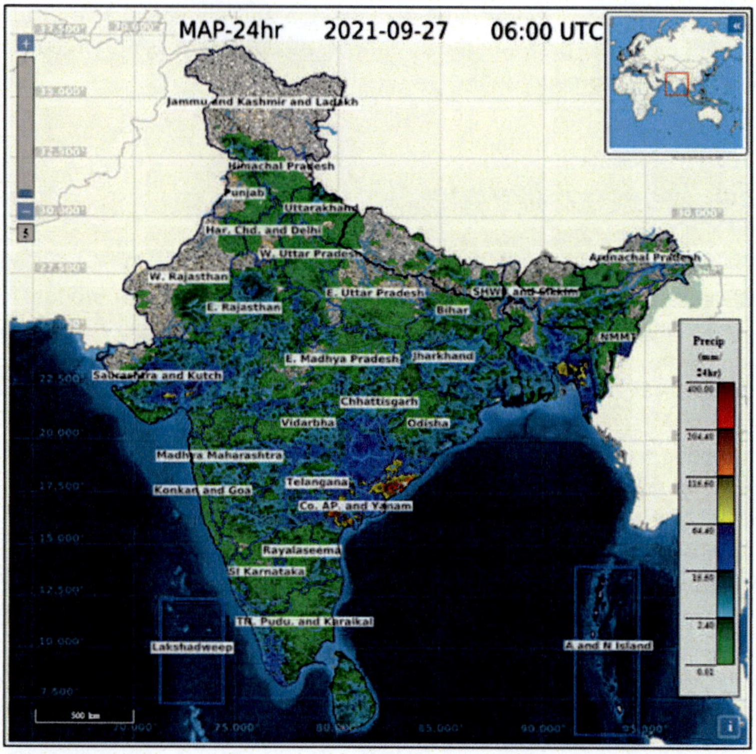

Fig. 8 Merged meal areal precipitation product

availability, the FFG System's auto-reprocessing window will incorporate the arrival of latent satellite or gauge data up to 24 h past the time of observation and optimize the reprocessing of merged MAP values (Fig. 8).

The diagnostic merged areal precipitation is computed by taking the weighted average using Inverse Distance Weighted technique based on IMD rain gauge sensors, quantitative satellite precipitation estimates of Infrared-based Global Hydro Estimator (GHE) from NOAA and Microwave-adjusted GHE using CMORPH algorithm from HRC, Real-time Doppler Weather Radar-Based estimates. Further, a lumped Sacramento soil moisture accounting model is used operationally to produce flash flood guidance estimates of a given duration from threshold runoff estimates at every watershed using these meteorological parameters.

4.1.2 SACSMA Average Soil Moisture (ASM)

The images and text provide soil water saturation fraction (dimensionless ratio of contents over capacity) for the upper zone (approximately 20–30 cm depth) of the

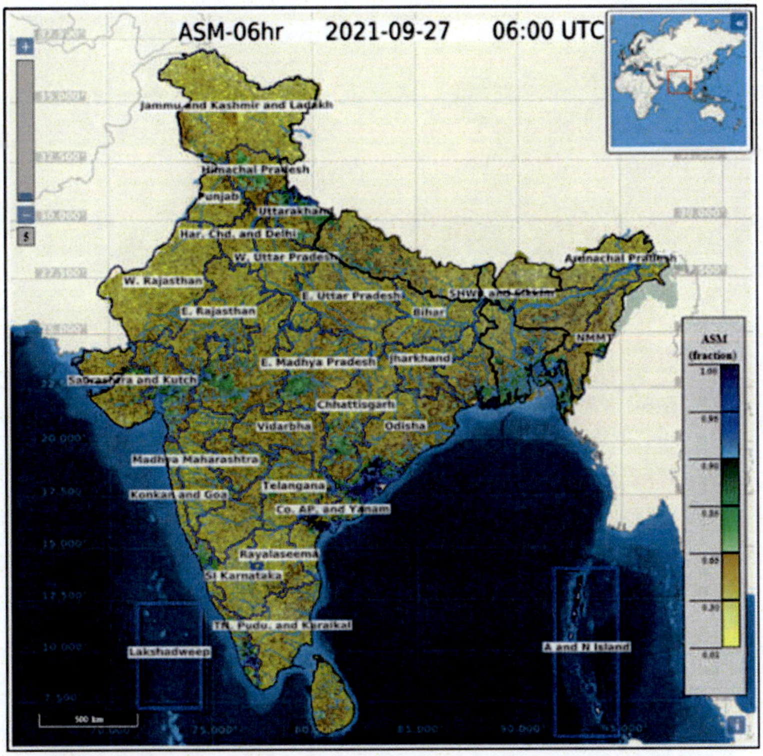

Fig. 9 Average soil moisture saturation fraction

Sacramento Soil Moisture Accounting Model for each of the sub-basins. The products are updated every hour.

- ASM 06-hr: Average soil water saturation (fraction of soil capacity in the upper zone) for model processing hour (Fig. 9).

4.1.3 Flash Flood Guidance (FFG)

Flash flood guidance of duration T for a small catchment is the volume of rainfall accumulated over a future period T that is just enough to cause bankfull flow at the outlet of the draining stream. The minimum this value over a particular catchment at T indicates the higher likelihood to cause bankfull runoff, as the saturation excess with emerging exceedance of rainfall in T plus times. The images and text provide 1-, 3- and 6-h Flash Flood Guidance (mm) for each sub-basin. For a given sub-basin

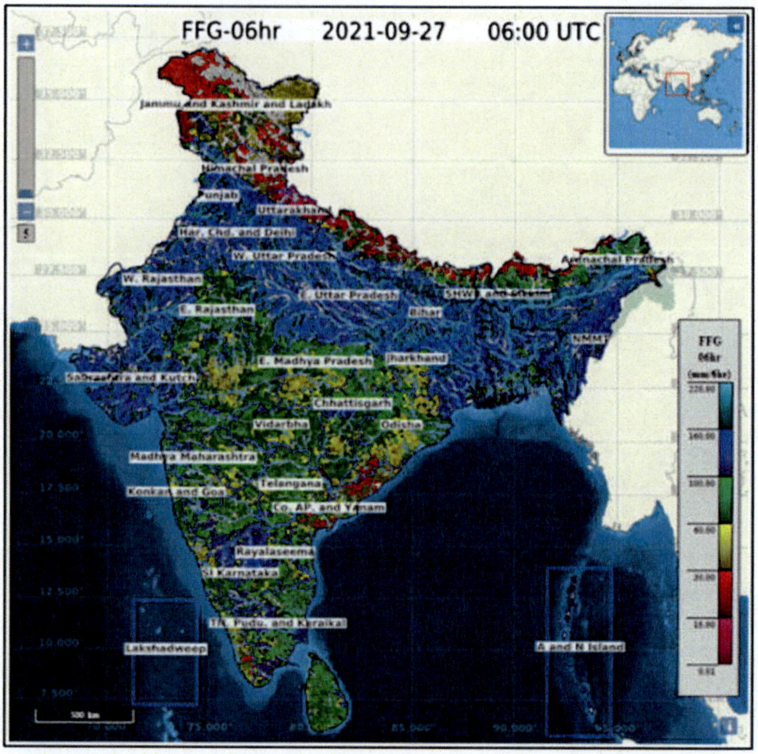

Fig. 10 Flash flood guidance (FFG) product (mm/h)

and duration (1-, 3- or 6-h), the FFG value indicates the total volume of rainfall over the given duration which is just enough to cause bankfull flow at the outlet of the draining stream. Consequently, rainfall volumes of the same duration that are greater than the FFG value indicate a likelihood of overbank flows at the draining stream outlet. Each of the FFG products is updated every hour. This product is appropriate to use in real time with nowcasts or forecasts of rainfall and other local information to estimate the risk of flash flooding in the sub-basins.

- FFG 01-hr: Required rainfall to cause bankfull flow over the next 1 h following the most recent (current) model processing hour. (mm/1 h)
- FFG 03-hr: Required rainfall to cause bankfull flow over the next 3 h following the most recent (current) model processing hour. (mm/3 h)
- FFG 06-hr: Required rainfall over the next 6 h to cause bankfull flow following the most recent (current) model processing hour. (mm/6 h) (Fig. 10).

4.2 *Threat Products*

See Figs. 11 and 12.

4.2.1 Imminent Flash Flood Threat (IFFT)

The images and text provide 1-, 3- and 6-h Imminent Flash Flood Threat (mm) for each sub-basin as in Fig. 12. The values indicate the difference of the Merged MAP of a given duration and the corresponding past model processing hour FFG of the same duration for a given sub-basin. In this sense, the IFFT value is considered a current "observation". The last 1-, 3- and 6-h durations of FFG are considered with current corresponding Merged MAP in the computation of IFFT.

- IFFT 01-hr: Difference of 01-hr FFG from a previous model processing hour and 01-hr MAP observed over the following 1 h (mm/1 h). The 1-h IFFT is valid at (01, 07, 13, and 19 UTC).

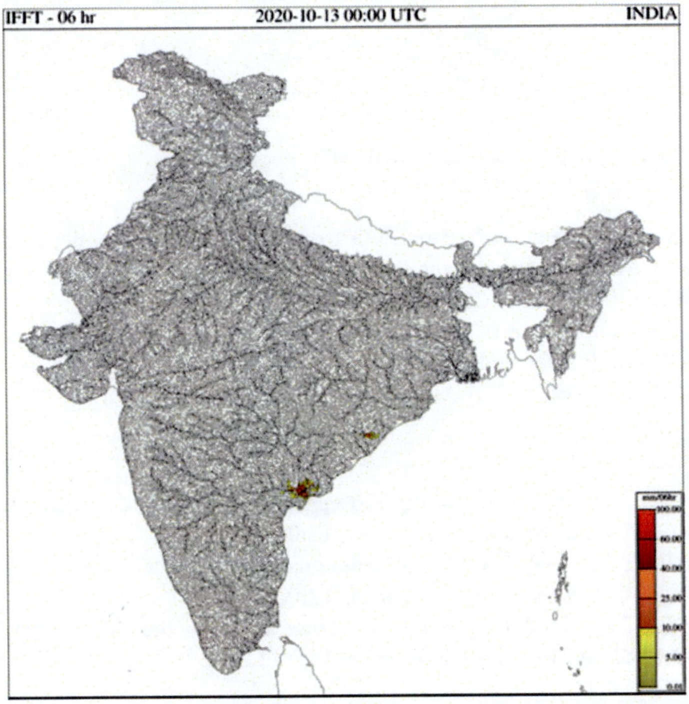

Fig. 11 Imminent flash flood threat product (mm/h)

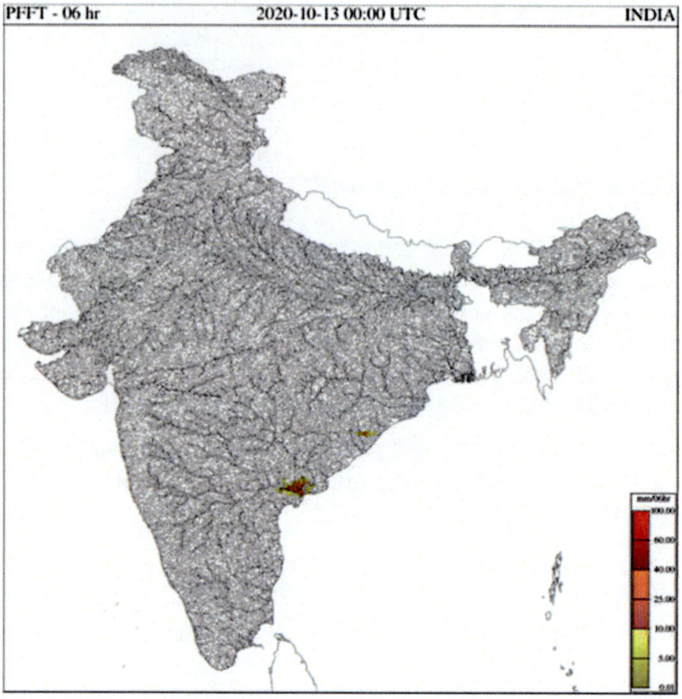

Fig. 12 Persistent flash flood threat product (mm/h)

- IFFT 03-hr: Difference of 03-hr FFG from a previous model processing hour and 03-hr MAP observed over the following 3 h (mm/3 h). The 3-h IFFT is valid at (03, 09, 15, and 21 UTC).
- IFFT 06-hr: Difference of 06-hr FFG from a previous model processing hour and 06-hr MAP observed over the following 6 h (mm/6 h). The 6-h IFFT is valid at (00, 06, 12, and 18 UTC).

For Example:

The 01-hr IFFT at 13:00 UTC = the difference between the 01-hr Merged MAP from 13:00 UTC and the 01-hr FFG from 12:00 UTC.
The 03-hr IFFT at 15:00 UTC = the difference between the 03-hr Merged MAP from 15:00 UTC and the 03-hr FFG from 12:00 UTC.
The 06-hr IFFT at 18:00 UTC = the difference between the 06-hr Merged MAP from 18:00 UTC and the 06-hr FFG from 12:00 UTC.

Additional examples (between FFG update cycles):

The 03-hr IFFT at 20:00 UTC = the difference between the 03-hr Merged MAP from 15:00 UTC and the 03-hr FFG from 12:00 UTC. Note that this 03-hr IFFT is

no longer valid at 20:00 UTC but will be updated with the 18:00 UTC FFG as noted in the following example.
The 03-hr IFFT at 21:00 UTC = the difference between the 03-hr Merged MAP from 21:00 UTC and the 03-hr FFG from 18:00 UTC, considered valid for 21:00 UTC.

In the images, an approximate measure of uncertainty in the IFFT estimates is indicated by the ranges in the color scale (with yellow indicating the range of values that are less likely to be of concern for flash flooding and with orange and red indicating progressively higher risk of flooding for the sub-basin of interest). The time stamp on each IFFT product represents the end of the valid period.

IFFT provides the forecaster with an idea of likely regions of imminent flash flood threats. Note that this set of products concerns the past rainfall and should be evaluated before using for warnings. IFFT is offered as a baseline product that must be carefully evaluated by the forecaster in real-time.

4.2.2 Persistence Flash Flood Threat (PFFT)

The images and text provide 1-, 3- and 6-h Persistence Flash Flood Threat (mm) for each sub-basin as in Fig. 12. The values indicate the difference of recent persisted Merged MAP of the given duration and the corresponding current FFG of the same duration for a given sub-basin. The last 1-, 3- and 6-h durations of Merged MAP are persisted and considered with current corresponding FFG in the computation of PFFT. In this sense, the PFFT is considered a “forecast” with persistence used as the rainfall forecast.

- PFFT 01-hr: Difference of 01-hr FFG for current model processing hour and current 01-hr MAP persisted for the next 1 h. (mm/1 h)
- PFFT 03-hr: Difference of 03-hr FFG for current model processing hour and current 03-hr MAP persisted for the next 3 h. (mm/3 h)
- PFFT 06-hr: Difference of 06-hr FFG for current model processing hour and current 06-hr MAP persisted for the next 6 h. (mm/6 h)

For Example:

The 01-hr PFFT at 12:00 UTC = the difference between the 01-hr Merged MAP from 12:00 UTC and the 01-hr FFG from 12:00 UTC, considered valid at 13:00 UTC.
The 03-hr PFFT at 12:00 UTC = the difference between the 03-hr Merged MAP from 12:00 UTC and the 03-hr FFG from 12:00 UTC, considered valid at 15:00 UTC.
The 06-hr PFFT at 12:00 UTC = the difference between the 06-hr Merged MAP from 12:00 UTC and the 06-hr FFG from 12:00 UTC, considered valid at 18:00 UTC.

Additional examples (between FFG update cycles):

The 03-hr PFFT at 17:00 UTC = the difference between the 03-hr Merged MAP from 12:00 UTC and the 03-hr FFG from 12:00 UTC, considered valid at 15:00 UTC. Note that this 03-hr PFFT is no longer valid at 17:00 UTC but will be updated with the 18:00 UTC FFG as noted in the following example.
The 03-hr PFFT at 18:00 UTC = the difference between the 03-hr Merged MAP from 18:00 UTC and the 03-hr FFG from 18:00 UTC, considered valid for 21:00 UTC.

In the images, an approximate measure of uncertainty in the PFFT estimates is indicated by the ranges in the color scale (with yellow indicating the range of values that are less likely to be of concern for flash flooding and with orange and red indicating progressively higher risk of flooding for the sub-basin of interest). Note that the time stamp on each PFFT product represents the beginning of the valid period.

Note that this set of products uses a rainfall forecast (persistence) and contains uncertainties. PFFT is offered as a baseline product that must be carefully evaluated by the forecaster in real time as illustrated in the Figs. 13 and 14.

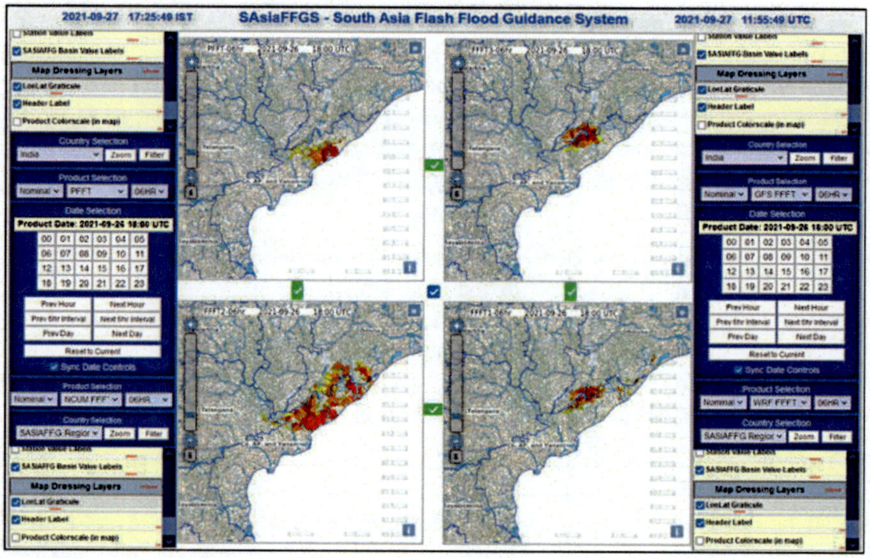

Fig. 13 Evaluation of all threat product—6 h over an area mm/hr)

Low		Can Occur with Uncertainty
Moderate		Moderate Confidence with substantial exeedance of RF
High		High Confidence with very substantial exeedance of RF

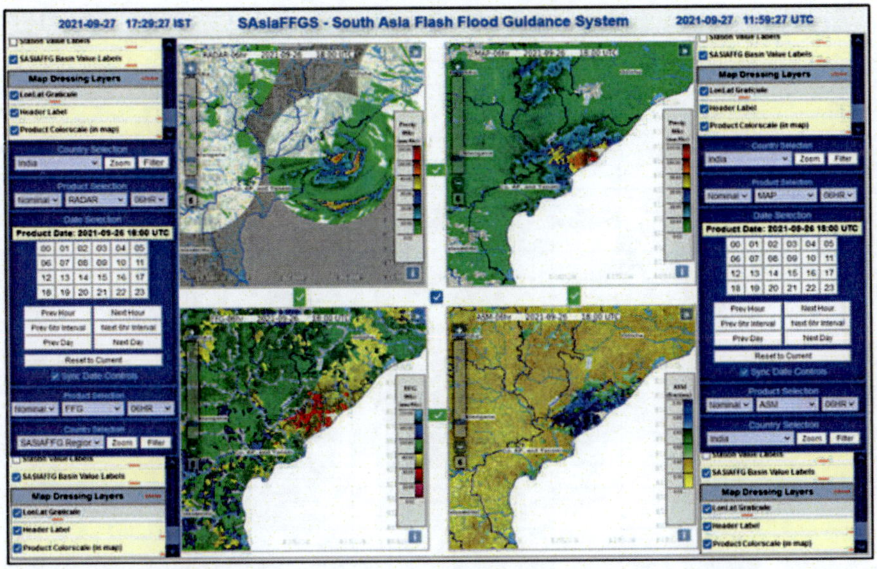

Fig. 14 Visualization of diagnostic products over an area (Top Left: DWR Rain Precipitations, Top Right: Mean Areal Precipitation observed during last 6 h, Bottom Left: Minimum Flash Flood Guidance Value shows high possibility of bankful runoff conditions, Bottom right: Average Soil Top shows saturation excess with 100% conditions

4.3 Prognostic Products

The prognostic products as in Fig. 15 the main product table of the Dissemination Interface Console contains Gridded Forecasted Precipitation (FCST), Forecasted Mean Areal Precipitation (FMAP), Forecasted Flash Flood Threat (FFFT) products for forecast periods of 3, 6 and 24 h based on 3 Numerical Weather Prediction Models, viz. IMD WRF, IMD GFS & NCUM model. Based on the precipitation forecast based on 0000 UTC & 1200 UTC, the system computed products every six hours and measure the occurrence of positive forecast flash flood threat (FFFT) over a given forecast period using forecast precipitation to drive the Soil and FFG models.

The following images, an approximate measure of uncertainty in the FFFT estimates is indicated by the ranges in the color scale (with yellow indicating the range of values that are less likely to be of concern for flash flooding and with orange and red indicating progressively higher risk of flooding for the sub-basin of interest).

The IMD's Global Forecast System GFS (gsm.v14), adopted from National Centre for Environmental Prediction (NCEP), at T1534L64 ($\sim$12 km) in horizontal resolution and in the vertical there are 64 hybrid sigma-pressure [7] layers with the top layer centered around 0.27 hPa ($\sim$55 km) is operationally used to estimate rainfall forecast at every watersheds for next 6, 12, 24 h as in Fig. 15.

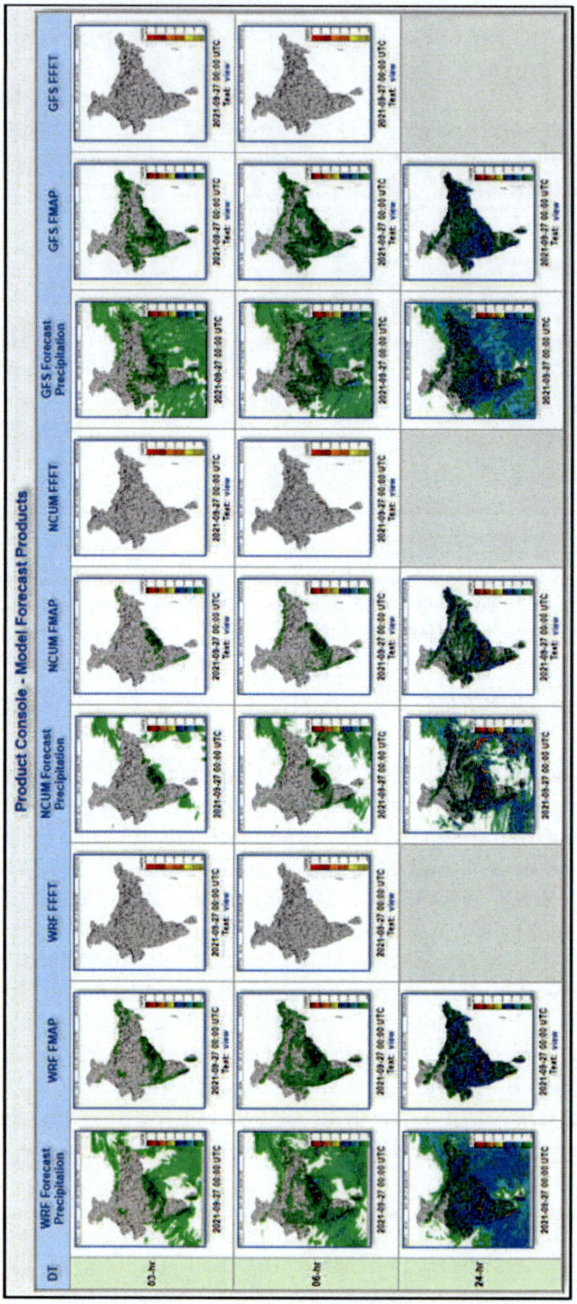

Fig. 15 Prognostic flash flood guidance (FFG) products based on IMD NWP models

The IMD's Weather Research and Forecast (WRF) model with its double nested operational configuration (3 km) is used for short-range forecasting of weather events with lead time of three days. The WRF rainfall forecast of 3 km is operationally used for computation of rainfall at every watershed for next 6, 12, 24 h as in Fig. 15.

The National Centre for Medium Range Forecasting (NCMRWF) Unified Model (NCUM) system is based on the Unified Model (UM) developed under the UM Partnership by Met Office, UK, BoM/CSIRO, Australia, KMA, South Korea, NIWA, New Zealand and MoES/NCMRWF, India [8]. The 4 km NCUM Regional rainfall forecast is operationally used for computation of rainfall at every watershed for next 6, 12, 24 h as in Fig. 15.

4.3.1 Forecast Mean Areal Precipitation (FMAP)

The images and text provide accumulations of mean areal precipitation (mm) for each sub-basin produced by using numerical forecasts from by the IMD WRF, IMD GFS & NCUM models. Each data product reflects basin-average precipitation accumulation forecasts from the navigation hour over the corresponding interval.

- **FMAP 03-hr:** Accumulated mean areal precipitation forecasted by the IMD WRF (FMAP1), IMD GFS (FMAP3) & NCUM (FMAP2) model to occur over 3 h beginning from the current navigation hour. (mm/3 h)
- **FMAP 06-hr:** Accumulated mean areal precipitation forecasted by the IMD WRF (FMAP1), IMD GFS (FMAP3) & NCUM (FMAP2) model to occur over 6 h beginning from the current navigation hour. (mm/6 h)
- **FMAP 24-hr:** Accumulated mean areal precipitation forecasted by the IMD WRF (FMAP1), IMD GFS (FMAP3) & NCUM (FMAP2) model to occur over 24 h beginning from the current navigation hour. (mm/24 h)

4.3.2 Forecast Flash Flood Threat (FFFT)

The images and text provide 1-, 3- and 6- Forecast Flash Flood Threat (mm) for each sub-basin. The values indicate the difference of the FMAP from the IMD WRF, IMD GFS & NCUM model of the given duration and the corresponding current FFG of the same duration. The current 3-h and 6-h FMAP products are considered with current corresponding FFG in the computation of FFFT.

- **FFFT1 03-hr:** Difference of 03-hr FFG from for current model processing hour and the corresponding IMD WF 03-hr FMAP1. (mm/3 h)
- **FFFT1 06-hr:** Difference of 06-hr FFG from for current model processing hour and the corresponding IMD WF 06-hr FMAP1. (mm/6 h)

For Example:

The 03-hr FFFT at 12:00 UTC = the difference between the 03-hr FMAP from 12:00 UTC and the 03-hr FFG from 12:00 UTC, considered valid at 15:00 UTC.

The 06-hr FFFT at 12:00 UTC = the difference between the 06-hr FMAP from 12:00 UTC and the 06-hr FFG from 12:00 UTC, considered valid at 18:00 UTC.

Additional examples (between FFG update cycles):

In the images, an approximate measure of uncertainty in the FFFT estimates is indicated by the ranges in the color scale (with yellow indicating the range of values that are less likely to be of concern for flash flooding and with orange and red indicating progressively higher risk of flooding for the sub-basin of interest). **FFFT carries significant uncertainty, and it is offered as a baseline product that must be carefully evaluated by the forecaster in real-time.**

4.3.3 Flash Flood Risk (FFR)

The Baseline Risk Product as in Fig. 16 below the main product table of the Dissemination Interface Console contains Flash Flood Risk (FFR) products for forecast periods of 12, 24 and 36 h. These products are computed every six hours and measure the occurrence of positive forecast flash flood threat (FFFT) over a given forecast period using forecast precipitation to drive the Soil and FFG models.

- **FFR112-hr:** Relative frequency of occurrence of positive flash flood threat over a forecast period of 12 h starting with the current IMD WRF, IMD GFS & NCUM model processing hour and using all the intervals (3 and 6 h) of threat computation (dimensionless).
- **FFR1 24-hr:** Relative frequency of occurrence of positive flash flood threat over a forecast period of 24 h starting with the current IMD WRF, IMD GFS & NCUM model processing hour and using all the intervals (3 and 6 h) of threat computation (dimensionless).
- **FFR1 36-hr:** Relative frequency of occurrence of positive flash flood threat over a forecast period of 36 h starting with the current IMD WRF, IMD GFS & NCUM model processing hour and using all the intervals (3 and 6 h) of threat computation (dimensionless).

5 Validation of the Operational Flash Flood Guidance System with a Case Study

5.1 Event Occurred

Heavy Rainfall-induced Pluvial Flash Flood over the urban watersheds and neighborhoods of Hyderabad, Telangana on October 13, 2020 (Fig. 17).

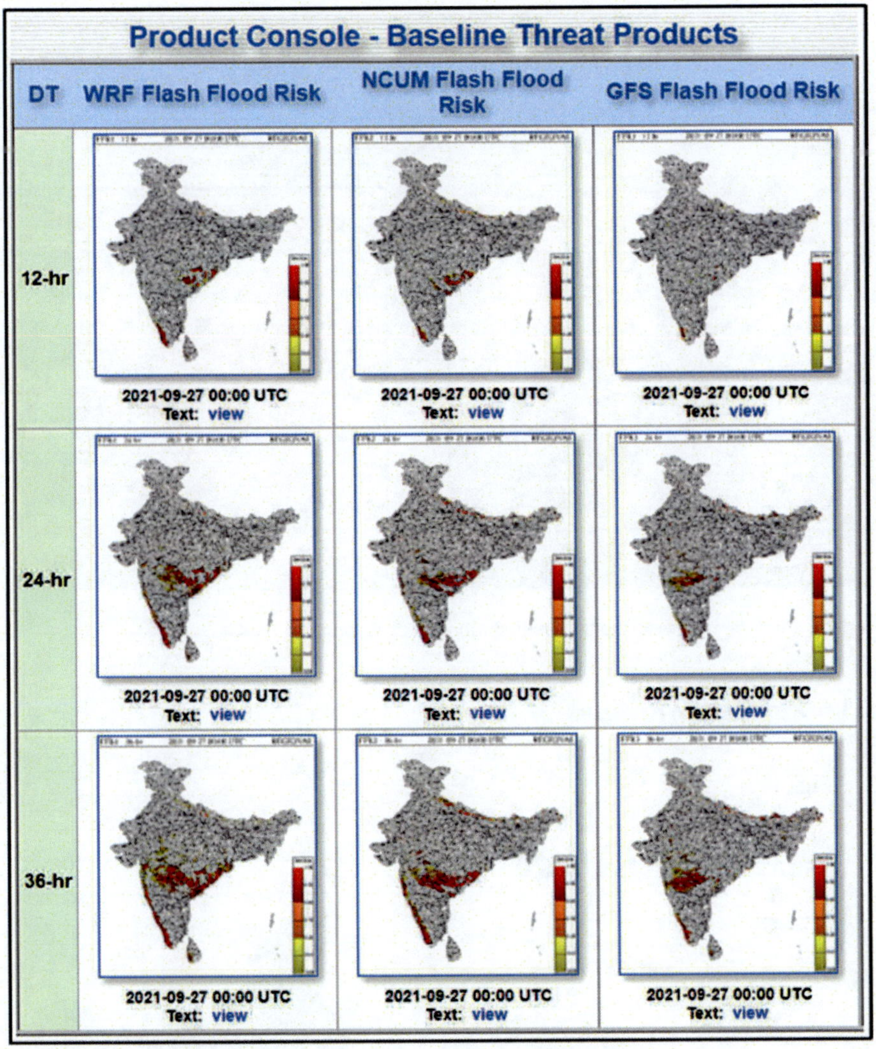

Fig. 16 WRF, NCUM, and GFS flash flood risk (FFR) product (mm/h)

5.2 *Synoptic Conditions*

During October 12–15, 2020, a cyclonic system crossed north Andhra Pradesh coast close to Kakinad a(near Lat. 17.0° N and long. 82.4°E) between 0630 and 0730 h IST on October 13, 2020, lay centered as a Deep Depression over the study area.

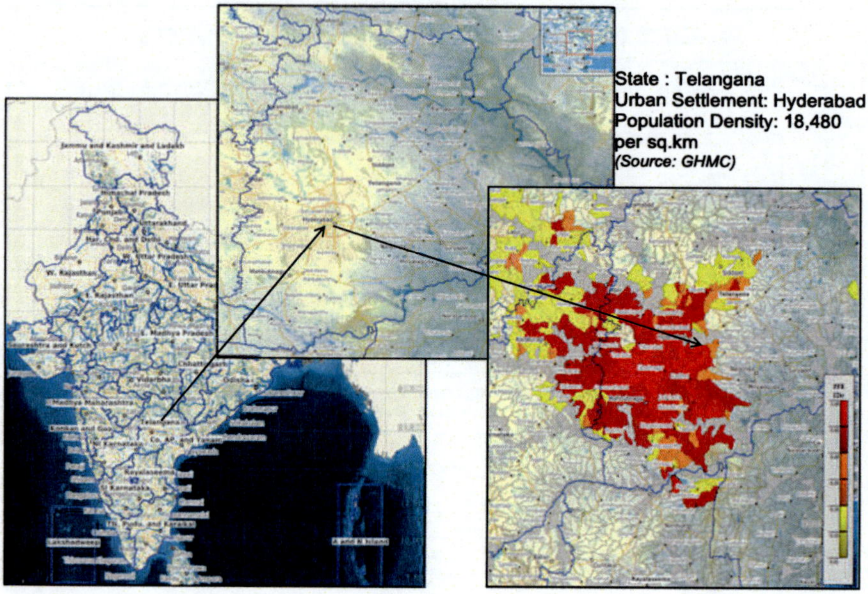

Fig. 17 Study area—watersheds & neighborhoods of Hyderabad (AoC)

5.3 *Rainfall Warnings, if Any*

See Fig. 18.

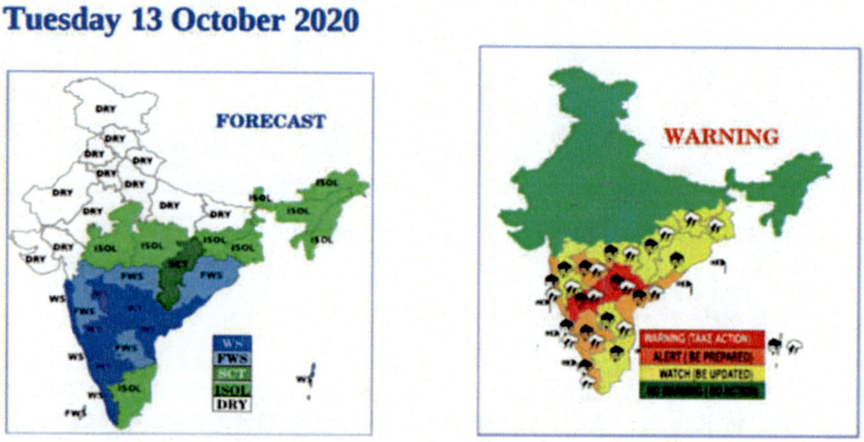

Fig. 18 IMD rainfall warnings and forecast issued

India Meteorological Department has issued “**Red Warning**” and **Widespread rainfall** activity due to synoptic conditions over Telangana State, India. District-wise heavy rainfall warning was given on October 13, 2020 over the study area [26].

5.4 *Diagnostic Conditions*

See Figs. 19 and 20.

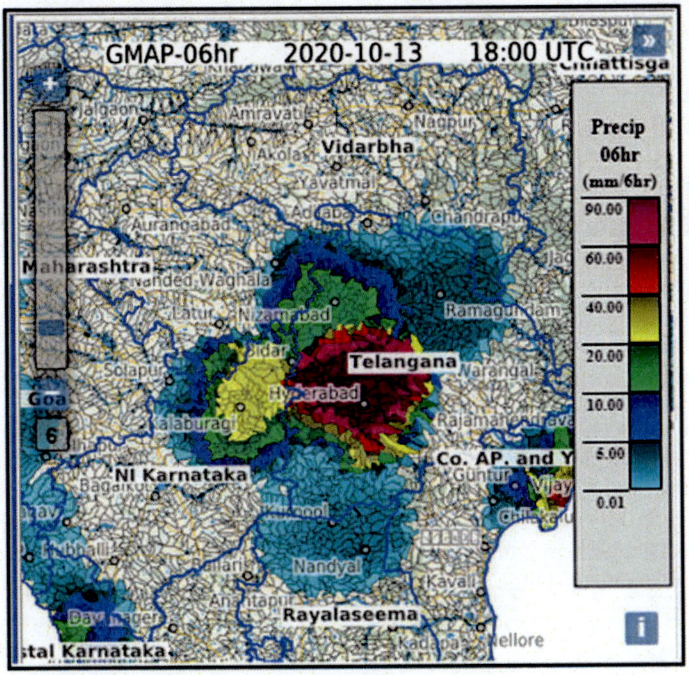

Fig. 19 Gauge mean precipitation (6 h accumulation)

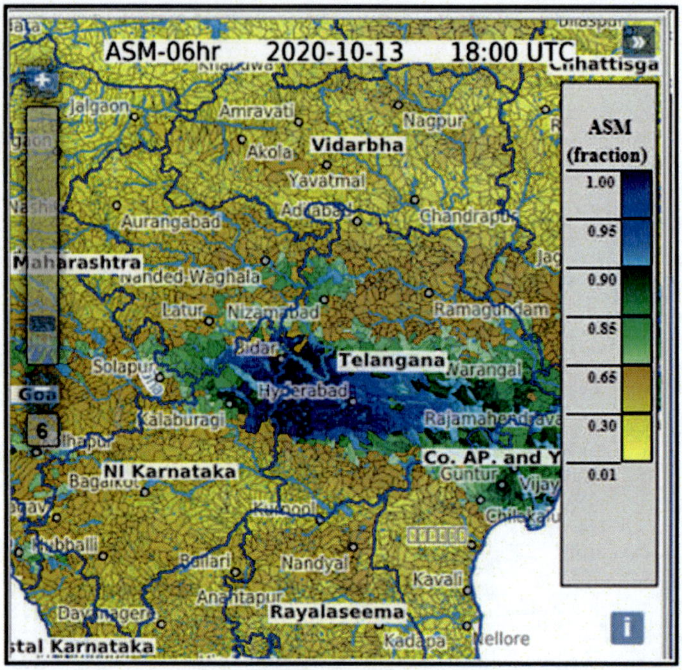

Fig. 20 Average soil moisture from SACSMA Model

(1) It was observed that the convective system became stationary over the AoC based on the Satellite Rainfall Estimate and high cloud top brightness temperature from INSAT 3D product.
(2) More than 11 cm rainfall reported during last 6 h as illustrated in Fig. 17.
(3) By persistence heavy rainfall, saturation excess became evident at the outlet of some watersheds with 100% fully saturated soil conditions as illustrated in Fig. 18.

5.5 *Hydrological Trigger Generation*

See Figs. 21 and 22.

(1) It is known that minimum the FFG value (mm/hr), maximum is the likelihood with high confidence of threat.
(2) Flash flood guidance value is minimum up to 10 mm/1 h indicates surface runoff at any moment with the persistent rainfall as illustrated in the figure.
(3) Moderate-to-high threat alert was issued through National Guidance Bulletin based on the indications of substantial exceedance of rainfall over watersheds & neighborhoods of the study area.

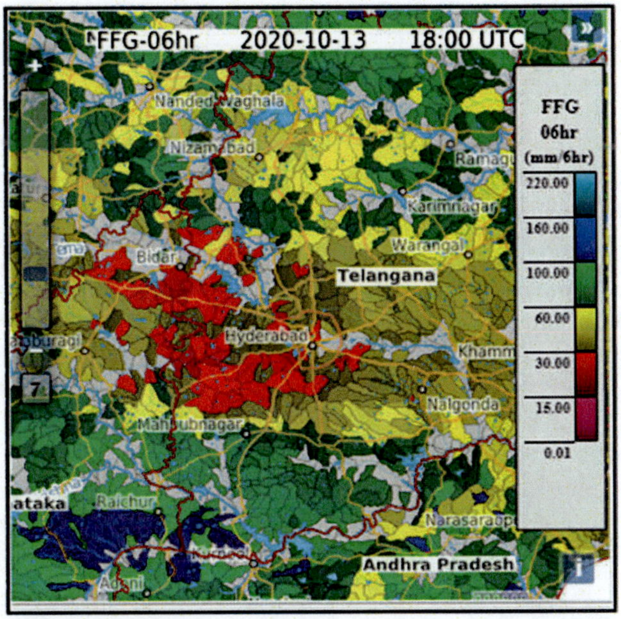

Fig. 21 Flash flood guidance value in mm/6 h

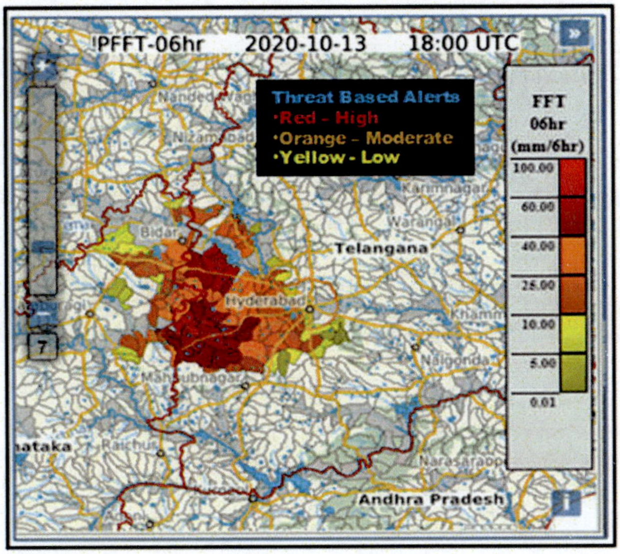

Fig. 22 Flash flood threat value in mm/6 h

Legend: Threat based Alerts		
Low		Can Occur with Uncertainty
Moderate		Moderate Confidence with substantial exeedance of RF
High		High Confidence with very substantial exeedance of RF

5.6 *Advisories/Alerts for Impact Based Flash Flood Guidance*

In addition to heavy rainfall warning, flash flood guidance alerts up to watershed level issued in the form of Threat & Risks issued from 18:15 h IST on 12.10.2020 (Approx. 24 h lead time) onward till the occurrence of the event as illustrated in Fig. 23. with impact assessment flowchart.

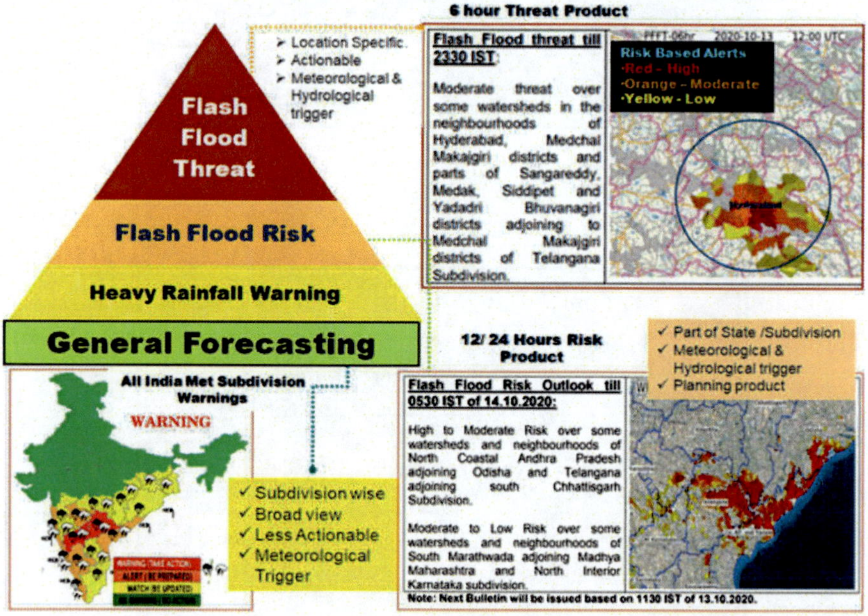

Fig. 23 Impact-based analysis for Hyderabad Flash Flood event dated 13.10.2020

5.7 *Observations*

See Figs. 24, 25 and 26.

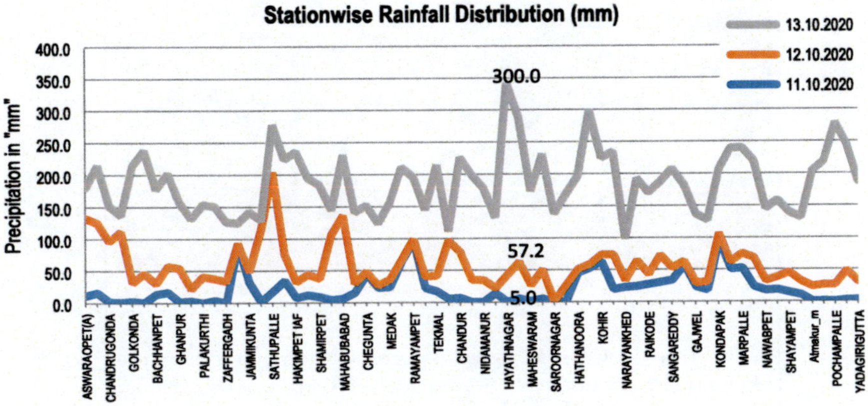

Fig. 24 In situ Gauge Data Observations of Hyderabad during October 11–13, 2020

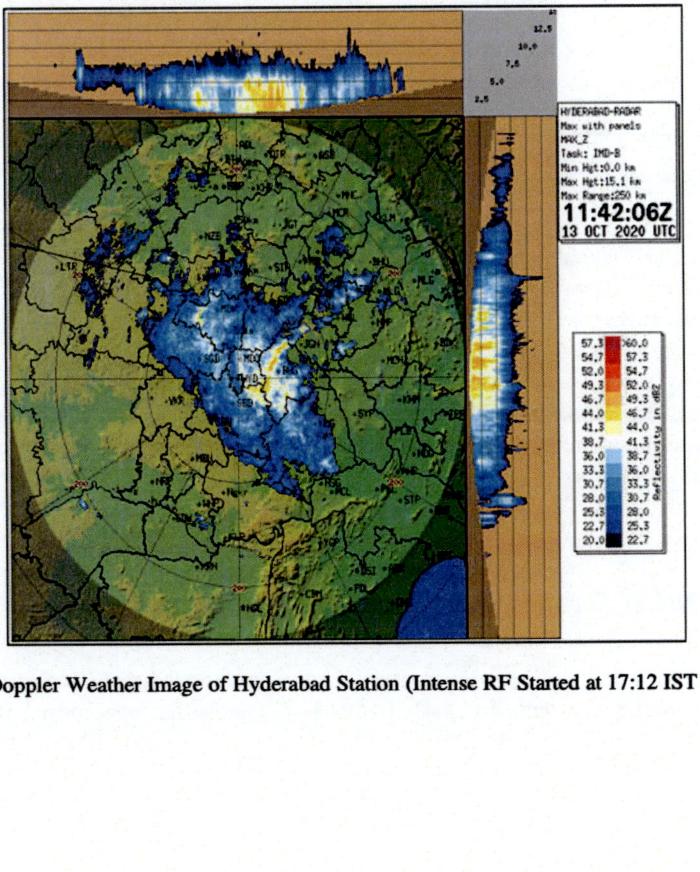

Fig. 25 Doppler Weather Image of Hyderabad Station (Intense RF Started at 17:12 IST of 13 Oct 2020)

By: Express Web Desk | Hyderabad | October 15, 2020 12:02:12 pm

Buildings and vehicles lie partially submerged in floodwater following heavy rains, at Falaknuma, in Hyderabad. (PTI Photo)

The heavy rains that lashed Hyderabad Tuesday plunged many parts of the city into darkness, with power supply yet to be restored in several areas even after 48 hours. Several residents took to Twitter on Thursday morning to complain about the lack of response from ground officials.

Following the unprecedented rainfall that resulted in the flooding of low-lying areas, deaths, and severe damages, the declared a two-day holiday for all government offices, private institutions, including banks and financial institutions, across the state for Wednesday and Thursday. In a tweet, informing of the holidays, principal secretary for Municipal Administration Arvind Kumar asked people to stay indoors and not venture out unless it was an emergency and work from home.

Fig. 26 Local Media/News reporting on incident (Express Web desk)

(1) It is clearly evident from the high reflectivity dbZ values of Doppler Weather Radar of Hyderabad indicating this intense short period rainfall over the saturation excess watersheds causing flash flood over the urban watersheds and neighborhoods of Hyderabad.
(2) Figure 23 shows sudden increase in rainfall (> 15 cm) at many stations
(3) It was also reported by local news and print media that the Hyderabad and neighborhoods received unprecedented flashy rains resulted in quick flooding at several low-lying areas.

5.8 *Skill of System for Pluvial Flash Floods*

(1) From the detailed discussion above, it is now understood that a platform and an early warning system for flash floods is available for operational purposes.

The SASIAFFGS was put into experimental mode during 2019 followed by pre-operational mode in 2020 and full operational mode since May 2021.

(2) One of the major difficulties faced during the validation is to determine the exact time and occurrence of the event case by case, where there exists no proper observational mechanism. It needs to be correlated with the available in situ data and saturation excess flow parameters indirectly.

(3) The results and performance evaluation of the FFGS products, NWP model derived risk products and adaptation of a homogeneous approach toward impact analysis gave way for its full operationalization.

6 Results and Concluding Remarks

i. This study reveals that the accuracy in predicting quantitative rainfall over watersheds, lead time computing the surface threshold runoff at operational level and is validated by many accurately forecasted Threat and Risk potentials of recent Flash Flood events.

ii. This study summarizes the operational schema and protocols adopted in India, evaluates the performance with a recent case study over South India. A standard operating procedure is adapted to monitor and enhance the predictability of short time hydro-meteorological hazards like flash floods. [*Refer, SOP on Hydro meteorological Services of India, 2020*]

iii. The impact-based alerts were issued 24 h prior to the occurrence of flash flood event in the watersheds of Hyderabad, Telangana and proper mitigations were taken by the local authorities.

Acknowledgements Implemented in South Asia by India Meteorological Department (IMD), Ministry of Earth Sciences in collaboration with World Meteorological Organization (**WMO**) supported by Hydrologic Research Center (**HRC**), **USAID** and **NOAA.** Authors wish to thank Numerical Weather Prediction, Radar Lab, National Weather Forecasting Centre of IMD and NCMRWF for their data support, guidance at operational level.

References

1. European Commission. Directorate General for the Environment. (2016). *European overview assessment of Member States' reports on preliminary flood risk assessment and identification of areas of potentially significant flood risk: final report.* LU: Publications Office. Accessed: September 30, 2021. [Online]. Available: https://data.europa.eu/doi/https://doi.org/10.2779/576456
2. Georgakakos, K. (2006). Analytical results for operational flash flood guidance. *Journal of Hydrology, 317*, 81–103. https://doi.org/10.1016/j.jhydrol.2005.05.009
3. Flood Forecasting—1st Edition. https://www.elsevier.com/books/flood-forecasting/adams/978-0-12-801884-2. Accessed September 30, 2021.

4. Gaume, E., et al. (2009). A compilation of data on European flash floods. *Journal of Hydrology, 367*(1/2), 70–78.
5. COMET / MetEd Lesson Viewer. https://www.meted.ucar.edu/education_training/lesson/958. Accessed September 30, 2021.
6. Yin, J., Yu, D., Yin, Z., Liu, M., & He, Q. (2016). Evaluating the impact and risk of pluvial flash flood on intra-urban road net- work: A case study in the city center of Shanghai, China. https://doi.org/10.1016/j.jhydrol.2016.03.037
7. Acosta-Coll, M., Ballester-Merelo, F., Martinez-Peiró, M., & De la Hoz-Franco, E. (2018). Real-time early warning system design for pluvial flash floods—A review. *Sensors, 18*(7), Art. no. 7. https://doi.org/10.3390/s18072255
8. Borga, M., Stoffel, M., Marchi, L., Marra, F., & Jakob, M. (2014). Hydrogeomorphic response to extreme rainfall in headwater systems: Flash floods and debris flows. *Journal of Hydrology, 518*, 194–205. https://doi.org/10.1016/j.jhydrol.2014.05.022
9. Cities and Flooding. https://elibrary.worldbank.org/doi/abs/https://doi.org/10.1596/978-0-8213-8866-2. Accessed September 30, 2021.
10. Terti, G., Ruin, I., Anquetin, S., & Gourley, J. J. (2017). A situation-based analysis of flash flood fatalities in the United States. *Bulletin of the American Meteorological Society, 98*(2), 333–345. https://doi.org/10.1175/BAMS-D-15-00276.1
11. David, Q. A., & Schmalz, B. (2020). Flood hazard analysis in small catchments: Comparison of hydrological and hydrodynamic approaches by the use of direct rainfall. *Journal of Flood Risk Management, 13*(4). https://doi.org/10.1111/jfr3.12639
12. AR5 Climate Change 2013: The Physical Science Basis—IPCC. https://www.ipcc.ch/report/ar5/wg1/. Accessed September 30, 2021.
13. Singh, P., Sinha, V. S. P., Vijhani, A., & Pahuja, N. (2018). Vulnerability assessment of urban road network from urban flood. *International Journal of Disaster and Risk Reduction, 28*, 237–250. https://doi.org/10.1016/j.ijdrr.2018.03.017
14. W. M. Organization (WMO) and World Meteorological Organization (WMO). (2011). *Manual on flood forecasting and warning*. WMO.
15. Yang, W.-Y., Li, D., Sun, T., & Ni, G.-H. (2015). Saturation-excess and infiltration-excess runoff on green roofs. *Ecological Engineering, 74*, 327–336. https://doi.org/10.1016/j.ecoleng.2014.10.023
16. Johnson, M. S., Coon, W. F., Mehta, V. K., Steenhuis, T. S., Brooks, E. S., & Boll, J. (2003). Application of two hydrologic models with different runoff mechanisms to a hillslope dominated watershed in the northeastern US: A comparison of HSPF and SMR. *Journal of Hydrology, 284*(1–4), 57–76. https://doi.org/10.1016/j.jhydrol.2003.07.005
17. Huffman, R., Fangmeier, D., Elliot, W., Workman, S., & Schwab, G. (2011). Infiltration and runoff. *Soil and Water Conservation Engineering*, 81–111.
18. Knapp, V., Durgunoglu, A., & Ortel,T. (1991). a review of rainfall-runoff modeling for stormwater management. U.S. Geologic Survey, Illinois District. Kokkonen, T., Koivusalo, H., & Karvonen, T. (2001). A semi-distributed approach to rainfallrunoff modelling—a case study in a snow affected catchment. *Environmental Modelling & Software, 16*(5), 481–493. https://doi.org/10.1016/S1364-8152(01)00028-7
19. Beven, K. J. (2012). *Rainfall-runoff modelling: The primer* (2nd ed.): Wiley-Blackwell.
20. Moradkhani, H., &Sorooshian, S. (2008). General review of rainfall-runoff modeling: Model calibration, data assimilation, and uncertainty analysis. In S. Sorooshian, K. L. Hsu, E. Coppola, B. Tomassetti, M. Verdecchia, & G. Visconti (Eds.), *Hydrological modelling and the water cycle: Coupling the atmospheric and hydrological models* (Vol. 63, pp. 12–35). Springer
21. Burnash, R. J. C., Ferral, R. L., & Maguire, R. A. (1973). *A generalized streamflow simulation system: Conceptual models for digital computers*. Sacramento, CA: Joint Federal State River Forecast Center. Chahine, M. T. (1992). The hydrological cycle and its influence on climate. *Nature, 359*(6394), 373–380. https://doi.org/10.1038/359373a0
22. Georgakakos, K. P. (1986). A generalized stochastic hydro-meteorological model for flood and flash flood forecasting, 1: Formulation. *Water Resource Research, 2213*, 2083–2095.

23. Shamir, E., Imam, B., Gupta, H. V., & Sorooshian, S. (2005). Application of temporal streamflow descriptors in hydrologic model parameter estimation. *Water Resources Research, 416*, W06021.
24. Carpenter, T. M., Sperfslage, J. A., Sweeney, T., & Fread, D. L. (1999). National threshold runoff estimation utilizing GIS in support of operational flash flood warning systems. *Journal of Hydrology, 224*, 21–44 [CrossRef]
25. https://www.ospo.noaa.gov/Products/atmosphere/ghe/
26. https://internal.imd.gov.in/pages/monsoon_main.php.
27. Standard Operating Procedures on Hydromet Services. https://mausam.imd.gov.in/

Evaluation of Effluent Quality and Operating Cost in Biological Wastewater Treatment Plants with Non-ideal Sensors in the Feedback Control

Abdul Gaffar Sheik, E. S. S. Tejaswini, Murali Mohan Seepana, and Seshagiri Rao Ambati

Abstract In the analysis of wastewater treatment plants (WWTPs), it is very general to use ideal sensors and actuators without noise and delay for the measurement of dissolved oxygen, flow rate, ammonia, and nitrate levels. However, to control WWTPs in practice, non-ideal sensors and actuators only exist, and hence evaluation by considering non-ideal sensors is required. It is important to assess the performance of the advanced control strategies under non-ideal conditions and it is addressed in this research. Biological WWTP model such as benchmark simulation model (BSM1-P) with activated sludge process of ASM3bioP is used in the present study. In this chapter, non-ideal sensors with delay and noise and different combinations of sensors are analyzed. Three control strategies are designed (Proportional integral (PI) controller, Model predictive controller (MPC), and Fuzzy logic controllers (FLC)) to check the effluent quality and operating costs. For the right dissolved oxygen and nitrite monitoring assessment, the significance of delay and noise in sensors are much higher, as for ideal sensors a good control performance is achieved by increasing the controller gain. A total of nine different sensor class combinations are studied. MPC has shown an improved effluent quality of 3.6% in the sensor class combination of DO–B_O and NO–B_O. Fuzzy is not responding well to the sensor class combinations and there are no improvements. Operational cost is improved by 1% in the sensor class combination of DO–D_O and NO–D_O in the application of PI controller.

Keywords Dissolved oxygen · Effluent quality · Operational cost · And Ideal sensor

A. G. Sheik · E. S. S. Tejaswini · M. M. Seepana · S. R. Ambati (✉)
Department of Chemical Engineering, National Institute of Technology, Warangal, Telangana 506004, India
e-mail: seshagiri@nitw.ac.in

A. K. Dikshit et al. (eds.), *Innovative Trends in Hydrological and Environmental Systems*, Lecture Notes in Civil Engineering 234,
https://doi.org/10.1007/978-981-19-0304-5_45

1 Introduction

Biological wastewater treatment plants (WWTP) are utilized all over the world to ensure water quality in receiving environments and to preserve aquatic biodiversity. The WWTP's generally keep many water quality parameters within permissible limits. Thus, the nutrient reduction is one of the most difficult goals to achieve and necessarily requires the use of advanced control design strategies. The secondary treatment or biological process is the most important among all the treatment processes in WWTPs. Biological processes play an important role in the reduction of nutrients and organic matter during this treatment. During nitrification, the amount of dissolved oxygen (DO) concentration plays a critical role in the reduction of ammonium and ammonia nitrogen (NH), orthophosphates (S_{PO}), and nitrate (NO) concentrations in the biological treatment. These are the nutrients that are more challenging to keep under permissible levels. As a result, DO and NO control has been the focus of numerous research investigations over the last few years. In WWTP simulation-based research, ideal sensors and actuators with no noise or delays are commonly used. They can be used to compare the performance of a proposed advanced control strategy to that of other well-established ones. Nevertheless, noise and delay in sensors and actuators are far more important for a proper DO and NO tracking evaluation, because, with optimal sensors, excellent control performance may be accomplished simply by increasing the controller's gain.

The earlier researchers noted that objectives of DO and NO tracking are all based on simulations of WWTPs, and the majority of them are developed by the International Association on Water Pollution Research and Control (IWA) using internationally recognized and accepted standards such as benchmark simulation model no. 1 (BSM1). Activated sludge models (ASM) such as ASM1, ASM2, ASM2d, and ASM3 have been proposed by the IWA task group [1]. Here, in this study, a version of the ASM3 model (i.e., ASM3bioP) [2, 3] was created largely to estimate biological phosphorus removal by incorporating modified processes from the ASM2d model but excluding the metal precipitation reactions [4]. WWTP systems are distinguished by the highly complex and nonlinear processes. These processes must be carried out under specific working conditions that must be maintained over time to ensure the proper operation of the WWTP. For the seven bioreactors in the A^2/O (Anaerobic, anoxic, and toxic) process, various control configurations are implemented [5, 6]. The optimized set-points are computed by the higher-level control loops [7–9]. Various control systems have been proposed to do this, ranging from the most basic Proportional Integral (PI) and Derivative Controllers (PID) to the most complicated model-based approaches such as a combination of Model Predictive Controllers (MPC) and Fuzzy Logic Controllers (FLC) [10]. Controlling of DO and NO is done by manipulating oxygen mass transfer coefficient and internal recycle, respectively. Different PI control combinations, MPC, and FLC strategies are applied in BSM1-P. The resultant outcomes show a tradeoff between operational cost and effluent quality [11]. Ammonia-based aeration control (ABAC) strategies are developed by using different combinations of MPC and Fuzzy. It was found that effluent quality is improved with increased operational cost [12]. Nested loop control based on feedforward is developed on BSM1-P. The outcomes are found that improved ammonia and nitrogen violations are archived with reduced effluent quality [13].

The noise and delays caused by the several non-ideal sensors located over the WWTP are the principal perturbations that influence these signals. When it comes to noise, the control strategies are defined by the need for precise control parameters and an identification process. As a result, when noise is present in the measurements, identifying it and properly tuning the controller becomes a difficult task [14]. This may cause actuation mechanical components to tear and wear, degrading performance and reducing instrumentation life span. The Benchmark Simulation Model No.1 (BSM1), a general-purpose model of a WWTP plant [15], provides a clear example of time-delays and noise effects, where the noise is introduced by the considered sensors. The resultant outcome shows the inaccurate measurement of the controlled variable and, as a result, an inaccurate actuation signal occurs. Although the control strategies suggested in the cited studies produce significant technological advances, ideal sensors and actuators are used in the majority of the works. However, some amount of research is done based on the delay [16], a sensor with noise without delay [17] is taken into consideration. The noise application in Ref. [18] does not adhere to the norms specified in BSM1, making any posterior comparison problematic. None of them use both noise, delay-affected sensors, and delay-affected actuators. The usage of sensor classifications is to define distinct types of sensors while limiting the number of sensor possibilities available. The objective of the chapter is to test the different combinations of sensor classifications with non-deal applications to check the corresponding plant performance by using the feedback control design.

In this chapter, a modeling-based configuration is employed to investigate the performance of WWTP. The chapter is articulated and structured as follows: WWTP is designed using ASM3bioP as an activated sludge model with the combination of anaerobic, anoxic, and oxic in a seven bioreactor process followed by non-reactive settler (BSM1-P). Moreover, plant effluent quality and operational cost performance are also assessed. The two control loops are created at lower levels for this working platform using PI, fuzzy logic, and model predictive controls. Non-ideal sensor classification combinations are tried for those control applications to find their corresponding plant performance are evaluated.

2 Materials and Methods

This section explains the BSM1-P working system, including detail on cost/effluent quality performance evaluation indices. MATLAB/SIMULINK is used to create the models and run the simulations (Mathwork, Inc).

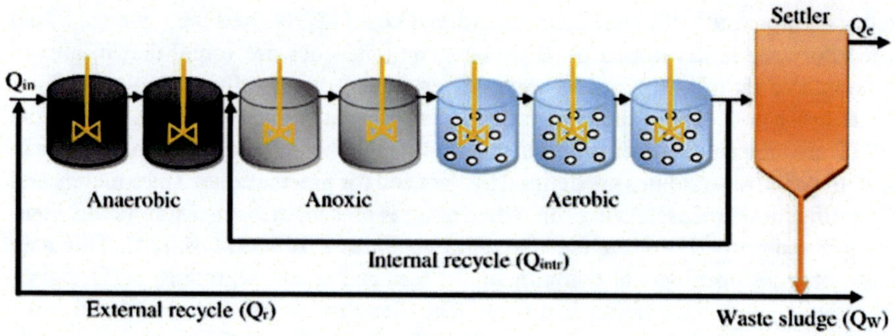

Fig. 1 WWTP framework for simultaneous removal of carbon, nitrogen, and phosphorous

2.1 WWTP Architecture

Figure 1 depicts the WWTP configuration, which consists of seven tanks connected in series, plus a secondary clarifier. WWTP framework has two anaerobic, two anoxic, and three aerobic bioreactors in the plant. The corresponding combined volume of bioreactors is 2000 m^3, 2000 $m^{3,}$ and 3999 m^3, respectively and fully mixed. The secondary clarifier tank volume is 6000 m^3. The WWTP framework consists of two recycle flow rates, first flow from the third aeration bioreactor (Q_{intr}) to the first anoxic bioreactor is 34500 m^3/d, and second flow from the secondary clarifier underflow (Q_r) to influent flow is 18,446 m^3/d. The WWTP is designed to have an average flow rate (Q_{in}) of 18,446 m^3/d and a sludge flow rate (Q_w) of 385 m^3/d. Even though fourteen days are available for analysis, only the last seven days are used for evaluation. The ASM2d influent data is used to modify the influent data values and concentration [4, 11]. The process rate reactions are very similar to ASM2d and it was integrated into ASM3bioP without any metal precipitation reactions [11]. In this study ASM3bioP [19] is selected for the biological process unit and double exponential settling velocity of the non-reactive secondary clarifier model is employed [20]. ASM3bioP model involves 17 state variables with 23 biological processes are used [11].

2.2 Plant Performance Criteria

EQI (kg polluts units/day) is determined as a weighted average sum of effluent concentrations. For dynamic simulation, fourteen days' data is available but the last seven days are considered as the plant performance assessment. [11, 12, 21]:

$$\text{EQI} = \frac{1}{100(t_f - t_0)} \int_{t_0}^{t_f} K U_{(t)} Q_{e(t)} \mathrm{d}t \tag{1}$$

$$\mathrm{KU}_{(t)} = \mathrm{KU}_{\mathrm{TSS}(t)} + \mathrm{KU}_{\mathrm{COD}(t)} + \mathrm{KU}_{\mathrm{BOD}(t)} + \mathrm{KU}_{\mathrm{TKN}(t)} + \mathrm{KU}_{\mathrm{NO_3}(t)} + \mathrm{KU}_{P_{\mathrm{tot}}(t)} \quad (2)$$

The t_o and t_f in Eq. (2) represents the starting and ending intervals of time for computing the EQI while the KU_t notify the average load of polluted concentrations in the influent and effluent data. Generally, it consists of TSS (total suspended solids), $\mathrm{BOD_5}$ (biological oxygen demand), COD (chemical oxygen demand), TKN (total Kjeldahl nitrogen), NO (nitrate), $\mathrm{S_{NH}}$ (ammonia), and TP in Eq. (2). Thus the corresponding expression for KU_t is given in Eq. (3).

$$\mathrm{KU}_t = \beta_t G_t \quad (3)$$

where β_t (g^{-1}) are weighting factors ascribe every component of the pollution. The weighting factor values are represented below. Moreover, the composition of different elements (G_t) is estimated by using Eq. (4)–(10).

The values of weighting factors are assigned each effluent component, the factors are considered as follows: $\beta_{\mathrm{ss}} = 2, \beta_{\mathrm{cod}} = 1$, $\beta_{\mathrm{TKN}} = 20$, $\beta_{\mathrm{NO}} = 10$, $\beta_{\mathrm{BOD_5}} = 2$, $\beta_p = 100$. Besides G_t, spontaneous concentrations of various nutrients are calculated corresponding to their state and particulate symbol with description is in the supplementary data and variables below:

$$G_{\mathrm{TSS}} = X_{\mathrm{TSS}} \quad (4)$$

$$G_{\mathrm{COD}} = S_S + S_I + X_I + X_S + X_H + X_{\mathrm{PAO}} + X_{\mathrm{PHA}} + X_{\mathrm{A}} \quad (5)$$

$$G_{\mathrm{BOD}} = 0.25\big(S_S + (1 - f_{S_i})X_S + (1 - f_{XI_H})X_H + (1 - f_{XI_P})(X_{\mathrm{PAO}} + X_{\mathrm{PHA}}) + (1 - f_{XI_A})X_A\big) \quad (6)$$

$$G_{\mathrm{TKN}} = S_{\mathrm{NH}} + i_{P,S_S}S_S + i_{N,S_I}S_I + i_{N,X_I}X_I + i_{N,X_S}X_S + i_{N,\mathrm{BM}}(X_H + X_{\mathrm{PAO}} + X_A) \quad (7)$$

$$G_{N_{\mathrm{tot}}} = G_{\mathrm{TKN}} + G_{\mathrm{NO_3}} \quad (8)$$

$$G_{\mathrm{NO_3}} = S_{\mathrm{NO_3}} \quad (9)$$

$$G_{P_{tot}} = S_{PO_4} + i_{P,S_I}S_I + i_{P,X_I}X_I + i_{P,X_S}X_S + i_{P,BM}(X_H + X_{PAO} + X_A) + X_{PP} \quad (10)$$

2.3 Operational Cost Index (OCI)

Further, to compute the cost of WWTP, operations were done with respect to the following energy basis: aeration energy (AE), mixing energy (ME), pumping energy (PE), and sludge production (SP) disposal. Therefore, the following sums of these energies are calculated as the operation cost (OC) per day. AE rate was utilized in the last three bioreactors by manipulating the oxygen mass transfer coefficient, PE consumption rate to keep wastewater violations based on the recycling rates. ME is the mechanical mixing with constant rate and SP represents the total suspended solids and waste flow in the sludge. However, all these energies (Kwh/d) were calculated in an identical way to BSM1 [15]. The carbon (CA) and metal (MA) addition (kg/d) are directly calculated based on the amount of dosages the calculations are given below [11, 12, 21]. In this study no carbon and metal dosages are added:

$$\mathrm{OCI} = \mathrm{AE} + \mathrm{PE} + 5 * \mathrm{SP} + 3 * \mathrm{CA} + 1.5 * \mathrm{MA} + \mathrm{ME}$$

Where individual components of OCI are calculated based on the Eq. (11)–(16)

$$\mathrm{AE} = \frac{S_O^{\mathrm{sat}}}{1800T} \int_{t_o}^{t_f} \sum_{i=5}^{7} V_i * K_L a_i(t)\mathrm{d}t \quad (11)$$

$$\mathrm{PE} = \frac{1}{T} \int_{t_o}^{t_f} (0.008 Q_r(t) + 0.004 Q_{\mathrm{intr}}(t) + 0.05 Q_w(t))\mathrm{d}t \quad (12)$$

$$\mathrm{SP} = \frac{1}{T} * \left(\mathrm{TSS}_a(t_f) - \mathrm{TSS}_a(t_o) + \mathrm{TSS}_s(t_f) - \mathrm{TSS}_s(t_o) + \int_{t_o}^{t_f} \mathrm{TSS}_w Q_w \mathrm{d}t \right) \quad (13)$$

$$\mathrm{ME} = \frac{1}{T} \int_{t_o}^{t_f} \mathrm{ME}(t)\mathrm{d}t \quad (14)$$

where

$$\mathrm{ME}(t) = 24 \sum_{i=1}^{5} \begin{cases} 0.005 * V_i & \text{if } K_L a_i \le 20d^{-1} \\ 0 & \text{if } K_L a_i \ge 20d^{-1} \end{cases}$$

Carbon addition (CA) (kg COD/d) is described as:

$$CA = \frac{CON_{CA}}{t_o * 1000} \int_{t_{start}}^{t_{end}} Q_{CA} * dt \tag{15}$$

Here, Q_{CA} is the sum of carbon flow rate added and CON_{CA} is the concentration of added carbon.

Metal addition (MA) is described as:

$$MA = \frac{CON_{MA}}{t_o * 1000} \int_{t_{start}}^{t_{end}} Q_{MA} * dt \tag{16}$$

Here, Q_{MA} is the sum of carbon flow rate added and CON_{MA} is the concentration of added metal.

The DO and NO in seventh and fourth reactors control performance is assessed with the error between the reference of setpoint and actual value of DO. The corresponding average values of the absolute error (mean $|e|$) the integral square error (ISE) and the integral absolute error (IAE) are calculated as follows:

$$\text{mean}|e| = \frac{\sum_{i=o}^{n} |e_i|}{n+1} \tag{17}$$

$$ISE = \int_{t=7\text{days}}^{t=14\text{days}} e_i^2 dt \tag{18}$$

$$IAE = \int_{t=7\text{ days}}^{t=14\text{ days}} |e_i| dt \tag{19}$$

Class Ao is essentially an ideal sensor. The one-minute response time prevents unrealistic control applications. Class Bo is intended to include traditional online analyzers with fast filtration and short sample loops. Analyzers with slow filtration or sedimentation are classified as Class Co sensors. Class Do contains all batch measurements such as respirometry and total component sensors [22, 23]. Further, four types of non-ideal sensors are considered to evaluate the performance as shown in Table 1. The four sensors have different response times and model orders.

3 Results and Discussions

Different combinations of non-ideal sensor classification for control applications are studied in the WWTP. In practical usage, the most frequently used control configurations using the specified control handles are:

Table 1 Different types of non-ideal sensors classes

Sensor class	Response time (min)	System order	The ratio of a time delay to response time
Ao	1	2	0.133
Bo	10	8	0.392
Co	20	8	0.392
Do	4	2	0.133

DO control in the last aeration tank: To improve the oxidation process, DO is added to the aeration basin by supplying oxygen to aerobic microbes so that they can successfully convert organic wastes into inorganic by-products. Theoretically, microbes must need at least 0.1 to 0.3 mg/L to metabolize and reproduce. The reason for considering that point of operation is as follows: In practice, the DO in the aeration basin concentration should be held at 1.5 to 4 gO_2/m^3 and the widely used value of DO 2 gO_2/m^3. The oxygen mass transfer coefficient (K_La) is accountable for the air flow rate; it depends on the dynamics of diffusion phenomena [11, 24].

NO control in the fourth anoxic tank: When an internal recirculation system is provided, the most optimal operating points for nitrate levels in the anoxic reactor should be kept in the range of 1–3 g N/m^3, with 1 g N/m^3 being the preferred value. In the anoxic bioreactors section, denitrification takes place. This is achieved by heterotrophs and poly accumulating organism's (PAO's) converting the nitrate delivered by the Q_{intr} from aerobic reactors to anoxic reactors into molecular nitrogen [4, 11].

3.1 Control Approaches

3.1.1 Proportional Integral Controller (PI)

The PI controllers can be implemented using various methods that are found in the literature. In this chapter, the Skogestad integral model control (SIMC) methodology is used to frame the PI controllers [25]. For using this method, transfer function models need to be identified around the operating point. Here, the prediction-error method (PEM) is used to identify the models based on the open-loop data for each loop. The first order plus time delay (FOPTD) models are implemented for individual control loops as expressed in Eq. (20).

$$G(s) = \frac{K_P e^{-d}}{\text{TS} + 1} \tag{20}$$

where, K_P denotes the process gain, d denotes the delay and T signifies the time constant of the system. The method of approach is represented in the flow diagram in Fig. 4 for further understanding. The method of identification and the controllers that have been designed are well explained. For both the DO and NO loops, open-loop data is produced, and control models are developed as a result. The oxygen mass transfer coefficient (K_La_7) and Q_{intr} are the manipulating variables of DO and NO, respectively in the seventh and fourth tank. The corresponding DO and NO values are 252 d^{-1}, 34,500 m^3/d in the last aerobic and anoxic reactors are reported in 2 gO_2/m^3 and 1 gN/m^3. A random input signal is given in K_La_7 and Q_{intr} by considering a variation of ±10% in the nominal value of 252 $d^{-1}$and 34500 m^3/d. the resultant output data for both the loops are used to implement FOPTD models using the prediction-error method. Each loop is designed with PI controllers based on these models and the SIMC [11]. The resultant PI controllers are designed using the SIMC method and the obtained controller parameters are given in below. The two PI control strategies are depicted in Fig. 2. DO loop: $K_c = 6.9256$ and $T_i = 0.0014262$ and NO loop: $K_c = 28{,}533.61$ and $T_i = 0.031488$ (Fig. 3).

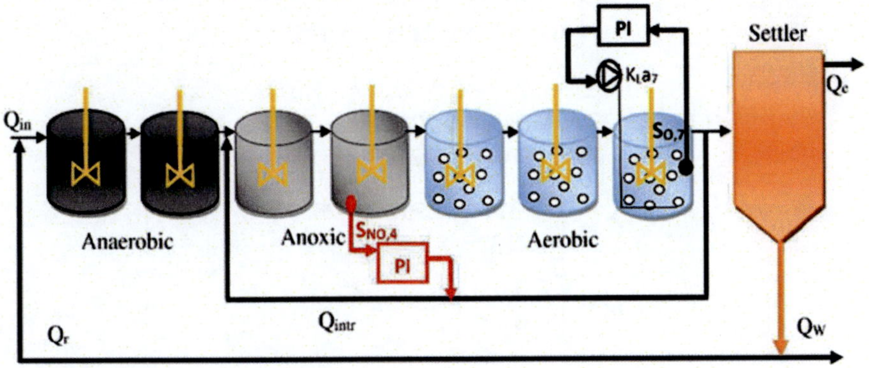

Fig. 2 PI control strategy in BWWTP

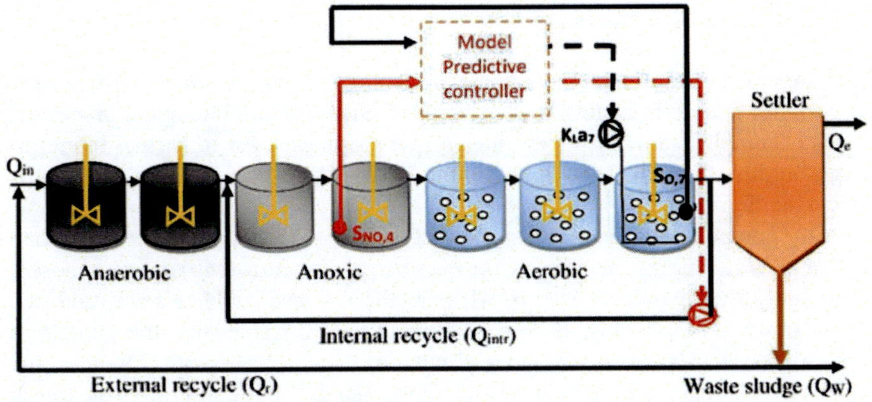

Fig. 3 MPC control strategy in BWWTP

3.1.2 Model Predictive Controller (MPC)

The model predictive controller is a well-known advanced controller that is used to predict the future movement of a significant variable and to deal with control issues by using optimized methods. MPC predicts the type of output variables over the prediction horizon at each unique control time interval. The method of approach is represented in the flow diagram in Fig. 4 for further understanding. The method of identification and the controllers that have been designed are well explained [11]. The MPC control strategy is depicted in Fig. 3. MPC tuning parameters for this model are $m = 2$ and $p = 10$, with $\Delta t = 0.0001$ days. For DO control,$\Gamma_G = 1$ and $\Gamma_{\Delta j} = 0.01$ are considered and for NO control, $\Gamma_G = 1$ and $\Gamma_{\Delta j} = 0.0001$ are utilized. MPC lower-level state-space model:

$$A = \begin{bmatrix} 0.3926 & -0.05 & 2.38e-5 \\ 0.1014 & 0.3318 & 0.2935 \\ 0.011339 & 0.5385 & 0.536 \end{bmatrix}$$

$$B = \begin{bmatrix} 1.005e-05 & -0.0002057 & -7.092e-05 \\ 1.775e-06 & -0.003394 & -3.07e-17 \\ -3.381e-06 & 0.002697 & 5.606e-17 \end{bmatrix}$$

$$C = \begin{bmatrix} 3.319 & -0.552 & -0.2939 \\ 0.4232 & -2.5 & 1.602 \end{bmatrix}$$

$$D = \begin{bmatrix} 0\,0\,0\,0 \\ 0\,0\,0\,0 \end{bmatrix}$$

3.1.3 Fuzzy Logic Control (FLC)

Fuzzy logic is a common technique in a wide range of control applications. Fuzzy logic control is used to control a wide range of chemical and biological processes. This is accomplished using fuzzy rules similar to those used in human inference design. In this study, FLC is applied to the WWTP. The Fuzzy control strategy is depicted in Fig. 5.

For the computation of control signals, FLC uses IF–THEN statement rules. When these outputs and inputs are coupled, the membership function (MF) must be chosen. In this study, the Mamdani fuzzy interface method is used, and MF's are used as a triangular shape functioning. Before developing the FC framework, the usage of a rules-based system is obtained based on simulation data. In the seventh bioreactor, the output variable scale of K_L a ranges from 200 to 280 d^{-1}. Furthermore, in the fourth bioreactor, the input variable scale of "E" is maintained between −30 and 30 $^{g}/m^3$, and the input variable scale of "ED" is maintained between −25 and 25 $^{g}/m^3$. The

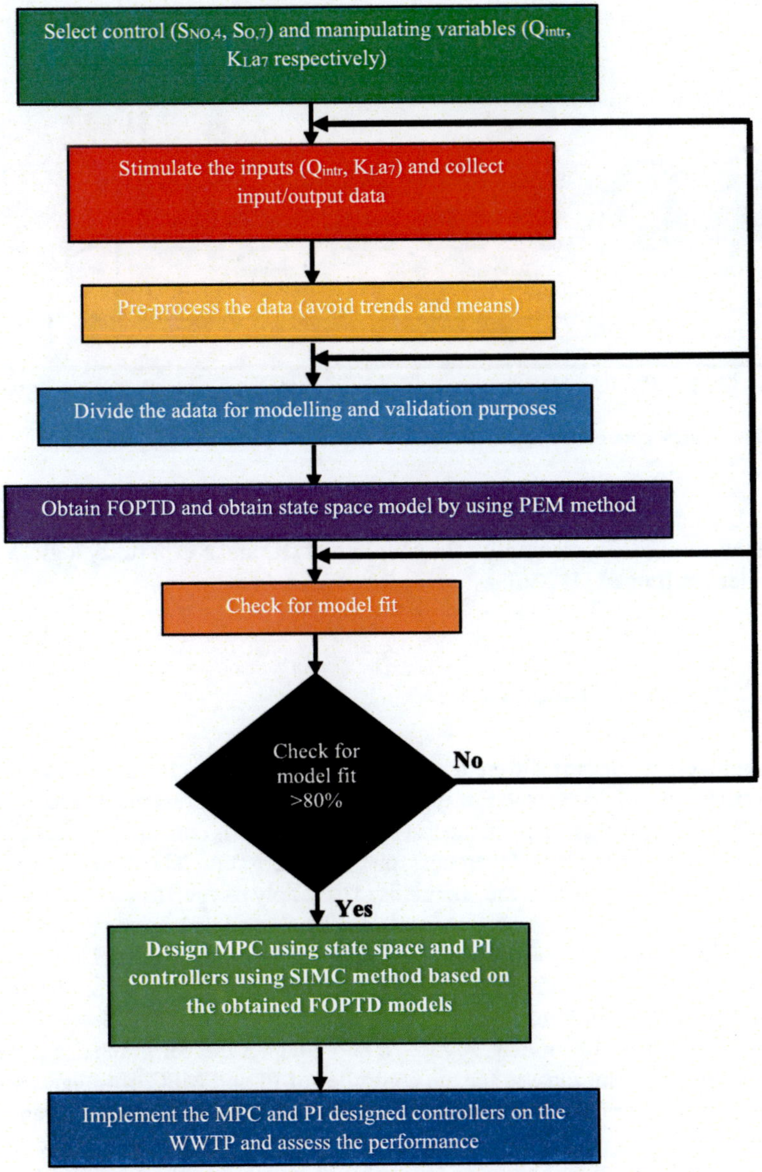

Fig. 4 A systematic approach for model identification and implementation of control

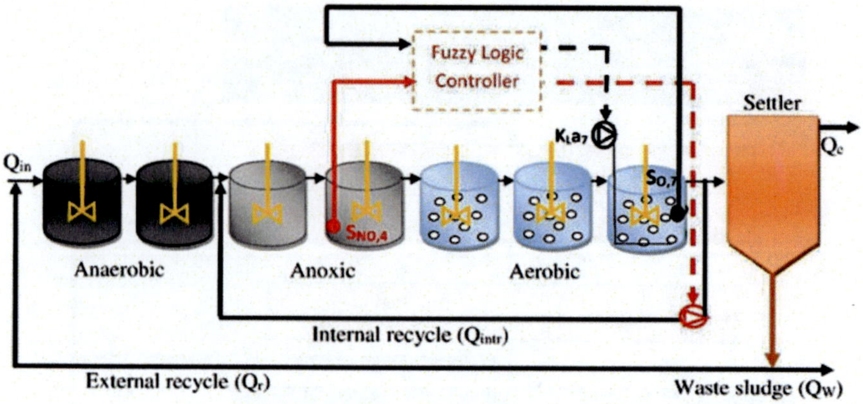

Fig. 5 Fuzzy logic control strategy in BWWTP

scale of Q_{int} output variable is 20100 to 45,000 1/d [11, 26]. The IF–THEN statement conditions are used. To accomplish the coupling of DO and NO in fuzzy logic, a total of 74 rules are framed [11, 26].

3.2 Simulation Outcomes

The simulation results for different combinations of sensor classes with feedback control strategies (PI, MPC, and Fuzzy) for fourteen day's data scenarios are reported in Table 3. During the assessment, the last seven day's data is considered to check EQI and OCI. To practice the WWTP process, one of the most crucial factors determining the operational expenditure is the energy used to supply oxygen to aerobic operation and the cost involved in pumping of materials. The non-ideal sensor combinations are compared with the ideal sensor that corresponds to the Ref. [11]. A total of nine sensor class combinations are selected. Each feedback control application of sensor combinations with resultant outcomes is reported from Table 3, where a, b, c are percentage of improvements. Figures 6 and 7 depict the bar graphs of different combinations of sensor classes for the controller of PI and MPC of integral square error (ISE) and integral absolute error (IAE). These bar graphs represent the outcomes of dissolved oxygen concentrations only. On seeing the bar graph of Fig. 6, it was noticed that MPC is showing higher IAE values compared to PI. It is almost double the values of PI. On seeing the bar graph of ISE in Fig. 7, it was reported that a similar trend is followed by the IAE pattern. Table 2 represents the nine combinations of sensor class of PI, MPC, with their average values of mean $|e|$, IAE and ISE. On seeing Table 2, the average mean $|e|$ vales are high in the case DO on comparing with NO. The percentage of improvements shown in Table 3 are compared with their corresponding ideal sensors of PI, MPC, and Fuzzy controllers [11]. From the Table

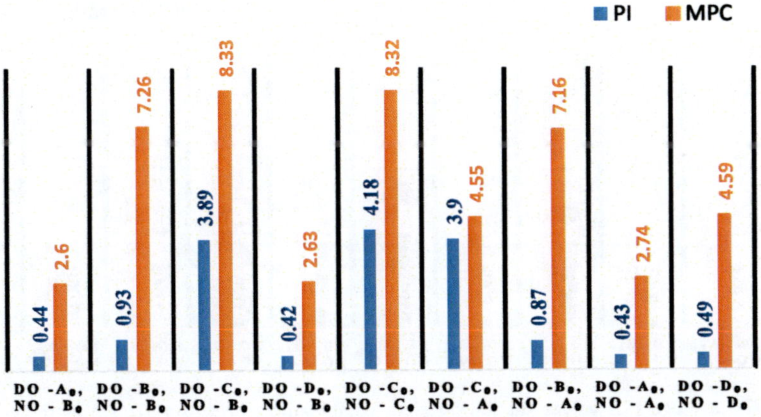

Fig. 6 The average IAE values for the different combinations of sensor classes in the control application of PI and MPC

3, it was observed that EQI is improved by 0.77% in the sensor class combination of DO-C_O, and NO-C_O of default PI. Moreover, OC is improved by 1% in the sensor class combination of DO-D_O and NO-D_O. In some other combinations of sensor classes, there is a tradeoff between OC and EQ in the default PI controller. In some cases, like DO-A_O and NO-A_O, DO-B_O and NO-C_O, and DO-A_O there is no improvement found and these results are not worthwhile in applying these sensor combinations. In another case, MPC controller is used and the resultant EQI is improved by 3.6% in the sensor class combination of DO–B_O and NO–B_O. Additionally, OC is improved by 0.11% in the sensor class combination of DO–D_O and NO–B_O. In most of the sensor combinations, a tradeoff is found between OCI and EQI in the MPC controller. In these combinations like DO-C_O and NO-A_O, and DO–D_O and NO-D_O has shown no improvement, while DO–B_O and NO–A_O, and DO–A_O and NO–A_O have shown the improved results in both the sections of OCI and EQI. In the last case, the Fuzzy controller is employed. It is found that there are no drastic changes in the improvement of EQI in all the sensor class combinations. Whereas, OCI is improved by 0.24% in almost all the sensor class combinations. A marginal improvement is observed in the sensor class combinations DO–B_O and NO–A_O, and DO–A_O and NO-A_O for both OCI and EQI. For all the combinations almost all the simulation results are similar in the Fuzzy controller.

4 Conclusions

Three control strategies (PI, MPC, and Fuzzy) are designed and the comparative results of three control combinations with different non-ideal sensors are evaluated. MPC showed optimal results when compared with the other two approaches. PI is

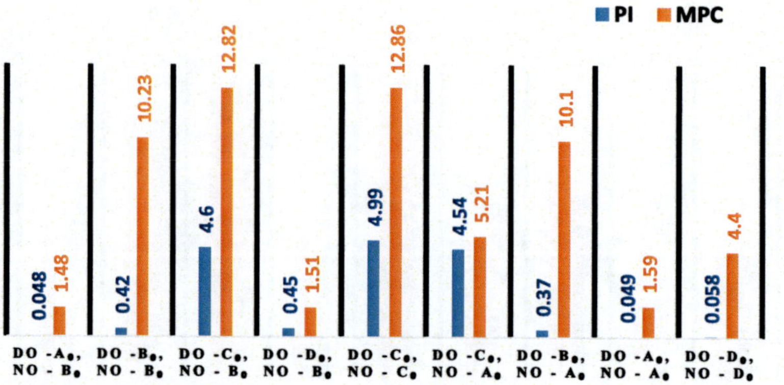

Fig. 7 The average ISE values for the different combinations of sensor classes in the control application of PI and MPC

Table 2 Nine combinations of sensor class of PI, MPC, with their average values of mean |*e*|, IAE and ISE

Sensor class	Evaluation criteria	Default PI		MPC	
		DO	NO	DO	NO
DO–A_O NO–B_O	Mean |*e*|	0.06	0.133	0.37	0.25
	IAE	0.44	0.9348	2.6	1.78
	ISE	0.048	0.208	1.48	0.72
DO–B_O NO–B_O	Mean |*e*|	0.13	0.133	1.03	0.22
	IAE	0.93	0.934	7.26	1.6
	ISE	0.42	0.2	10.23	0.55
DO–C_O NO–B_O	Mean |*e*|	0.55	0.13	1.19	0.23
	IAE	3.89	0.94	8.33	1.65
	ISE	4.6	0.21	12.82	0.56
DO–D_O NO–B_O	Mean |*e*|	0.060	0.13	0.37	0.25
	IAE	0.42	0.93	2.63	1.80
	ISE	0.45	0.20	1.51	0.73
DO–C_O NO–C_O	Mean |*e*|	0.59	0.24	1.18	0.37
	IAE	4.18	1.72	8.31	2.63
	ISE	4.99	0.78	12.86	1.51
DO–C_O NO–A_O	Mean |*e*|	0.55	0.084	0.65	0.53
	IAE	3.90	0.58	4.55	3.76

(continued)

Table 2 (continued)

Sensor class	Evaluation criteria	Default PI		MPC	
		DO	NO	DO	NO
	ISE	4.54	0.082	5.21	2.59
DO–Bo NO–Ao	Mean |e|	0.12	0.07	1.02	0.16
	IAE	0.87	0.54	7.16	1.12
	ISE	0.37	0.069	10.10	0.25
DO–Ao NO–Ao	Mean |e|	0.062	0.07	0.39	0.14
	IAE	0.43	0.54	2.74	1.00
	ISE	0.049	0.068	1.59	0.23
DO–Bo NO–Ao	Mean |e|	0.070	0.11	0.65	0.15
	IAE	0.49	0.80	4.59	1.10
	ISE	0.058	0.21	4.40	0.28

Table 3 Combinations of sensor class with PI, MPC, and Fuzzy with comparative improvements

Non-ideal sensor class	Performance of plant	Default PI	a	MPC	b	Fuzzy	c
DO–Ao NO–Bo	EQI	13,239	–	12,899	2.6	13,383	−0.01
	OCI	18,684	−0.016	18,630	−0.06	18,694	0.24
DO–Bo NO–Bo	EQI	13,233	0.045	12,763	3.6	13,384	−0.02
	OCI	18,689	−0.042	18,886	−1.4	18,694	0.24
DO–Co NO–Bo	EQI	13,156	0.62	13,183	1.8	13,383	−0.01
	OCI	18,759	−0.41	19,006	−2.0	18,694	0.24
DO–Do NO–Bo	EQI	13,237	0.015	13,244	−0.03	13,383	−0.01
	OCI	18,685	−0.02	18,597	0.11	18,694	0.24
DO–Co NO–Co	EQI	13,137	0.77	12,998	1.85	13,383	−0.01
	OCI	18,783	−0.39	19,050	−2.3	18,694	0.24
DO–Co NO–Ao	EQI	13,167	0.54	13,832	−4.44	13,380	0.007
	OCI	18,754	−0.39	19,010	−2.1	18,694	0.24
DO–Bo NO–Ao	EQI	13,245	−0.04	13,085	1.19	13,381	–
	OCI	18,682	−0.005	18,616	0.016	18,694	0.24
DO–Ao NO–Ao	EQI	13,245	-0.04	13,085	1.19	13,381	–
	OCI	18,682	−0.005	18,616	0.016	18,694	0.24
DO–Do NO–Do	EQI	15,590	−17.7	13,361	−0.89	13,380	0.007
	OCI	18,493	1.0	18,648	−0.15	18,694	0.24

compared under nine different sensor classes and it is observed that the operational cost is low at Do and Do, high at Co and Co combination. Whereas, the effluent quality showed improvement at Co and Co and worsened at Do and Do combination. The comparison is done with ideal sensors of PI. MPC is compared with nine different sensor classes and the operational cost is low at Do and Bo, high at Co and Co combination, whereas the effluent quality showed improvement at Bo and Bo and worsened at Bo and Ao combination. The comparison is done with ideal sensors of MPC. Fuzzy is not well responding to changing sensors with the same cost for all sensors combinations, and minor changes are observed in EQI.

References

1. Henze, M., Gujer, W., Mino, T, & van Loosdrecht, M. C. (2000) Activated sludge models ASM1, ASM2, ASM2d, and ASM3. IWA Scientific and Technical Report No. 9, IWA Publishing.
2. Rieger, L., Koch, G., Kühni, M., Gujer, W., & Siegrist, H. (2001). The EAWAG bio-P module for activated ludge model no. 3. *Water Research, 35*(16), 3887–3903. https://doi.org/10.1016/S0043-1354(01)00110-5
3. Solon, K. (2015). Activated sludge model no. 3 with bioP module (ASMbioP) implemented within the benchmark simulation model no.1, Technical Report. Division of Industrial Electrical Engineering and Automation Faculty of Engineering, Lund University, Sweden.
4. Gernaey, K. V., & Jørgensen, S. B. (2004). Benchmarking combined biological phosphorus and nitrogen removal wastewater treatment processes. *Control Engineering Practice, 12*(3), 357–373. https://doi.org/10.1016/S0967-0661(03)00080-7
5. Ostace, G. S., Baeza, J. A., Guerrero, J., Guisasola, A., Cristea, V. M., Agachi, P. Ş, & Lafuente, F. J. (2013). Development and economic assessment of different WWTP control strategies for optimal simultaneous removal of carbon, nitrogen and phosphorus. *Computers & Chemical Engineering, 53*, 164–177. https://doi.org/10.1016/j.compchemeng.2013.03.007
6. Stare, A., Vrečko, D., Hvala, N., & Strmčnik, S. (2007). Comparison of control strategies for nitrogen removal in an activated sludge process in terms of operating costs: A simulation study. *Water Research, 41*, 2004–2014. https://doi.org/10.1016/j.watres.2007.01.029
7. Guerrero, J., Guisasola, A., Comas, J., Rodríguez-Roda, I., & Baeza, J. A. (2012). Multicriteria selection of optimum WWTP control setpoints based on microbiology-related failures, effluent quality and operating costs. *Chemical Engineering Journal, 188*, 23–29. https://doi.org/10.1016/j.envsoft.2010.10.012
8. Guerrero, J., Guisasola, A., Vilanova, R., & Baeza, J. A. (2011). Improving the performance of a WWTP control system by model-based setpoint optimisation. *Environmental Modelling and Software, 26*, 492–497. https://doi.org/10.1016/j.cej.2012.01.115
9. Rojas, J. D., Baeza, J. A., & Vilanova, R. (2011). Effect of the controller tuning on the performance of the BSM1 using a data driven approach. In *Watermatex. 8th IWA Symposium on Systems Analysis and Integrated Assessment* (pp. 785–792)
10. Ogata, K. (2010). *Modern control engineering*. Prentice Hall.
11. Shiek, A. G., Machavolu, V. S. R. K., Seepana, M. M., & Ambati, S. R. (2020). Design of control strategies for nutrient removal in a biological wastewater treatment process. *Environmental Science and Pollution Research, 28*, 12092–12106. https://doi.org/10.1007/s11356-020-09347-2
12. Sheik, A. G., Seepana, M. M., & Ambati, S. R. (2021). Supervisory control configurations design for nitrogen and phosphorus removal in wastewater treatment plants. *Water Environment Research*. https://doi.org/10.1002/wer.1512

13. Maheswari, P., Sheik, A. G., Tejaswini, E. S. S., & Ambati, S. R. (2020). Nested control loop configuration for a three stage biological wastewater treatment process. *Chemical Pro and Pro Modelling, 16*(2), 87–100. https://doi.org/10.1515/cppm-2020-0035
14. Zhang, X., Ji, J., & Xu, J. (2019). Parameter identification of time-delayed nonlinear systems: An integrated method with adaptive noise correction. *Journal of Franklin Institute, 356*, 5858–5880. https://doi.org/10.1016/j.jfranklin.2019.03.023
15. Copp, J. B. (2002). *The cost simulation benchmark: description and simulator manual (Cost action 624 and action 682)*. Office for Official Publications of the European Union.
16. Sanchez, A., & Katebi, M. (2003). Predictive control of dissolved oxygen in an activated sludge wastewater treatment plant. In *European Control Conference (ECC)* (pp. 2424–2429)
17. Man, Y., Shen, W., Chen, X., Long, Z., & Corriou, J. P. (2018) Dissolved oxygen control strategies for the industrial sequencing batch reactor of the wastewater treatment process in the papermaking industry. *Environmental Science Water Research Technology, 4*, 654–662. https://doi.org/10.1039/c8ew00035b
18. Lin, M. J., & Luo, F. (2016). Adaptive neural control of the dissolved oxygen concentration in WWTPs based on disturbance observer. *Neurocomputing, 185*, 133–141. https://doi.org/10.1016/j.neucom.2015.12.045
19. Rieger, L., Koch, G., Kühni, M., Gujer, W., & Siegrist, H. (2001). The EAWAG Bio-P module for activated sludge model No. 3. *Water Research, 35*(16), 3887–3903. https://doi.org/10.1016/S0043-1354(01)00110-5
20. Takács, I., Patry, G. G., & Nolasco, D. (1991). A dynamic model of the clarification-thickening process. *Water Research, 25*(10), 1263–1271. https://doi.org/10.1016/0043-1354(91)90066-Y
21. Hongyang, X., Pedret, C., Santin, I., & Vilanova, R. (2018). Decentralized model predictive control for N and P removal in wastewater treatment plants. In *22nd International Conference on System Theory, Control and Computing (ICSTCC)* (pp. 224–230). IEEE. https://doi.org/10.1109/ICSTCC.2018.8540675
22. Gernaey, K. V., & Jeppsson, U. (2014). *Benchmarking of control strategies for wastewater treatment plants*. IWA Publishing.
23. Rieger, L., Alex, J., Gujer, W., & Siegrist, H. (2006). Modelling of aeration systems at wastewater treatment plants. *Water Science and Technology, 53*(4–5), 439–447. https://doi.org/10.2166/wst.2006.100
24. Amand, L., Olsson, G., & Carlsson, B. (2013). Aeration control—A review. *Water Science & Technology, 67*(11), 2374–2398. https://doi.org/10.2166/wst.2013.139
25. Grimholt, C., & Skogestad, S. (2018). Optimal PI and PID control of first-order plus delay processes andevaluation of the original and improved SIMC rules. *Journal of Process Control, 70*, 36–46. https://doi.org/10.1016/j.jprocont.2018.06.011
26. Mohammad, H. D., Gobinath, R., & Ramarao, K. (2021). *Soft computing techniques in solid waste and wastewater management*. Elsevier Science.

Climate Change Impact on the Groundwater Resources of Nagavali Basin Using QSWAT Model

M. Jothi Jeya Sri and K. Venkata Reddy

Abstract Groundwater is one of the major freshwater resources, which has been under stress nowadays. The rate depletion of groundwater resources is more than its recharge rate due to the pumping of groundwater for industrial and agricultural use. Global warming causes a change in precipitation and temperature rate. Nagavali basin is a tropical river basin that lies between Mahanadi and the Godavari river basins of South India. It is a major source for agriculture, industrial and domestic purposes. This report assesses the possible future climatic condition and changes in the aquifer recharge with climactic parameters under two Representative Concentration Pathway (RCP) scenarios of 4.5 and 8.5. The hydrological balance was studied with the SWAT model using the climate parameters. The groundwater recharge and evapotranspiration rate with the change of temperature and precipitation for the calibration period 1991–2018 and the future period of 2021–2080 were carried out. The QSWAT model was calibrated for the observed streamflow data. From the calibration, it is clear that the observed and simulated streamflow data have a significant correlation with an R^2 value of 0.85. The future prediction of rainfall and temperature for the period of 2021–2080 shows the pattern of future weather. There is an increase in temperature in the future period of 2% for RCP 4.5 and 4% for RCP8.5 scenarios. In the future period, the amount of recharge is increased compared to the present status. The average aquifer recharge was 9.62 and 11.64 m under RCP 4.5 and 8.5. The average evapotranspiration is decreased in the future period by 16 and 11% under RCP 4.5 and RCP 8.5 scenarios compared to base period. Due to increased percentage of precipitation and decreased evapotranspiration under the RCP 8.5 scenario, there is an increase in recharge. Methodology presented in this paper is useful for groundwater studies of the basins in semi-arid regions under climate change scenarios.

M. Jothi Jeya Sri (✉) · K. Venkata Reddy
Department of Civil Engineering, National Institute of Technology, Warangal, Telangana, India
e-mail: jothi_191856@student.nitw.ac.in

K. Venkata Reddy
e-mail: kvreddy@nitw.ac.in

A. K. Dikshit et al. (eds.), *Innovative Trends in Hydrological and Environmental Systems*, Lecture Notes in Civil Engineering 234,
https://doi.org/10.1007/978-981-19-0304-5_46

Keywords Climate change · Climate model · Evapotranspiration · Groundwater recharge · Representative concentration pathway (RCP)

1 Introduction

Groundwater is an integral part of the hydrological cycle. Groundwater is a key resource for human development. On a global scale, one-third of the population depends on groundwater for their drinking water, in urban as well as rural areas. Groundwater also plays a pivotal role in agriculture, and an increasing portion of groundwater is being used for agriculture. Central Ground Water Board mentioned that contribution of groundwater for irrigation was 48.19% in their 2009 report (Kambale et al. 2007). The IPCC (2013) has reported that changes in climatic factors will affect the groundwater availability and quality. Presently in most of the places in India groundwater is being over-exploited. This over-exploitation of water resources may result in future famine. Hence, it is necessary to maintain the groundwater availability for the future period.

The climate parameters temperature, precipitation, wind speed, and solar radiation have major effects on the groundwater recharge capacity. The change in trend pattern of climate variables gives the idea of the change of climate over the region. Many previous assessments had been focused on trend analysis of climate variables and change in groundwater level with climate change. Kumar [14] reviewed coupling of climate data with the specific hydrological model used for the study of water balance components. The spatiotemporal change of climate variables has its impact on the water balance component of the watershed and reported that the highest rainfall accelerated the recharge over the watershed [17]. Marhaento et al. [16] studied the change of hydrological components in response to future landuse change and climate change. With their study, it was concluded that the fraction of changes is more under the climate change scenario, so the impact of land-use change is slower compared to climate change. Mogaji et al. [18] assessed the recharge rate and PET with the empirical relationship equation. The change of climatic conditions concerning economic and environmental development scenarios such as A1B, A2, B1. Alam et al. [1] assessed the groundwater storage of California central valley using distributed variable infiltration capacity model (VIC), where the change in groundwater was assessed for the different cropping patterns and agriculture practice.

Shrestha et al. [30] assessed the change of groundwater level for the period of 2011–2100 using the WETPASS and MODLFOW model. They used average precipitation and temperature from five different GCM under two different RCP scenarios of 4.5 and 8.5. In their study, the recharge rate gets decreased mainly during the dry season. Kahsay et al. [12] assessed future groundwater recharge and base flow of Tekeze basin, Ethiopia using the WETSPA model. They concluded that the rate of precipitation has impact on variation of recharge rate. Kambale et al. [13] predicted the climate change impact on the northern part of India for the future period of 2030. The groundwater level variation was studied using the hydrological models

HYDRUS-1D and MODFLOW. They reported that the recharge rate might vary with the number of climate parameters considered for analysis and soil types over the region. Shrestha et al. [30] assessed the climatic factor change for the period of 2006–2100 and its influence on recharge for four Asian cities. The climate parameters from five different RCM were coupled with the model WETSPASS and SWAT model. Based on the size of the area the model was selected and concluded that recharge is mainly due to amount of rainfall and temperature. More decrease in recharge occurs during the summer season due to an increase in temperature. Loukika et al. [15] estimated the groundwater recharge of chintalapudi village, West Godavari for the period of 1990–2000. For the groundwater recharge estimation coupled SWAT-MODFLOW were used, where they calculated the groundwater recharge on monthly basis in wells based on aquifer parameters.

In this paper, groundwater recharge variation for Nagavali basin is studied under climate change scenarios over 60 years from 2021 to 2080. The monthly groundwater recharge has been estimated using the SWAT model. The climate models namely ACCESS, CNRM, and MPI have been used for the climate data retrieved, and the groundwater recharge has been estimated using a model. The seasonal and annual variation of climate variables and groundwater recharge has been studied over the entire study basin.

2 Study Area

The study area was the Nagavali river basin shown in Fig. 1 is an imperative basin among the east-flowing river. An independent and interstate river lies in between the Mahanadi and Godavari basins. The Nagavali originates near the Lakhabahal in the Kalahandi district of Odisha at an elevation of about 1300 m (source: http://www.dowrodisha.gov.in/BasinMaps/IndexofBasins.htm). It drains parts of the districts of Odisha and Andhra Pradesh State. Basin lies on the geographical coordinates of 18° 17′ to 19° 44′ latitude and 82° 53′ to 83° 54′ longitude. The total basin area is 9510 Sq. Km. This river plays a major role in supporting irrigation demand and water supply demand for the district of Koraput, Rayagada, Vizianagaram, and Srikakulam. The total population of the basin as per census 2011 was 6, 76, 678. The top region of basin was covered with forest so that the lower portion of 5559Sq. Km was analyzed in the study.

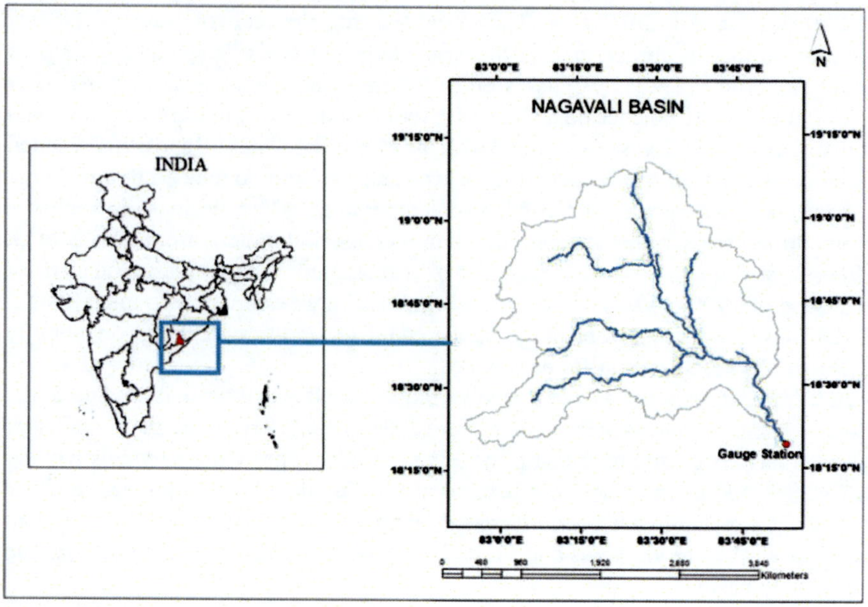

Fig. 1 Location map of Nagavali river basin

3 Data and Methods

3.1 Data

For calculation of recharge of aquifer using SWAT model, the SWAT model requires weather parameters (precipitation and temperature), streamflow data, the digital elevation model, landuse and landcover, and soil properties.

The climate parameters (rainfall and temperature) for the historic period were downloaded from the Indian Meteorological Department (IMD), Pune. The rainfall data is available with the resolution of 0.25° × 0.25°. For the analysis, eight grid points data within the basin were downloaded for the period of 1990–2018. The temperature of 1°×1° grid was also collected for the period of 1990–2018. For the future rainfall and temperature, three climate models were selected data has been downloaded and bias-corrected for the future period of 2021–2080.

An Aster DEM of 30 m × 30 m shown in Fig. 2 was used to delineate watershed and stream network generation. The landuse/landcover (LULC) data and soil map of the study area was clipped from SWAT Indian datasets. LULC shown in Fig. 3 is useful for the generation of hydrological response units (HRUs). The soil map shown in Fig. 4 shows that the entire area has three types of soil namely loam, sandy loam, and sandy clay loam. The majority of the upstream portion is covered with sandy

Fig. 2 Digital elevation model

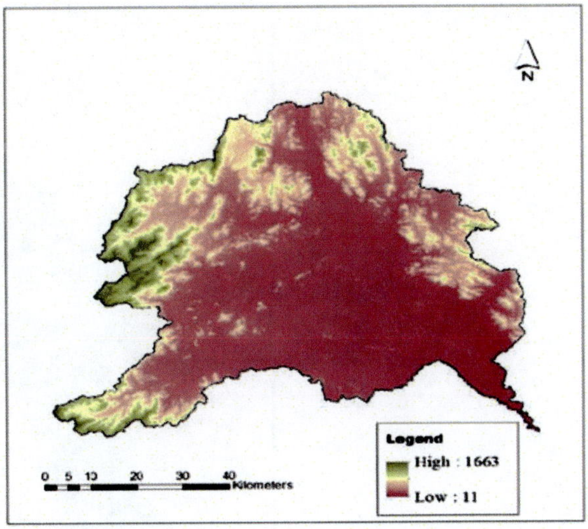

Fig. 3 Landuse/landcover map

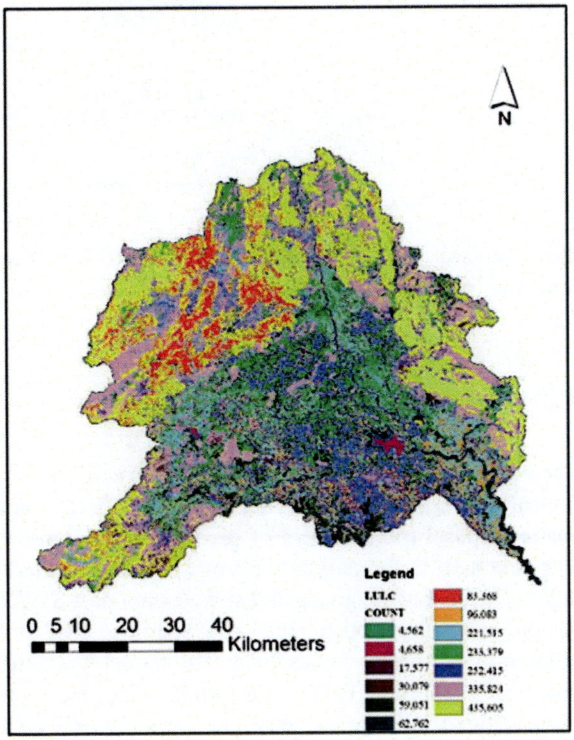

Fig. 4 Soil map

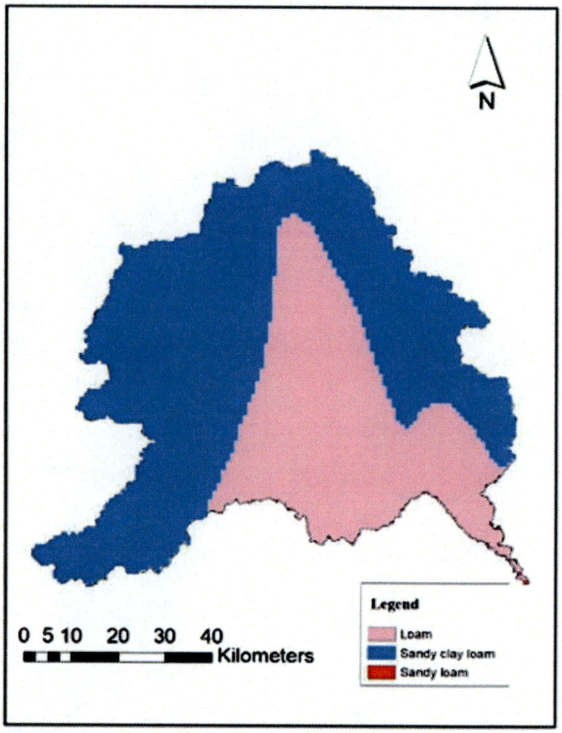

clay loam soil. The stream flow data obtained from the Srikakulam gauge station is essential for the calibration of the SWAT model.

3.2 Methodology

The methodology for the work includes the coupling of climate data with the hydrological model SWAT that is stated in Fig. 5. The study involves analysis of the spatiotemporal trend pattern of precipitation, temperature, and groundwater level. The first step is to analyze the trend of measured and downloaded climatic variables. Then using the geospatial and weather data SWAT model was initially running for the historic period of 1991–2018. In the SWAT model, the basin was subdivided into sub-basins for analysis. The model was calibrated and validated with the observed streamflow data of the Srikakulam gauge station using SWAT CUP. The climate model data from three models namely ACCESS, MPI, CNRM were used for future precipitation and temperature. The data has been downloaded under 2 RCP Scenarios RCP 4.5 and RCP 8.5 and the downloaded model data were bias-corrected using Quantile mapping. The bias-corrected climate variable was given as input for

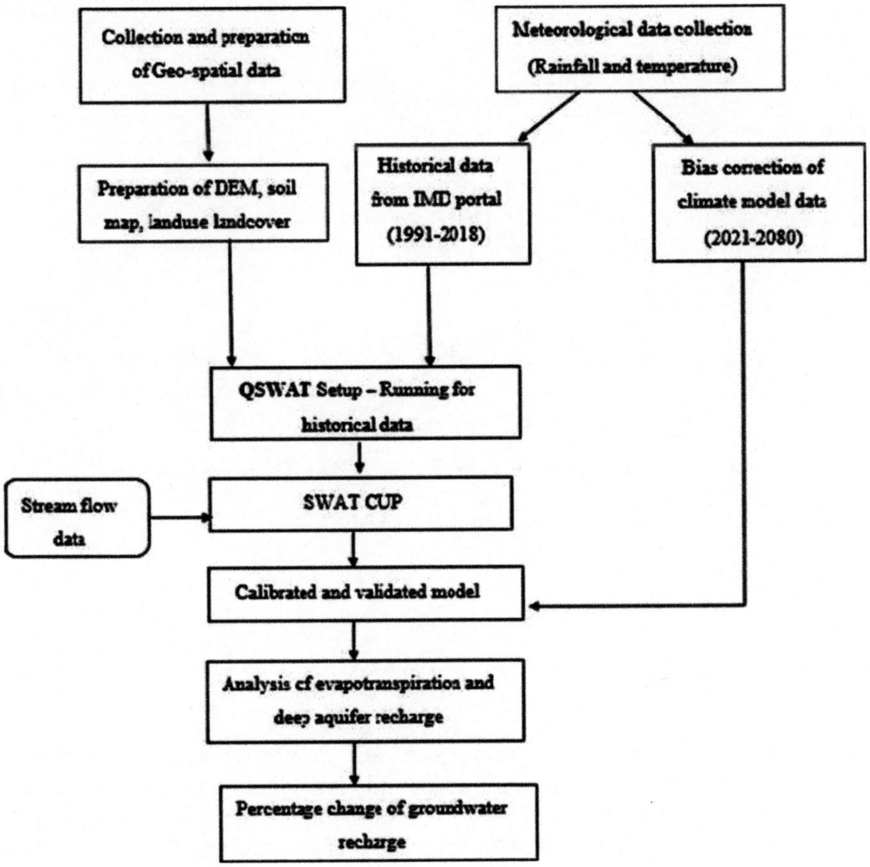

Fig. 5 Conceptual flowchart for assessing the groundwater recharge

calibrated SWAT model. Using the SWAT HRU output file the deep percolation value, evapotranspiration amount for each sub-basin and each land-use type were calculated. Finally, the percentage change in evapotranspiration and groundwater recharge was calculated over 60 years for the prediction of future possibilities of recharge.

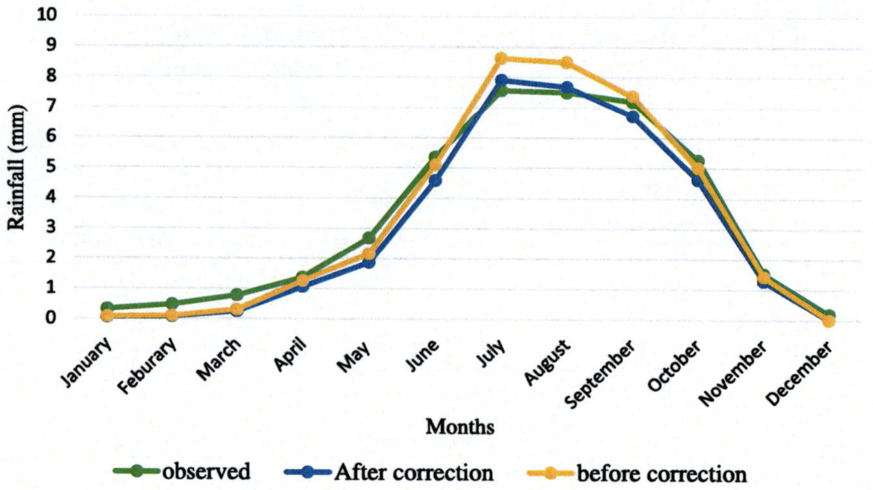

Fig. 6 Correlation between observed and simulated rainfall

4 Result and Discussion

4.1 Bias Correction

The climate data has been corrected for systematic error in the data using Quantile mapping method. The bias corrections of monthly precipitation for the base period of 1950–2005 are presented in Fig. 6. From the figure, corrected precipitation is found to be consistent with observed precipitation. Hence, the corrected data shows close agreement with the observed value in the historic period. So with the best fit transformation function, the future period (2021- 2080) rainfall has been corrected.

4.2 Trend Analysis

4.2.1 Precipitation

MK test statistic (S) indicates a significant increasing trend in future precipitation. Rainfall is predicted to increase gradually in both the scenarios. From Fig. 7 the rainfall variation compared to base period shows that there is 6% increase in RCP 4.5 scenario and 85% increase in RCP 8.5 scenario. The average annual precipitation was 2192.93 and 1186.54 mm. There is a decrease in precipitation for the years 2024, 2025, and 2026 compared to the present average precipitation. Hence, in the future, these years may receive less rainfall.

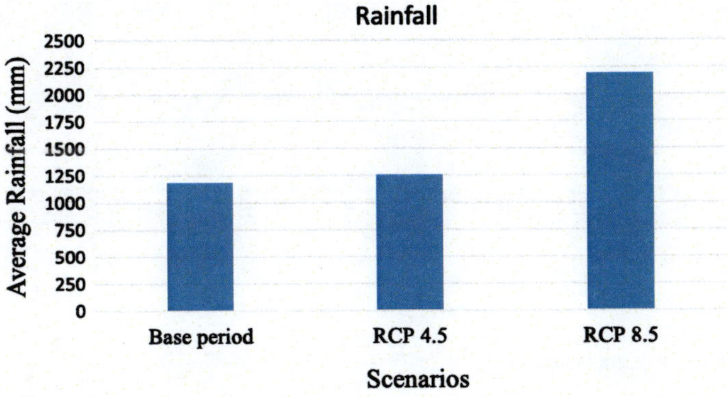

Fig. 7 Rainfall variation

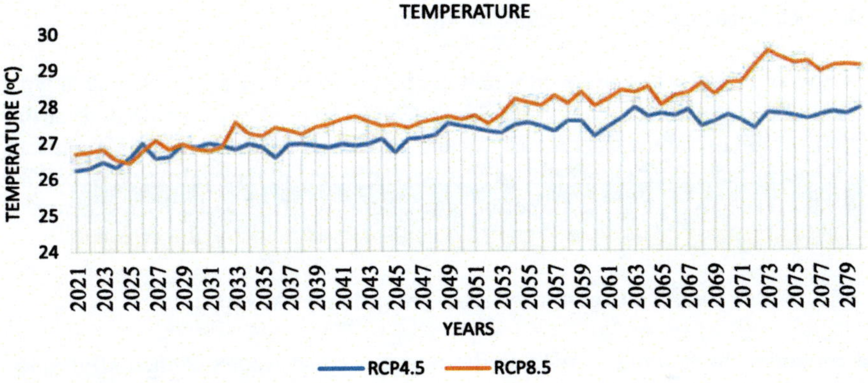

Fig. 8 Trend plot of future temperature

4.2.2 Temperature

For the base period, the maximum average temperature was 32.26 °C and the minimum average temperature was 20.92 °C. The average temperature increases by 2% and 4% under RCP 4.5 and RCP 8.5 scenarios. The temperature plot shown in Fig. 8 depicts that the temperature is more under RCP 8.5 scenario.

4.2.3 Evapotranspiration

The average evapotranspiration is 119.39 and 126.91 mm under RCP 4.5 and RCP 8.5 scenarios, respectively. The increase in temperature by 4% under RCP 8.5 is causing more evapotranspiration compared to RCP 4.5 scenario (Fig. 9).

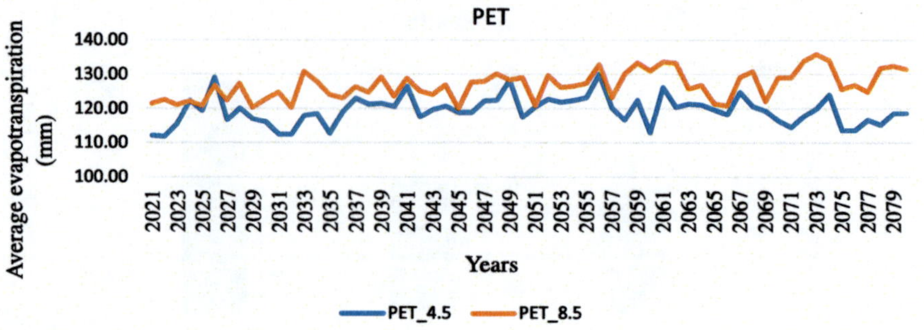

Fig. 9 Potential evapotranspiration

4.3 Estimation of Groundwater Recharge

4.3.1 Calibration and Validation of Model

The SWAT model was calibrated with the observed monthly streamflow data with the simulated streamflow data using SWAT CUP. The model was calibrated for the period of 1991–2005 and validated for the period for 2006–2014. The statistical model performance indicators R^2 and NSE for the period of the calibration were 0.85 and 0.80, respectively and for the period of validation was 0.73 and 0.70, respectively.

4.3.2 Water Balance Component and Groundwater Recharge

After running the calibrated SWAT model, the water balance components and groundwater recharge have been simulated for the future period of 2021–2080. The rainfall, evapotranspiration, and recharge variation have been presented in the table show that there is an increase in future precipitation with the increase of temperature by 0.46 and 1.08 °C under climate change scenarios (RCP 4.5 and RCP 8.5). However, there is a possibility of decrease in the rate of evaporation due to an increase in rainfall amount (Table 1).

Therefore, the decrease in PET by 16% to 11% under RCP 4.5 and RCP 8.5 leads to an increase of recharge rate by 57–90%. The average groundwater recharge (Amount of water entering the deep aquifer) is 9.62 m of water and 11.64 m of water under scenarios 4.5 and 8.5. The reason for the increase in recharge is due to the

Table 1 Percentage change of water balance components

	Rainfall (mm)	% change	AET (mm)	% change	Recharge (m)	% change
Base period	1187.4		142.04		6.13	
RCP4.5	1255.84	6	119.45	−16	9.62	57
RCP8.5	2192.93	85	126.91	−11	11.64	90

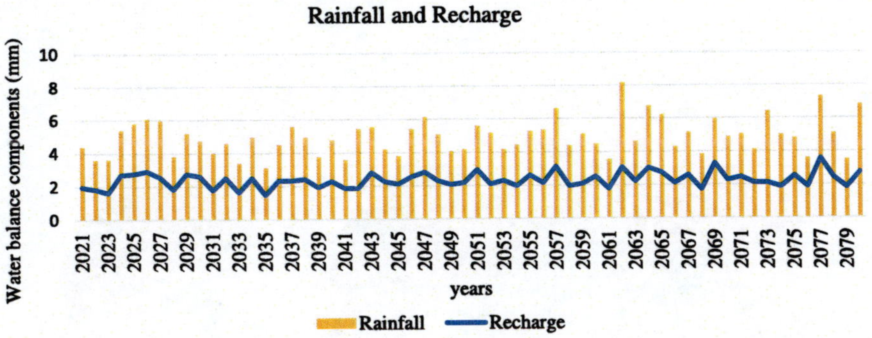

(i)

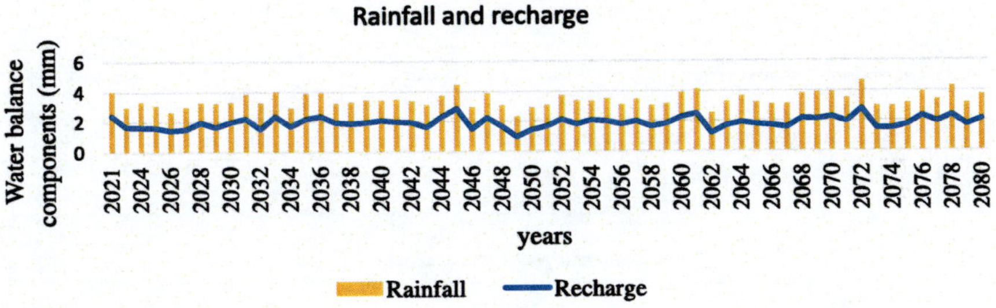

(ii)

Fig. 10 Trend plot of recharge and rainfall under **a** RCP 4.5, **b** RCP 8.5

permeability of the soil and the increase in precipitation rate. The combination plot of rainfall and recharge is shown in Fig. 10a, b indicates that the rainfall has direct influence on the recharge. Therefore, the year with higher rainfall has higher recharge capacity.

The spatial distribution map of recharge and evapotranspiration for both the scenarios was prepared in GIS environment under both the scenarios. The recharge map is shown in Fig. 11. It is seen that the future period (2021–2080) has possibility of higher recharge capacity over the basin under both scenarios. The evapotranspiration plot is shown in Fig. 12.

4.3.3 Temporal Analysis of Water Balance

The temporal analysis refers to the seasonal analysis of weather and water balance components. For seasonal analysis, the entire year was divided into three seasons which is presented in Table 2. Season I refers to the monsoon season (June to September), Season II refers to the winter season (October to January), and Season III is termed as the summer season (February to May).

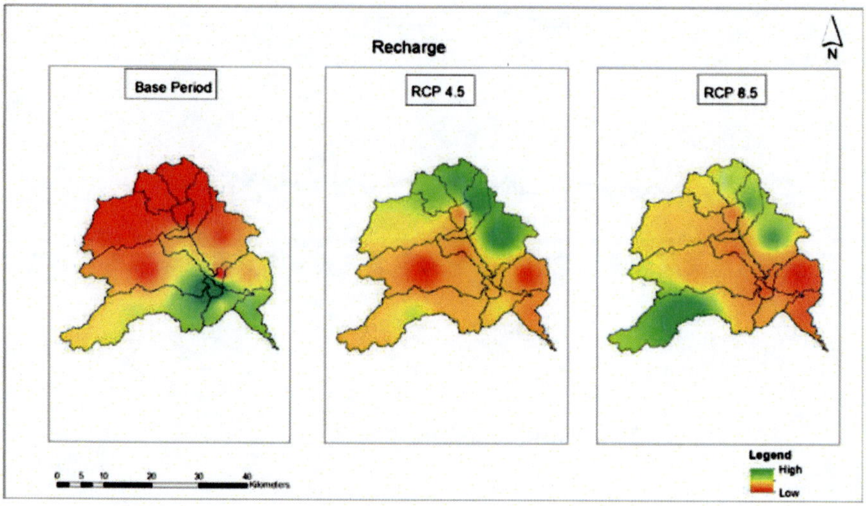

Fig. 11 Spatial distribution of recharge in the basin

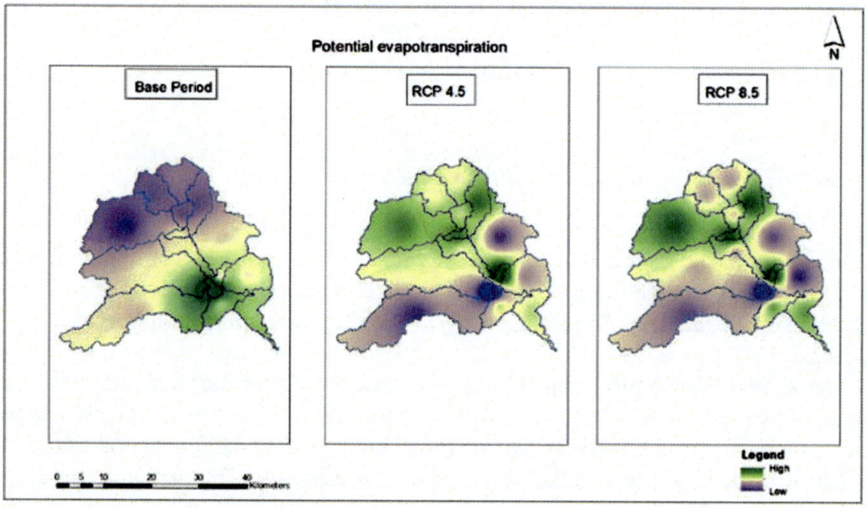

Fig. 12 Spatial distribution of evapotranspiration in the basin

The season-wise analysis of the water balance is presented in Table 2. Under RCP 4.5 scenario, the monsoon season receives 76% of total rainfall, contributes 56% of total recharge, and 25% of evapotranspiration. However, in the summer season, the increase of temperature contributes to 40% of total evapotranspiration, 1% of total recharge. Hence, in the summer season, the recharge rate is limited with increased evapotranspiration with increased temperature.

Table 2 Season-wise water balance

Scenario and period	Season	Precipitation		Evapotranspiration		Recharge	
		P (mm)	%	ET (mm)	%	R (mm)	%
RCP 4.5 (2021–2080)	Monsoon	952.55	76	365.48	25	5389.17	56
	Post monsoon	193.86	15	432.00	30	4126.71	43
	Pre-monsoon	109.43	9	635.87	44	101.09	1
	Total	1255.84		1433.35		9616.97	
RCP 8.5 (2021–2080)	Monsoon	1527.19	70	386.56	25	6918.46	59%
	Post monsoon	584.82	27	448.93	29	4620.68	40%
	Pre-monsoon	80.92	4	687.5	45	103.82	1%
	Total	2192.93		1522.99		11,643	

Under RCP 8.5 Scenario, the monsoon season receives 70% of total rainfall, contributes 59% of total recharge with 25% of evapotranspiration. The rate of recharge is less and potential evapotranspiration is more during the summer season. Hence the increase in temperature is affecting the recharge condition. In the RCP 8.5 scenario, the percentage of winter rainfall is 27% that is more than RCP 4.5 winter rainfall and it indicates that there is, possibility of a shift in the rainfall if the likely emission scenario of the atmosphere is 8.5.

5 Conclusions

The study involves the change in recharge with the influence of climatic conditions. This study analyses the spatiotemporal patterns of weather, groundwater recharge, and evapotranspiration during the period of 2021–2080 under RCP scenarios of 4.5 and 8.5. Trend pattern of rainfall and temperature shows that it increases gradually in the upcoming years. There is a decrease in the future evapotranspiration by 16 and 11% under RCP 4.5 and 8.5 scenarios compared to the present status. From the simulation output, it is seen that the groundwater recharge is increasing in the future period. The average groundwater recharge for the study area in the base period was 6.13 m. In the future period, the average recharge over the basin was 9.62 and 11.64 m under RCP 4.5 and RCP 8.5, respectively and average evapotranspiration was 119.45 mm and 126.91 m. There is decrease in recharge during summer season with the increase of evapotranspiration rate. The increase in precipitation by 6–85% is causing the recharge to increase by 57–90% in the future period. Overall, in the Nagavali basin, there is increase in groundwater recharge in the upcoming years. The reason for the increase of recharge is increase in rainfall rate and soil with higher permeability.

References

1. Alam, S., Gebremichael, M., Dozier, J., & Lettenmaier, D. P. (2019). Climate change impacts on groundwater storage in the Central Valley, California. *Climatic Change, 157*(3–4), 387–406. https://doi.org/10.1007/s10584-019-02585-5g
2. Batelaan, O., & De Smedt, F. (2007). GIS-based recharge estimation by coupling surface—subsurface water balances. *Journal of Hydrology, 337*(3–4), 337–355. https://doi.org/10.1016/j.jhydrol.2007.02.001
3. Chung, I. M., Kim, N. W., Lee, J., & Sophocleous, M. (2010). Assessing distributed groundwater recharge rate using integrated surface water-groundwater modelling: Application to Mihocheon watershed, South Korea. *Hydrogeology Journal, 18*(5), 1253–1264. https://doi.org/10.1007/s10040-010-0593
4. Chunn, D., Faramarzi, M., Smerdon, B., & Alessi, D. S. (2019). Application of an integrated SWAT-MODFLOW model to evaluate potential impacts of climate change and water withdrawals on groundwater-surface water interactions in west-central Alberta. *Water (Switzerland), 11*(1). https://doi.org/10.3390/w11010110
5. Feng, W., Shum, C. K., Zhong, M., & Pan, Y. (2018). Groundwater storage changes in China from satellite gravity: An overview. *Remote Sensing, 10*(5), 1–25. https://doi.org/10.3390/rs10050674
6. Gajbhiye, S., Meshram, C., Mirabbasi, R., & Sharma, S. K. (2016). Trend analysis of rainfall time series for Sindh river basin in India. *Theoretical and Applied Climatology, 125*(3–4), 593–608. https://doi.org/10.1007/s00704-015-1529-4
7. Ghazavi, R., & Ebrahimi, H. (2019). Predicting the impacts of climate change on groundwater recharge in an arid environment using modeling approach. *International Journal of Climate Change Strategies and Management, 11*(1), 88–99. https://doi.org/10.1108/IJCCSM-04-2017-0085
8. Goderniaux, P., Brouyére, S., Blenkinsop, S., Burton, A., Fowler, H. J., Orban, P., & Dassargues, A. (2011). Modeling climate change impacts on groundwater resources using transient stochastic climatic scenarios. *Water Resources Research, 47*(12). https://doi.org/10.1029/2010WR010082
9. Gosain, A. K., Rao, S., & Basuray, D. (2006). Climate change impact assessment on hydrology of Indian river basins. *Current Science, 90*(3), 346–353.
10. Jalota, S. K., Vashisht, B. B., Sharma, S., & Kaur, S. (2018). Climate change projections. *Understanding Climate Change Impacts on Crop Productivity and Water Balance*. https://doi.org/10.1016/b978-0-12-809520-1.00002
11. Jana, S., Das, M., Roy, D., Das, S., & Mazumdar, A. (2015). Simulation of climate change impact in a river basin in Eastern India. *International Journal of Hydrology Science and Technology, 5*(4), 314–332. https://doi.org/10.1504/IJHST.2015.072631
12. Kahsay, K. D., Pingale, S. M., Hatiye, S. D. (2018). Impact of climate change on groundwater recharge and base flow in the sub-catchment of Tekeze basin, Ethiopia. *Groundwater for Sustainable Development, 6*, 121–133.1 https://doi.org/10.1016/j.gsd.2017.12.002
13. Kambale, J. B., Singh, D. K., & Sarangi, A. (2017). Impact of climate change on groundwater recharge in a semi-arid region of northern India. *Applied Ecology and Environmental Research, 15*(1), 335–362. https://doi.org/10.15666/aeer/1501_335362
14. Kumar, C. P. (2016). Impact of climate change on groundwater resources. *Natural Resources Management: Concepts, Methodologies, Tools, and Applications*, 3–2 (January 2010), 1094–1120. https://doi.org/10.4018/978-1-5225-0803-8.ch052
15. Loukika, K. N., Reddy, K. V., Rao, K. H., & Singh, A. (2020). Estimation of Groundwater recharge rate using SWAT MODFLOW model. *Springer Singapore*. https://doi.org/10.1007/978-981-13-7067-0
16. Marhaento, H., Booij, M. J., & Hoekstra, A. Y. (2018). Hydrological response to future land-use change and climate change in a tropical catchment. *Hydrological Sciences Journal, 63*(9), 1368–1385. https://doi.org/10.1080/02626667.2018.1511054

17. Meresa, E., & Taye, G. (2019). Estimation of groundwater recharge using GIS-based WetSpass model for Birki watershed, the eastern zone of Tigray, Northern Ethiopia. *Sustainable Water Resources Management, 5*(4), 1555–1566. https://doi.org/10.1007/s40899-018-0282-0
18. Mogaji, K. A., Lim, H. S., & Abdullah, K. (2013). Regional modelling of climate change impacts on groundwater resources sustainability in peninsular Malaysia. *Modern Applied Science, 7*(5), 72–97. https://doi.org/10.5539/mas.v7n5p72
19. Ng, G. H. C., McLaughlin, D., Entekhabi, D., & Scanlon, B. R. (2010). Probabilistic analysis of the effects of climate change on groundwater recharge. *Water Resources Research, 46*(7), 1–18. https://doi.org/10.1029/2009wr007904
20. Niraula, R., Meixner, T., Dominguez, F., Bhattarai, N., Rodell, M., Ajami, H., Gochis, D., & Castro, C. (2017). How might recharge change under projected climate change in the Western U.S.? *Geophysical Research Letters, 44*(20), 10407–10418. https://doi.org/10.1002/2017GL07542
21. Nistor, M. M., Dezsi, Ş, Cheval, S., & Baciu, M. (2016). Climate change effects on groundwater resources: A new assessment method through climate indices and effective precipitation in Beliş district, Western Carpathians. *Meteorological Applications, 23*(3), 554–561. https://doi.org/10.1002/met.1578
22. Nkwasa, A., Chawanda, C. J., Msigwa, A., Komakech, H. C., Verbeiren, B., & van Griensven, A. (2020). How can we represent seasonal land use dynamics in SWAT and SWAT+ models for African cultivated catchments. *Water (Switzerland), 12*(6). https://doi.org/10.3390/W12061541
23. Nyenje, P. M., & Batelaan, O. (2009). Estimating the effects of climate change on groundwater recharge and baseflow in the upper Ssezibwa catchment, Uganda. *Hydrological Sciences Journal, 54*(4), 713–726. https://doi.org/10.1623/hysj.54.4.713
24. Pisani, B., Samper, J., & Marques, J. E. (2019). Climate change impact on groundwater resources of a hard rock mountain region (Serra da Estrela, Central Portugal). *Sustainable Water Resources Management, 5*(1), 289–304. https://doi.org/10.1007/s40899-017-0129-0
25. Reszler, C., Harum, T., Poltnig, W., Saccon, P., Reichl, P., Ruch, C., Kopeinig, C., & Freundl, G. (2012). Impact of climate change on groundwater resources in Southern Austria. 14(October), 7871. https://doi.org/10.13140/RG.2.2.25462.45125
26. Sahoo, S, Dhar A, Debsarkar A, Pradhan B, & Alamri AM (2020) Future Water Use Planning by Water Evaluation and Planning System Model. *Water Resources Management, 34*(15), 4649–4664. https://doi.org/10.1007/s11269-020-02680-8
27. Setti, S. (2018). Nagavali paper assessment of water balance for a forest dominated coastal river basin in India using a semi distributed hydrological model. *Modeling Earth Systems and Environment*. https://doi.org/10.1007/s40808-017-0402-0
28. Setti, S., Maheswaran, R., Radha, D., Sridhar, V., Barik, K. K., & Narasimham, M. L. (2020). Attribution of hydrologic changes in a tropical river basin to rainfall variability and land-use change: Case study from India. *Journal of Hydrologic Engineering, 25*(8), 05020015. https://doi.org/10.1061/(asce)he.1943-5584.0001937
29. Setti, S., Rathinasamy, M., & Chandramouli, S. (2018). Assessment of water balance for a forest dominated coastal river basin in India using a semi distributed hydrological model. *Modeling Earth Systems and Environment, 4*(1), 127–140. https://doi.org/10.1007/s40808-017-0402-0
30. Shrestha, S., Bach, T. V., & Pandey, V. P. (2016). Climate change impacts on groundwater resources in Mekong Delta under representative concentration pathways (RCPs) scenarios. *Environmental Science and Policy, 61*, 1–13. https://doi.org/10.1016/j.envsci.2016.03.010
31. Singh, A. K., Tripathi, J. N., Kotlia, B. S., Singh, K. K., & Kumar, A. (2019). Monitoring groundwater fluctuations over India during Indian Summer Monsoon (ISM) and Northeast monsoon using GRACE satellite: Impact on agriculture. *Quaternary International, 507*(October), 342–351. https://doi.org/10.1016/j.quaint.2018.10.036
32. Trzaska, S., & Schnarr, E. (2014). A review of downscaling methods for climate change projections. United States Agency for International Development by Tetra Tech ARD, September (pp. 1–42).

Hydrological Appraisal of Munneru River Basin Using SWAT Model

Eswar Sai Buri, Keesara Venkata Reddy, and Kotapati Narayana Loukika

Abstract Precipitation, temperature and remaining meteorological and hydrological variables are affecting the stream flow in a catchment. Hydrological appraisal for any river basin using hydrological models is necessary in the view of changing climate. In the present study, changes in the stream flow for the observed period are assessed for Munneru River basin, which comes under lower Krishna River region, India, using SWAT model. The goal of the study is to calibrate and validate the SWAT model for the chosen period using geospatial, meteorological and hydrological data, which will aid in calculating stream flows and understanding the study area's hydrological parameters. For model set-up, observed data from 1980 to 2017 is considered, and data from 1980 to 1982 is taken as the warm-up period. The model is calibrated for the period1983–2002 and validated for 2003–2017 streamflow data. The calibration, validation and sensitivity analyses are carried out in SWAT-CUP using SUFI-2 algorithm. Coefficient of determination (R^2) and Nash–Sutcliffe Efficiency (NSE) are used to evaluate the model performance. Results concluded that SWAT model simulates the stream flow at gauge station for the entire period of run with the correlation coefficient of 0.93 after updating the best-fitted parameters.

Keywords Munneru · Precipitation · SWAT · SWAT-CUP · SUFI-2 · Sensitivity analysis · Temperature

E. Sai Buri (✉) · K. Venkata Reddy · K. N. Loukika
Department of Civil Engineering, National Institute of Technology Warangal, Warangal, India
e-mail: beswar@student.nitw.ac.in

K. Venkata Reddy
e-mail: kvreddy@nitw.ac.in

K. N. Loukika
e-mail: loukikak@student.nitw.ac.in

A. K. Dikshit et al. (eds.), *Innovative Trends in Hydrological and Environmental Systems*, Lecture Notes in Civil Engineering 234,
https://doi.org/10.1007/978-981-19-0304-5_47

1 Introduction

Water plays a prominent role for the survival of human race and other living organisms. Due to the rapid growth of population and industrialization, it has become a scarce resource [10]. Hydrological models are increasingly being utilized to predict and understand the hydrological processes. By developing simplified conceptual representation of the complicated hydrological cycle using different parameters and multiple sets of mathematical equations, models are the effective tools for understanding and quantifying the components of the water balance and hydrological fluxes within the area of a catchment. Because of the size of the catchment and the dominating processes present, the performance and appropriateness of any hydrological model can vary between the catchments. These models are frequently developed for various reasons such as water demand estimation, hydrological processes comprehension, flood risk assessments and so on and also for distant geographic regions. For assessing the impact of current and future climate, estimation of various factors and selection of hydrological models are critical [8]. SWAT model was used for quantification of hydrological parameters at basin scale for current and future periods [4–6, 11–14].

SWAT model simulated the hydrological aspects in the long term for Purna River basin, India [9]. Calibrated SWAT model can be used to determine the current and potential impacts of climate and land-use change for the Bani catchment and also the catchment's water supplies, which is becoming more important knowledge for water managers [3]. The capacity of integrating gridded precipitation products into hydrological assessment models within the Atlantic coastal plain is represented by outcomes from long-term SWAT model calibration and validation [7]. In present study, the monthly observed streamflow values of CWC gauge station, present at Keesara village of Munneru River basin, are used to analyse the performance of SWAT model for the observed period (1980–2017) of 38 years.

2 Materials and Methods

2.1 Study Area

The Munneru River basin is one of the important and vulnerable sub-basins of Krishna River basin, India. It extends its drainage area across the districts of Khammam, Warangal (Telangana State), and Krishna (Andhra Pradesh State) comes under the lower Krishna basin, India. The boundaries of Munneru basin lie between 16.6–18.1° N latitudes and 79.2–80.8° E longitudes. The total drainage area of the basin is 9854 km^2. The gauge station is maintained by the Central Water Commission (CWC) at Keesara village for measuring gauge, discharge, sedimentation and qualitative aspects of the water (http://cwc.gov.in/kgbo/home/). After delineation, location map of the basin with stream network is shown in Fig. 1.

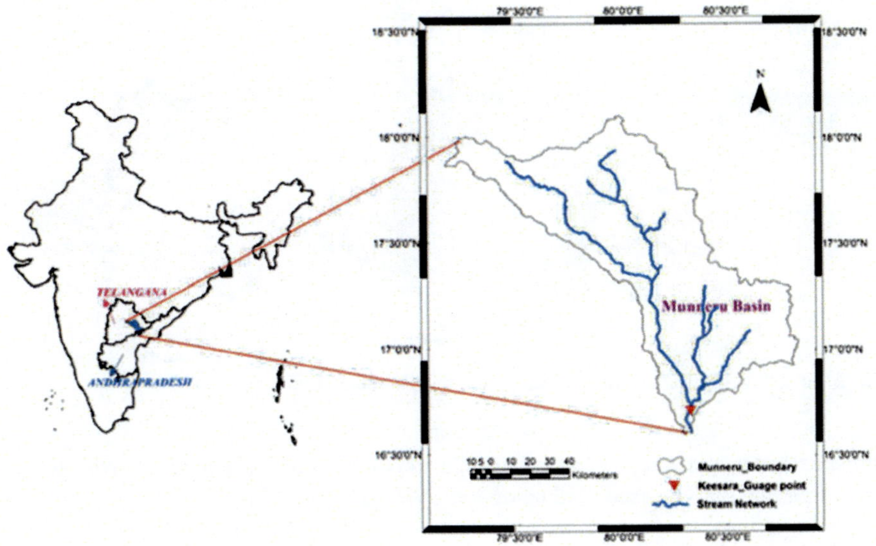

Fig. 1 Location map of Munneru basin

2.2 Data Used

Digital elevation model (DEM) prepared from CARTOSAT satellite data with 30 m resolution is used in the present study. Land use and land cover map (LULC) is obtained from National Remote Sensing Centre (NRSC) which is of 1:250 K scale. Soil map is taken from International Soil Reference and Information Centre (ISRIC) which is at 1 km resolution. Precipitation data for the period of 1901–2020 with 0.250 × 0.250 and maximum and minimum temperature for the period of 1951–2020 with 1.00 × 1.00 resolution are obtained from Indian Meteorological Department (IMD). Gauge, discharge, sediment and water quality data from 1965 to 2017 are obtained from Central Water Commission (CWC) at Keesara gauge point of the basin. DEM, LULC and soil map for the study are shown in Fig. 2.

2.3 SWAT

SWAT is a basin-scale, semi-distributed, physically oriented, open-source software that is widely used around the globe to examine the influence of sediment transport, fertilizer load and various water management strategies (http://swat.tamu.edu/docs/). QSWAT, a QGIS-interfaced version of SWAT, may be used to run a continuous simulation of a catchment model at various time increments and spatial scales. SWAT can forecast flow volumes on an hourly, daily, monthly and annual basis. Surface runoff, lateral subsurface flow, groundwater flow, evapotranspiration, snowmelt, stream

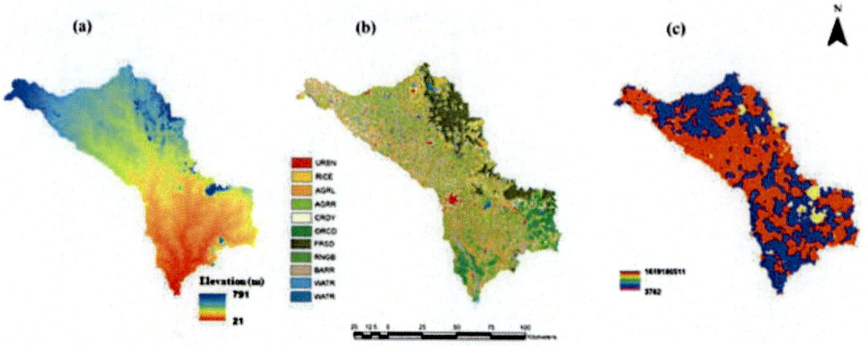

Fig. 2 (a) DEM, (b) LULC and (c) soil map of the Munneru basin

transmission losses and pond water storage and losses are all examples of hydrologic processes that may be modelled using the SWAT model.

2.4 SWAT CUP

For calibration and validation of SWAT model, SWAT-CUP (SWAT-Calibration Uncertainty Programme) algorithm was developed by [1]. To reduce the uncertainty associated with model predictions, the SUFI-2 method in the SWAT-CUP was used to calibrate and validate the model based on observed streamflow data [2, 11].

The sensitivity and uncertainty analyses were also carried out in this study. From the available observed streamflow data at gauge station, the model was calibrated for a period of 20 years (1983–2002) and validated for a period of 15 years (2003–2017). The flow chart of the detailed methodology of the study is represented in Fig. 3.

3 Results and Discussions

The monthly streamflow values from 1983 to 2017 are simulated and quantified for the study area. A SUFI-2 method within SWAT-CUP is used for calibrating and validating the model. The performance level of the SWAT model is determined using the coefficient of determination (R^2) and the Nash–Sutcliffe Efficiency (NSE) parameter for both calibration and validation, after thousands of simulations. The R^2 and NSE values for the calibration period are found to be 0.89 and 0.87, and for the validation period, the values are 0.86, 0.85 respectively. The results seem satisfactory for monthly time step. Percentage bias values also come under 15 for both calibration and validation. The 95PPU plot for observed and simulated flows in calibration period is shown in Fig. 4 and for validation period is shown in Fig. 5.

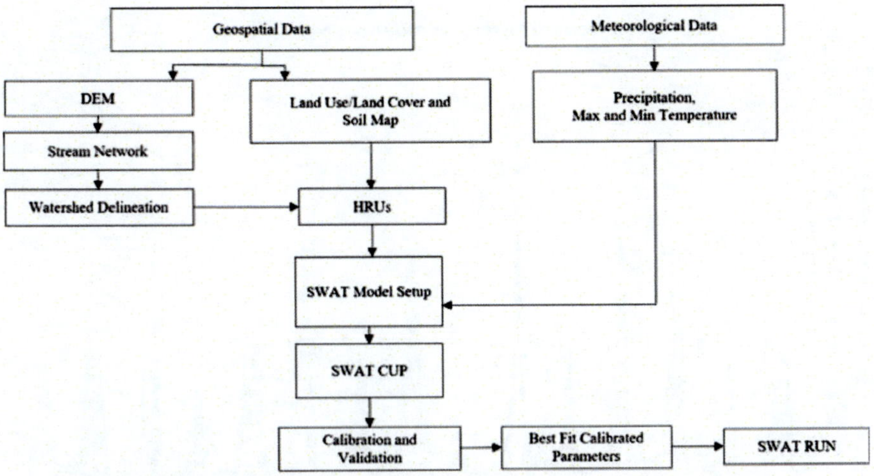

Fig. 3 Flow chart for proposed methodology

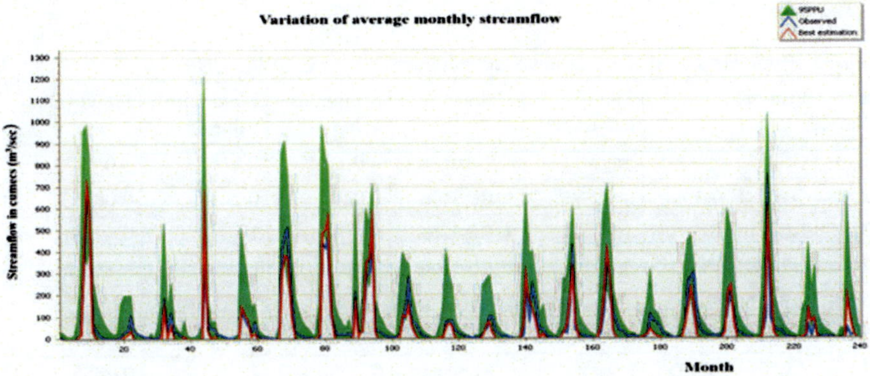

Fig. 4 95PPU plot for observed and simulated flows in cumecs for calibration period

The maximum and minimum range of the parameters used and best-fitted values are shown in Table 1 after calibration.

Out of 12 parameters which are considered in this study, nine parameters are effective. The results of global sensitivity analysis by *P*-value and t-Stat are shown in Fig. 6. After getting best-fit parameters, manual updation of parameters done for the already ran SWAT model and the observed and simulated streamflows in cumecs for the entire period (1983–2017) are shown in Fig. 7. The correlation between these two flows appears to be satisfactory. There are six times out of thirty-eight years of study, the maximum discharge crosses 4000 m^3/s. For entire period of study, the variation in annual maximum streamflow is represented in Fig. 8.

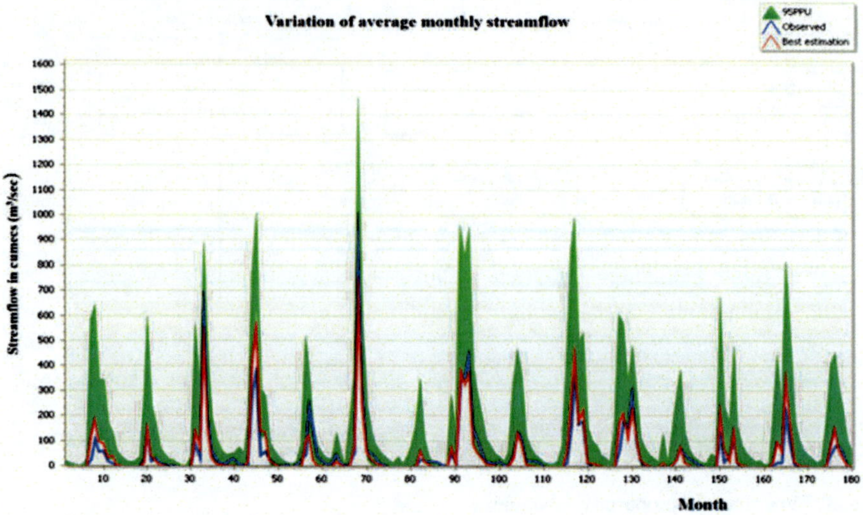

Fig. 5 95PPU plot for observed and simulated flows in cumecs for validation period

Table 1 Parameters used and maximum and minimum ranges in SWAT-CUP

S. No.	Parameter	File extension	Method	Minimum value	Maximum value	Fitted value
1	CN2	.mgt	r_Relative	−0.1	0.1	−0.0879
2	GW_REVAP	.gw	v_Replace	0.02	0.2	0.08453
3	ALPHA_BF	.gw	v_Replace	0	1	0.4905
4	GW_DELAY	.gw	a_Absolute	−30	60	−15.915
5	GWQMN	.gw	a_Absolute	−1000	1000	205
6	RCHRG_DP	.gw	a_Absolute	−0.05	0.05	−0.03105
7	ALPHA_BF_D	.gw	v_Replace	0	1	0.0855
8	SOL_AWC	.sol	r_Relative	−0.1	0.1	0.0547
9	EPCO	.hru	v_Replace	0	1	0.7655
10	ESCO	.hru	v_Replace	0	1	0.3715
11	SLSUBBSN	.hru	r_Relative	−0.2	0.2	−0.053
12	SURLAG	.bsn	r_Relative	−0.2	0.2	0.1894

4 Conclusions

SWAT model simulates the outflow at the guage station for the entire period of run with correlation coefficient of 0.935 after updation of best-fitted parameters. Model is capable of simulating peak discharges for the study area. The study found that the SWAT model simulated the hydrological processes for long-term periods for the

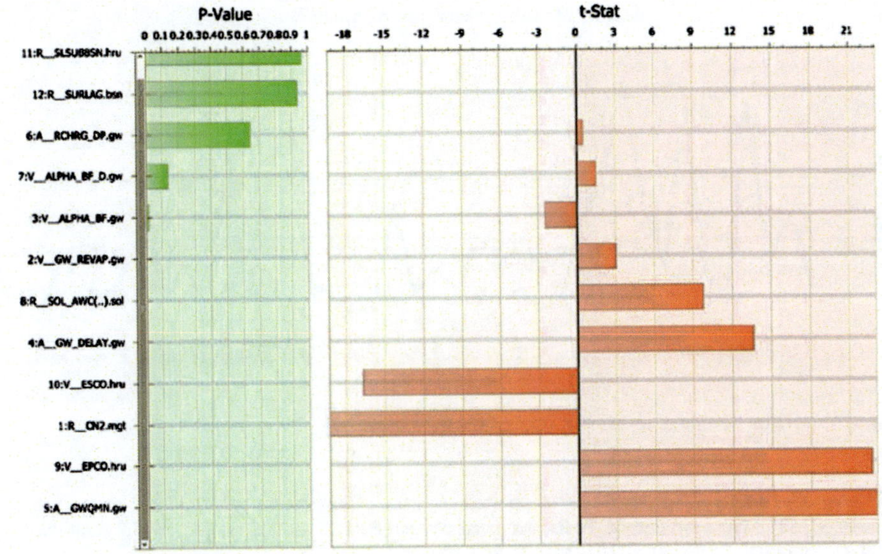

Fig. 6 Results of parameter sensitivity analysis (*P*-value and t-stat)

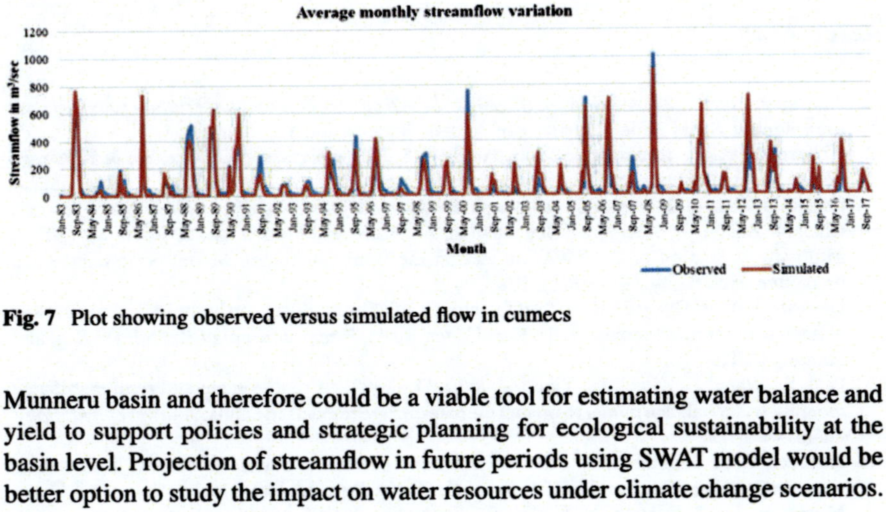

Fig. 7 Plot showing observed versus simulated flow in cumecs

Munneru basin and therefore could be a viable tool for estimating water balance and yield to support policies and strategic planning for ecological sustainability at the basin level. Projection of streamflow in future periods using SWAT model would be better option to study the impact on water resources under climate change scenarios.

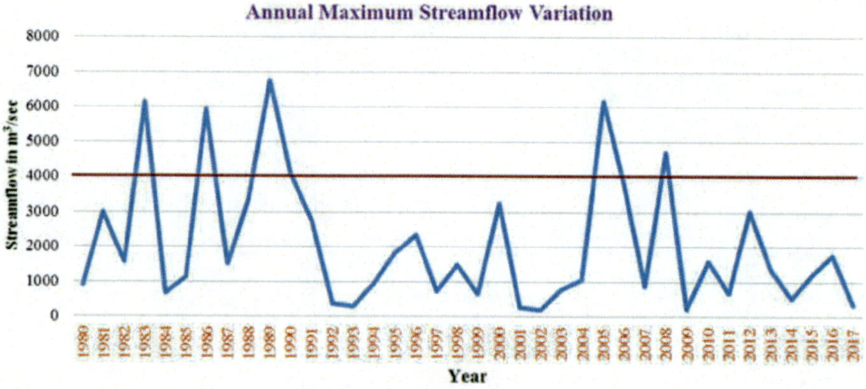

Fig. 8 Variation of annual maximum streamflow at Keesara

Acknowledgements This work is supported by the Department of Science and Technology (DST), Government of India, as part of the BRICS—DST project with grant No: DST/IMRCD/BRICS/PilotCall2/IWMM-BIS/2018 (G).

References

1. Abbaspour, K. C. (2014). *SWAT-CUP 2012: SWAT calibration and uncertainty programs—A user manual*. Swiss Federal Institute of Aquatic Science and Technology.
2. Abbaspour, K. C., Rouholahnejad, E., Vaghefi, S. R., Srinivasan, R., Yang, H., & Kløve, B. (2015). A continental-scale hydrology and water quality model for Europe: Calibration and uncertainty of a high-resolution large-scale SWAT model. *Journal of hydrology, 524*, 733–752.
3. Chaibou Begou, J., Jomaa, S., Benabdallah, S., Bazie, P., Afouda, A., & Rode, M. (2016). Multi-site validation of the SWAT model on the Bani catchment: Model performance and predictive uncertainty. *Water, 8*(5), 178.
4. Devkota, L. P., & Gyawali, D. R. (2015). Impacts of climate change on hydrological regime and water resources management of the Koshi River Basin, Nepal. *Journal of Hydrology: Regional Studies, 4*, 502–515.
5. Guo, S., Wang, J., Xiong, L., Ying, A., & Li, D. (2002). A macro-scale and semi-distributed monthly water balance model to predict climate change impacts in China. *Journal of hydrology, 268*(1–4), 1–15.
6. Iskender, I., & Sajikumar, N. (2016). Evaluation of surface runoff estimation in ungauged watersheds using SWAT and GIUH. *Procedia Technology, 24*, 109–115.
7. Mapes, K. L., & Pricope, N. G. (2020). Evaluating SWAT model performance for runoff, percolation, and sediment loss estimation in low-gradient watersheds of the Atlantic coastal plain. *Hydrology, 7*(2), 21.
8. Mishra, N., Aggarwal, S. P., & Dadhwal, V. K. (2008). *Macroscale hydrological modelling and impact of land cover change on stream flows of the Mahanadi River basin*. A Master thesis submitted to Andhra University, Indian Institute of Remote Sensing (National Remote Sensing Agency) Dept. of Space, Govt. of India.
9. Nilawar, A. P., & Waikar, M. L. (2018). Use of SWAT to determine the effects of climate and land use changes on streamflow and sediment concentration in the Purna River basin, India. *Environmental Earth Sciences, 77*(23), 1–13.

10. Prajnya, V. H., & Tuppad, P. (2016). Study of hydrologic response of Kabini Basin using soil and water assessment tool (SWAT). *International Journal of Engineering Science, 6*(8), 2373–2380.
11. Yang, J., Reichert, P., Abbaspour, K. C., Xia, J., & Yang, H. (2008). Comparing uncertainty analysis techniques for a SWAT application to the Chaohe Basin in China. *Journal of hydrology, 358*(1–2), 1–23.
12. Zhang, D., Chen, X., Yao, H., & Lin, B. (2015). Improved calibration scheme of SWAT by separating wet and dry seasons. *Ecological Modelling, 301*, 54–61.
13. Zhang, G. H., Nearing, M. A., & Liu, B. Y. (2005). Potential effects of climate change on rainfall erosivity in the Yellow River basin of China. *Transactions of the ASAE, 48*(2), 511–517.
14. Zierl, B., & Bugmann, H. (2005). Global change impacts on hydrological processes in Alpine catchments. *Water Resources Research, 41*(2).

Hydrological Modelling of Ungauged Tanks in SWAT Using BHUVAN Data

K. Ramabrahmam and K. Venkata Reddy

Abstract Tanks are the small water storage reservoirs colonized in arid and semi-arid regions of southern India. Tanks were constructed to collect the local runoff and have been an important source of irrigation in Andhra Pradesh, Karnataka, Tamilnadu and Telangana states. The irrigation tanks are accounted for more than one-third of the irrigated area in South India. Accordingly, there is a need to address hydrological issues of the irrigation tanks for effective utilization and management of water within the tank system and river basin. The semi-arid region of Telangana has numerous ungauged tanks. Hence, in this paper, an attempt was made to model the hydrologic processes of the ungauged tanks in the selected study area Salivagu watershed, Telangana, India. This study used geospatially extracted tank area and volumes from the Water Body Information System (WBIS) provided by Bhuvan, National Remote Sensing Centre (NRSC), India. The SWAT hydrological model was used to simulate the flow in the Salivagu river by considering the tanks in the watershed. The model performance statistics of NSE, R2 and PBIAS are 0.84, 0.78 and −5.3 during calibration and 0.92, 0.93 and −10.8 during validation, respectively. The study evidenced that; SWAT model can be effectively used for flow simulation of the watershed with tanks using WBIS data of NRSC-Bhuvan.

Keywords Hydrological · Irrigation · NRSC · SWAT · Tanks · Ungauged · WBIS

1 Introduction

Tanks are the major surface water storage structures and play a significant role in addressing challenges of limited water availability, by both increasing groundwater recharge as well as the water available for irrigation [1]. Tank irrigation systems of

K. Ramabrahmam (✉) · K. Venkata Reddy
Department of Civil Engineering, National Institute of Technology, Warangal, Telangana 506004, India
e-mail: rambrahmam@student.nitw.ac.in

K. Venkata Reddy
e-mail: kvreddy@nitw.ac.in

A. K. Dikshit et al. (eds.), *Innovative Trends in Hydrological and Environmental Systems*, Lecture Notes in Civil Engineering 234,
https://doi.org/10.1007/978-981-19-0304-5_48

India are a century old but it is significant in semi-arid regions due to their contribution to water resources management. The tanks account for more than one-third of the total irrigated area in south India, in the middle of the nineteenth century, the tank irrigated area was reduced and groundwater irrigated area was increased due to the green revolution in India [2]. As time passes, these tanks were silted, partly deteriorated due to the negligence of local authorities as a result, the tank capacities were reduced. At present many of state governments are started renovating these tanks to improve water availability at local scale. This restoration of local scale tanks has an impact on watershed hydrology. The hydrology of basins with tanks in the semi-arid region is unique as it consists of intervening storage structures [3]. Therefore, it is necessary to know the hydrological impact due to restored irrigation tanks in the semi-arid hot regions. Significant rainfall-runoff modelling studies have been carried out for flow simulations with single tank and cascade tanks in the basin. Jayakody et al. [4] and Bishop et al. [5] were modelled the tanks conceptually for estimating hydrological components at tank scale. The model results were in close agreement with monthly runoff rather than daily runoff. Kanagaraj et al. [6] modelled the irrigation tank surface runoff hydrologically using USDA-Natural Resources Conservation Service (NRCS) curve number method of the small ungauged watershed. The model simulated result shown a 10–20% difference with actual observations. Jayatilaka et al. [7] used the ROSES model for simulating the flow in cascade tanks in the small catchment but the predictive capacity of the model was not evaluated. Krishnaveni and Rajeswari [3] the physical distributed model (MIKE SHE/MIKE 11) used to model the physical process of tank cascades in the stream network of the small watershed. The study evidenced that tanks have a great influence on the basins water balancing. Bucak et al. [8] was used the SWAT model to predict the future water availability in the lake using climate model data. The results projected a decrease in future water availability. Jayanthi and Keesara [9] applied the SWAT model for the ungauged watershed with tank systems by transfer the parameters from the gauged watershed. Study results with a decrease in future flow of the basin. Gal et al. [10] developed height-volume-area (HVA) model to estimate the inflow volume of the lake in an ungauged basin using a remote sensing approach. The study shows an increase in the flow but results were statically not significant. Ogilvie et al. [11] forecasted the runoff using the conceptual hydrological model and the observed volume was estimated using remote sensing data and field observations. The study result with improved NSE from 0.64 to 0.94 on daily values. In the above literature, the conceptual models required more observed data and were poor in spatial variability. The remote sensing approach is the better solution in the absence of in-situ data of lakes and gives better results when combine with hydrological modelling data [11]. The physical-based semi-distributed model of SWAT can accomplish both spatial variability and remote sensing data usage in the modelling process of ungauged lakes. The SWAT model performed best in the simulation of high, medium and low flows of dry and moist conditions [12]. In the earlier studies, rainfall-runoff modelling of the single tank and cascade tanks were considered for flow simulation and water balance within the small catchment. The catchment with numerous small and medium tanks was not considered for the effect of flow simulation in the catchment. In this study, an

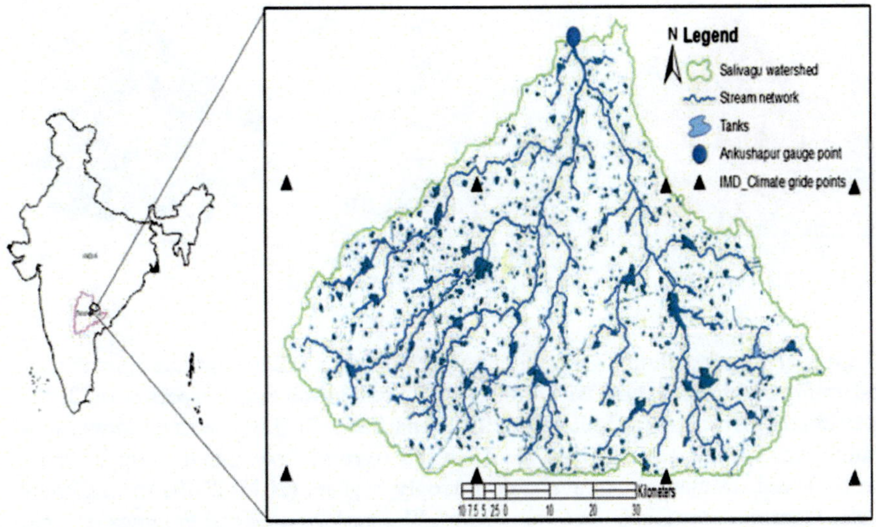

Fig. 1 Location map of Salivagu watershed with tanks

attempt was made to model the hydrologic process of ungauged tanks in the Salivagu watershed, Telangana in the SWAT model using the WBIS-NRSC-Bhuvan data. The purpose of the study is to understand the importance of tanks in the catchment flow.

2 Study Area

The Salivagu river is one of the tributaries of the Maner subbasin of the Godavari River basin. The boundaries of the Salivagu watershed lies between 18° 0′ 8″ N to 18° 23′ 3″ N latitude and 79° 14′ 25″ E to 79° 57′ 41″ E longitude and is located in Warangal and Hanamkonda districts in the state of Telangana, India. The total drainage area of the basin is 1879 km^2. The gauge station was maintained by the Central Water Commission (CWC) at Ankushapur village for measuring gauge, discharge, of the water (http://www.cwc.gov.in). The delineated river network of the basin with tanks are shown in Fig. 1.

3 Data Used

Digital Elevation Model (DEM) extracted from Shuttle Radar Topography Mission (SRTM) with 30 m resolution. Land Use and Land Cover map (LULC) obtained from National Remote Sensing Centre (NRSC) which is of 1:250 K scale. Soil Map prepared from International Soil Reference and Information Centre (ISRIC)

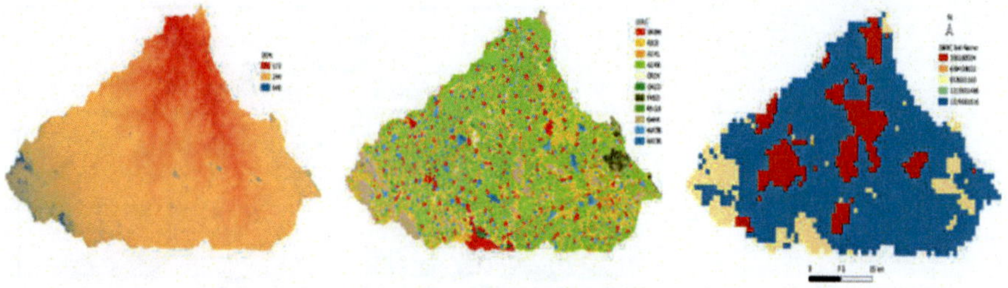

Fig. 2 **a** DEM, **b** LULC and **c** soil map for the Salivagu watershed

which is at 1 km resolution. Precipitation with 0.250 × 0.250 resolution, maximum and minimum temperature with 1.00 × 1.00 resolution for the period of 2017–2019 obtained from Indian Meteorological Department (IMD). The streamflow gauge station was installed in 2017 on the Salivagu river at Ankushapur village, hence the observed discharge data available for only 3 years (2017–2019) from Central Water Commission (CWC) at Gauge point. The surface area and volumes of tanks extracted from the satellite imageries are served as the best substitute in the absence of observed data. In this study, Waterbody Information System (WBIS) (https://bhuvanwbis.nrsc.gov.in) was used for tank area and volume time series data which is available in the Bhuvan platform, developed by National Remote Sensing Centre (NRSC). The WBIS has data pertaining to water surface area and volume of water bodies for the entire country from 2012. DEM, LULC and soil map for the study are shown in Fig. 2.

4 Methodology

SWAT is a physically based semi-distributed basin-scale hydrological model [13] widely used around the globe to assess the impact of sediment transport, fertilizer load and different water management practices. QSWAT, a version of SWAT interfaced with QGIS, used for continuous simulation of a catchment model operating at different time steps and different spatial scales. The tank surface area, volume data at normal and the emergency level was used to represent the tanks in the watershed. The hydrologic processes was simulated using the SWAT model, including surface runoff, lateral subsurface flow, groundwater flow, evapotranspiration and pond water storage and losses.

The SWAT-Calibration Uncertainty Programme (SWAT-CUP) [14, 15] was used for calibration and validation of SWAT simulated flow. To minimize the uncertainty associated with model predictions, the calibration and validation of the model were carried out based on the observed streamflow data, by using the SUFI-2 algorithm in the SWAT-CUP [14, 16]. The sensitivity and uncertainty analyses were also carried out in this study. The available observed streamflow data at the gauge stations of Ankushapur, the model was calibrated for a period of 2 years (2017–2018) and

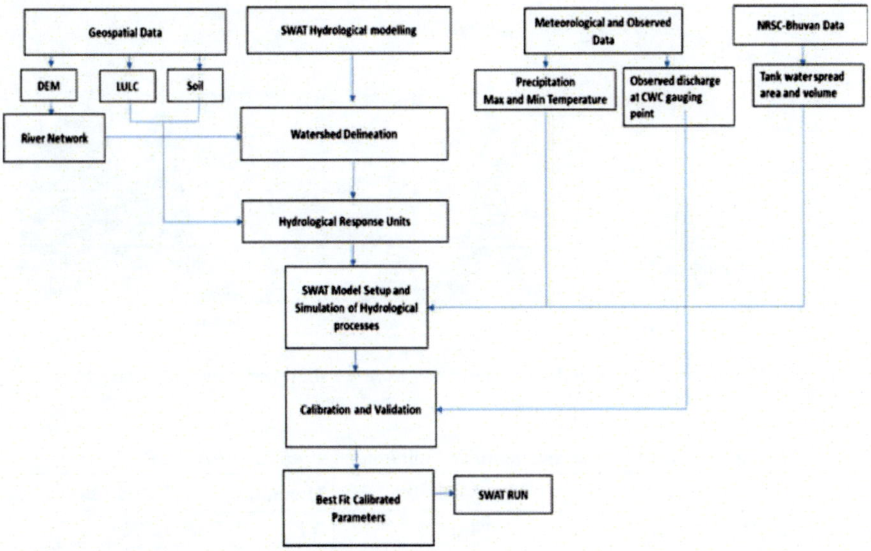

Fig. 3 Flowchart of the proposed methodology

validated for 1 year (2019). The flowchart is shown the detailed methodology of the study represented in Fig. 3.

5 Results and Discussion

The mean monthly streamflow from 2017 to 2019 is simulated and quantified for the study area. A SUFI-2 method within SWAT-CUP is used for calibrating and validating the model. The performance of the SWAT model is determined using the coefficient of determination (R2) and the Nash–Sutcliffe Efficiency parameter (NSE) for both calibration and validation. The NSE, R2, PBIAS are 0.87, 0.89, −5.3 during calibration and 0.92, 0.93, −10.8 validation, respectively. The 95PPU plot for observed and simulated flows in the calibration and validation period is shown in Fig. 4. The selected SWAT parameters of maximum, minimum range values used in the simulation and best-fitted values after calibration are shown in Table 1.

The global sensitivity analysis results by *P*-value and t-stat are shown in Fig. 5 for the selected parameters in the study. The best-fitted parameters in the calibration were manually updated and calibrated in the initially simulated SWAT model.

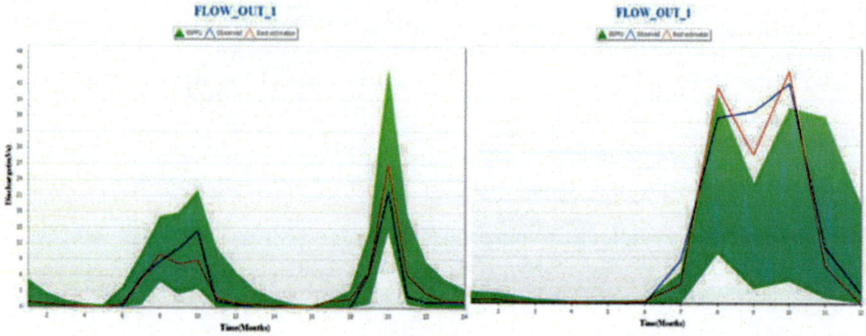

Fig. 4 95PPU plot for observed and simulated flows during calibration and validation period

Table 1 The parameters fitted and maximum and minimum ranges in SWAT-CUP

S. No.	Parameter	Minimum	Maximum	Fitted value
1	R__CN2.mgt	−0.1	0.1	0.097
2	V__ALPHA_BF.gw	0.3	1	0.593
3	A__GW_DELAY.gw	−30	90	63.66
4	A__GWQMN.gw	−200	1000	145
5	V__GW_REVAP.gw	0.08	0.2	0.1716
6	V__ESCO.hru	0.3	0.75	0.549
7	R__SOL_AWC(..).sol	−0.05	0.05	0.033
8	A__RCHRG_DP.gw	−0.05	0.05	0.00185
9	A__REVAPMN.gw	−500	1000	570.25
10	V__CH_N2.rte	0.01	0.1	0.0986
11	V__CH_K2.rte	0.01	50	36.22
12	V__CH_K1.sub	0.01	50	28.73
13	V__CH_N1.sub	0.01	0.1	0.0525
14	V__LAT_TTIME.hru	0	60	43.169
15	V__SLSOIL.hru	40	70	64.915
16	V__ALPHA_BF_D.gw	0	1.0	0.760

A-Absolute, R-Relative, V-Replace are parameter range change methods

6 Conclusions

The flow simulation through tanks in the SWAT model shown good results using WBIS-NRSC-Bhuvan tank data. The medium and large size tanks have considerable influence on the streamflow generation at the local scale. The model is capable of simulating peak discharges and base flows for the current study. The study evidenced that NRSC- Bhuvan provided tank area and volumes series are useful in absence of

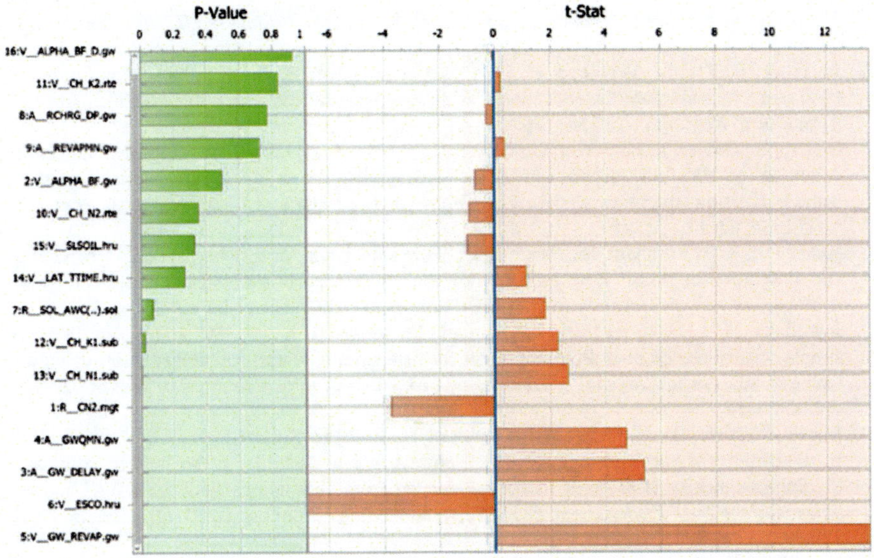

Fig. 5 SWAT parameters of sensitivity analysis of P-value and t-stat during calibration

observed data. The SWAT model can be used for hydrological flow simulation of watersheds with enriched traditional irrigation tanks. This study can be used for estimating water balance and yield to support policies and strategic planning for ecological sustainability at the local and basin level. The water balance analysis at tank scale and basin-scale using observed or NRSC-Bhuvan data with SWAT model would be useful for local scale water management.

Acknowledgements The research described in this paper was carried out with funds provided by the Science and Engineering Research Board (SERB), DST India through project EMR/2017/004691. We are thankful to CWC (Central Water Commission Hyderabad), NRSC (National Remote Sensing Centre, Balnagar, Hyderabad), and the Irrigation department of Telangana state for providing necessary data.

References

1. Van Meter, K. J., Steiff, M., McLaughlin, D. L., & Basu, N. B. (2016). The socioecohydrology of rainwater harvesting in India: Understanding water storage and release dynamics across spatial scales. *Hydrology and Earth System Sciences, 20*, 2629–2647. https://doi.org/10.5194/hess-20-2629-2016
2. Palanisami, K. (2006). Sustainable management of tank irrigation systems in India. *Journal of Developments in Sustainable Agriculture, 1*, 34–40. https://doi.org/10.11178/jdsa.1.34
3. Krishnaveni, M., & Rajeswari, A. (2019). Flow routing and surface water balance simulation of a tank cascaded catchment using coupled MIKE SHE/MIKE 11 modeling. *Journal of The*

Institution of Engineers (India): Series A, 100, 65–74. https://doi.org/10.1007/s40030-018-0335-2
4. Jayakody, B. V. P. L., Mowjood, M. I. M., & Gunawardena, E. R. N. (2004). Modifying and testing of a cascade water balance model using a mini cascade system with two tanks. *Tropical Agricultural Research, 16*, 159–170.
5. Bishop, E., Calder, I., Batchelor, C., Garratt, J., & Gosain, A. (2009). Development and application of the HYLUC-Cascade model for investigating the water resource impacts of rainwater harvesting structures in India. *The Centre for Land Use and Water Resources Research. Newcastle, 61*, 11.
6. Kanagaraj, R. P., Abraham, M., Perumal, M., Panneerselvam, V., & Sekar, S. (2021). Modeling and simulation of runoff from an irrigation tank watershed to evaluate the utilizable water. *Arabian Journal of Geosciences, 14*. https://doi.org/10.1007/s12517-020-06267-w
7. Jayatilaka, C. J., Sakthivadivel, R., Shinogi, Y., Makin, I. W., & Witharana, P. (2003). A simple water balance modelling approach for determining water availability in an irrigation tank cascade system. *Journal of Hydrology, 273*, 81–102. https://doi.org/10.1016/S0022-1694(02)00360-8
8. Bucak, T., Trolle, D., Andersen, H. E., Thodsen, H., Erdoğan, Ş, Levi, E. E., Filiz, N., Jeppesen, E., Beklioğlu, M., & Bucak, T. (2017). Future water availability in the largest freshwater Mediterranean lake is at great risk as evidenced from simulations with the SWAT model. *Science of the Total Environment, 581–582*, 413–425. https://doi.org/10.1016/j.scitotenv.2016.12.149
9. Jayanthi, S. L. S. V., & Keesara, V. R. (2019). Climate change impact on water resources of medium irrigation tank. *ISH Journal of Hydraulic Engineering, 00*, 1–12. https://doi.org/10.1080/09715010.2019.1649605
10. Gal, L., Grippa, M., Hiernaux, P., Peugeot, C., Mougin, E., & Kergoat, L. (2016). Changes in lakes water volume and runoff over ungauged Sahelian watersheds. *Journal of Hydrology, 540*, 1176–1188. https://doi.org/10.1016/j.jhydrol.2016.07.035
11. Ogilvie, A., Belaud, G., Massuel, S., Mulligan, M., Le Goulven, P., Malaterre, P. O., & Calvez, R. (2018). Combining Landsat observations with hydrological modelling for improved surface water monitoring of small lakes. *Journal of Hydrology, 566*, 109–121. https://doi.org/10.1016/j.jhydrol.2018.08.076
12. Tegegne, G., Park, D. K., & Kim, Y. O. (2017). Comparison of hydrological models for the assessment of water resources in a data-scarce region, the Upper Blue Nile River Basin. *Journal of Hydrology: Regional Studies, 14*, 49–66. https://doi.org/10.1016/j.ejrh.2017.10.002
13. Arnold, J. G., Srinivasan, R., Muttiah, R. S., & Williams, J. R. (1998). *Basin scale model called SWAT (Soil and Water speed and storage, advanced software debugging policy to meet the needs, and the management to the tank model*. 34, 73–89.
14. Abbaspour, K. C., Rouholahnejad, E., Vaghefi, S., Srinivasan, R., Yang, H., & Kløve, B. (2015). A continental-scale hydrology and water quality model for Europe: Calibration and uncertainty of a high-resolution large-scale SWAT model. *Journal of Hydrology, 524*, 733–752. https://doi.org/10.1016/j.jhydrol.2015.03.027
15. Abbaspour, K. C. (2014). *SWAT-CUP 2012: SWAT Calibration and uncertainty programs—A user manual: Swiss federal institute of aquatic science and technology*.
16. Yang, J., Reichert, P., Abbaspour, K. C., Xia, J., & Yang, H. (2008). Comparing uncertainty analysis techniques for a SWAT application to the Chaohe Basin in China. *Journal of hy-drology, 358*(1–2), 1–23.

Cost-Effective Methodology to Quantify the Submarine Groundwater Discharge from Coastal Unconfined Aquifers

Yovan Lino, Athira S. Rajan, Santhosh Prabhu, Harikripa Narayana Udayashankar, R. S. Sudharsanan, and Keshava Balakrishna

Abstract Submarine groundwater discharge is a pathway of fresh and recirculated groundwater to the sea. It is recently understood that this phenomenon delivers thirteen times more nutrients to the sea than the river water. This is due to the high residence time of groundwater in the coastal aquifers when compared with the river water. Studies to estimate the discharge in the field is very scarce; therefore, the real magnitude of discharge and the nutrient budget through SGD is still inconclusive. On the other hand, quantifying the SGD rates in a location would help calculate the optimal pumping rate from the coastal aquifer without leading to seawater ingress. Methods available currently to estimate SGD rates are resource intensive and needs dedicated instruments. We suggest a cost-effective methodology using the basic civil engineering survey instruments and fabricated tools to estimate the SGD rates from the coastal aquifer. This method not only estimates the discharge, but it also helps delineate the seawater-freshwater interface in the coastal aquifer and could differentiate between the fresh and recirculated SGD. The limitation of this methodology is that the SGD rates can only be estimated from the coastal unconfined aquifers, not the deeper aquifers.

1 Introduction

1.1 What Is Submarine Groundwater Discharge?

The discharge of groundwater into the sea is a natural process that acts as a conduit for the nutrients, pollutants, and trace metals to the sea. Submarine groundwater discharge (SGD) is defined as 'the flow of water through continental margins from the seabed to the coastal ocean, with scale lengths of meters to kilometers, regardless

Y. Lino · A. S. Rajan · S. Prabhu · H. N. Udayashankar · R. S. Sudharsanan · K. Balakrishna (✉)
Department of Civil Engineering, Manipal Institute of Technology, Manipal Academy of Higher Education, Manipal 576104, India
e-mail: k.balakrishna@manipal.edu

A. K. Dikshit et al. (eds.), *Innovative Trends in Hydrological and Environmental Systems*, Lecture Notes in Civil Engineering 234,
https://doi.org/10.1007/978-981-19-0304-5_49

of fluid composition or driving force' [1]. SGD is an integral part of the hydrological cycle. According to traditional hydrologists, the term groundwater represents the water that originated from the terrestrial aquifer, and it does not include recirculated seawater as groundwater, but in ocean research, any subsurface water is considered groundwater. However, some researchers consider both freshwaters and recirculated seawater that seeps into the ocean as SGD [2]. Seawater recirculation is controlled by subtidal pumping, wave set up, thermal gradient, and density-driven circulation [1]. Freshwater discharge commonly occurs in the form of direct seepage, as submarine springs or even as tide-controlled discharge from the aquifers [3]. SGD typically decreases as we move toward the ocean from the coast, but a reverse condition (increase in the groundwater discharge toward the ocean) can be observed at certain places. It is site dependent. SGD is primarily controlled by tidal variation. During low tide, the inland hydraulic gradient is higher; thus, the groundwater discharge rate will also be higher; during the high tide, a gradient reversal occurs due to seawater level. Freshwater discharge is inversely proportional to tidal height. During spring tide, total SGD flux is higher due to the pumping action of recirculated seawater. While at neap tide, the freshwater discharge will be higher [4]. Thus, SGD is spatially and temporally varied. Over the years, scientists have been adopting several methods to quantify the freshwater discharging into the sea. Groundwater seepage is patchy, intermittent, momentarily unpredictable, and several aquifers may be involved.

1.2 Methods to Detect and Quantify SGD

SGD can be estimated through analytical or numerical modeling, by tracer techniques or through direct methods.

Seepage meter is an efficient method to estimate SGD since they provide the variabilities of discharge in a minimal area. It was developed for the first time by Israelson and Reeve [5] to calculate the water lost from canals. Lee [6] designed the most commonly used seepage meter, which uses a steel drum with an opening on the top to fit a plastic bag to collect the seepage water. The seepage meter is fixed with the open end down in the sediment, and the seepage water in the inner surface area of the seepage meter will be collected in the plastic bag. The discharge is calculated from the surface area of the seepage meter and the quantity of water collected. Since seepage along most of the coastline is patchy, diffusive, and temporally variable, seepage meter readings are taken in closed intervals, and measurements must be done at different times to detect temporal variabilities. Usually, seepage meters are synchronous with piezometer nests to accurately determine hydraulic conductivity [7].

Regional aerial Thermal Infrared (TIR) mapping carried out in a large volcanic island of Hawaii has found more than 30 sources of groundwater plumes rich in nutrients on the dry western half of the island [8]. These nutrient-rich plumes were cold and less dense (a mixture of groundwater and seawater). Hence, the thermal infrared study can be used at high latitudes where groundwaters are warm relative to seawater [9–11] and at lower latitudes where groundwaters are cool [12, 13]. This is because of the contrasting difference in temperature of groundwater and seawater for the thermal infrared camera to pick up the signals.

Numerical models are used to model the aquifer to determine the groundwater flow, mixing zone, the flow of fluids with varied densities, etc. However, most models only provide the fresh groundwater component of SGD. Difficulties in numerical modeling include modeling the interaction zone between seawater and bottom sediment and difficulty calculating the dynamism in other factors. Collecting data required creating a model and calibrating it makes this method time consuming and difficult. GIS techniques have been proven effective to identify SGD zones [14], where they have used a digital elevation model (DEM) to locate terrestrially sourced SGD and used radon activity to verify the results. They have found out that the delineated flow accumulation from DEM strongly correlates with observed radon activity. Hence, they suggested that with the help of high-resolution DEM, it may be possible to quantify SGD without using radon measurements.

Isotopes are a commonly used tool in groundwater studies to estimate the movement and recharge of groundwater. This can be an excellent tool to evaluate the hydrological sources contributing to SGD to understand better the recharge sources, leaky surface water bodies, fracture zones leading to the recharge of coastal aquifers, etc. Naturally occurring geochemical tracers can be used to determine SGD on a local scale. Radon is one such tracer that occurs during the decay of radioactive isotopes in sediments. They are enriched in fresh groundwater and recirculated seawater [15]. The difference in their abundance in groundwater and seawater makes it an efficient method to analyze.

On the other hand, radium is absorbed by sediments and is immobile in freshwater. When the seawater replaces the fresh groundwater in the sediments, radium is released from sediments, mixed with brackish water, and discharged as SGD [16]. Therefore, the analysis of radium gives an account of the recirculated seawater component of SGD [15]. Thus, both radium and radon measurements are needed to differentiate between the fresh groundwater and recirculated seawater component of SGD. Natural radium and radon isotopes are excellent tracers for SGD because of their conservative state, higher concentration in groundwater and seawater, and relative ease to measure [17]. It has been pointed out that ^{222}Rn inputs to the groundwater inland may also be a point source from deep-seated fractures. Thus, care should be taken during the calculations of SGD flux through ^{222}Rn activity in groundwater inland and coastal water.

1.3 Why Is SGD Significant?

Although it is a natural process, the freshwater being discharged into the sea carries a large quantity of nutrients and other contaminants into the sea [1]. Sometimes, the abundant delivery of these nutrients may harm marine organisms and can also affect marine ecology and biogeochemistry. Groundwater contains a higher concentration of nutrients (NO^3, Si, PO^4, NH^4 and C) and trace metals (Co, Ni, Zn, Pb, Cd, Li, Cu, Fe, Al, Hg, etc.) than any other water body. Direct flow of this groundwater into the sea may lead to abundant primary productivity, which will further lead to harmful algal bloom. Algal bloom can reduce the dissolved oxygen (DO) content in the water, which may affect the survival of other macro aquatic organisms. Submarine groundwater can also affect coral reef growth. A recent study about coral reefs showed that nutrients derived from SGD could increase the rates of bioerosion and coral growth. At the highest SGD sites, it is found that salinity stress induces a decrease in coral growth and finally causes its mortality [18].

1.4 Effects of SGD in Coastal Ecosystem

Marsh [19] observed that SGD is a source of nutrients for coral reefs in Guam Island. In 1980, Johannes [20] quantified the nitrate flux through SGD along the northern coast of Perth, Australia, which was three times that of the river. This paper further emphasized the importance of this flux to coastal productivity. Lee and Hu attributed the formations of harmful algal blooms and red tides to SGD [21, 22]. In 2009, McCoy and Corbett [23] pointed out that the excessive flux of nutrients in the continental shelf can also be attributed to SGD. Other than that, SGD is a source of substantial inputs of contaminants, nutrients (infiltrated from domestic, municipal, agricultural, and industrial practices) to the coastal ocean [24]. In a 2004 study done by Slomp and Capellen [23], they found the sources of nutrients in groundwater and found out that groundwater's redox conditions and storage time in the subterranean estuary prominently affect the water chemistry. Studies have estimated that the nutrient flux from SGD is equal or greater than the riverine flux to the ocean [25, 26]. This excess flux to the coastal ocean could play a significant role in coastal ecology [27, 28]. Nutrient flux through SGD is the reason for primary biological productivity in Hampyeong Bay, Yellow Sea [26]. Gonneea [29] suggested that the longer residence time of groundwater in the subterranean estuary would end up in depleting groundwater oxygen and favoring nitrification, denitrification, and anaerobic ammonium oxidation. Also, analysis done in four embayments at Majorca, Spain, with different land-use characteristics indicated a link between human activities and compounds arriving at the coast [30].

Shaw and Prepas [31] estimated the concentration of Ba through SGD to the southeast coast of North America and found that its flux is about four times that of riverine inputs. In 2007, Bone et al. [32] calculated the flux of Hg through SGD to Waquoit Bay, Massachusetts, which was ten times higher than that of atmospheric inputs. The atmospheric deposition was an essential source of Fe to the sea [33], which is an essential source of nutrients for planktons [34]. However, Windom et al. [35] argued that a significant portion of Fe on the southeast coast of Brazil is through the SGD.

Most of the SGD studies have been quantified using these isotopes in the past couple of decades. A study conducted at coastal waters of Majorca found a relationship between SGD derived ^{224}Ra and phytoplankton biomass which impacts coastal productivity. Samples from elevated radium levels showed increased pCO_2 concentration, which is enhanced in groundwater than in the atmosphere. Moore [1] noted that some of the metals like uranium are removed in the subterranean estuary because of changing redox conditions when seawater enters the aquifers. Cai et al. [36] have calculated the dissolved inorganic carbon (DIC) flux through SGD as 170 mmol per square meter per day at the South Carolina coast, which is higher than riverine inputs. In the same area, Goni and Gardner [37] have estimated the dissolved organic carbon (DOC) flux and found out to be 50 mmol per square meter per day. In 2006, Moore [38] calculated Ra flux and DIC and DOC in Okatee, South Carolina. For DIC and DOC, they estimated the fluxes to be 2000 mmol per square meter per day and 170 mmol per square meter per day. They pointed out that carbon and radium have a strong correlation with each other, and a simple stoichiometric ratio determines the release of these elements during diagenesis. Santos et al. [39] estimated the flux of DOC through SGD to the west coast of Florida as 19–27 mmol per square meter per day.

1.5 SGD Versus Saltwater Intrusion

Seawater intrusion or saltwater intrusion (SWI) is a counterpart of SGD, where the saline seawater infiltrates into the coastal aquifer. This occurs due to a variety of reasons including lower inland hydraulic head, over-exploitation of groundwater, storms etc. Saline water intrusion at the coastal aquifer interface can be modeled using the Ghyben-Herzberg relation. This relation was proposed by Ghyben [40] and Herzberg [41], and it is stated as 'the depth of the interface is approximately 40 times the height of the water table above the sea level' [42]. Ghyben-Herzberg principle works based on the following assumption: the aquifer is homogenous, hydrodynamic dispersion is negligible, and the saline groundwater is at rest. Many scientists consider it an unrealistic hydrostatic situation because the assumptions used in computing the relationship are ideal and barely exist in the field. Although SGD and SWI are counterparts, methods to quantify SGD are entirely different from SWI assessment methods.

1.6 Why Is a Cost-Effective Methodology Needed to Quantify SGD?

Groundwater forms a part of the water cycle on earth, and it plays a pivotal role in the existence of human beings. Because of the integral and essential part of the hydrological cycle, their availability depends on the conditions of recharge and rainfall. It has until recently been regarded as a reliable source of uncontaminated water. Over the years, water demand has increased, contributing to water scarcity in many parts of the world. The condition could worsen in the future due to the steadily increasing population. India is heading toward a freshwater crisis due to unsuitable water resource management, particularly groundwater [43]. Quantification of SGD rates is usually done using radon measurements which is expensive and need sophisticated tools. Therefore, cost-effective analysis aids are needed to quantify the SGD rates. This would aid more research and, therefore, more SGD estimates. Countries like India, where the freshwater resource is depleting, could use methodologies like these to estimate the SGD rates from its coastline accurately.

2 Materials and Methods

2.1 Ground Elevation Profile

A detailed ground profile is needed in the study area between the groundwater well to the lowest of the low tide line (LLTL) Fig. 1. The selection criteria for the ground elevation profile site should include close proximity of the groundwater well from the

Fig. 1 Positioning of autolevel, levelling staff and the porewater sampling transect in the field

beach and the visibility to each point from one another (for sighting the instrument). The closer the groundwater well is to the coast, the more accurate the hydraulic conductivity measurements. GPS measurements should be noted for latitude, longitude, and altitude. The GPS should be held at the ground level and allowed to stabilize before noting down the readings. Then the autolevel equipment should be set up (to see horizontal plane using bulls' eye and spirit levels) beside the well, and reference elevation should be taken at a fixed point near the groundwater well. Then the measuring chain of 30 m in length should be laid in the transect facing perpendicular to the sea. The water table elevation from the ground level and the well depth measurements should be noted down at the field. Leveling staff (4 m height or higher preferred) should be used to determine the elevation of the ground at every five meters interval up to the LLTL. Care should be taken when moving the leveling staff in the field. After 30 m, the chain should be extended further toward the beach and measurements were sighted. If at any instances where the instrument becomes invisible for sighting through the level staff, the instrument should be shifted toward the beach and taken 'back-sighting' (BS) to fall back to the previous plane of measurement.

2.2 *Hydraulic Head Measurements*

Hydraulic head differences between the seawater and the coastal groundwater can be delineated through the beach profile and the groundwater. The hydraulic head measurement can be made by the parameters measured in the field (water level in the groundwater well, seepage face and the beach profile) using the equation [44] given below:

$$i \propto \frac{h_a - h_b}{L} \propto \frac{dh}{dl}$$

Here,

i—Hydraulic gradient

h_a—Groundwater level in the well

h_b—Seepage face

L—distance between seepage face and groundwater well.

The relative head difference between the mean sea level and the groundwater in the coastal aquifer is acquired purely based on the auto-level measurements. Therefore, the relative sea level during the time of beach profile should be adjusted with the GPS measurements and matched with the profile for plotting. For the sites with no seepage seen in the intertidal zone, the porewater measurements should be considered. If a substantial dip in EC was seen along the intertidal zone, which can be a signature of seepage, it could be taken as a seepage face. For sites with no substantial EC variations and no seepage seen, the LLTL height can be taken as seepage face.

2.3 Porewater Profile

Porewater profile can be measured using a fabricated stainless steel (SS) porewater sampler. This porewater sampler is generally a 0.5 mm diameter tube with a SS rod within the tube. However, the size and length can be customized depending on the usage and the depth requirements. The bottom end of the tube should have perforations to allow porewater to enter the tube. The tube's top end and the rod are attached with a handle for easy installation-removal of the sampler. The tube fixed with the rod is lowered in the sediment column to the desired depth for sampling. Then the inner SS rod is removed, and flexible tubing is attached to the top end of the tube which is in the sediment. A 50 ml syringe is attached to the other end of the flexible tubing to sample the porewater. This porewater is then analyzed for basic physicochemical parameters. This data, along with the beach profile, can then be plotted to get the porewater profile of the intertidal zone.

Porewater samples (Fig. 2) should be collected during low tide to ensure the maximum coverage of the intertidal zone for porewater measurements. The high tide line should be marked in the field, and the measuring chain should be laid perpendicular from the highest of the high tide line (HHTL) toward the sea, up to the low tide mark. The porewater sampler should then be inserted into the sediment, and the sample is drawn at 0.5 and 1 m depth at every five-meter interval from HHTL. Before taking the sample for the measurements of physicochemical parameters, a small quantity of porewater should be taken out to release the air column in the tubing and the sampler. Then, ~50 ml of the sample can be taken for flushing the tubing, syringe, sample measuring container and the probes (multiparameter meter or electrical conductivity (EC) meter). After that, 50 ml of porewater can be taken and measured for pH, EC and temperature (*T*). Samples should be placed in a thick plastic container and measured immediately to get the appropriate temperature and avoid atmospheric influence. The temperature probe should be given enough time (10 s) to stabilize before performing temperature measurements.

If samples are required for analysis: the porewater should be filtered using a 0.45 μm glass fiber filter in the field immediately after collection using a vacuum-operated glass filtration kit or syringe filters. Then the samples should be stored in labeled bottles, sealed with parafilm. This sample can be analyzed for major anions

Fig. 2 Seepage meter design and deployment in the intertidal zone

and cations in the laboratory. Standard methods recommended for porewater sample collection can be used [45]s.

Note: Ambient air temperature can be taken to clear the atmospheric heating of porewater. It can be done by stabilizing the probe for ~30 s under shade. This allows the user to measure accurate ambient air temperature measurements and avoid temperature sensor getting heated by direct solar radiation.

2.4 Seepage Meter Measurements

Fabricated seepage meters can be made by slicing 200 liters plastic drum (75 cm in diameter) into four pieces (Fig. 2 and 3). The slices portions can then be used as 'open-end' seepage meter to isolate a portion of the exposed area in the intertidal zone to measure seepage. Ziplock bags/plastic bags (2 liters) are to be fastened to a plastic tubing (2.5 m) used as a collection bag and connected to the seepage meter. The plastic bags should be tested to calculate the storage capacity without exerting pressure on the tubing and seepage meter. The seepage meter is then to be placed at a 5-m interval from each other, starting from the seepage face. The seepage bags should then be monitored carefully with time intervals noted, and the seepage water should be collected at regular intervals. Eight seepage meters (four seepage meters at transect 1 and four at transect 2) are recommended during the field over a complete tidal cycle. However, one porewater measurement per tidal cycle, preferably during the low tide condition, would also be sufficient.

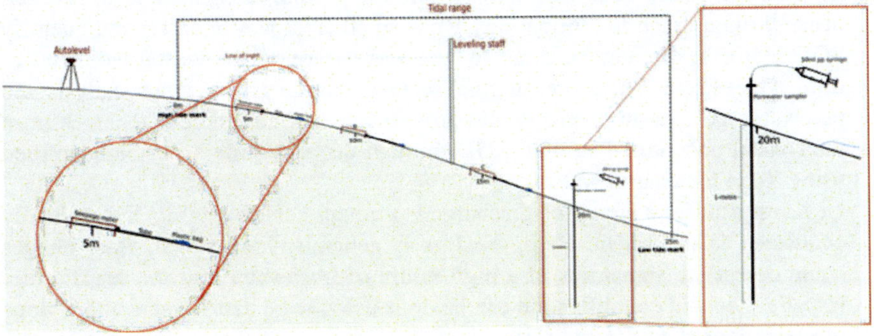

Fig. 3 Cross section of the intertidal zone showing the position of seepage meter, porewater sampler and the beach profiling transect

3 Coastal aquifers and SGD

The kind of geological formation in the selected study area adversely affects the kind of discharge happening on the coast. The recommended measurements for a different kind of aquifer set-up are given below:

In porous unconsolidated coastal aquifers: Which are hydraulically connected to the ocean, beach profiling from the groundwater well to the intertidal zone would provide calculations of the inland hydraulic gradient to determine the presence of SGD. This procedure can be done in the initial stages of the study to determine the probable SGD zones, and the sites having high inland hydraulic gradients can be earmarked for detailed studies. The detailed study includes monthly/frequent porewater sampling in the confirmed sites to monitor the movement of seepage face from the monsoon season to the next monsoon season to calculate the amount of groundwater being discharged as SGD. This can be done using seepage meters and porewater samplers for a complete tidal cycle for a full hydrologic year. The SGD discharge data, along with the rainfall and recharge data of the coastal aquifer, would help determine the water budget of the coastal aquifer. Furthermore, Rn and Ra measurements can also be done to determine SGD fluxes.

In hard-rock aquifers with no fractures: Where SGD is expected to be null to feeble, satellite thermal images analysis and porewater analysis for EC could be done to investigate SGD from the deep and shallow aquifers respectively. These methods are cost and resource-efficient. Rn measurements in the coastal water could be done in these areas but not recommended.

In fractured hard-rock aquifers: Where the groundwater flows to the coastal ocean through fractured formations, Rn is an efficient way to detect and quantify SGD. Here, the discharge is direct to the ocean or through the coastal sediments as diffusive seepage. Therefore, thermal images, coastal groundwater analysis and porewater EC measurements would provide clues of nearshore discharges from the coastal unconfined aquifer. SGD through deeper aquifers can be quantified using Rn in the coastal waters.

In Karst aquifers: Here, the discharge is through karstic aquifers hydraulically connected to the ocean. Here, the flux is generally high and thermal images could detect the signatures of a high influx of freshwater into the sea. Further investigation and quantification can be done using fixed data loggers and isotope investigations (^{18}O, ^{2}H and ^{3}H).

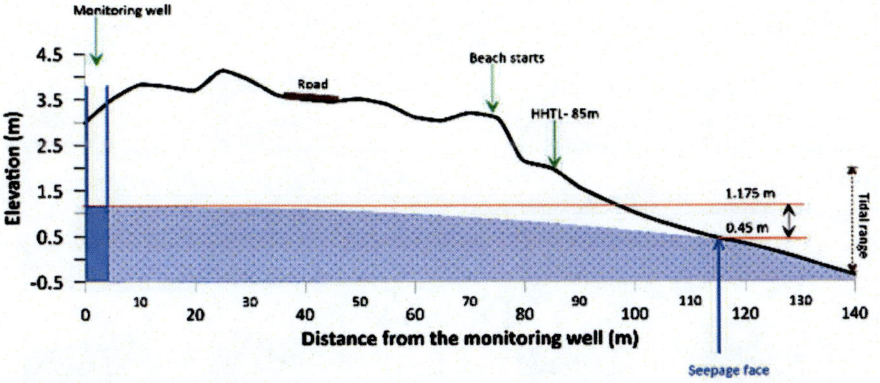

Fig. 4 Beach profile showing the position of groundwater well, seepage face, ground profile and hydraulic head differences

4 Results and Discussion

4.1 Beach Profile

The beach profile plot is adjusted to the mean sea level through the tide levels and the GPS altitude measurements. The beach aspect is calculated by the ratio between the HHTL and LLTL divided by the distance of the intertidal zone. The beach profile contains information about the groundwater level in the well, established seepage face by porewater, and seepage water measurements and their head difference is also plotted (Fig. 4). The plot shows the ground profile plotted against elevation (referenced to MSL) in the 'y' axis and the distance from the monitoring well to the beach on the 'x' axis. The groundwater level in the well is illustrated in the figure, and the water level is shown (red line) wherever there is seepage in the intertidal zone. The tidal range is also shown in the graph. Other markers like HTL, starting of the beach, is also shown. From the figure, the beach aspect and hydraulic gradient can be measured for use later in seepage measurements.

4.2 Porewater Variations

Probable SGD zones may show any one of the discharge patterns in Fig. 5 where the freshwater discharges slightly above the high tide line in monsoon and moves toward the low tide mark toward post-monsoon. When the water level in the aquifer drops during the post-monsoon, the seepage face centralizes between HHTL and LLTL. The portion between the seepage face and high tide line gets infiltrated with seawater during high tide and gets flushed out during low tide. This seawater pocket

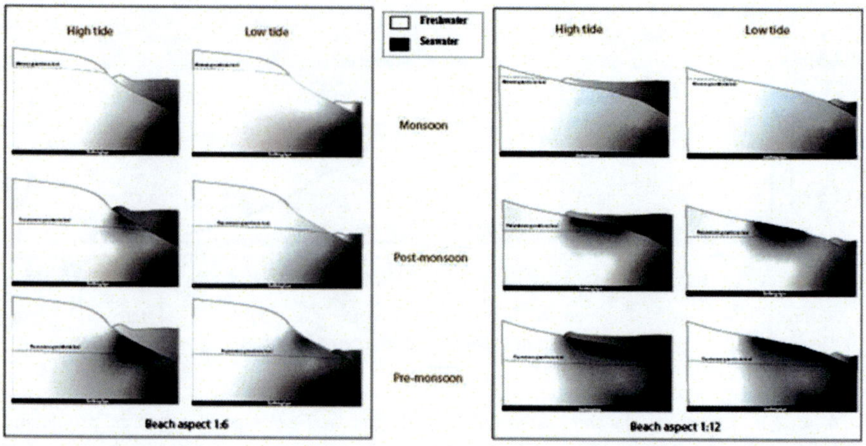

Fig. 5 Illustration showing the porewater electrical conductivity variations in the intertidal zone at different beach aspect

present above the seepage face is called upper saline plume (USP), and the water flushed out from this portion is recirculated SGD (rSGD). Between monsoon to the next post-monsoon period, this USP increases in size as the seepage face is lowered due to inland hydraulic gradient. The illustration shows the position of USP and seepage for steep and gently sloping beaches over monsoon to post-monsoon to the pre-monsoon season.

4.3 Seepage Rates Measurement

Seepage rates at the sampling point can be measured by calculating the area inside the seepage meter and converting the seepage rates recorded during the monitoring period. This value is then multiplied by the total length and breadth of the beach to estimate the seepage rates for the specific study site. The average seepage rates at every seepage meter should then be converted to 'measured seepage volume' for the entire tidal cycle.

Figure 6 shows the 3D cross-section of the intertidal zone and the position of seepage meters and porewater sampling transects. The zoomed-out portion shows the seepage face where the seepage meter measurements were carried out. Generally, the seepage face continues toward and beyond the low tide mark. However, because the wave disturbances are closer to the low tide mark, the installation of the seepage meter was limited to the point where the wave disturbances are not present in the intertidal zone.

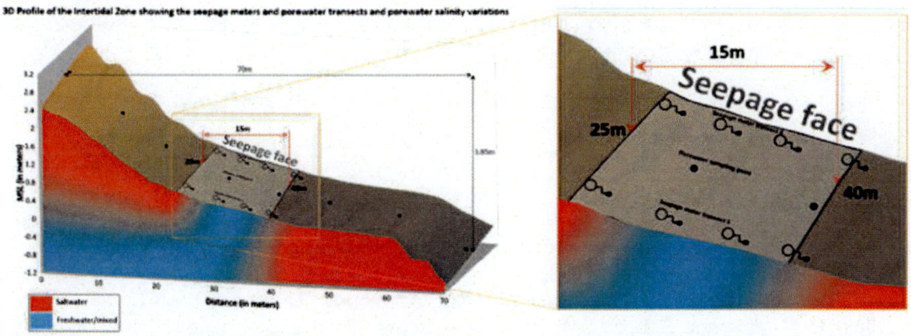

Fig. 6 3D profile showing the position of seepage meters, porewater samplers and the porewater EC profile

5 Limitations

Although seepage meter measurement is the most reliable mode to quantify SGD, using these in high energy sea conditions would not provide realistic estimates. The influence of waves, wind and tidal variations during the seepage measurements affects the seepage meter measurements. Seepage meters deployed underwater may get affected by currents, pressure built up in the seepage bags and the density difference between the seepage water and seawater itself. Seepage meters deployed in the intertidal zone is devoid of all the disturbances and can provide accurate estimates. However, it can only provide the nearshore SGD rates from the coastal unconfined aquifer. Porewater monitoring near the low tide mark cannot be done during the low tide conditions. This might lead to a data gap during the high tide period of the sampling and, therefore, cannot be used. However, the data gap will not affect the seepage estimates.

6 Conclusion

Quantifying the SGD rates are commonly done using radon mass balance estimates, which is not cost-efficient. Estimation of freshwater discharge to the coastal ocean was done at only a few portions of coastlines around the world. In India, > 1% of the 7500 km coastline is accounted for SGD. Therefore, developing a cost-efficient methodology is essential to get estimates for SGD throughout the coastline. Our methodology uses basic survey equipment and fabricated meters and tools to quantify the discharge and monitor the seepage face. The limitation of this methodology is that it only accounts for seepage from coastal unconfined aquifers, not from the deeper aquifers. Quantification of SGD rates throughout the complete coastline is crucial, especially in countries like India, where there is an increasing demand for freshwater resources. Estimations are also required for oceanographers to estimate the nutrient budgets from inland to the coastal ocean.

References

1. Moore, W. S. (2010). The effect of submarine groundwater discharge on the ocean. *Annual Review of Marine Science, 2*, 59–88.
2. Jiao, J., & Post, V. (2019). *Coastal hydrogeology*. Cambridge University Press.
3. Finkl, C. V., & Krupa, S. L. (2003). Environmental impacts of coastal-plain activities on sandy beach systems: Hazards, perception and mitigation. *Journal of Coastal Research, 35*, 132–150.
4. Taniguchi, M., Burnett, W. C., Cable, J. E., & Turner, J. V. (2003). Assessment methodologies of submarine groundwater discharge. *Land and Marine Hydrogeology, 1*.
5. Israelson, O. W., & Reeve, R. C. (1944). Canal lining experiments in the Delta Area. *Utah. Utah Agricultural Experiment Station Bulletin, 313*, 52.
6. Lee, D. R. (1977). A device for measuring seepage flux in lakes and estuaries. *Limnology and Oceanography, 22*, 140–147.
7. Burnett, W. C., Aggarwal, P. K., Aureli, A., Bokuniewicz, H., Cable, J. E., Charette, M. A., Kontar, E., Krupa, S., Kulkarni, K. M., Loveless, A., & Moore, W. S. (2006). Quantifying submarine groundwater discharge in the coastal zone via multiple methods. *Science of the Total Environment, 367*(2–3), 498–543.
8. Johnson, A. G., Glenn, C. R., Burnett, W. C., Peterson, R. N., & Lucey, P. G. (2008). Aerial infrared imaging reveals large nutrient-rich groundwater inputs to the ocean. *Geophysical Research Letters, 35*(15).
9. Roxburgh, I. S. (1985). Thermal infrared detection of submarine spring associa-ted with the Plymouth Limestone. *Hydrological Sciences Journal, 30*, 185–196.
10. Banks, W., Paylor, R., & Hughes, W. (1996). Using thermal infrared imagery to delineate groundwater discharge. *Groundwater, 34*, 434–444.
11. Miller, D., & Ullman, W. (2004). Ecological consequences of ground water discharge to Delaware Bay, United States. *Ground Water, 42*, 959–970.
12. Shaban, A. M., Khawlie, M., Abdallah, C., & Faour, G. (2005). Geologic controls of submarine groundwater discharge: Application of remote sen-sing to north Lebanon. *Environmental Geology, 47*, 512–522.
13. Duarte, T., Hemond, H., Frankel, D., & Frankel, S. (2006). Assessment of submarine groundwater discharge by handheld aerial infrared imagery: Case study of Kaloko fishpond and bay, Hawai'i. *Limnology and Oceanography-Methods, 4*, 227–236. https://doi.org/10.4319/lom.2006.4.227
14. Rapaglia, J., Grant, C., Bokuniewicz, H., Pick, T., & Scholten, J. (2015). A GIS typology to locate sites of submarine groundwater discharge. *Journal of environmental radioactivity, 145*, 10–18.
15. Oberdorfer, J. A. (2003). Hydrogeologic modeling of submarine groundwater discharge: Comparison to other quantitative methods. *Biogeochemistry, 66*, 159–169.
16. Moore, W. S. (1996). Large groundwater inputs to coastal waters revealed by 226 Ra enrichments. *Nature, 380* (6575), 612–614.
17. Cable, J. E., Burnett, W. C., & Chanton, J. P. (1997). Magnitude and variations of groundwater along a Florida marine shoreline. *Biogeochemistry, 38*, 189–205.
18. Lubarsky, K. A., Silbiger, N. J., & Donahue, M. J. (2018). Effects of submarine groundwater discharge on coral accretion and bioerosion on two shallow reef flats. *Limnology and Oceanography, 63*(4), 1660–1676.
19. Jr Marsh, J. A. (1977). Terrestrial inputs of nitrogen and phosphates on fringing reefs on Guam. In *Proceedings: Third International Coral Reef Symposium*, pp. 331–336, University of Miami, Miami, Florida.

20. Johannes, R. (1980). The Ecological Significance of the Submarine Discharge of Groundwater. *Marine Ecology Progress Series*. https://doi.org/10.3354/meps003365
21. Hu, C., Muller-Karger, F. E., & Swarzenski, P. W. (2006). Hurricanes, submarine groundwater discharge, and Florida's red tides, *Geophysical Research Letters, 33*, L11601https://doi.org/10.1029/2005GL025449
22. Kim, H. C., & Lee, K. (2009). Significant contribution of dissolved organic matter to seawater alkalinity. *Geophysical Research Letters, 36*(20).
23. Mccoy, C. A., & Corbett, D. R. (2009). Review of submarine groundwater discharge (SGD) in coastal zones of the Southeast and Gulf Coast regions of the United States with management implications. *Journal of Environmental Management, 90*(1), 644–651. https://doi.org/10.1016/j.jenvman.2008.03.002
24. Hays, R. L., & Ullman, W. J. (2007). Direct determination of total and fresh groundwater discharge and nutrient loads from a sandy beachface at low tide (Cape Henlopen, Delaware). *Limnology and Oceanography, 52*(1), 240–247.
25. Leote, C., Ibanez, J. S., & Rocha, C. (2008). Submarine groundwater discharge as a nitrogen source to the Ria Formosa studied with seepage meters. *Biogeochemistry, 88*, 185–194.
26. Kim, G., Kim, J. S., & Hwang, D. W. (2011). Submarine groundwater discharge from oceanic islands standing in oligotrophic oceans: Implications for global biological production and organic carbon fluxes. *Limnology and Oceanography, 56*(2), 673–682.
27. Burnett, W. C., et al. (2003). Burnett_2003_Biogeochemistry.pdf. *Biogeochemistry, 66*, 3–33. https://doi.org/10.1023/B:BIOG.0000006066.21240.53
28. Barlow, P.M. (2003). Ground water in freshwater–saltwater envi-ronments of the Atlantic coast. *US Geological Survey Circular 1262*.
29. Gonneea, M. E., Mulligan, A. E., & Charette, M. A. (2013). Seasonal cycles in radium and barium within a subterranean estuary: Implications for groundwater derived chemical fluxes to surface waters. *Geochimica et Cosmochimica Acta, 119*, 164–177. https://doi.org/10.1016/j.gca.2013.05.034
30. Basterretxea, G., et al. (2010). Submarine groundwater discharge to the coastal environment of a Mediterranean Island (Majorca, Spain): Ecosystem and biogeochemical significance. *Ecosystems, 13*(5), 629–643. https://doi.org/10.1007/s10021-010-9334-5
31. Shaw, R. D., & Prepas, E. E. (1989). Anomalous, short-term influx of water into seepage meters. *Limnology and Oceanography, 34*, 1343–1351.
32. Bone, S. E., Gonneea, M. E., & Charette, M. A. (2007). Geochemical cycling ofarsenic in a coastal aquifer. *Environmental Science and Technology, 40*, 3273–3278.
33. Duce, R. A., & Tindale, N. W. (1991). Atmosheric transport of iron and its deposition in the ocean. *Limnology and Oceanography, 36*, 1715–1726.
34. Martin, J. H., & Fitzwater, S. (1988). Iron deficiency limits phytolankton growth in the northeast Pacific subarctic. *Nature, 331*, 341–343.
35. Windom, H. L., Moore, W. S., Niencheski, L. F. H., & Jahrike, R. A. (2006). Submarine groundwater discharge: A large previously unrecognized source of dissolved iron to the South Atlantic Ocean. *Marine Chemistry, 102*, 252–266.
36. Cai, W.-J., Wang, Y., Krest, J., & Moore, W. S. (2003). The geochemistry of dissolved inorganic carbon in a surficial groundwater aquifer in North Inlet, South Carolina, and the carbon fluxes to the coastal ocean. *Geochimica et Cosmochimica Acta, 67*, 631–639.
37. Goni, M. A., & Gardner, L. R. (2003). Seasonal dynamics in dissolved organic carbon concentrations in a coastal water-table aquifer at the forest-marsh interface. *Aquatic Geochemistry, 9*, 209–232.
38. Moore, W.S., Blanton, J.O., & Joye, S.B., (2006). Estimates of flushing times, submarine groundwater discharge, and nutrient fluxes to Okatee Estuary, South Carolina. *Journal of Geophysical Research: Oceans, 111*(C9).
39. Santos, I. R., et al. (2009). Tidal pumping drives nutrient and dissolved organic matter dynamics in a Gulf of Mexico subterranean estuary. *Geochimica et Cosmochimica Acta, 73*(5), 1325–1339. https://doi.org/10.1016/j.gca.2008.11.029

40. Ghyben, W.B. (1899). Notes in verband met voorgenomen put boring Nabji Amsterdam. In Tijdschrift van let koninklijk Institut van Ingenieurs. The Hague, Netherlands.
41. Herzberg, A. (1901). Die Wasserversorgung einiger Nordseebader. Zeitschriftf ̈ur Gasbeleuchtung und Wasserversorgung 44: 815–819, 842–844.
42. Todd, D. K. (2005). Ground-Water-Hydrology-By-Dk-Tood.Pdf, pp. 206–225.
43. Babu, D. S., Padmalal, D., & Rao, N. P. (2018). Submarine groundwater discharge in the Indian context. *Current Science, 115*(12), 2197–2198.
44. Freeze, R. A., & Cherry, J. A. (1979). *Groundwater* (p. 604). Prentice-Hall Inc.
45. Rosenberry, D. O. (2008). A seepage meter designed for use in flowing water. *Journal of Hydrology, 359*(1–2), 118–130. https://doi.org/10.1016/j.jhydrol.2008.06.029

Effect of Grid Type on the Flow Field Predictions in Baffled Circular Surface Aeration Tanks

K. Devarajan and Ajey Kumar Patel

Abstract The performance of a new hybrid grid over the conventional unstructured tetrahedral grid in modelling the flow field predictions in the circular baffled aeration tank was analyzed in the present study. The RANS approach with standard $k-\varepsilon$ model was used for modelling the various flow features. The hybrid grid was developed by dividing the entire tank domain into various blocks and providing tetrahedral elements in the blocks close to the impeller and sweeping with hexahedral elements in the remaining blocks. Unstructured tetrahedral grid consists of tetrahedral elements in the complete domain of the vessel. The numerical convergence was assessed by evaluating the Grid Convergence Index (GCI) of the mean and turbulent flow fields. Both the grid types have provided accurate and reliable predictions of N_{pt} while the hybrid grid has developed the reliable and superior predictions of $N_{p\varepsilon}$, mean radial velocity and turbulent kinetic energy as compared to the unstructured tetrahedral grid at lesser computational cost. The energy imbalance (relative difference between N_{pt} and $N_{p\varepsilon}$) associated with the hybrid grid is 5.3% which increases to 15% with the unstructured tetrahedral grid. Therefore, the hybrid grid can be employed for effectively modelling the flow features in the surface aeration tanks.

Keywords Surface aeration tank · CFD · Grid type · GCI · Mixing · Turbulence

1 Introduction

Surface aeration tanks are widely employed for the biological oxidation of organic wastes, suspended solids and nutrients present in the wastewater. The surface aerators require steady supply of energy for continuous working of turbines employed for mixing operations as well as for pumps used for re-circulating the sludge and mixed liquor [1]. Surface aerators consume 45–75% of the total energy required for the

K. Devarajan (✉) · A. K. Patel
National Institute of Technology Warangal, Warangal, Telangana, India

A. K. Patel
e-mail: akpcivil@nitw.ac.in

A. K. Dikshit et al. (eds.), *Innovative Trends in Hydrological and Environmental Systems*, Lecture Notes in Civil Engineering 234,
https://doi.org/10.1007/978-981-19-0304-5_50

wastewater treatment plant [2]. Thus, proper configuration and operating conditions of the surface aeration tanks need to be designed to achieve higher treatment efficiency at lesser cost by maintaining the required effluent quality criteria. The detailed characterization of mean and turbulent flow fields is necessary for optimal design of the surface aeration tanks as the underlying physical and chemical phenomena are closely connected with the flow features [3]. The Computational Fluid Dynamics (CFD) approach is widely adopted for hydraulic modelling of the surface aerators due to the scale-up issues and difficulties in measuring the fluctuating quantities near the impeller surface by the experimental techniques [4]. The Reynolds Averaged Navier-Stokes (RANS) approach is regarded as the major tool for modelling the flow field characteristics due to appropriate balance between the accuracy of predictions and computational cost [5]. The RANS approach was able to accurately predict the mean and turbulent quantities in the bulk circulation region while the same near the impeller were under/over predicted [6]. However, Wechsler et al. [4], Aubin et al. [7], Deglon and Meyer [8] and Coroneo et al. [9] concluded that the mean and turbulent flow field predictions near the impeller surface can be improved by properly verifying the numerical errors associated with the modelling process. But these authors have considered the grid resolution and numerical discretization scheme as the major sources of numerical error in the respective CFD modelling works.

The review article of Liu [10] specifies that the type of grid used for discretizing the reactor domain also affects the accuracy of predictions from the CFD models. But this aspect hasn't received wide attention in the earlier studies related with the CFD modelling of reactor vessels. The structured grid types provide accurate flow field predictions while extensive time is required for the grid generation and modelling processes. On the other hand, the unstructured grids can be generated in a faster manner while the accuracy of predictions is affected by the numerical diffusion error [11]. Longest and Vinchurkar [11, 12] have analysed the impact of various types of grids such as structured hexahedral, unstructured hexahedral, unstructured tetrahedral, flow adaptive tetrahedral and prismatic grid types on the velocity and particle deposition profiles of the respiratory airways. It was recommended in those studies regarding the necessity of the development of new hybrid grid types which can maintain the advantages of structured and unstructured grid types. Garcia et al. [13] has compared the performance of two unstructured grid schemes viz. cut-cell scheme and tetrahedral scheme in modelling the mean and turbulent flow fields associated with the reactor vessels. Although, the power number computed from impeller torque (N_{pt}) and pumping number were independent of the type of grid, the power number from turbulent dissipation rate ($N_{p\varepsilon}$) predicted from the cut-cell scheme was better than that obtained from the tetrahedral scheme. Further, it was suggested to develop new grid schemes for accurately modelling the various flow features associated with the reactor vessels. Therefore, it is necessary to develop a new hybrid grid by combining the structured and unstructured grid elements in a proper fashion for effectively modelling the flow features related with the surface aeration tank.

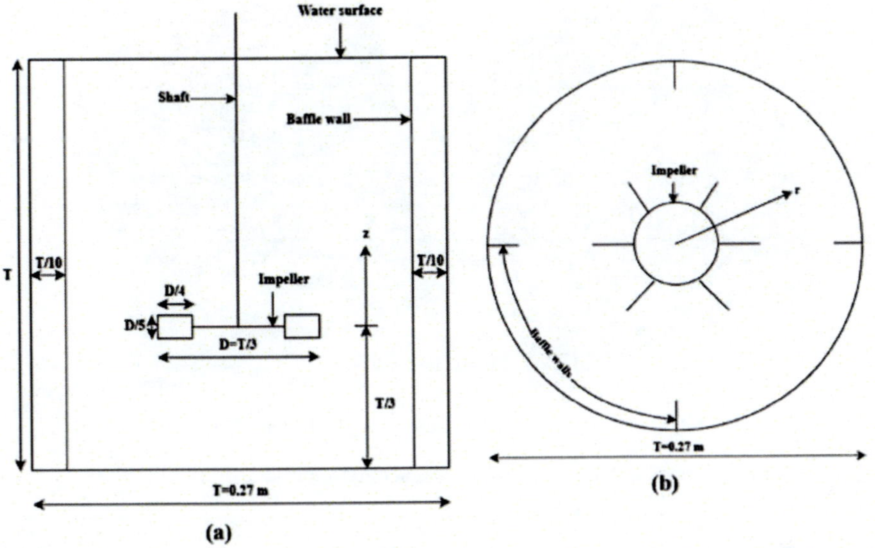

Fig. 1 **a** Sectional elevation and **b** Plan of stirred tank reactor of standard configuration. *Source* Wu and Patterson [15]

2 Surface Aeration Tank Configuration

The standard configuration of fully baffled surface aeration tank having circular cross-section was considered for modelling purposes in the present study. The standard configuration of the reactor vessel was widely investigated in the earlier studies using advanced experimental and computational approaches [14, 15] so that enormous data regarding the mean and turbulent quantities at various locations in the vessel is available for the validation process. The Rushton Turbine (RT) impeller with six blades was concentrically mounted at a clearance of 9.3 cm from the bottom of the vessel and used for agitating the fluid within the reactor vessel. The impeller was rotated at a speed of 200 rpm and the Reynolds number of flow was 29,000. The working fluid considered for the investigations comprises of pure water. The plan and sectional elevation of the surface aeration tank with detailed dimensions are illustrated in the Figs. 1a, b respectively.

3 Numerical Grid Development

The performance of two grid types namely, hybrid grid and unstructured tetrahedral grid on the flow field predictions was compared in this particular study. The three dimensional geometry of the reactor vessel was developed using ANSYS Design Modeler tool and the associated dimensions were accurately represented in the

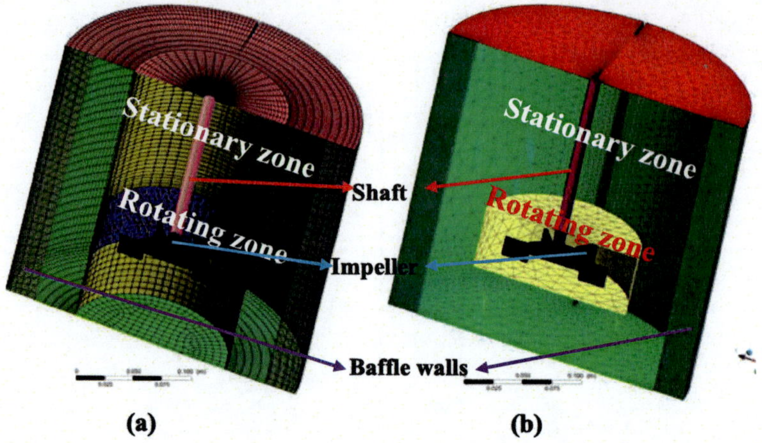

Fig. 2 Isometric views of **a** Hybrid grid and **b** Unstructured tetrahedral grid

computational domain. The finite volume grids were generated using the ANSYS Meshing tool for properly defining the geometric features of the vessel and the grid refinements were provided close to the impeller for resolving the sharp velocity gradients. The hybrid grid was developed by splitting the computational domain of the stirred tank into various blocks. The block around the impeller was discretized using the tetrahedral elements so as to properly define the complex shape of the RT impeller. In the other blocks generated around this zone, the hexahedral elements were developed by sweeping a surface mesh consisting of quadrilateral dominant elements along the axial direction of the tank. The tetrahedral elements were distributed based on the Patch Confirming Algorithm [16] and the hexahedral elements were provided according to the General Sweeping Algorithm available in ANSYS [16]. On the other hand, the generation of unstructured tetrahedral grid doesn't requires any block wise treatment as that of the hybrid grid and the entire tank geometry was discretized using the tetrahedral elements based on the Patch Confirming Algorithm. The three dimensional geometry of the RT impeller as well as the blade and disc thicknesses were properly resolved in both the grid types. Moreover, the inflation layers were provided around the impeller so as to resolve the boundary layer gradients by refining the grid elements in a direction perpendicular to the impeller walls. The isometric views of hybrid and unstructured tetrahedral grids are illustrated in the Figs. 2a, b respectively.

4 Governing Equations

The steady state Reynolds Averaged Navier Stokes (RANS) equations were used for modelling the flow features associated with the surface aeration tank and the same are provided in the Eqs. (1)–(2). The standard $k - \varepsilon$ model was used for resolving the Reynolds stresses arising from the ensemble averaging process.

Continuity equation:

$$\nabla.(\rho\vec{\mathbf{u}}) = 0 \tag{1}$$

Momentum equation:

$$\nabla.(\rho\vec{u}\vec{\mathbf{u}}) = -\nabla P + \nabla.\left(\overline{\overline{\tau}} + \overline{\overline{\tau^R}}\right) + \rho\vec{g} + \overrightarrow{F^{\mathrm{MRF}}} \tag{2}$$

In the above equations, ρ, P and $\mathbf{u}$ are the time averaged density, static pressure and velocity of the fluid. The terms $\overline{\overline{\tau}}$ and $\overline{\overline{\tau^R}}$ indicate the viscous stress tensor and Reynolds stress tensor respectively. The term $\rho\vec{g}$ represents the gravitational body force and the term $\overrightarrow{F^{\mathrm{MRF}}}$ includes the combined effects of centrifugal and coriolis forces which is considered during the modelling of impeller rotation. The turbulent viscosity (μ_t) was determined according to the Eq. (3):

$$\mu_t = C_\mu \rho \frac{k^2}{\varepsilon} \tag{3}$$

The transport equations for determining the turbulent kinetic energy (k) and turbulent dissipation rate (ε) are specified as per the Eqs. (4) and (5) respectively.

$$\nabla.(\rho\vec{\mathbf{u}}k) = \nabla.\left(\frac{\mu_t}{\sigma_k}\nabla k\right) + G_k - \rho\varepsilon \tag{4}$$

$$\nabla.(\rho\vec{\mathbf{u}}\varepsilon) = \nabla.\left(\frac{\mu_t}{\sigma_\varepsilon}\nabla \varepsilon\right) + \frac{\varepsilon}{k}(C_{1\varepsilon}G_k - C_{2\varepsilon}\rho\varepsilon) \tag{5}$$

In the above equations, G_k represents the turbulence generation rate which is calculated from mean velocity gradients, σ_k and σ_ε indicate the turbulent Prandtl numbers, $C_{1\varepsilon}$ and $C_{2\varepsilon}$ are the constants. The empirical constants associated with the standard $k - \varepsilon$ model are given below.

$$C_\mu = 0.09,\ C_{1\varepsilon} = 1.44,\ C_{2\varepsilon} = 1.92,\ \sigma_k = 1,\ \sigma_\varepsilon = 1.3$$

5 Computational Methodology

The three dimensional steady state CFD models were developed by simultaneously solving the ensemble averaged continuity and momentum equations. The Reynolds stresses obtained during the ensemble averaging procedure were closed using the standard $k-\varepsilon$ model. The impeller rotation was modelled using the pseudo steady state Multiple Reference Frame (MRF) technique. The SIMPLE scheme was used for coupling the continuity and momentum equations and the coupled equations were numerically discretized using the second order upwind scheme. The peripheral surfaces of the reactor vessel and the parts of the impeller were represented using no slip boundary condition whereas the top surface of the tank was simulated using the symmetry boundary condition. The standard wall function [17] was used to describe the viscous flow associated with the solid surfaces. The CFD simulations were performed till the monitoring parameters such as N_{pt} and $N_{p\varepsilon}$ attain constant values and the residuals allied with the governing equations fall below the level of 10^{-6}. An advanced workstation having double precision 64 bit Intel (R) Xeon (R) E5-1620 3.6 GHz processor with 12 cores was used for developing the CFD models.

6 Results and Discussion

6.1 Numerical Convergence

The variations of N_{pt}, $N_{p\varepsilon}$ and axial profiles of normalized mean radial velocity with the grid resolution for both the grid types are shown in the Figs. 3, 4, 5 and 6 respectively. The flow field predictions from the five grids of both the grid types have been used for the preliminary grid independence analysis. The size of the elements comprising the impeller was successively reduced to resolve the sharp velocity gradients, trailing vortices and high turbulence regions near the impeller. The hybrid grid exhibits considerable variations in the predictions of N_{pt} and mean radial velocity up to Grid 2 while the unstructured tetrahedral grid attains the convergence of the same after the Grid 3. The $N_{p\varepsilon}$ has shown continuous increase up to the Grid 4 for both the grid types and provides negligible improvements with the Grid 5. Therefore, the proper grid convergence of turbulent flow quantities requires high resolution grids as compared to the mean flow fields. The N_{pt}, peak mean radial velocity and $N_{p\varepsilon}$ increases by 4.7%, 11% and 33.4% from Grid 1 to Grid 5 for hybrid grid while the same increases by 6.4%, 19% and 38% from Grid 1 to Grid 5 for the unstructured tetrahedral grid. Thus, the unstructured tetrahedral grid is more sensitive to the grid resolution as compared to the hybrid grid.

The systematic and scientific grid independence study was performed based on the recommendations provided by the American Society of Mechanical Engineers (ASME) [18]. The flow field results from three grids among the five grids adopted for the preliminary grid independence analysis were used for this purpose. Numerical discretization error was computed in terms of Grid Convergence Index (GCI) whereby the relative error from a grid refinement process is converted to an equivalent uncertainty estimate corresponding to an asymptotic convergence level (i.e., grid doubling with second order accuracy). The recommendations provided by Roy [19] were also adopted to avoid the unrealistic larger or smaller GCI magnitudes

Table 1 Features of the grids generated under hybrid and unstructured tetrahedral grid schemes

Grid designation	Grid size	r	GCI (%)				Computational time (Hrs.)
			N_{pt}	$N_{p\varepsilon}$	u_{rp}	k_p	
Hybrid grid							
Grid 1 (Coarse)	300,573	–					4
Grid 2 (Medium)	996,072	1.49	3.6	24.24	10.35	35.76	20
Grid 5 (Fine)	7,418,360	1.95	1	5.44	1.27	1.73	40
Unstructured tetrahedral grid							
Grid 1 (Coarse)	266,733	–					5
Grid 3 (Medium)	1,001,202	1.55	6.1	31.7	17.27	37.11	27
Grid 5 (Fine)	7,142,519	1.93	0.35	10	7.07	3.19	60

which can arise due to unstructured nature of grids, non-uniform refinement factor and turbulent flow conditions in the vessel. The details of the grids used for the calculation of numerical discretization error are given in the Table 1.

The refinement factor (r) between the coarse, medium and fine grids used for the analysis of both the grid types is more than 1.3 as per the recommendations of Celik et al. [18]. The conventional grid doubling was almost achieved between the medium and fine grids of both the grid types. It can be seen from the Table 1 that the GCI of N_{pt}, $N_{p\varepsilon}$, peak mean radial velocity and turbulent kinetic energy decreases with the increase in the grid resolution for both the grid types. The N_{pt}, $N_{p\varepsilon}$, peak radial velocity and turbulent kinetic energy have achieved monotonic grid convergence for both the grid types. The hybrid and unstructured tetrahedral grid types provide reliable predictions of N_{pt} with GCI less than 1%. The hybrid grid type produces superior convergence of $N_{p\varepsilon}$ with a GCI of 5.44% which increases to 10% for the unstructured tetrahedral grid type. The GCI associated with the peak radial velocity and turbulent kinetic energy from the hybrid grid are 1.27% and 1.73% respectively which drastically increases to 7.07% and 3.19% for the unstructured tetrahedral grid type. The higher GCI of various flow fields obtained from the unstructured tetrahedral grid type may be due to the development of higher magnitudes of numerical diffusion error from the underlying grid elements [12]. The asymptotic or near asymptotic convergence of the various flow quantities is also an important consideration while analyzing the reliability of the CFD predictions [19]. The difference between formal order of the numerical scheme $\left(p_f\right)$ and the observed order of accuracy $\left(\hat{p}\right)$ associated with the predictions from fine, medium and coarse grids provides a clear idea regarding the asymptotic convergence of various flow quantities. The hybrid and unstructured tetrahedral grid types generate asymptotic convergence of N_{pt} as the $\hat{p}$ is close to p_f. The hybrid grid type develops asymptotic grid convergence for $N_{p\varepsilon}$ as the relative difference between $\hat{p}$ and p_f is less than 4% while the unstructured tetrahedral grid type exhibits non-asymptotic grid convergence of $N_{p\varepsilon}$ as this relative deviation increases to 33%. Moreover, the relative deviation between the $\hat{p}$ and p_f for the peak radial velocity and turbulent kinetic energy associated with the hybrid grid type are 5.3% and 7% respectively which increases to 33.33% and 37.50% related with the unstructured tetrahedral grid type. Thus, the hybrid grid type develops reliable predictions of the mean and turbulent flow fields due to the lesser magnitudes of GCI and near asymptotic convergence of the same as compared to the unstructured tetrahedral grid type. Further, the computational time increases with the increase in the grid resolution for both the grid types. Also, the computational time associated with the fine grid of hybrid grid type is much lesser than that related with the fine grid of unstructured tetrahedral grid type.

6.2 *Accuracy of Flow Field Predictions*

The mean and turbulent flow field predictions from the hybrid and unstructured tetrahedral grids were compared with the corresponding experimental results of Bates et al. [20] and Wu and Patterson [15] so as to evaluate the accuracy of the flow field predictions. As shown in the Figs. 3 and 4, both the grids accurately predict the N_{pt} with errors less than 5%. The hybrid grid provides superior prediction of $N_{p\varepsilon}$ with an error of 10% while the error associated with the unstructured tetrahedral grid is 18%. The energy input from the impeller should be effectively transported to the entire domain of the vessel so as to achieve necessary mixing characteristics in the reactor vessel [5]. The term 'energy imbalance' is defined as the relative difference between the N_{pt} and $N_{p\varepsilon}$ which indirectly provides a clear insight regarding the energy input into the vessel from the impeller and dissipation of the same in the entire domain of the reactor vessel [5]. The energy imbalance needs to be minimized to obtain appropriate transfer of energy from the impeller to the entire volume of the reactor vessel [5]. The energy imbalance associated with the hybrid grid is 5.3% which

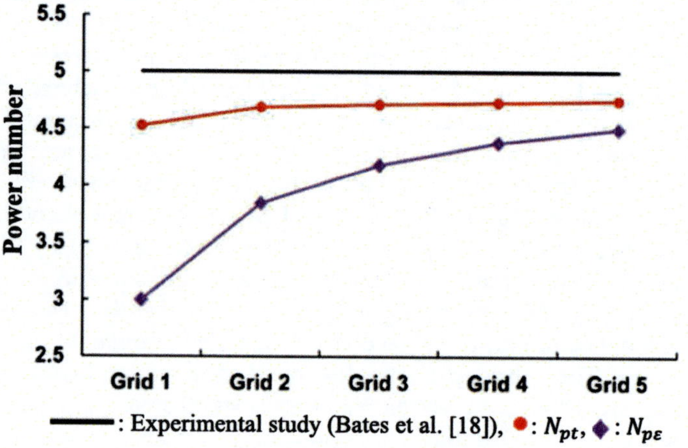

Fig. 3 Variations of N_{pt} and $N_{p\varepsilon}$ with grid resolution for Hybrid grid

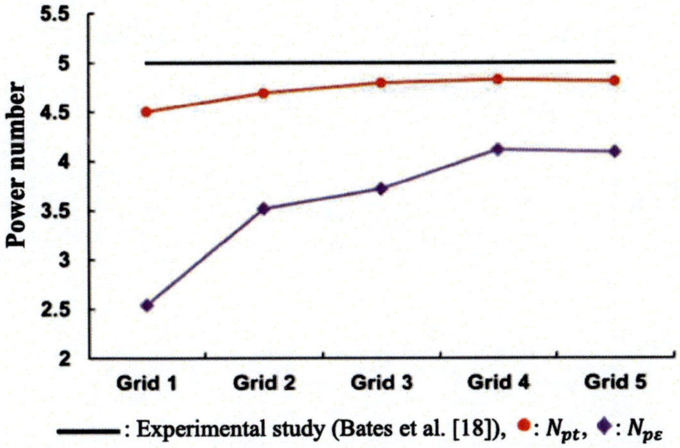

Fig. 4 Variations of N_{pt} and $N_{p\varepsilon}$ with grid resolution for Unstructured tetrahedral grid

considerably increases to 15% for the unstructured tetrahedral grid (Fig. 7). The axial profiles of mean radial velocity from both the grid types confirm to single peak curves representing the discharge stream jet emerging from the impeller. The predictions of axial profile of mean radial velocity from both the grid types are highly similar and close to that from the experimental profile near the impeller blade as illustrated in the Figs. 5 and 6 respectively. However, the predictions of mean radial velocity away from the impeller from the hybrid grid are much better than that from the unstructured tetrahedral grid as shown in the Fig. 8. Thus, the hybrid grid outperforms the unstructured tetrahedral grid in terms of accuracy of mean and turbulent flow field predictions, numerical convergence as well as the computational cost. The new hybrid grid will also be beneficial in accurately modelling the flow features associated with the complex geometries such as pumps, turbines, pipe networks and rotor-stator mixers at lesser computational cost.

7 Conclusions

The mean and turbulent flow fields associated with the baffled circular surface aeration tank agitated using the RT impeller were modelled using the RANS approach with standard $k-\varepsilon$ model closure conditions. The performance of a new hybrid grid over the conventional unstructured tetrahedral grid was evaluated in terms of accuracy of predictions, numerical convergence and computational cost. Both the grids provide superior numerical convergence and accurate predictions of N_{pt}. The numerical convergence and accuracy of prediction of mean radial velocity from the hybrid grid were better than that from the unstructured tetrahedral grid. Also, the hybrid grid has developed excellent numerical convergence and accurate predictions of turbulent

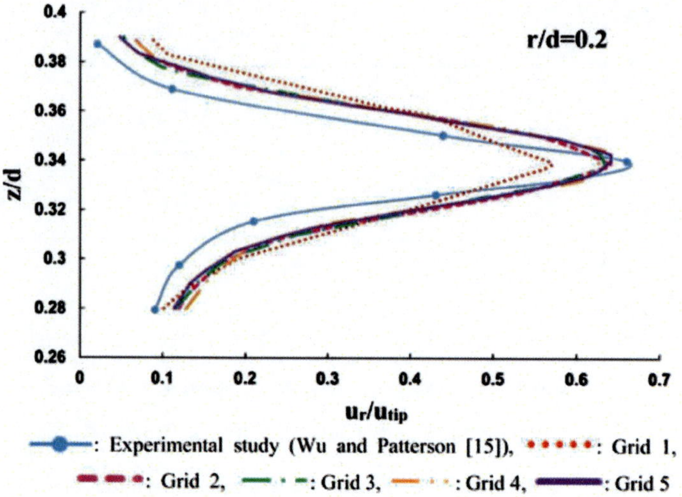

Fig. 5 Variation of axial profiles of $\frac{u_r}{u_{\text{tip}}}$ close to the impeller with grid resolution for Hybrid grid

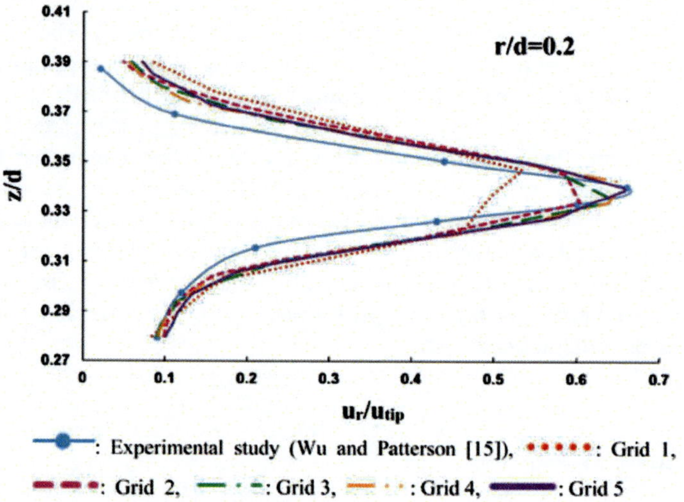

Fig. 6 Variation of axial profiles of $\frac{u_r}{u_{\text{tip}}}$ close to the impeller with grid resolution for Unstructured tetrahedral grid

flow fields as compared to the unstructured tetrahedral grid. Moreover, the computational time associated with the hybrid grid was lesser than that of the unstructured tetrahedral grid. Therefore, the hybrid grid can be adopted for modelling the surface aeration tanks due to reliable and accurate predictions of mean and turbulent flow fields at lesser computational cost.

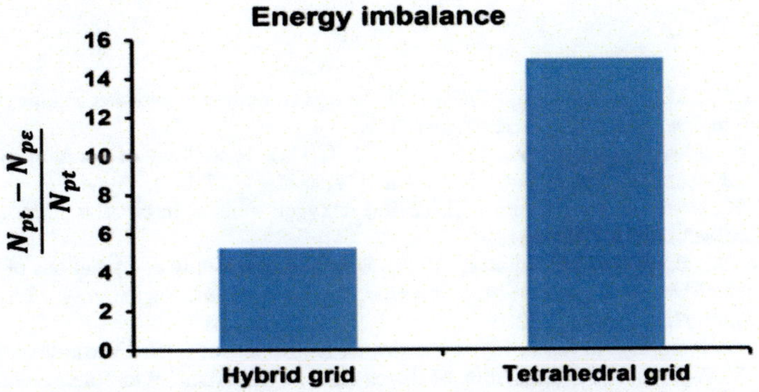

Fig. 7 Energy imbalance related with the hybrid and unstructured tetrahedral grids

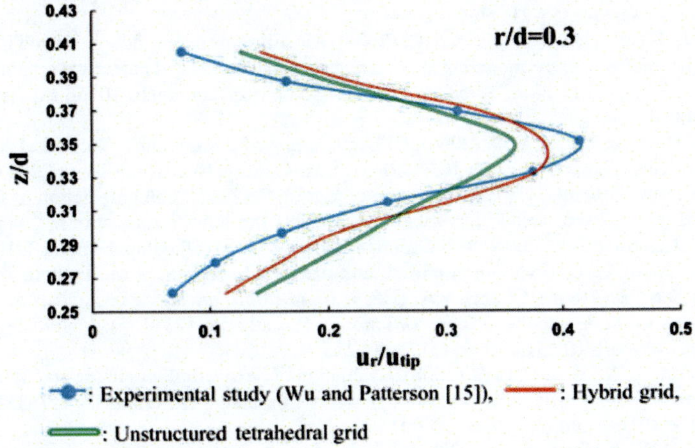

Fig. 8 Comparison of axial profiles of $\frac{u_r}{u_{tip}}$ between the hybrid and unstructured tetrahedral grids away from the impeller

Acknowledgements This work was supported by the Science and Engineering Research Board (SERB) of Government of India (Grant No. SB/S3/CEE/0057/2013).

References

1. Karpinska, A. M., & Bridgeman, J. (2016). CFD aided modelling of activated sludge systems—A critical review. *Water Research, 88*, 861–879.
2. Rieger, L., Alex, J., Gujer, W., & Siegrist, H. (2006). Modelling of aeration systems at wastewater treatment plants. *Water Science and Technology, 53*(4–5), 439–447.
3. Rao, A. R., Patel, A. K., & Kumar, B. (2009). Oxygen transfer in circular surface aeration tanks. *EnvTech, 30*(7), 747–753.
4. Wechsler, K., Breuer, M., & Durst, F. (1999). Steady and unsteady computations of turbulent flows induced by a 4/45° pitched-blade impeller. *Journal of Fluids Engineering, 121*, 318–329. https://doi.org/10.1115/1.2822210
5. Joshi, J. B., Nere, N. K., Rane, C. V., Murthy, B. N., Mathpati, C. S., Patwardhan, A. W., & Ranade, V. V. (2011). CFD simulation of stirred tanks: Comparison of turbulence models. Part I: Radial flow impellers. *The Canadian Journal of Chemical Engineering, 89*, 23–82. https://doi.org/10.1002/cjce.20446
6. Brucato, A., Ciofalo, M., Grisafi, F., & Micale, G. (1998). numerical prediction of flow fields in baffled stirred vessels: A comparison of alternative modelling approaches. *Chemical Engineering Science, 53*(21), 3653–3684.
7. Aubin, J., Fletcher, D. F., & Xuereb, C. (2004). Modeling turbulent flow in stirred tanks with CFD: The influence of the modeling approach, turbulence model and numerical scheme. *Experimental Thermal and Fluid Science, 28*, 431–445. https://doi.org/10.1016/j.expthermflusci.2003.04.001
8. Deglon, D. A., & Meyer, C. J. (2006). CFD modelling of stirred tanks: Numerical considerations. *Minerals Engineering, 19*, 1059–1068. https://doi.org/10.1016/j.mineng.2006.04.001
9. Coroneo, M., Montante, G., Paglianti, A., & Magelli, F. (2011). CFD prediction of fluid flow and mixing in stirred tanks: Numerical issues about the RANS simulations. *Computers & Chemical Engineering, 35*, 1959–1968. https://doi.org/10.1016/j.compchemeng.2010.12.007
10. Liu, M. (2016). CFD Modeling of stirred tank reactors. In Kresta, S. M., Etchells III, A. W., Dickey, D.S., & Atiemo Obeng, V.A. (Eds.), *Chapter 5b, advances in industrial mixing, a companion to the handbook of industrial mixing* (1st ed., pp. 123–145). John Wiley & Sons Inc. https://doi.org/10.1595/205651317X696225
11. Vinchurkar, S., & Longest, P. W. (2008). Evaluation of hexahedral, prismatic and hybrid mesh styles for simulating respiratory aerosol dynamics. *Computers & Fluids, 37*, 317–331. https://doi.org/10.1016/j.compfluid.2007.05.001
12. Longest, P. W., & Vinchurkar, S. (2007). Effects of mesh style and grid convergence on particle deposition in bifurcating airway models with comparisons to experimental data. *Medical Engineering & Physics, 29*(3), 350–366.
13. Garcia, A.A., Mendoza Escamilla, V. X., Martinez Delgadillo, S. A., Gonzalez Neria, I., Gutierrez Torres, C., & Jimenez Bernal, J.A. (2019). On the performance of different rans based models to describe the turbulent flow in an agitated vessel using non-structured grids and PIV validation. *Brazilian Journal of Chemical Engineering, 36*(1), 361–382. https://doi.org/10.1590/0104-6632.20190361s20180091
14. Zhang, Y., Gao, Z., & Li, Z. (2017). Transitional flow in a Rushton turbine stirred tank. *AIChE Journal, 63*(8), 3610–3623. https://doi.org/10.1002/aic
15. Wu, H., & Patterson, G. K. (1989). Laser Doppler measurements of turbulent flow parameters in a stirred mixer. *Chemical Engineering Science, 44*, 2207–2221. https://doi.org/10.1016/0009-2509(89)85155-3
16. ANSYS. (2013). *ANSYS fluent theory guide release 15.0*. ANSYS Inc, USA.
17. Launder, B. E., & Spalding, D. B. (1974). The numerical computation of turbulent flows. *Computer Methods in Applied Mechanics and Engineering, 3*, 269–289. https://doi.org/10.1016/0045-7825(74)90029-2
18. Celik, I.B., Ghia, U., Roache, P.J., Freitas, C.J., Coleman, H., & Raad, P.E. (2008). Procedure for estimation and reporting of uncertainty due to discretization in CFD applications. *Journal of fluids Engineering, 130*, 078001(1–4).

19. Roy, C. J. (2010). Review of discretization error estimators in scientific computing. In *48th AIAA Aerospace science meeting Incl. New horizons forum & aerospace exposition.* American Institute of Aeronautics and Astronautics, Orlando, Florida, pp. 1–29. https://doi.org/10.2514/6.2010-126
20. Bates, R. L., Fondy, P. L., & Corpstein, R. R. (1963). An examination of some geometric parameters of impeller power. *Industrial & Engineering Chemistry Process Design and Development,* 2(4), 310–314. https://doi.org/10.1021/i260008a011

Assessment of Environmental Flows of Tungabhadra River Using Global Environmental Flow Calculator

P. N. Chandi Priya, Rehana Shaik, and Rahul Kumar Singh

Abstract Maintaining the full scale of naturally occurring river flow is usually impossible due to the development of the water resources and variations of land and soil usage in the catchment. These developed resources can create differences in the balance of the ecosystem and socio-economic activities. These designed structures also cause a decrease in the minimum flow regime in downstream. Environmental flows (EFs) are the medium that help maintain river flow in healthy or ecological conditions. The river's hydrologic (river mapping), hydraulic (cross-section, water depth, and velocity), and environmental conditions (riparian flora and fauna) are significant considerations for estimating environmental flows (EF's). There are various desktop assessment methods for calculating the environmental flows. In this present study, the global environmental flow calculator (GEFC) method is used to estimate the ecological flows by using the flow duration curves (FDC) generated from the given monthly discharge data of the river. The FDC in this system contains 17 fixed percentile points concerning the discharge. In the current study, we analyze the environmental flows of the Tungabhadra River basin by considering the different discharge stations, which are Balehonuur, Haralahalli, Hosaritti, Shivamogga, Honalli, Rattihalli, and Tungabhadra Dam, with a mean annual flow (MAF) of 36%, 24.8%, 27.2%, 16.2%, 23.3%, 21.1%, and 12.2%, respectively, to maintain the ecological conditions of the river. The monthly discharge data from 1995 to 2017 for those stations are obtain from the Advance Center for Integrated Water Resource

P. N. Chandi Priya (✉)
Lab for Spatial Informatics, International Institute of Information Technology, Hyderabad 500032, India
e-mail: nagachandi.p@research.iiit.ac.in

R. Shaik
Department of Civil Engineering, International Institute of Information Technology, Hyderabad, India
e-mail: rehana.s@iiit.ac.in

R. K. Singh
Department of Civil Engineering, International Institute of Technology, Roorkee 247667, India
e-mail: rsingh@hy.iitr.ac.in

A. K. Dikshit et al. (eds.), *Innovative Trends in Hydrological and Environmental Systems*, Lecture Notes in Civil Engineering 234,
https://doi.org/10.1007/978-981-19-0304-5_51

(ACIWR) Bengaluru, India. The river flow health is a study which helps in understanding the environmental variables that effects the habitat structure, flow regime, water quality, and biological conditions of the river. To estimate the Flow Health of Tungabhadra River, we used a tool called Flow Health which uses nine indicators to represent the Flow Health (FH) score for the stations Balehonnur, Haralahalli, Hosaritti, Shivamogga, Honalli, Rattihalli, and Tungabhadra. This tool uses the gauge discharge data in the form of reference (1995–2005) and test periods (2006–2017), with Flow Health score of 0.72, 0.4, 0.72,0.70, 0.58, 0.73, 0.71 and 0.72, 0.63, 0.63, 0.7, 0.66, 0.67, 0.66 for test and reference period with respect to stations. The study noted that majority of the discharge stations along the Tungabhadra River show a moderate to low flow variations for the reference and test periods. Overall, Tungabhadra river health, measured by the flow indices, had declined from 1995–2005 to 2006–2017.

Keywords GEFC · Environmental flows · MAR · Discharge · EMC · Flow health

1 Introduction

An environmental flow is a water supply within a river, wetland, and many to maintain marginal ecosystems. The flows can be regulated according to the requirements using approachable methods. Environmental flows (EFs) offer a homogeneous way to manage river flows to meet the necessities of people, agriculture, industry, and ecosystems undergoing climate changes and using limited sources [1]. Managing the full spectrum of natural flows of the river is not possible due to the development of water resources and land and soil changes in a catchment. Hence, EFs can be considered as an agreement connecting the development of the watershed on one hand and maintaining the river ecosystem on other [2, 3].

For maintaining the river's natural regime, the river flow characteristics are necessary for the possibility of the river's self-purification, preserving aquatic life and vegetation, increasing groundwater level, encouraging subsistence, and upholding waterway conditions [3, 4]. Operating the rivers requires environmental flow charges having certain timing, frequency, magnitude, and duration. These conditions make sure the flow regime potential for carrying the processes of the ecosystem and are considered as EFs or environmental water requirements (EWR) [5, 6].

Many environmental flow assessment methods were developed in recent years, namely range of variability approach (RVA), desktop reserve model (DRM), and global environmental flow calculator (GEFC). In this study, we used the desktop environmental flow assessment method known as a global environmental flow calculator using flow duration curves to estimate the EFs. All the FDC values in this system are represented from top to bottom using seventeen fixed percentile points 0.01, 0.1, 1, 5, 10, 20, 30, 40, 50, 60, 70, 80, 90, 95, 99, 99.9, and 99.99 [7, 8]. EF aims to maintain an ecosystem in the desired feature known as environmental management class or ecological management category classified into six classes from A to F. This study follows the terminology environmental management classes. The larger value of environmental management classes (EMC), the more water needs to assign for ecological maintenance [9, 10]. Although various researchers have performed studies on various river basins of India for estimating the environmental flows and managing rivers, the concept of EFs is still in its infancy stages. Smakthin and Anputhas [2] set up a global environmental flow calculator to calculate the EFs and generalize the environmental classes of more than ten river basins in India by considering the single discharge station for each river [3]. Uday Kumar and Jayakumar [11] analyzed the minimum flow regime of the Krishna River downstream by comparing the three desktop models, viz., RVA, DRM, and GEFC [12, 13].

This study considers seven different discharge stations for calculating EFs of one particular river basin, Tungabhadra River, using one desktop method, GEFC. At the same time, the literature review of previous works revealed that they were done on different rivers with a single discharge location or used various EF methods.

2 Study Area and Data Used

Tungabhadra River is a major tributary of Krishna River, India. Tungabhadra River is the convergence of the Tunga and Bhadra rivers at Koodli. Tungabhadra flows from the eastern slope of the Western Ghats, Karnataka (Fig. 1). The area of catchment is 71,417 km^2, length 531 km, and located at a latitude of 14° 0′ 30″ N, and longitude of 75° 40′ 27″ E. In the present study, we consider seven discharge stations along the Tungabhadra River stretch to estimate environmental flows. Balehonuur, Haralahalli, Hosaritti, Shivamogga, Honalli, Rattihalli, and Tungabhadra Dam are the discharge stations considered in the current study.

For calculation of environmental flows of Tungabhadra River, monthly discharge data from 1995 to 2017 was obtained from the Advance Center for Integrated Water Resource (ACIWR), Bengaluru, Karnataka, India. In the current study, we considered the different discharge stations along the river to calculate environmental flows. The locations of selected gauging stations Balehonuur, Haralahalli, Hosaritti, Shivamogga, Honalli, Rattihalli, and Tungabhadra Dam are shown in Fig. 1.

3 Methodology

Environmental flows (EF) are described as the medium that helps maintain river flow in healthy conditions or some specific ecological scenario. For maintaining a fair ecological condition of rivers, EF in the range of 20%–30% is needed. Environmental flows can be analyzed and calculated in two ways: detailed assessment methods and desktop environmental flow assessment methods. In precise assessment methods, we need a sufficient amount of fieldwork, time, and resource data of

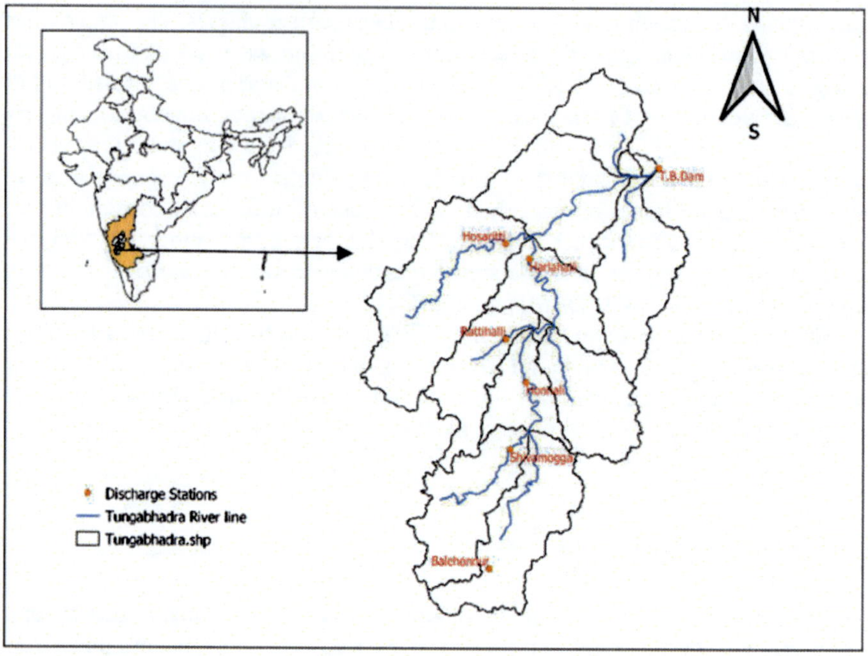

Fig. 1 Map showing discharge stations location, Tungabhadra catchment area

the river and its ecosystem. In contrast, desktop methods do not require any fieldwork, and the necessary amount of time is also significantly less. Currently, there are various types of desktop environmental flow assessment methods for calculating and analyzing environmental flows of rivers globally. Some of the ways are the range of variability approach (RVA), catchment abstraction management strategies (CAMS) [14], desktop reserve model (DRM), global environmental flow calculator (GEFC), etc. In this present study, we use the GEFC to calculate the Tungabhadra River's environmental flows, India.

GEFC uses either the globally defined or user-defined time series data. Here, GEFC uses the user-defined monthly discharge data of a river to analyze the EFs of the river using flow duration curves (FDCs). FDCs in the GEFC represents the seventeen fixed percentiles, which are 0.01, 0.1, 1, 5, 10, 20, 30, 40, 50, 60, 70, 80, 90, 95, 99, 99.9, and 99.99%, with one FDC shift for every change in the flow as described in the Smakthin and Anputhas [2] and Hughes and Münster [15]. The FDC shift in GEFC helps us to identify the suggested streamflow in a revised stream by shifting the natural flow curve within fixed percentile places. The GEFC uses the seventeen fixed percentile points, which helps shift of natural flow curve as explained above [16].

The flow duration curve is cumulative curve indicates the percentile time framed discharges equal or went above during certain time interval. Region below the curve

Table 1 Environmental management classes (EMC) used in GEFC

EMC	Nature	Description
Class A	Natural	Normal state or minor changes in stream
Class B	Slightly modified	Large obscure of habitats regardless the growth of water resources
Class C	Moderately modified	Habitats are disrupted, and ecosystem roles are unharmed. Obscure of sensitive species
Class D	Largely modified	Large variations in natural habitat, and ecosystem roles are disturbed
Class E	Seriously modified	Habitat variety and number are reduced. The species abundance dropped
Class F	Critically modified	Critical state and ecology changed with near deterioration of natural habitat

offers average daily flow and intermediate daily flow as 50% value, Q50% [17]. Flow duration curve distinguishes the ability of basin to provide flows in different magnitudes, Q75%, Q80%, and Q90%. Other main objective of environmental flows is to maintain the proper riverine ecosystem, representing various environmental management classes. GEFC uses six different management classes that directly correspond to the river's environmental flows in the given time series data. These six classes range from A to F, referring to the river's natures in terms of hydrologic alteration and the natural percentile mean annual runoff (MAR) for all class (Table 1).

After determining the flow duration curves (FDC) and environmental management classes (EMC), GEFC gives simulated monthly time series of environmental flows for the selected management class. From EMC, we learn it proved that: "Higher the EMC, higher the water to assign to the ecosystem" [3, 15].

3.1 Flow Health

The hydrological adjustments generated in the Tungabhadra River can be estimated using the application called Flow Health. Flow Health assists in computing, designing, and managing of the flow regimes [18]. The main aim of Flow Health software is to allocate a score for hydrological criteria in river health assessments. Also, it plays a role in environmental flow assessment. There are four major purposes of the Flow Health tool; first is to assist and provide annual score for hydrology indicator in river health, second to recommend the minimum monthly environmental flow regime, third to test the hydrological health of flow regime in monthly means, and last is to give monthly flow series based on designed flow regime [18].

The flow health score depends on the nine indicator value generated using the reference and test period datasets. The indicator values obtained will be in the range of 1 and 0. The values of one represents that the indicators are close to reference period, whereas the value of zero indicates that it is contrast from reference period

[19]. There are nine indicators which Flow Health depends on are shown in Tables 3 and 4 [18].

The flow duration curve helps in determining the hydrological conditions and the flow regime of the rivers, using discharge values with fixed percentile exceedance values. This is achieved by using global environmental flow calculator tool, which uses different environmental classes (EMC) as explained above.

4 Results and Discussion

The mean annual runoff (MAR) during the pre-impact or pre-altered period for the Balehonuur discharge station was 2140 MCM, Haralahalli discharge station was 7025 MCM, Hosaritti discharge station was 1006 MCM. Shivamogga discharge station was 5058 MCM, Honalli discharge station was 4888 MCM, Rattihalli discharge station was 363.43 MCM, and Tungabhadra Dam discharge station was 5734 MCM. The GEFC analyzed mean annual runoff (MAR) percentile according to the environmental management classes is explained in Table 2. The flow health obtained from reference period and test period for seven discharge stations is shown in the Tables 3 and 4.

Flow Duration Curves

- The comparison of environmental flows obtained from EMC class C for seven discharge stations of Tungabhadra River is shown in Fig. 10. Class C is the moderately modified condition and the MAR values obtained to justify maintaining the river ecological conditions in the selected river basin. As the EF range should maintain between 10% and 30% for maintaining the environmental conditions of the global rivers; in the current study, the values from class C satisfy the requirement by keeping the MAR values between 12 and 36% (Figs. 2, 3, 4, 5, 6, 7 and 8).

Table 2 Estimated environmental flows for different EMC at seven discharge locations of Tungabhadra River

Discharge stations	Environmental flows MAR (%)						
	Natural MAR (MCM)	Class A	Class B	Class C	Class D	Class E	Class F
Balehonuur	2140	72.3	51.5	36	25	17.1	11.6
Haralahalli	7025	64.2	40.5	24.8	15.4	9.8	6.5
Hosaritti	1006	61.4	40.2	27.2	18	11.6	7.2
Shivamogga	5058	59.5	32.6	16.2	7.7	3.6	1.7
Honalli	4888	62.8	38.7	23.3	14.5	9.5	6.7
Rattihalli	363.4	51	31.2	21.1	14.9	10.7	7.7
Tungabhadra dam	5734	51.4	27	12.2	5.5	2.9	1.7

Table 3 Flow Health Scores for reference period (1995–2005) for discharge stations of Tungabhadra River

Station name	High flow (HF)	Highest month (HM)	Low flow (LF)	Lowest month (LM)	Persistently higher (PH)	Persistently lower (PL)	Persistently very low (PVL)	Seasonality flow shift (SFS)	Flood flow interval (FFI)	Flow Health score (FH)
Balehonnur	0.92	0.92	0.75	0.75	0	1	1	0.44	0.83	0.72
Haralahalli	0.81	0.81	0.75	0.75	0	1	1	0	0.83	0.63
Hosaritti	0.81	0.81	0.76	0.74	0	1	1	0	0.97	0.63
Shivamogga	0.92	0.93	0.75	0.75	0	1	1	0	1	0.70
Honalli	0.81	0.81	0.75	0.75	0	1	1	0	0.91	0.66
Rattihalli	0.8	0.8	0.73	0.73	0	1	1	1	1	0.67
Tungabhadra Dam	0.81	0.81	0.75	0.75	0	1	1	0	0.91	0.66

Table 4 Flow Health Scores for test period (2006–2017) for discharge stations of Tungabhadra River

Station name	High flow (HF)	Highest month (HM)	Low flow (LF)	Lowest month (LM)	Persistently higher (PH)	Persistently lower (PL)	Persistently very low (PVL)	Seasonality flow shift (SFS)	Flood flow interval (FFI)	Flow Health score (FH)
Balehonnur	1	1	0.75	0.75	0	1	1	0	1	0.72
Haralahalli	0	0	0.75	0.75	0	1	1	0	0.45	0.4
Hosaritti	1	1	0.75	0.75	0	1	1	0	1	0.72
Shivamogga	1	1	0.75	0.75	0	1	1	0	1	0.70
Honalli	0.5	0.5	0.75	0.75	0	1	1	0	0.92	0.58
Rattihalli	1	1	0.73	0.75	0	1	1	0	1	0.73
Tungabhadra Dam	0.91	1	0.75	0.75	0	1	1	0	1	0.71

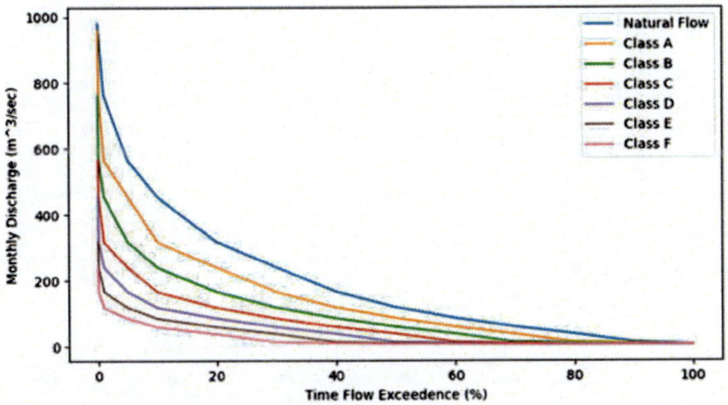

Fig. 2 FDCs of EMC comparing with natural flow for Balehonnur station

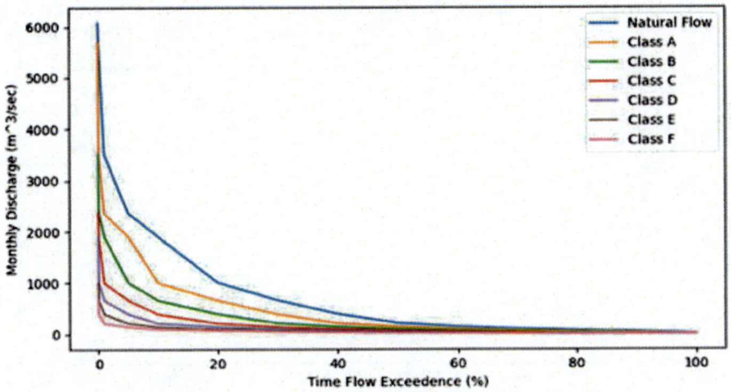

Fig. 3 FDCs of EMC comparing with natural flow for Haralahalli station

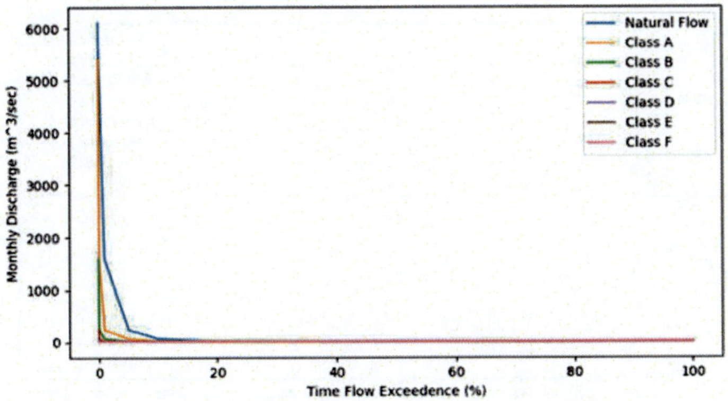

Fig. 4 FDCs of EMC comparing with natural flow for Hosaritti station

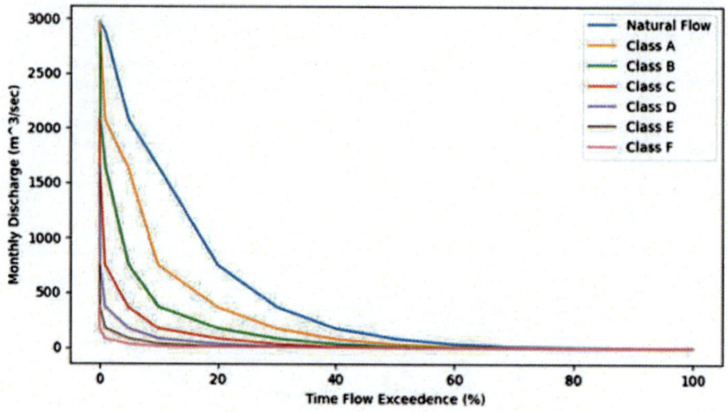

Fig. 5 FDCs of EMC comparing with natural flow for Shivamogga station

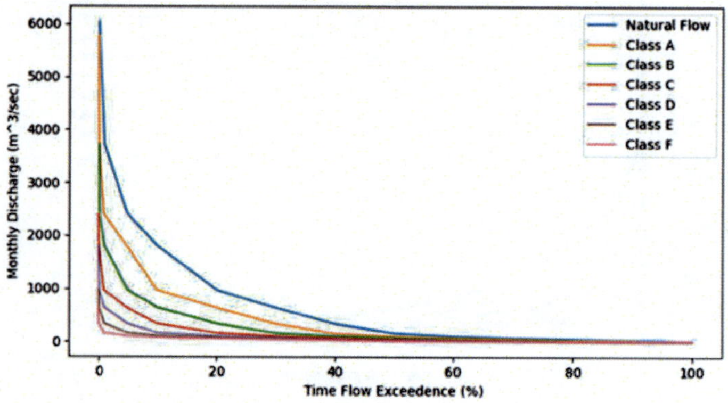

Fig. 6 FDCs of EMC comparing with natural flow for Honalli station

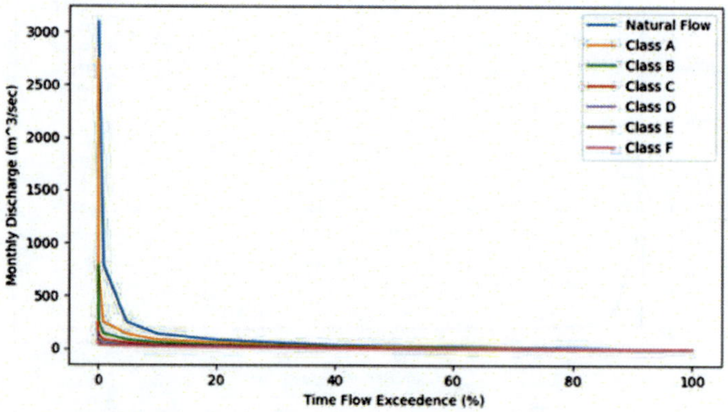

Fig. 7 FDCs of EMC comparing with natural flow for Rattihalli station

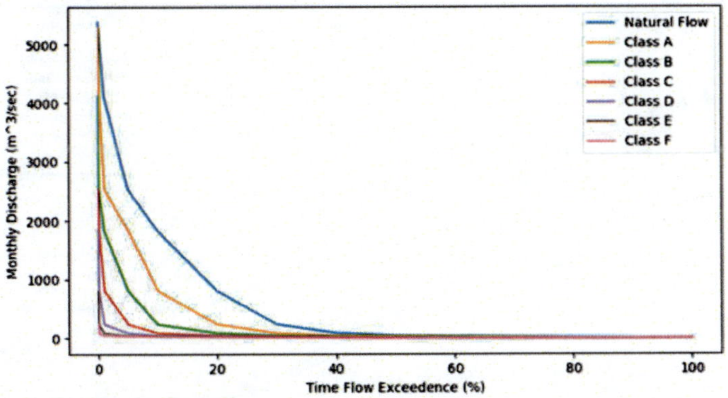

Fig. 8 FDCs of EMC comparing with natural flow for Tungabhadra Dam station

- The flow duration curves of all the six classes concerning natural flow have been shown as the considered model output. The graphs obtained observe that each class has a shift of one toward left from the natural flow. Maintaining a river between Class A and Class B, which means to maintain a river condition in natural or slightly modified conditions, requires a mean annual runoff (MAR) of above 40% to 50%. At the same time, the GEFC estimated 24% of annual MAR (6291 mcm) in order to maintain the downstream reach of Tungabhadra River at reasonable ecological condition (Class C). This study helps to understand that maintenance entails an amount of water to transmit from upstream of the water resources and structures to the downstream for the healthy river condition and better ecosystem conditions. As MAR falling below 30%, the shifting of the curve falls from the natural flow region, which shows the deteriorated conditions of the flow of the river, which causes a significant imbalance in the ecosystem (Fig. 9).

Flow Health

The flow health application was successfully run for the Flow Health flow score values using reference and test periods. Figures 10, 11, 12, 13, 14, 15, and 16, show the mosaic flow scores of the nine indicators for Balehonnur, Haralahalli, Hosaritti, Shivamogga, Honalli, Ratihalli, and Tungabhadra stations, respectively. FH with five different classes with different colors using flow deviation ranges are plotted. From Tables 3 and 4, the estimated score for the Flow Health considering reference and test periods are shown.

Here, from Tables 3 and 4, the HF (high flow) values for the stations Balehonnur, Hosaritti, Shimoga, and Rattihalli are 1 which represents a very small deviation of flow during test period compared to the reference period. The LF (low flow) remains almost similar (between 0.73 and 0.76) during reference and test years. There is high seasonal shift flows observed during test period, and same is observed for the reference years near Balehonnur and Rattihalli stations. The Flow Health (FH) score estimated for the Haralahalli drop from 0.63 to 0.4, and for Honalli, the score drops

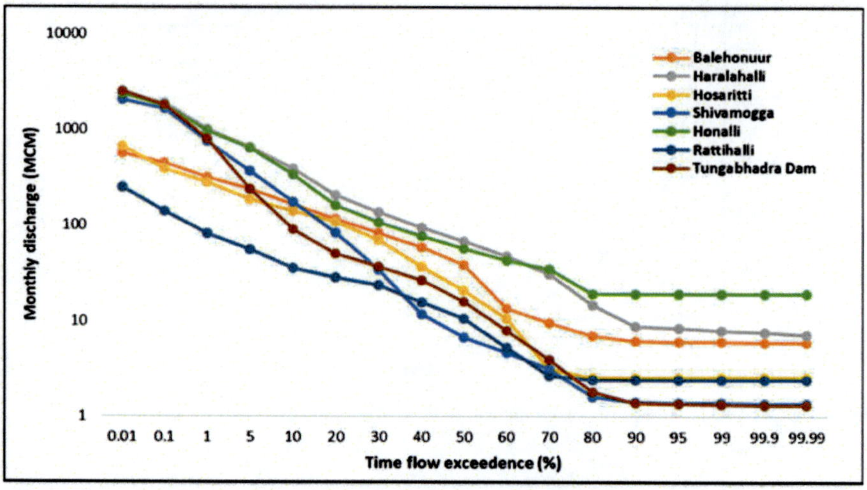

Fig. 9 FDCs for EMC of Class C comparing with all 7 discharge stations

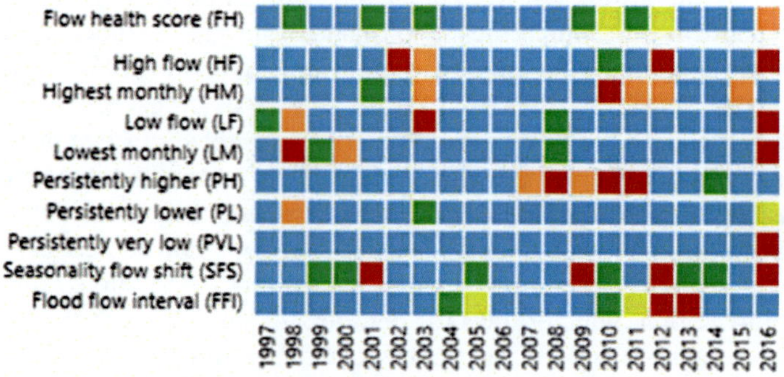

Fig. 10 Flow health scores for Balehonnur station

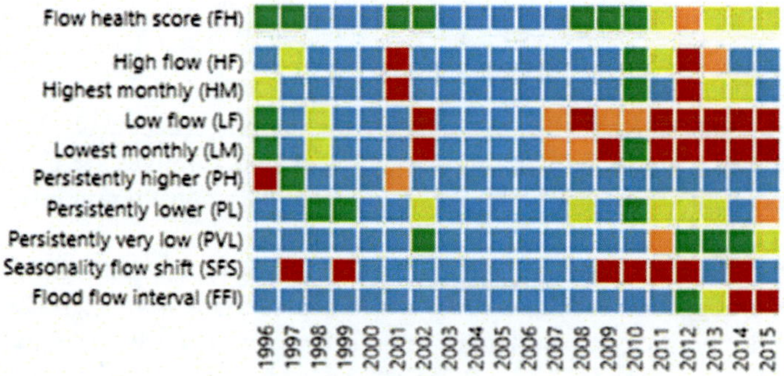

Fig. 11 Flow health scores for Haralahalli station

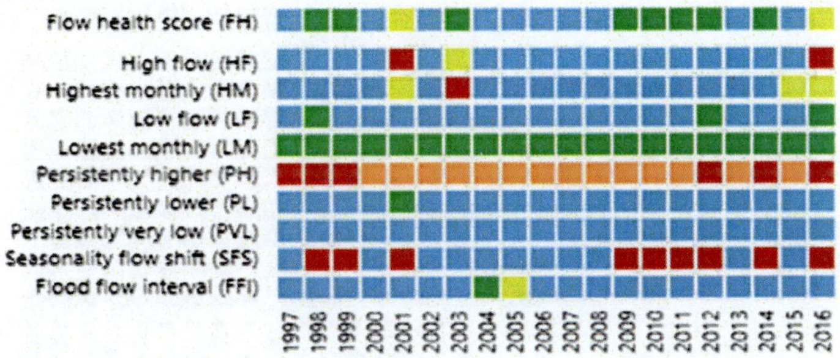

Fig. 12 Flow health scores for Hosaritti station

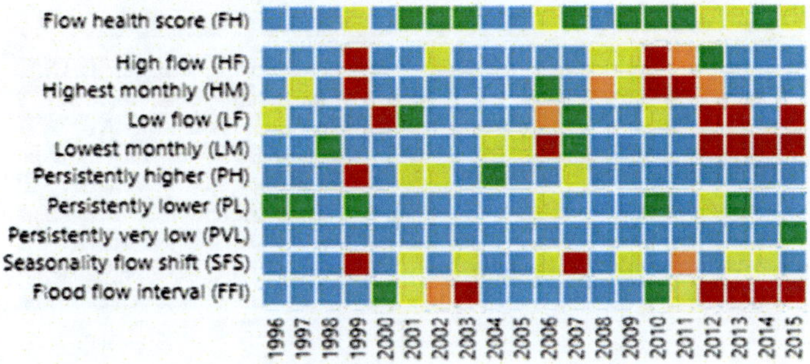

Fig. 13 Flow health scores for Shivamogga station

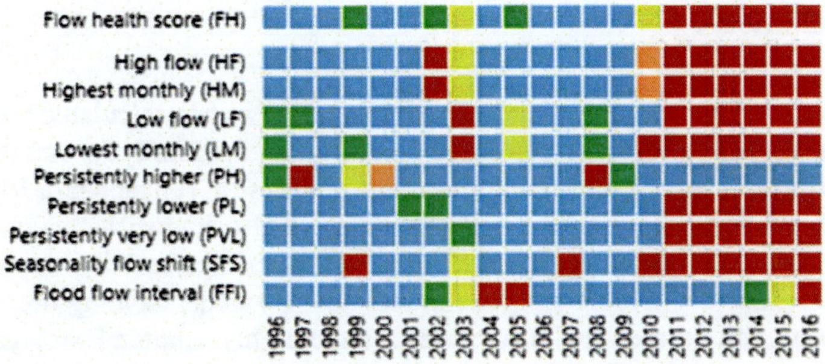

Fig. 14 Flow health scores for Honalli station

from 0.66 to 0.58 from reference period to test period which represents the moderate flow changes. And for the stations Balehonnur, Hosaritti, Rattihalli, Shivamogga, and Tungabhadra, the flow health (FH) score remains almost same ranging from 0.63 to 0.73, only with a slight increase in test period compared to reference years. None of the stations reach the Flow Health (FH) score to 1 in either cases.

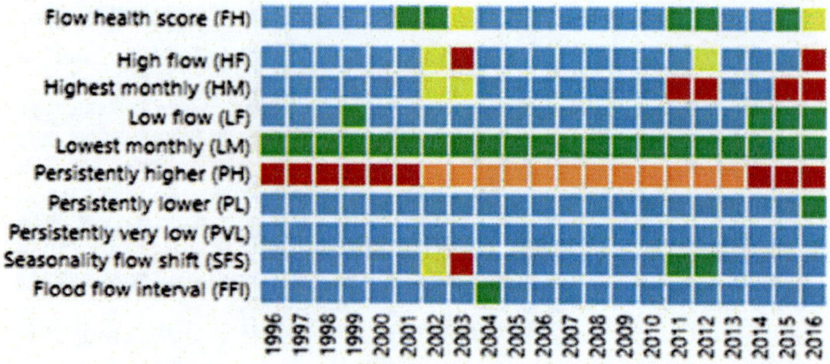

Fig. 15 Flow health scores for Rattihalli station

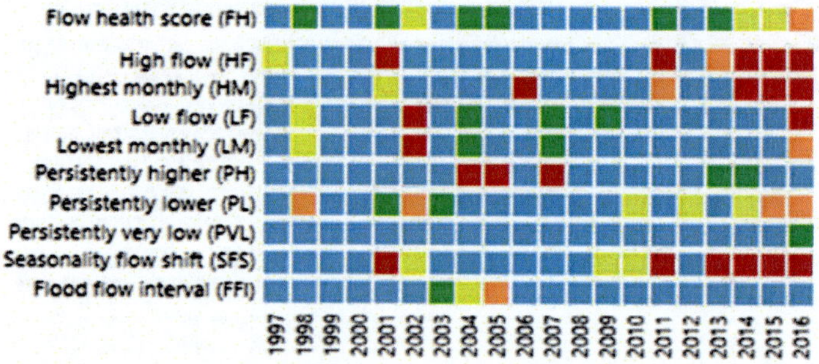

Fig. 16 Flow health scores for Tungabhadra station

5 Conclusion

Environmental flows of natural and environmental flows are calculated from 1995 to 2017 using the GEFC software, which comes under the family for rapid desktop assessment of EFs at seven different discharge stations of Tungabhadra River. The environmental water requirement, (EWR) which is also known as environmental flows, is estimated for six environmental management classes, and then the best class which falls under the given conditions is selected. By understanding the available ecological conditions, the river is positioned in the highly suitable EMC (i.e., Class C). The main characteristic feature of evaluated environmental water requirement is that higher the flow variability of a river, less environmental water requirement in all classes. With respect to simulated flow values and the obtained flow curves, Balehonnur and Haralahalli have the least variable flow patterns, thus have the high EWR. Stations with a high variability of flow patterns like Rattihalli or Tungabhadra Dam have the low EWR in most classes. The results imply that river ecosystems, in general, need to maintain an appropriate state even with limited EF allocations 10%–30% of natural MAR depending on hydrological variability. As of now, Tungabhadra River Basin has no water policy to determine EFR in order to protect the downstream riverine Ecosystem. Thus, this study helps to understand the changes in the flow and helps to observe the required amount of flow to be maintained in the downstream regions to balance the ecosystem and socioeconomic activities.

The Flow Health (FH) score for reference (1995–2005) and testing years (2006–2017) from Flow Health tool are estimated for the seven discharge stations of the Tungabhadra River. From the obtained results, flow deviations are observed to be maintained in the moderate state for the stations Balehonnur, Hossaritti, Shivamogga, Ratihalli, and Tungabhadra. Large deviation of FH score is observed in the Haralahalli and Honalli stations. As we know, there is no possibility to get a very small or no variations of flow in a river due to various natural and anthropogenic activities. Thus, from the results, we observed that majority of the discharge stations along the Tungabhadra River show a moderate to low flow variations for the reference and test years. These flow variations help in assessing the environmental water requirement using flow duration curves which we performed with the help of GEFC. Overall, Tungabhadra river health, measured by the flow indices, had declined from 1995–2005 to 2006–2017.

References

1. Environmental Flows—IUCN. https://www.iucn.org/theme/water/our-work/past-projects/environmental-flows
2. Smakhtin, V. U., & Anputhas, M. (2006). *An assessment of EF requirements of Indian river basins* [Research Report 107]. International Water Management Institute, Colombo.
3. Iyer, R. R. (2005). *The notion of environmental flows: A caution NIE/IWMI workshop on environmental flows*. New Delhi, March 23–24, 2005.
4. Jhingran, V. G. (1991). *Fish and fisheries of India* (3rd ed.). Hindustan Publishing Corporation.
5. Knights, P. (2002). Environmental flows: Lessons from an Australian experience. In *Proceedings of International Conference: Dialog on Water, Food and Environment* (18 pp.). Hanoi, Vietnam.
6. Jain, S. K., Sharma, K. D., & Singh, V. P. (2005). Interlinking of Indian Rivers—Issues and analytical framework. In *Proceedings of the IWRA XII World Water Congress* (Vol. 1, pp. 1.25–1.34). New Delhi.
7. Jowett, I. G. (1997). Instream flow methods: A comparison of approaches. *Regulated Rivers: Research and Management, 13*, 115–127.
8. Tharme, R. E. (2003). A global perspective on environmental flow assessment: Emerging trends in the development and application of environmental flow methodologies for rivers. *River Research and Applications, 19*, 397–441.
9. Acreman, M., & Dunbar, M. J. (2004). Defining environmental river flow requirements—A review. *Hydrology and Earth System Sciences, 8*, 861–876.
10. Hughes, D. A., & Hannart, P. (2003). A desktop model used to provide an initial estimate of the ecological instream flow requirements of rivers in South Africa. *Journal of Hydrology, 270*, 167–181.
11. Uday Kumar, A., & Jayakumara, K.V. (2018). Assessment of environmental flows using hydrological methods for Krishna River, India.
12. Smakhtin, V. U., Revenga, C., Döll, P. (2004). Taking into account environmental water requirements in global-scale water resources assessments. In *Research report of the CGIAR comprehensive assessment programme of water use in agriculture*. IWMI comprehensive assessment research report (Vol. 2, 24pp.). International Water Management Institute, Colombo, Sri Lanka.

13. Smakhtin, V., Revenga, C., & Döll, P. (2004). A pilot global assessment of environmental water requirements and scarcity. *Water International, 29*, 307–317.
14. Environment Agency. (2001). *Managing water abstraction: The catchment abstraction management strategy process*. Environment Agency.
15. Hughes, D. A., & Münster, F. (2000). Hydrological information and techniques to support the determination of the water quantity component of the ecological reserve. *Water Research Commission Report TT 137/ 00* (91 pp.). Pretoria, South Africa.
16. Tharme, R. E., Smakhtin, V. U. (2003). Environmental flow assessment in Asia: Capitalizing on existing momentum. In *Proceedings of the First Southeast Asia Water Forum, November 17–21, 2003, Chiang Mai, Thailand* (Vol. 2, pp. 301–313). Thailand Water Resources Association.
17. Hughes, D. A., & Smakhtin, V. U. (1996). Daily flow time series patching or extension: A spatial interpolation approach based on flow duration curves. *Journal of Hydrological Sciences, 41*(6), 851–871.
18. Flow Health: *Software to assess deviation of River flows from reference and to design a monthly environmental flow regime technical manual and user guide version 2.0.*
19. Uday Kumar, A., & Jayakumar, K. V. (2018). *Assessment of hydrological alteration and environmental flow requirements for Srisailam dam on Krishna River*, India

Geographic Information System-Based Appraisal of Groundwater Quality Index of the Dimoria Block, Assam, India

Priyanka Kotoky, Bimlesh Kumar, and Ajay S. Kalamdhad

Abstract The Dimoria block of the state of Assam, India, has been taken up to appraise the groundwater quality index (GWQI) and to obtain the variability of the water quality parameters affecting the block's groundwater quality over time and space. 5 physico-chemical water quality parameters (pH, TDS, turbidity, TH, and TA), 8 main ions (Na^+, Ca^{2+}, Mg^{2+}, K^+, F^-, NO_3^-, Cl^-, and SO_4^{2-}), and 5 heavy metals (Mn, Fe, Pb, As, and Zn) were analysed for the groundwater samples that were amassed from 10 sampling sites casing the entire study area in pre-monsoon (PrM) season, monsoon (Mon) season, and post-monsoon (PoM) season in the 2019–2020 year phase. The spatio-temporal variations of GWQI are studied using a geographic information system (GIS). Based on the I.S. 10500:2012 standards and GWQI, the results revealed that the more prominent parts of the block's groundwater are not apt for consumption rationale. Amongst the 5 physico-chemical water quality parameters, pH, turbidity, and TH exceeded their respective permissible limits in all the seasons. Na^+, Ca^{2+}, and K^+ were found to be the dominant cations. Amongst the major anions, F^- and NO_3^- played the most dominant role. The most influential heavy metals are Mn, Fe, and As in all the seasons. Based on GWQI and GIS-based spatio-temporal maps, it was monitored that the eminence of groundwater decreased from Mon to PoM to PrM seasons due to the augment of concentration of the varied physico-chemical parameters, cations, anions, and heavy metals.

Keywords Anions · Cations · Geographic information system · Groundwater quality index · Heavy metals

---

P. Kotoky (✉) · B. Kumar · A. S. Kalamdhad
Department of Civil Engineering, Indian Institute of Technology Guwahati, Guwahati 781039, India

B. Kumar
e-mail: bimk@iitg.ac.in

A. S. Kalamdhad
e-mail: kajay@iitg.ac.in

A. K. Dikshit et al. (eds.), *Innovative Trends in Hydrological and Environmental Systems*, Lecture Notes in Civil Engineering 234,
https://doi.org/10.1007/978-981-19-0304-5_52

1 Introduction

One of the most vital inherent water resources is groundwater. Huge quantities of groundwater are devoured daily for diverse purposes like drinking, irrigation, industrial, and commercial uses [1]. Due to such uses, the physical, chemical, and organic compositions of groundwater are being amended [2, 3]. The contamination of groundwater sources has become a global environmental concern due to its undeviating impact on the health and environment. For preserving and protecting the groundwater resource, long-term conservation, development, and management of this natural resource are critical [4, 5]. This includes groundwater quality assessment and geographic information system (GIS)-based mapping that takes into account constant monitoring of the groundwater, thereby providing information about its quality. GIS over traditional methods can serve up as an incredibly functional tool not only for modelling of groundwater but also for the appraisal of spatio-temporal deviations in groundwater quality [5–7]. Incorporation of GIS with groundwater quality assists the appraising and supervising the spatio-temporal-based details of groundwater quality and its contamination and thus, is very useful for decoding multifaceted scheduling and supervision of groundwater problems [8, 9].

On the other hand, the appraisal of groundwater eminence is imperative for the socio-economic progress of any section of the world. The fortitude of groundwater quality for human utilization is fundamental for the welfare of the ever-escalating populace. Superior quality groundwater will make sure the sustainability of socio-economic growth, because the government precedence is shifted to other segments of the economy, rather than fighting eruptions of water-borne ailments owing to using up of contaminated groundwater. The assessed groundwater quality data (measures or concentrations of the water quality parameters of the groundwater) can be integrated with the GIS to get the overall groundwater quality in terms of the groundwater quality index (GWQI). Horton [10] in the mid of the precedent century was the foremost researcher to recommend the profits of appraising a water quality index (WQI), and thereafter, numerous studies regarding water quality indexes were accounted in different places for lakes, rivers, coastal areas, and groundwater. These indices' values are vital when taking into consideration water use by individuals and appraisals for consumers or collaborators. A WQI signifying any water resource can be affected by chemical, physical, and biological factors [11, 12]. The WQI provides a solitary figure that conveys the comprehensive water quality of a natural water resource at a specific place and time derived from numerous water quality parameters. It is a mathematical tool functional to transmute a huge measure of groundwater quality data into a sole number that elucidates the intensity of groundwater contamination. GWQI serves as a constructive tool for monitoring the groundwater quality and for appraising its potential to fit a particular purpose [10, 11, 13]. The employment of a GWQI may not be the finest method to comprehend large-scale groundwater quality circumstances, but for explicit use, it would be the solitary way. GWQI can trounce communication predicaments between scientists, water resource managers, policymakers, governmental agencies, and the local public [14]. Thus, the GIS-based

GWQI maps are effortlessly interpretable and can serve as a communication tool to inform concerning the groundwater quality eminence for improved use in future.

For the current study, the Dimoria block of the state of Assam is selected as the study area. Despite being a very significant block, it has been denied of apposite measures for season-wise appraisal and monitoring of groundwater quality. From the reports of the Central Ground Water Board [15] and the Public Health Engineering Department (Assam), it is found that there is the problem of iron, arsenic, and other heavy metals in groundwater causing associated health vulnerabilities. The groundwater reliability is much in this block because of the surface water pollution and the dearth of appropriate and well-sustained water delivering networks. All through the rainy period (monsoon season), the surface water becomes unusable for consumption, and during the dry periods (post-monsoon and pre-monsoon seasons), water dries up. So the barely dependable water supply comes from groundwater. Also, an unswerving incessant scrutinizing of various water quality parameters of groundwater is not available in this block. Even the topmost government organizations responsible for carrying out groundwater quality assessment take into account only a small quantity of water quality parameters, and there is an absence of seasonal and continuous appraisal and monitoring of groundwater. Thus, there is a requirement for an exhaustive scientific study with precise appraisal and monitoring of the groundwater of the block. Therefore, the rationales of this study are: (1) to gauge and endow with an overview of current groundwater quality of the Dimoria block with special reference to 5 physico-chemical parameters, viz. negative logarithm of hydrogen ion concentration (pH), total dissolved solids (TDS), turbidity, total alkalinity (TA), and total hardness (TH), 8 major ions concentration, viz. sodium (Na^+), calcium (Ca^{2+}), magnesium (Mg^{2+}), potassium (K^+), fluoride (F^-), chloride (Cl^-), nitrate ($NO_3{}^-$), and sulphate ($SO_4{}^{2-}$), and 5 heavy metals, viz. manganese (Mn), iron (Fe), lead (Pb), Arsenic (As), and zinc (Zn); (2) estimation of GWQI spatio-temporal variability to distinguish potable and non-potable zones.

2 Materials and Methodology

2.1 *Study Area*

The Dimoria block is positioned in the Kamrup Metropolitan district of the state of Assam, India, between north latitudes 25° 43′ N and 26° 51′ N, and amid east longitudes 90° 36′ E and 92° 12′ E. It has an area of 261.64 km^2 covering mostly forests and agricultural areas. It is bordered on the west by the Guwahati block of the Kamrup Metropolitan district, on the north by the Chandrapur Block of the Kamrup Metropolitan district and Marigaon district, and on the east by the Marigaon district. On the south lies the state of Meghalaya. Taking into account the exploration survey of the study area, 10 sampling locations were selected that were evenly allocated casing the whole block. The samples were collected from

Amsing (26.157043° N and 91.896918° E), Gomoria Gaon (26.143862° N and 91.972853° E), Lumsum Pathar (26.122466° N and 92.013537° E), Thengabhanga (26.159509° N and 92.038844° E), Murkata (26.091313° N and 92.099548° E), Maloibari (26.169411° N and 92.094049° E), Kamarkuchi (26.121182° N and 91.905108° E), Khetri (26.126053° N and 92.083430° E), Topatali-Umsiang (26.102297° N and 92.165109° E), and Guripathar (26.138816° N and 92.159106° E) areas that represent the most inhabited areas in the block.

In this block, a chief segment of the population takes the assistance of groundwater as a resource of drinking water and for other household and agricultural purposes. The key groundwater sources in this block are ring wells (RW), hand tube wells (HTW), hand bore wells (HBW), tara pumps (TP), and deep tube wells (DTW). So, the sampling from these groundwater sources is adequate as the sampling locations have sources that replicate and characterize the authentic groundwater. Based on reconnaissance survey in the block, the 10 sampling locations that were selected were based on factors such as accessibility of the sampling location and presence of wells; wells nearer to infecting sources like industries, city wastewater drains, trash dumpsites, agricultural lands etc.; wells inferred for contaminants like iron, nitrate, or such contaminants. Thus, the current study is implemented for the appraisal of the water quality parameters in the Dimoria block and also to study whether the groundwater is apposite for domestic usages devoid of any health risk or not.

2.2 *Data Acquisition*

The coordinates of the selected 10 sampling points were obtained by using the GARMIN GPS-60 receiver. The importation of the sampling points data to ArcGIS (v. 10.5) was done using the WGS_1984_UTM_Zone_46N projection. The boundary of the Dimoria block was delineated using ArcMap (v. 10.5). Using ArcGIS's point feature tool, the places in the block were digitized. The Survey of India topographic sheet no. ng46 was also utilized to create the ultimate study area map. The obtained map and the GPS data build the spatial database. Figure 1 shows the Dimoria block map depicting the sampling sites.

Seasonal analysis was carried out by sampling in the block for 3 seasons: pre-monsoon (PrM) season, monsoon (Mon) season, and post-monsoon (PoM) season during the year 2019–2020. Assortment of the samples, hauling, storage, and investigation were made per APHA guiding principles [16]. The collected groundwater specimens were examined for a totality of 18 water quality parameters, viz. pH, TDS, turbidity, TH, TA, Na^+, Ca^{2+}, Mg^{2+}, K+, F^-, NO_3^-, Cl^-, SO_4^{2-}, Mn, Fe, Pb, As, and Zn. The analyses were implemented in triplicates.

The pH of the groundwater samples was gauged via a digital pH meter. The TDS (mg/l) of the samples was measured using a digital electrical conductivity meter. The turbidity (NTU) of the samples was measured using a nephelometric turbidimeter. The TH (mg/l as $CaCO_3$) and TA (mg/l as $CaCO_3$) of the samples were measured using APHA titrimetric methods [15]. The Na^+ (mg/l), Ca^{2+} (mg/l), and

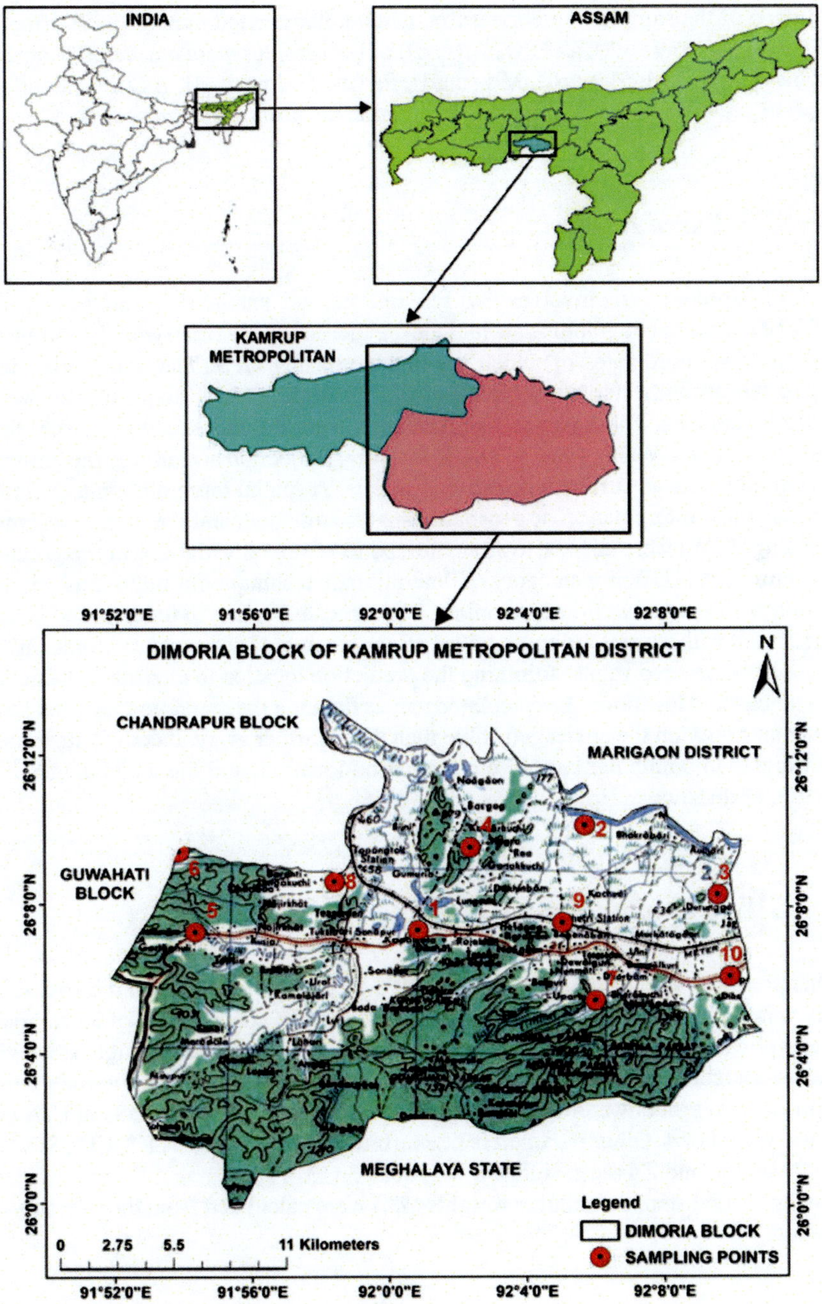

Fig. 1 Dimoria block depicting the sampling points

K^+ (mg/l) of the samples were measured using a flame photometer. The F^- (mg/l), NO_3^- (mg/l), Cl^- (mg/l), and SO_4^{2-} (mg/l) of the samples were measured using ion chromatography. Mg^{2+} (mg/l), Mn (mg/l), Fe (mg/l), Pb (mg/l), As (mg/l), and Zn (mg/l) of the samples were gauged using atomic absorption spectroscopy.

2.3 GIS Based Analysis

The GIS-based analysis involves two sets of data, viz. spatial data and non-spatial data. The spatial or geographic data include the final study area map with the sampling locations, the coordinates of which are obtained using GPS. The non-spatial data include the groundwater quality data of each sampling point about all the water quality parameters. These two sets of data are integrated into ArcGIS (v. 10.5) GIS platform using the WGS_1984_UTM_Zone_46N projection and adding the groundwater quality data as attributes to the spatial data. A spatial interpolation technique, a geostatistical interpolation or geospatial analysis method, and the inverse distance weighting (IDW) were utilized to attain the spatio-temporal variations of the groundwater eminence. IDW is a category of deterministic technique for multivariate interpolation with a set of well-known points. The allocated values to unidentified points are intended with a weighted mean of the values obtainable at the known points. IDW utilizes the measured values adjoining the prediction location to envisage a value for any unmeasured location. The calculated values flanking the prediction location have further authority on the envisaged value than those farther away. It contributes superior weights to points flanking to the prediction location, and the weights ebb as a function of distance.

2.4 GWQI Computation

In this present study, for calculating GWQI, 3 steps were pursued. In the first step, every water quality parameter was assigned a weight (w_i) on the basis of its comparative implication on the overall quality of groundwater (Table 1). The highest weight of 5 was assigned to the parameters like F^-, NO_3^-, Pb, and As owing to their chief significance in groundwater quality appraisal. Parameters like pH, TDS, and TA are given a weight of 4. Other parameters like turbidity, TH, Na^+, Ca^{2+}, K^+, Cl^-, SO_4^{2-}, Mg^{2+}, Mn, Fe, and Zn are assigned a weight of 3 [13, 17].

In the second step, the relative weights (W_i) were calculated from the subsequent equation.

$$W_i = w_i / \sum_{i=1}^{n} w_i \tag{1}$$

Table 1 Water quality parameters with the relative weights

Parameters	Weight (w_i)	Relative weight (W_i)
pH	4	0.062
TDS	4	0.062
Turbidity	3	0.046
TH	3	0.046
TA	4	0.062
Na^+	3	0.046
Ca^{2+}	3	0.046
K^+	3	0.046
F^-	5	0.077
NO_3^-	5	0.077
Cl^-	3	0.046
SO_4^{2-}	3	0.046
Mg^{2+}	3	0.046
Mn	3	0.046
Fe	3	0.046
Pb	5	0.077
As	5	0.077
Zn	3	0.046
Sum	65	1.000

where w_i is the weight of each parameter, W_i is the relative weight, and n is the parameter quantity.

In the third step, a quality rating (q_i) for every parameter is allocated by dividing its concentration in each groundwater sample by its corresponding standard in accordance with the principles laid down in the I.S. 10500:2012 [18], and the result is multiplied by 100.

$$q_i = (C_i/S_i) \times 100 \tag{2}$$

where q_i is the quality rating; C_i is the parameter concentration in each groundwater sample in mg/l, and S_i is the Indian drinking water quality standard for each parameter in mg/l according to the guidelines of the BIS 10500:2012 [18].

For calculating the GWQI, the SI is foremost established for each parameter. Subsequently, the GWQI is determined as stated by the next equation.

$$\mathrm{SI}_i = W_i \times q_i \tag{3}$$

$$\mathrm{GWQI} = \sum \mathrm{SI}_i \tag{4}$$

Table 2 Range of GWQI and type of groundwater

Range	Type of groundwater
<50	Excellent
50–100	Good
100.01–200	Poor
200.01–300	Very poor
>300	Unsuitable for drinking purpose

where SI_i is the sub-index of the *i*th parameter; q_i is the rating based on the *i*th parameter's concentration.

Table 1 shows the water quality parameters and their relative weights, and Table 2 shows the range of GWQI and type of groundwater [17].

3 Results and Discussion

The levels of the physico-chemical entities of the groundwater tasters were statistically examined, and the outcomes, for instance maximum, minimum, mean, and standard deviation, are furnished in Tables 3 and 4 for the 3 seasons in the Dimoria block. Table 5 demonstrates the percentage of groundwater samples surpassing the permissible limit in the block based on BIS 10500:2012 standards [18].

The levels of pH in the groundwater samples range from 5.59 to 6.76, 5.59 to 7.66, and 6.01 to 7.27 during PrM, Mon, and PoM seasons correspondingly, which indicates acidic nature. 80%, 60%, and 80% of the samples surpassed the permissible limits of I.S. 10500:2012 standards during PrM, Mon, and PoM seasons correspondingly. This decrease in pH of the groundwater samples during the PrM and PoM seasons compared to the Mon season may be owing to the increasing levels of various salts and minerals in the groundwater of the block [19].

The values of TDS vary from 198.4 to 712.96 mg/l, 57.6 to 486.4 mg/l, and 327.68 to 821.12 mg/l through PrM, Mon, and PoM seasons correspondingly. 50%, 0%, and 60% of the samples surpassed the permissible limits of I.S. 10500:2012 standards during PrM, Mon, and PoM seasons correspondingly. Elevated values of TDS through PoM and PrM seasons are owing to the decline of groundwater during lean periods or non-monsoon seasons (PoM and PrM seasons) and leaching and dissolution of salts from rocks and soil and furthermore by human-induced exploits [20]. The runoff with households' sewage that permeates into the groundwater may also be one of the chief reasons for the augmentation in TDS levels during PoM and PrM seasons [21].

The values of turbidity vary from 0.35 to 7.12 NTU, 0.6 to 15.4 NTU, and 0.15 to 12.17 NTU during PrM, Mon, and PoM seasons correspondingly. 60%, 90%, and 80% of the samples exceeded the allowable limits of I.S. 10500:2012 standards during PrM, Mon, and PoM seasons correspondingly. During PrM and PoM seasons (lean periods), due to very low infiltration of precipitation into groundwater, there is a

Table 3 Statistics of the parameters

Parameter	Minimum			Maximum		
	PrM	Mon	PoM	PrM	Mon	PoM
pH	5.59	5.59	6.01	6.76	7.66	7.27
TDS	198.40	57.60	327.68	712.96	486.40	821.12
Turbidity	0.35	0.60	0.15	7.12	15.40	12.17
TH	83.00	44.00	60.00	480.00	476.00	390.00
TA	60.00	90.00	60.00	156.00	202.00	178.00
Na^+	13.50	0.90	7.20	149.00	91.00	119.60
Ca^{2+}	11.90	3.90	15.40	91.60	32.90	116.90
Mg^{2+}	4.03	0.21	3.36	24.13	17.73	20.11
K^+	2.10	0.10	1.20	59.20	21.90	39.50
F^-	0.29	0.10	0.44	1.92	1.46	1.69
NO_3^-	5.69	0.26	11.80	70.31	15.57	53.80
Cl^-	18.71	1.42	12.50	44.24	11.20	29.49
SO_4^{2-}	10.26	1.00	6.84	153.00	56.67	102.00
Mn	0.05	0.01	0.01	2.80	1.97	2.17
Fe	0.05	0.01	0.01	8.39	6.06	8.73
Pb	0.000	0.000	0.000	0.000	0.000	0.000
As	0.000	0.000	0.000	0.012	0.011	0.016
Zn	0.052	0.000	0.035	5.476	1.478	2.660

*The units of all concentrations are in mg/l excluding pH and turbidity. The unit of turbidity is NTU and pH is unitless

low turbid water input to groundwater. Therefore, there is a low turbid water input to the groundwater during the non-monsoon seasons. This declining trend of turbidity in the lean periods can be attributable to the fact that turbidity in groundwater reduces with the augment of deposition time [22].

The values of TH differ from 83 to 480 mg/l, 44 to 476 mg/l, and 60 to 390 mg/l through PrM, Mon, and PoM seasons correspondingly. 40%, 30%, and 40% of the samples surpassed the permissible limits of I.S. 10500:2012 standards through PrM, Mon, and PoM seasons correspondingly. Although the mean Mg^{2+} concentration of Mon season samples is found lower than samples of PoM season, the high TH that is found in the Mon season in contrast to PoM season can be point to the fact that the TH in groundwater is primarily due to calcium carbonates and bicarbonates, and secondarily due to magnesium carbonates and bicarbonates; and can consist of other compounds such as magnesium sulphate, magnesium chloride, calcium sulphate, and calcium chloride [23]. Moreover, in the block, the values of Ca^{2+} ions are found to be more; thus, the TH levels are also found to be high.

The values of TA differ from 60 to 156 mg/l, 90 to 202 mg/l, and 60 to 178 mg/l in PrM, Mon, and PoM seasons correspondingly. 0%, 10%, and 0% of the samples

Table 4 Statistics of the parameters

Parameter	Mean			Standard deviation		
	PrM	Mon	PoM	PrM	Mon	PoM
pH	6.06	6.52	6.45	0.37	0.66	0.41
TDS	501.31	210.56	507.46	162.48	131.68	146.00
Turbidity	2.37	4.08	2.36	2.10	4.20	3.52
TH	213.70	168.20	166.80	135.78	134.46	103.15
TA	104.40	144.00	110.40	31.46	43.32	39.44
Na^+	65.51	34.92	49.61	45.96	32.10	38.16
Ca^{2+}	43.01	12.55	46.33	25.16	9.15	36.02
Mg^{2+}	11.10	4.62	9.25	6.00	5.08	5.00
K^+	16.44	2.66	10.05	18.77	6.78	12.83
F^-	0.78	0.55	0.91	0.60	0.51	0.42
NO_3^-	33.73	3.06	21.19	21.72	4.51	12.90
Cl^-	29.06	5.01	19.39	7.10	2.90	4.74
SO_4^{2-}	48.81	12.73	32.54	43.86	17.33	29.24
Mn	0.80	0.43	0.64	0.96	0.64	0.75
Fe	2.04	1.16	1.94	3.02	1.93	3.23
Pb	0.000	0.000	0.000	0.000	0.000	0.000
As	0.001	0.002	0.003	0.004	0.004	0.006
Zn	0.995	0.181	0.468	1.634	0.457	0.823

[a]The units of all concentrations are in mg/l excluding pH and turbidity. The unit of turbidity is NTU and pH is unitless

exceeded the permitted limits of I.S. 10500:2012 standards in PrM, Mon, and PoM seasons correspondingly. As the pH values of the groundwater are found to be in the acidic range, therefore TA also falls low in level.

The values of Na^+ vary from 13.5 to 149 mg/l, 0.9 to 91 mg/l, and 7.2 to 119.6 mg/l during PrM, Mon, and PoM seasons correspondingly. 80%, 50%, and 70% of the samples went over the permissible limits of I.S. 10500:2012 standards during PrM, Mon, and PoM seasons correspondingly.

The values of Ca^{2+} differ from 11.9 to 91.6 mg/l, 3.9 to 32.9 mg/l, and 15.4 to 116.9 mg/l in PrM, Mon, and PoM seasons correspondingly. 20%, 0%, and 30% of the samples surpassed the acceptable limits of I.S. 10500:2012 standards in PrM, Mon, and PoM seasons correspondingly. Municipality potential sources like sewages, domestic wastes, and industrial wastes contribute to the complex amount of Ca^{2+} ions in groundwater leading to the origin of ionic pollutants. The concentration of Ca^{2+} ions in groundwater might be increasing during PrM and PoM seasons compared to the Mon season due to the leaching process [24].

The values of Mg^{2+} vary from 4.03 to 24.13 mg/l, 0.21 to 17.73 mg/l, and 3.36 to 20.11 mg/l during PrM, Mon, and PoM seasons correspondingly. 0% of the samples

Table 5 Percentage of groundwater samples surpassing the permissible limit in the block

Parameter	Permissible limit	Maximum permissible limit	% of samples exceeding the permissible limit		
			PrM	Mon	PoM
pH	6.5–8.5	6.5–8.5	80	60	80
TDS	500	2000	50	0	60
Turbidity	1	5	60	90	80
TH	200	600	40	30	40
TA	200	600	0	10	0
Na^+	20	40	80	50	70
Ca^{2+}	75	200	20	0	30
Mg^{2+}	30	100	0	0	0
K^+	10	10	60	10	30
F^-	1	1.5	30	20	30
NO_3^-	45	45	40	0	10
Cl^-	250	1000	0	0	0
SO_4^{2-}	200	400	0	0	0
Mn	0.1	0.3	80	40	80
Fe	0.3	1	50	40	40
Pb	0.01	0.01	0	0	0
As	0.01	0.01	10	10	10
Zn	5	15	10	0	0

[a]The units of all concentrations are in mg/l excluding pH and turbidity. The unit of turbidity is NTU and pH is unitless

surpassed the permissible limits of I.S. 10500:2012 standards during PrM, Mon, and PoM seasons. The major resource of Mg^{2+} in the groundwater of this area is because of Mg^{2+} bearing minerals, ion exchange of minerals from rocks, and also animal, household, and industrial squander.

The values of K^+ vary from 2.1 to 59.2 mg/l, 0.1 to 21.9 mg/l, and 1.2 to 39.5 mg/l during PrM, Mon, and PoM seasons correspondingly. 60%, 10%, and 30% of the samples beat the acceptable limits of I.S. 10500:2012 standards during PrM, Mon, and PoM seasons correspondingly. K^+ concentrations at a few sites are bizarrely very soaring. This may be attributable to salt scrapes existing eugenically and fertilizer percolating/leaching through the sub-surface.

The values of F^- range from 0.29 to 1.92 mg/l, 0.1 to 1.46 mg/l, and 0.44 to 1.69 mg/l during PrM, Mon, and PoM seasons correspondingly. 30%, 20%, and 30% of the samples went beyond the acceptable limits of I.S. 10500:2012 standards during PrM, Mon, and PoM seasons correspondingly. In India, it is recognized that elevated levels of F^- in groundwater are generally characteristic of geogenic causes: primarily from rocks holding rich fluorine having minerals [25]. The Dimoria block region is occupied by granitic and gneisses rocks, and these hold a relatively large quantity of high fluorine minerals [5].

The values of NO_3^- vary from 5.69 to 70.31 mg/l, 0.26 to 15.57 mg/l, and 11.8 to 53.8 mg/l during PrM, Mon, and PoM seasons correspondingly. 40%, 0%, and 10% of the samples went beyond the permissible limits of I.S. 10500:2012 standards during PrM, Mon, and PoM seasons correspondingly. Preceding studies on NO_3^- contamination in groundwater have affirmed that the impending occurrence of NO_3^- and its allocation in groundwater is principally due to the appliance of inorganic nitrogenous manures as well as fertilizers, cesspit leakage from the septic cisterns, brisk urbanization, industry-laden, and community sewage, landfill leachates, and atmospheric desiccated and sodden depositions [26–28]. The presence of NO_3^- may be from farming areas due to leaching progression from plants' nutrients and NO_3^- fertilizers [29, 30].

The values of Cl^- vary from 18.71 to 44.24 mg/l, 1.42 to 11.2 mg/l, and 12.5 to 21.49 mg/l in PrM, Mon, and PoM seasons correspondingly. 0% of the samples surpassed the permissible limits of I.S. 10500:2012 standards during PrM, Mon, and PoM seasons. Very low Cl^- levels in groundwater could be characteristic of leaching from gangrenous cisterns, familial and animal wastes, community sewages, farming, and fertilizers [31–33].

The values of SO_4^{2-} differ from 10.26 to 153 mg/l, 1 to 56.67 mg/l, and 6.84 to 102 mg/l during PrM, Mon, and PoM seasons correspondingly. 0% of the samples exceeded the permissible limits of I.S. 10500:2012 standards in PrM, Mon, and PoM seasons.

The concentrations of Mn differ from 0.05 to 2.8 mg/l, 0.01 to 1.97 mg/l, and 0.01 to 2.17 mg/l during PrM, Mon, and PoM seasons correspondingly. 80%, 40%, and 80% of the samples went beyond the permissible limits of I.S. 10500:2012 standards during PrM, Mon, and PoM seasons correspondingly. Mn in groundwater comes from rainfall, dissolution of Mn in minerals from surrounding rocks, and leaching of Mn in percolating through soils. Greater Mn values are found in groundwater that is acidic (low pH) and are in an anaerobic condition. Mn can be leached from above lying soils, minerals in underlying rocks, and aquifer. The well depth, reducing states, residence time, and salinity are the main factors contributing to the disbanding and immigration of Fe and Mn to groundwater [34]. With the augment in residence time, the groundwater progressively alters from oxic to reductive. Then, the reductive dissolution of Mn oxides augments the Mn concentration in groundwater [35].

The values of Fe diverge from 0.05 to 8.39 mg/l, 0.01 to 6.06 mg/l, and 0.01 to 8.73 mg/l during PrM, Mon, and PoM seasons correspondingly. 50%, 40%, and 40% of the samples went beyond the permissible limits of I.S. 10500:2012 standards during PrM, Mon, and PoM seasons correspondingly. Fe is one of the dominant metals present in the groundwater. Elevated Fe levels in the groundwater may be for the interactions of oxidized Fe minerals and organic matter, and successive disbanding of Fe_2CO_3 at a rationally lower pH. Such groundwater is lucid when initially drawn from the well, but rapidly becomes murky and later brown by the precipitation of $Fe(OH)_3$, which is a familiar trouble in groundwater [36]. An additional cause for the elevated Fe concentration may be the elimination of dissolved oxygen by organic matter, leading to reduced conditions. In reducing conditions, the dispensability of Fe-bearing minerals (siderite, marcasite, etc.) amplifies, leading to the fortification of dissolved Fe in the aquifer water [37].

Pb in groundwater is found to be 0 mg/l during PrM, Mon, and PoM seasons. No samples went beyond the permissible limits of I.S. 10500:2012 standards in PrM, Mon, and PoM seasons. Pb is the most unwanted heavy metal in groundwater. The incidence of Pb in groundwater may be due to Pb availability in worldwide contamination, the liberation of Pb adsorbed on sediments, dust transmuted via the atmosphere and earth's crust erosion, and lastly due to precipitation and deposition of airborne aerosols [38, 39].

The values of As vary from 0 to 0.012 mg/l, 0 to 0.011 mg/l, and 0 to 0.016 mg/l during PrM, Mon, and PoM seasons correspondingly. 10%, 10%, and 10% of the samples went beyond the permissible limits of I.S. 10500:2012 standards in PrM, Mon, and PoM seasons correspondingly. Anthropogenic sources of As are in the form of inorganic ions and organic metal composites [40]. The occurrence of As in groundwater is generally connected with sedimentary rocks of marine sources, fossil fuels, disintegrating of volcanic rocks, mineral deposits, mining wastes, irrigation practices, and agricultural use. As is also adsorbed onto clay colloids bound to organic matter to form water-soluble complexes (mostly with Al, Fe, Ca, and Mg) [36, 41].

The values of Zn vary from 0.052 to 5.476 mg/l, 0 to 1.478 mg/l, and 0.035 to 2.66 mg/l during PrM, Mon, and PoM seasons correspondingly. 10%, 0%, and 0% of the samples went beyond the permissible limits of I.S. 10500:2012 standards during PrM, Mon, and PoM seasons correspondingly. Zn is utilized as an anti-corrosion agent, encrusted on Fe pipelines to shield them against corrosion. In borehole construction, galvanized pipelines are utilized. Corrosion of pump parts could also direct to Zn being released into the groundwater [36]. Zn is present in a very low concentration which is due to its curbed portability from the place of rock weathering or from natural sources [18]. Human beings need Zn at a required level, and if not, it distresses the metabolism and immune system resulting in higher susceptibility to infections. A higher concentration of Zn is also responsible for birth defects in pregnant women [42, 43].

Based on the IS 10500:2012 standards, the results revealed that a major portion of the groundwater samples is not appropriate for consumption in the Dimoria block. In the PrM season, except for TA, Mg^{2+}, Cl^-, and $SO_4{}^{2-}$, every other parameter surpassed their individual permissible limits. In the Mon season, except for TDS, Ca^{2+}, Mg^{2+}, $NO_3{}^-$, Cl^-, and $SO_4{}^{2-}$, all other parameters went beyond their respective permissible limits. In the PoM season, except for TA, Mg^{2+}, Cl^-, and $SO_4{}^{2-}$, every other parameter went above their respective permissible limits.

Amongst the 5 physico-chemical parameters, pH, TDS, turbidity, TH, and TA exceeded their respective permissible limits during one or more seasons, the most prominent being pH, turbidity, and TH. Dominant cations found in the block are Na^+, Ca^{2+}, and K^+. Amongst the major cations, Na^+ and K^+ played a governing role. The levels of Mg^{2+} are relatively very low in contrast to other major cations. Amongst the major anions, F^- and $NO_3{}^-$ played the most dominant role. The levels of Cl^- and $SO_4{}^{2-}$ are relatively very low in contrast to other key anions. The levels of the major ions are found to be soaring signifying leaching and human-laden activities preponderate over groundwater dilution. Amongst the metals, the most influential ones are Mn, Fe, and As in all the seasons. The Zn concentrations are found to be relatively low. Therefore, it can be signified that the majority of the groundwater samples are not fulfilling the drinking water quality standards prearranged by the I.S. 10500:2012 standards.

In this study, to establish the spatio-temporal variation outlines of the GWQI values and to differentiate elevated and non-elevated GWQI values, spatio-temporal variation maps for the GWQI values were engendered through ArcGIS (v. 10.5) software. The original water quality parameters of every groundwater well were amalgamated into a solitary index in a comprehensible and proficient system. Using the computed GWQI value for each well, the spatio-temporal GIS-based GWQI maps for the Dimoria block for all the seasons are plotted (Figs. 2, 3 and 4). The GWQI values were categorized according to the prescribed classification (Table 2). For the GWQI maps, the blue-coloured areas denote the GWQI values not more than 50 (excellent quality groundwater), green-coloured areas show the GWQI values that are within 50–100 (good quality groundwater), yellow-coloured areas show the

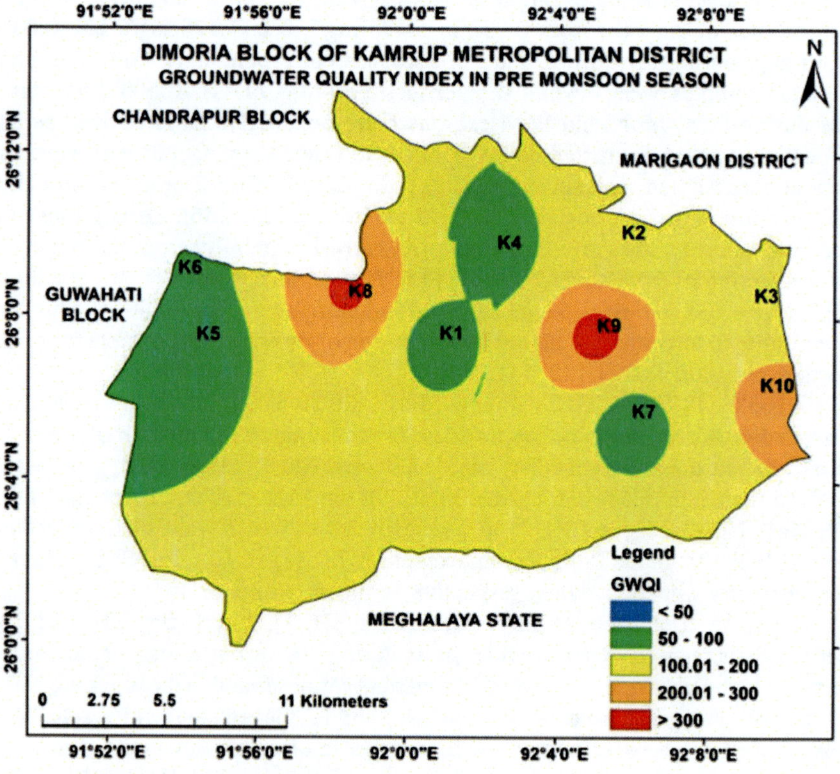

Fig. 2 GWQI for Dimoria block for the pre-monsoon season

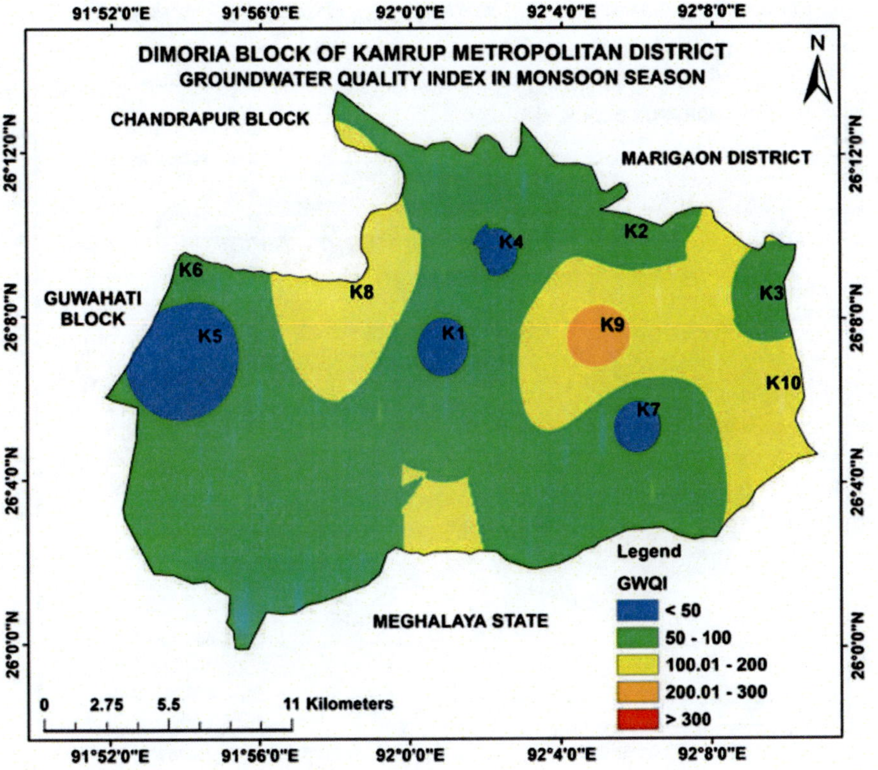

Fig. 3 GWQI for Dimoria block for the monsoon season

GWQI values within 100.01–200 (poor quality groundwater), orange-coloured areas show the GWQI values within 200.01–300 (very poor quality groundwater), and red-coloured areas show the GWQI values more than 300 (groundwater unsuitable for drinking purposes).

In the pre-monsoon season, 0% of the groundwater samples have GWQI values less than 50 (excellent quality), 50% of the groundwater samples have GWQI values between 50 and 100 (good quality), 20% of the groundwater samples have GWQI values between 100.01 and 200 (poor quality), 10% of the groundwater samples have GWQI values between 200.01 and 300 (very poor quality), and 20% of the groundwater samples have GWQI values more than 300 (unsuitable). In the monsoon season, 40%, 30%, 20%, 10%, and 0% of the groundwater samples have excellent, good, poor, very poor, and unsuitable groundwater quality, respectively. In the post-monsoon season, 30%, 30%, 20%, 0%, and 20% of the groundwater samples have excellent, good, poor, very poor, and unsuitable groundwater quality correspondingly.

From the GIS-based spatio-temporal GWQI maps of the Dimoria block, it was monitored that the eminence of groundwater decreased from Mon to PoM to PrM

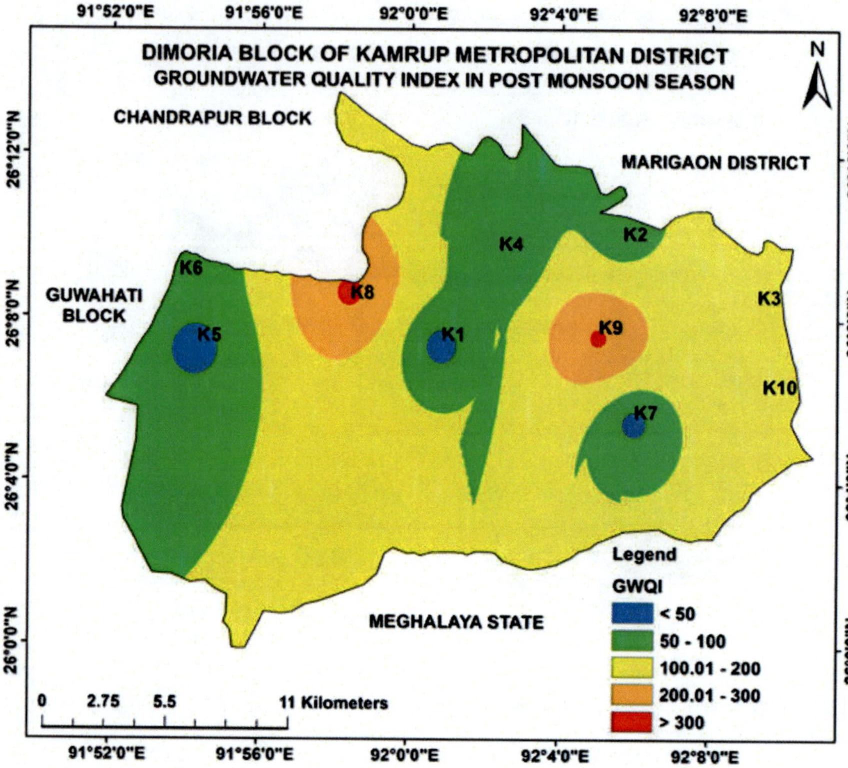

Fig. 4 GWQI for Dimoria block for the post-monsoon season

seasons due to the augment of levels of the varied physico-chemical parameters, cations, anions, and metals during the non-monsoon seasons compared to the monsoon season. The GWQI maps also designated that meagre quality of groundwater was found the utmost in the PrM and PoM seasons as compared to the Mon season in the block. It can be perceived that the groundwater excellence is anticipated to perk up in the Mon season as compared to the PrM and PoM seasons. This is generally because of the dilution of contaminants during the rainy season. This indicates that the Dimoria block's groundwater is not apposite for straight consumption and it entails sustainable processing before its consumption.

4 Conclusions

The current study put on views that the employment of GIS-based techniques could endow productive information for groundwater quality appraisal. From the GIS-based spatio-temporal GWQI maps of the Dimoria block, it can be perceived that the groundwater superiority is anticipated to perk up in the Mon season in contrast to the PrM and PoM seasons. This is usually because of the dilution of contaminants during the rainy season than in the dry seasons. The spatio-temporal variation maps of the block's groundwater quality also indicated that greater parts of the groundwater samples are not satisfying the I.S. 10500:2012 drinking water quality standards. The consequential GWQI maps, being effortlessly comprehensible, could provide a communication contrivance to enlighten the residents as well as the government organizations about the groundwater quality prominence for enhanced use in upcoming years. The graphical and statistical groundwater quality appraisals in the block divulge that the groundwater quality in several fractions of the block is not suitable to be used for direct consumption. The spatial database and GWQI developed in this study will thus be necessitated for the monitoring and regulation of groundwater contamination remedies in the block.

References

1. Darnley, A. G., Bjorklund, A., Bolviken, B., Gustavsson, N., Koval, P. V., Uk, J.P., Xuejing, X. (1995). *A global geochemical database. Recommendations for international geochemical mapping* [Final report of IGCP project, 259].
2. Gibbs, R. J. (1970). Mechanisms controlling world water chemistry. *Science, 170*(3962), 1088–1090.
3. Butler, M., Wallace, J., & Lowe, M. (2002). Ground-water quality classification using GIS contouring methods for Cedar Valley, Iron County, Utah. In *Digital mapping techniques* (Vol. 207).
4. Machiwal, D., & Jha, M. K. (2015). Identifying sources of groundwater contamination in a hard-rock aquifer system using multivariate statistical analyses and GIS-based geostatistical modeling techniques. *Journal of Hydrology: Regional Studies, 4*, 80–110.
5. Kotoky, P., Sarma, B., & Kotoky, E. D. (2017). Assessment and mapping of fluoride contamination of groundwater of the Hatigaon Area of Assam, India using geographic information system. *International Journal of Science and Research (IJSR), 6*(4), 1692–1698.
6. Negewo, T. F., & Sarma, A. K. (2021). Estimation of water yield under baseline and future climate change scenarios in Genale watershed, Genale Dawa River basin, Ethiopia, using SWAT model. *Journal of Hydrologic Engineering, 26*(3), 05020051.
7. Negewo, T. F., & Sarma. A. K. (2021). *Evaluation of climate change-induced impact on streamflow and sediment yield of Genale watershed*. Ethiopia.
8. Hajigholizadeh, M., & Melesse, A. M. (2017). Assortment and spatiotemporal analysis of surface water quality using cluster and discriminant analyses. *CATENA, 151*, 247–258.
9. Iscen, C. F., Emiroglu, Ö., Ilhan, S., Arslan, N., Yilmaz, V., & Ahiska, S. (2008). Application of multivariate statistical techniques in the assessment of surface water quality in Uluabat Lake, Turkey. *Environmental Monitoring and Assessment, 144*(1), 269–276.
10. Horton, R. K. (1965). An index number system for rating water quality. *Journal of Water Pollution Control Federation, 37*(3), 300–306.
11. Nasirian, M. (2007). *A new water quality index for environmental contamination contributed by mineral processing: A case study of Amang (Tin Tailing) processing activity.*
12. dos Santos, S. F., Moreira, A. B., Bisinoti, M. C., Gimenez, S. M. N., & Yabe, M. J. S. (2008). Water quality index as a simple indicator of aquaculture effects on aquatic bodies. *Ecological Indicators, 8*(5), 476–484.

13. Kotoky, P., & Sarma, B. (2017). Assessment of water quality index of the Brahmaputra River of Guwahati City of Kamrup District of Assam, India. *International Journal of Engineering Research and Technology (IJERT), 6*(03).
14. Abessi, O., & Saeedi, M. (2011). Groundwater quality index development for Qazvin Province.
15. CGWB. (2010). *Groundwater quality in shallow aquifers of India.* Ministry of Water Resources, Government of India, Faridabad.
16. APHA/AWWA/WEF. (2012) *Standard methods for the examination of water and wastewater* (20th ed.). American Public Health Association/American Water Works Association/Water Environment Federation.
17. Vasanthavigar, M., Srinivasamoorthy, K., Vijayaragavan, K., Ganthi, R. R., Chidambaram, S., Anandhan, P., & Vasudevan, S. (2010). Application of water quality index for groundwater quality assessment: Thirumanimuttar sub-basin, Tamilnadu, India. *Environmental Monitoring and Assessment, 171*(1), 595–609.
18. BIS I. (2012). *10500 Indian standard drinking water–specification* (2nd rev.). Bureau of Indian Standards.
19. Anwar, K. M., & Aggarwal, V. (2016). Studies on seasonal variation in ground water quality: A statistical approach. *Journal of Environmental Research and Development, 11*(1), 123–131.
20. Sharma, D. A., Rishi, M. S., & Keesari, T. (2017). Evaluation of groundwater quality and suitability for irrigation and drinking purposes in southwest Punjab, India using hydrochemical approach. *Applied Water Science, 7*(6), 3137–3150.
21. Sharma, D. A., Rishi, M. S., Keesari, T., & Sinha, U. K. (2016). Assessment of groundwater quality of Bathinda district, Punjab with reference to nitrate contamination. *Journal of Applied Geochemistry, 18*(4), 480–489.
22. Azis, A., Yusuf, H., Faisal, Z., & Suradi, M. (2015). Water turbidity impact on discharge decrease of groundwater recharge in recharge reservoir. *Procedia Engineering, 125*, 199–206.
23. Pal, P. (2017). *Industrial water treatment process technology.* Butterworth-Heinemann.
24. Baysal, A., Ozbek, N., & Akman, S. (2013). Determination of trace metals in waste water and their removal processes. In *Waste water-treatment technologies and recent analytical developments* (pp. 145–171).
25. Mukherjee, I., & Singh, U. K. (2018). Groundwater fluoride contamination, probable release, and containment mechanisms: A review on Indian context. *Environmental Geochemistry and Health, 40*(6), 2259–2301.
26. Chen, J., Wu, H., Qian, H., & Gao, Y. (2017). Assessing nitrate and fluoride contaminants in drinking water and their health risk of rural residents living in a semiarid region of northwest China. *Exposure and Health, 9*(3), 183–195.
27. Wei, Y. N., Fan, W., Wang, W., & Deng, L. (2017). Identification of nitrate pollution sources of groundwater and analysis of potential pollution paths in loess regions: A case study in Tongchuan region, China. *Environmental Earth Sciences, 76*(12), 1–13.
28. Adimalla, N., Li, P., & Qian, H. (2018). Evaluation of groundwater contamination for fluoride and nitrate in semi-arid region of Nirmal Province, South India: a special emphasis on human health risk assessment (HHRA). *Human and Ecological Risk Assessment: An International Journal.*
29. Beaglehole, R., Bonita, R., & Kjellström, T. (1993). *Basic epidemiology* (pp. 133–142). World Health Organization.
30. WHO, World Health Organization, World Health Organisation Staff. (2004). *Guidelines for drinking-water quality* (Vol. 1). World Health Organization.
31. Appelo, C. A. J., & Postma, D. (2004). *Geochemistry, groundwater and pollution.* CRC Press.
32. Narsimha, A., & Sudarshan, V. (2013). Hydrogeochemistry of groundwater in Basara area, adilabad district, Andhra Pradesh, India. *Journal of Applied Geochemistry, 15*(2), 224–237.
33. Adimalla, N., & Venkatayogi, S. (2018). Geochemical characterization and evaluation of groundwater suitability for domestic and agricultural utility in semi-arid region of Basara, Telangana State, South India. *Applied Water Science, 8*(1), 1–14.
34. Weng, H. X., Qin, Y. C., & Chen, X. H. (2007). Elevated iron and manganese concentrations in groundwater derived from the Holocene transgression in the Hang-Jia-Hu Plain, China. *Hydrogeology Journal, 15*(4), 715–726.

35. Luzati, S., Beqiraj, A., Goga, E. B., & Jaupaj, O. (2016). Iron and Manganese in Groundwater of Rrogozhina Aquifer, Western Albania. *Journal of Environmental Science and Engineering, 5*, 276–285.
36. Mondal, N. C., Singh, V. S., Puranik, S. C., & Singh, V. P. (2010). Trace element concentration in groundwater of Pesarlanka Island, Krishna Delta, India. *Environmental Monitoring and Assessment, 163*(1), 215–227.
37. Applin, K. R., & Zhao, N. (1989). The kinetics of Fe (II) oxidation and well screen encrustation. *Groundwater, 27*(2), 168–174.
38. Dodge, R. E., & Brass, G. W. (1984). Skeletal extension, density and calcification of the reef coral, Montastrea annularis: St. Croix, US Virgin Islands. *Bulletin of Marine Science, 34*(2), 288–307.
39. Erel, Y., & Patterson, C. C. (1994). Leakage of industrial lead into the hydrocycle. *Geochimica et Cosmochimica Acta, 58*(15), 3289–3296.
40. Pal, P., Sen, M., Manna, A., Pal, J., Pal, P., Roy, S., & Roy, P. (2009). Contamination of groundwater by arsenic: A review of occurrence, causes, impacts, remedies and membrane-based purification. *Journal of Integrative Environmental Sciences, 6*(4), 295–316.
41. Ramesh, R., Kumar, K. S., Eswaramoorthi, S., & Purvaja, G. R. (1995). Migration and contamination of major and trace elements in groundwater of Madras City, India. *Environmental Geology, 25*(2), 126–136.
42. ATSDR. (2005). *Pediatric environmental health specialty units (PEHSU)*. Agency for Toxic Substances and Disease Registry.
43. Landrigan, P. J., & Tamburlini, G. (2005). *Children's health and the environment: A transatlantic dialogue.*

Rising Temperatures and Its Impacts on Thermal Comfort and Productivity—A Case Study from Select Workplaces in Southern India

P. K. Latha, Rekha Shanmugam, Manikandan Krishnamoorthy, and Vidhya Venugopal

Abstract *Background*: Exposure to extreme heat due to climate change can result in occupational illnesses, thermal discomfort, productivity losses, as well as increasing the risk of injury. With this background, the present study aims to fill this gap by assessing three occupational settings with different heat levels to find the association between the heat exposures related to worker's perception of indoor thermal comfort and the Productivity Loss (PL). *Methodology*: A cross-sectional study for 856 workers from [high ($n = 507$), medium ($n = 219$), and low-process heat ($n = 130$)] workplaces located in Southern India was done. We evaluated the indoor heat profile, the wet bulb globe temperature (WBGT) index, using a portable WBGT monitor. We recorded workers' perceived thermal comfort using a questionnaire from The American Society of Heating, Refrigerating, and Air Conditioning Engineers (ASHRAE Standard Project Committee in Thermal environmental conditions for human occupancy. American Society of Heating, Refrigerating, and Airconditioning Engineers, Inc., [1]) and the PL using a validated high occupational temperature health and productivity suppression (HOTHAPS) questionnaire. *Results*: The average WBGT in high, medium, and low-process heat industries was 29.6 °C ± 3.7, 27.8 °C ± 1.7, and 27.1 °C ± 2.3. Exposure to heat above the safe limit was 64% ($n = 548$), with a maximum of 71% ($n = 362$) in the high-process heat industry. Perception of thermal discomfort was by 90%, 64%, and 48% in high, medium, and low-process heat industries, respectively, and maximum perception on PL was by workers in medium-heat workplace (36%). A significant association between perceived PL and heat exposure was reported among the high and medium process workers, and higher risk of

P. K. Latha
Academy of Maritime Education and Training (AMET) University, Chennai, India
e-mail: latha@ehe.org.in

P. K. Latha · R. Shanmugam · M. Krishnamoorthy · V. Venugopal (✉)
Faculty of Public Health, Ramachandra Institute of Higher Education and Research, Chennai, India
e-mail: vvidhya@ehe.org.in

R. Shanmugam
e-mail: Rekha@ehe.org.in

A. K. Dikshit et al. (eds.), *Innovative Trends in Hydrological and Environmental Systems*, Lecture Notes in Civil Engineering 234,
https://doi.org/10.1007/978-981-19-0304-5_53

thermal discomfort (OR = 26.9, CI-10.97–68.6) was among the medium-process heat exposed workers. *Conclusion*: A passive cooling interventions are needed to avert health and productivity risks and manage increasing energy needs to cool the workplaces in the changing climate scenario.

Keywords Thermal comfort · Climate change · Productivity loss · Heat stress

1 Introduction

Globally, a significant increase in daily average temperatures per decade and climate change scenarios project a high probability of more frequent heatwaves [2, 3]. The ambient temperature influences indoor and outdoor thermal comfort [4, 5]. The impact of ambient temperature on thermal comfort has been studied quite extensively [6, 7]. This temperature rise is expected to increase work-related heat stress and damage workers' productivity.

Thermal comfort is defined in an international standard organization (ISO) 7730 (1994) as "the condition of mind that which expresses satisfaction with the thermal environment." American Society of Heating, Refrigerating, and Air Conditioning Engineers (ASHRAE) Standard 55 defines thermal comfort as "that state of mind which expresses satisfaction with the thermal environment." It involves the well-being of the occupants in a particular environment for a particular climate about their capacity to adapt to thermal equilibrium, physiological, psychological, and behavioral changes [1]. Studies show that temperature rise is likely to subject nearly 60% of India's working populations to thermal discomfort in their workplaces [8], and the need to provide cooling interventions in the workplace is increasing.

In climate-vulnerable regions and developing countries that rely heavily on manual work, most residential buildings, and work still rely on natural ventilation for cooling, and thermal discomfort can be significant in terms of adverse impacts on health and productivity. Productivity loss (PL) in the present study is defined as reduced performance or productivity at work. Several literature works have inferred that exposure to heat decreases the physical capacity and productive working time [9, 10]. According to ISO 7243, for a working individual, human work capacity starts to decrease at a wet bulb globe temperature (WBGT) a heat index of 25 °C and becomes very challenging to carry out any physical work when the WBGT value surpasses 40 °C [11]. Heat stress exposures beyond human endurance result in reducing physical activity, a natural defense mechanism to avoid heat gain [11] that has direct implications on workers' ability and performance [10] and therefore has direct impacts on the workplace productivity and economic loss [12].

Evidence on thermal discomfort with high baseline temperatures in workplaces lacks and remains a critical research gap. With this background, the present study aims to fill this gap by assessing three occupational settings with different heat levels to find the association between the heat exposures related to worker's perception of indoor thermal comfort and the PL.

2 Materials and Method

2.1 Study Design

We did our present cross-sectional study in three types of workplaces with varying levels of heat, high-process heat (steel manufacturing and automobile industry), medium process heat (healthcare center and auto spare parts), and low-process heat (garment industry) spread across Kanchipuram, Thiruvallur, and Salem districts of Southern India. We obtained ethical clearance from the Institutional Ethics Committee (IEC) and took prior workplace approval to conduct the study. We explained the risks and benefits of participating in the study and obtained a signed informed consent for the qualitative data on different days in each location when work was in progress. We did not administer the perception questionnaires to the same workers necessarily in both seasons. We recruited the workers' for the study based on the inclusion criteria of age (18–60 years) and with heat exposures at the same workplace for at least one year. We excluded workers with pre-existing medical conditions like diabetes, hypertension, cardiac illnesses, thyroid diseases, or any co-morbid conditions. We recruited a total of 856 workers [high-process heat workplace ($n = 507$), medium-process heat workplace ($n = 219$), and low-process heat workplace ($n = 130$)] for the study based on their willingness to take part in the study.

Quantitative Data
We evaluated the indoor heat profile, the Wet Bulb Globe Temperature (WBGT) index, of each workplace using a portable WBGT heat stress monitor (QuesTemp 34; QUEST. Technologies, Oconomowoc, WI, USA) with an accuracy level of ±0.5 °C between 0 and 120 °C of dry bulb temperature and ±5% relative humidity. We took measurements after allowing the equipment to stabilize for at least 15 min in each work location during regular shift hours in most workplaces (9.00 am to 3.00 pm), as per the protocols recommended by NIOSH. The workers' work intensity (heavy or moderate) and each worker's "Clo" factor are rated by a Certified Industrial Hygienist according to the ACGIH guidelines [13], added the "Clo" value to the measured WBGT for each worker accordingly. Based on the WBGT values, we categorized the workers' heat exposures into unexposed (<27.5 °C) and exposed (≥27.5 °C). Then, we used the WBGT permissible heat exposure TLV to evaluate the risk of heat stress and the corresponding WBGT under which continuous work during an hour could be safely undertaken [13]. In some cases, we attributed the WBGT measured in each workplace location for other workers working close by based on their workload.

Qualitative Data
We recorded workers' perceived thermal comfort data using a questionnaire adapted from ASHRAE 2004 [1] that included demographic details like age, gender, education status, and other details like type of work as per ACGIH [13], workers' exposure to heat and thermal responses like indoor humidity and the ventilation status.

We recorded the PL perceived by the workers using a validated High Occupational Temperature Health and Productivity Suppression (HOTHAPS) questionnaire [14].

Data Analysis

We did all data analysis using SPSS version 15.0. We used the bivariate analysis to identify associations between heat and productivity for workers using the chi-square test. We used the Crude Odds Ratio (COR) to measure the association with the cut-off of 0.05 to interpret the significance of the *p*-values.

3 Results

Demographic Characteristics of Participants

Of 856 workers, 71% were males, and 29% were females, and we interviewed an equal number of workers in both seasons. Table 1 shows the descriptive characteristics of the study population. The study population's mean age was 33.8 ± 9.3 years, and the maximum number of workers was between 18 and 40 years ~86.0% had some basic education level, 84% of workers were non-smokers and did not consume alcohol ($n = 409$, 48%). The study population was engaged in jobs with heavy work, and the rest had moderate workloads ($n = 447$, 52%).

Workplace Heat Stress Profiles

The average WBGT in high, medium, and low-process heat industries was measured to be 29.6 °C ± 3.7, 27.8 °C ± 1.7, and 27.1 °C ± 2.3 (Fig. 1). Exposure to heat above the safe limit [13] among the 856 workers was 64% ($n = 548$), with a maximum of

Table 1 Demographic characteristics of the study participants from various work sectors ($n = 856$)

S. No.	Variables		*n*	Percentage (%)
1	Age	18–40 years	624	73
		41–60 years	230	27
		>60 years	2	0.2
2	Gender	Male	604	71
		Female	252	29
3	Smoking/alcoholic status	Yes	138	16
		No	718	84
4	Literacy status	Illiterate	118	14
		Literate	738	86
5	Years of exposure	≥3 years	632	74
		<3 years	224	26
6	Work category	Heavy	409	48
		Moderate	447	52

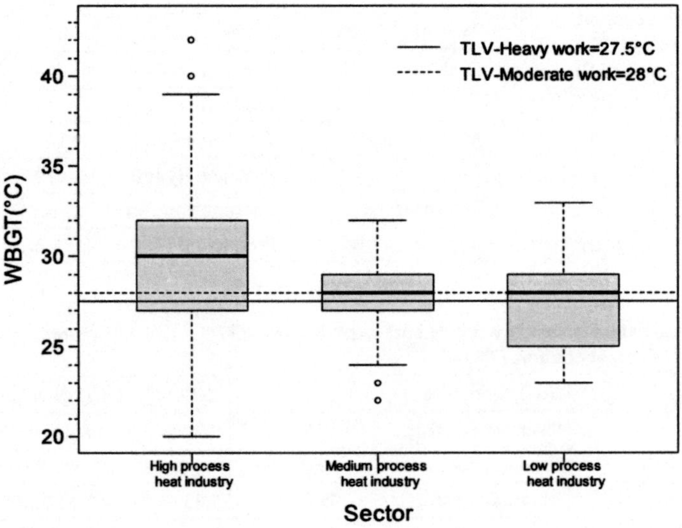

Fig. 1 Wet bulb globe temperature (WBGT) profiles across various workplaces during 2013–2019 in Southern India

71% ($n = 362$) in the high-process heat industry. Based on the WBGT (attributed) profiles, in high-process heat industries, the workers engaged with heavy workload had a 1.2 times higher risk of heat exposures (WBGTs) than workers doing moderate work; however, the risk was less than that 1 odds in moderate and light-process heat industry.

Workers Perception on Thermal Comfort and Productivity Loss

Perception of thermal discomfort was by 90%, 64%, and 48% of the workers in high, medium, and low-process heat industries, respectively, and maximum perception on PL was by workers in medium-heat workplace (36%) followed by high-heat (5%) and low-heat (2%) (Table 2). The statistical results showed a significant association between heat exposure (WBGT) and perceived thermal discomfort in high, medium, and low-process; however, a significant association between perceived PL and heat exposure was reported among the high-process heat and medium-process heat workers only. We observed a higher risk of thermal discomfort (OR = 26.9, CI-10.97–68.6) among the medium-process heat exposed workers than the low-heat exposed workers (Table 3).

Table 2 Self-reported Productivity Loss (PL) and Thermal Comfort (TC)

Type of industry	Study variables	*n*	%
High-heat	Thermal comfort	458	90
	Productivity loss	23	5
Medium-heat	Thermal comfort	141	64
	Productivity loss	79	36
Low-heat	Thermal comfort	62	48
	Productivity loss	2	2

Table 3 Association between workers heat exposure (WBGT°C) and self-reported productivity loss (PL) and thermal comfort (TC)

Type of industry	Study variables	X^2	COR[a] (95% CI); *p*-value[b]
High-heat	Thermal comfort	15.8	3.22 (1.770–5.860); 0.0001
	Productivity loss	4.60	4.4 (1.019–1 9.02); 0.03
Medium-heat	Thermal comfort	35.0	5.83 (3.177–10.707); 0.0001
	Productivity loss	8.6	0.433 (0.247–0.760); 0.003
Low-heat	Thermal comfort	59.5	**26.94 (10.97–68.6); 0.0001**
	Productivity loss	0.002	1.06 (0.065–17.39); 0.965

Note
[a]*COR* crude odds ratio more than 1 denotes the presence of odds
[b]*p*-value < 0.05 is significant

4 Discussion

4.1 Heat Stress Profiles and Thermal Discomfort

Tropical settings, high ambient temperatures, and humidity levels significantly influence indoor temperatures, which, when combined with heat-generating processes expose workers to high stress with consequent health implications [15, 16].

Heat stress profiles prevailing in the workplace evaluated via area WBGT monitors in the industry showed high occupational heat stress exposures (Fig. 1) that exceeded the recommended TLVs, a scenario similar to other studies conducted in India [8, 17, 18]. Apart from the environmental heat imposed on the workers that subjects them to thermal discomfort, the high metabolic workload on workers also adds to the additional burden of heat stress to the workers' comfort, proven by experimental studies [19]. Similar to the finding in the present study, workers engaged in heavy physical work viz., trunk work, carrying, pushing, and pulling heavy loads throughout their shifts were at high risk of thermal discomfort [20] and heat stress that is commonly reported in many Indian studies [21, 22].

4.2 Workers Perception of Thermal Comfort on Productivity Loss

The thermal environment is one of the main factors that influence thermal comfort and consequently, the productivity of occupants inside any occupational setting. Research has described the connection between thermal comfort and productivity [23]. Thermal comfort depends not only on environmental factors but also on the physical, physiological and psychological aspects [24], and the exposure to indoor process heat. In the present study, thermal discomfort was high among the high-process heat industry, followed by the medium and low-process heat industry. Similar findings are reported in few Indian studies [17, 25]. Similarly, we observed a significant association between heat exposure (WBGT) and the workers' perception of thermal comfort irrespective of the level of heat they expos to [17, 25].

Rising temperature and higher WBGT in workplaces would have significant productivity and economic implications for most regions in Tamil Nadu, South India, especially during the hot season (May). Researches show an average relationship of 2% decrement in work performance per degree °C when the temperature exceeds 25 °C [26]. In the present study, PL is reported by a higher percentage of workers in the medium-heat industry (8.6%) and has a significant association with heat exposure in high and medium-process heat industries. The previous studies observed that high-process heat industry workers are exposed to high-heat environments [27]. In the coming decades, there is a risk of heat-induced decline in the workers' workability in occupational settings and impacts on occupational health globally, especially in developing country settings like India [18, 28]. This study results evidence that the worker's productivity declines among heat exposed workers with manual workload would further weaken them economically that could also affect them physically and psychologically [9, 29].

5 Conclusion

Our study examined the present WBGT and related productivity losses among indoor workers with moderate/heavy work quantitatively. The study findings show that Indian workers are subjected to heat stress, thermal discomfort, and PL in low ventilated occupational workplaces, which are predicted to increase with the predicted rising temperature scenario due to climate change. Interventions are needed to avert health and productivity risks among the working population and manage increasing energy needs to cool the workplaces. Passive cooling technologies could be an effective and sustainable solution to avert occupational health and high-energy risks in the changing climate scenario.

References

1. ASHRAE Standard Project Committee. (2004). *Thermal environmental conditions for human occupancy*. American Society of Heating, Refrigerating, and Airconditioning Engineers, Inc.
2. IPCC. (2014). Mitigation of climate change. In *Contribution of Working Group III to the Fifth Assessment Report of the Intergovernmental Panel on Climate Change* (Vol. 1454).
3. Angeles-Malaspina, M., González-Cruz, J. E., & Ramírez-Beltran, N. (2018) Projections of heat waves events in the intra-Americas region using multimodel ensemble. In *Advances in Meteorology*. https://doi.org/10.1155/2018/7827984
4. Santamouris, M. (2018). *Minimizing energy consumption, energy poverty and global and local climate change in the built environment: innovating to zero: Causalities and impacts in a zero concept world*. Elsevier. ISBN: 9780128114179.
5. Huang, C.-H., Tsai, H.-H., & Chen, H. (2020). Influence of weather factors on thermal comfort in subtropical urban environments. *Sustainability, 12*, 2001. https://doi.org/10.3390/su1205 2001
6. Gwak, J., Shino, M., Ueda, K., & Kamata, M. (2019). An investigation of the effects of changes in the indoor ambient temperature on arousal level, thermal comfort, and physiological indices. *Applied Sciences, 9*, 899. https://doi.org/10.3390/app9050899
7. Yao, Y., Lian, Z., Liu, W., & Shen, Q. (2008). Experimental study on physiological responses and thermal comfort under various ambient temperatures. *Physiology and Behavior, 93*, 310–321. https://doi.org/10.1016/j.physbeh.2007.09.012 Epub 2007 Oct 22.
8. Venugopal, V., Chinnadurai, J. S., Lucas, R. A., & Kjellstrom, T. (2016). Occupational heat stress profiles in selected workplaces in India. *International Journal of Environmental Research and Public Health, 13*, 89. https://doi.org/10.3390/ijerph13010089
9. Parson, K. (2014). Human thermal environments: The effects of hot, moderate, and cold environments on human health, comfort, and performance (3rd ed.). ASIN: B00L2EBI9A.
10. Kjellstrom, T., Kovats, R. S., Lloyd, S. J., et al. (2009). The direct impact of climate change on regional labor productivity. *Archives of Environmental and Occupational Health, 64*, 217–227. https://doi.org/10.1080/19338240903352776
11. Kjellstrom, T., Holmer, I., & Lemke, B. (2009). Workplace heat stress, health and productivity–an increasing challenge for low and middle-income countries during climate change. *Global Health Action, 2*, 2047. https://doi.org/10.3402/gha.v2i0.2047
12. Kjellstrom, T., Briggs, D., Freyberg, C., et al. (2016). Heat, human performance, and occupational health: A key issue for the assessment of global climate change impacts. *Annual Review of Public Health, 37*, 97–112. https://doi.org/10.1146/annurev-publhealth-032315-021740
13. ACGIH. (2010). *Threshold limit values for chemical substances and physical agents and biological exposure indices* (pp. 200–219).
14. Kjellström, T., Lucas, R., Lemke, B., et al. (2014). Measuring and estimating occupational heat exposure and effects in relation to climate change: "Hothaps" tools for impact assessments and prevention approaches. *Climate Change and Health, 1*, 45–53. https://doi.org/10.1079/978178 0642659.0045
15. Kjellstrom, T., & Crowe, J. (2011). Climate change, workplace heat exposure, and occupational health and productivity in Central America. *International Journal of Occupational and Environmental Health, 17*, 270–281. https://doi.org/10.1179/107735211799041931
16. Kjellstrom, T., & McMichael, A. J. (2013). Climate change threats to population health and well-being: The imperative of protective solutions that will last. *Global Health Action, 6*, 20816. https://doi.org/10.3402/gha.v6i0.20816
17. Latha, P., Venugopal, V., Elangovan, R., et al. (2020). Thermal discomfort and health symptoms in Indian occupational settings in the climate change scenario. *Pollution Research, 39*(2), 479–486. ISSN 0257-8050.
18. Venugopal, V., Latha, P., Shanmugam, R., et al. (2020). Risk of kidney stone among workers exposed to high occupational heat stress—A case study from southern Indian steel industry. *Science of the Total Environment, 722*, 137619. https://doi.org/10.1016/j.scitotenv.2020.137619

19. Haneda, M., Wargocki, P., Dalewski, M., & Tanabe, S. (2009). *The effects of thermal discomfort on task performance, fatigue and mental work load examined in a subjective experiment*. https://www.researchgate.net/publication/287642446
20. Lan, L., Lian, Z., & Pan, L. (2010). The effects of air temperature on office workers' well-being, workload and productivity-evaluated with subjective ratings. *Applied Ergonomics, 42*, 29–36. https://doi.org/10.1016/j.apergo.2010.04.003
21. Tawatsupa, B., Yiengprugsawan, V., Kjellstrom, T., et al. (2013). Association between heat stress and occupational injury among Thai workers: Findings of the Thai Cohort study. *Industrial Health, 51*, 34–46. https://doi.org/10.2486/indhealth.2012-0138
22. Nag, P., Nag, A., Sekhar, P., & Pandit, S. (2009). *Vulnerability to heat stress: Scenario in western India.* National Institute of Occupational Health.
23. Bueno, A. M., de Paula Xavier, A. A., & Broday, E. E. (2021). Evaluating the connection between thermal comfort and productivity in buildings: A systematic literature review. *Buildings, 11*, 244. https://doi.org/10.3390/buildings11060244
24. Nematchoua, M. K., Tchinda, R., Ricciardi, P., & Djongyang, N. (2014). A field study on thermal comfort in naturally-ventilated buildings located in the equatorial climatic region of Cameroon. *Renewable and Sustainable Energy Reviews, 39*, 381–393. https://doi.org/10.1016/j.rser.2014.07.010
25. Latha, P., & Venugopal, V. (2020). Rethinking workplace thermal comfort in climate change context. *Journal of Indian Association for Environmental Management (JIAEM), 40*, 48–52.
26. Seppanen, O., Fisk, W. J., & Faulkner, D. (2004). *Control of temperature for health and productivity in offices.* https://escholarship.org/uc/item/39s1m92c
27. Krishnamurthy, M., Ramalingam, P., Perumal, K., et al. (2017). Occupational heat stress impacts on health and productivity in a steel industry in Southern India. *Safety and Health at Work, 8*, 99–104. https://doi.org/10.1016/j.shaw.2016.08.005
28. Sanders, M. L. (2015). Being nonprofit-like in a market economy: Understanding the mission-market tension in nonprofit organizing. *Nonprofit and Voluntary Sector Quarterly, 44*, 205–222. https://doi.org/10.1177/0899764013508606
29. Langkulsen, U., Vichit-Vadakan, N., & Taptagaporn, S. (2010). Health impact of climate change on occupational health and productivity in Thailand. *Global Health Action, 3*, 5607. https://doi.org/10.3402/gha.v3i0.5607

SWAT Modeling and Water Quality Analysis of Meenachil River, Kerala

P. V. Sujana and Anupama Surenjan

Abstract The Meenachil River is an important river serving Central Kerala. The rapid growth of population and change in land use have had an adverse effect on the river basin. A hydrological research in Meenachil River Basin (MRB) would aid in implementing an enhanced management program to prevent degradation of soil and water resources in the area. Soil and Water Assessment Tool (SWAT) modeling for streamflow has done in MRB for analyzing the water quality of the river. Using the SuFi2 algorithm in SWAT-CUP, the model was successfully calibrated and validated for Kidangoor gauging station. The model was calibrated during a ten-year period (1997–2006) and verified over a nine-year period (2007–2015). As reflected by coefficient of determination (R^2) and Nash–Sutcliffe simulation efficiency (NSE) values, model predictions done remarkably efficient on monthly basis during the calibration and validation periods. The R^2 and NSE for the simulated monthly streamflow are 0.79 and 0.77 for the calibration period, and 0.82 and 0.75 for the validation period. The calibrated model was used to study the effects of landscape, climate condition, and fertilizer application on sediment and nutrient loadings. From the investigation, it was observed that hydrological processes play an important role in the transport of pollutants and fertilizer which led to excessive nutrient loadings in the river. The model performs effectively, indicating that it can be used to estimate streamflow in MRB and provide knowledge about the water quality under various land scape and climate conditions.

Keywords Meenachil River · SWAT 2012 · SWAT-CUP · SuFi2 algorithm · Streamflow · Surface water quality

P. V. Sujana (✉) · A. Surenjan
Department of Civil Engineering, National Institute of Technology Karnataka, Surathkal, India

A. K. Dikshit et al. (eds.), *Innovative Trends in Hydrological and Environmental Systems*, Lecture Notes in Civil Engineering 234,
https://doi.org/10.1007/978-981-19-0304-5_54

1 Introduction

Meenachil River is the only river in Kerala, which has human settlement from origin to end of its flow. It originates at Vagamon in Western Ghats and flows though Kottayam District of Kerala state, India [1]. The rapid growth of population and change in land use have adversely affected the river basin. During six months of the year, the river basin experiences water scarcity. The increased cultivation of rubber plant in this particular area has severely impacted the environment due to higher runoff rate than the mixed crop, increased groundwater dependence during the summer, allowing the erosion cycle and the overflow of pesticides and fertilizers from the plantation [2]. The rising demand for water affects the natural water system and is considered as the primary cause of the degradation of water quality in the river. Agricultural activities have resulted in increased salinity, alkalinity, and water-logging problems in the area. Run-off from agricultural fields, industrial effluents, sewage from urban settlements, mining activities and over exploitation of groundwater, domestic and industrial pollution, combined with deforestation, use of pesticides and fertilizers have affected the river water quality extensively making water unfit for drinking. It was reported that the use of fertilizers has also affected groundwater quality in the basin [3]. A hydrological method is important to address and resolve the environmental issues occurring in this area.

Modeling tools help in understanding the complex interactions between the pollution source, the landscape, the nutrient cycle, climate, and the surface water quality. Numerous models have been developed to estimate streamflow and pollutant concentrations in streams and other water bodies. The current investigation is concentrated on the use of Soil and Water Assessment Tool (SWAT) modeling a physically based distributed model for nutrient simulation, sediment yields, and to analyze the impact of different land management practices on water quality [1]. This paper is focused in accessing the ability of SWAT version 2012 to simulate streamflow and associated nutrients and suggest alternative management practices in controlling pollution in Meenachil River of Kidangoor station, Kerala.

2 Materials and Methods

2.1 Soil and Water Assessment Tool (SWAT) Model

The SWAT model, a comprehensive, semi-distributed, continuous time, processed-based model [4–6], was developed in the1990s by the United States Department of Agriculture (USDA). SWAT software is based on spatially semi-dispersed hydrological and water quality model, which is intended to find out runoff water, sediments, and nutrients from the mainstream watersheds to its outlet. The model is a particularly flexible software that has been utilized in various parts of the world to check and predict the impact of land use pattern and practices on water, fertilizer yield, and sediments from farming activities in small to large basins with different land use, soils, and management scenarios, over an amount of time. The SWAT model can be used to create models to examine the effects of alternative decision making on water resource management and non-point source pollution in large river basins. It uses the concept of excess runoff mechanism and infiltration. It assumes that the runoff occurs every occasion the rainfall exceeds the rate of infiltration. Main components of the model include soil temperature and properties, weather, pesticides, plant growth, land management, nutrients, hydrology, and bacteria and pathogens. In SWAT, ArcGIS tool defines hydrological characteristics of the watershed and store-relevant spatial and tabular data, which can divide watershed into sub-basins for high-level spatial detail simulation. These small units called Hydrologic Response Units (HRUs) consist of homogeneous slope, soil characteristics, land use, and management. The HRUs represent percentages of the sub-watershed area [4].

The SWAT model simulates the hydrological cycle as indicated by Eq. (1) of water balance [6].

$$\mathrm{SW}_t = \mathrm{SW}_o + \sum \left(R_{\mathrm{day}} - Q_{\mathrm{surf}} - E_{\mathrm{a}} - W_{\mathrm{seep}} - Q_{\mathrm{gw}} \right) \tag{1}$$

SW_t stands for soil water content (final) (mm), SW_o for soil water content (initial) in day I (mm), and t for time (days). The proportion of precipitation on a given day is denoted by R_{day} I (mm), Q_{surf} indicates the surface runoff in day i (mm), E_{a} indicates the amount of evapotranspiration in day i (mm), W_{seep} represents the water percolation to the bottom of the soil profile on day i (mm), and Q_{gw} represents the groundwater discharge in day i (mm). For sustainable management of water resources, the important parameter for the evaluation of the study area is the water yield. Sum of water leaving the HRU and entering the channel during the time step is defined as the water yield in that area [5]. Water yield within a watershed is evaluated by the model based on Eq. (2):

$$W_{\mathrm{yield}} = Q_{\mathrm{surf}} + Q_{\mathrm{gw}} + Q_{\mathrm{lat}} - T_{\mathrm{loss}} \tag{2}$$

Here, W_{yield} represents the value of water yield (mm), Q_{gw} represents the contribution of groundwater to streamflow (mm), Q_{surf} indicates the surface runoff (mm), T_{loss} indicates the value of transmission losses (mm), and Q_{lat} represents the lateral flow contribution to stream (mm) and through bed transmission from tributary in the HRU.

The SWAT-simulated hydrological processes include infiltration, canopy storage, and surface. The processes in the soil involve drainage, return flow from shallow aquifers, lateral flow which transport water to the river, shallow aquifer recharge, and capillary rise from shallow aquifer into the root zone aquifer recharge which removes water from the system and evapotranspiration. Nutrient cycle for nitrogen and phosphorus is generated in SWAT. Two inorganic forms such as ammonium and nitrate and three organic forms of nitrogen are simulated. Similarly, three inorganic and three organic forms of phosphorous are simulated by SWAT [7].

Nutrient simulation in SWAT
The transport and transformation of nitrogen (N) and phosphorus (P) simulated in SWAT within an HRU are a function of nutrient cycles with inorganic and organic forms. N and P loss from the soil is assumed by plant absorption and surface runoff in SWAT. Simulated losses of N also occur in percolation below the root zone, in lateral subsurface flow, and by volatilization to the atmosphere. Movement of nitrate (NO_3–N) in surface runoff, lateral subsurface flow, and percolation are computed as the product of the average soil layer NO_3–N concentration and the volume of water in each flow pathway. Soluble P lost through surface runoff is determined as a function of the P solution concentration in 10 mm of top soil, the surface runoff volume, and a partitioning factor [8].

2.2 *Study Area*

Meenachil River is one among the significant rivers of Kottayam district in Kerala. It emerges from Western Ghats and flows through the taluks of Meenachil, Vaikom, and Kottayam. This then splits into number of tributaries before ending into Vembanad Lake. The river has a total length of 78 km and has a catchment area of 1272 km^2, extending from Vagamon in the east at an elevation of 1195 m. The entire Meenachil watershed area geographically lies between 09° 26′ 24″ and 09° 51′ 00″ N latitude and longitude of 76° 22′ 12″ and 76° 55′ 12″ E. Total annual yield of the river is 2349 million m^3 out of which 1110 million m^3 has been used in every year [9]. The river has 47 sub-watersheds, 114 micro–watersheds, and 38 tributaries including major and minor ones. The Meenachil River Basin falls within a tropical climate region; high variations in elevation and proximity to the sea influence the climatic parameters. The temperature of the area varies in between 24 and 32 °C throughout the year [9]. The basin's average annual rainfall is 3510 mm. It is most available in the southwest (June–August) and northeast (October–December) monsoon seasons. Only 10% occurring throughout January to May (summer months) [1]. The MRB mainly comprises of precambrian metamorphic rocks which form a hilly background. Well-drained laterite soils are the most common soil type in the area. Riverine alluvium soils and forest soils can also be found in some areas. Rubber trees are cultivated throughout large areas of the river basin. Other crops such as spices and paddy are also grown in the river basin [9].

Eazon [10] studied the flow pattern of Meenachil River using Indicators of Hydrologic Alteration (IHA) software reflects the basins, ecological distress, quality issues, siltation, and weed growth. In the study, an alteration in flow pattern was observed. This indicates large-scale anthropogenic influence in the river which may be causing a slow death to the river by the same society which is very much dependent on it. Narayanan [11] studied the Ichthyofauna of Aymanam Panchayat, in Vembanad wetland in Kerala. The Meenachil River feeds the Vembanad wetland. Out of 34 freshwater species, nine were threatened. One exotic species also noted. Abhilash [12] conducted a study on an invasive weed using Geographical Information System at Kumarakom, the tributary system of the Meenachil River. The study reflected that nutrients, water depth, land use patterns, and associated species were the major factors responsible for the growth and spreading of this exotic weed. Meenachil River has an accumulation of toxic metals right before the monsoon season [9]. This is because the continuous discharge of sewage from the banking town areas of the river and erosion from the plantations along the basin are causing problems to the river. According to George [13], significant rise of total phosphorous in water and sediment indicates the contribution of leachate from widespread rubber plantations in the upland catchment area. Joseph [14, 15] found that the water in Pennar River, a branch of Meenachil is highly contaminated and not safe for drinking. Uncontrolled use of chemical fertilizers and pesticides, and unethical dumping of domestic wastes are the major causes of deterioration of water. Smitha et al. studied the algal diversity in Meenachil River [16]. Sixteen algal species were observed in the study. Vincy [17] studied the toxic microorganisms and compared the microbiological quality of the Meenachil River water. The pathogenic bacteria present in the river indicates that the bacteriological quality is poor in the river. It also indicates that raw sewage is being dumped into the river. The location map of the Meenachil River is shown in Fig. 1.

3 Data Collection

SWAT requires a variety of information. The spatially distributed data required for Arc SWAT include the Digital Elevation Model (DEM), soil data, and land use data layers either as shape files or grid data. For calibration and prediction purposes, weather data and observed streamflow data are also necessary.

3.1 DEM (Digital Elevation Model)

Topography data are used to delineate a watershed into multiple sub-watersheds and also to calculate watershed/sub-watershed parameters such as slope and slope length. Digital Elevation model (DEM) represents the topography of an area. At a certain spatial resolution, it describes the elevation of any point in a particular area. As it contains the elevation information, it determines the possible directions of flow and physical characteristics of the watershed [18]. The DEM is extracted from the Shuttle Radar Topography Mission (SRTM) data which are of 90 m resolution, year 2014 downloaded from United State Geological Survey (USGS) (https://earthexplorer.usgs.gov/). This DEM shown in Fig. 2 was used to delineate the watershed using delineation tool in SWAT.

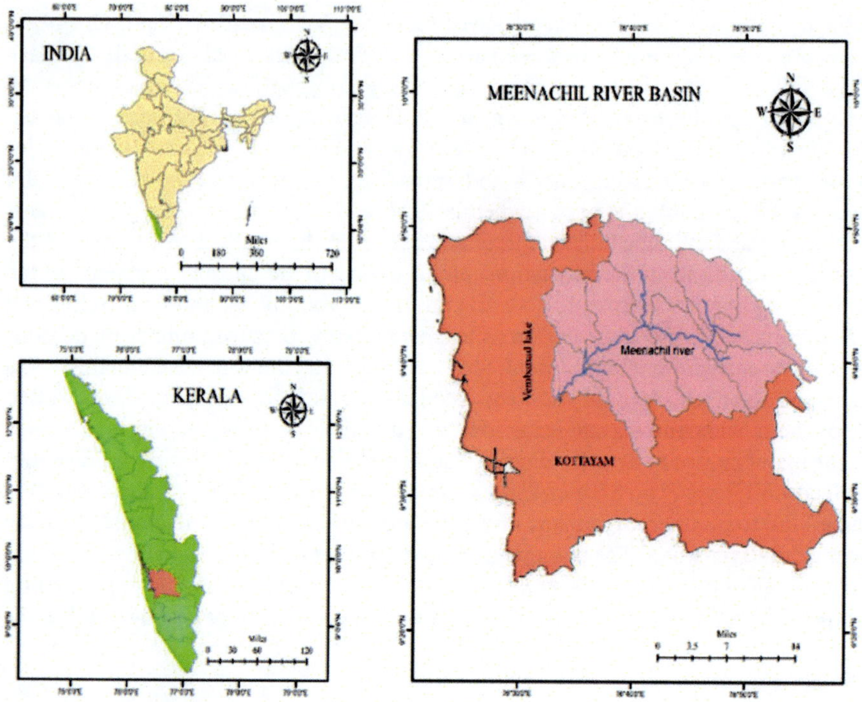

Fig. 1 Location map of study area

3.2 *Climate Data*

The climate data required are precipitation and maximum/minimum air temperature. Daily observed data for precipitation, maximum or minimum temperature (1997–2015) were used for the hydrologic simulation. The temperature data and rainfall data of high resolution (0.25° × 0.25°) gridded dataset are collected from India Meteorological Department (IMD) https://mausam.imd.gov.in/ for this study.

3.3 *Land Use/Land Cover*

The land use map of watershed is clipped from the land use map of India, spatial resolution 100 m [19]. Land use map gives the purpose of usage of each portion of study area. LULC classification is done in ArcGIS 10.3. The distribution of various land use/land cover classes in the study area is provided in Fig. 3. The study area comprises of nine different land use categories (Table 1).

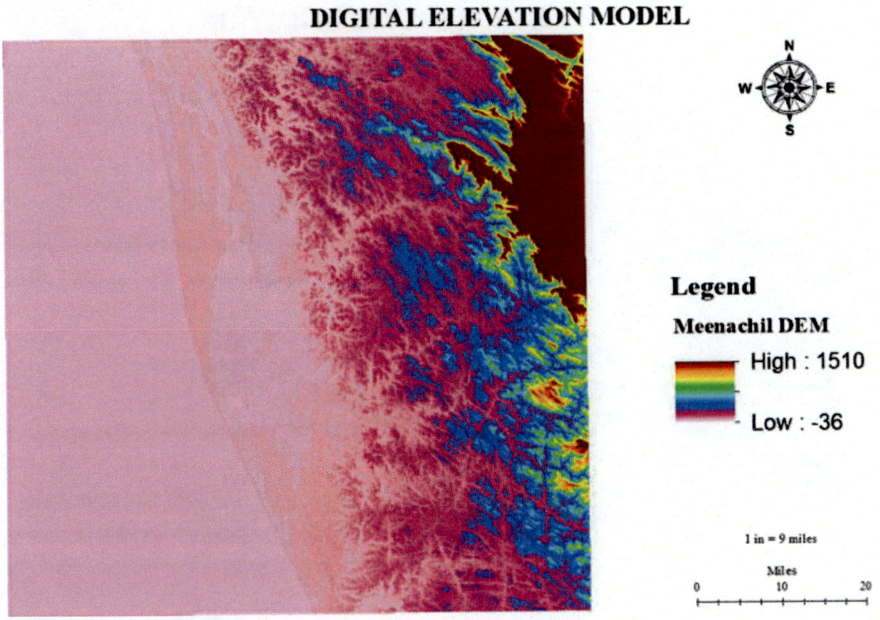

Fig. 2 Digital elevation model

3.4 Soil Data

The soil data are obtained (2005) from NBSS (National Bureau of soil survey and Land use planning). The shapefile of the study area extracted, clipped, and classified in Arc GIS 10.3. Meenachil basin consists of mainly three different types of soils (Table 2). Different soils having different properties like hydraulic conductivity, bulk density, available water capacity, structure, etc. Figure 4 represents the soil classification.

3.5 Streamflow Data

The discharge data are collected from the Central Water Commission, India (CWC), via India-WRIS (Water Resources Information System) portal. The discharge data for Kidangoor gauge station for the time period 1997 to 2015 in monthly scale were collected for this study. https://indiawris.gov.in/wris/#/.

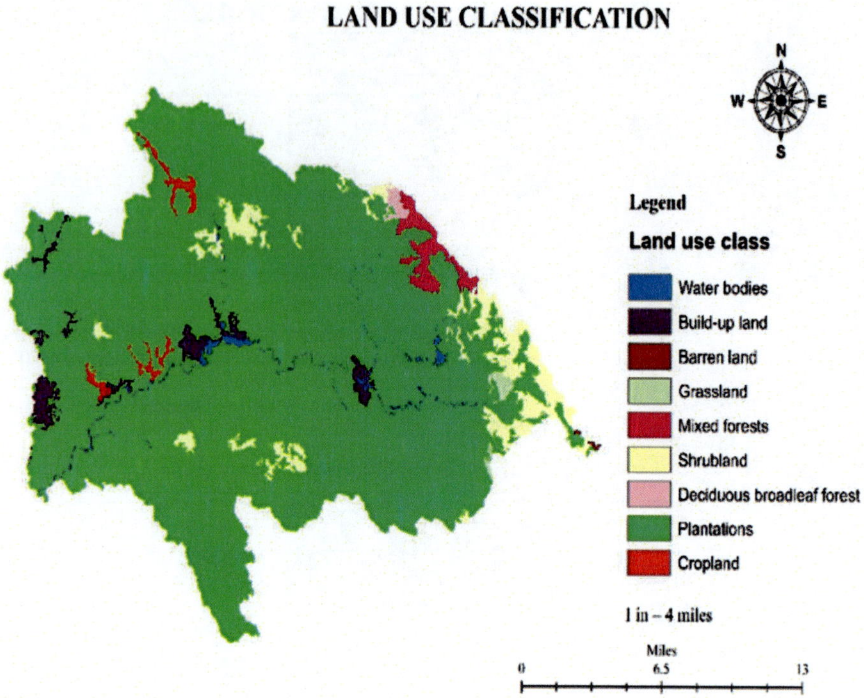

Fig. 3 Land use map of Meenachil River Basin-2005

Table 1 Land use/land cover categories

Land use type	Area (%) cover
Deciduous broadleaf forest	0.32
Crop land	0.87
Build-up land	2.06
Mixed forests	1.38
Shrubland	5.52
Barren land	0.04
Water bodies	0.91
Plantations	88.55
Grassland	0.35

Table 2 Soil classification

Soil type	% Area
Sandy-clay-loam	46.007
Sandy-loam	0.004
Clay-loam	53.989

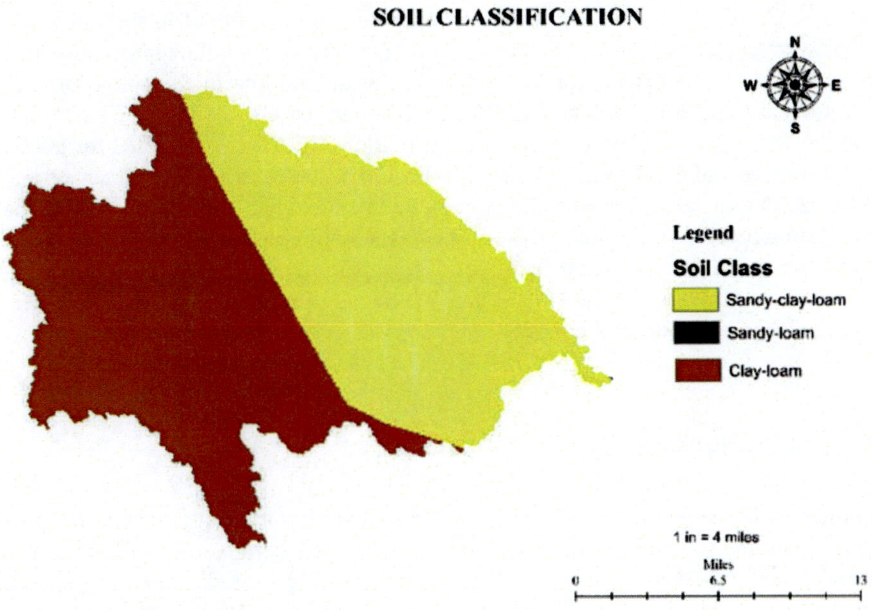

Fig. 4 Major soils of Meenachil River Basin

3.6 Water Quality Data

Monthly measured values of various water quality parameters of Meenachil River at Kidangoor station have been collected from Kerala Pollution Control Board (KPCB). Water quality parameters have been collected from India-WRIS (Water Resources Information System) portal also.

4 Model Application

The major steps involved for the application of SWAT model to the Meenachil River basin are (1) data preparation, (2) watershed delineation, (3) HRU definition, (4) sensitivity analysis, and (5) model calibration and validation. The precipitation and temperature data files for the calibration period (1997–2006) and validation period (2007–2015) were created for the observed data in the format specified in SWAT. The spatial data sets required were projected to the same projection, WGS_1984_UTM_ZONE43N using ArcGIS 10.3. Project setup has done to store necessary folders and databases during modeling. DEM was inserted for the delineation of the watershed and to understand the directions of flow. The land use/land cover spatial data were reclassified into SWAT land cover/plant types. User-defined soil types were added to the soil database. Multiple HRU definition suggested by the

ArcSWAT user's manual [20], threshold value of 5% land use, 20% soil, and 20% slope were applied for this study. Nine hydrologic parameters influencing water flow were tested for sensitivity for the simulation of streamflow in the study area. In order to calibrate the model, SWAT-CUP software was used. Sequential Uncertainty Fitting (SuFi-2) algorithm was used for calibration. SWAT was executed for a total simulation period of 19 years, which includes 1997–2006 as a calibration period and 2007–2015 as a validation period. Parameter adjustment has carried out only during calibration period. The validation process was accomplished for different time period by simply providing previously calibrated input parameters. Precipitation, temperature, and streamflow data set (1997–2015) and land use map of the year 2005 were used for the validation of the model.

5 Evaluation of Model Performance

Simulated data from the SWAT model can be compared statistically to observed data to evaluate the predictive capability of the model. The correlation coefficient (R^2) together with the Nash–Sutcliffe model efficiency (NSE) coefficient as a method to evaluate and analyze simulated monthly data. The R^2 value is a measure of the strength of the linear correlation between the predicted and observed values. The NSE value, which is a measure of the predictive power of the model, is defined as (Eq. (3)):

$$\mathrm{NSE} = \frac{1 - \sum_i^n (\mathrm{obs}_{I^-} - \mathrm{sim}_i)^2}{\sum_i^n (\mathrm{obs}_i - \mathrm{sim}_{\mathrm{mean}})^2} \tag{3}$$

where NSE is the Nash–Sutcliffe model efficiency, i = time step, n = total number of simulated time steps, obs_i and sim_i are the observed and simulated values of streamflow, respectively, and $\mathrm{sim}_{\mathrm{mean}}$ is the mean of the observation for the simulated time period [1].

R^2 and NSE values close to zero cannot be acceptable for model prediction. Generally, R^2 and NSE values greater than 0.5 are considered acceptable. Those parameter values equal to one show a perfect match between simulated and observed data values. It also indicates that predicted variable can estimate the observed variable.

6 Results and Discussions

6.1 Watershed Delineation

The watershed was delineated by using the Digital Elevation Model; the model will find out the different possible streams that can form based on the relative elevations

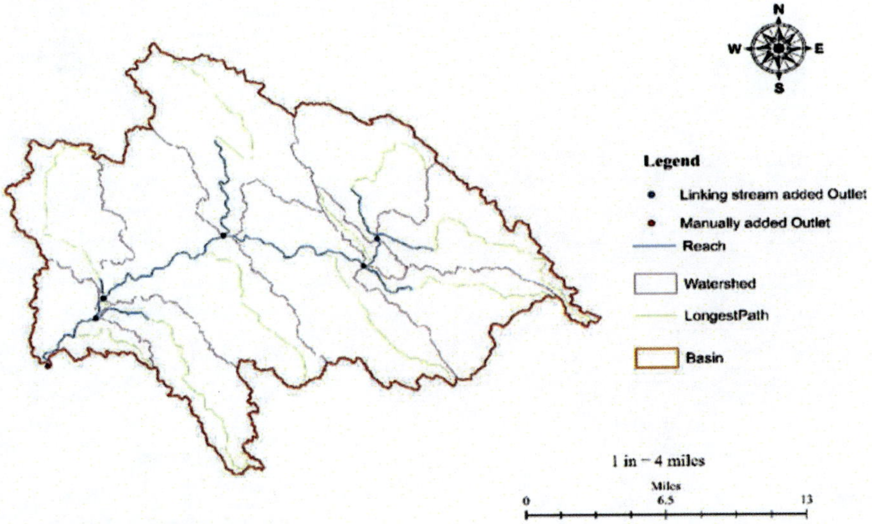

Fig. 5 Delineated watershed

of various points in the study area. Once the streams were formed, one outlet was designated at the end of the stream networks, and then, the delineation process was carried out. The delineation is based on the topography of the area around each stream and reach of each stream. After the delineation process, the entire study area was divided into 11 sub-basins. The elevation of the entire sub-basin area is −3 to 1188 m with a mean elevation of 145.62 m. These sub-basins were divided further in the HRU generation stage. The delineated watershed is shown in Fig. 5.

6.2 HRU (Hydrologic Response Unit) Generation

The large-scale diversity of the study area was illustrated by dividing the entire watershed into sub-basins and further sub-dividing it into series of hydrologic response units, which are unique soil-land use combinations. Land use data, soil class, and slope class were given as inputs for the generation of HRUs, and as a result, 97 HRUs were generated. All the components such as soil water content, surface runoff, lateral flow, and evapotranspiration were simulated for all the HRUs (Fig. 6).

6.3 Calibration and Validation of the Model

For calibrating the model, a preliminary sensitivity analysis based on the available climatic hydrologic data for the period 1997–2006 (10 years) was performed. The

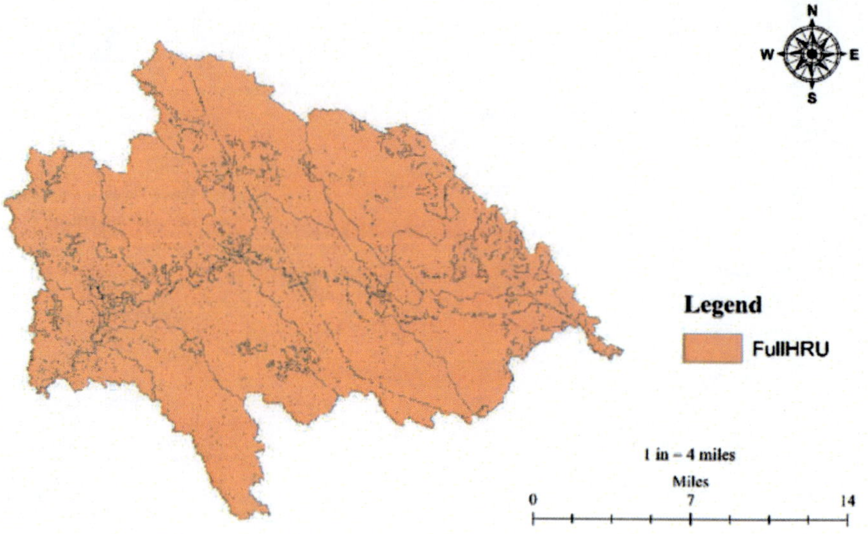

Fig. 6 HRUs

sensitivity analysis has been carried out for identifying the key parameters required for model calibration. There are more than 40 parameters influencing surface water in SWAT among which 9 parameters were shortlisted depending on the sensitiveness of the parameters toward the objective of the study for model calibration [1]. In this study, global sensitivity analysis was performed using SWAT-CUP. SuFi2 algorithm in SWAT-CUP is used for calibration. Streamflow collected during 2007–2015 (9 years) for the Kidangoor station is used for validating the predictive capability of the SWAT model applied to Meenachil River Basin. In this study, R^2 and NSE are considered as evaluation criteria. The outlet of the drainage area, i.e., Kidangoor gauging station, is used in calibration for streamflow.

The streamflow calibration process was completed by varying several SWAT hydrologic calibration parameters within their acceptable ranges (Table 3), to match the model predicted monthly streamflow time series with corresponding measured values.

The most sensitive parameter in the calibrated parameter is RCHRG_DP. A larger *p*-factor (around 1) and a smaller r-factor (around 0) should be achieved to have a better calibration and uncertainty results. The model performance has been assessed by comparing observed versus simulated monthly flows during calibration as well validation periods. The calibration after 3 iterations gave a comparably good fitting for the simulated streamflow with respect to the observed streamflow. The SWAT model shows good performance in simulating streamflow in the Meenachil River. The 95PPU (95 Percentage Prediction Uncertainty) *P* factor 0.61 and *R* factor or thickness coefficient equal to 0.65 during the calibration. Figure 7 shows the 95PPU band obtained after uncertainty analysis during calibration period.

Table 3 Sensitive parameters and fitted values after calibration using SuFi2

S. No.	Parameter name	Description	Fitted value	Minimum value	Maximum value
1	V__ALFA_BF.gw	Base flow alpha factor (days)	0.9	0	1
2	V__GW_DELAY.gw	Groundwater delay	50	0	500
3	R__GWQMN.gw	Threshold depth of water in the shallow aquifer required for return flow to occur (mm)	300	0	5000
4	R__GW_REVAP.gw	Groundwater revap coefficient	0.1964	0.02	0.2
5	R__REVAPMN.gw	Threshold depth of water in the shallow aquifer required for revap to occur (mm)	130	0	500
6	R__RCHRG_DP.gw	Deep aquifer percolation fraction	0.94	0	1
7	R__ESCO. Bsn	Soil evaporation compensation factor (unit less)	0.06	0	1
8	R__CH_N2.rte	Manning's n for the main channels (unit less)	0.162	0	0.3
9	R__CH_K2.rte	Effective hydraulic conductivity in main channel alluvium (mm/h)	147	0	150

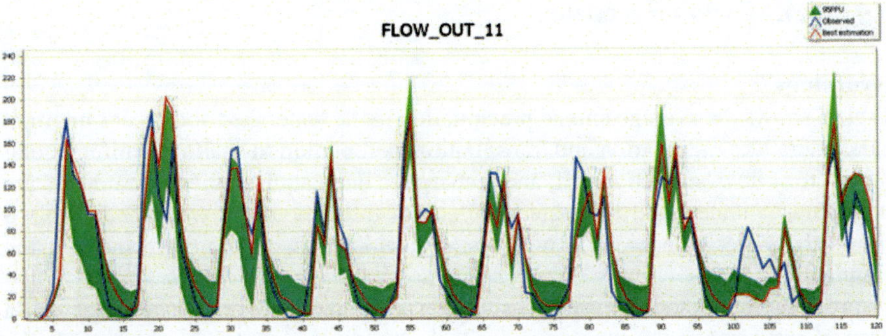

Fig. 7 Hydrograph for simulated and observation flows with 95PPU band during calibration period

Table 4 Statistical indicators

	Calibration	Validation	Remarks
R^2	0.79	0.82	Good
NSE	0.77	0.75	Good

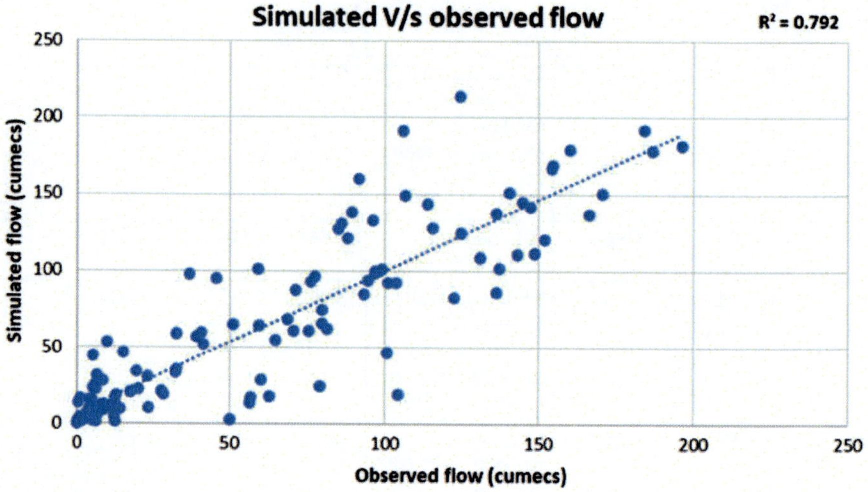

Fig. 8 Scatter plot for monthly simulated streamflow versus observed streamflow during calibration period (1997–2006)

The measured and simulated monthly flow values matched well and showed a strong correlation with reasonable accuracy, as reflected by the strong R^2 and NSE values (Table 4) for both monthly results. Figures 8 and 9 show the graphical representation of the calibration and validation results on monthly basis.

6.4 *SWAT Model Results*

Hydrology

Table 5 shows the average annual precipitation in the watershed and other important parameters like evaporation and transpiration, percolation to shallow aquifer, lateral flow, return flow, surface runoff, and recharge to deep aquifer; these parameters are important to understand the water availability throughout the watershed. Table 6 shows the water balance ratio of the model which reflects the correctness of the simulation.

From Table 7, potential evapotranspiration is the amount of evaporation that would occur if sufficient water source was available, and as expected, it is maximum in the month of March because high solar insolation along with this low precipitation may create water scarce situation during this month.

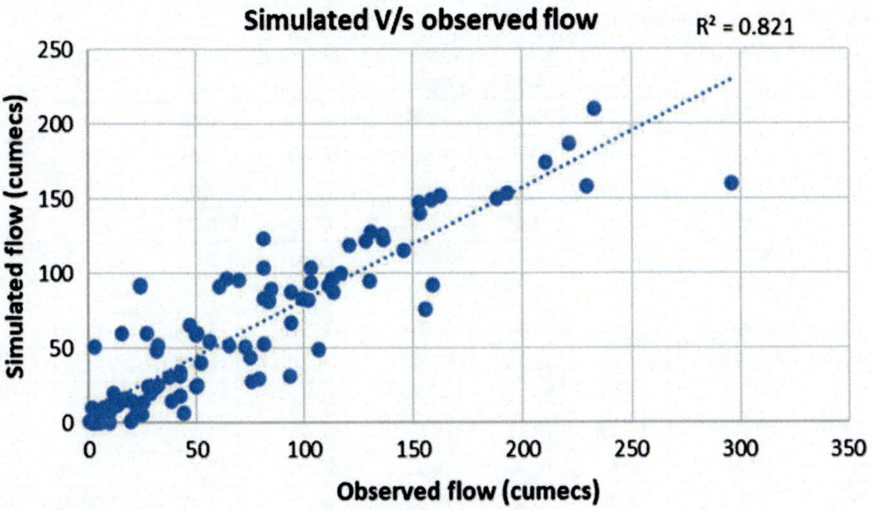

Fig. 9 Scatter plot for monthly simulated streamflow versus observed streamflow during validation period (2007–2015)

Table 5 Model check details

Parameter	Value (mm H_2O)
Precipitation	3364.5
Evaporation and transpiration	699.4
Percolation to shallow aquifer	1487.06
Return flow	1383.34
Lateral flow	153.3
Surface runoff	1017.46
Recharge to deep aquifer	74.35

Table 6 Water balance ratio details

Ratios	Value
Streamflow/precipitation	0.76
Baseflow/total flow	0.6
Surface runoff/total flow	0.4
Percolation/precipitation	0.44
Deep recharge/precipitation	0.02
ET/precipitation	0.21

Table 7 Average monthly basin values

Month	Rain (mm)	SURF Q (mm)	LAT Q (mm)	Water yield (mm)	ET (mm)	Sed. yield (mm)	PET (mm)
1	9.62	0.21	1.65	76.91	24.01	0.03	127.28
2	28.18	1.50	0.93	30.43	42.33	0.61	122.75
3	64.34	1.87	1.45	12.22	71.15	0.03	175.41
4	170.94	12.15	4.44	24.04	73.90	0.07	113.77
5	287.94	86.76	8.46	116.35	79.57	0.64	118.92
6	623.59	230.56	20.07	329.78	68.89	6.47	89.81
7	652.04	230.65	29.57	453.80	69.75	13.63	87.33
8	496.75	166.87	26.48	445.94	68.50	16.92	92.59
9	368.13	110.27	20.21	366.37	58.61	20.39	91.60
10	396.87	117.19	19.11	355.06	55.25	28.02	77.45
11	207.62	49.73	14.72	256.22	50.35	17.20	86.16
12	58.22	9.68	6.20	159.70	36.76	4.27	109.44

High precipitation was observed in the month of July due to the effect of monsoon. The river basin experiences southwest monsoon starting from June and lasts till August. The northeast monsoon starts from October to the end of November [1]. Heavy precipitation during monsoon rain events leads to transport of pollutants to the river. The study area consists of 3 soil class which are sandy clay loam 46.007%, clay loam 53.389%, and sandy loam 0.004%. The study area is dominated by lateritic soils (clay). During high intensity rain in the monsoon months, the soil column gets saturated and the contact zone between the hard rock and the over burden is liquefied, resulting in soil slip.

Nitrogen Cycle

Figure 10 shows the nitrogen cycle in the river basin. A high amount of fertilizer application was found. There was an increase in nitrogen (N) pollution of surface water. The application of nitrogenous fertilizers and straw decaying could be a possible source. Drainage from manure storage sites, livestock husbandry operations, and drainage from landfill sites are major causes of nitrogen pollution [13]. These point sources severely affect the quality of stream water. This causes eutrophication in the Meenachil River. The increased amount of NO_3 concentration is an indication of anthropogenic pollution in the study area. It is mainly due to influences of poor sanitary conditions and over usage of higher fertilizers for higher crop yields in the study area. The NO_3 originates from ammonium and NO_3 fertilizers and aerobic decomposition of organic matter in the soils. Sewages generated from domestic and industrial activities and septic tanks are also remote sources of NO_3 in the area. Controlled application of chemical fertilizers can reduce the nitrate loadings at the watershed outlet without affecting crop yield.

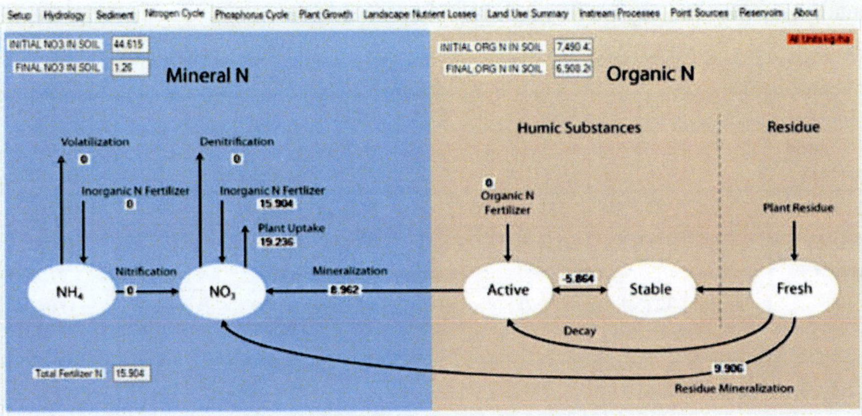

Fig. 10 Nitrogen cycle simulated in SWAT

Landscape Nutrient Losses

The loss of nutrients and sediment from land originates from both diffuse, i.e., non-point sources (fertilizers, agricultural land) and point sources (industrial plants/factories, hospitals). Source type, emissions, and transport processes will determine the overall nutrient loss to stream. Generally, river basins, which have abundance of agricultural activities, have a major role in contribution of N and P into streams. Significant rise of phosphorous simulated (2.861 kg/ha) suggests the contribution of leachate from widespread rubber plantations in the upland catchment area. Phosphorus recovery from wastewaters and its reuse for agricultural purposes could bring practical solutions for both the intrusion of phosphate ores and the reduction in P-polluted water bodies.

Rubber trees are extensively cultivated in vast areas in the entire river basin (88.55%–720.15 km^2) shown in Table 8. Crops like spices, paddies, etc., are also another agricultural land use in the basin area. The water body is polluted from the nearby paddy field, coconut, and oil palm plantations. As the percentage area under rubber plantation increases, the surface runoff decreases in the river basin. From Table 8, it is clear that more water is lost by evapotranspiration from the rubber plantation compared to other crop-cultivated area. This is maximum in the month of March due to high solar insolation along with low rainfall which may create water scarce situation in this month. Rubber is planted after removal of traditional crops or vegetations and building terraces. Due to the removal of top soils, the subsurface soils are exposed, which absorb water poorly. This can lead to accelerated soil erosion, disruption of natural streamflow, elevated stream sediments, and greater risk of landslides. Due to complete coverage, penetration of sunlight is restricted under canopy rubber plantation. This may affect water quality of the rivers flowing under rubber plantation for a long stretch.

Table 8 Land use summary

LULC	AREA (km^2)	PREC (mm)	SURFQ (mm)	GWQ (mm)	ET (mm)	SED (mm)	NO_3 (kg/h)	ORGN (kg/h)
Grasses	4.53	3215.65	1104.59	1206.04	801.04	15.66	3.83	22.40
Plantation (rubber)	720.15	3363.88	1019.74	1530.80	701.85	89.15	0.52	29.73
Shrubland	34.9	3388.58	952.30	1682.06	650.48	487.80	0.26	20.00
Mixed forests	11.14	3388.58	1034.86	1607.87	652.80	191.62	0.31	19.18
Water	0.06	3215.65	0.00	0.00	1126.45	0.00	0.00	0.00

7 Conclusion

SWAT2012 model was used to simulate streamflow in Meenachil River Basin of Kerala State, India. SWAT model was successfully calibrated and validated using SuFi2 algorithm in SWAT-CUP. The model performance evaluation was conducted with the prescribed statistical coefficients. A decent correlation was obtained agreement between the measured and predicted values on a monthly scale, as indicated by statistical parameters such as R^2 and NSE. This good performance of the model makes it feasible for predicting streamflow in Meenachil River Basin. The calibrated model was used to study the effects of landscape, climate condition, and fertilizer application on sediment and nutrient loadings. With a monthly time step, this study has provided knowledge and insights into the availability of water at a sub-basin level, which helps to locate regions where a detailed analysis may be necessary. SWAT is used to simulate the nutrient cycle and provide useful insight into its application on similarly impacted agricultural watershed in central Kerala. The application of fertilizers in agricultural land, the drainage from composting sites and landfills, and animal husbandry operations have led to the presence of excess nitrogen and phosphorus in the water bodies. This in turn degrades the water quality, and the excess nutrients lead to increased algal growth. This phenomenon known as eutrophication affects the aquatic biota and eventually renders the water source unusable. From the study area, it was observed that there is a lack of awareness among the agricultural community regarding the environmental implications of the excessive use of chemical fertilizers and pesticides. From various studies, it is clear that the non-sanitary latrines, open defecation, poor drainage facilities, water-logging environment, and unethical dumping of domestic wastes have led to the deterioration of water quality in the study area. Consequently, the river is subjected to natural contamination by human activities. Hence, a general change in land use pattern, the reduced cultivation of rubber plants to reduce surface runoff and erosion, regular monitoring, removal of macrophytes in downstream, and an increased awareness among the community would help in implementing a good management strategy for reviving and protecting the water quality of Meenachil River and its tributaries.

Acknowledgements The Kerala State Pollution Control Board is thanked for providing the water quality data of Meenachil River at Kidangoor station.

References

1. George, C., & James, E. J. (2013). Simulation of streamflow using soil and water assessment tool (SWAT) in Meenachil river basin of Kerala, India. *Scholars Journal of Engineering and Technology, 1*, 68–77.
2. Celine, G., & James, E. J. (2015). Assessing the implications of extension of rubber plantation on the hydrology of humid tropical river basin. *International Journal of Environmental Research, 9*, 841–852. https://doi.org/10.22059/ijer.2015.971
3. Vincy, M. V., Brilliant, R., & Pradeepkumar, A. P. (2015). Hydrochemical characterization and quality assessment of groundwater for drinking and irrigation purposes: A case study of Meenachil River Basin, Western Ghats, Kerala, India. *Environmental Monitoring and Assessment, 187*, 1–19. https://doi.org/10.1007/s10661-014-4217-4
4. Gassman, P. W., Reyes, M. R., Green, C. H., & Arnold, J. G. (2007). The soil and water assessment tool: Historical development, applications, and future research directions. *Transactions of the ASABE, 50*, 1211–1250. https://doi.org/10.13031/2013.23637
5. Arnold, J. G., Moriasi, D. N., Gassman, P. W., Abbaspour, K. C., White, M. J., Srinivasan, R., Santhi, C., Harmel, R. D., Van Griensven, A., Van Liew, M. W., Kannan, N., & Jha, M. K. (2012). SWAT: Model use, calibration, and validation. *Transactions of the ASABE, 55*, 1491–1508.
6. Neitsch, S. L., Arnold, J. G., Kiniry, J. R., & Williams, J. (2005). *Soil and water assessment tool theoretical documentation version.*
7. Jha, M. K., Gassman, P. W., & Arnold, J. G. (2007). Water quality modeling for the Raccoon River watershed using SWAT. *Transactions of the ASABE, 50*, 479–493.
8. Geoffrey, D., Gooch, A. R.-C., & Inacke, P. S. (2010). Integrating water resources management: interdisciplinary methodologies and strategies in practice. *Water Intelligence Online, 9*. https://doi.org/10.2166/9781780401461
9. Nair, I. V., Singh, K., Arumugam, M., & Clarson, D. (2011). Monitoring of trace metal pollution in meenachil river at Kottayam, Kerala (India). *E-Journal of Chemistry, 8*, 257–263. https://doi.org/10.1155/2011/952040
10. Biju Eazon, D., Gajendran, C., Sivan, P. P., & Mahendran, C. (2021). Ecological impact assessment of South Indian river basins by water graphics. *Materials Today: Proceedings*. https://doi.org/10.1016/j.matpr.2020.12.267
11. Narayanan, S. P., Thapanjith, T., & Thomas, A. P. (2005). A study on the icthyofauna of Aymanam Panchayath, in Vembanad wetland, Kerala. *Zoos' Print Journal, 20*, 1980–1982. https://doi.org/10.11609/jott.zpj.1329.1980-2
12. Abhilash, P. C., Singh, N., Sylas, V. P., Ajay Kumar, B., Mathew, J. C., Satheesh, R., & Thomas, A. P. (2008). Eco-distribution mapping of invasive weed *Limnocharis flava* (L.) Buchenau using geographical information system: Implications for containment and integrated weed management for ecosystem conservation. *Taiwania, 53*, 30–41. https://doi.org/10.6165/tai.2008.53(1).30
13. George, P., & Joseph, S. (2017). Appraisal of nutrient distribution in the surface water and bed sediments of a small mountainous river. *Environmental Monitoring and Assessment, 189*, 1–10. https://doi.org/10.1007/s10661-017-5874-x
14. Joseph, P. V., & Jacob, C. (2010). Physicochemical characteristics of Pennar River, A Fresh Water Wetland in Kerala, India. *E-Journal of Chemistry, 7*, 1266–1273.
15. Joseph, P. V., & Jacob, C. (2010). Bacterial, fungal and algal population of pennar river: A fresh water wetland in Kottayam district, Kerala. *Asian Journal of Chemistry, 22*, 4286–4290.

16. Sebastian, S. (2016). Algal diversity of River Meenachil in Kerala, India. *Indian Journal of Applied Research, 6*, 203–204.
17. Vincy, M. V., Brilliant, R., & Pradeepkumar, A. P. (2017). Prevalence of indicator and pathogenic bacteria in a tropical river of Western Ghats, India. *Applied Water Science, 7*, 833–844. https://doi.org/10.1007/s13201-015-0296-9
18. Barkved, L., Fazi, S., & Lo Porto, A. (Eds.). *Scientific report on pollution source assessment, including source apportionment results and pollution prevention measures* (pp. 1–91).
19. Roy, P. S., Meiyappan, P., Joshi, P. K., Kale, M. P., Srivastav, V. K., Srivasatava, S. K., Behera, M. D., Roy, A., Sharma, Y., Ramachandran, R. M., Bhavani, P., Jain, A. K., & Krishnamurthy, Y. V. N. (2016). *Decadel landuse and land cover classifications across India, 1985, 1995, 2005.*
20. Frankenberge, J., *Instructional videos | SWAT | soil & water assessment tool.* https://swat.tamu.edu/workshops/instructional-videos/

Physical and Hydro Geochemical Impact Assessment of Inactive Mines in and Around Southwestern Part of Cuddapah Basin Using a Conceptual Site Model (CSM)

Y. Sudharshan Reddy, V. Sunitha, and B. Suvarna

Abstract Inactive mines have a tremendous influence on the environment and human health, and are frequently the result of a lack of attention and planning. As a result, this report outlines all of the inactive mine site characteristics in terms of physical and chemical consequences, community safety, and human and environmental health threats. To that purpose, this work develops a conceptual site model that may be used to locate, describe, and map inactive mining locations. Inactive mines are creating haphazard dangers to local communities and the environment in the research area, based on site characteristics such as vertical and horizontal openings, toxic impoundments, and steep waste dumps. Water chemistry studies have been undertaken near inactive mines in terms of chemical impact evaluation. Groundwater in the vicinity of inactive mining zones has a high pH and alkalinity, and the dissolved ion concentration (TDS) of 2% is more than the permitted limit, according to the analytical data. Sulfate and chloride are two anions that are within the acceptable range. The nitrate concentrations in the research area ranged from 0.14 to 589 mg/L, and anthropogenic input from inactive mining areas caused 45% of the groundwater samples to exceed the permissible range. As, Ni, Pb, and Sr all have greater heavy metal concentrations than the permissible limit. This suggests that the greater concentration of heavy metals in the research area is due to human input from inactive mine disposal sites. As a result, the rehabilitation program must be carried out in order to eliminate haphazard difficulties in the research region caused by inactive problems.

Keywords Conceptual site model (CSM) · Groundwater quality · Inactive mines · Southwestern part of Cuddapah basin

Y. S. Reddy · V. Sunitha (✉) · B. Suvarna
Department of Geology, Yogi Vemana University, Kadapa 516005, India

A. K. Dikshit et al. (eds.), *Innovative Trends in Hydrological and Environmental Systems*, Lecture Notes in Civil Engineering 234,
https://doi.org/10.1007/978-981-19-0304-5_55

1 Introduction

In general, mining operations results in inactive or abandoned mine sites, which have a physical and chemical influence on the environment and human communities. Mining is the sole route for toxins to reach the ecosystem [1–4]. As a result, the majority of industrialized and developing countries are dealing with substantial environmental issues caused by inactive mines, which is primarily a human health issue. In the case of mines that are inactive or abandoned, the environmental concern associated with mining operations is more acute and visible. Physical and groundwater quality issues are important concerns when determining whether environmental issues associated to inactive mines are applicable. For example, several researchers [2] determined that environmental pollution from abandoned barite mines had negative consequences for global health and economic development. The principal physical difficulties faced by inactive mining sites include vertical and horizontal openings, dangerous impoundments, dangerous high walls and steep slopes, trash piles, and so on. Groundwater quality surrounding inactive mining locations is a big challenge in terms of chemical impact [5–7].

Inactive mines are damaging groundwater supplies because they are not being properly remedied. Mining activities typically have an impact on water resources at all phases of the mine's life cycle, including after it has ceased operations. When a mine closes, the dewatering facility closes as well. Metals diffuse extensively and reach shallow aquifers through openings such as adits, fractures, and fissures in dormant mines during the monsoon season, when considerable rains occur. Heavy metals are present in high concentrations in this sort of groundwater, causing a plethora of health issues for humans [8–11]. When groundwater and surface water enter mine impoundments or sites and come into contact with primary or secondary minerals under hazardous conditions, groundwater problems with mine drainage arise. As a result, under particular E^H–pH conditions, those minerals come into contact with and dissolve in groundwater, polluting the aquifer system. Inactive mine impoundments damage groundwater resources in inactive mining sites, which are owing to the widespread use of chemical fertilizers to boost crop output in farm land, with excess fertilizer being deposited in the soil. Due to high rainfall during the monsoon, these fertilizers breakdown in rainwater and are carried to an inactive mine impoundment, where they are kept for a long time. Heavy metals and toxic components were found in this mine water, which will easily infiltrate the groundwater table and damage the entire aquifer system beneath the earth's surface. This contaminated groundwater will be returned to the surface via tube wells, and local residents will use it for drinking, cooking, and cultivation [12, 13]. However, environmental pollution studies on inactive mines in various areas, such as those conducted by several researchers, revealed that the problem in mining sites is an inventory of acid-producing sulfides and acid-neutralizing carbonates, such as the typical representative of the Mangampeta barite mine area in Andhra Pradesh, India [14]. Lack of mine dewatering facilities, the leakage of toxic leachate, and the discharge of untreated water leading to mine floods are some of the environmental issues connected with dormant mining sites [15].

In this sense, the current study project aims to relate to physical and hydro geochemical impact, which refers to environmental factors that can harm (i.e., cause injury or death) to the human body. Several physical hazards, such as dangerous impoundments, mine shafts, vertical and horizontal mine openings, sinkholes, mine dumps, and glory holes, can be found at inactive mine sites. Unprotected inactive mine openings are create negative impact on human and animal health. Pollution of water, air, and soil by contaminants from inactive mines are causing to threat local inhabitants. However, the physical and chemical impact assessment of inactive mine sites has yet to be investigated in the study area. However, keep in mind that the purpose of this project is to develop a conceptual site model that illustrates the environmental and physical impacts of the inactive mining site in the study area. As a result, the main goal of the study were to (1) evaluate the physical characteristics of inactive mines, (2) estimate the level of heavy metal concentrations in groundwater around inactive mine sites, and (3) investigate the socioeconomic implications of inactive mine sites in the study area. The model will be used as a decision-making framework or guidance for the restoration of dormant mining sites in the Cuddapah Basin, A.P., and southern India.

2 Study Area

Cuddapah Basin is one of India's largest and most complicated sedimentary basins, covering 44,500 km^2. The research area is located in the southwestern part of the Cuddapah basin and includes four mandals: Vempalli, Vemula, Lingala, and Pulivendla. It is located between latitude 14° 18′ 0″ N and 14° 28′ 0″; longitude 78° 0′ 00″ E 78° 30′ 0″ and is covered by toposheet no. 57 J/02, 57 J/03, 57 J/04, 57 J/07 (Fig. 1). The studied area is stratigraphically part of the lower Cuddapah group consist of Papaghni and Chitravati sub groups [16]. Lithologically the study area consist of 6 km thick succession of quartzite, sandstone, conglomerate, shale, slate, carbonate, and igneous intrusive formations (Fig. 1). Residual hill, pediment, inter mountain valley, pediplain, structural hill or linear ridges, and flood plain are the principal geomorphic units in the study area. The study area climate is classed as semi-arid, with an average yearly temperature of 29.60 °C and maximum temperatures of around 40 °C in April and May. The average lowest temperature during winter is around 19 °C, with some days falling below 16 °C. The majority of the cultivation in the study area is carried out in dry conditions, but a small percentage of the cropped area in the mandals of Pulivendla, Vempalli, Vemula, and Lingala has access to irrigation facilities through the Pulivendla branch canal, Papaghni flood plain, and KC canal, respectively. Irrigation is mostly provided by tube wells and bore wells, followed by canals and tanks. The water level in the Cuddapah basin southern section ranges from 10 to 20 m deep [17].

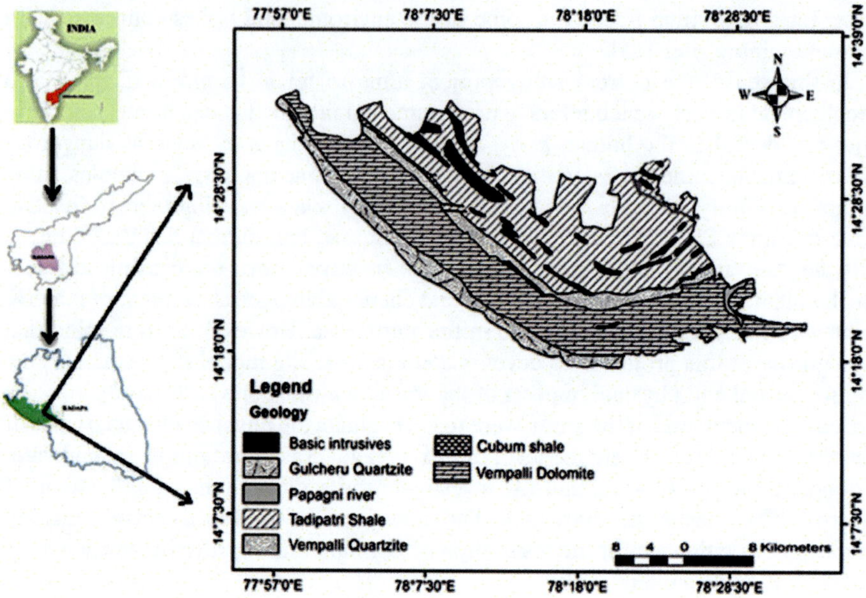

Fig. 1 Location and geology map of the study area

3 Methodology

Conceptual site modeling (CSM) was created to make assessing the physical and environmental implications of inactive mines in the southwestern region of the Cuddapah basin; it involves the identification and location of all inactive mines in the study area. The inactive mine site and its accompanying features are identified using Landsat photos (15 m spatial resolution), Toposheets, Google Earth images, and aerial photography and under the part of the field work, traveling through inactive mine sites to physically define the mine characteristics found in the southwestern region of the Cuddapah basin. The present status of mine features is considered in the site description, as well as how they affect different aspects of the environment and pose a concern to public safety and health. Site descriptions of abandoned mine features can establish trails and/or exposure routes based on photographs and actual site assessments.

At September 2019, 50 groundwater and 10 soil samples were collected in and near inactive mining sites in the Cuddapah basin in Andhra Pradesh, and required steps were taken to avoid pollution [18]. Groundwater samples were taken in two-liter pre-cleaned and dried polyethylene bottles, while soil samples were taken in a clean plastic bag. During field work, the pH value, EC, and total dissolved solids were evaluated using the water analyzer 371 field kit; Ca^{2+}, Mg^{2+}, Na^{+}, and K^{+}, as well as other significant cations, were determined using titration and flame photometry (Systronic Model No. 128), mostly anions like SO_4^{2-} are measured with a flame photometer, and Cl^- is measured with a titration; fluoride was measured with an Orion 4 star ion meter (model: pH/ISE), and nitrate (NO_3^-) was measured with a spectrophotometer. The study's analysis and estimates are based on APHA and other industry-standard procedures [19, 20]. The analysis and calculations of the study is based on APHA [18] and other standard methods [19, 20]. In order to predict the asbestos fibers in the soils, SEM analysis was performed.

4 Conceptual Site Modeling (CSM)

The analytical framework model of conceptual site modeling is extremely useful. It gives a graphic representation of the current physical characteristics of inactive mines, as well as known potential pollution sources associated with inactive mines in the research area. The CSM model depicts the various chemical migration paths that have been identified. Based on the physical and chemical dangers existing in inactive mining regions, it illustrates the human and ecological pollution pathways. Exposure scenarios are created using conceptual site model elements relating to exposure medium, exposure pathways, and populations of concern [21, 22]. The physical and environmental challenges of dormant mining sites in the Cuddapah basin's southwest corner are depicted schematically in Fig. 2.

5 Inactive Mines Sites

In the study region (southwest section of the Cuddapah basin), numerous inactive open pit and underground mines (barite, yellow ochre, asbestos, and other minerals) have been discovered, causing aggravation to local residents. If a mine is dormant but has a large environmental impact, it poses a considerable risk to the health and safety of local human communities if it is not adequately decommissioned. High walls, vertical openings, dangerous inactive mine impoundments, waste piles, and tailing ponds, among other things, have been exposed by physically inactive mines, posing a serious threat to the health and safety of the surrounding human communities [23, 24]. Mining waste piles and inactive mine impoundments were disrupted in the environment, blocking streams and contaminating surface and groundwater sources. The majority of the abandoned mines are in residential areas, with the exception of a small patch of agricultural land in the southern corner of the Cuddapah basin. The spoil dumps have the highest percentage of silica oxide and the second highest percentage of aluminum oxide in their mineral composition. The spoil dump covers a large area, and the rubbish disperses throughout the area and surrounding agricultural lands, posing a risk to crop productivity [24].

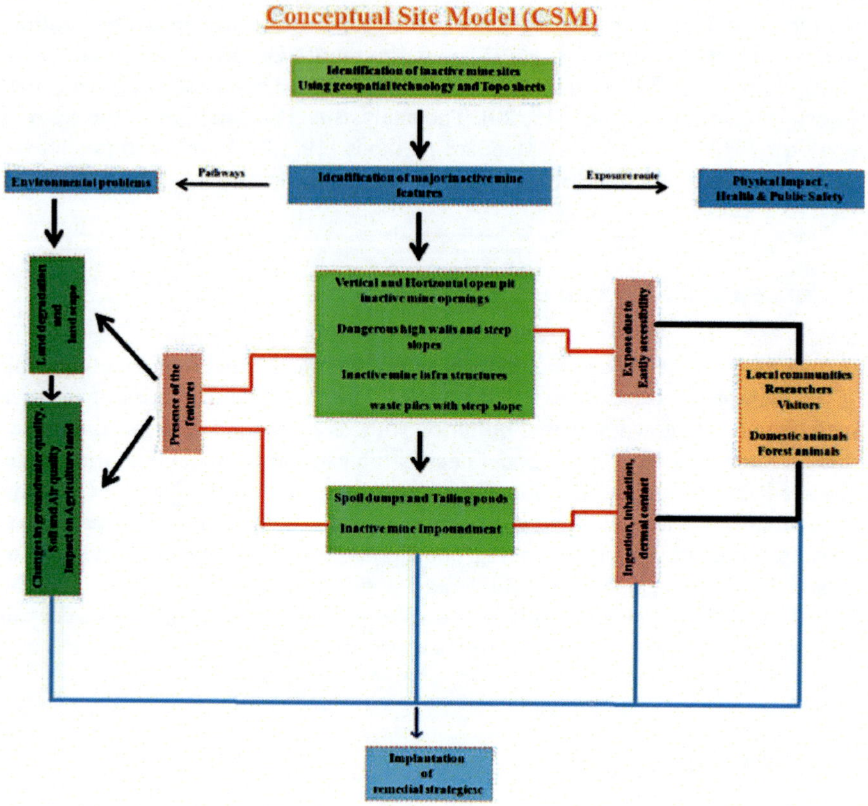

Fig. 2 Conceptual site model of physical and environmental problems of inactive mines in the study area [3]

6 Results and Discussion

6.1 *Environmental Impact of Inactive Mine Lands*

A long history of active mining in the southwest corner of the Cuddapah basin has resulted in inactive mine sites and a huge number of unprotected open pit mines, posing a threat to humans and livestock and these mines can be detected on both private and public lands. Physical and chemical risks may arise from inactive mine sites and related mining waste. Many of these dangers are the consequences such as past mining activities; ore processing techniques, inappropriate closer procedures, or the deposit are exposure to the elements on the surface, changed landscapes, disused open pits and shafts, land that is no longer useable owing to soil erosion, inactive mine waste, subsidence, and changes in vegetation cover are all frequent physical and environmental challenges in most inactive mine sites. The next sections and Table 1 detail the many characteristics of inactive mining sites.

Physical impact and the Characterization of Inactive Mine Sites

Falling into the mine opening is the leading cause of injury and death at inactive mine sites. An inactive mine lands can contain a variety of physical risks. After that are inactive mine openings and impoundments.

Table 1 Concentration of major oxides of different inactive mine site in the SW part of the Cuddapah basin

Inactive/abandoned mine	Village	Latitude	Longitude	Effects	
				Physical	Chemical
Barite	Votlapalli	N 14° 21′ 11.7″	E 78° 18′ 00.2″	High risk due to contain land subsidence, dangerous impoundment, vertical mine openings (no barricaded)	Effect on groundwater quality and soil quality due to presence of heavy metals
	Vemula	N 14° 21′ 19.3″	E 78° 18′ 39.3″		
	Vempalli	N 14° 22′ 54.3″	E 78° 27′ 13.4″		
	Mugguraikona	N 14° 20′ 35″	E 78° 19′ 38″		
Yellow ochre	Alavalapadu	N 14° 23′ 43.6″	E 78° 13′ 27.8″	High risk due to contain steep slope high wall with loose surface, dangerous impoundment	Effect on groundwater quality and soil quality due to presence of heavy metals
	Ammayagari palli	N 14° 19′ 52.8″	E 78° 22′ 11.7″		
	Chagaleru	N 14° 3′ 37.9″	E 78° 21′ 29.8″		
	Ramanuthala palli	N 14° 21′ 11.7″	E 78° 18′ 00.2″		
Asbestos	Brahmanapalli	N 14° 25′ 0.08″	E 78° 11′ 08.7″	High risk due to contain steep slope high wall with loose surface, steep portal mine openings, dangerous processing equipment facilities air pollution	Effect on groundwater quality and soil quality, acid mine drainage
White clay	Lingala	N 14° 25′ 12.1″	E 78° 10′ 45.6″	High risk due to contain steep slope high wall with loose surface, dangerous impoundment	Effect on groundwater quality and soil quality

Fig. 3 **a, b** Inactive vertical open pit barite mine openings at Vempalli and Vemula villages

Inactive Mine Openings

In inactive mine sites, mine openings (such as vertical and horizontal open pits) can have substantial physical effects on humans. In the SW section of the Cuddapah basin, near Vemula and Vempalli villages, vertical and horizontal open pit mine holes of inactive barite mines are displayed in Fig. 3. These openings are often well-known, posing a threat to local communities as adventurers seek to enter and explore them. These mines may have loose soil with unstable earth that could collapse, as well as a lack of oxygen, which could hurt or kill humans. Other physical repercussions of mine openings include vegetation-covered apertures or fallen old structures, which may pose a risk of collapsing from a height of a hundred feet if they come too close to unstable covered mine openings.

Inactive Mine Impoundments

Mine impoundments are underground mines where water is entered and filled by the surface flow during occurring of heavy floods, groundwater seepage, and direct precipitation are the main sources of this water. This water may contain toxic metal ions and it may have a negative impact on human health. Solid waste and polluted mining water enter the environment and are absorbed by humans and animals by ingestion, inhalation, and direct and dermal contact. Contamination of groundwater resources disturbs inactive mine impoundments in the environment. The majority of the inactive mine impoundments are situated on gentle slopes with nearby residential and agricultural areas. A large amount of residential and agricultural waste (chemical fertilizers, pesticides, etc.) is washed away and flows into the upper to lower slope due to the significant rainfall in the rainy season following the monsoon. This water will be deposited into the inactive mine impoundment because it is in the low gradient category (Fig. 4). The majority of mine activities take place below the water table, in either underground or open pit workings. As a result, filthy impound water seeps into the groundwater and is used for irrigation and drinking. Any strip pit with impounded water that is not barricaded, regardless of depth, is regarded an enticing nuisance by nearby inhabitants who desire to access them and fall into the mine, causing injury or death.

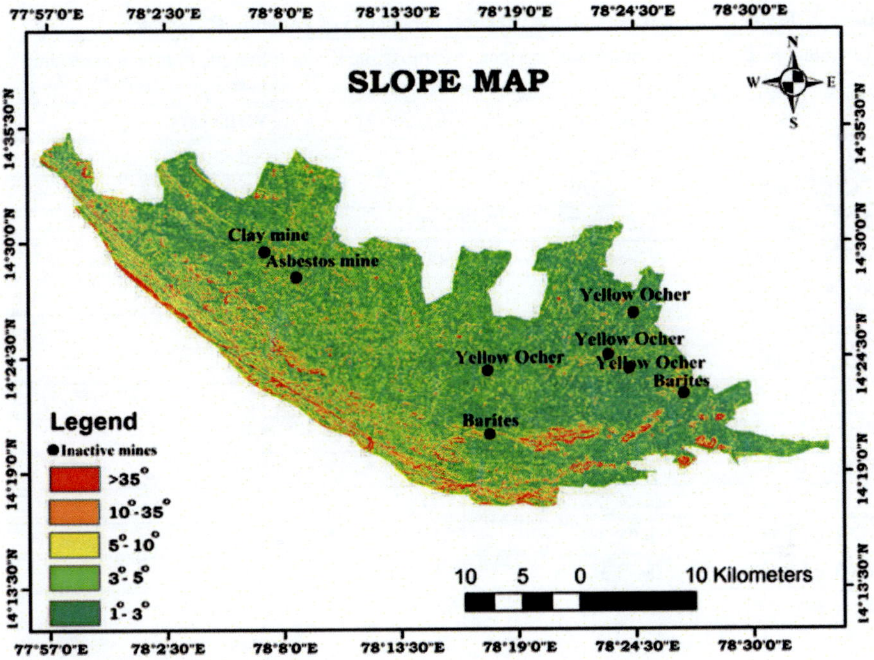

Fig. 4 Slope map of the study area

7 Chemical Impact of Inactive Mine Sites

7.1 Chemical Contamination of Underground Aquifers

Inactive mines can have a variety of effects on groundwater quality. Metals are the most common dissolved contaminants in the mining area, although they might also include SO_4^{2-}, NO_3^-, Pd, Cu, Mn, Cd, Fe, and Zn. Once dissolved, these contaminants will travel to local surface water to groundwater (surface water pollution may also occur when contaminated soil or waste is eroded and washed into water bodies). These mine streams may contain considerable amounts of heavy metals and total dissolved solids, as well as elevated temperatures and altered pH values, depending on the nature of the ore body and local geochemical settings [25, 26]. Metals that surpass water quality regulations in neutral and alkaline mine water can have a major impact on human health and the environment. The geochemical analysis results of water samples collected around inactive mine sites in the SW part of Cuddapah basin is given in Table 2.

Groundwater Chemistry

The pH of groundwater samples ranges from 7.3 to 8.9, indicating that it is alkaline. The observed groundwater samples' electrical conductivity (EC) ranged from 115

Table 2 Range and comparison of the chemical parameters of groundwater samples

Chemical parameters	Concentrations ions	Permissible limits	Sample numbers exceeding permissible limit (%)
	Range	WHO [27]	WHO [27]
pH	7.4–8.9	9.2	Nil
EC (μS/cm)	115–2920	–	Nil
TDS (mg/L)	110–1987	1500	2
HCO_3^- (mg/L)	24–366	600	Nil
SO_4^{2-} (mg/L)	10.6–224	600	Nil
Cl^- (mg/L)	0.03–322	600	Nil
NO_3^- (mg/L)	0.14–589	45	42
As (μg/L)	0–56.3	10	17
Cd (μg/L)	0–25.6	30	Nil
Cr (μg/L)	0–42	50	Nil
Fe (μg/L)	0–49	300	Nil
Ni (μg/L)	0–98.3	70	9
Pb (μg/L)	0–92.3	10	8
Sr (μg/L)	0–1140	30	46

to 220 S. The total dissolved ionic concentrations range from 110 to 197 mg/L, with only 2% of samples exceeding the allowed limit [27]. Sulfates and chlorides were identified in the research region in concentrations ranging from 10.6 to 224 mg/L and 0.03 to 322 mg/L, respectively, both of which were below the permissible limit [27]. Due to the exchange ion process and ion dissolution process, sulfates and chloride are presence in the study region [28]. The concentrations of NO_3^- in the studied area range from 0.14 to 589 mg/L. According to the World Health Organization (WHO), 42% of samples surpass the permitted level (Table 2). Nitrates are commonly found in groundwater as a result intensive usage of unregulated N-fertilizer, septic tanks, and poor waste management [29]. NO_3^- pollution in drinking water is more harmful to adults and children. Methemoglobinemia is a common health condition caused by an excessive intake of nitrate in the human body (blue baby syndrome). Fifteen forms of cancer, birth defects, high blood pressure, and other major chronic health issues are among the other serious chronic health issues [30]. Higher nitrate levels in groundwater in the research area are mostly due to human activities such as extensive irrigation and the use of restricted nitrate fertilizers for boosting agricultural yields, animal waste, and septic tank leakage [31, 32]. Clay and gravelly loam soils predominate in the research area, it has a low permeability. Hence, the nitrate infiltration through the soils is very low [17]. Apart from that an inactive mine impoundment is one of the causes of nitrate contamination in groundwater. Inactive mines are found near resident communities in the study region, and they have a mild slope. Due to high rains, the majority of residential and agricultural waste (fertilizer, pesticides, etc.) will be flown in to inactive mine impoundments and stored for a long time; as

a result, nitrates will easily infiltrate groundwater through infiltration and return for irrigation and drinking.

Heavy Metals in Ground Waters

Heavy metal pollution causes serious human health concerns, biodiversity loss, and environmental degradation by passing down the food chain. Heavy metals in ground water can come from either natural (geogenic) or anthropogenic sources. Due to a lack of monitoring and poor waste management at dormant mine sites, heavy metals are released into the environment and water supplies. Sr > Ni > Pb > As > Fe > Cr > Cd is the order of ionic dominance pattern of heavy metals in the research area. Expect As, Ni, Pb, and Sr from the analytical data; the remaining metals, such as Cd, Cr, and Fe, are discovered within the allowed limit (Table 2).

Cadmium is found in abundance in the environment as a result of natural sources and processes including rock and soil erosion and abrasion. Cadmium levels in groundwater range from 0 to 25.6 µg/L, which is well within the permissible limit. Cadmium is frequently detected in the study as a result of geogenic (Ca^{2+} and Mg^{2+} bearing minerals) and anthropogenic (phosphate fertilizers and containing Cd agro-chemicals) sources [33, 34]. Iron is the second most frequent metal in the earth's crust, and it is also found in soil and groundwater systems [33]. The presence of iron in groundwater samples ranged from 0 to 49 µg/L, and it was within the allowable limit. The weathering of Fe_2O_3 containing carbonate rocks, as well as the oxidation of pyrite in the carbonate rocks, are the primary sources of Fe. Chromium levels in groundwater range from 0 to 42 µg/L, with the permitted limit being discovered [27]. The research area's principal anthropogenic sources of Chromium in groundwater include sewage sludge, wastewater, and mining activities [35, 36].

Strontium is a prevalent component of carbonate minerals, and it is most likely to arise via their disintegration, accounting for 15–28% of the entire chemical intake of Strontium into the ocean to karst waters [37]. Strontium levels in groundwater samples ranged from 0 to 1140 µg /L, with 46% of samples above the permissible limit. Alkaline magma rock and silty sedimentary rock both contain nickel. Nickel is clearly associated with alkaline magmatic rocks and silty sedimentary rocks. Magnesium-iron silicates are frequently found with nickel. Nickel is easily activated and coupled with iron and manganese during the weathering process. It can move a great distance with its hydroxides in the cation form of Ni^{2+}. Nickel makes its way into the environment in large quantities. Nickel concentrations in groundwater samples varied from 0 to 98.3 µg/L, with 9% of samples exceeding the allowed limit. Human activities such as mining, smelter emissions, coal combustion, sewage, and phosphate fertilizer use are all boosting nickel concentrations in specific locations [38]. Arsenic (As) is a naturally occurring element found in soil, the atmosphere, rocks, natural streams, and living beings. Arsenic remains stably linked to sediments in geochemical settings, and dissolved levels remain very low until the parent rock releases it. The indications of chronic arsenic exposure have been described all throughout the world, including skin lesions such as hyper and hypo pigmentation, peripheral neuropathy, and vesicular diseases. The amount of arsenic in groundwater samples ranged from 0 to 56.3 µg/L, with 17% of samples above the allowed limit. Arsenic acid compounds collected from the subsurface that surrounds aquifers resulted in a higher concentration of Arsenic in the research area [39–41]. Anthropogenic activities such as mining, extensive use of arsenical pesticides, herbicides, and the use of arsenic as an addition in livestock feed, notably for chicken, are other key causes of increasing arsenic concentrations in groundwater [42]. The most dangerous heavy metal is lead. Water, food, and breathing can all be used to introduce it into living creatures. Lead has serious health consequences. The lead concentration in groundwater samples ranged from 0 to 92.3 μg/L, with 8% of samples exceeding the allowed limit. Anthropogenic activities are the primary cause of rising lead levels in natural streams.

8 Conclusions and Recommendations

The conceptual site model for inactive mines was created to better understanding that how the inactive mines harm the ecosystem and endanger the health and safety of humans and animals. The characteristics of inactive mining sites in the southern region of the Cuddapah basin are linked to many physical dangers, according to the conceptual model. Vertical open pit mine openings, filthy inactive mine impoundments, perilous steep high walls, and trash heaps are the principal safety risks of inactive mining sites in the study area. Heavy metals are abundant in inactive mine waste piles, and they have the potential to harm human health and the environment. Groundwater pollution is mostly caused by anthropogenic activities and inactive mine impoundments. According to hydrochemical analysis, 42% of groundwater samples in the study area had high nitrate concentrations, owing to the widespread use of nitrogen fertilizer in the area. Excess fertilizers and other pollution indicators (animal dung, poultry waste, and residential waste) were dumped into inactive mining impoundments, where they were kept for a long time, adding pollutants to shallow aquifers via infiltration. In terms of heavy metal content, As, Ni, Pb, and Sr are present in higher concentrations in groundwater samples from the research area, with 17%, 9%, and 46% of the samples above the W.H.O. recommended permitted limits, respectively. This suggests that the greater concentration of heavy metals in the research area is due to human input from inactive mine disposal sites. Agricultural land pollution and groundwater contamination, as well as the dispersion of fugitive dust from inactive mine sites are major cause to polluting the groundwater.

Mine closure models and geochemical models to avoid environmental pollution from inactive mines, as well as the installation of sedimentation tanks to store water and enlarge the water storage area, are among the possible restoration strategies to be followed, according to analytical data. Setting up barriers around the open pit mine's vertical and horizontal mining zones, and then revegetating the entire mining area with metal-tolerant plants and restoring the polluted soil are some of the suggested procedures for inactive mine restoration.

Acknowledgements This research was funded by the Department of Science and Technology of the Government of India (DST-INSPIRE), sanctioned to first author which is gratefully acknowledged.

References

1. Muralidhara Reddy, B., Sunitha, V., Prasad, M., Sudharshan Reddy, Y., & Ramakrishna Reddy, M. (2019). Evaluation of groundwater suitability for domestic and agricultural utility in semi-arid region of Anantapur, Andhra Pradesh State, South India. *Groundwater for Sustainable Development, 9*, 100262. https://doi.org/10.1016/j.gsd.2019.100262
2. Younger, P., Banwart, S., & Hedin, R. (2002). *Mine water hydrology, pollution and remediation*. Kluwer.
3. Sudharshan Reddy, Y., Sunitha, V., & Suvarna, B. (2020). Monitoring of groundwater quality for drinking purposes using the WQI method and its health implications around inactive mines in Vemula-Vempalli region, Kadapa District, South India. *Applied Water Science, 10*, 202. https://doi.org/10.1007/s13201-020-01284-2
4. USEPA. (1993). *Well land protection: A guide for small community's office of research and development office of water* (EPA/625/R-93/002). Washington, DC.
5. Smith, S. (2009). *Mining and the environment—Briefing paper no. 6/09*. NSW Parliamentary Library Research Service. Sydney, State of New South Wales.
6. Sahu, H. B., & Dash, S. (2011). Land degradation due to mining in India and its mitigation measures. In *Proceedings of Second International Conference on Environmental Science and Technology*.
7. Chapman, A. (2011). *Acid mine drainage in South Africa: An emerging environmental problem*. Institute for Futures Research, University of Stellenbosch, Republic of South Africa.
8. Sudharshan Reddy, Y., Sunitha, V., Suvarna, B., Prasad, M., Muralidhara Reddy, M., & Ramakrishna Reddy, M. (2020). Data on physical impacts and hydro geochemical assessment of inactive/abandoned mines in and around Southwestern parts of the Cuddapah basin using a conceptual site model (CSM). *Data in Brief, 29*, 105187. https://doi.org/10.1016/j.dib.2020.105187
9. Ziarkowski, D. V. (1998). *Abandoned mine lands preliminary assessment handbook*. United States Environmental Protection Agency (U. S. EPA).
10. Nagaraju, A., Suresh, S., Killam, K., & Hudson-Edwards, K. (2006). Hydrogeochemistry of waters of Mangompeta barite mining area, Cuddapah Basin, Andhra Pradesh, India. *Turkey Environmental Science, 30*, 203–219.
11. Franciskovic-Bilinski, S. (2006). Barium anomaly in Kupa River Drainage Basin. *Journal of Geochemical Exploration, 88*, 106–109.
12. Sudharshan Reddy, Y., & Sunitha, V. (2020). Development of water quality index for groundwater quality evaluation in and around Vemula Mandal, YSR District (A.P., India). *Journal of Indian Geophysical Union, 24*(1), 68–77.
13. Xiao, Y., Shao, J., Cui, Y., Zhang, G., & Zhang, Q. (2017). Groundwater circulation and hydrogeochemical evolution in Nomhon of Qaidam Basin, northwest China. *Journal of Earth System Science, 126*(2).
14. Prasad, M., Sudharshan Reddy, Y., Balaji, E., Sunitha, V., & Rama Krishna Reddy, M. (2019). Application of geospatial technology in evaluating the impact of mining associated urbanization on agricultural lands. *Nature Environment and Pollution Technology, 18*(3), 1041–1044.
15. Adamu, C. I., Nganje, T. N., & Aniekan, E. (2013). Hydrochemical assessment of pond and stream water near abandoned barite mine sites in parts of Oban massif and Mamfe Embayment, Southeastern Nigeria. *Environment and Earth Science, 71*, 3793–4381. https://doi.org/10.1007/s12665-013-2757-5

16. Nagaraja Rao, B. K., Rajurkar, S. T., Ramalingaswamy, G., & Ravindra Babu, B. (1987). Stratigraphy, structure and evolution of the Cuddapah Basis. *Geological Society of India, 6*, 33–86.
17. Central Ground Water Board. (2013). *Ministry of water resources Government of India.* Ground Water Brochure YSR District (Kadapa), Andhra Pradesh.
18. APHA. (2012). *Standard methods for the examination of water and wastewater* (22nd ed.).
19. Hem, J. D. (1991). *Study and interpretation of the chemical characteristics of natural water: USGS.*
20. Raghunath, H. M. (1987). *Groundwater* (2nd ed., p. 563). Wiley Eastern Limited.
21. Mhlongo, S., & Amponsah-Dacosta, F. (2015). Environmental and physical hazards of historic magnesite mining sites in the Tshipise magnesite field: A conceptual site model (CSM) and implications. In *Environment 2015*.
22. Vande Graaff, S., Unger, C., & Evans, R. (2012). *Aus IMM abandoned mines survey report*. The Aus IMM Sustainability Committee, Carlton Victoria Australia.
23. Sudharshan Reddy, Y., Sunitha, V., & Suvarna, B. (2020). Groundwater quality evaluation using GIS and water quality index in and around inactive mines, Southwestern parts of Cuddapah basin, Andhra Pradesh, South India. *HydroResearch, 3*, 146–157. https://doi.org/10.1016/j.hydres.2020.11.001
24. Adamu, C. I., Nganje, T. N., & Aniekan, E. (2014). Heavy metal contamination and health risk assessment associated with abandoned barite mines in Cross River State, southeastern Nigeria. *Environmental Nanotechnology, Monitoring and Management*. https://doi.org/10.1016/j.enmm.2014.11.001
25. Yin, H., Zhao, H., Xie, D., Sang, S., Shi, Y., & Tian, M. (2019), Mechanism of mine water inrush from overlying porous aquifer in quaternary: A case study in Xinhe Coal Mine of Shandong Province, China. *Arabian Journal of Geosciences, 12*(5).
26. Vasant, M. W., Dipak, B. P., Shrikant, V. M., Manesh, L. A., & Uday, L. S. (2019). Nitrate associated health risks from groundwater of Kadava River Basin Nashik, Maharashtra, India. *Human and Ecological Risk Assessment*. https://doi.org/10.1080/10807039.2018.1528861
27. WHO. (2011). *WHO guidelines for drinking-water quality* (4th ed.). World Health Organization.
28. Kumar, S., Rajesh, V., & Khan, N. (2020). Evaluation of groundwater quality in Ramanathapuram district, using water quality index (WQI). *Modeling Earth Systems and Environment*. https://doi.org/10.1007/s40808-020-01025-z
29. Canter, L. W. (1996). *Nitrates in groundwater*. CRC Press. ISBN 0873715691.
30. Forman, D., Al-Dabbagh, S., & Doll, R. (1985). Nitrates, nitrites and gastric cancer in Great Britain. *Nature, 313*, 620–625.
31. Li, P., Wu, J., Tian, R., et al. (2018). Geochemistry, hydraulic connectivity and quality appraisal of multilayered groundwater in the Hongdunzi Coal Mine Northwest China. *Mine Water and the Environment, 37*(2), 222–237.
32. Wagh, V., Shrikant, M., Aniket, M., Ajaykumar, K., Dipak, P., & Abhay, V. (2020). Study of groundwater contamination and drinking suitability in basaltic terrain of Maharashtra, India through PIG and multivariate statistical techniques. *Journal of water supply: Research and Technology AQUA*. https://doi.org/10.2166/aqua.2020.108/651699/jws2020108
33. Deverel, S. J., Goldberg, S., & Fujii, R. (2012). Chemistry of trace elements in soils and groundwater. ASCE manual and reports on engineering practice no. 71. *Agricultural Salinity Assessment and Management, 4*(2), 89–137.
34. Markovic, T., Miko, S., KapeljBuljan, R., Larva, O., & Peh, Z. (2006). Behaviour of metals and nutrients in soils and groundwater of a karst polje. *Journal of Geochemical Exploration, 88*, 124–129.
35. Adriano, D. C. (2001). *Trace elements in terrestrial environments biogeochemistry, bioavailability, and risks of metals* (p. 865). Springer. ISBN 978-1-4684-9505-8.
36. Hu, H. (2002). Human health and heavy metals exposure. In M. McCally (Ed.) *Life support: The environment and human health*. MIT Press (Chapter 4).

37. Gonneea, M. E., Charette, M. A., Liu, Q., Herrera-Silveira, J. A., Morales-Ojeda, S. M. (2014). Trace element geochemistry of groundwater in a karst subterranean estuary (Yucatan Peninsula, Mexico). *Geochimica et Cosmochimica Acta, 132*, 31–49.
38. IARC. (1990). Nickel and nickel compounds. In *Chromium, nickel and welding. IARC monographs on the evaluation of carcinogenic risks to humans* (Vol. 49, pp. 257–445). International Agency for Research on Cancer Lyon.
39. Wagh, V. M., Panaskar, D. B., Mukate, S. V., Gaikwad, S. K., Muley, A. A., & Varade, A. M. (2018). *Health risk assessment of heavy metal contamination in groundwater of Kadava River.*
40. Prasad, B., & Kumari, S. (2008). Heavy metal pollution index of ground water of an abandoned open cast mine filled with fly ash: A case study. *Mine Water and the Environment, 27*, 265–267.
41. Newton, G., Reynolds, S., Newton-Reed, S., Tuffly, M., Miller, M., Reeves, S., Mistchenko, J., & Bailey, J. (2000). *California's abandoned mines, a report on the magnitude and scope of the issue in the state* (Vol. I). Department of Conservation Office of Mine Reclamation Abandoned Mine Lands Unit, Sacramento, California.
42. Yousaf, B., Liu, G., Wang, R., Imtiaz, M., Rizwan, M. S., Zia-ur-Rehman, M., Qadir, A., & Si, Y. (2016). The importance of evaluating metal exposure and predicting human health risks in urban-periurban environments influenced by emerging industry. *Chemosphere, 150*, 79.

Analysis of Groundwater Level Fluctuation Near the Rejuvenated Tank Under Mission Kakatiya

Chakilam Sree Laxmi Pavani and Keesara Venkata Reddy

Abstract Groundwater is an essential natural resource and plays a substantial role in irrigation, agriculture, public and domestic water supply. Thus, conserving water through various measures would help to preserve the groundwater. The government of India has launched Jal Shakthi Abhiyan which focuses on water conservation and rainwater harvesting to solve the problem of water scarcity. Restoring the minor irrigation tanks by desilting process to their original capacity under the "Mission Kakatiya" program is taken up by government of Telangana. In the present study, spatiotemporal variation and trend analysis of groundwater fluctuation levels were analyzed using spatial analyst tool with Inverse Distance Weighted (IDW) and clustered column method trend analysis. A real-world scenario for the study area Padamati Cheruvu tank, Warangal District, Telangana, India is considered to check the efficiency of the model. Groundwater level data for 5 years (2016–2020) of seven observation wells of type Dug well (DW) were considered. To check the fluctuations of groundwater recharge, out of seven observation wells, five were considered in the Influence Zone (IZ) and the remaining two in the Non-Influence Zone (NIZ). The study was performed to analyze the impact of groundwater fluctuation levels annually and seasonally (Pre and Post Monsoon) near the study area. The obtained results highlight the variations of groundwater recharge of seven observation wells, which shows that there is a need to model and analyze the fluctuations of groundwater recharge before and after rejuvenation of tanks.

Keywords Groundwater level fluctuation · Geospatial techniques · Tank rejuvenation · Warangal

---

C. Sree Laxmi Pavani (✉)
Kakatiya Institute of Technology and Science, Warangal, Telangana 506 015, India
e-mail: slpavani@student.nitwac.in

K. V. Reddy
National Institute of Technology, Warangal, Warangal, Telangana 506 004, India

A. K. Dikshit et al. (eds.), *Innovative Trends in Hydrological and Environmental Systems*, Lecture Notes in Civil Engineering 234,
https://doi.org/10.1007/978-981-19-0304-5_56

1 Introduction

In several parts of the world, groundwater resources are used to meet industrial, drinking water supply, and irrigational practices. The groundwater resources are refilled purely by rainfall infiltration. As growing population and rapid industrialization have led groundwater resources to often over-exploit in various segments of the world, especially hard rock areas lead to hasty deterioration in the levels of groundwater. To preserve water resources for sustainable management, with respect to space and time groundwater fluctuation reigns attention throughout the world [1]. Formation of Jal Shakthi Abhiyan was made by Ministry of Jal shakti in May 2019 to reduce the issues related to water scarcity in India. Jal shakti Abhiyan launched 21 districts in Telangana state, under which various water conservation measures were taken to ease the problem of water scarcity in the Warangal District of Telangana. The measures of restoration of tanks involve desilting of tank beds, so as to expand the irrigated area and to cover the command area gap for adaptation of resource conservation-cum-production technologies [2]. Mission Kakatiya is one of the leading programs of the government of Telangana, for rejuvenating the irrigation tanks. The state government claims that it would raise the groundwater level to the range of consumption of its present allocation from the two river basins, namely Krishna and Godavari, and will support the area to become drought-proofed. These tanks have substantial degradation, in the storage of water and reduce the area irrigated over the decades [3]. Even though these systems provide various services such as protective irrigation, drinking water, etc. their involvement and importance declined during the post-independence India. Recognizing the numerous benefits from these traditional systems, tank rehabilitations have been one of the policy priorities at the central as well as in some states [4]. In the shallow aquifers, the groundwater gets restocked annually, and therefore, water levels and their fluctuation play a crucial role in the assessment of groundwater. The degree of saturation and extent of recharge in the hard rock aquifers can be indicated with the analysis of pre-monsoon and post-monsoon groundwater levels [5]. Using GIS spatial analysis, the trend of water table and water level fluctuation can easily be understood by anyone [6]. Geospatial maps are evidenced to be a substantial asset for the analysis of groundwater level fluctuation [7]. In the present study, seven observation (OB) wells near Padamati Cheruvu tank, of Mission Kakatiya (Non-Pilot Phase-II), Devannapet Village, Hasanparthy Mandal, Warangal District, Telangana State were considered to analyze the groundwater fluctuations after restoration of tank. The registered Padamati cheruvu tank command area is 42.72 ha with a desilted quantity of 32,561 m^3. Out of 7 OB wells, five are taken in Influence Zones (IZ) below command area and the other two in Non-Influence Zone (NIZ) away from the command area. The analysis of OB wells is carried out using geospatial techniques and trend analysis to detect the groundwater recharge for the period 2016–2020.

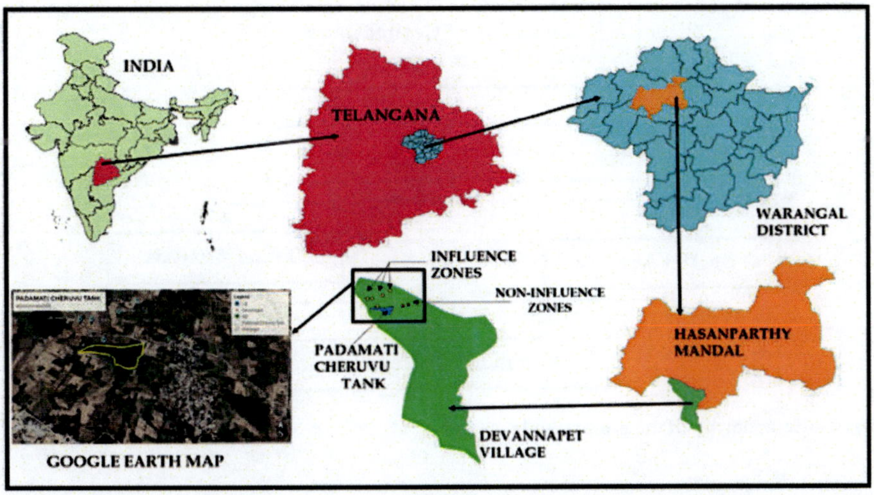

Fig. 1 Study area location map showing Country, State, District, Mandal, and Village Boundaries with Padamati Cheruvu tank and seven observation wells in IZ and NIZ along with Google Earth map

2 Study Area

Padamati cheruvu tank and its nearby observation dug wells are located in Devannapet Village, Hasanparthy Mandal, Warangal District, Telangana State (Fig. 1). The study area Devannapet village lies between 18° 2′ 13.03″ to 18° 1′ 57.97″ N and 79° 29′ 7.97″ to 79° 29′ 41.03″ E with an areal extent of 6.77 km^2. The study area experiences a hot, seasonally wet, tropical savannah climate. The maximum temperature recorded 44 °C in summer and may fall up to 15 °C in winter. Padamati cheruvu tank along with nearby seven observation dug wells, i.e., five in the Zone of Influence and two in the Zone of Non-Influence are shown in Google Earth map. Warangal District falls in the drainage basins of both Godavari and Krishna Rivers. The average annual precipitation of Warangal district is 955 mm. The percentage distribution of rainfall, season-wise is 83% in southwest monsoon, 12% in northeast monsoon, nil percentage in winter, and 5% in summer [8].

3 Methodology

Aquifer data of observation dug wells is collected from Groundwater Department of Warangal Urban area. Monthly groundwater level data, Depth to Water levels (DWL)—below ground level (bgl) of observation (OB) wells of type dug well, under Mission Kakatiya (MK), Non-Pilot Phase-II, from May 2016 to December 2020 is considered for the study area. Five OB wells in the zone of influence and two in

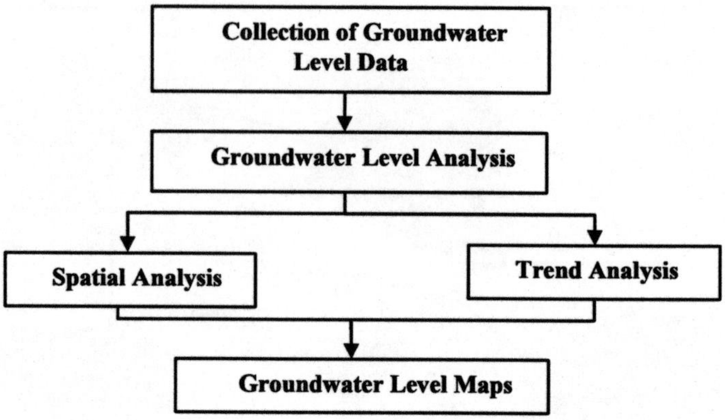

Fig. 2 Methodology of the present study

the zone of non-influence are studied for the years 2016–2020. The depth to water level fluctuation data was arranged annually, for the years 2016–2020 and also for the Post-Monsoon and Pre-Monsoon seasons, to represent and analyze the data. The spatiotemporal data were analyzed annually using spatial and trend analysis. The detailed methodology implemented in the present research work is shown in Fig. 2.

3.1 Spatial Analysis of Groundwater Level Fluctuation

Geographical Information system (GIS) Spatial analysis tool was used to study the spatiotemporal variations of groundwater level fluctuations for a period of 5 years. Using the Inverse Distance method (IDW) analyst tool, seasonal and annual groundwater level fluctuations were analyzed for the mean annual groundwater level data of depth to water level (DWL). IDW method is an interpolation technique which is the simplest method among all other interpolation methods such as krigging, natural neighbor, spline, and trend. IDW method is quite easy to define and understand the results therefore chosen for the analysis of present research work. For interpolation studies, this is the better method for observing the space and time-related fluctuations in groundwater levels. For the years 2016 to 2020 Spatial analysis maps were created for annual and monsoon seasons (pre-monsoon and post-monsoon). The sample points of seven influence and non-influence zones were used to analyze the consequences of adjoining points with respect to the mean values and the depth to water level (DWL) was categorized in the five ranges viz., Very low, low, medium, high, and very high. Very low, low DWLs show that the GW levels are high with sufficient water, medium DWLs show average water availability, and high and very high DWLs show that the water levels are low in this range.

3.2 Trend Analysis of Groundwater Level Fluctuation

Trend analysis is performed using clustered column method in excel to analyze the groundwater level fluctuations of the study area. Using this method graphs are generated to show monthly annual and seasonal (pre-monsoon and post-monsoon) variations of 7 observation wells in the influence and non-influence zones for the years 2016–2020. A clustered column graph presents more than one data series in vertical columns. Each data series shares the same axis labels, so vertical bars are grouped by category of 7 observation wells. Clustered column allows the direct comparison of multiple data series, but they become complex visually. They work best in situations where data points are limited. As the observations wells are limited to 7, this method best suits to analyze the groundwater level variations of the study area near Padamati cheruvu tank. This method allows direct comparison of multiple data series per category and also can display the change over time. The clustered column method is inverted to show the results, as the data was collected for depth to water levels from ground.

4 Results and Discussion

4.1 Annual Groundwater Level Fluctuations Using Spatial Analysis

Geospatial analysis of mean DWL-bgl, during the years 2016 to 2020 were performed using Spatial analyst tool in GIS and are shown for the 7 OB wells data. In the year 2016, mean water level varies from 6.13 m in IZ-2 to 9.8 m in NIZ-1. In 2017, DWL varies from 6.59 m in IZ-2 and 9.94 m in NIZ-2. In 2018, DWL varies from 9.81 m in IZ-2 to 13.55 m in NIZ-2. In 2019, DWL varies from 9.94 m in IZ-4 to 12.41 m in NIZ-1. In 2020, DWL is 4.38 m in IZ-1 and 5.38 m in NIZ-1. The spatial analysis indicates that the DWL in m. bgl lies between 4.38 m and 13.55 m during the years 2016 to 2020 (Table 1). Mean annual and seasonal variation shows the DWL value varying from 4.57 to 13.34 m bgl. Which indicated a large difference in GWL fluctuations. Table-1 shows the data of Warangal District, Dharmasagar basin, Hasanparthy Mandal, Devannapet village, Padamati cheruvu tank of type observation dug well, Depth to water level (DWL) below ground level (bgl) for influence and non-influence zones.

Annual mean Spatial Analysis Map

Annual mean spatial analysis map shows that the DWL has minimum of 8.01 in IZ-3 and is having maximum of 9.97 in NIZ-1. Which states that, there is sufficient groundwater available in the zones of influence, and is low in the zones of non-influence. This is primarily due to recharge of aquifer by the rejuvenation of Padamati cheruvu tank which is nearer to the influence zones 1, 2, and 3. Influence zones 4 and

Table 1 Depth to water level (DWL) below ground level (bgl) for IZ and NIZ OB wells near Padamati cheruvu tank

OB wells in the zone of influence/non-influence	Longitude (DD)	Latitude (DD)	MSL (m)	Total depth (m)	2016 DWL mean	2017 DWL mean	2018 DWL mean	2019 DWL mean	2020 DWL mean	Mean annual DWL	Mean annual pre-monsoon	Mean annual post-monsoon
IZ-1	79.48806	18.03278	284	17.5	6.29	8.09	10.58	12.01	4.38	8.27	11.34	4.60
IZ-2	79.48917	18.03583	276	11	6.13	6.59	9.81	11.3	8.17	8.40	11.43	4.57
IZ-3	79.49083	18.03694	276	16.5	7.02	7.36	10.89	10.05	4.75	8.01	10.93	5.24
IZ-4	79.48667	18.035	272	15.5	8.36	9.08	9.88	9.94	4.92	8.44	11.51	4.98
IZ-5	79.48556	18.03472	279	14	8.85	9.14	10.04	10.18	4.85	8.61	11.10	5.48
NIZ-1	79.49472	18.03333	279	18	9.8	9.85	12.44	12.41	5.38	9.98	12.82	8.50
NIZ-2	79.49361	18.03306	286	15.5	9.18	9.94	13.55	11.15	5.06	9.78	13.34	7.51

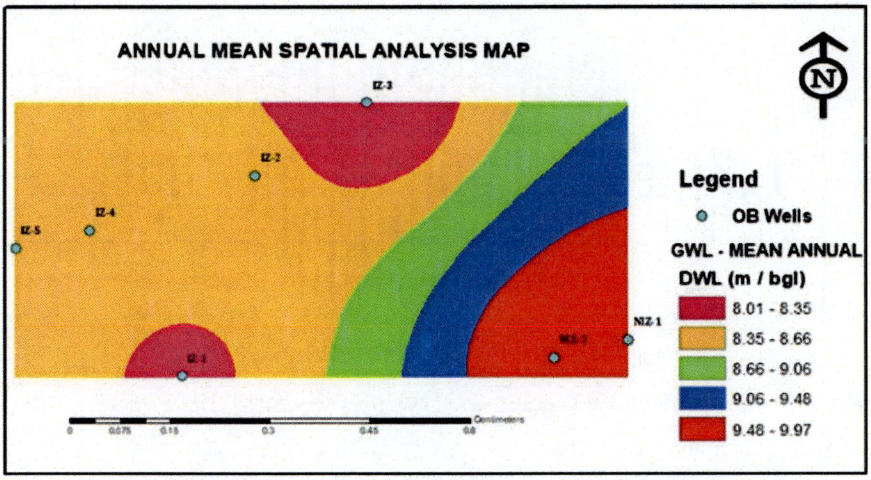

Fig. 3 Annual mean spatial analysis map showing IZ and NIZ of 7 OB wells

5 have moderate recharge and non-influence zones 1 and 2 have low recharge in the aquifer as it is occupied by built-up area (Fig. 3). The spatial variation map clearly indicates that groundwater level is more in influence zones when compared to non-Influence zones. In Non-Influence zones, the GWL is low as there is built-up area and village nearby for which the groundwater usage is more whereas in influence zones groundwater is been utilized for irrigation purposes, which is less and this can be improved by utilizing the water for other purposes, such as public, domestic and agriculture purposes of the study area.

4.2 Annual Groundwater Level Variations Using Trend Analysis

Government of Telangana has taken up Mission Kakatiya tank rejuvenation desilting process in May 2016 as Non-Pilot Phase-II and is continued till November 2016 and was having a total desilted quantity of 32,561 m^3. DWL of 7 observation wells were collected by groundwater department Warangal from November 2016. After the process of desilting, it is observed that in the trend analysis map groundwater recharge has been improved from the year 2016 to 2020.

Monthly Groundwater Level Fluctuations

From the results of monthly groundwater fluctuations, it is observed that, in May 2016, before the process of desilting of soil from observation wells, the wells were almost dry. After desilting process, in November 2016, the groundwater recharge is increased and started fluctuating for different years from 2016 to 2020 (Fig. 4).

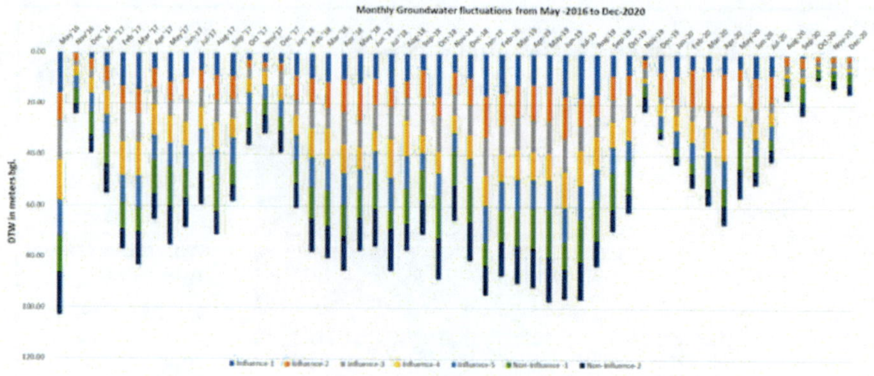

Fig. 4 Monthly groundwater level fluctuation of DWL from May 2016 to Dec 2020

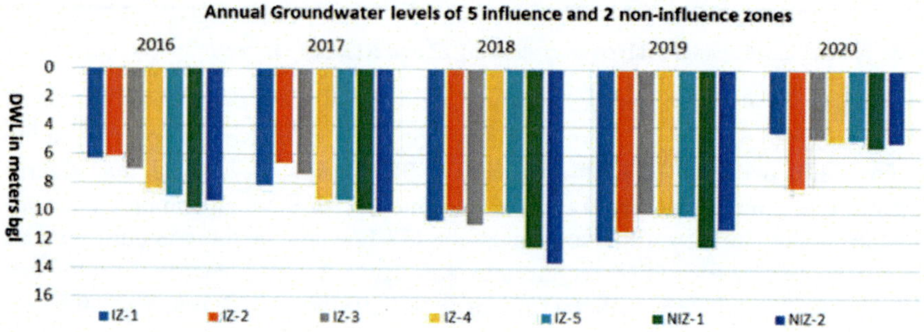

Fig. 5 Annual trend analysis of DWL-bgl (2016–2020)

Annual Mean Trend Analysis

From the Annual mean trend analysis, it is observed that the groundwater recharge was increased in the year 2016, later in the years 2017, 2018, and 2019 it was decreased due to decrease in rainfall. And it has been slowly improved in the year 2020 due to heavy rainfall and floods that occurred in the Warangal region during the period of 2019 and 2020 (Fig. 5).

4.3 *Pre-monsoon Spatial Analysis*

Pre-Monsoon spatial analysis shows that the zones of influence have sufficient groundwater recharge, the recharge is moderate between influence and non-influence zones and is low in the zones of non-influence (Fig. 6).

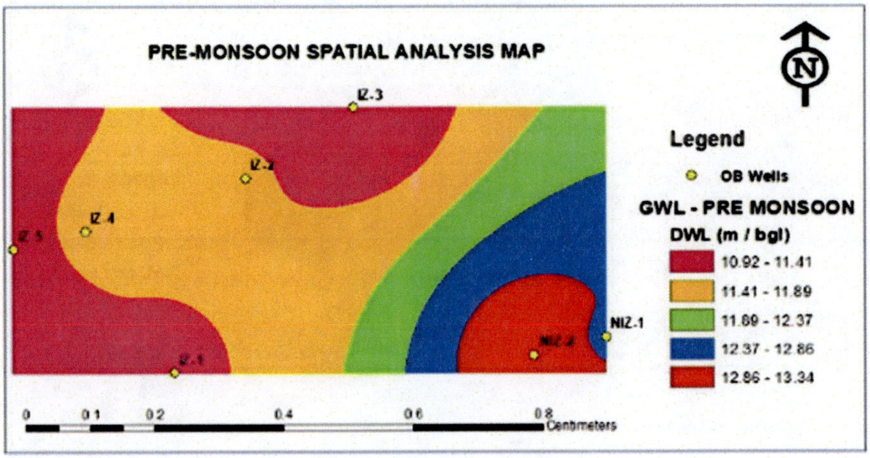

Fig. 6 Pre-monsoon spatial analysis of DWL-bgl (2016–2020)

4.4 Pre-monsoon Trend Analysis

Pre-monsoon trend analysis shows that the groundwater recharge is very low in the years 2016 and 2019. Whereas in the years 2017, 2018, and 2020 there is sufficient groundwater recharge. Therefore, the fluctuation of pre-monsoon season needs to be taken care for improving the groundwater recharge (Fig. 7).

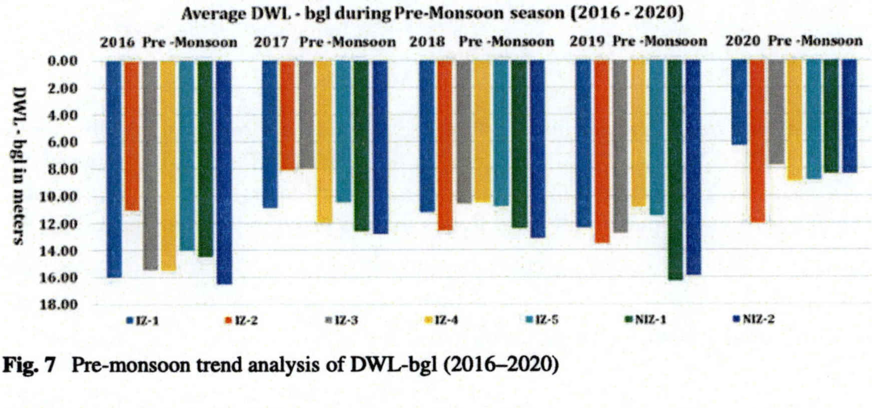

Fig. 7 Pre-monsoon trend analysis of DWL-bgl (2016–2020)

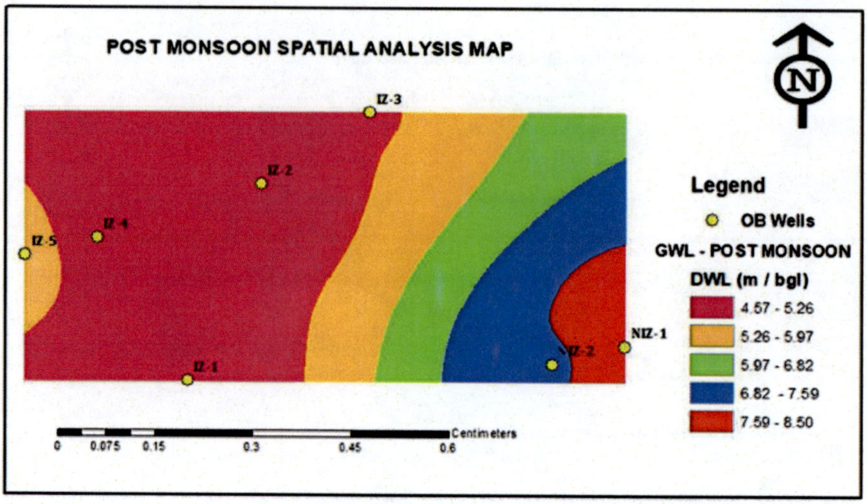

Fig. 8 Post-monsoon spatial analysis of GWL (2016–2020)

4.5 Post-monsoon Spatial Analysis

Post-monsoon spatial analysis also map shows that the zones of influence have sufficient groundwater recharge, the recharge is moderate between influence and non-influence zones and is low in the zones of non-influence (Fig. 8).

4.6 Post-monsoon Trend Analysis

Post-monsoon trend analysis displays that the groundwater recharge is sufficient in the years 2016, 2017, and 2020. Whereas in the years 2018 and 2019 groundwater recharge is less. This indicates that in the post-monsoon season sufficient groundwater recharge is available due to rains and floods that occurred in the study area (Fig. 9).

5 Summary and Conclusions

In the present study, the analysis of groundwater level fluctuations for OB dug wells in the IZ and NIZ near the Padamati Cheruvu rejuvenated tank under Mission Kakatiya was analyzed using spatial and trend analysis. From the results, it is observed that in the years 2016, 2017, and 2020 the months of November, December, and January have adequate groundwater recharge, due to sufficient rainfall. Whereas in remaining

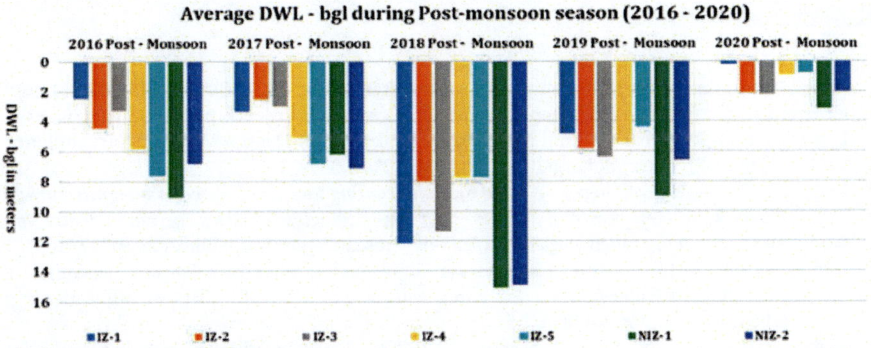

Fig. 9 Post-monsoon trend analysis of DWL (2016–2020)

months of the years 2018, 2019, the groundwater recharge is low due to less precipitation occurred during the year. It is also observed that in the months of winters, i.e., monsoon and post-monsoon the groundwater recharge is high. Whereas in the months of summer and pre-monsoon period the groundwater recharge is less, due to hot climate and high humidity in the study area of Warangal region. From Pre-Monsoon trend analysis, it can be stated that the groundwater is very low in this season and from Post-monsoon trend analysis it is witnessed that, the groundwater recharge is more. Spatial analysis maps indicate that the zones of influence have more groundwater recharge and are less in the zones of non-influence for the various years from 2016 to 2020. As there are many fluctuations in groundwater recharge of the study area, it is necessary to analyze the groundwater recharge of various observation wells continuously near rejuvenation tank, Mission Kakatiya. Remote sensing and GIS play an important role in obtaining useful information for monitoring and managing groundwater resources, conducting spatiotemporal studies, organizing and analyzing the available data to extract meaningful conclusions. Measures are to be taken that, over-exploitation of GW is not taken place, flow of runoff and infiltration is to be taken care of. The significant reasons for high groundwater fluctuation are more runoff, less infiltration, and over-exploitation of groundwater in OB wells. Whenever the groundwater recharge is more, GW can be supplied to nearby irrigation and agricultural fields and also can be supplied to the nearby village people for domestic and industrial purposes for efficient use of groundwater. The analysis of water table fluctuation has been proved to be efficient for assessing groundwater recharge in OB wells near tank rejuvenation.

Acknowledgements The authors are grateful to the Deputy Director, Ground Water Department, Warangal Urban District for their continuous support in providing the groundwater level fluctuations data of observation wells near Padamati Cheruvu tank for the study.

References

1. Anand, B., Karunanidhi, D., Subramani, T., Srinivasamoorthy, K., & Suresh, M. (2020). Long-term trend detection and spatiotemporal analysis of groundwater levels using GIS techniques in Lower Bhavani River basin, Tamil Nadu, India. *Environment, Development and Sustainability, 22*(4), 2779–2800. https://doi.org/10.1007/s10668-019-00318-3
2. Kumar, D., Bassi, N., & Ganguly, A. (2016). *Rejuvenating tanks in Telangana integrated urban water management view project development of methodologies for assessing global water and food security challenges view project*. https://www.researchgate.net/publication/307173207
3. Kumar, M. D. (2018). Mission Kakatiya for Rejuvenating tanks in Telangana: Making it a mission possible. In *Water policy science and politics* (pp. 115–128). Elsevier. https://doi.org/10.1016/b978-0-12-814903-4.00007-5
4. Reddy, V. R., Reddy, M. S., & Palanisami, K. (2018). Tank rehabilitation in India: Review of experiences and strategies. In *Agricultural water management* (Vol. 209, pp. 32–43). Elsevier B. V. https://doi.org/10.1016/j.agwat.2018.07.013
5. Maggirwar, B. C., & Umrikar, B. N. (2011). Influence of various factors on the fluctuation of groundwater level in hard rock terrain and its importance in the assessment of groundwater. *Journal of Geology and Mining Research, 3*(11), 305–317. http://www.academicjournals.org/JGMR
6. Chandra, S., Singh, P. K., Tiwari, A. K., Panigrahy, B. P., & Kumar, A. (2015). Evaluation of hydrogeological factors and their relationship with seasonal water table fluctuation in Dhanbad district, Jharkhand, India. *ISH Journal of Hydraulic Engineering, 21*(2), 193–206. https://doi.org/10.1080/09715010.2014.1002542
7. Tiwari, A. K., Singh, P. K., Chandra, S., & Ghosh, A. (2016). Assessment of groundwater level fluctuation by using remote sensing and GIS in West Bokaro coalfield, Jharkhand, India. *ISH Journal of Hydraulic Engineering, 22*(1), 59–67. https://doi.org/10.1080/09715010.2015.1067575
8. Sagar, R. V. V. (2013). *For official use only Central Ground Water Board, Ministry of Water Resources, Government of India Ground Water Brochure, Warangal, Andhra Pradesh (AAP-2012–13)*. Last viewed on July 07, 2021.

Decolorization of Azo Dye-Contaminated Water using Microbes: A Review

Vamshi Krishna Mukkera and Srivani Katuri

Abstract One of the heavy users of water is the textile industry, which also yields huge volumes of tainted water. The most important contaminants are azo dyes. This tainted effluent water when released into the freshwater will result in the increase of BOD, COD, TOC values of water thereby making it difficult for the aquatic beings in breathing. The toxicity of the effluent may also affect organisms when the water is ingested. Many physical and chemical techniques such as adsorption, coagulation-flocculation, oxidation, filtration, and electrochemical methods may be used for color removal from wastewater among which microbial processes for the treatment of textile wastewater have the advantage of being cost-effective and environmentally friendly. Moreover, in most cases, the final products of the physical or chemical techniques are either highly toxic than their parent toxins or produce more sludge, which is the reason microbial degradation is opted, that produce fewer toxic products and less sludge. Reductive cleavage of Azo bond, leading to the formation of aromatic amines, is the initial reaction during the biological metabolism of Azo dyes. Aromatic types of amines are mineralized under methanogenic conditions. But sulfonated aromatic amines are resistant, require aerobic microbial consortia for their clarification or degradation. Studies that used algae, fungi, bacteria and other microbial consortia along with the application of Advanced Oxidation Processes (AOPs) for the degradation of azo dyes for better results are discussed in this review. It is also discussed how the different conditions in degradation can influence the removal of azo dyes.

Keywords Azo dyes · Effluents · Decolorization · Microbes · Enzymes · Degradation · Analysis

V. K. Mukkera (✉) · S. Katuri
Department of Chemical Engineering, National Institute of Technology, Warangal, Warangal, Telangana 506004, India

A. K. Dikshit et al. (eds.), *Innovative Trends in Hydrological and Environmental Systems*, Lecture Notes in Civil Engineering 234,
https://doi.org/10.1007/978-981-19-0304-5_57

Nomenclature

AD	Azo Dye
AOP	Advanced Oxidation Process
BOD	Biochemical Oxygen Demand
COD	Chemical Oxygen Demand
DO	Dissolved Oxygen
MFC	Microbial Fuel Cell
NADH	Nicotinamide Adenine Dinucleotide
NADH-DCIP	Nicotinamide Adenine Dinucleotide Dichloroindophenol NADPH-Nicotinamide Adenine Dinucleotide Phosphate
TOC	Total Organic Content

1 Introduction

According to [38], intake of water and production of effluents in the textile industry relied on the processes followed in turning fibre into textile. According to [7] the stages of processing and types of dyes applied during this turning determine the characteristics of wastewater like pH, dissolved oxygen, organic and inorganic chemical content. Ubiquitously, 2.8×10^8 kg of dyes used in textile industries are freed in industrial effluents every year [25], in one of the cases of its representation is nearly 0.7 kg of NaCl, 35–65 g of dye and 75–150 L of water are required to dye 1 kg of cotton where tainted effluents produced has 25% of the applied unfixed dyes, their concentration reaching 2000 ppm [6] and its result includes the undesirable effects of azo dyes on surrounding location, such as their inhibitory effect on marine photosynthesis, their ability to lessen dissolved oxygen. They are toxic to flora, fauna. They also affect TOC, BOD, COD, suspended solids, salinity, color, values of pH. The dyes Acid Red 18 and Acid Red 27 not only cause mutations but are also carcinogenic due to the presence of the methyl group. Normally, it is significant to evaluate the toxicity of compounds formed after the process of decolorization of dyes [24]. If the ratio of BOD to COD ranges from 0.3 to 0.6, it denotes that the runoffs have a major portion of non-biologically degradable organic matter [39]. Toxicity is classified based on various methodologies and it is shown in Fig. 1.

Physically, azo dyes are branded of the azo bond whereas chemically, they have gained a property of sucking up the light when rays of the visible spectrum are projected on it [2, 13]. So as to create the diversity of azo dyes, the phenyl groups in the structure shown in Fig. 2 can be replaced with those of the naphthalene groups that also may carry various groups like methyl ($-CH_3$), chloro (–Cl), nitro ($-NO_2$), hydroxyl (–OH), amino ($-NH_2$), and carboxyl (–COOH). During routine aerobic wastewater treatment, azo dyes are not degraded [40]. The azo dyes are electron acceptors. They receive electrons from reduced flavin nucleotides. Redox mediators enhanced the reduction of azo dyes [26]. A Ds reduction results in the genesis

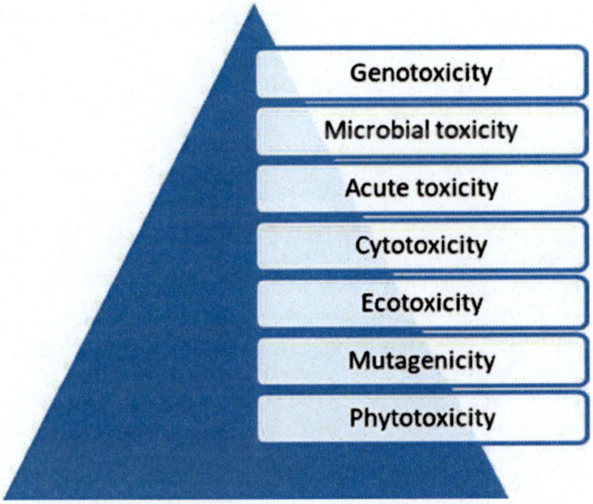

Fig. 1 Classification of toxicity

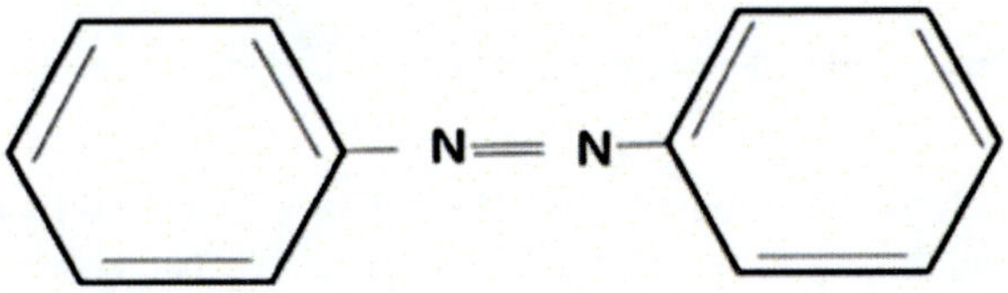

Fig. 2 Assembly of a typical azo dye

of aromatic amines with the exception of a few aromatic amines substituted with hydroxyl and carboxyl groups. They can be degraded under methanogenic or anaerobic conditions. Whereas, the exceptions can be readily degraded, aerobically [27]. Hence a combination of both the aerobic as well as anaerobic processes of degradation can be used for the reduction of azo dyes. There are a number of studies on the biological degradation of azo dyes in the effluent wastewater. Many researchers focused mainly on the usage of bacteria and fungi among microbes, out of which bacteria were adopted because of their high activity of degradation, their ability to degrade a large number of azo dyes and high flexibility of usage [15].

The present review focus on the technologies involving microbes that are used for the decolorization of azo dyes from the textile effluents. Many numbers of investigations indicated that the dyes used in textile industries and their runoffs have toxic effects on the growth rates and phytology of many plants which have vital ecological responsibilities, like setting up a natural territory for wildlife, prevention of soil from getting eroded and providing organic matter that is required for soil fertility [15]. Hence it is essential for the remediation of the effluents before being released into the environment, which can be carried out by using microbes. Usage of microbes is an appealing substitute for bioremediation processes in the treatment of textile industry effluents. Methods that involve the biotic species are ecologically friendly, yield lesser sludge than physicochemical processes and are cost-effective. The efficacy of

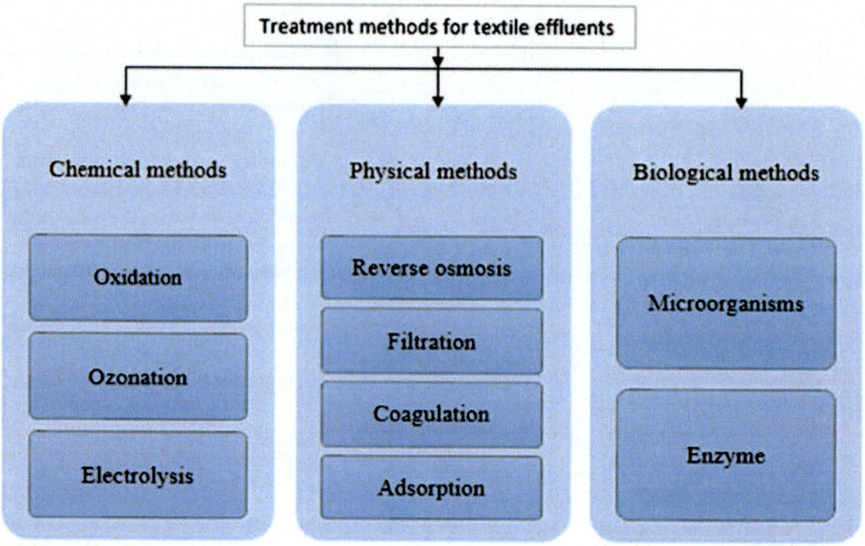

Fig. 3 Various treating methods for the degradation of dyes in tainted water effluent

different procedures in the decoloration relies on its numerous substituent groups and the physical–chemical features of runoffs, pH, temperature, and the existence of other pollutants [41].

1.1 Methods of Degradation

To remove the dyes, a number of physical and chemical methods were used. Nonetheless, their application has got many obstacles such as cost, energy. These methods cannot degrade some of the resisting azo dyes and produce serious ecological problems [17]. Azo dyes are complicated aromatic composites with significant architectural variety. Decolorization of synthetic dyes by traditional biological processes is not so effective [20]. Due to strict laws, many investigations are developed and efficient methods for handling wastewater are found, where various microbes were also adopted for developing the methods of treatment of textile effluents. Different methods of treatment are shown in Fig. 3.

1.2 Physical Azo Dye Reduction

The dyes having sulfur group can be removed by the physical methods like coagulation and flocculation but it has less effect on acid and direct dyes. Other methods like adsorption (which is used for the higher amount of removal and for large number of azo dyes), the process in which the substance used for adsorption must have a high affinity to azo dyes and the chance of adsorption to happen again should be high, i.e., for adsorbent regeneration [34]. Various adsorbents are used, one of which is powdered activated carbon (PAC) [33]. But it has a disadvantage, as it is highly expensive. Some other adsorbents like Kaolin, Bentonite, polymeric resins, peat which are less costly are also used. Some carbon-rich materials like corn, wheat, maize stalks were also used. But the application of the above materials have drawbacks with the regeneration and high amount of sludge production [3]. There are many methods of filtration like ultrafiltration, reverse osmosis. Membranes are used in textile industries for the removal of dyestuff and to reduce the color, BOD and COD. External condition like temperature is to be considered for the selection of filter. But usage of membranes has disadvantages like the high cost of investment, membrane fouling, production of secondary waste [15].

1.3 Chemical Azo Dye Reduction

Use of Advanced Oxidation Process (AOP)

In AOP hydroxyl radicals are generated so as to destroy the chemical compounds by oxidation of pollutants rapidly [30]. There are two routes to generate radicals,

1. The usage of permanganate (MnO_4), ozone (O_3), hydrogen peroxide (H_2O_2) along with the use of UV rays.
2. The usage of oxygen at moderate to high temperatures as found in Wet Air Oxidation (WAO) processes.

In Advanced Oxidation Processes (AOP), some of the catalysts like TiO_2, ZnO_2, Mn and Fe are employed along with oxidizing agents like O_3 and H_2O_2. Ozone is used because of its effect in reacting with a numerous azo dyes, and also better decolorization efficiencies [1]. Application of ozone process is limited because of its short life span, not effective to dyes which are insoluble in water, low COD reduction and its high cost [3]. Fenton's oxidation has got many advantages like high degradation efficiency, benign process, less costly materials [5]. But this process has a drawback of production of high sludge content [33]. The AOPs can either be in the homogenous phase or in heterogenous phase. Where in heterogenous phase, catalysts which are used are in solid phase and contaminants in the liquid phase [5]. Catalysts produce the OH^- to destroy the dye compounds [3]. AOPs do not produce many secondary pollutants that need additional detoxification. AOPs consume less amount of energy. But some of its drawbacks are higher operating costs, scaling up of processes to the industrial level is not [5].

1.4 Biotic Methods

The use of microbes to deal with the biodegradation is important in the treatment of wastewater. Various biotechnological methods are used to treat the azo dye-contaminated water effluents in an environmentally friendly way. Even though the dyes are very obstinate to the degradation, these methods are used for complete degradation and also have the advantages like ecologically supportive end products, less costly, generating less sludge, non-toxic end products. The following microbes can be used in degradation processes. They are Bacteria, fungi, Yeasts, Actinomycetes, Algae. The degradation products such as aromatic amines can inhibit bacterial degradation [32]. However, such inhibition action is not possible for fungi. Moreover, fungi can be used for degrading complex molecules along with some ligninolytic enzymes which include laccase, manganese peroxidase (MnP) and lignin peroxidase (LiP) [21] and Tyrosinases [4] riboflavin reductases and NADH-DCIP reductases [42]. Fungi can also be used along with Catalysis processes.

Decolorization by Bacteria

The decolorization using bacteria proceeds in two steps. The first step is cleaving of the dye linkages reductively, producing the colorless, toxic aromatic amines whereas the second step is the degradation of the aromatic amines. Generally, Bacteria are used for degradation under aerobic conditions. If the conditions are anoxic, the strains like *Pseudomonas* sp. and *Proteus mirabilis* are used for degradation [29]. A bacterial microbe *Sphingomonas xenophaga* is used to decolorize the dye Mordant yellow 3. The effect is increased using glucose whereas, it is decreased when *Pseudomonas leuteola, Aeromonas hydrophila* and some additional mixed cultures were used [29]. The decolorization is analyzed using HPLC and mass spectrometry after the degradation of reactive red 22 by *Pseudomonas leuteola*, which indicated the presence of two aromatic amines. Following Table 1 gives information about the work done by some of the researchers.

Decolorization by Fungi

The fungi are present everywhere in the universe. It can be found in the soil, plants, organic wastes etc. Their endurance is based on the adaptation of their metabolism to the ever-changing content of carbon and nitrogen sources. Many of the reports have focused on the degradation using bacteria. But decolorizing of aromatic amines inhibit the action of bacteria [42]. Hence fungi are used for decolorization. Many works on the treatment of azo dye-contaminated water are concentrated on the white-rot fungi [35] among which *Phanerochaete chrysosporium* is studied widely, whereas some other white-rot fungi like *Trametes (Coriolus) versicolor*, *Bjerkandera adusta, Aspergillus ochraceus*, species of Pleurotus, Phlebia, and different other fungi [42] are also used. *Pleurotus ostreatus*, *Pichia* sp., *Penicillium* sp., and *Candida tropicalis* are used for the decolorization of azo dyes by adsorption and/or degradation [42]. However, it has some drawbacks like nitrogen limiting conditions, lengthy growth

Table 1 Degradation using bacteria

Microbe	Dye	References
Pseudomonas sp.	Reactive Blue 13	[42]
Corynebacterium glutamicum	Reactive Green 19 A	[35]
Enterobacter EC3	Reactive Black 5	[42]
Mutant Bacillus ACT2	Congo Red	[42]
Lactobacillus acidophilus and *Lactobacillus fermentum*	Water and oil soluble azo dyes	[35]
Aeromonas hydrophila	Reactive Red 141	[42]
Bacillus VUS	Navy Blue 2G	[42]
Citrobacter sp. *CK3*	Reactive Red 180	[35]
Acinetobacter calcoaceticus	Direct Brown MR	[42]
Pseudomonas sp. *SU-EBT*	Congo red	[42]
Rhizobium radiobacter	Reactive Red 141	[35]
Brevibacillus latarosporus	Golden Yellow ER	[42]

cycles. It is difficult to preserve fungi in bioreactors for longer time [8]. Following Table 2 gives information about the work done by some of the researchers.

Decolorization by Yeast

The yeasts are used since they are powerful adsorbents and can be used for a larger quantity of dye presence [35]. The treatment of azo dye-contaminated water using yeast is concerned mainly with biosorption. Yeast cultures like *Issatchenkia occidentalis* [43] are utilized generally for the degradation of ADs. Moreover, the above yeast also shows the bioaccumulation of reactive dyes like Remazol Blue, Black B and Red RB during the growth of it [9]. Moreover, some of the works also showed

Table 2 Decolorization of azo dyes using fungi

Microbe	Dye	References
Trametes versicolor	Reactive Black 5	[37]
Trametes pubiscens	Remazol Brilliant Blue	[35]
Trametes hirsute	Acid Blue, Reactive Green	[44]
Phanerochaete chrysosporium	Direct Red 80	[44]
Phanerochaete chrysosporium	Astrazon Red FBL	[35]
Aspergillus flavus	Acid Red 151	[44]
Trametes trogii	Remazol Brilliant Blue R	[44]

that yeast can act as the important dye adsorbent and can uptake high amount of dye concentration.

Decolorization by Algae

Algae, which are photosynthetic are present everywhere. A study suggests that the degradation of azo dyes by algae is possible through a form of azoreductase [11]. The color reduction of dye-contaminated wastewater effluent by algae follows the different mechanisms like complete use of chromophores for the production of algae and the adsorption of chromophores. It was concluded from the studies that nearly 30 of the azo dyes can be degraded using the algae Chlorella. The future outcomes might also intend that algae play a vital part in the degradation in stabilizing ponds, and this process can be less costly [37].

2 Conditions Affecting Decolorization

The dyes are utilized for several purposes like coloring, processing of fibre. The runoffs of water from dyeing processes are of different chemical compositions which include various organic compounds, Sulfurous compounds, nutrients, salts, toxic substances [19, 41]. Various physicochemical parameters like the level of mixing, addition or removal of oxygen, variation of temperature, variation in acidic conditions, the structure of dye, concentration of dye, supplementing different carbon and nitrogen sources affect the microbial degradation. The presence of electron donors also affects microbial degradation.

2.1 Effect of Oxygen and Mixing

The degradation of azo dye effluent occurs in anaerobic, facultatively anaerobic, aerobic conditions. It is studied that the decolorization process of azo dye-contaminated water takes place in perfectly anaerobic conditions [32]. The same results were observed when the microbial strains such as *Pseudomonas luteola, Proteus mirabilis, Pseudomonas* sp. *SUK1, Micrococcus glutamicus* NCIM-2168 [37] were used. It was studied that, as the mixing is augmented, the degradation of azo dyes is reduced as there is an increase in oxygen content thereby increasing DO [12]. Moreover, the impact of the presence of oxygen on the reducing process can be irreversible [9] and the intermediates obtained through decolorization of simple dyes lead to the opening of their aromatic chains [12]. It was also stated that, in the process of removal of color from Reactive Red 22 using Escherichia coli, when the agitation speed is zero, the DO content in the culture dropped to nearly zero, but as the speed is increased to 200 rpm, DO content increased to 0.5 mg per L. No notable decolorization was observed.

2.2 Effect of Carbon and Nitrogen Sources Addition/removal

The decolorization process of azo dyes is very difficult when the amount of carbon is less. AD contaminated water has less carbon content and henceforth the addition of sources of carbon or nitrogen should be done for the degradation process [14]. It was stated that anaerobic consortia, acidogenic bacteria convert the soluble substrates, such as carbohydrates, to volatile organic acids or alcohols [38]. Whereas, the addition of nitrogen sources like beef extract, peptone, urea, yeast extract generates NADH that is used as an electron donor for the effective reduction of azo dyes to amines [13]. In some reports, it was found that the presence of lignocellulose substrate increases decolorization. It was observed that the presence of lignocellulosic substrates enhances decolorization by effective production of lignolytic enzymes. Similar results are obtained when microbes like *Comamonas* sp. *UVS*, showed active decolorization of the dye Direct Blue GLL along with the generation of lignolytic enzymes.

2.3 Effect of Temperature

The activity of the microorganism is dependent on the temperature of the environment. Hence temperature plays an important role in the treatment of wastewater. A few studies are on the activation energy required for the microbial degradation [15]. It is reported that the physiological temperature changes bring about a change in activation energy [44]. It was observed that the rate of degradation was increased with increase in temperature up to a particular temperature and later it decreased [38]. But, some of the enzymes produced by the bacteria can be thermally stable even at high temperatures [31].

2.4 Effect of pH

The optimum value for the pH (since it has a large effect on degradation of azo dyes) for degradation of azo dyes is in the range of 6.5 and 10 for the microorganism *Proteus mirabilis*. The values of the pH should be neutral for the decolorization to be high whereas decolorization decreases when in highly acidic medium or highly alkaline medium [38]. Arica and Bayramoglu [4] stated that, as the value of pH was reduced, the decolorization of Reactive Red 12 dye by fungi sajor-caju was improved. In the process of degradation of azo dyes, the amines are formed which results in an increase of the pH value which showed a negative effect decolorization. But [12] obtained that the rate of reduction of dyes into amines increased 2.5 times with the increase of pH from 5 to 7 for *Pseudomonas luteola*. For the degradation of dye

Reactive Red 190 by isolated *Citrobacter* sp. *C3*, the reaction was effective at acidic conditions [44].

2.5 Effect of Dye Concentration

Many studies state that with the rise in dye concentration, there is a reduction in the decolorization rate. The reason given for it is the poisonous consequence of dyes on microbes, blocking of active sites of azoreductase by excess dye molecules with different structures [37]. Saratale et al. [36] in their study stated that azo dyes having the sulphonic group on their rings inhibited the growth of microbes. But, [37] stated that the use of cocultures of the microbes can reduce the inhibition of growth due to the dye concentrations.

2.6 Effect of Dye Structure

The structures of the dyes are very diverse because of the presence of several functional groups. It was identified that the degradation rate is higher in dyes of simpler structures than those of the ones having $-SO_3H$, $-SO_2NH_2$ in the phenyl ring [23]. The functional group of azo dyes with a higher electronic density might be not favorable to the second electron transfer forming the dianion, leading to low or no capability for decolorization [31]. Even the chemical structure of the compound may affect the decolorization rate [42].

2.7 Effect of Electron Donor

For decolorization, the azo dyes have to be reduced. But if the organic content in the effluent is low that complete decolorization process is not possible by the microbes, there will be a need for the external addition of electron donors so that reduction increases and it enhances the complete process of decolorization [19]. Hence there is a need to study how the external addition of electron donors affect the rate of degradation [22]. For the dye Reactive Orange 16 using isolated *Bacillus* sp. *ADR*. Ghodake et al. [19] reported that some of the electron contributors like sodium formate, sodium acetate, sodium pyruvate, sodium succinate, sodium citrate are added to improve the rate. Brás et al. [10] stated that the addition of glucose or acetate ions enhances the reduction. Methanol is a perfect electron donor [18]. Moreover, the addition of electron donors not only favor the reduction but also enhances enzymatic action for the decolorization process [31]. The enzymes can be either reductive or oxidative. One of the major used reductive enzymes is azoreductase and the oxidative enzyme is

Table 3 Effect of reducing enzymes

Theory	References
They can be used for the studies of environmental biotechnology	[43]
Bacillus subtilis is induced with electrophiles (catechol, diamide)	[43]
Enzymes like Xenophilus azovoranss K46F, Pigmentiphaga kulle K24, Enterococcus faecalis, Staphylococcus aureus, Esherichia colii, *Bacillus* sp. OY1 to 2, and Rhodobacter sphaeroids are studied	[43]
NADH-DCIP reductases are enzymes that are formed in the bacterial and fungi cultures and are used for the detoxification	[36]
Riboflavins role for the degradation was found out for the dyes Mordant Yellow 10 with anaerobic granulated sludge and Brilliant Blue G by a microbial conglomerate	[24]
Dye Scarlet Ruby by *P. vulgaris* and *Micrococcus glutamicus*, Dye Reactive Green 19A by M. glutamicus Dye Navy Blue Gl by isolated *Bacillus* sp.	[36]

lignin peroxidase. Table 3 gives the compilation of various works showing the effect of reducing enzymes.

The mechanism of azo dye oxidation by peroxidases such as LiP probably involves the oxidation of the phenolic group to produce a radical at the carbon bearing the azo linkages. The water attacks this phenolic carbon to cleave the molecule producing phenyldiazene, and the phenyldiazene can then be oxidized by a one-electron reaction generating N_2 [18]. Laccases are the compounds that are made of copper on the aromatic rings and hence named based on copper position. They act as the catalyst for the metabolism and degradation of various phenols and azo dyes [18]. Table 4 gives the compilation of various works showing the effect of reducing enzymes.

Table 4 Effect of oxidizing enzymes

Theory	References
LiP is used in the degradation of the multiaromatic hydrocarbons, ADs	[16]
It was detected that decontaminated enzymes have a better capability to degrade various ADs	[19]
Bacteria *Micrococcus glutamicus, Pseudomonas desmolyticum* can catalyze the azo dye oxidation	[36]
Enzyme laccase is used in the decolorization of dye CI reactive Orange 16	[19]
Moreover, refinement of the enzyme "veratryl alcohol oxidase" from *Commamonas* sp. *UVS* in decolorization of Red HE7B and Direct Blue GL has been described	[19]

3 Methods of Analysis

UV–vis spectroscopy is the method of analysis that gets the information of decolorization based on the appearance of peaks. The color level in the effluents of industries or the mixture of dyes can be found using American Dye Manufacturers Institute tristimulus filter method (ADMI 3WL). Another analytical procedure is "thin layer chromatography (TLC)" which is used to distinguish closely related dyes and also to find out the decolorization [28]. Another technique used for the analysis is "high performance liquid chromatography (HPLC)" which utilizes high pressure and tight packing [12]. The degradation can be found using this analysis. And the widely used analytical method is Fourier transform infrared spectroscopy (FTIR) which explains the interactions of various functional groups in ADs after degradation and also determine decolorization rate. The mass spectrometry is used for finding the spectra of the mixture, molecular weights and structural information. Based on the state of the mixture, it can be termed liquid chromatography-mass spectrometry (LC–MS) or gas chromatography-mass spectrometry (GC–MS) [37]. Moreover, nuclear magnetic resonance (NMR) analytical method is used for the study to get the structural information of the compounds. It is also possible to find out the decolorization by calculating values of BOD, COD and TOC [37].

4 Future Work

In some situations, the final products of degradation are more toxic than the parent dyes. Hence there is a need for dye mineralization after the dye degradation, a process is to be evolved for a maximum number of dyes. Moreover, sophisticated methods are to be used for the degradation. This can be obtained by combining the AOPs along with the microbial degradation. Novel Azo dyes have to be designed by physical or chemical restructuring so that more efficient degradation is possible and less effluent waste is released from the textile industries.

5 Conclusions

Since the rules and regulations in the treatment of wastewater have become strict, there should be an efficient way to degrade azo dyes. This can be achieved using microbial degradation since it is less costly and generate less sludge. The effluents when released into the freshwater, result in the increase of BOD, COD, TOC values of water thereby making it difficult for the aquatic beings in breathing and the toxicity of the effluent may also affect organisms when the water is ingested. When there is bio metabolism of azo dyes, aromatic amines are formed after the reduction of azo bonds. The bacteria are used in the degradation since they have a quick growth rate, elevated hydraulic retention stretch. Not all azo dyes are biodegradable. Hence a new type of processes of AOPs and microbial Fuel Cells combined with microbial degradation is to be established. The literature presented in this paper explained the various conditions affecting the degradation process using microbes like the presence or absence of oxygen or the oxidation or reduction processes used to degrade azo dyes. It is stated that the degradation is done by reducing the azo bonds of dyes or cleaving reactions of azo dyes. Work can be done by, (a) Choosing appropriate enzymes for biodegradation, (b) industrial effluents. Optimization studies can be done for microbial degradation of different azo dyes. From many research studies, it is evident that microbial degradation made that they are very promising to conventional treatment methods.

References

1. Alaton, I., Balcioglu, I. A., & Bahnemann, D. (2002). Advanced oxidation of a reactive dyebath effluent: Comparison of O_3, H_2O_2/UV-C and TiO_2/UV-A processes. *Water Research, 36*. https://doi.org/10.1016/s0043-1354(01)00335-9
2. Ali, N., Hameed, A., & Ahmed, S. (2010). Role of brown-rot fungi in the bioremoval of azo dyes under different conditions. *Brazilian Journal of Microbiology, 41*(4). https://doi.org/10.1590/S1517-83822010000400009
3. Anjaneyulu, Y., Sreedhara Chary, N., & Samuel Suman Raj, D. (2005). Decolourization of industrial effluents—Available methods and emerging technologies—A review. *Reviews in Environmental Science and Biotechnology, 4*(4). https://doi.org/10.1007/s11157-005-1246-z
4. Arica, M. Y., & Bayramoglu, G. (2007). Biosorption of reactive Red-120 dye from aqueous solution by native and modified fungus biomass preparations of Lentinus sajor-caju. *Journal of Hazardous Materials, 149*(2). https://doi.org/10.1016/j.jhazmat.2007.04.021
5. Atalay, S., & Ersöz, G. (2016). *Novel catalysts in advanced oxidation of organic pollutants*. https://doi.org/10.1007/978-3-319-28950-2
6. Babu, B. R., Parande, A. K., Raghu, S., & Kumar, T. P. (2007). *Cotton textile processing: Waste generation and effluent treatment*. https://www.cotton.org//journal/2007-11/3/141.cfm
7. Banat, I. M., Nigam, P., Singh, D., & Marchant, R. (1996). Microbial decolourization of textile-dye-containing effluents: A review. *Bioresource Technology, 58*(3). https://doi.org/10.1016/S0960-8524(96)00113-7
8. Bhosale, S., Saratale, G., & Govindwar, S. (2006). Biotransformation enzymes in Cunninghamella blakesleeana (NCIM-687). *Journal of Basic Microbiology, 46*(6). https://doi.org/10.1002/jobm.200510117
9. Bragger, J. L., Lloyd, A. W., Soozandehfar, S. H., Bloomfield, S. F., Marriott, C., & Martin, G. P. (1997). Investigations into the azo reducing activity of a common colonic microorganism. *International Journal of Pharmaceutics, 157*(1). https://doi.org/10.1016/S0378-5173(97)00214-7
10. Brás, R., Isabel A. Ferra, M., Pinheiro, H. M., & Gonçalves, I. C. (2001). Batch tests for assessing decolourisation of azo dyes by methanogenic and mixed cultures. *Journal of Biotechnology, 89*(2–3). https://doi.org/10.1016/S0168-1656(01)00312-1
11. Casieri, L., Varese, G. C., Anastasi, A., Prigione, V., Svobodová, K., Filippelo Marchisio, V., & Novotný, Č. (2008). Decolourization and detoxication of reactive industrial dyes by immobilized fungi Trametes pubescens and Pleurotus ostreatus. *Folia Microbiologica, 53*(1). https://doi.org/10.1007/s12223-008-0006-1

12. Chang, J. S., Chou, C., Lin, Y. C., Lin, P. J., Ho, J. Y., & Lee Hu, T. (2001). Kinetic characteristics of bacterial azo-dye decolourization by Pseudomonas luteola. *Water Research, 35*(12). https://doi.org/10.1016/S0043-1354(00)00581-9
13. Chang, J. S., Kuo, T. S., Chao, Y. P., Ho, J. Y., & Lin, P. J. (2000). Azo dye decolourization with a mutant Escherichia coli strain. *Biotechnology Letters, 22*(9). https://doi.org/10.1023/A:1005624707777
14. Dhanve, R. S., Shedbalkar, U. U., & Jadhav, J. P. (2008). Biodegradation of diazo reactive dye Navy blue HE2R (reactive blue 172) by an isolated *Exiguobacterium* sp. RD3. *Biotechnology and Bioprocess Engineering, 13*(1). https://doi.org/10.1007/s12257-007-0165-y
15. dos Santos, A. B., Cervantes, F. J., & van Lier, J. B. (2007). Review paper on current technologies for decolourisation of textile wastewaters: Perspectives for anaerobic biotechnology. *Bioresource Technology, 98*(12). https://doi.org/10.1016/j.biortech.2006.11.013
16. Durán, N., & Esposito, E. (2000). Potential applications of oxidative enzymes and phenoloxidase-like compounds in wastewater and soil treatment: A review. *Applied Catalysis B: Environmental, 28*(2). https://doi.org/10.1016/S0926-3373(00)00168-5
17. Forgacs, E., Cserháti, T., & Oros, G. (2004). Removal of synthetic dyes from wastewaters: a review. *Environment International, 30*(7). https://doi.org/10.1016/j.envint.2004.02.001
18. Georgiou, D., Metallinou, C., Aivasidis, A., Voudrias, E., & Gimouhopoulos, K. (2004). Decolourization of azo-reactive dyes and cotton-textile wastewater using anaerobic digestion and acetate-consuming bacteria. *Biochemical Engineering Journal, 19*(1). https://doi.org/10.1016/j.bej.2003.11.003
19. Ghodake, G. S., Talke, A. A., Jadhav, J. P., & Govindwar, S. P. (2009). Potential of Brassica juncea in order to treat textile-effluent-contaminated sites. *International Journal of Phytoremediation, 11*(4). https://doi.org/10.1080/15226510802429518
20. Ghoreishi, S. M., & Haghighi, R. (2003). Chemical catalytic reaction and biological oxidation for treatment of non-biodegradable textile effluent. *Chemical Engineering Journal, 95*(1–3). https://doi.org/10.1016/S1385-8947(03)00100-1
21. Gomi, N., Yoshida, S., Matsumoto, K., Okudomi, M., Konno, H., Hisabori, T., & Sugano, Y. (2011). Degradation of the synthetic dye amaranth by the fungus Bjerkandera adusta Dec 1: Inference of the degradation pathway from an analysis of decolourized products. *Biodegradation, 22*(6). https://doi.org/10.1007/s10532-011-9478-9
22. He, Q., & Sanford, R. A. (2003). Characterization of Fe (III) reduction by chlororespiring Anaeromxyobacter dehalogenans. *Applied and Environmental Microbiology, 69*(5). https://doi.org/10.1128/AEM.69.5.2712-2718.2003
23. Hsueh, C. C., Chen, B. Y., & Yen, C. Y. (2009). Understanding effects of chemical structure on azo dye decolourization characteristics by Aeromonas hydrophila. *Journal of Hazardous Materials, 167*(1–3). https://doi.org/10.1016/j.jhazmat.2009.01.077
24. Jadhav, S. B., Phugare, S. S., Patil, P. S., & Jadhav, J. P. (2011). Biochemical degradation pathway of textile dye Remazol red and subsequent toxicological evaluation by cytotoxicity, genotoxicity and oxidative stress studies. *International Biodeterioration and Biodegradation, 65*(6). https://doi.org/10.1016/j.ibiod.2011.04.003
25. Jin, X. C., Liu, G. Q., Xu, Z. H., & Tao, W. Y. (2007). Decolourization of a dye industry effluent by Aspergillus fumigatus XC6. *Applied Microbiology and Biotechnology, 74*(1). https://doi.org/10.1007/s00253-006-0658-1
26. Kudlich, M., Keck, A., Klein, J., & Stolz, A. (1997). Localization of the enzyme system involved in anaerobic reduction of azo dyes by *Sphingomonas* sp. Strain BN6 and effect of artificial redox mediators on the rate of azo dye reduction. *Applied and Environmental Microbiology*. https://doi.org/10.1128/aem.63.9.3691-3694.1997
27. Loidl, M., Hinteregger, C., Ditzelmiiller, G., Ferschl, A., & Streichsbier, F. (1990). *Hicrnbiology*.
28. Mohana, S., Shrivastava, S., Divecha, J., & Madamwar, D. (2008). Response surface methodology for optimization of medium for decolourization of textile dye Direct Black 22 by a novel bacterial consortium. *Bioresource Technology, 99*(3). https://doi.org/10.1016/j.biortech.2006.12.033

29. Moosvi, S., Keharia, H., & Madamwar, D. (2005). Decolourization of textile dye Reactive Violet 5 by a newly isolated bacterial consortium RVM 11.1. *World Journal of Microbiology and Biotechnology, 21*(5). https://doi.org/10.1007/s11274-004-3612-3
30. Muruganandham, M., Suri, R. P. S., Jafari, S., Sillanpää, M., Lee, G.-J., Wu, J. J., & Swaminathan, M. (2014). Recent developments in homogeneous advanced oxidation processes for water and wastewater treatment. *International Journal of Photoenergy*. https://doi.org/10.1155/2014/821674
31. Pearce, C. I., Lloyd, J. R., & Guthrie, J. T. (2003). The removal of colour from textile wastewater using whole bacterial cells: A review. *Dyes and Pigments, 58*(3). https://doi.org/10.1016/S0143-7208(03)00064-0
32. Qu, Y., Shi, S., Ma, F., & Yan, B. (2010). Decolourization of reactive dark blue K-R by the synergism of fungus and bacterium using response surface methodology. *Bioresource Technology, 101*(21). https://doi.org/10.1016/j.biortech.2010.05.025. Ramakrishna, K. R., & Viraraghavan, T. (1997). Dye removal using low-cost adsorbents. *Water Science and Technology, 36*(2–3). https://doi.org/10.1016/S0273-1223(97)00387-9
33. Robinson, T., McMullan, G., Marchant, R., & Nigam, P. (2001). Remediation of dyes in textile effluent: A critical review on current treatment technologies with a proposed alternative. *Bioresource Technology, 77*(3). https://doi.org/10.1016/S0960-8524(00)00080-8
34. Sadhasivam, S., Savitha, S., Swaminathan, K., & Lin, F. H. (2009). Metabolically inactive Trichoderma harzianum mediated adsorption of synthetic dyes: Equilibrium and kinetic studies. *Journal of the Taiwan Institute of Chemical Engineers, 40*(4). https://doi.org/10.1016/j.jtice.2009.01.002
35. Safarikova, M., Ptáčková, L., Kibriková, I., & Šafařík, I. (2005). Biosorption of water-soluble dyes on magnetically modified Saccharomyces cerevisiae subsp. uvarum cells. *Chemosphere, 59*(6). https://doi.org/10.1016/j.chemosphere.2004.10.062
36. Saratale, R. G., Saratale, G. D., Chang, J. S., & Govindwar, S. P. (2009). Ecofriendly degradation of sulfonated diazo dye C.I. reactive green 19A using micrococcus glutamicus NCIM-2168. *Bioresource Technology, 100*(17). https://doi.org/10.1016/j.biortech.2009.03.051
37. Saratale, R. G., Saratale, G. D., Kalyani, D. C., Chang, J. S., & Govindwar, S. P. (2009). Enhanced decolourization and biodegradation of textile azo dye Scarlet R by using developed microbial consortium-GR. *Bioresource Technology, 100*(9). https://doi.org/10.1016/j.biortech.2008.12.013
38. Saratale, R. G., Saratale, G. D., Chang, J. S., & Govindwar, S. P. (2011). Bacterial decolourization and degradation of azo dyes: A review. *Journal of the Taiwan Institute of Chemical Engineers, 42*(1). https://doi.org/10.1016/j.jtice.2010.06.006
39. Savin, I. I., & Butnaru, R. (2008). Wastewater characteristics in textile finishing mills. *Environmental Engineering and Management Journal, 7*(6).
40. Shaul, G. M., Holdsworth, T. J., Dempsey, C. R., & Dostal, K. A. (1991). Fate of water-soluble azo dyes in the activated sludge process. *Chemosphere, 22*(1–2). https://doi.org/10.1016/0045-6535(91)90269-J
41. Solís, M., Solís, A., Pérez, H. I., Manjarrez, N., & Flores, M. (2012). Microbial decolouration of azo dyes: A review. *Process Biochemistry, 47*(12). https://doi.org/10.1016/j.procbio.2012.08.014
42. Telke, A., Kalyani, D., Jadhav, J., & Govindwar, S. (2008). Kinetics and mechanism of reactive red 141 degradation by a bacterial isolate Rhizobium radiobacter MTCC 8161. *Acta Chimica Slovenica, 55*(2).
43. Wang, H., Su, J. Q., Zheng, X. W., Tian, Y., Xiong, X. J., & Zheng, T. L. (2009). Bacterial decolourization and degradation of the reactive dye Reactive Red 180 by *Citrobacter* sp. CK3. *International Biodeterioration and Biodegradation, 63*(4). https://doi.org/10.1016/j.ibiod.2008.11.006
44. Wang, H., Zheng, X. W., Su, J. Q., Tian, Y., Xiong, X. J., & Zheng, T. L. (2009). Biological decolourization of the reactive dyes Reactive Black 5 by a novel isolated bacterial strain *Enterobacter* sp. EC3. *Journal of Hazardous Materials, 171*(1–3). https://doi.org/10.1016/j.jhazmat.2009.06.050